Climbing The Gamma Ladder To Light-Speed And Beyond.

Volume 2.

To order additional copies of this book, contact:
Xlibris
844-714-8691
www.Xlibris.com
Orders@Xlibris.com

ISBN: Softcover 978-1-6698-6198-0
EBook 978-1-6698-6197-3

Print information available on the last page

Rev. date: 01/11/2023

This volume includes a whimsical tour of Light-Speed and ever so slightly faster than light impulse travel and associated infinite and super-infinite spacecraft Lorentz factors.

We cover content related to leaving a universe of origin based on the possibility that a light-speed spacecraft may over-run universal future time to enter a larger realm. Analogs for multiverses, forests, biospheres, and the like and hyperspaces, bulks, hyper-bulks and the like are also included.

Of special note, I make use of recourse to GOD and immensely infinite possible values for Lorentz factors to argue the plausibility of some of the concepts presented herein.

I justify invocation of the Divine under a simple assumption that there either is or is not a GOD and reality either does or does not have meaning.

Additionally, I provide whimsical content on how Lorentz invariant and Lorentz variant effects may manifest on human ontology including the soul and its faculties and accidental aspects.

ESSAY 1) Imprinting Or Quantum Entanglement Of A Spacecraft And Crew With A Low Frequency Photon

A fascinating method of super-luminal travel would involve imprinting or quantum entanglement of a spacecraft and crew with a low frequency photon.

In cases of quantum entanglement, the photon and classical feedback channel photon would travel through an expanding spacetime and become cosmically red-shifted as a host universe or hyperspace expanded.

The redshifted photons ideally would be collected almost instantly at a receiver located near the boundary of the cosmic light-cone relative to the origin.

Ideally, the collected photons would decohere as the leading edge of the photon interacted with special optical equipment that statistically skews the photon decoherence and thus arrival after less than the leading 1 percent of the photon wave-length has arrived at the receiver optics.

Now, the velocity of the mid-point of both waves in free space is c = distance/time.

So, in first order, the effective velocity of the waves is about [C + [(1/2)ʎ](Photon Frequency)]].

In second order, the effective velocity of the waves is about,

[C + [[(1/2)[1 – [(2)(Fraction of photon wavelength having arrived at the point in time of decoherence)]]dʎ](Photon Frequency)]]

More specifically for the above specific example, the photon velocity is about,

[C + [[(1/2)[1 – [(2)(0.01)]]ʎ](Photon Frequency)]]

Here, ʎ is the photon wavelength.

Now, there will be a processing time for photon reception. So, the combined process effective photon travel speed will be in 3rd order,

[C + [[(1/2)[1 – [(2)(0.01)]]ʎ](Photon Frequency)]]

-[[C + [[(1/2)[1 – [(2)(0.01)]]ʎ](Photon Frequency)]] [(Photon reception process time)/[(Effective Photon Travel Time Before Not Counting The Decoherence Process Set-Up Time) + (Photon reception process time)]]]

More generally, the photon travel time including time to set up and process decoherence at the receiver is,

[C + [[(1/2)[1 – [(2)(Fraction of photon wavelength having arrived at the point in time of decoherence)]]dʎ](Photon Frequency)]]-

[[C + [[(1/2)[1 – [(2)(Fraction of photon wavelength having arrived at the point in time of decoherence)]]dʎ](Photon Frequency)]] [(Photon reception process time)/[(Effective Photon Travel Time Before Not Counting The Decoherence Process Set-Up Time) + (Photon reception process time)]]]

For hyperspatial travel in N-D-Space-1-D-Time, the formula for travel time in 3rd order will be,

[(C in N-D-Space-1-D-Time) + [[(1/2)[1 – [(2)(Fraction of photon wavelength having arrived at the point in time of decoherence)]]dʎ](Photon Frequency)]]-

[[(C in N-D-Space-1-D-Time) + [[(1/2)[1 – [(2)(Fraction of photon wavelength having arrived at the point in time of decoherence)]]dʎ](Photon Frequency)]] [(Photon reception process time)/[(Effective Photon Travel Time Before Not Counting The Decoherence Process Set-Up Time) + (Photon reception process time)]]]

Here, we assume the speed of light is constant in a given realm but can vary from one hyperspace to another.

Additionally, we assume that the photons remain in the same light-cone for our universe, and also in any hyperspaces of travel.

Regarding the above travel methods of photon red-shift skewing, consider time derivatives and integrals related to change of position, velocity, acceleration, mass, kinetic energy, momentum and the like for hyperspaces or universes of more than one temporal dimension.

Each of the following roster elements can be the basis of expressions of differentiated functions for power, energy, velocity, momentum, and gamma and the like by differentiation in the following manners.

d f/dt,1 d(d f/dt,1)/dt,2 d(d f/dt,2)/dt,1 d[d(d f/dt,1)/dt,2]/dt,3 d[d(d f/dt,1)/dt,3]/dt,2

d[d(d f/dt,2)/dt,1]/dt,3 d[d(d f/dt,2)/dt,3]/dt,1 d[d(d f/dt,3)/dt,1]/dt,2 d[d(d f/dt,3)/dt,2]/dt,1 d f/dT,1 d(d f/dT,1)/dT,2 d(d f/dT,2)/dT,1 d[d(d f/dT,1)/dT,2]/dT,3 d[d(d f/dT,1)/dT,3]/dT,2

d[d(d f/dT,2)/dT,1]/dT,3 d[d(d f/dT,2)/dT,3]/dT,1 d[d(d f/dT,3)/dT,1]/dT,2 d[d(d f/dT,3)/dT,2]/dT,1

T in each case is a time-like blend of various sorts of multiple time dimensions that appropriately and usefully enables derivation in one iteration.

We can go further to consider.

d f/dƮ,1 d(d f/dƮ,1)/dƮ,2 d(d f/dƮ,2)/dƮ,1 d[d(d f/dƮ,1)/dƮ,2]/dƮ,3 d[d(d f/dƮ,1)/dƮ,3]/dƮ,2

d[d(d f/dƮ,2)/dƮ,1]/dƮ,3 d[d(d f/dƮ,2)/dƮ,3]/dƮ,1 d[d(d f/dƮ,3)/dƮ,1]/dƮ,2 d[d(d f/dƮ,3)/dƮ,2]/dƮ,1

Ʈ in each case is a time-like blend of time-like blends of various sorts of multiple time dimensions that appropriately and usefully enables derivation in one iteration.

By the same token, we can also integrate functions for power, energy, velocity, momentum, and gamma and the like by respect to multiple time dimensions as follows.

∫f(dt,1) ∫∫d f(dt,1)(dt,2) ∫∫d f(dt,2)(dt,1) ∭f(dt,1)(dt,2)(dt,3) ∭f(dt,1)(dt,3)(dt,2)

∭f(dt,2)(dt,1)(dt,3) ∭f(dt,2)(dt,3)(dt,1) ∭f(dt,3)(dt,1)(dt,2) ∭f(dt,3)(dt,2)(dt,1)

∫f(dT,1) ∬f(dT,1) (dT,2) ∬f(dT,2) ((dT,1)) ∭f(dT,1) (dT,2) (dT,3) ∭f(dT,1) (dT,3) (dT,2)

∭f(dT,2) (dT,1) (dT,3) ∭f(dT,2) (dT,3) (dT,1) ∭f(dT,3) (dT,1) (dT,2) ∭f(dT,3) (dT,2) (dT,1)

We can go further to consider.

∫ f(dƮ,1) ∫ ∫ f(dƮ,1)(dƮ,2) ∫ ∫ f(dƮ,2)(dƮ,1) ∫ ∫ ∫ f(dƮ,1)(dƮ,2)(dƮ,3) ∫ ∫ ∫ f(dƮ,1)(dƮ,3)(dƮ,2)

∫ ∫ ∫ f(dƮ,2)(dƮ,1)(dƮ,3) ∫ ∫ ∫ f(dƮ,2)(dƮ,3)(dƮ,1) ∫ ∫ ∫ f(dƮ,3)(dƮ,1)(dƮ,2) ∫ ∫ ∫ f(dƮ,3)(dƮ,2)(dƮ,1)

We can continue to yet higher order derivatives and integrals suffice it to say that the general patterns continue for higher numbers of dimensions.

We can also consider higher order blends of time and thus are not limited to the blends defined by T and Ʈ. We chose these two symbols for their resemblance of the letter t.

ESSAY 2) Light-Speed Craft Travel Out Of Physical Order.

Now, we generally divide the ontology of created reality into two main primary classes.

The first class of ontology is matter also sometimes referred more generally to as mass-energy-space-time.

The second class is spirit. We accordingly refer to the soul as spiritual and angels as pure spirits.

However, we can lexicographically intuit a level above that of mass-energy-space-time that is as different from mass-energy-space-time and created spirit as the latter two are from each other.

Now, again, we can intuit that a spacecraft having attained the velocity of light may have infinite Lorentz factors and have run out of room to travel forward in time in for a given host universe, multiverse, etc. As such, the spacecraft may leave the host realm and enter other realms.

Upon still greater infinite Lorentz factors, the craft may leave a host cosmic order all together to in a sense travel out of a host created order.

Having traveled out of said host cosmic (physical) order, the craft may travel into or amongst a level above that of mass-energy-space-time which we will refer to as Above-mass-energy-space-time-1.

We can lexicographically intuit another level we will refer to as, Above-mass-energy-space-time-2 that is above that of Above-mass-energy-space-time-1 and mass-energy-space-time and which is as different from Above-mass-energy-space-time-1, mass-energy-space-time, and created spirit as the latter three are from each other.

We do not know how many levels as such as there may be so we will refer to the set of higher levels by the notation, [Σ(k = 1; k = ↑):(Above-mass-energy-space-time-k)]. Since there may be any finite or infinite numbers of such higher levels and progressively more as the process of creation ex-nihilo continues, we may use the following operator to denote the mechanisms for such higher levels and their application for and related to Lorentz factors and gamma. We imply affixing the operator to formulas for gamma and related parameters to denote the number of and nature of levels.

[r[Σ(k = 1; k = ↑):(Above-mass-energy-space-time-k where k can be any finite or infinite positive integer on either a standard or hyper-extended number line)]].

The upward pointing arrow as the series limit simply stands for open ended indefinite ordinals which can and likely would increase as the process of creation ex-nihilo continues forever.

ESSAY 3) Unlimited Numbers Of Particle Masses And Strengths Of Forces Considering Multiverses, And So On.

Now, we are familiar with the matter-antimatter duality, as well as the Standard Model distinctions of and specific species of bosons and fermions. We even consider the supersymmetry models and extended levels of supersymmetry. We further consider other universes for which the relative masses and strengths of fundamental particles and forces can differ in perhaps unlimited numbers of values.

So, each Above-mass-energy-space-time-k may likewise have unlimited numbers of embodiments.

Each specific embodiment may lie on a coordinate system defined by R EXP R or a space defined by as many dimensions as there are real numbers where each coordinate axis is a real number line.

We can continue to even more extremes by considering specific embodiments that lie in a coordinate system defined as (hR) EXP (hR) where each coordinate axis is hyper-extended into the realm of hyper-real numbers and where the number of dimensions is equal to the number of real and hyper-real numbers on the hyper-extended number line.

We imply affixing the operator to formulas for gamma and related parameters to denote the number of and nature of the elements in the set (hR) EXP (hR) . We do not actually append the operator in iterated notation for the sake of brevity.

[r[Σ(k = 1; k = ↑):[Set of (hR) EXP (hR) Above-mass-energy-space-time-k where k can be any finite or infinite positive integer on either a standard or hyper-extended number line]]]

ESSAY 4) Breakdown Of Special Relativistic Rules For Spacecraft Kinetic Energy For Sufficiently Relativistic Spacecraft.

Relativistic kinetic energy is fascinating. Accordingly, the translational relativistic kinetic energy is proportional to the spacecraft Lorentz factor which equals {1/{1 – [(v/c) EXP 2]}} EXP (1/2).

However, for light-speed inertial impulse spacecraft with non-modified inertial invariant mass, the relativistic spacecraft kinetic energy can grow to [(∞)(M)(c EXP 2)].

Perhaps at the speed of light, the rules for relativistic kinetic energy begin to break down and are supplanted with other formulas.

A fascinating notion for inertial spacecraft that achieve velocities ever so slightly greater than c in inertial motion is the concept that these scenarios might be associated with super-infinite kinetic energies.

Accordingly, for inertial impulse spacecraft with non-modified inertial invariant mass, the relativistic spacecraft kinetic energy grows to [(◯)(M)(c EXP 2)]. Here, the symbol, ◯, denotes the least super-infinite value.

A fascinating scenario would involve a system of complicated rotors, actuators for which the super-posed motions would result in portions of the complex system attaining a velocity of zero for a brief period of time then moving back up to light-speed in infinite Lorentz factors. Thus, the temporarily stationary element would become at rest then all of the sudden be boosted to light-speed.

Such a system would need to be fabricated from extreme materials perhaps of types such as naked singularitium (a conjectural material more dense than the Planck Density Level), wormhole wall material, or perhaps raw non-quantized or continuous mass-energy-space-time.

Such a system might be scaled to cosmic light-cone or super-cosmic light-cone extensities and may be manufactured in adjacent duplicity so that passengers can be conveyed from one place to another while merely needed to do multiple transfers.

These systems may also be accelerated to a translational velocity of light-speed with infinite Lorentz factors.

Oddly enough, when such systems are traveling at light-speed, passengers can be conveyed across a large multiplicity of such co-traveling systems for which effective travel at near the speed of light can be accomplished across a cosmic habitat comprised of such light-cone scale or super-light-cone scale systems.

In cases for which such systems can have periods of travel at speeds finitely but vanishingly small multiples of light-speed, the passengers can attain effectively light-speed travel.

Another possibility would include systems that have cars of slow motion for very brief periods of time but where the passengers are quantum mechanically teleported into the car, classically beamed into the car, or otherwise exotically transported into the car at or near transport speeds of light.

The systems described above and below can in many cases serve as co-translational light-speed motion systems allowing passengers to travel within said system.

Once uploaded into the system, the passengers may also be quantum or classically mechanically teleported within the system at the speed of light or classically beamed within the system.

Given sufficient elaboration of the systems described above, the systems as a whole may self-teleport or self-tunnel ahead of a precursor previous position. Oddly enough, the teleportation can occur relative to reference frames already traveling at the speed of light.

Ideally, the massive quadrupole or higher ordered moments of inertia for massive systems would have zero time dependent changes allowing the systems to work without gravitational radiation leakage.

Antigravity or gravity wave generators can be switched on to cancel gravitational wave energy otherwise bled from the systems.

These systems can operate in ordinary 4-D space-time or N-D-Space-M-D-Time where N is any integer greater that two and N is any counting number.

These systems can include any craft of multiple degrees of light-speed motion presented in this book as well as any types of quantum mechanical, classical mechanical, classically beamed, or quantum tunneling transport mechanisms presented in this book as non-limiting systems.

ESSAY 5) Upsweep Of Infinite And Super-Infinite Energy Reserves From Hyperspaces.

Consider mechanisms of traveling in ordinary 4-D spacetime and within a portion of an extension of the three spatial dimensions that protrudes or extends into a 4-D space in a dimensional extension of [1 – (ε,4)]-D where ε,4 is a vanishingly infinitesimal or vanishingly small finite fraction less than one.

Such a craft may be able to extract untold finite, infinite, and perhaps even super-infinite reserves of hyperspatial or 4-D space-time hidden or potential energy.

These scenarios may also enable exotic space-time transport effects perhaps such as wormhole travel, teleportation of quantum or classical types, warp drives, and other completely different exotic transport methods not yet considered.

Assuming that the spacecraft extracts some of its propulsion energy via its extension into 4-D-Space and the 4-D-Space is quantized at the Planck Length level, the spacecraft Lorentz factor will become:

{{{[(Average Energy Of Planck Length Edged 4-D Hypercube Elements Swept Of Energy)(Number of Planck Length Edged 4-D Hypercube Elements Swept Of Energy)](Fraction Of Hyperspatial Upswept Energy Converted To Spacecraft Kinetic Energy)}

+ {(Upswept Electromagnetic Energy In Ordinary 4-D Space-Time By The Spacecraft)(Dimensionless Fraction Of Ordinary 4-D Space-Time Upswept Electromagnetic Energy Converted To Spacecraft Kinetic Energy)}

+ {(Upswept Magnetic Field Energy In Ordinary 4-D Space-Time By The Spacecraft)(Dimensionless Fraction Of Ordinary 4-D Space-Time Upswept Magnetic Field Energy Converted To Spacecraft Kinetic Energy)}

+ {(Upswept Background Neutrino Energy In Ordinary 4-D Space-Time By The Spacecraft)(Dimensionless Fraction Of Ordinary 4-D Space-Time Upswept Background Neutrino Energy Converted To Spacecraft Kinetic Energy)}

+ {(Upswept Background Ionic Latent Energy In Ordinary 4-D Space-Time By The Spacecraft)(Dimensionless Fraction Of Ordinary 4-D Space-Time Upswept Background Ionic Latent Energy Converted To Spacecraft Kinetic Energy)}

+ {(Upswept Background Gravitational Wave Energy In Ordinary 4-D Space-Time By The Spacecraft)(Dimensionless Fraction Of Ordinary 4-D Space-Time Upswept Background Gravitational Wave Energy Converted To Spacecraft Kinetic Energy)}

+ (Quantity Of Spacecraft Kinetic Energy Derived From Upswept Nuclear Fusion Fuel In Ordinary 4-D Space-Time By The Spacecraft)

+ (Quantity Of Spacecraft Kinetic Energy Derived From Upswept Chemical Fuel In Ordinary 4-D Space-Time By The Spacecraft)

+ (Quantity Of Spacecraft Kinetic Energy Derived From Upswept Cold Dark Matter Fuel In Ordinary 4-D Space-Time By The Spacecraft)} + [(M)(c EXP 2)]}/ [(M)(c EXP 2)]

Consider mechanisms of traveling in ordinary 4-D spacetime and within a portion of an extension of the three spatial dimensions that protrudes or extends into a five dimensional hyper-space in a dimensional extension of [1 – (ε,5)]-D where ε,5 is a vanishingly infinitesimal or vanishingly small finite fraction less than one.

Such a craft may be able to extract untold finite, infinite, and perhaps even super-infinite reserves of hyperspatial or 4-D space-time hidden or potential energy.

These scenarios may also enable exotic space-time transport effects perhaps such as wormhole travel, teleportation of quantum or classical types, warp drives, and other completely different exotic transport methods not yet considered.

Assuming that the spacecraft extracts some of its propulsion energy via its extension into 5-D-Space and the 5-D-Space is quantized at the Planck Length level, the spacecraft Lorentz factor will become:

{{{[(Average Energy Of Planck Length Edged 5-D Hypercube Elements Swept Of Energy)(Number of Planck Length Edged 5-D Hypercube Elements Swept Of Energy)](Fraction Of Hyperspatial Upswept Energy Converted To Spacecraft Kinetic Energy)}

+ {(Upswept Electromagnetic Energy In Ordinary 4-D Space-Time By The Spacecraft)(Dimensionless Fraction Of Ordinary 4-D Space-Time Upswept Electromagnetic Energy Converted To Spacecraft Kinetic Energy)}

+ {(Upswept Magnetic Field Energy In Ordinary 4-D Space-Time By The Spacecraft)(Dimensionless Fraction Of Ordinary 4-D Space-Time Upswept Magnetic Field Energy Converted To Spacecraft Kinetic Energy)}

+ {(Upswept Background Neutrino Energy In Ordinary 4-D Space-Time By The Spacecraft)(Dimensionless Fraction Of Ordinary 4-D Space-Time Upswept Background Neutrino Energy Converted To Spacecraft Kinetic Energy)}

+ {(Upswept Background Ionic Latent Energy In Ordinary 4-D Space-Time By The Spacecraft)(Dimensionless Fraction Of Ordinary 4-D Space-Time Upswept Background Ionic Latent Energy Converted To Spacecraft Kinetic Energy)}

+ {(Upswept Background Gravitational Wave Energy In Ordinary 4-D Space-Time By The Spacecraft)(Dimensionless Fraction Of Ordinary 4-D Space-Time Upswept Background Gravitational Wave Energy Converted To Spacecraft Kinetic Energy)}

+ (Quantity Of Spacecraft Kinetic Energy Derived From Upswept Nuclear Fusion Fuel In Ordinary 4-D Space-Time By The Spacecraft)

+ (Quantity Of Spacecraft Kinetic Energy Derived From Upswept Chemical Fuel In Ordinary 4-D Space-Time By The Spacecraft)

+ (Quantity Of Spacecraft Kinetic Energy Derived From Upswept Cold Dark Matter Fuel In Ordinary 4-D Space-Time By The Spacecraft)} + [(M)(c EXP 2)]}/ [(M)(c EXP 2)]

Consider mechanisms of traveling in ordinary 4-D space-time and within a portion of an extension of the three spatial dimensions that protrudes or extends into a six dimensional hyper-space in a dimensional extension of [1 – (ε,6)]-D where ε,6 is a vanishingly infinitesimal or vanishingly small finite fraction less than one.

Such a craft may be able to extract untold finite, infinite, and perhaps even super-infinite reserves of either hyperspatial or 4-D space-time hidden or potential energy.

These scenarios may also enable exotic space-time transport effects perhaps such as wormhole travel, teleportation of quantum or classical types, warp drives, and other completely different exotic transport methods not yet considered.

Assuming that the spacecraft extracts some of its propulsion energy via its extension into 6-D-Space and the 6-D-Space is quantized at the Planck Length level, the spacecraft Lorentz factor will become:

{{{[(Average Energy Of Planck Length Edged 6-D Hypercube Elements Swept Of Energy)(Number of Planck Length Edged 6-D Hypercube Elements Swept Of Energy)](Fraction Of Hyperspatial Upswept Energy Converted To Spacecraft Kinetic Energy)}

+ {(Upswept Electromagnetic Energy In Ordinary 4-D Space-Time By The Spacecraft)(Dimensionless Fraction Of Ordinary 4-D Space-Time Upswept Electromagnetic Energy Converted To Spacecraft Kinetic Energy)}

+ {(Upswept Magnetic Field Energy In Ordinary 4-D Space-Time By The Spacecraft)(Dimensionless Fraction Of Ordinary 4-D Space-Time Upswept Magnetic Field Energy Converted To Spacecraft Kinetic Energy)}

+ {(Upswept Background Neutrino Energy In Ordinary 4-D Space-Time By The Spacecraft)(Dimensionless Fraction Of Ordinary 4-D Space-Time Upswept Background Neutrino Energy Converted To Spacecraft Kinetic Energy)}

+ {(Upswept Background Ionic Latent Energy In Ordinary 4-D Space-Time By The Spacecraft)(Dimensionless Fraction Of Ordinary 4-D Space-Time Upswept Background Ionic Latent Energy Converted To Spacecraft Kinetic Energy)}

+ {(Upswept Background Gravitational Wave Energy In Ordinary 4-D Space-Time By The Spacecraft)(Dimensionless Fraction Of Ordinary 4-D Space-Time Upswept Background Gravitational Wave Energy Converted To Spacecraft Kinetic Energy)}

+ (Quantity Of Spacecraft Kinetic Energy Derived From Upswept Nuclear Fusion Fuel In Ordinary 4-D Space-Time By The Spacecraft)

+ (Quantity Of Spacecraft Kinetic Energy Derived From Upswept Chemical Fuel In Ordinary 4-D Space-Time By The Spacecraft)

+ (Quantity Of Spacecraft Kinetic Energy Derived From Upswept Cold Dark Matter Fuel In Ordinary 4-D Space-Time By The Spacecraft)} + [(M)(c EXP 2)]}/ [(M)(c EXP 2)]

Consider mechanisms of traveling in ordinary 4-D space-time and within a portion of an extension of the three spatial dimensions that protrudes or extends into an N dimensional hyper-space in a dimensional extension of [1 – (ε,N)]-D where ε,N is a vanishingly infinitesimal or vanishingly small finite fraction less than one.

Such a craft may be able to extract untold finite, infinite, and perhaps even super-infinite reserves of either hyperspatial or 4-D space-time hidden or potential energy.

These scenarios may also enable exotic space-time transport effects perhaps such as wormhole travel, teleportation of quantum or classical types, warp drives, and other completely different exotic transport methods not yet considered.

Assuming that the spacecraft extracts some of its propulsion energy via its extension into N-D-Space and the N-D-Space is quantized at the Planck Length level, the spacecraft Lorentz factor will become:

{{{[(Average Energy Of Planck Length Edged N-D Hypercube Elements Swept Of Energy)(Number of Planck Length Edged N-D Hypercube Elements Swept Of Energy)](Fraction Of Hyperspatial Upswept Energy Converted To Spacecraft Kinetic Energy)}

+ {(Upswept Electromagnetic Energy In Ordinary 4-D Space-Time By The Spacecraft)(Dimensionless Fraction Of Ordinary 4-D Space-Time Upswept Electromagnetic Energy Converted To Spacecraft Kinetic Energy)}

+ {(Upswept Magnetic Field Energy In Ordinary 4-D Space-Time By The Spacecraft)(Dimensionless Fraction Of Ordinary 4-D Space-Time Upswept Magnetic Field Energy Converted To Spacecraft Kinetic Energy)}

+ {(Upswept Background Neutrino Energy In Ordinary 4-D Space-Time By The Spacecraft)(Dimensionless Fraction Of Ordinary 4-D Space-Time Upswept Background Neutrino Energy Converted To Spacecraft Kinetic Energy)}

+ {(Upswept Background Ionic Latent Energy In Ordinary 4-D Space-Time By The Spacecraft)(Dimensionless Fraction Of Ordinary 4-D Space-Time Upswept Background Ionic Latent Energy Converted To Spacecraft Kinetic Energy)}

+ {(Upswept Background Gravitational Wave Energy In Ordinary 4-D Space-Time By The Spacecraft)(Dimensionless Fraction Of Ordinary 4-D Space-Time Upswept Background Gravitational Wave Energy Converted To Spacecraft Kinetic Energy)}

+ (Quantity Of Spacecraft Kinetic Energy Derived From Upswept Nuclear Fusion Fuel In Ordinary 4-D Space-Time By The Spacecraft)

+ (Quantity Of Spacecraft Kinetic Energy Derived From Upswept Chemical Fuel In Ordinary 4-D Space-Time By The Spacecraft)

+ (Quantity Of Spacecraft Kinetic Energy Derived From Upswept Cold Dark Matter Fuel In Ordinary 4-D Space-Time By The Spacecraft)} + [(M)(c EXP 2)]}/ [(M)(c EXP 2)]

We can now consider scenarios where the space-time protrusions occur in multiple higher dimensional realms of a given dimensionality.

Consider mechanisms of traveling in ordinary 4-D spacetime and within a portion of an extension of the three spatial dimensions that protrudes or extends into a multiple of, k, 4 dimensional hyperspaces in dimensional extensions of Σ [1 – (ε,4,k)]-D where ε,4,k is a vanishingly infinitesimal or vanishingly small finite fraction less than one.

Such a craft may be able to extract untold finite, infinite, and perhaps even super-infinite reserves of hyperspatial or 4-D space-time hidden or potential energy.

These scenarios may also enable exotic space-time transport effects perhaps such as wormhole travel, teleportation of quantum or classical types, warp drives, and other completely different exotic transport methods not yet considered.

Assuming that the spacecraft extracts some of its propulsion energy via its extension into the multiple of 4-D-Spaces and the 4-D-Spaces are quantized at the Planck Length level, the spacecraft Lorentz factor will become:

{{{Σ(k = 1; k = ↑∞):{{[(Average Energy Of Planck Length Edged 4-D Hypercube Elements Swept Of Energy)(Number of Planck Length Edged 4-D Hypercube Elements Swept Of Energy)](Fraction Of Hyperspatial Upswept Energy Converted To Spacecraft Kinetic Energy)},k}}

+ {(Upswept Electromagnetic Energy In Ordinary 4-D Space-Time By The Spacecraft)(Dimensionless Fraction Of Ordinary 4-D Space-Time Upswept Electromagnetic Energy Converted To Spacecraft Kinetic Energy)}

+ {(Upswept Magnetic Field Energy In Ordinary 4-D Space-Time By The Spacecraft)(Dimensionless Fraction Of Ordinary 4-D Space-Time Upswept Magnetic Field Energy Converted To Spacecraft Kinetic Energy)}

+ {(Upswept Background Neutrino Energy In Ordinary 4-D Space-Time By The Spacecraft)(Dimensionless Fraction Of Ordinary 4-D Space-Time Upswept Background Neutrino Energy Converted To Spacecraft Kinetic Energy)}

+ {(Upswept Background Ionic Latent Energy In Ordinary 4-D Space-Time By The Spacecraft)(Dimensionless Fraction Of Ordinary 4-D Space-Time Upswept Background Ionic Latent Energy Converted To Spacecraft Kinetic Energy)}

+ {(Upswept Background Gravitational Wave Energy In Ordinary 4-D Space-Time By The Spacecraft)(Dimensionless Fraction Of Ordinary 4-D Space-Time Upswept Background Gravitational Wave Energy Converted To Spacecraft Kinetic Energy)}

+ (Quantity Of Spacecraft Kinetic Energy Derived From Upswept Nuclear Fusion Fuel In Ordinary 4-D Space-Time By The Spacecraft)

+ (Quantity Of Spacecraft Kinetic Energy Derived From Upswept Chemical Fuel In Ordinary 4-D Space-Time By The Spacecraft)

+ (Quantity Of Spacecraft Kinetic Energy Derived From Upswept Cold Dark Matter Fuel In Ordinary 4-D Space-Time By The Spacecraft)} + [(M)(c EXP 2)]}/ [(M)(c EXP 2)]

Consider mechanisms of traveling in ordinary 4-D spacetime and within a portion of an extension of the three spatial dimensions that protrudes or extends into a multiple of, k, 5 dimensional hyperspaces in dimensional extensions of Σ [1 – (ε,5,k)]-D where ε,5,k is a vanishingly infinitesimal or vanishingly small finite fraction less than one.

Such a craft may be able to extract untold finite, infinite, and perhaps even super-infinite reserves of hyperspatial or 4-D space-time hidden or potential energy.

These scenarios may also enable exotic space-time transport effects perhaps such as wormhole travel, teleportation of quantum or classical types, warp drives, and other completely different exotic transport methods not yet considered.

Assuming that the spacecraft extracts some of its propulsion energy via its extension into the multiple of 5-D-Spaces and the 5-D-Spaces are quantized at the Planck Length level, the spacecraft Lorentz factor will become:

{{{Σ(k = 1; k = ↑∞):{{[(Average Energy Of Planck Length Edged 5-D Hypercube Elements Swept Of Energy)(Number of Planck Length Edged 5-D Hypercube Elements Swept Of Energy)](Fraction Of Hyperspatial Upswept Energy Converted To Spacecraft Kinetic Energy)},k}}

+ {(Upswept Electromagnetic Energy In Ordinary 4-D Space-Time By The Spacecraft)(Dimensionless Fraction Of Ordinary 4-D Space-Time Upswept Electromagnetic Energy Converted To Spacecraft Kinetic Energy)}

+ {(Upswept Magnetic Field Energy In Ordinary 4-D Space-Time By The Spacecraft)(Dimensionless Fraction Of Ordinary 4-D Space-Time Upswept Magnetic Field Energy Converted To Spacecraft Kinetic Energy)}

+ {(Upswept Background Neutrino Energy In Ordinary 4-D Space-Time By The Spacecraft)(Dimensionless Fraction Of Ordinary 4-D Space-Time Upswept Background Neutrino Energy Converted To Spacecraft Kinetic Energy)}

+ {(Upswept Background Ionic Latent Energy In Ordinary 4-D Space-Time By The Spacecraft)(Dimensionless Fraction Of Ordinary 4-D Space-Time Upswept Background Ionic Latent Energy Converted To Spacecraft Kinetic Energy)}

+ {(Upswept Background Gravitational Wave Energy In Ordinary 4-D Space-Time By The Spacecraft)(Dimensionless Fraction Of Ordinary 4-D Space-Time Upswept Background Gravitational Wave Energy Converted To Spacecraft Kinetic Energy)}

+ (Quantity Of Spacecraft Kinetic Energy Derived From Upswept Nuclear Fusion Fuel In Ordinary 4-D Space-Time By The Spacecraft)

+ (Quantity Of Spacecraft Kinetic Energy Derived From Upswept Chemical Fuel In Ordinary 4-D Space-Time By The Spacecraft)

+ (Quantity Of Spacecraft Kinetic Energy Derived From Upswept Cold Dark Matter Fuel In Ordinary 4-D Space-Time By The Spacecraft)} + [(M)(c EXP 2)]}/ [(M)(c EXP 2)]

Consider mechanisms of traveling in ordinary 4-D spacetime and within a portion of an extension of the three spatial dimensions that protrudes or extends into a multiple of, k, 6 dimensional hyperspaces in dimensional extensions of Σ [1 – (ε,6,k)]-D where ε,6,k is a vanishingly infinitesimal or vanishingly small finite fraction less than one.

Such a craft may be able to extract untold finite, infinite, and perhaps even super-infinite reserves of either hyperspatial or 4-D space-time hidden or potential energy.

These scenarios may also enable exotic space-time transport effects perhaps such as wormhole travel, teleportation of quantum or classical types, warp drives, and other completely different exotic transport methods not yet considered.

Assuming that the spacecraft extracts some of its propulsion energy via its extension into the multiple of 6-D-Spaces and the 6-D-Spaces are quantized at the Planck Length level, the spacecraft Lorentz factor will become:

{{{Σ(k = 1; k = ↑∞):{{[(Average Energy Of Planck Length Edged 6-D Hypercube Elements Swept Of Energy)(Number of Planck Length Edged 6-D Hypercube Elements Swept Of Energy)](Fraction Of Hyperspatial Upswept Energy Converted To Spacecraft Kinetic Energy)},k}}

+ {(Upswept Electromagnetic Energy In Ordinary 4-D Space-Time By The Spacecraft)(Dimensionless Fraction Of Ordinary 4-D Space-Time Upswept Electromagnetic Energy Converted To Spacecraft Kinetic Energy)}

+ {(Upswept Magnetic Field Energy In Ordinary 4-D Space-Time By The Spacecraft)(Dimensionless Fraction Of Ordinary 4-D Space-Time Upswept Magnetic Field Energy Converted To Spacecraft Kinetic Energy)}

+ {(Upswept Background Neutrino Energy In Ordinary 4-D Space-Time By The Spacecraft)(Dimensionless Fraction Of Ordinary 4-D Space-Time Upswept Background Neutrino Energy Converted To Spacecraft Kinetic Energy)}

+ {(Upswept Background Ionic Latent Energy In Ordinary 4-D Space-Time By The Spacecraft)(Dimensionless Fraction Of Ordinary 4-D Space-Time Upswept Background Ionic Latent Energy Converted To Spacecraft Kinetic Energy)}

+ {(Upswept Background Gravitational Wave Energy In Ordinary 4-D Space-Time By The Spacecraft)(Dimensionless Fraction Of Ordinary 4-D Space-Time Upswept Background Gravitational Wave Energy Converted To Spacecraft Kinetic Energy)}

+ (Quantity Of Spacecraft Kinetic Energy Derived From Upswept Nuclear Fusion Fuel In Ordinary 4-D Space-Time By The Spacecraft)

+ (Quantity Of Spacecraft Kinetic Energy Derived From Upswept Chemical Fuel In Ordinary 4-D Space-Time By The Spacecraft)

+ (Quantity Of Spacecraft Kinetic Energy Derived From Upswept Cold Dark Matter Fuel In Ordinary 4-D Space-Time By The Spacecraft)} + [(M)(c EXP 2)]}/ [(M)(c EXP 2)]

Consider mechanisms of traveling in ordinary 4-D spacetime and within a portion of an extension of the three spatial dimensions that protrudes or extends into a multiple of, k, N dimensional hyperspaces in dimensional extensions of Σ [1 – (ε,5,k)]-D where ε,5,k is a vanishingly infinitesimal or vanishingly small finite fraction less than one.

Such a craft may be able to extract untold finite, infinite, and perhaps even super-infinite reserves of either hyperspatial or 4-D space-time hidden or potential energy.

These scenarios may also enable exotic space-time transport effects perhaps such as wormhole travel, teleportation of quantum or classical types, warp drives, and other completely different exotic transport methods not yet considered.

Assuming that the spacecraft extracts some of its propulsion energy via its extension into the multiple of N-D-Spaces and the N-D-Spaces are quantized at the Planck Length level, the spacecraft Lorentz factor will become:

{{{Σ(k = 1; k = ↑∞):{{[(Average Energy Of Planck Length Edged N-D Hypercube Elements Swept Of Energy)(Number of Planck Length Edged N-D Hypercube Elements Swept Of Energy)](Fraction Of Hyperspatial Upswept Energy Converted To Spacecraft Kinetic Energy)},k}}

+ {(Upswept Electromagnetic Energy In Ordinary 4-D Space-Time By The Spacecraft)(Dimensionless Fraction Of Ordinary 4-D Space-Time Upswept Electromagnetic Energy Converted To Spacecraft Kinetic Energy)}

+ {(Upswept Magnetic Field Energy In Ordinary 4-D Space-Time By The Spacecraft)(Dimensionless Fraction Of Ordinary 4-D Space-Time Upswept Magnetic Field Energy Converted To Spacecraft Kinetic Energy)}

+ {(Upswept Background Neutrino Energy In Ordinary 4-D Space-Time By The Spacecraft)(Dimensionless Fraction Of Ordinary 4-D Space-Time Upswept Background Neutrino Energy Converted To Spacecraft Kinetic Energy)}

+ {(Upswept Background Ionic Latent Energy In Ordinary 4-D Space-Time By The Spacecraft)(Dimensionless Fraction Of Ordinary 4-D Space-Time Upswept Background Ionic Latent Energy Converted To Spacecraft Kinetic Energy)}

+ {(Upswept Background Gravitational Wave Energy In Ordinary 4-D Space-Time By The Spacecraft)(Dimensionless Fraction Of Ordinary 4-D Space-Time Upswept Background Gravitational Wave Energy Converted To Spacecraft Kinetic Energy)}

+ (Quantity Of Spacecraft Kinetic Energy Derived From Upswept Nuclear Fusion Fuel In Ordinary 4-D Space-Time By The Spacecraft)

+ (Quantity Of Spacecraft Kinetic Energy Derived From Upswept Chemical Fuel In Ordinary 4-D Space-Time By The Spacecraft)

+ (Quantity Of Spacecraft Kinetic Energy Derived From Upswept Cold Dark Matter Fuel In Ordinary 4-D Space-Time By The Spacecraft)} + [(M)(c EXP 2)]}/ [(M)(c EXP 2)]

We now consider scenarios of hyperspatial protrusions into sets of hyperspaces of different dimensionalities all else remaining the same.

{{{Σ(k = 1; k = ↑∞):{{[(Average Energy Of Planck Length Edged 4-D Hypercube Elements Swept Of Energy)(Number of Planck Length Edged 4-D Hypercube Elements Swept Of Energy)](Fraction Of Hyperspatial Upswept Energy Converted To Spacecraft Kinetic Energy)},k}}

+

{Σ(k = 1; k = ↑∞):{{[(Average Energy Of Planck Length Edged 5-D Hypercube Elements Swept Of Energy)(Number of Planck Length Edged 5-D Hypercube Elements Swept Of Energy)](Fraction Of Hyperspatial Upswept Energy Converted To Spacecraft Kinetic Energy)},k}}

+

{Σ(k = 1; k = ↑∞):{{[(Average Energy Of Planck Length Edged 6-D Hypercube Elements Swept Of Energy)(Number of Planck Length Edged 6-D Hypercube Elements Swept Of Energy)](Fraction Of Hyperspatial Upswept Energy Converted To Spacecraft Kinetic Energy)},k}}

+ ... + {Σ(k = 1; k = ↑∞):{{[(Average Energy Of Planck Length Edged N-D Hypercube Elements Swept Of Energy)(Number of Planck Length Edged N-D Hypercube Elements Swept Of Energy)](Fraction Of Hyperspatial Upswept Energy Converted To Spacecraft Kinetic Energy)},k}}

+ {(Upswept Electromagnetic Energy In Ordinary 4-D Space-Time By The Spacecraft)(Dimensionless Fraction Of Ordinary 4-D Space-Time Upswept Electromagnetic Energy Converted To Spacecraft Kinetic Energy)}

+ {(Upswept Magnetic Field Energy In Ordinary 4-D Space-Time By The Spacecraft)(Dimensionless Fraction Of Ordinary 4-D Space-Time Upswept Magnetic Field Energy Converted To Spacecraft Kinetic Energy)}

+ {(Upswept Background Neutrino Energy In Ordinary 4-D Space-Time By The Spacecraft)(Dimensionless Fraction Of Ordinary 4-D Space-Time Upswept Background Neutrino Energy Converted To Spacecraft Kinetic Energy)}

+ {(Upswept Background Ionic Latent Energy In Ordinary 4-D Space-Time By The Spacecraft)(Dimensionless Fraction Of Ordinary 4-D Space-Time Upswept Background Ionic Latent Energy Converted To Spacecraft Kinetic Energy)}

+ {(Upswept Background Gravitational Wave Energy In Ordinary 4-D Space-Time By The Spacecraft)(Dimensionless Fraction Of Ordinary 4-D Space-Time Upswept Background Gravitational Wave Energy Converted To Spacecraft Kinetic Energy)}

+ (Quantity Of Spacecraft Kinetic Energy Derived From Upswept Nuclear Fusion Fuel In Ordinary 4-D Space-Time By The Spacecraft)

+ (Quantity Of Spacecraft Kinetic Energy Derived From Upswept Chemical Fuel In Ordinary 4-D Space-Time By The Spacecraft)

+ (Quantity Of Spacecraft Kinetic Energy Derived From Upswept Cold Dark Matter Fuel In Ordinary 4-D Space-Time By The Spacecraft)} + [(M)(c EXP 2)]}/ [(M)(c EXP 2)]

We now consider scenarios for which the hyperspaces are positively curved, negatively curved, positively curved and torsioned at one or more scales in arbitrary patterns including but not limited to fractals, negatively curved, or negatively curved and torsioned at one or more scales in arbitrary patterns including but not limited to fractals.

We denote all possibilities of curvature geometries by the term, (kth hyperspace Curvature Factor, p,n,pt,nt)

Gamma = {{{Σ(k = 1; k = ↑∞):{{{[(Average Energy Of Planck Length Edged 4-D Hypercube Elements Swept Of Energy)(Number of Planck Length Edged 4-D Hypercube Elements Swept Of Energy)](Fraction Of Hyperspatial Upswept Energy Converted To Spacecraft Kinetic Energy)},k}(kth hyperspace Curvature Factor, p,n,pt,nt)}}

+

{Σ(k = 1; k = ↑∞):{{{[(Average Energy Of Planck Length Edged 5-D Hypercube Elements Swept Of Energy)(Number of Planck Length Edged 5-D Hypercube Elements Swept Of Energy)](Fraction Of Hyperspatial Upswept Energy Converted To Spacecraft Kinetic Energy)},k}(kth hyperspace Curvature Factor, p,n,pt,nt)}}

+

{Σ(k = 1; k = ↑∞):{{{[(Average Energy Of Planck Length Edged 6-D Hypercube Elements Swept Of Energy)(Number of Planck Length Edged 6-D Hypercube Elements Swept Of Energy)](Fraction Of Hyperspatial Upswept Energy Converted To Spacecraft Kinetic Energy)},k}(kth hyperspace Curvature Factor, p,n,pt,nt)}}

+ ... + {Σ(k = 1; k = ↑∞):{{{[(Average Energy Of Planck Length Edged N-D Hypercube Elements Swept Of Energy)(Number of Planck Length Edged N-D Hypercube Elements Swept Of Energy)](Fraction Of Hyperspatial Upswept Energy Converted To Spacecraft Kinetic Energy)},k}(kth hyperspace Curvature Factor, p,n,pt,nt)}}

+ {(Upswept Electromagnetic Energy In Ordinary 4-D Space-Time By The Spacecraft)(Dimensionless Fraction Of Ordinary 4-D Space-Time Upswept Electromagnetic Energy Converted To Spacecraft Kinetic Energy)}

+ {(Upswept Magnetic Field Energy In Ordinary 4-D Space-Time By The Spacecraft)(Dimensionless Fraction Of Ordinary 4-D Space-Time Upswept Magnetic Field Energy Converted To Spacecraft Kinetic Energy)}

+ {(Upswept Background Neutrino Energy In Ordinary 4-D Space-Time By The Spacecraft)(Dimensionless Fraction Of Ordinary 4-D Space-Time Upswept Background Neutrino Energy Converted To Spacecraft Kinetic Energy)}

+ {(Upswept Background Ionic Latent Energy In Ordinary 4-D Space-Time By The Spacecraft)(Dimensionless Fraction Of Ordinary 4-D Space-Time Upswept Background Ionic Latent Energy Converted To Spacecraft Kinetic Energy)}

+ {(Upswept Background Gravitational Wave Energy In Ordinary 4-D Space-Time By The Spacecraft)(Dimensionless Fraction Of Ordinary 4-D Space-Time Upswept Background Gravitational Wave Energy Converted To Spacecraft Kinetic Energy)}

+ (Quantity Of Spacecraft Kinetic Energy Derived From Upswept Nuclear Fusion Fuel In Ordinary 4-D Space-Time By The Spacecraft)

+ (Quantity Of Spacecraft Kinetic Energy Derived From Upswept Chemical Fuel In Ordinary 4-D Space-Time By The Spacecraft)

+ (Quantity Of Spacecraft Kinetic Energy Derived From Upswept Cold Dark Matter Fuel In Ordinary 4-D Space-Time By The Spacecraft)} + [(M)(c EXP 2)]}/ [(M)(c EXP 2)]

ESSAY 6) Arbitrary Edge With Of Cubes.

The edge width of cubes can be arbitrary.

For example, the edge width of the cubes can be a sub-unitary finite fraction of the Planck Length Unit for hyperspaces correspondingly quantized.

As another example, the edge width of the cubes can be (Planck Length Unit)/[f(Aleph 0)] for continuous hyperspace.

As another example, the edge width of the cubes can be (Planck Length Unit)/[f(Aleph 1)] for super-continuous hyperspace at a first level.

As another example, the edge width of the cubes can be (Planck Length Unit)/[f(Aleph 2)] for super-continuous hyperspace at a second level.

As another example, the edge width of the cubes can be (Planck Length Unit)/[f(Aleph 3)] for super-continuous hyperspace at a third level.

In general, the edge width of the cubes can be (Planck Length Unit)/[f(Aleph H)] for super-continuous hyperspace at an Hth level.

Note that for any Standard Model bosons and fermions in the hyperspaces, perhaps the hyperspaces must be quantized at the Planck Length scale just so that the fundamental standard model particles can fit in the hyperspaces under the condition that supersymmetry models additionally apply to both our universe and said hyperspaces of mass-energy acquisition.

However, in cases where only continuous or super-continuous embodiments of energy are present or present and extracted, the edge width of the hypercubic elements swept of energy can be arbitrarily but commensurately infinitesimal.

ESSAY 7) Various Levels Of Hyper-Information Related To Infinite Lorentz Factor Spacecraft.

Here, we consider spacecraft that have traveled to such great infinite Lorentz factors such that they run out of future room to travel forward in time within the universe, multiverse, forest, biosphere, etc., and even the hyperspaces or hyper-space-times of travel.

Thus, these spacecraft effectively acquire or out-run all of the information of the original host universe, multiverse, forest, biosphere, etc., and even the hyperspaces or hyper-space-times of travel because the spacecraft have exhausted all thermodynamic states to travel forward in time within.

So, in a sense, these spacecraft enter a realm of hyper-information which is to information as information is to more or less lack of information in the empty set context. We refer hyper-information as such by the variable hyper-1-information.

Eventually, as the spacecraft Lorentz factor increased to even greater infinite values, perhaps the craft would enter the realm of hyper-hyper-information. We refer hyper-hyper-information as such by the variable hyper-2-information.

As the spacecraft Lorentz factor increased to even greater infinite values, perhaps the craft would enter the realm of hyper-hyper-hyper-information. We refer hyper-hyper-hyper-information as such by the variable hyper-3-information.

We can go onward to consider hyper-4-information, hyper-5-information, hyper-6-information, and so-on.

Perhaps what is transmitted in quantum mechanical teleportation before the classical channel feedback loop is closed is hyper-1-information or analog thereof.

Accordingly, nature may be extremely strict at preventing real information from traveling at faster than light.

We imply affixing the following operator to formulas for gamma and related parameters to denote the above hyper-k-information realm entered by spacecraft and its lowest ordinated level and perhaps having relevance to the establishment of quantum entanglement links before a classical feed-back signal can be sent to actually deposit information.

[Σ(k = 1; k = ◯):[q(Establishment of kth realm of travel associated with hyper-k-information and for which hyper-1-information may be related to quantum entanglement links set-up before information or objects can be teleported at the limit of light-speed via before the required classical feedback channel is completed and for which at least the next few higher ordinated terms such hyper-2-information, and hyper-3-information, and so-on would be related to progressively even more sublime and exotic links than those associated with quantum-link establishments in quantum-mechanical teleportation processes)]]

The symbol, ◯, stands for a limit that is at least as great as the lowest super-finite ordinal such as perhaps present on hyper-extended number lines of arbitrary degrees of hyper-extension.

In short, as I like to mention, light-speed travel offers unbounded mystical potentials with relevance to the very ontology of the created cosmic order. So, I urge you to not fret any light-speed travel limits imposed strictly by Mother Nature.

We may forever discover new mysteries and new opportunities in light-speed travel.

As for travel at very finite Lorentz factors, say perhaps at a value of 8, even such travel can open up our cosmic light-cone for travel, exploration, and settlement.

Within our observable portion of our universe, there are an estimate 10 EXP 24 stars. This is about equal to the number of fine grains of table sugar that would cover the entire United States 100 meters deep. The number of planets orbiting stars in our cosmic light-cone is likely 10 times greater yet or equal to the number of fine grains of table sugar that would cover the entire United States 1,000 meters deep. The number of moons orbiting planets orbiting stars in our cosmic light-cone is likely ten times greater yet or equal to the number of fine grains of table sugar that would cover the entire United States 10,000 meters deep.

Our cosmic light-cone is most likely just a tiny portion of our universe. Our universe may be just one of untold infinite numbers of universes in a multiverse which may be just one of untold infinite numbers of multiverses in a so-called forest, which may be just one of untold infinite numbers of forests in a so-called biosphere and so-on. The hierarchy may extend to untold infinite numbers of levels.

A Future Star-farers Story Yet To Be Told

It is sometime in the early 23rd Century, Earth time. You have entered orbit around an exo-gas-giant-planet within the habitable zone of a star about 50 light-years from Earth. As you glance at the beautiful blue, green, and white ribbons in a cacophony of shades, and tones, making up the planet's atmosphere, you also catch a glimpse of a moon that resembles Earth in tone and color but not in land mass.

After traveling about 10 years ship time having reached a gamma factor of about 8, you muse that you and your crew members as mere human persons have arrived as the ETs visiting this planetary system which has all of the biochemical signatures of infusement with life. Your beam sail and very high mass ratio fusion rocket, dual mode starship, has made this all possible.

You also notice the star studded background and realize that humanity now collectively has in theory the capacity to send explorers and colonists beyond an Earth centered cosmic light cone. You muse at the technical hurdles needed to be overcome to bridge the gaps between galaxies, and even more so, between galaxy clusters. However, you realize that self-contained world ships with large crews in theory are up to the task even if manned star travel is limited to roughly the gamma factor achieved by the mothership from which you now gaze upon this lovely planetary system.

Being an explorer at heart and never content to stay still, you begin to day-dream with a sense of exuberance at the ensemble of similar planetary systems that have long been demonstrated to exist within the observable universe, perhaps upwards of 10^{20} such systems in total just within your cosmic light-cone.

As you look out a carbonaceous super-material spacecraft window, you start to contemplate the number of similar possible scenes perceivable by a typical human. First, you consider a typical human vision of angular resolution roughly equal to 10,000 by 10,000 pixels. Then you recall a childhood present handed down through your familial lineage from your many times great uncle, back in the days when he worked along with numerous other space-heads on the basic thermodynamics and kinematics of the type of ship you now crew. You then recall reading in this gift of a copy of the Guinness Book Of World Records dating back to the 20th Century about the ability of humans to sense about 10 million different colors if placed side by side.

Now you make a mental estimate of the number of possible planetary sights that can in theory be viewed and sensed by a human person as (10 EXP 4)(10 EXP 4) EXP 10,000,000 or 10 EXP 80,000,000. Next you assume the capacity to detect roughly 10,000 tones for each color to derive a number of possible sights equal to (10 EXP 4)(10 EXP 4) EXP [(10,000,000)(10,000)] or 10 EXP 800,000,000,000. You now become satisfied from a purely artistic standpoint of the huge rough estimate of the number of possible vistas.

As the nature of human emotions, attitudes, feelings, and thoughts in terms of the still unconfirmed continuity or super-continuity of such subjective experiences is brought into the estimate, a sense of exhilaration begins to flow within your head and body along with a mild general relaxed euphoria of the perhaps infinite numbers of individual subjective experiences which are possible in consideration of the coupling of these psychodynamic variables and the mere perceptual field of visual field pixilations.

Now that my brief fictional account is complete, day dreams similar to the above story help keep me grounded in my life's calling to help pave the theoretical path to the stars.

This is not simply a Jim show, but is motivated for the benefit of all those yet to be born, in a great hope of future communion in an ever growing cosmic civilization of love for which Humanity will live in fraternal love and in lasting peace amongst the perhaps untold numbers of any existent extra-terrestrial and ultra-terrestrial peoples currently unknown to humanity.

Such extreme gamma factors permit extreme real time travel into the future. This future is not merely some extrapolation of the present nor is it the kind of holographic future according to conspiracy theorists that can be traveled into to find out what folks are going to do so that dangerous actions can be stopped before they occur. Such an appearant future would not really be the true future but would instead be a mere latent or potential projection from the present. Extreme gamma factor space travel at below the velocity of light permits crew members to travel into the cosmically distant future as it will actually unfold in an unalterable manner. Now, according to Stephen Hawking, the future of our universe if closed may be indistinguishable from its birth. Perhaps traveling at suitably extreme sub-light speeds would enable future distant descendants or emissaries from our cosmic era to travel into the past of our universe as it actually unfolded or perhaps into the future of another Big Bang incarnation of our current universe. The caveat, is well, the ability to pass through the transitional energy state of our collapsing universe into our past and/or the future incarnation of our universe upon its eventual collapse.

A suitably cloaked space craft may in some poorly understood, but valid sense, travel within a margin of 4-D space-time or within a boundary seam of 4-D space-time. Such travel may thus occur near or at the boundary transition between 4-D space-time and not 4-D space-time. It is thus conceivable that such travel may be properly viewed as more exotic than travel in higher dimensional space, hyperspace, other universes, or parallel space-times of the Many Worlds Interpretation of Quantum Theory forms. It may even be reasonable to suggest that such travel marginally occurs in space-time in an absolute sense regardless of the space-time considered. Can you imagine traveling but not so in space nor in space-time of any form!

With the passing near the time of this writing of the 67th anniversary of the destruction of Hiroshima and Nagasaki, I have been contemplating the terrible power of the atomic nucleus, which if unchecked in its applications for warfare, can certainly lead to our destruction. However, this cosmic energy source can be applied for great and profound good, in its enablement of the unlimited extension of the human presence into the eternal blackness of space-time. Atoms for Peace must replace Atoms for War. Atoms for peace can be an ultimate charitable mantra by promoting untold future living space for humanity. Atoms for War can certainly lead to the death of Humanity. The choice of paths to follow is profoundly ours, here and now, in this generation and the next one or two to follow. Choose wisely. With Peace, all things are possible including eternal progress.

I am especially interested in nuclear fusion because this is the dominant real energy source within our universe. Since the mean lifetime of the proton has been verified to be greater than about 10^{35} years and may in fact be eternally stable. Nuclear fusion fuel will be available for commensurately long time periods.

Nuclear fusion fuels can enable achievement of virtually unlimited special relativistic Lorentz transformation factors for cases where the fuel is sequestered by space craft in situ as it travels in interstellar or intergalactic space such as may be possible with the latest interstellar ramjet concepts. Alternatively, the nuclear fuels can be embodied in fusion fuel runways thus permitting extreme gamma factors with mass ratios of essentially one.

Antimatter may well be producible in large bulk quantities and safely stored as rocket fuel at some future time. However, most of the Standard Model baryonic matter within our universe is hydrogen and helium isotopes and thus may be more effectively harnessed for its direct latent energy as opposed to using nuclear fusion power sources to produce antimatter fuels.

As someone who is a nuclear energy freak, I see great potential for Humanity to harness nuclear energy in the decades, centuries, and millennia that lie ahead.

Nuclear fusion will always be available to turn the light on, whether for night-time illumination, or the light of nuclear energy for energizing our near term and distant future starships.

In a sense, nuclear energy is the most natural form of energy in our present cosmic era. For nuclear energy runs the stars, and heats the cores of Earth-like planets to thus assist with the provision of a thermodynamic gradient and planetary chemistry necessary for the development and evolution of any extraterrestrial life forms.

However, protonic and neutronic nuclear energy may not be the only useful form of nuclear energy. Science fiction is familiar with the concept of a quark bomb where protons would be induced to fission thus yielding energy virtually identical in quantity to the reaction of an equal mass consisting of half matter and half antimatter of the same species. Strangelets, charmedlets, and bottomlets when produced in any stable or metastable varieties in a responsible manner may be useful for catalyzing exotic quarkonium transformative reactions within rocket fuel or within captured interstellar Standard Model baryonic matter. Perhaps additional nuclear forces will be discovered that provide a higher mass specific yield in reactants than the best known nuclear fusion fuels.

This book is a glance at some plausible and possible outcomes that can evolve from the mantra, “Atoms For Peace”. This book is just one glimpse of a possible future full of excitement, pioneering exploration, and the expansion of human civilization, and stories yet to be told. This is the story of The Galactic Explorer!

ESSAY 8) The Infinitely Many Modes Of Travel.

Here, we still again consider spacecraft that have traveled to such great infinite Lorentz factors such that they run out of future room to travel forward in time within the universe, multiverse, forest, biosphere, etc., and even the hyperspaces or hyper-space-times of travel.

Thus, these spacecraft effectively acquire or out-run all of the information of the original host universe, multiverse, forest, biosphere, etc., and even the hyperspaces or hyper-space-times of travel because the spacecraft have exhausted all thermodynamic states to travel forward in time in.

So, in a sense, a sufficiently light-speed craft having sufficiently great infinite Lorentz factors must go somewhere else other than the space and time of the previous realm(s) traveled.

Now, we come to a fascinating notion. What if, because a spacecraft can not fit into the future of a given realm of travel, the craft ends up conveyancing in some other way other than the two known ways of travel, through space and forward in time.

In a sense, travel in space and in time can be represented abstractly in a two dimensional Cartesian plane. Here, the x-axis is space and the y-axis is time.

For a third mode of conyence, we add the z-axis which is orthogonal to the x and y axis.

Upon running out of room in the z-axis of conveyance, perhaps the craft will need to begin a conveyance defined in a fourth axis.

Upon running out of room in the 4th axis of conveyance, perhaps the craft will need to begin a conveyance defined in a fifth axis.

We can go on in the depths of eternity to produce a coordinate system of conveyance of the form R EXP R where R is the complete set of real numbers. As such, the number of axes of conveyance is equal the number of real numbers and the number of point-ticks on each axis is equal to the number of real numbers.

But hold on a minute! We can conjecture to even more extreme scenarios.

For example, consider prosects for which each axis is a hyper-extended number line to any arbitrary extent.

Thus, we develop scenarios for which ordinals that are super-integral are paired with a point on each axis. These abstract ordinals would not be real numbers nor even integers, but yet higher ordinals.

Note that we are not even talking about infinite integers which in theory exist on any ordinary number line. Instead, we are implying somethings loosely described as hyper-quantities. These super-integral ordinals are not qualities either. So, we can further develop the following tuple, for lack of a better word, to denote the extent of the conveyance coordinate system.

(R → super-integral ordinals) EXP (R → super-integral ordinals)

This loosely translates into the number of conveyance possibilities by axis and by position along axis, in manner analogous to systems of coordinate axis commonly considered in linear algebra.

So, I must urge you not to fret any light-speed travel limits imposed by creation as a unity.

Back in the days in the 1980's when I was going through some grave psychological difficulties very often punctuated by despair and depression, one of my family members would console me by calmly but emphatically stating to me that "This life is a mere stepping stone in eternity!". To this very day, I live this mantra enthusiastically and plan for a wonderful afterlife and hope and work so that many others can say the same thing.

ESSAY 9) Conjectures On Infinite Levels Of Sub-Quantum Hidden Variables.

Now, most quantum mechanicists do not believe in hidden variables.

Accordingly, the probabilistic explanations of quantum level phenomenon are the most simple related mechanisms.

However, I must say that I am a "GOD does not play dice!" kind of guy. I relish the above quote made by Einstein regarding the process of quantum entanglement and non-local correlations in entangled particles.

Accordingly, I believe in quantum hidden variables and am open to prospects that at some future time we may be able to observe them in a laboratory setting.

Additionally, I am open minded to prospects for hidden sub-quantum variables which I will refer to as hidden-sub-quantum-1 variables.

Furthermore, I can envision hidden-sub-sub-quantum variables which I will refer to as hidden-sub-quantum-2 variables.

Furthermore, I can envision hidden-sub-sub-sub-quantum variables which I will refer to as hidden-sub-quantum-3 variables.

I am even open minded to hidden-sub-quantum-4 variables, hidden-sub-quantum-4 variables, and so on all the way up to and perhaps beyond hidden-sub-quantum- $\bigcirc$ variables.

Here, $\bigcirc$, stand for the least super-infinite ordinal.

By super-infinite ordinal, I essentially mean the first hyper-integral ordinal such as would be abstracted as paired with a point on a hyper-extended number line.

I also do not believe that faster than light travel is possible except to such small degrees such that the spirit of Special Relativity is not violated and no true backward time travel would occur, at least to any extent which would violate causality in accordance with the chronological protection theorem.

One way we might be able to observe quantum hidden variables is to observe particles having non-zero invariant mass traveling at speeds associated with extreme finite Lorentz factors if not particles having infinite kinetic energy and thus infinite Lorentz factors. Such particles would in theory have magnified structures to such an extent that quantum hidden variables may be observable directly or indirectly.

Additionally, particle Lorentz factors which are still greater infinite numbers may reveal hidden-sub-quantum-1 variables, then hidden-sub-quantum-2 variables, then hidden-sub-quantum-3 variables and so on as the particle Lorentz factors become ever more infinite. Super-infinite Lorentz factors may even enable observation of hidden-sub-quantum-$\bigcirc$ variables. Perhaps we can even continue on to yet deeper levels of hidden compositions or structures.

Light-speed travel of massive systems may even enable violation of Lorentz invariance in such a manner that new realms of travel or conveyance manifest as commensurately associated with progressively ordinated terms of hidden-sub-quantum-k variables where k would even supersede $\bigcirc$.

So, do not at all fret any light-speed limits. Light-speed travel will open untold realms of exploration for us that will make concepts of warp drive, wormhole transport, and the like seem like a young pre-school aged child's coloring books in comparison.

I am picking up on Special and General Relativity where Einstein left off.

Regarding finite Lorentz factors, imagine stepping into a completely sealed compartment about the size of a typical closet in a house or apartment. Further consider that the compartment is a fiat having an unfueled mass of perhaps ten metric tons and is made of special 0.1 femtometer thick quarkonium. One femtometer is a quadrillionth of a meter or one millionth of a nanometer or 0.000001 nanometer.

Now, assume the booth is fueled with antihydrogen of mass equal to about 10 billion metric tons compressed to the density 10 billion metric tons per cubic meter. When the booth travels, it combines the antihydrogen with background hydrogen while recycling essentially all drag energy. Thus, the booth is an antimatter relativistic rocket ramjet with an effective specific impulse equal to 2 c or two times that of light-speed.

Now, after switching on the console and drive mode, you enter your location and desired destination. When you are ready to say good-by to all of your loves ones back on Earth and tell them you will meet them again in the next life, you press the go icon on the screen and hear a few clicks, buzz noises, and clanking much like a modern high capacity office photocopy machine.

Next, an automated female style voice announces that you have arrived at your destination. Your fiat had quickly ship-frame accelerated to a Lorentz factor of about 10 EXP 18 upon which reactionary breaking via linear induction drag introducing coils then other deceleration mechanisms where deployed by the fiat.

Within a mere few seconds ship time, you traveled 10 billion light-years and 10 billion years into the future.

The Sun and Earth you left behind are now dead. The Sun is a white dwarf now almost a black dwarf corpse and the Earth was long ago evaporated by the Sun during its red giant phase.

On your new planet of habitation, you hope to stick around for about one year before doing your next 10 billion light-year trip yet still further way from the Milky Way Galaxy.

The spacecraft has an antigravity system that is precisely controlled and cancels out the acceleration induced g-forces experienced from within.

Now, such scenarios may actually be possible within the realm of the Standard Model and advances in applications of known physics.

Just imagine what infinite Lorentz factor travel will enable. Truly, right here and now as you read this text, you have begun the initial stages of eventually making all of this a dream come true. OH! WHAT DREAMS WILL FOLLOW.

ESSAY 10) Serial Deployment Of Light-Speed Scout Ships From Light-Speed Host Ships.

As always, do not fret any light-speed limits Mother Nature imposes. It works out better for us in the long run.

Now, we can imagine inertial massive reference frames such as a human crewed starship traveling at the speed of light in variously infinite Lorentz factors.

Once a massive impulse powered spacecraft would attain light speed, perhaps it can launch a much smaller scout craft that can attain light-speed velocities and infinite Lorentz factors relative to the mothership.

However, we contemplate that the relative light-speed of the scout craft relative to the mothership might be greater, perhaps much, much, greater than the constant c.

The scout craft may optionally be crewed or have a teleportive quantum entanglement link such that the mothership could self-quantum-teleport to the scout craft.

We can also consider scenarios where the scout ship itself launches a second scout ship where the velocity of the second scout ship may be a light-speed that is perhaps much, much, greater than the constant c relative to the first scout ship.

The second scout craft may optionally be crewed or have a teleportive quantum entanglement link such that the mothership could self-quantum-teleport to the second scout craft.

Alternatively, the second scout craft may optionally be crewed or have a teleportive quantum entanglement link such that the first scout ship could self-quantum-teleport to the second scout craft.

We can also consider scenarios where the second scout ship itself launches a third scout ship where the velocity of the third scout ship may be a light-speed that is perhaps much, much, greater than the constant c relative to the second scout ship.

The third scout craft may optionally be crewed or have a teleportive quantum entanglement link such that the mothership could self-quantum-teleport to the third scout craft.

Alternatively, the third scout craft may optionally be crewed or have a teleportive quantum entanglement link such that the first scout ship could self-quantum-teleport to the third scout craft.

Alternatively, the third scout craft may optionally be crewed or have a teleportive quantum entanglement link such that the second scout ship could self-quantum-teleport to the third craft.

We can consider analog scenarios for 4th, 5th, 6th, and so-on scout ships.

So, all you folks who fret any light-speed limits, fear not! Light-speed travel can become really farout and exotic.

A mothership traveling at a standard light-speed in a vacuum having infinite Lorentz factors can travel an infinite number of light-years through space and an infinite number of years into the future in one tiny Planck Time Unit which equal {{[h/(2π)] G/(c EXP 5)} EXP (1/2)}. Here, h, G, and c are the Planck Constant, gravitational constant, and speed of light respectively in MKS units. The Planck time is [5.39 x (10 EXP – 44)] second or 0.0000000000000000000000000000000000000000000539 second.

The conjectural increases in light-speeds associated with various light-speed scout ships may manifest as mother nature makes first attempts to limit infinite spacecraft gamma factors to certain infinite value ranges. The increase in light speeds may be a way that nature fills out travel opportunities at velocities of light-speed greater than standard c.

Another argument for the increased values of c may be derived from the fact that the Lorentz factors of the scout ships relative to the previous scout ships and mothership become so intense so as to foster breakthrough kinematics and space-time topologies that would manifest in said increased velocities of light.

ESSAY 11) Onset Of Acausality In Light-Speed Travel.

A fascinating aspect of light speed travel would be the onset of acausality.

Philosophically speaking, travel of massive bodies at light speed would result in a lack of distinction between cause and effect in the traveling reference frame and might even lead to a roiling throth of cause and effect similar to the space-time foam that theoretically manifests at distance scales and time scales less than then Planck Length and Planck Time units. Thus, travel of massive bodies at light speed may result in the production of big bangs much as our universe seems to have resulted from a precursor state of space-time foam.

Now, a fascinating result would be the conversion of a spacecraft and her crew upon achieving light speed into space-time foam and new big bangs relative to a given exotic reference frame, all the while the spacecraft and crew experience normal integrity at light speed. Thus, the spacecraft having become infinitely relativistic might express a different kind of relativity. This relativity would be the distinction between full integrity of the spacecraft and crew in their own reference frame but an acausal frothy mess in the background reference frame.

A lack of distinction between cause and effect for a spacecraft traveling at light speed might in a sense result in a state of acausality of the spacecraft and her crew in an existential and ontological manner. In a limited sense, the natural substantial and accidental properties of the spacecraft and crew would transcend the limits of created nature to such an extent that at light speed, the craft and crew members would be self-existent.

We can also consider the onset of more than one level of acausality. The number of levels of acausality and also the number of commensurate levels of self-existence may be finite or arbitrarily infinite with levels never previously reached by created persons and their spacecraft. The size of the infinite numbers of acausality and self-existence can be any positive infinite integer on the Cartesian number line or even on hyper-extended number lines.

We may affix the following operator to denote the physics and philosophy underlying the light speed acausality and self-existent states of the spacecraft and its crew to formulas for gamma and related parameters.

[r(Physics and philosophy underlying the light speed acausality and self-existent states of the spacecraft and its crew where the infinite numbers of acausality and self-existence can be any positive infinite integer on the Cartesian number line or even on hyper-extended number lines)]

We can intuit a spacecraft traveling at the speed of light of the least infinite Lorentz factor.

Additionally, we can consider reference frames traveling along the same parallel at said least infinite Lorentz factor ahead of the spacecraft thus traveling at the speed of light so compositionally summed with respect to the background and light-speed travel with respect to the spacecraft. We will refer to this second light-speed reference frame as the second light-speed craft or light-speed-craft-2.

Likewise, we can consider a third light-speed craft traveling along the same parallel at said least infinite Lorentz factor ahead of the light-speed-craft-2 thus traveling at the speed of light so composition summed with respect to the background and light-speed travel with respect to light-speed-craft-2. We will refer to this third light-speed reference frame as light-speed-craft-3.

We can continue the pattern light-speed-craft-4, light-speed-craft-5, and so-on.

However, we can also consider relative speeds of light of craft with respect to backwardly adjacent craft that are associated with such relatively great Lorentz factors such that the Lorentz factors are functions of ◯. The latter symbol stands for the least super-infinite Lorentz factor and thus implies that all standard ways of defining ever greater infinities have been exhausted in developing approximations equal to said super-infinite Lorentz factors. By functions of super-infinite values we mean any functions that amplify ◯.

Now, we can consider scenarios defined by a vector along an axis representing the infinite variety of speeds of light, all of which may be placed within a vanishingly small light-speed velocity spectrum.

The vector as such may serve as a gradient vector that extends from the first light-speed and undefinably beyond the first light-speed such as associated with the least infinite number.

We can take the divergence of the light-speeds vector field **C** in n abstracted spatial dimensions.

The divergence is:

div **C** = $\partial \mathbf{c_1}/\partial x_1 + \partial \mathbf{c_2}/\partial x_2 + \ldots + \partial \mathbf{c_n}/\partial x_n$

The gradient of a field is its direction of steepest change.

For more n abstracted spatial dimensions of the light-speed ascension field, the gradient is:

Grad C = $(\partial c/\partial x_1)\mathbf{i_1} + (\partial c/\partial x_2)\mathbf{i_2}, \ldots, + (\partial c/\partial x_n)\mathbf{i_n}$

The Laplacian is the divergence of the gradients in the light-speed field and is defined abstractly in n spatial dimensions in Cartesian coordinates as:

Dell grad C = $(\partial^2 c/\partial x_1^2) + (\partial^2 c/\partial x_2^2) + \ldots + (\partial^2 c/\partial x_n^2)$

Now we can consider the realm of light-speed values as if an n-dimensional space in the abstract sense but not in the literal sense.

The gradient of the light-speed ascension field may be infinite in magnitude and perhaps even super-infinite in magnitude.

We can also consider the divergence of the gradient of the fields of ascending light-speed.

The divergence of the gradients or the associated Laplacian may plausibly have a value of less than zero but be non-negative. In other words, the Laplacian may have a modulus of less than zero but non-negative as a non-limiting option.

We can also consider analogs of the light-speed ascension field for Lorentz factors, spacecraft momentum, spacecraft kinetic energy, spacecraft acceleration in the ship frame; and ironically, acceleration in the background reference frame.

In cases for which the travel upward to ever greater light-speeds of travel develops into an infinite modulus vector, the climb to ever greater speeds of light may runway not only due to a momentum-style analog, but also in a manner analogous to the acceleration of a negative refractive index light-sail for which the incident radiative power would scale as the fourth power of the temperature of a black body source of incident light used to accelerate the light-sail.

We may affix of the following operator to formulas for gamma and related parameters to denote the above light-speed field concepts and the associated physics especially but not only as related to gamma. However, we do not in actuality insert said operator into said lengthy formulas for gamma and related parameters so as to save file space.

{f{{light-speed-craft-k | $1 \leq k \leq$ ◯, or k > ◯};{div **C** = $\partial \mathbf{c_1}/\partial x_1 + \partial \mathbf{c_2}/\partial x_2 + \ldots + \partial \mathbf{c_n}/\partial x_n$ | mod div **C** ≤ 0};{Grad C = $(\partial c/\partial x_1)\mathbf{i_1} + (\partial c/\partial x_2)\mathbf{i_2},\ldots, + (\partial c/\partial x_n)\mathbf{i_n}$ | $\infty <$ mod Grad C $\leq$ ◯, or mod Grad C > ◯}}}

We can take light-speed, gamma, velocity, acceleration, momentum, kinetic energy, and so on derivatives of the above operator as functionally modified to represent ship and background time, light-speed, gamma, velocity, acceleration, momentum, kinetic energy, and so on. We may affix the resulting derivative operators to formulas for gamma and related parameters do denote the operative mechanisms without actually including these operators in the formulas.

{d … {d{d {c{f{{light-speed-craft-k | $1 \leq k \leq$ ◯, or k >◯};{div **C** = $\partial \mathbf{c_1}/\partial x_1 + \partial \mathbf{c_2}/\partial x_2 + \ldots + \partial \mathbf{c_n}/\partial x_n$ | mod div **C** ≤ 0};{Grad C = $(\partial c/\partial x_1)\mathbf{i_1} + (\partial c/\partial x_2)i_2,\ldots, + (\partial c/\partial x_n)\mathbf{i_n}$ | $\infty <$ mod Grad C $\leq$ ◯, or mod Grad C > ◯}}}}/dc}/dc}/ … /d c}

{d … {d{d {c{f{{light-speed-craft-k | $1 \leq k \leq$ ◯, or k >◯};{div **C** = $\partial \mathbf{c_1}/\partial x_1 + \partial \mathbf{c_2}/\partial x_2 + \ldots + \partial \mathbf{c_n}/\partial x_n$ | mod div **C** ≤ 0};{Grad C = $(\partial c/\partial x_1)\mathbf{i_1} + (\partial c/\partial x_2)i_2,\ldots, + (\partial c/\partial x_n)\mathbf{i_n}$ | $\infty <$ mod Grad C $\leq$ ◯, or mod Grad C > ◯}}}}/dγ}/dγ}/ … /d γ}

{d … {d{d {c{f{{light-speed-craft-k | $1 \leq k \leq$ ◯, or k >◯};{div **C** = $\partial \mathbf{c_1}/\partial x_1 + \partial \mathbf{c_2}/\partial x_2 + \ldots + \partial \mathbf{c_n}/\partial x_n$ | mod div **C** ≤ 0};{Grad C = $(\partial c/\partial x_1)\mathbf{i_1} + (\partial c/\partial x_2)i_2,\ldots, + (\partial c/\partial x_n)\mathbf{i_n}$ | $\infty <$ mod Grad C $\leq$ ◯, or mod Grad C > ◯}}}}/dv}/dv}/ … /d v} {d … {d{d {c{f{{light-speed-craft-k | 1 ≤ k ≤ ◯, or k >◯};{div **C** = $\partial \mathbf{c}_1/\partial x_1 + \partial \mathbf{c}_2/\partial x_2 + \ldots + \partial \mathbf{c}_n/\partial x_n$ | mod div **C** ≤ 0};{Grad C = $(\partial c/\partial x_1)\mathbf{i}_1 + (\partial c/\partial x_2)i_2, \ldots, + (\partial c/\partial x_n)\mathbf{i}_n$ | ∞ < mod Grad C ≤ ◯, or mod Grad C > ◯}}}}/da}/da}/ … /d a}

{d … {d{d {c{f{{light-speed-craft-k | 1 ≤ k ≤ ◯, or k >◯};{div **C** = $\partial \mathbf{c}_1/\partial x_1 + \partial \mathbf{c}_2/\partial x_2 + \ldots + \partial \mathbf{c}_n/\partial x_n$ | mod div **C** ≤ 0};{Grad C = $(\partial c/\partial x_1)\mathbf{i}_1 + (\partial c/\partial x_2)i_2, \ldots, + (\partial c/\partial x_n)\mathbf{i}_n$ | ∞ < mod Grad C ≤ ◯, or mod Grad C > ◯}}}}/dP}/dP}/ … /d P}

{d … {d{d {c{f{{light-speed-craft-k | 1 ≤ k ≤ ◯, or k >◯};{div **C** = $\partial \mathbf{c}_1/\partial x_1 + \partial \mathbf{c}_2/\partial x_2 + \ldots + \partial \mathbf{c}_n/\partial x_n$ | mod div **C** ≤ 0};{Grad C = $(\partial c/\partial x_1)\mathbf{i}_1 + (\partial c/\partial x_2)i_2, \ldots, + (\partial c/\partial x_n)\mathbf{i}_n$ | ∞ < mod Grad C ≤ ◯, or mod Grad C > ◯}}}}/dKE}/dKE}/ … /d KE}

{d … {d{d {c{f{{light-speed-craft-k | 1 ≤ k ≤ ◯, or k >◯};{div **C** = $\partial \mathbf{c}_1/\partial x_1 + \partial \mathbf{c}_2/\partial x_2 + \ldots + \partial \mathbf{c}_n/\partial x_n$ | mod div **C** ≤ 0};{Grad C = $(\partial c/\partial x_1)\mathbf{i}_1 + (\partial c/\partial x_2)i_2, \ldots, + (\partial c/\partial x_n)\mathbf{i}_n$ | ∞ < mod Grad C ≤ ◯, or mod Grad C > ◯}}}}/dt}/dt}/ … /dt}

{d … {d{d {c{f{{light-speed-craft-k | 1 ≤ k ≤ ◯, or k >◯};{div **C** = $\partial \mathbf{c}_1/\partial x_1 + \partial \mathbf{c}_2/\partial x_2 + \ldots + \partial \mathbf{c}_n/\partial x_n$ | mod div **C** ≤ 0};{Grad C = $(\partial c/\partial x_1)\mathbf{i}_1 + (\partial c/\partial x_2)i_2, \ldots, + (\partial c/\partial x_n)\mathbf{i}_n$ | ∞ < mod Grad C ≤ ◯, or mod Grad C > ◯}}}}/dT}/dT}/ … /d T}

{d … {d{d {γ{f{{light-speed-craft-k | 1 ≤ k ≤ ◯, or k >◯};{div **C** = $\partial \mathbf{c}_1/\partial x_1 + \partial \mathbf{c}_2/\partial x_2 + \ldots + \partial \mathbf{c}_n/\partial x_n$ | mod div **C** ≤ 0};{Grad C = $(\partial c/\partial x_1)\mathbf{i}_1 + (\partial c/\partial x_2)i_2, \ldots, + (\partial c/\partial x_n)\mathbf{i}_n$ | ∞ < mod Grad C ≤ ◯, or mod Grad C > ◯}}}}/dc}/dc}/ … /d c}

{d … {d{d {γ{f{{light-speed-craft-k | 1 ≤ k ≤ ◯, or k >◯};{div **C** = $\partial \mathbf{c}_1/\partial x_1 + \partial \mathbf{c}_2/\partial x_2 + \ldots + \partial \mathbf{c}_n/\partial x_n$ | mod div **C** ≤ 0};{Grad C = $(\partial c/\partial x_1)\mathbf{i}_1 + (\partial c/\partial x_2)i_2, \ldots, + (\partial c/\partial x_n)\mathbf{i}_n$ | ∞ < mod Grad C ≤ ◯, or mod Grad C > ◯}}}}/dγ}/dγ}/ … /d γ}

{d … {d{d {γ{f{{light-speed-craft-k | 1 ≤ k ≤ ◯, or k >◯};{div **C** = $\partial \mathbf{c}_1/\partial x_1 + \partial \mathbf{c}_2/\partial x_2 + \ldots + \partial \mathbf{c}_n/\partial x_n$ | mod div **C** ≤ 0};{Grad C = $(\partial c/\partial x_1)\mathbf{i}_1 + (\partial c/\partial x_2)i_2, \ldots, + (\partial c/\partial x_n)\mathbf{i}_n$ | ∞ < mod Grad C ≤ ◯, or mod Grad C > ◯}}}}/dv}/dv}/ … /d v}

{d … {d{d {γ{f{{light-speed-craft-k | 1 ≤ k ≤ ◯, or k >◯};{div **C** = $\partial \mathbf{c}_1/\partial x_1 + \partial \mathbf{c}_2/\partial x_2 + \ldots + \partial \mathbf{c}_n/\partial x_n$ | mod div **C** ≤ 0};{Grad C = $(\partial c/\partial x_1)\mathbf{i}_1 + (\partial c/\partial x_2)i_2, \ldots, + (\partial c/\partial x_n)\mathbf{i}_n$ | ∞ < mod Grad C ≤ ◯, or mod Grad C > ◯}}}}/da}/da}/ … /d a}

{d … {d{d {γ{f{{light-speed-craft-k | 1 ≤ k ≤ ◯, or k >◯};{div **C** = $\partial \mathbf{c}_1/\partial x_1 + \partial \mathbf{c}_2/\partial x_2 + \ldots + \partial \mathbf{c}_n/\partial x_n$ | mod div **C** ≤ 0};{Grad C = $(\partial c/\partial x_1)\mathbf{i}_1 + (\partial c/\partial x_2)i_2, \ldots, + (\partial c/\partial x_n)\mathbf{i}_n$ | ∞ < mod Grad C ≤ ◯, or mod Grad C > ◯}}}}/dP}/dP}/ … /d P}

{d … {d{d {γ{f{{light-speed-craft-k | 1 ≤ k ≤ ◯, or k >◯};{div **C** = $\partial \mathbf{c}_1/\partial x_1 + \partial \mathbf{c}_2/\partial x_2 + \ldots + \partial \mathbf{c}_n/\partial x_n$ | mod div **C** ≤ 0};{Grad C = $(\partial c/\partial x_1)\mathbf{i}_1 + (\partial c/\partial x_2)i_2, \ldots, + (\partial c/\partial x_n)\mathbf{i}_n$ | ∞ < mod Grad C ≤ ◯, or mod Grad C > ◯}}}}/dKE}/dKE}/ … /d KE} {d … {d{d {γ{f{{light-speed-craft-k | 1 ≤ k ≤ ◯, or k >◯};{div **C** = $\partial \mathbf{c_1}/\partial x_1 + \partial \mathbf{c_2}/\partial x_2 + \dots +$ $\partial \mathbf{c_n}/\partial x_n$ | mod div **C** ≤ 0};{Grad C = $(\partial c/\partial x_1)\mathbf{i_1}$ + $(\partial c/\partial x_2)i_2$,…, + $(\partial c/\partial x_n)\mathbf{i_n}$ | ∞ < mod Grad C ≤ ◯, or mod Grad C > ◯}}}}/dt}/dt}/ … /dt}

{d … {d{d {γ{f{{light-speed-craft-k | 1 ≤ k ≤ ◯, or k >◯};{div **C** = $\partial \mathbf{c_1}/\partial x_1 + \partial \mathbf{c_2}/\partial x_2 + \dots +$ $\partial \mathbf{c_n}/\partial x_n$ | mod div **C** ≤ 0};{Grad C = $(\partial c/\partial x_1)\mathbf{i_1}$ + $(\partial c/\partial x_2)i_2$,…, + $(\partial c/\partial x_n)\mathbf{i_n}$ | ∞ < mod Grad C ≤ ◯, or mod Grad C > ◯}}}}/dT}/dT}/ … /d T}

{d … {d{d {v{f{{light-speed-craft-k | 1 ≤ k ≤ ◯, or k >◯};{div **C** = $\partial \mathbf{c_1}/\partial x_1 + \partial \mathbf{c_2}/\partial x_2 + \dots +$ $\partial \mathbf{c_n}/\partial x_n$ | mod div **C** ≤ 0};{Grad C = $(\partial c/\partial x_1)\mathbf{i_1}$ + $(\partial c/\partial x_2)i_2$,…, + $(\partial c/\partial x_n)\mathbf{i_n}$ | ∞ < mod Grad C ≤ ◯, or mod Grad C > ◯}}}}/dc}/dc}/ … /d c}

{d … {d{d {v{f{{light-speed-craft-k | 1 ≤ k ≤ ◯, or k >◯};{div **C** = $\partial \mathbf{c_1}/\partial x_1 + \partial \mathbf{c_2}/\partial x_2 + \dots +$ $\partial \mathbf{c_n}/\partial x_n$ | mod div **C** ≤ 0};{Grad C = $(\partial c/\partial x_1)\mathbf{i_1}$ + $(\partial c/\partial x_2)i_2$,…, + $(\partial c/\partial x_n)\mathbf{i_n}$ | ∞ < mod Grad C ≤ ◯, or mod Grad C > ◯}}}}/dγ}/dγ}/ … /d γ}

{d … {d{d {v{f{{light-speed-craft-k | 1 ≤ k ≤ ◯, or k >◯};{div **C** = $\partial \mathbf{c_1}/\partial x_1 + \partial \mathbf{c_2}/\partial x_2 + \dots +$ $\partial \mathbf{c_n}/\partial x_n$ | mod div **C** ≤ 0};{Grad C = $(\partial c/\partial x_1)\mathbf{i_1}$ + $(\partial c/\partial x_2)i_2$,…, + $(\partial c/\partial x_n)\mathbf{i_n}$ | ∞ < mod Grad C ≤ ◯, or mod Grad C > ◯}}}}/dv}/dv}/ … /d v}

{d … {d{d {v{f{{light-speed-craft-k | 1 ≤ k ≤ ◯, or k >◯};{div **C** = $\partial \mathbf{c_1}/\partial x_1 + \partial \mathbf{c_2}/\partial x_2 + \dots +$ $\partial \mathbf{c_n}/\partial x_n$ | mod div **C** ≤ 0};{Grad C = $(\partial c/\partial x_1)\mathbf{i_1}$ + $(\partial c/\partial x_2)i_2$,…, + $(\partial c/\partial x_n)\mathbf{i_n}$ | ∞ < mod Grad C ≤ ◯, or mod Grad C > ◯}}}}/da}/da}/ … /d a}

{d … {d{d {v{f{{light-speed-craft-k | 1 ≤ k ≤ ◯, or k >◯};{div **C** = $\partial \mathbf{c_1}/\partial x_1 + \partial \mathbf{c_2}/\partial x_2 + \dots +$ $\partial \mathbf{c_n}/\partial x_n$ | mod div **C** ≤ 0};{Grad C = $(\partial c/\partial x_1)\mathbf{i_1}$ + $(\partial c/\partial x_2)i_2$,…, + $(\partial c/\partial x_n)\mathbf{i_n}$ | ∞ < mod Grad C ≤ ◯, or mod Grad C > ◯}}}}/dP}/dP}/ … /d P}

{d … {d{d {v{f{{light-speed-craft-k | 1 ≤ k ≤ ◯, or k >◯};{div **C** = $\partial \mathbf{c_1}/\partial x_1 + \partial \mathbf{c_2}/\partial x_2 + \dots +$ $\partial \mathbf{c_n}/\partial x_n$ | mod div **C** ≤ 0};{Grad C = $(\partial c/\partial x_1)\mathbf{i_1}$ + $(\partial c/\partial x_2)i_2$,…, + $(\partial c/\partial x_n)\mathbf{i_n}$ | ∞ < mod Grad C ≤ ◯, or mod Grad C > ◯}}}}/dKE}/dKE}/ … /d KE}

{d … {d{d {v{f{{light-speed-craft-k | 1 ≤ k ≤ ◯, or k >◯};{div **C** = $\partial \mathbf{c_1}/\partial x_1 + \partial \mathbf{c_2}/\partial x_2 + \dots +$ $\partial \mathbf{c_n}/\partial x_n$ | mod div **C** ≤ 0};{Grad C = $(\partial c/\partial x_1)\mathbf{i_1}$ + $(\partial c/\partial x_2)i_2$,…, + $(\partial c/\partial x_n)\mathbf{i_n}$ | ∞ < mod Grad C ≤ ◯, or mod Grad C > ◯}}}}/dt}/dt}/ … /dt}

{d … {d{d {v{f{{light-speed-craft-k | 1 ≤ k ≤ ◯, or k >◯};{div **C** = $\partial \mathbf{c_1}/\partial x_1 + \partial \mathbf{c_2}/\partial x_2 + \dots +$ $\partial \mathbf{c_n}/\partial x_n$ | mod div **C** ≤ 0};{Grad C = $(\partial c/\partial x_1)\mathbf{i_1}$ + $(\partial c/\partial x_2)i_2$,…, + $(\partial c/\partial x_n)\mathbf{i_n}$ | ∞ < mod Grad C ≤ ◯, or mod Grad C > ◯}}}}/dT}/dT}/ … /d T} {d … {d{d {a{f{{light-speed-craft-k | $1 \le k \le \bigcirc$, or $k > \bigcirc$};{div **C** = $\partial \mathbf{c_1}/\partial x_1 + \partial \mathbf{c_2}/\partial x_2 + \ldots + \partial \mathbf{c_n}/\partial x_n$ | mod div **C** ≤ 0};{Grad C = $(\partial c/\partial x_1)\mathbf{i_1} + (\partial c/\partial x_2)i_2, \ldots, + (\partial c/\partial x_n)\mathbf{i_n}$ | $\infty <$ mod Grad C $\le \bigcirc$, or mod Grad C $> \bigcirc$}}}}/dc}/dc}/ … /d c}

{d … {d{d {a{f{{light-speed-craft-k | $1 \le k \le \bigcirc$, or $k > \bigcirc$};{div **C** = $\partial \mathbf{c_1}/\partial x_1 + \partial \mathbf{c_2}/\partial x_2 + \ldots + \partial \mathbf{c_n}/\partial x_n$ | mod div **C** ≤ 0};{Grad C = $(\partial c/\partial x_1)\mathbf{i_1} + (\partial c/\partial x_2)i_2, \ldots, + (\partial c/\partial x_n)\mathbf{i_n}$ | $\infty <$ mod Grad C $\le \bigcirc$, or mod Grad C $> \bigcirc$}}}}/dγ}/dγ}/ … /d γ}

{d … {d{d {a{f{{light-speed-craft-k | $1 \le k \le \bigcirc$, or $k > \bigcirc$};{div **C** = $\partial \mathbf{c_1}/\partial x_1 + \partial \mathbf{c_2}/\partial x_2 + \ldots + \partial \mathbf{c_n}/\partial x_n$ | mod div **C** ≤ 0};{Grad C = $(\partial c/\partial x_1)\mathbf{i_1} + (\partial c/\partial x_2)i_2, \ldots, + (\partial c/\partial x_n)\mathbf{i_n}$ | $\infty <$ mod Grad C $\le \bigcirc$, or mod Grad C $> \bigcirc$}}}}/dv}/dv}/ … /d v}

{d … {d{d {a{f{{light-speed-craft-k | $1 \le k \le \bigcirc$, or $k > \bigcirc$};{div **C** = $\partial \mathbf{c_1}/\partial x_1 + \partial \mathbf{c_2}/\partial x_2 + \ldots + \partial \mathbf{c_n}/\partial x_n$ | mod div **C** ≤ 0};{Grad C = $(\partial c/\partial x_1)\mathbf{i_1} + (\partial c/\partial x_2)i_2, \ldots, + (\partial c/\partial x_n)\mathbf{i_n}$ | $\infty <$ mod Grad C $\le \bigcirc$, or mod Grad C $> \bigcirc$}}}}/da}/da}/ … /d a}

{d … {d{d {a{f{{light-speed-craft-k | $1 \le k \le \bigcirc$, or $k > \bigcirc$};{div **C** = $\partial \mathbf{c_1}/\partial x_1 + \partial \mathbf{c_2}/\partial x_2 + \ldots + \partial \mathbf{c_n}/\partial x_n$ | mod div **C** ≤ 0};{Grad C = $(\partial c/\partial x_1)\mathbf{i_1} + (\partial c/\partial x_2)i_2, \ldots, + (\partial c/\partial x_n)\mathbf{i_n}$ | $\infty <$ mod Grad C $\le \bigcirc$, or mod Grad C $> \bigcirc$}}}}/dP}/dP}/ … /d P}

{d … {d{d {a{f{{light-speed-craft-k | $1 \le k \le \bigcirc$, or $k > \bigcirc$};{div **C** = $\partial \mathbf{c_1}/\partial x_1 + \partial \mathbf{c_2}/\partial x_2 + \ldots + \partial \mathbf{c_n}/\partial x_n$ | mod div **C** ≤ 0};{Grad C = $(\partial c/\partial x_1)\mathbf{i_1} + (\partial c/\partial x_2)i_2, \ldots, + (\partial c/\partial x_n)\mathbf{i_n}$ | $\infty <$ mod Grad C $\le \bigcirc$, or mod Grad C $> \bigcirc$}}}}/dKE}/dKE}/ … /d KE}

{d … {d{d {a{f{{light-speed-craft-k | $1 \le k \le \bigcirc$, or $k > \bigcirc$};{div **C** = $\partial \mathbf{c_1}/\partial x_1 + \partial \mathbf{c_2}/\partial x_2 + \ldots + \partial \mathbf{c_n}/\partial x_n$ | mod div **C** ≤ 0};{Grad C = $(\partial c/\partial x_1)\mathbf{i_1} + (\partial c/\partial x_2)i_2, \ldots, + (\partial c/\partial x_n)\mathbf{i_n}$ | $\infty <$ mod Grad C $\le \bigcirc$, or mod Grad C $> \bigcirc$}}}}/dt}/dt}/ … /dt}

{d … {d{d {a{f{{light-speed-craft-k | $1 \le k \le \bigcirc$, or $k > \bigcirc$};{div **C** = $\partial \mathbf{c_1}/\partial x_1 + \partial \mathbf{c_2}/\partial x_2 + \ldots + \partial \mathbf{c_n}/\partial x_n$ | mod div **C** ≤ 0};{Grad C = $(\partial c/\partial x_1)\mathbf{i_1} + (\partial c/\partial x_2)i_2, \ldots, + (\partial c/\partial x_n)\mathbf{i_n}$ | $\infty <$ mod Grad C $\le \bigcirc$, or mod Grad C $> \bigcirc$}}}}/dT}/dT}/ … /d T}

{d … {d{d {P{f{{light-speed-craft-k | $1 \le k \le \bigcirc$, or $k > \bigcirc$};{div **C** = $\partial \mathbf{c_1}/\partial x_1 + \partial \mathbf{c_2}/\partial x_2 + \ldots + \partial \mathbf{c_n}/\partial x_n$ | mod div **C** ≤ 0};{Grad C = $(\partial c/\partial x_1)\mathbf{i_1} + (\partial c/\partial x_2)i_2, \ldots, + (\partial c/\partial x_n)\mathbf{i_n}$ | $\infty <$ mod Grad C $\le \bigcirc$, or mod Grad C $> \bigcirc$}}}}/dc}/dc}/ … /d c}

{d … {d{d {P{f{{light-speed-craft-k | $1 \le k \le \bigcirc$, or $k > \bigcirc$};{div **C** = $\partial \mathbf{c_1}/\partial x_1 + \partial \mathbf{c_2}/\partial x_2 + \ldots + \partial \mathbf{c_n}/\partial x_n$ | mod div **C** ≤ 0};{Grad C = $(\partial c/\partial x_1)\mathbf{i_1} + (\partial c/\partial x_2)i_2, \ldots, + (\partial c/\partial x_n)\mathbf{i_n}$ | $\infty <$ mod Grad C $\le \bigcirc$, or mod Grad C $> \bigcirc$}}}}/dγ}/dγ}/ … /d γ} {d … {d{d {P{f{{light-speed-craft-k | 1 ≤ k ≤ ◯, or k >◯};{div **C** = $\partial \mathbf{c_1}/\partial x_1 + \partial \mathbf{c_2}/\partial x_2 + \dots + \partial \mathbf{c_n}/\partial x_n$ | mod div **C** ≤ 0};{Grad C = $(\partial c/\partial x_1)\mathbf{i_1}$ + $(\partial c/\partial x_2)i_2$,…, + $(\partial c/\partial x_n)\mathbf{i_n}$ | ∞ < mod Grad C ≤ ◯, or mod Grad C > ◯}}}}/dv}/dv}/ … /d v}

{d … {d{d {P{f{{light-speed-craft-k | 1 ≤ k ≤ ◯, or k >◯};{div **C** = $\partial \mathbf{c_1}/\partial x_1 + \partial \mathbf{c_2}/\partial x_2 + \dots + \partial \mathbf{c_n}/\partial x_n$ | mod div **C** ≤ 0};{Grad C = $(\partial c/\partial x_1)\mathbf{i_1}$ + $(\partial c/\partial x_2)i_2$,…, + $(\partial c/\partial x_n)\mathbf{i_n}$ | ∞ < mod Grad C ≤ ◯, or mod Grad C > ◯}}}}/da}/da}/ … /d a}

{d … {d{d {P{f{{light-speed-craft-k | 1 ≤ k ≤ ◯, or k >◯};{div **C** = $\partial \mathbf{c_1}/\partial x_1 + \partial \mathbf{c_2}/\partial x_2 + \dots + \partial \mathbf{c_n}/\partial x_n$ | mod div **C** ≤ 0};{Grad C = $(\partial c/\partial x_1)\mathbf{i_1}$ + $(\partial c/\partial x_2)i_2$,…, + $(\partial c/\partial x_n)\mathbf{i_n}$ | ∞ < mod Grad C ≤ ◯, or mod Grad C > ◯}}}}/dP}/dP}/ … /d P}

{d … {d{d {P{f{{light-speed-craft-k | 1 ≤ k ≤ ◯, or k >◯};{div **C** = $\partial \mathbf{c_1}/\partial x_1 + \partial \mathbf{c_2}/\partial x_2 + \dots + \partial \mathbf{c_n}/\partial x_n$ | mod div **C** ≤ 0};{Grad C = $(\partial c/\partial x_1)\mathbf{i_1}$ + $(\partial c/\partial x_2)i_2$,…, + $(\partial c/\partial x_n)\mathbf{i_n}$ | ∞ < mod Grad C ≤ ◯, or mod Grad C > ◯}}}}/dKE}/dKE}/ … /d KE}

{d … {d{d {P{f{{light-speed-craft-k | 1 ≤ k ≤ ◯, or k >◯};{div **C** = $\partial \mathbf{c_1}/\partial x_1 + \partial \mathbf{c_2}/\partial x_2 + \dots + \partial \mathbf{c_n}/\partial x_n$ | mod div **C** ≤ 0};{Grad C = $(\partial c/\partial x_1)\mathbf{i_1}$ + $(\partial c/\partial x_2)i_2$,…, + $(\partial c/\partial x_n)\mathbf{i_n}$ | ∞ < mod Grad C ≤ ◯, or mod Grad C > ◯}}}}/dt}/dt}/ … /dt}

{d … {d{d {P{f{{light-speed-craft-k | 1 ≤ k ≤ ◯, or k >◯};{div **C** = $\partial \mathbf{c_1}/\partial x_1 + \partial \mathbf{c_2}/\partial x_2 + \dots + \partial \mathbf{c_n}/\partial x_n$ | mod div **C** ≤ 0};{Grad C = $(\partial c/\partial x_1)\mathbf{i_1}$ + $(\partial c/\partial x_2)i_2$,…, + $(\partial c/\partial x_n)\mathbf{i_n}$ | ∞ < mod Grad C ≤ ◯, or mod Grad C > ◯}}}}/dT}/dT}/ … /d T}

{d … {d{d {KE{f{{light-speed-craft-k | 1 ≤ k ≤ ◯, or k >◯};{div **C** = $\partial \mathbf{c_1}/\partial x_1 + \partial \mathbf{c_2}/\partial x_2 + \dots + \partial \mathbf{c_n}/\partial x_n$ | mod div **C** ≤ 0};{Grad C = $(\partial c/\partial x_1)\mathbf{i_1}$ + $(\partial c/\partial x_2)i_2$,…, + $(\partial c/\partial x_n)\mathbf{i_n}$ | ∞ < mod Grad C ≤ ◯, or mod Grad C > ◯}}}}/dc}/dc}/ … /d c}

{d … {d{d {KE {f{{light-speed-craft-k | 1 ≤ k ≤ ◯, or k >◯};{div **C** = $\partial \mathbf{c_1}/\partial x_1 + \partial \mathbf{c_2}/\partial x_2 + \dots + \partial \mathbf{c_n}/\partial x_n$ | mod div **C** ≤ 0};{Grad C = $(\partial c/\partial x_1)\mathbf{i_1}$ + $(\partial c/\partial x_2)i_2$,…, + $(\partial c/\partial x_n)\mathbf{i_n}$ | ∞ < mod Grad C ≤ ◯, or mod Grad C > ◯}}}}/dγ}/dγ}/ … /d γ}

{d … {d{d {KE {f{{light-speed-craft-k | 1 ≤ k ≤ ◯, or k >◯};{div **C** = $\partial \mathbf{c_1}/\partial x_1 + \partial \mathbf{c_2}/\partial x_2 + \dots + \partial \mathbf{c_n}/\partial x_n$ | mod div **C** ≤ 0};{Grad C = $(\partial c/\partial x_1)\mathbf{i_1}$ + $(\partial c/\partial x_2)i_2$,…, + $(\partial c/\partial x_n)\mathbf{i_n}$ | ∞ < mod Grad C ≤ ◯, or mod Grad C > ◯}}}}/dv}/dv}/ … /d v}

{d … {d{d {KE {f{{light-speed-craft-k | 1 ≤ k ≤ ◯, or k >◯};{div **C** = $\partial \mathbf{c_1}/\partial x_1 + \partial \mathbf{c_2}/\partial x_2 + \dots + \partial \mathbf{c_n}/\partial x_n$ | mod div **C** ≤ 0};{Grad C = $(\partial c/\partial x_1)\mathbf{i_1}$ + $(\partial c/\partial x_2)i_2$,…, + $(\partial c/\partial x_n)\mathbf{i_n}$ | ∞ < mod Grad C ≤ ◯, or mod Grad C > ◯}}}}/da}/da}/ … /d a} {d … {d{d {KE {f{{light-speed-craft-k | 1 ≤ k ≤ ◯, or k >◯};{div **C** = $\partial \mathbf{c_1}/\partial x_1 + \partial \mathbf{c_2}/\partial x_2 + \ldots + \partial \mathbf{c_n}/\partial x_n$ | mod div **C** ≤ 0};{Grad C = $(\partial c/\partial x_1)\mathbf{i_1} + (\partial c/\partial x_2)i_2,\ldots, + (\partial c/\partial x_n)\mathbf{i_n}$ **|** ∞ < mod Grad C ≤ ◯, or mod Grad C > ◯}}}}/dP}/dP}/ … /d P}

{d … {d{d {KE {f{{light-speed-craft-k | 1 ≤ k ≤ ◯, or k >◯};{div **C** = $\partial \mathbf{c_1}/\partial x_1 + \partial \mathbf{c_2}/\partial x_2 + \ldots + \partial \mathbf{c_n}/\partial x_n$ | mod div **C** ≤ 0};{Grad C = $(\partial c/\partial x_1)\mathbf{i_1} + (\partial c/\partial x_2)i_2,\ldots, + (\partial c/\partial x_n)\mathbf{i_n}$ **|** ∞ < mod Grad C ≤ ◯, or mod Grad C > ◯}}}}/dKE}/dKE}/ … /d KE}

{d … {d{d {KE {f{{light-speed-craft-k | 1 ≤ k ≤ ◯, or k >◯};{div **C** = $\partial \mathbf{c_1}/\partial x_1 + \partial \mathbf{c_2}/\partial x_2 + \ldots + \partial \mathbf{c_n}/\partial x_n$ | mod div **C** ≤ 0};{Grad C = $(\partial c/\partial x_1)\mathbf{i_1} + (\partial c/\partial x_2)i_2,\ldots, + (\partial c/\partial x_n)\mathbf{i_n}$ **|** ∞ < mod Grad C ≤ ◯, or mod Grad C > ◯}}}}/dt}/dt}/ … /dt}

{d … {d{d {KE {f{{light-speed-craft-k | 1 ≤ k ≤ ◯, or k >◯};{div **C** = $\partial \mathbf{c_1}/\partial x_1 + \partial \mathbf{c_2}/\partial x_2 + \ldots + \partial \mathbf{c_n}/\partial x_n$ | mod div **C** ≤ 0};{Grad C = $(\partial c/\partial x_1)\mathbf{i_1} + (\partial c/\partial x_2)i_2,\ldots, + (\partial c/\partial x_n)\mathbf{i_n}$ **|** ∞ < mod Grad C ≤ ◯, or mod Grad C > ◯}}}}/dT}/dT}/ … /d T}

{d … {d{d {t{f{{light-speed-craft-k | 1 ≤ k ≤ ◯, or k >◯};{div **C** = $\partial \mathbf{c_1}/\partial x_1 + \partial \mathbf{c_2}/\partial x_2 + \ldots + \partial \mathbf{c_n}/\partial x_n$ | mod div **C** ≤ 0};{Grad C = $(\partial c/\partial x_1)\mathbf{i_1} + (\partial c/\partial x_2)i_2,\ldots, + (\partial c/\partial x_n)\mathbf{i_n}$ **|** ∞ < mod Grad C ≤ ◯, or mod Grad C > ◯}}}}/dc}/dc}/ … /d c}

{d … {d{d {t{f{{light-speed-craft-k | 1 ≤ k ≤ ◯, or k >◯};{div **C** = $\partial \mathbf{c_1}/\partial x_1 + \partial \mathbf{c_2}/\partial x_2 + \ldots + \partial \mathbf{c_n}/\partial x_n$ | mod div **C** ≤ 0};{Grad C = $(\partial c/\partial x_1)\mathbf{i_1} + (\partial c/\partial x_2)i_2,\ldots, + (\partial c/\partial x_n)\mathbf{i_n}$ **|** ∞ < mod Grad C ≤ ◯, or mod Grad C > ◯}}}}/dγ}/dγ}/ … /d γ}

{d … {d{d {t{f{{light-speed-craft-k | 1 ≤ k ≤ ◯, or k >◯};{div **C** = $\partial \mathbf{c_1}/\partial x_1 + \partial \mathbf{c_2}/\partial x_2 + \ldots + \partial \mathbf{c_n}/\partial x_n$ | mod div **C** ≤ 0};{Grad C = $(\partial c/\partial x_1)\mathbf{i_1} + (\partial c/\partial x_2)i_2,\ldots, + (\partial c/\partial x_n)\mathbf{i_n}$ **|** ∞ < mod Grad C ≤ ◯, or mod Grad C > ◯}}}}/dv}/dv}/ … /d v}

{d … {d{d {t{f{{light-speed-craft-k | 1 ≤ k ≤ ◯, or k >◯};{div **C** = $\partial \mathbf{c_1}/\partial x_1 + \partial \mathbf{c_2}/\partial x_2 + \ldots + \partial \mathbf{c_n}/\partial x_n$ | mod div **C** ≤ 0};{Grad C = $(\partial c/\partial x_1)\mathbf{i_1} + (\partial c/\partial x_2)i_2,\ldots, + (\partial c/\partial x_n)\mathbf{i_n}$ **|** ∞ < mod Grad C ≤ ◯, or mod Grad C > ◯}}}}/da}/da}/ … /d a}

{d … {d{d {t{f{{light-speed-craft-k | 1 ≤ k ≤ ◯, or k >◯};{div **C** = $\partial \mathbf{c_1}/\partial x_1 + \partial \mathbf{c_2}/\partial x_2 + \ldots + \partial \mathbf{c_n}/\partial x_n$ | mod div **C** ≤ 0};{Grad C = $(\partial c/\partial x_1)\mathbf{i_1} + (\partial c/\partial x_2)i_2,\ldots, + (\partial c/\partial x_n)\mathbf{i_n}$ **|** ∞ < mod Grad C ≤ ◯, or mod Grad C > ◯}}}}/dP}/dP}/ … /d P}

{d … {d{d {t{f{{light-speed-craft-k | 1 ≤ k ≤ ◯, or k >◯};{div **C** = $\partial \mathbf{c_1}/\partial x_1 + \partial \mathbf{c_2}/\partial x_2 + \ldots + \partial \mathbf{c_n}/\partial x_n$ | mod div **C** ≤ 0};{Grad C = $(\partial c/\partial x_1)\mathbf{i_1} + (\partial c/\partial x_2)i_2,\ldots, + (\partial c/\partial x_n)\mathbf{i_n}$ **|** ∞ < mod Grad C ≤ ◯, or mod Grad C > ◯}}}}/dKE}/dKE}/ … /d KE} {d … {d{d {t{f{{light-speed-craft-k | 1 ≤ k ≤ ◯, or k >◯};{div **C** = $\partial \mathbf{c_1}/\partial x_1 + \partial \mathbf{c_2}/\partial x_2 + \ldots + \partial \mathbf{c_n}/\partial x_n$ | mod div **C** ≤ 0};{Grad C = $(\partial c/\partial x_1)\mathbf{i_1} + (\partial c/\partial x_2)i_2, \ldots, + (\partial c/\partial x_n)\mathbf{i_n}$ | ∞ < mod Grad C ≤ ◯, or mod Grad C > ◯}}}}/dt}/dt}/ … /dt}

{d … {d{d {t{f{{light-speed-craft-k | 1 ≤ k ≤ ◯, or k >◯};{div **C** = $\partial \mathbf{c_1}/\partial x_1 + \partial \mathbf{c_2}/\partial x_2 + \ldots + \partial \mathbf{c_n}/\partial x_n$ | mod div **C** ≤ 0};{Grad C = $(\partial c/\partial x_1)\mathbf{i_1} + (\partial c/\partial x_2)i_2, \ldots, + (\partial c/\partial x_n)\mathbf{i_n}$ | ∞ < mod Grad C ≤ ◯, or mod Grad C > ◯}}}}/dT}/dT}/ … /d T}

{d … {d{d {T{f{{light-speed-craft-k | 1 ≤ k ≤ ◯, or k >◯};{div **C** = $\partial \mathbf{c_1}/\partial x_1 + \partial \mathbf{c_2}/\partial x_2 + \ldots + \partial \mathbf{c_n}/\partial x_n$ | mod div **C** ≤ 0};{Grad C = $(\partial c/\partial x_1)\mathbf{i_1} + (\partial c/\partial x_2)i_2, \ldots, + (\partial c/\partial x_n)\mathbf{i_n}$ | ∞ < mod Grad C ≤ ◯, or mod Grad C > ◯}}}}/dc}/dc}/ … /d c}

{d … {d{d {T{f{{light-speed-craft-k | 1 ≤ k ≤ ◯, or k >◯};{div **C** = $\partial \mathbf{c_1}/\partial x_1 + \partial \mathbf{c_2}/\partial x_2 + \ldots + \partial \mathbf{c_n}/\partial x_n$ | mod div **C** ≤ 0};{Grad C = $(\partial c/\partial x_1)\mathbf{i_1} + (\partial c/\partial x_2)i_2, \ldots, + (\partial c/\partial x_n)\mathbf{i_n}$ | ∞ < mod Grad C ≤ ◯, or mod Grad C > ◯}}}}/dγ}/dγ}/ … /d γ}

{d … {d{d {T{f{{light-speed-craft-k | 1 ≤ k ≤ ◯, or k >◯};{div **C** = $\partial \mathbf{c_1}/\partial x_1 + \partial \mathbf{c_2}/\partial x_2 + \ldots + \partial \mathbf{c_n}/\partial x_n$ | mod div **C** ≤ 0};{Grad C = $(\partial c/\partial x_1)\mathbf{i_1} + (\partial c/\partial x_2)i_2, \ldots, + (\partial c/\partial x_n)\mathbf{i_n}$ | ∞ < mod Grad C ≤ ◯, or mod Grad C > ◯}}}}/dv}/dv}/ … /d v}

{d … {d{d {T{f{{light-speed-craft-k | 1 ≤ k ≤ ◯, or k >◯};{div **C** = $\partial \mathbf{c_1}/\partial x_1 + \partial \mathbf{c_2}/\partial x_2 + \ldots + \partial \mathbf{c_n}/\partial x_n$ | mod div **C** ≤ 0};{Grad C = $(\partial c/\partial x_1)\mathbf{i_1} + (\partial c/\partial x_2)i_2, \ldots, + (\partial c/\partial x_n)\mathbf{i_n}$ | ∞ < mod Grad C ≤ ◯, or mod Grad C > ◯}}}}/da}/da}/ … /d a}

{d … {d{d {T{f{{light-speed-craft-k | 1 ≤ k ≤ ◯, or k >◯};{div **C** = $\partial \mathbf{c_1}/\partial x_1 + \partial \mathbf{c_2}/\partial x_2 + \ldots + \partial \mathbf{c_n}/\partial x_n$ | mod div **C** ≤ 0};{Grad C = $(\partial c/\partial x_1)\mathbf{i_1} + (\partial c/\partial x_2)i_2, \ldots, + (\partial c/\partial x_n)\mathbf{i_n}$ | ∞ < mod Grad C ≤ ◯, or mod Grad C > ◯}}}}/dP}/dP}/ … /d P}

{d … {d{d {T{f{{light-speed-craft-k | 1 ≤ k ≤ ◯, or k >◯};{div **C** = $\partial \mathbf{c_1}/\partial x_1 + \partial \mathbf{c_2}/\partial x_2 + \ldots + \partial \mathbf{c_n}/\partial x_n$ | mod div **C** ≤ 0};{Grad C = $(\partial c/\partial x_1)\mathbf{i_1} + (\partial c/\partial x_2)i_2, \ldots, + (\partial c/\partial x_n)\mathbf{i_n}$ | ∞ < mod Grad C ≤ ◯, or mod Grad C > ◯}}}}/dKE}/dKE}/ … /d KE}

{d … {d{d {T{f{{light-speed-craft-k | 1 ≤ k ≤ ◯, or k >◯};{div **C** = $\partial \mathbf{c_1}/\partial x_1 + \partial \mathbf{c_2}/\partial x_2 + \ldots + \partial \mathbf{c_n}/\partial x_n$ | mod div **C** ≤ 0};{Grad C = $(\partial c/\partial x_1)\mathbf{i_1} + (\partial c/\partial x_2)i_2, \ldots, + (\partial c/\partial x_n)\mathbf{i_n}$ | ∞ < mod Grad C ≤ ◯, or mod Grad C > ◯}}}}/dt}/dt}/ … /dt}

{d … {d{d {T{f{{light-speed-craft-k | 1 ≤ k ≤ ◯, or k >◯};{div **C** = $\partial \mathbf{c_1}/\partial x_1 + \partial \mathbf{c_2}/\partial x_2 + \ldots + \partial \mathbf{c_n}/\partial x_n$ | mod div **C** ≤ 0};{Grad C = $(\partial c/\partial x_1)\mathbf{i_1} + (\partial c/\partial x_2)i_2, \ldots, + (\partial c/\partial x_n)\mathbf{i_n}$ | ∞ < mod Grad C ≤ ◯, or mod Grad C > ◯}}}}/dT}/dT}/ … /d T}

and so on.

We can take light-speed, gamma, velocity, acceleration, momentum, kinetic energy, and so on integrals of the basic first operator as functionally modified to represent ship and background time, light-speed, gamma, velocity, acceleration, momentum, kinetic energy, and so on. We may affix the resulting integral operators to formulas for gamma and related parameters to denote the operative mechanisms without actually including these operators in the formulas.

{∫ … {∫∫∫ {c{f{{light-speed-craft-k | $1 \le k \le \bigcirc$, or $k > \bigcirc$};{div **C** = $\partial \mathbf{c}_1/\partial x_1 + \partial \mathbf{c}_2/\partial x_2 + \ldots + \partial \mathbf{c}_n/\partial x_n$ | mod div **C** ≤ 0};{Grad C = $(\partial c/\partial x_1)\mathbf{i}_1 + (\partial c/\partial x_2)i_2, \ldots, + (\partial c/\partial x_n)\mathbf{i}_n$ | $\infty <$ mod Grad C $\le \bigcirc$, or mod Grad C $> \bigcirc$}}}}dc}dc} … d c}

{∫ … {∫∫∫ {c{f{{light-speed-craft-k | $1 \le k \le \bigcirc$, or $k > \bigcirc$};{div **C** = $\partial \mathbf{c}_1/\partial x_1 + \partial \mathbf{c}_2/\partial x_2 + \ldots + \partial \mathbf{c}_n/\partial x_n$ | mod div **C** ≤ 0};{Grad C = $(\partial c/\partial x_1)\mathbf{i}_1 + (\partial c/\partial x_2)i_2, \ldots, + (\partial c/\partial x_n)\mathbf{i}_n$ | $\infty <$ mod Grad C $\le \bigcirc$, or mod Grad C $> \bigcirc$}}}}dγ}dγ} … d γ}

{∫ … {∫∫∫ {c{f{{light-speed-craft-k | $1 \le k \le \bigcirc$, or $k > \bigcirc$};{div **C** = $\partial \mathbf{c}_1/\partial x_1 + \partial \mathbf{c}_2/\partial x_2 + \ldots + \partial \mathbf{c}_n/\partial x_n$ | mod div **C** ≤ 0};{Grad C = $(\partial c/\partial x_1)\mathbf{i}_1 + (\partial c/\partial x_2)i_2, \ldots, + (\partial c/\partial x_n)\mathbf{i}_n$ | $\infty <$ mod Grad C $\le \bigcirc$, or mod Grad C $> \bigcirc$}}}}dv}dv} … d v}

{∫ … {∫∫∫ {c{f{{light-speed-craft-k | $1 \le k \le \bigcirc$, or $k > \bigcirc$};{div **C** = $\partial \mathbf{c}_1/\partial x_1 + \partial \mathbf{c}_2/\partial x_2 + \ldots + \partial \mathbf{c}_n/\partial x_n$ | mod div **C** ≤ 0};{Grad C = $(\partial c/\partial x_1)\mathbf{i}_1 + (\partial c/\partial x_2)i_2, \ldots, + (\partial c/\partial x_n)\mathbf{i}_n$ | $\infty <$ mod Grad C $\le \bigcirc$, or mod Grad C $> \bigcirc$}}}}da}da} … d a}

{∫ … {∫∫∫ {c{f{{light-speed-craft-k | $1 \le k \le \bigcirc$, or $k > \bigcirc$};{div **C** = $\partial \mathbf{c}_1/\partial x_1 + \partial \mathbf{c}_2/\partial x_2 + \ldots + \partial \mathbf{c}_n/\partial x_n$ | mod div **C** ≤ 0};{Grad C = $(\partial c/\partial x_1)\mathbf{i}_1 + (\partial c/\partial x_2)i_2, \ldots, + (\partial c/\partial x_n)\mathbf{i}_n$ | $\infty <$ mod Grad C $\le \bigcirc$, or mod Grad C $> \bigcirc$}}}}dP}dP} … d P}

{∫ … {∫∫∫ {c{f{{light-speed-craft-k | $1 \le k \le \bigcirc$, or $k > \bigcirc$};{div **C** = $\partial \mathbf{c}_1/\partial x_1 + \partial \mathbf{c}_2/\partial x_2 + \ldots + \partial \mathbf{c}_n/\partial x_n$ | mod div **C** ≤ 0};{Grad C = $(\partial c/\partial x_1)\mathbf{i}_1 + (\partial c/\partial x_2)i_2, \ldots, + (\partial c/\partial x_n)\mathbf{i}_n$ | $\infty <$ mod Grad C $\le \bigcirc$, or mod Grad C $> \bigcirc$}}}}dKE}dKE} … d KE}

{∫ … {∫∫∫ {c{f{{light-speed-craft-k | $1 \le k \le \bigcirc$, or $k > \bigcirc$};{div **C** = $\partial \mathbf{c}_1/\partial x_1 + \partial \mathbf{c}_2/\partial x_2 + \ldots + \partial \mathbf{c}_n/\partial x_n$ | mod div **C** ≤ 0};{Grad C = $(\partial c/\partial x_1)\mathbf{i}_1 + (\partial c/\partial x_2)i_2, \ldots, + (\partial c/\partial x_n)\mathbf{i}_n$ | $\infty <$ mod Grad C $\le \bigcirc$, or mod Grad C $> \bigcirc$}}}}dt}dt} ... dt}

{∫ … {∫∫∫ {c{f{{light-speed-craft-k | $1 \le k \le \bigcirc$, or $k > \bigcirc$};{div **C** = $\partial \mathbf{c}_1/\partial x_1 + \partial \mathbf{c}_2/\partial x_2 + \ldots + \partial \mathbf{c}_n/\partial x_n$ | mod div **C** ≤ 0};{Grad C = $(\partial c/\partial x_1)\mathbf{i}_1 + (\partial c/\partial x_2)i_2, \ldots, + (\partial c/\partial x_n)\mathbf{i}_n$ | $\infty <$ mod Grad C $\le \bigcirc$, or mod Grad C $> \bigcirc$}}}}dT}dT} … d T}

{∫ … {∫∫∫ {γ{f{{light-speed-craft-k | $1 \le k \le \bigcirc$, or $k > \bigcirc$};{div **C** = $\partial \mathbf{c}_1/\partial x_1 + \partial \mathbf{c}_2/\partial x_2 + \ldots + \partial \mathbf{c}_n/\partial x_n$ | mod div **C** ≤ 0};{Grad C = $(\partial c/\partial x_1)\mathbf{i}_1 + (\partial c/\partial x_2)i_2, \ldots, + (\partial c/\partial x_n)\mathbf{i}_n$ | $\infty <$ mod Grad C $\le \bigcirc$, or mod Grad C $> \bigcirc$}}}}dc}dc} … d c}

{∫ … {∫∫∫ {γ{f{{light-speed-craft-k | 1 ≤ k ≤ ◯, or k > ◯};{div **C** = $\partial \mathbf{c_1}/\partial x_1 + \partial \mathbf{c_2}/\partial x_2 + \ldots + \partial \mathbf{c_n}/\partial x_n$ | mod div **C** ≤ 0};{Grad C = $(\partial c/\partial x_1)\mathbf{i_1} + (\partial c/\partial x_2)i_2, \ldots, + (\partial c/\partial x_n)\mathbf{i_n}$ | ∞ < mod Grad C ≤ ◯, or mod Grad C > ◯}}}}dγ}dγ} … d γ}

{∫ … {∫∫∫ {γ{f{{light-speed-craft-k | 1 ≤ k ≤ ◯, or k > ◯};{div **C** = $\partial \mathbf{c_1}/\partial x_1 + \partial \mathbf{c_2}/\partial x_2 + \ldots + \partial \mathbf{c_n}/\partial x_n$ | mod div **C** ≤ 0};{Grad C = $(\partial c/\partial x_1)\mathbf{i_1} + (\partial c/\partial x_2)i_2, \ldots, + (\partial c/\partial x_n)\mathbf{i_n}$ | ∞ < mod Grad C ≤ ◯, or mod Grad C > ◯}}}}dv}dv} … d v}

{∫ … {∫∫∫ {γ{f{{light-speed-craft-k | 1 ≤ k ≤ ◯, or k > ◯};{div **C** = $\partial \mathbf{c_1}/\partial x_1 + \partial \mathbf{c_2}/\partial x_2 + \ldots + \partial \mathbf{c_n}/\partial x_n$ | mod div **C** ≤ 0};{Grad C = $(\partial c/\partial x_1)\mathbf{i_1} + (\partial c/\partial x_2)i_2, \ldots, + (\partial c/\partial x_n)\mathbf{i_n}$ | ∞ < mod Grad C ≤ ◯, or mod Grad C > ◯}}}}da}da} … d a}

{∫ … {∫∫∫ {γ{f{{light-speed-craft-k | 1 ≤ k ≤ ◯, or k > ◯};{div **C** = $\partial \mathbf{c_1}/\partial x_1 + \partial \mathbf{c_2}/\partial x_2 + \ldots + \partial \mathbf{c_n}/\partial x_n$ | mod div **C** ≤ 0};{Grad C = $(\partial c/\partial x_1)\mathbf{i_1} + (\partial c/\partial x_2)i_2, \ldots, + (\partial c/\partial x_n)\mathbf{i_n}$ | ∞ < mod Grad C ≤ ◯, or mod Grad C > ◯}}}}dP}dP} … d P}

{∫ … {∫∫∫ {γ{f{{light-speed-craft-k | 1 ≤ k ≤ ◯, or k > ◯};{div **C** = $\partial \mathbf{c_1}/\partial x_1 + \partial \mathbf{c_2}/\partial x_2 + \ldots + \partial \mathbf{c_n}/\partial x_n$ | mod div **C** ≤ 0};{Grad C = $(\partial c/\partial x_1)\mathbf{i_1} + (\partial c/\partial x_2)i_2, \ldots, + (\partial c/\partial x_n)\mathbf{i_n}$ | ∞ < mod Grad C ≤ ◯, or mod Grad C > ◯}}}}dKE}dKE} … d KE}

{∫ … {∫∫∫ {γ{f{{light-speed-craft-k | 1 ≤ k ≤ ◯, or k > ◯};{div **C** = $\partial \mathbf{c_1}/\partial x_1 + \partial \mathbf{c_2}/\partial x_2 + \ldots + \partial \mathbf{c_n}/\partial x_n$ | mod div **C** ≤ 0};{Grad C = $(\partial c/\partial x_1)\mathbf{i_1} + (\partial c/\partial x_2)i_2, \ldots, + (\partial c/\partial x_n)\mathbf{i_n}$ | ∞ < mod Grad C ≤ ◯, or mod Grad C > ◯}}}}dt}dt} ... dt}

{∫ … {∫∫∫ {γ{f{{light-speed-craft-k | 1 ≤ k ≤ ◯, or k > ◯};{div **C** = $\partial \mathbf{c_1}/\partial x_1 + \partial \mathbf{c_2}/\partial x_2 + \ldots + \partial \mathbf{c_n}/\partial x_n$ | mod div **C** ≤ 0};{Grad C = $(\partial c/\partial x_1)\mathbf{i_1} + (\partial c/\partial x_2)i_2, \ldots, + (\partial c/\partial x_n)\mathbf{i_n}$ | ∞ < mod Grad C ≤ ◯, or mod Grad C > ◯}}}}dT}dT} … d T}

{∫ … {∫∫∫ {v{f{{light-speed-craft-k | 1 ≤ k ≤ ◯, or k > ◯};{div **C** = $\partial \mathbf{c_1}/\partial x_1 + \partial \mathbf{c_2}/\partial x_2 + \ldots + \partial \mathbf{c_n}/\partial x_n$ | mod div **C** ≤ 0};{Grad C = $(\partial c/\partial x_1)\mathbf{i_1} + (\partial c/\partial x_2)i_2, \ldots, + (\partial c/\partial x_n)\mathbf{i_n}$ | ∞ < mod Grad C ≤ ◯, or mod Grad C > ◯}}}}dc}dc} … d c}

{∫ … {∫∫∫ {v{f{{light-speed-craft-k | 1 ≤ k ≤ ◯, or k > ◯};{div **C** = $\partial \mathbf{c_1}/\partial x_1 + \partial \mathbf{c_2}/\partial x_2 + \ldots + \partial \mathbf{c_n}/\partial x_n$ | mod div **C** ≤ 0};{Grad C = $(\partial c/\partial x_1)\mathbf{i_1} + (\partial c/\partial x_2)i_2, \ldots, + (\partial c/\partial x_n)\mathbf{i_n}$ | ∞ < mod Grad C ≤ ◯, or mod Grad C > ◯}}}}dγ}dγ} … d γ}

{∫ … {∫∫∫ {v{f{{light-speed-craft-k | 1 ≤ k ≤ ◯, or k > ◯};{div **C** = $\partial \mathbf{c_1}/\partial x_1 + \partial \mathbf{c_2}/\partial x_2 + \ldots + \partial \mathbf{c_n}/\partial x_n$ | mod div **C** ≤ 0};{Grad C = $(\partial c/\partial x_1)\mathbf{i_1} + (\partial c/\partial x_2)i_2, \ldots, + (\partial c/\partial x_n)\mathbf{i_n}$ | ∞ < mod Grad C ≤ ◯, or mod Grad C > ◯}}}}dv}dv} … d v}

{∫ … {∫{∫ {v{f{{light-speed-craft-k | $1 \leq k \leq \bigcirc$, or $k > \bigcirc$};{div **C** = $\partial \mathbf{c_1}/\partial x_1 + \partial \mathbf{c_2}/\partial x_2 + \ldots + \partial \mathbf{c_n}/\partial x_n$ | mod div **C** ≤ 0};{Grad C = $(\partial c/\partial x_1)\mathbf{i_1} + (\partial c/\partial x_2)i_2, \ldots, + (\partial c/\partial x_n)\mathbf{i_n}$ | $\infty <$ mod Grad C ≤ $\bigcirc$, or mod Grad C > $\bigcirc$}}}}da}da} … d a}

{∫ … {∫{∫ {v{f{{light-speed-craft-k | $1 \leq k \leq \bigcirc$, or $k > \bigcirc$};{div **C** = $\partial \mathbf{c_1}/\partial x_1 + \partial \mathbf{c_2}/\partial x_2 + \ldots + \partial \mathbf{c_n}/\partial x_n$ | mod div **C** ≤ 0};{Grad C = $(\partial c/\partial x_1)\mathbf{i_1} + (\partial c/\partial x_2)i_2, \ldots, + (\partial c/\partial x_n)\mathbf{i_n}$ | $\infty <$ mod Grad C ≤ $\bigcirc$, or mod Grad C > $\bigcirc$}}}}dP}dP} … d P}

{∫ … {∫{∫ {v{f{{light-speed-craft-k | $1 \leq k \leq \bigcirc$, or $k > \bigcirc$};{div **C** = $\partial \mathbf{c_1}/\partial x_1 + \partial \mathbf{c_2}/\partial x_2 + \ldots + \partial \mathbf{c_n}/\partial x_n$ | mod div **C** ≤ 0};{Grad C = $(\partial c/\partial x_1)\mathbf{i_1} + (\partial c/\partial x_2)i_2, \ldots, + (\partial c/\partial x_n)\mathbf{i_n}$ | $\infty <$ mod Grad C ≤ $\bigcirc$, or mod Grad C > $\bigcirc$}}}}dKE}dKE} … d KE}

{∫ … {∫{∫ {v{f{{light-speed-craft-k | $1 \leq k \leq \bigcirc$, or $k > \bigcirc$};{div **C** = $\partial \mathbf{c_1}/\partial x_1 + \partial \mathbf{c_2}/\partial x_2 + \ldots + \partial \mathbf{c_n}/\partial x_n$ | mod div **C** ≤ 0};{Grad C = $(\partial c/\partial x_1)\mathbf{i_1} + (\partial c/\partial x_2)i_2, \ldots, + (\partial c/\partial x_n)\mathbf{i_n}$ | $\infty <$ mod Grad C ≤ $\bigcirc$, or mod Grad C > $\bigcirc$}}}}dt}dt} ... dt}

{∫ … {∫{∫ {v{f{{light-speed-craft-k | $1 \leq k \leq \bigcirc$, or $k > \bigcirc$};{div **C** = $\partial \mathbf{c_1}/\partial x_1 + \partial \mathbf{c_2}/\partial x_2 + \ldots + \partial \mathbf{c_n}/\partial x_n$ | mod div **C** ≤ 0};{Grad C = $(\partial c/\partial x_1)\mathbf{i_1} + (\partial c/\partial x_2)i_2, \ldots, + (\partial c/\partial x_n)\mathbf{i_n}$ | $\infty <$ mod Grad C ≤ $\bigcirc$, or mod Grad C > $\bigcirc$}}}}dT}dT} … d T}

{∫ … {∫{∫ {a{f{{light-speed-craft-k | $1 \leq k \leq \bigcirc$, or $k > \bigcirc$};{div **C** = $\partial \mathbf{c_1}/\partial x_1 + \partial \mathbf{c_2}/\partial x_2 + \ldots + \partial \mathbf{c_n}/\partial x_n$ | mod div **C** ≤ 0};{Grad C = $(\partial c/\partial x_1)\mathbf{i_1} + (\partial c/\partial x_2)i_2, \ldots, + (\partial c/\partial x_n)\mathbf{i_n}$ | $\infty <$ mod Grad C ≤ $\bigcirc$, or mod Grad C > $\bigcirc$}}}}dc}dc} … d c}

{∫ … {∫{∫ {a{f{{light-speed-craft-k | $1 \leq k \leq \bigcirc$, or $k > \bigcirc$};{div **C** = $\partial \mathbf{c_1}/\partial x_1 + \partial \mathbf{c_2}/\partial x_2 + \ldots + \partial \mathbf{c_n}/\partial x_n$ | mod div **C** ≤ 0};{Grad C = $(\partial c/\partial x_1)\mathbf{i_1} + (\partial c/\partial x_2)i_2, \ldots, + (\partial c/\partial x_n)\mathbf{i_n}$ | $\infty <$ mod Grad C ≤ $\bigcirc$, or mod Grad C > $\bigcirc$}}}}dγ}dγ} … d γ}

{∫ … {∫{∫ {a{f{{light-speed-craft-k | $1 \leq k \leq \bigcirc$, or $k > \bigcirc$};{div **C** = $\partial \mathbf{c_1}/\partial x_1 + \partial \mathbf{c_2}/\partial x_2 + \ldots + \partial \mathbf{c_n}/\partial x_n$ | mod div **C** ≤ 0};{Grad C = $(\partial c/\partial x_1)\mathbf{i_1} + (\partial c/\partial x_2)i_2, \ldots, + (\partial c/\partial x_n)\mathbf{i_n}$ | $\infty <$ mod Grad C ≤ $\bigcirc$, or mod Grad C > $\bigcirc$}}}}dv}dv} … d v}

{∫ … {∫{∫ {a{f{{light-speed-craft-k | $1 \leq k \leq \bigcirc$, or $k > \bigcirc$};{div **C** = $\partial \mathbf{c_1}/\partial x_1 + \partial \mathbf{c_2}/\partial x_2 + \ldots + \partial \mathbf{c_n}/\partial x_n$ | mod div **C** ≤ 0};{Grad C = $(\partial c/\partial x_1)\mathbf{i_1} + (\partial c/\partial x_2)i_2, \ldots, + (\partial c/\partial x_n)\mathbf{i_n}$ | $\infty <$ mod Grad C ≤ $\bigcirc$, or mod Grad C > $\bigcirc$}}}}da}da} … d a}

{∫ … {∫{∫ {a{f{{light-speed-craft-k | $1 \leq k \leq \bigcirc$, or $k > \bigcirc$};{div **C** = $\partial \mathbf{c_1}/\partial x_1 + \partial \mathbf{c_2}/\partial x_2 + \ldots + \partial \mathbf{c_n}/\partial x_n$ | mod div **C** ≤ 0};{Grad C = $(\partial c/\partial x_1)\mathbf{i_1} + (\partial c/\partial x_2)i_2, \ldots, + (\partial c/\partial x_n)\mathbf{i_n}$ | $\infty <$ mod Grad C ≤ $\bigcirc$, or mod Grad C > $\bigcirc$}}}}dP}dP} … d P}

{∫ … {∫{∫ {a{f{{light-speed-craft-k | 1 ≤ k ≤ ◯, or k > ◯};{div **C** = $\partial \mathbf{c_1}/\partial x_1 + \partial \mathbf{c_2}/\partial x_2 + \ldots + \partial \mathbf{c_n}/\partial x_n$ | mod div **C** ≤ 0};{Grad C = $(\partial c/\partial x_1)\mathbf{i_1} + (\partial c/\partial x_2)i_2, \ldots, + (\partial c/\partial x_n)\mathbf{i_n}$ | ∞ < mod Grad C ≤ ◯, or mod Grad C > ◯}}}}dKE}dKE} … d KE}

{∫ … {∫{∫ {a{f{{light-speed-craft-k | 1 ≤ k ≤ ◯, or k > ◯};{div **C** = $\partial \mathbf{c_1}/\partial x_1 + \partial \mathbf{c_2}/\partial x_2 + \ldots + \partial \mathbf{c_n}/\partial x_n$ | mod div **C** ≤ 0};{Grad C = $(\partial c/\partial x_1)\mathbf{i_1} + (\partial c/\partial x_2)i_2, \ldots, + (\partial c/\partial x_n)\mathbf{i_n}$ | ∞ < mod Grad C ≤ ◯, or mod Grad C > ◯}}}}dt}dt} ... dt}

{∫ … {∫{∫ {a{f{{light-speed-craft-k | 1 ≤ k ≤ ◯, or k > ◯};{div **C** = $\partial \mathbf{c_1}/\partial x_1 + \partial \mathbf{c_2}/\partial x_2 + \ldots + \partial \mathbf{c_n}/\partial x_n$ | mod div **C** ≤ 0};{Grad C = $(\partial c/\partial x_1)\mathbf{i_1} + (\partial c/\partial x_2)i_2, \ldots, + (\partial c/\partial x_n)\mathbf{i_n}$ | ∞ < mod Grad C ≤ ◯, or mod Grad C > ◯}}}}dT}dT} … d T}

{∫ … {∫{∫ {P{f{{light-speed-craft-k | 1 ≤ k ≤ ◯, or k > ◯};{div **C** = $\partial \mathbf{c_1}/\partial x_1 + \partial \mathbf{c_2}/\partial x_2 + \ldots + \partial \mathbf{c_n}/\partial x_n$ | mod div **C** ≤ 0};{Grad C = $(\partial c/\partial x_1)\mathbf{i_1} + (\partial c/\partial x_2)i_2, \ldots, + (\partial c/\partial x_n)\mathbf{i_n}$ | ∞ < mod Grad C ≤ ◯, or mod Grad C > ◯}}}}dc}dc} … d c}

{∫ … {∫{∫ {P{f{{light-speed-craft-k | 1 ≤ k ≤ ◯, or k > ◯};{div **C** = $\partial \mathbf{c_1}/\partial x_1 + \partial \mathbf{c_2}/\partial x_2 + \ldots + \partial \mathbf{c_n}/\partial x_n$ | mod div **C** ≤ 0};{Grad C = $(\partial c/\partial x_1)\mathbf{i_1} + (\partial c/\partial x_2)i_2, \ldots, + (\partial c/\partial x_n)\mathbf{i_n}$ | ∞ < mod Grad C ≤ ◯, or mod Grad C > ◯}}}}dγ}dγ} … d γ}

{∫ … {∫{∫ {P{f{{light-speed-craft-k | 1 ≤ k ≤ ◯, or k > ◯};{div **C** = $\partial \mathbf{c_1}/\partial x_1 + \partial \mathbf{c_2}/\partial x_2 + \ldots + \partial \mathbf{c_n}/\partial x_n$ | mod div **C** ≤ 0};{Grad C = $(\partial c/\partial x_1)\mathbf{i_1} + (\partial c/\partial x_2)i_2, \ldots, + (\partial c/\partial x_n)\mathbf{i_n}$ | ∞ < mod Grad C ≤ ◯, or mod Grad C > ◯}}}}dv}dv} … d v}

{∫ … {∫{∫ {P{f{{light-speed-craft-k | 1 ≤ k ≤ ◯, or k > ◯};{div **C** = $\partial \mathbf{c_1}/\partial x_1 + \partial \mathbf{c_2}/\partial x_2 + \ldots + \partial \mathbf{c_n}/\partial x_n$ | mod div **C** ≤ 0};{Grad C = $(\partial c/\partial x_1)\mathbf{i_1} + (\partial c/\partial x_2)i_2, \ldots, + (\partial c/\partial x_n)\mathbf{i_n}$ | ∞ < mod Grad C ≤ ◯, or mod Grad C > ◯}}}}da}da} … d a}

{∫ … {∫{∫ {P{f{{light-speed-craft-k | 1 ≤ k ≤ ◯, or k > ◯};{div **C** = $\partial \mathbf{c_1}/\partial x_1 + \partial \mathbf{c_2}/\partial x_2 + \ldots + \partial \mathbf{c_n}/\partial x_n$ | mod div **C** ≤ 0};{Grad C = $(\partial c/\partial x_1)\mathbf{i_1} + (\partial c/\partial x_2)i_2, \ldots, + (\partial c/\partial x_n)\mathbf{i_n}$ | ∞ < mod Grad C ≤ ◯, or mod Grad C > ◯}}}}dP}dP} … d P}

{∫ … {∫{∫ {P{f{{light-speed-craft-k | 1 ≤ k ≤ ◯, or k > ◯};{div **C** = $\partial \mathbf{c_1}/\partial x_1 + \partial \mathbf{c_2}/\partial x_2 + \ldots + \partial \mathbf{c_n}/\partial x_n$ | mod div **C** ≤ 0};{Grad C = $(\partial c/\partial x_1)\mathbf{i_1} + (\partial c/\partial x_2)i_2, \ldots, + (\partial c/\partial x_n)\mathbf{i_n}$ | ∞ < mod Grad C ≤ ◯, or mod Grad C > ◯}}}}dKE}dKE} … d KE}

{∫ … {∫{∫ {P{f{{light-speed-craft-k | 1 ≤ k ≤ ◯, or k > ◯};{div **C** = $\partial \mathbf{c_1}/\partial x_1 + \partial \mathbf{c_2}/\partial x_2 + \ldots + \partial \mathbf{c_n}/\partial x_n$ | mod div **C** ≤ 0};{Grad C = $(\partial c/\partial x_1)\mathbf{i_1} + (\partial c/\partial x_2)i_2, \ldots, + (\partial c/\partial x_n)\mathbf{i_n}$ | ∞ < mod Grad C ≤ ◯, or mod Grad C > ◯}}}}dt}dt} ... dt}

{∫ … {∫∫∫ {P{f{{light-speed-craft-k | 1 ≤ k ≤ ◯, or k > ◯};{div **C** = $\partial \mathbf{c}_1/\partial x_1 + \partial \mathbf{c}_2/\partial x_2 + \dots + \partial \mathbf{c}_n/\partial x_n$ | mod div **C** ≤ 0};{Grad C = $(\partial c/\partial x_1)\mathbf{i}_1 + (\partial c/\partial x_2)i_2, \dots, + (\partial c/\partial x_n)\mathbf{i}_n$ | ∞ < mod Grad C ≤ ◯, or mod Grad C > ◯}}}}dT}dT} … d T}

{∫ … {∫∫∫ {KE{f{{light-speed-craft-k | 1 ≤ k ≤ ◯, or k > ◯};{div **C** = $\partial \mathbf{c}_1/\partial x_1 + \partial \mathbf{c}_2/\partial x_2 + \dots + \partial \mathbf{c}_n/\partial x_n$ | mod div **C** ≤ 0};{Grad C = $(\partial c/\partial x_1)\mathbf{i}_1 + (\partial c/\partial x_2)i_2, \dots, + (\partial c/\partial x_n)\mathbf{i}_n$ | ∞ < mod Grad C ≤ ◯, or mod Grad C > ◯}}}}dc}dc} … d c}

{∫ … {∫∫∫ {KE {f{{light-speed-craft-k | 1 ≤ k ≤ ◯, or k > ◯};{div **C** = $\partial \mathbf{c}_1/\partial x_1 + \partial \mathbf{c}_2/\partial x_2 + \dots + \partial \mathbf{c}_n/\partial x_n$ | mod div **C** ≤ 0};{Grad C = $(\partial c/\partial x_1)\mathbf{i}_1 + (\partial c/\partial x_2)i_2, \dots, + (\partial c/\partial x_n)\mathbf{i}_n$ | ∞ < mod Grad C ≤ ◯, or mod Grad C > ◯}}}}dγ}dγ} … d γ}

{∫ … {∫∫∫ {KE {f{{light-speed-craft-k | 1 ≤ k ≤ ◯, or k > ◯};{div **C** = $\partial \mathbf{c}_1/\partial x_1 + \partial \mathbf{c}_2/\partial x_2 + \dots + \partial \mathbf{c}_n/\partial x_n$ | mod div **C** ≤ 0};{Grad C = $(\partial c/\partial x_1)\mathbf{i}_1 + (\partial c/\partial x_2)i_2, \dots, + (\partial c/\partial x_n)\mathbf{i}_n$ | ∞ < mod Grad C ≤ ◯, or mod Grad C > ◯}}}}dv}dv} … d v}

{∫ … {∫∫∫ {KE {f{{light-speed-craft-k | 1 ≤ k ≤ ◯, or k > ◯};{div **C** = $\partial \mathbf{c}_1/\partial x_1 + \partial \mathbf{c}_2/\partial x_2 + \dots + \partial \mathbf{c}_n/\partial x_n$ | mod div **C** ≤ 0};{Grad C = $(\partial c/\partial x_1)\mathbf{i}_1 + (\partial c/\partial x_2)i_2, \dots, + (\partial c/\partial x_n)\mathbf{i}_n$ | ∞ < mod Grad C ≤ ◯, or mod Grad C > ◯}}}}da}da} … d a}

{∫ … {∫∫∫ {KE {f{{light-speed-craft-k | 1 ≤ k ≤ ◯, or k > ◯};{div **C** = $\partial \mathbf{c}_1/\partial x_1 + \partial \mathbf{c}_2/\partial x_2 + \dots + \partial \mathbf{c}_n/\partial x_n$ | mod div **C** ≤ 0};{Grad C = $(\partial c/\partial x_1)\mathbf{i}_1 + (\partial c/\partial x_2)i_2, \dots, + (\partial c/\partial x_n)\mathbf{i}_n$ | ∞ < mod Grad C ≤ ◯, or mod Grad C > ◯}}}}dP}dP} … d P}

{∫ … {∫∫∫ {KE {f{{light-speed-craft-k | 1 ≤ k ≤ ◯, or k > ◯};{div **C** = $\partial \mathbf{c}_1/\partial x_1 + \partial \mathbf{c}_2/\partial x_2 + \dots + \partial \mathbf{c}_n/\partial x_n$ | mod div **C** ≤ 0};{Grad C = $(\partial c/\partial x_1)\mathbf{i}_1 + (\partial c/\partial x_2)i_2, \dots, + (\partial c/\partial x_n)\mathbf{i}_n$ | ∞ < mod Grad C ≤ ◯, or mod Grad C > ◯}}}}dKE}dKE} … d KE}

{∫ … {∫∫∫ {KE {f{{light-speed-craft-k | 1 ≤ k ≤ ◯, or k > ◯};{div **C** = $\partial \mathbf{c}_1/\partial x_1 + \partial \mathbf{c}_2/\partial x_2 + \dots + \partial \mathbf{c}_n/\partial x_n$ | mod div **C** ≤ 0};{Grad C = $(\partial c/\partial x_1)\mathbf{i}_1 + (\partial c/\partial x_2)i_2, \dots, + (\partial c/\partial x_n)\mathbf{i}_n$ | ∞ < mod Grad C ≤ ◯, or mod Grad C > ◯}}}}dt}dt} ... dt}

{∫ … {∫∫∫ {KE {f{{light-speed-craft-k | 1 ≤ k ≤ ◯, or k > ◯};{div **C** = $\partial \mathbf{c}_1/\partial x_1 + \partial \mathbf{c}_2/\partial x_2 + \dots + \partial \mathbf{c}_n/\partial x_n$ | mod div **C** ≤ 0};{Grad C = $(\partial c/\partial x_1)\mathbf{i}_1 + (\partial c/\partial x_2)i_2, \dots, + (\partial c/\partial x_n)\mathbf{i}_n$ | ∞ < mod Grad C ≤ ◯, or mod Grad C > ◯}}}}dT}dT} … d T}

{∫ … {∫∫∫ {t{f{{light-speed-craft-k | 1 ≤ k ≤ ◯, or k > ◯};{div **C** = $\partial \mathbf{c}_1/\partial x_1 + \partial \mathbf{c}_2/\partial x_2 + \dots + \partial \mathbf{c}_n/\partial x_n$ | mod div **C** ≤ 0};{Grad C = $(\partial c/\partial x_1)\mathbf{i}_1 + (\partial c/\partial x_2)i_2, \dots, + (\partial c/\partial x_n)\mathbf{i}_n$ | ∞ < mod Grad C ≤ ◯, or mod Grad C > ◯}}}}dc}dc} … d c}

{∫ … {∫{∫ {t{f{{light-speed-craft-k | 1 ≤ k ≤ ◯, or k > ◯};{div **C** = $\partial \mathbf{c}_1/\partial x_1 + \partial \mathbf{c}_2/\partial x_2 + \dots + \partial \mathbf{c}_n/\partial x_n$ | mod div **C** ≤ 0};{Grad C = $(\partial c/\partial x_1)\mathbf{i}_1 + (\partial c/\partial x_2)i_2,\dots, + (\partial c/\partial x_n)\mathbf{i}_n$ | ∞ < mod Grad C ≤ ◯, or mod Grad C > ◯}}}}dγ}dγ} … d γ}

{∫ … {∫{∫ {t{f{{light-speed-craft-k | 1 ≤ k ≤ ◯, or k > ◯};{div **C** = $\partial \mathbf{c}_1/\partial x_1 + \partial \mathbf{c}_2/\partial x_2 + \dots + \partial \mathbf{c}_n/\partial x_n$ | mod div **C** ≤ 0};{Grad C = $(\partial c/\partial x_1)\mathbf{i}_1 + (\partial c/\partial x_2)i_2,\dots, + (\partial c/\partial x_n)\mathbf{i}_n$ | ∞ < mod Grad C ≤ ◯, or mod Grad C > ◯}}}}dv}dv} … d v}

{∫ … {∫{∫ {t{f{{light-speed-craft-k | 1 ≤ k ≤ ◯, or k > ◯};{div **C** = $\partial \mathbf{c}_1/\partial x_1 + \partial \mathbf{c}_2/\partial x_2 + \dots + \partial \mathbf{c}_n/\partial x_n$ | mod div **C** ≤ 0};{Grad C = $(\partial c/\partial x_1)\mathbf{i}_1 + (\partial c/\partial x_2)i_2,\dots, + (\partial c/\partial x_n)\mathbf{i}_n$ | ∞ < mod Grad C ≤ ◯, or mod Grad C > ◯}}}}da}da} … d a}

{∫ … {∫{∫ {t{f{{light-speed-craft-k | 1 ≤ k ≤ ◯, or k > ◯};{div **C** = $\partial \mathbf{c}_1/\partial x_1 + \partial \mathbf{c}_2/\partial x_2 + \dots + \partial \mathbf{c}_n/\partial x_n$ | mod div **C** ≤ 0};{Grad C = $(\partial c/\partial x_1)\mathbf{i}_1 + (\partial c/\partial x_2)i_2,\dots, + (\partial c/\partial x_n)\mathbf{i}_n$ | ∞ < mod Grad C ≤ ◯, or mod Grad C > ◯}}}}dP}dP} … d P}

{∫ … {∫{∫ {t{f{{light-speed-craft-k | 1 ≤ k ≤ ◯, or k > ◯};{div **C** = $\partial \mathbf{c}_1/\partial x_1 + \partial \mathbf{c}_2/\partial x_2 + \dots + \partial \mathbf{c}_n/\partial x_n$ | mod div **C** ≤ 0};{Grad C = $(\partial c/\partial x_1)\mathbf{i}_1 + (\partial c/\partial x_2)i_2,\dots, + (\partial c/\partial x_n)\mathbf{i}_n$ | ∞ < mod Grad C ≤ ◯, or mod Grad C > ◯}}}}dKE}dKE} … d KE}

{∫ … {∫{∫ {t{f{{light-speed-craft-k | 1 ≤ k ≤ ◯, or k > ◯};{div **C** = $\partial \mathbf{c}_1/\partial x_1 + \partial \mathbf{c}_2/\partial x_2 + \dots + \partial \mathbf{c}_n/\partial x_n$ | mod div **C** ≤ 0};{Grad C = $(\partial c/\partial x_1)\mathbf{i}_1 + (\partial c/\partial x_2)i_2,\dots, + (\partial c/\partial x_n)\mathbf{i}_n$ | ∞ < mod Grad C ≤ ◯, or mod Grad C > ◯}}}}dt}dt} ... dt}

{∫ … {∫{∫ {t{f{{light-speed-craft-k | 1 ≤ k ≤ ◯, or k > ◯};{div **C** = $\partial \mathbf{c}_1/\partial x_1 + \partial \mathbf{c}_2/\partial x_2 + \dots + \partial \mathbf{c}_n/\partial x_n$ | mod div **C** ≤ 0};{Grad C = $(\partial c/\partial x_1)\mathbf{i}_1 + (\partial c/\partial x_2)i_2,\dots, + (\partial c/\partial x_n)\mathbf{i}_n$ | ∞ < mod Grad C ≤ ◯, or mod Grad C > ◯}}}}dT}dT} … d T}

{∫ … {∫{∫ {T{f{{light-speed-craft-k | 1 ≤ k ≤ ◯, or k > ◯};{div **C** = $\partial \mathbf{c}_1/\partial x_1 + \partial \mathbf{c}_2/\partial x_2 + \dots + \partial \mathbf{c}_n/\partial x_n$ | mod div **C** ≤ 0};{Grad C = $(\partial c/\partial x_1)\mathbf{i}_1 + (\partial c/\partial x_2)i_2,\dots, + (\partial c/\partial x_n)\mathbf{i}_n$ | ∞ < mod Grad C ≤ ◯, or mod Grad C > ◯}}}}dc}dc} … d c}

{∫ … {∫{∫ {T{f{{light-speed-craft-k | 1 ≤ k ≤ ◯, or k > ◯};{div **C** = $\partial \mathbf{c}_1/\partial x_1 + \partial \mathbf{c}_2/\partial x_2 + \dots + \partial \mathbf{c}_n/\partial x_n$ | mod div **C** ≤ 0};{Grad C = $(\partial c/\partial x_1)\mathbf{i}_1 + (\partial c/\partial x_2)i_2,\dots, + (\partial c/\partial x_n)\mathbf{i}_n$ | ∞ < mod Grad C ≤ ◯, or mod Grad C > ◯}}}}dγ}dγ} … d γ}

{∫ … {∫{∫ {T{f{{light-speed-craft-k | 1 ≤ k ≤ ◯, or k > ◯};{div **C** = $\partial \mathbf{c}_1/\partial x_1 + \partial \mathbf{c}_2/\partial x_2 + \dots + \partial \mathbf{c}_n/\partial x_n$ | mod div **C** ≤ 0};{Grad C = $(\partial c/\partial x_1)\mathbf{i}_1 + (\partial c/\partial x_2)i_2,\dots, + (\partial c/\partial x_n)\mathbf{i}_n$ | ∞ < mod Grad C ≤ ◯, or mod Grad C > ◯}}}}dv}dv} … d v}

{∫ … {∫{∫ {T{f{{light-speed-craft-k | 1 ≤ k ≤ ○, or k > ○};{div **C** = $\partial \mathbf{c}_1/\partial x_1 + \partial \mathbf{c}_2/\partial x_2 + \dots +$ $\partial \mathbf{c}_n/\partial x_n$ | mod div **C** ≤ 0};{Grad C = $(\partial c/\partial x_1)\mathbf{i}_1 + (\partial c/\partial x_2)i_2, \dots, + (\partial c/\partial x_n)\mathbf{i}_n$ | ∞ < mod Grad C ≤ ○, or mod Grad C > ○}}}}da}da} … d a}

{∫ … {∫{∫ {T{f{{light-speed-craft-k | 1 ≤ k ≤ ○, or k > ○};{div **C** = $\partial \mathbf{c}_1/\partial x_1 + \partial \mathbf{c}_2/\partial x_2 + \dots +$ $\partial \mathbf{c}_n/\partial x_n$ | mod div **C** ≤ 0};{Grad C = $(\partial c/\partial x_1)\mathbf{i}_1 + (\partial c/\partial x_2)i_2, \dots, + (\partial c/\partial x_n)\mathbf{i}_n$ | ∞ < mod Grad C ≤ ○, or mod Grad C > ○}}}}dP}dP} … d P}

{∫ … {∫{∫ {T{f{{light-speed-craft-k | 1 ≤ k ≤ ○, or k > ○};{div **C** = $\partial \mathbf{c}_1/\partial x_1 + \partial \mathbf{c}_2/\partial x_2 + \dots +$ $\partial \mathbf{c}_n/\partial x_n$ | mod div **C** ≤ 0};{Grad C = $(\partial c/\partial x_1)\mathbf{i}_1 + (\partial c/\partial x_2)i_2, \dots, + (\partial c/\partial x_n)\mathbf{i}_n$ | ∞ < mod Grad C ≤ ○, or mod Grad C > ○}}}}dKE}dKE} … d KE}

{∫ … {∫{∫ {T{f{{light-speed-craft-k | 1 ≤ k ≤ ○, or k > ○};{div **C** = $\partial \mathbf{c}_1/\partial x_1 + \partial \mathbf{c}_2/\partial x_2 + \dots +$ $\partial \mathbf{c}_n/\partial x_n$ | mod div **C** ≤ 0};{Grad C = $(\partial c/\partial x_1)\mathbf{i}_1 + (\partial c/\partial x_2)i_2, \dots, + (\partial c/\partial x_n)\mathbf{i}_n$ | ∞ < mod Grad C ≤ ○, or mod Grad C > ○}}}}dt}dt} ... dt}

{∫ … {∫{∫ {T{f{{light-speed-craft-k | 1 ≤ k ≤ ○, or k > ○};{div **C** = $\partial \mathbf{c}_1/\partial x_1 + \partial \mathbf{c}_2/\partial x_2 + \dots +$ $\partial \mathbf{c}_n/\partial x_n$ | mod div **C** ≤ 0};{Grad C = $(\partial c/\partial x_1)\mathbf{i}_1 + (\partial c/\partial x_2)i_2, \dots, + (\partial c/\partial x_n)\mathbf{i}_n$ | ∞ < mod Grad C ≤ ○, or mod Grad C > ○}}}}dT}dT} … d T}

and so on.

We can also take multi-variable derivatives and integrals involving the following variables as non-limiting options; light-speed, gamma, velocity, acceleration, momentum, kinetic energy, and so on.

Each of the following roster elements can be the basis of expressions of differentiated functions for power, energy, velocity, momentum, and gamma and the like by differentiation in the following manners.

d f/dt,1 d(d f/dt,1)/dt,2 d(d f/dt,2)/dt,1 d[d(d f/dt,1)/dt,2]/dt,3 d[d(d f/dt,1)/dt,3]/dt,2

d[d(d f/dt,2)/dt,1]/dt,3 d[d(d f/dt,2)/dt,3]/dt,1 d[d(d f/dt,3)/dt,1]/dt,2 d[d(d f/dt,3)/dt,2]/dt,1

d f/dT,1 d(d f/dT,1)/dT,2 d(d f/dT,2)/dT,1 d[d(d f/dT,1)/dT,2]/dT,3 d[d(d f/dT,1)/dT,3]/dT,2

d[d(d f/dT,2)/dT,1]/dT,3 d[d(d f/dT,2)/dT,3]/dT,1 d[d(d f/dT,3)/dT,1]/dT,2 d[d(d f/dT,3)/dT,2]/dT,1

T in each case is a time-like blend of various sorts of multiple time dimensions that appropriately and usefully enables derivation in one iteration.

We can go further to consider.

d f/dƮ,1 d(d f/dƮ,1)/dƮ,2 d(d f/dƮ,2)/dƮ,1 d[d(d f/dƮ,1)/dƮ,2]/dƮ,3 d[d(d f/dƮ,1)/dƮ,3]/dƮ,2

d[d(d f/dƮ,2)/dƮ,1]/dƮ,3 d[d(d f/dƮ,2)/dƮ,3]/dƮ,1 d[d(d f/dƮ,3)/dƮ,1]/dƮ,2 d[d(d f/dƮ,3)/dƮ,2]/dƮ,1

Ʈ in each case is a time-like blend of time-like blends of various sorts of multiple time dimensions that appropriately and usefully enables derivation in one iteration.

By the same token, we can also integrate functions for power, energy, velocity, momentum, and gamma and the like by respect to multiple time dimensions as follows.

∫ f(dt,1) ∫∫d f(dt,1)(dt,2) ∫∫d f(dt,2)(dt,1) ∭ f(dt,1)(dt,2)(dt,3) ∭ f(dt,1)(dt,3)(dt,2)

∭ f(dt,2)(dt,1)(dt,3) ∭ f(dt,2)(dt,3)(dt,1) ∭ f(dt,3)(dt,1)(dt,2) ∭ f(dt,3)(dt,2)(dt,1)

∫ f(dT,1) ∬ f(dT,1) (dT,2) ∬ f(dT,2) ((dT,1)) ∭ f(dT,1) (dT,2) (dT,3) ∭ f(dT,1) (dT,3) (dT,2)

∭ f(dT,2) (dT,1) (dT,3) ∭ f(dT,2) (dT,3) (dT,1) ∭ f(dT,3) (dT,1) (dT,2) ∭ f(dT,3) (dT,2) (dT,1)

We can go further to consider.

∫ f(dƮ,1) ∫ ∫ f(dƮ,1)(dƮ,2) ∫ ∫ f(dƮ,2)(dƮ,1) ∫ ∫ ∫ f(dƮ,1)(dƮ,2)(dƮ,3) ∫ ∫ ∫ f(dƮ,1)(dƮ,3)(dƮ,2)

∫ ∫ ∫ f(dƮ,2)(dƮ,1)(dƮ,3) ∫ ∫ ∫ f(dƮ,2)(dƮ,3)(dƮ,1) ∫ ∫ ∫ f(dƮ,3)(dƮ,1)(dƮ,2) ∫ ∫ ∫ f(dƮ,3)(dƮ,2)(dƮ,1)

We can continue to yet higher order derivatives and integrals suffice it to say that the general patterns continue for higher numbers of dimensions.

We can also consider higher order blends of time and thus are not limited to the blends defined by T and Ʈ. We chose these two symbols for their resemblance of the letter t.

ESSAY 12) Hugely Infinite Lorentz Factors For Spacecraft And Levels Of Hyper-Relativity-Of-Simultaneity.

Now, as we covered previously, we considered light-speed reference frames at increasing infinite Lorentz factors as being more and more relativistic in the sense of increasingly exuberant existential expression of the relativity of simultaneity.

Now, we consider spacecraft having Lorentz factors that are so infinite that the spacecraft and her crew take on a hyper-relativity-of-simultaneity in the sense that such increased relativity of simultaneity is of a whole qualitative level above relativity of simultaneity. We refer to said hyper-relativity-of-simultaneity as hyper-relativity-of-simultaneity-1.

We can go to even further extremes to consider hyper-hyper-relativity-of-simultaneity which is a whole qualitative level above hyper-relativity-of-simultaneity or hyper-relativity-of-simultaneity-1. We refer to said hyper-hyper-relativity-of-simultaneity as hyper-relativity-of-simultaneity-2.

We can go to even further extremes to consider hyper-hyper-hyper-relativity-of-simultaneity which is a whole qualitative level above hyper-hyper-relativity-of-simultaneity or hyper-relativity-of-simultaneity-2. We refer to said hyper-hyper-hyper-relativity-of-simultaneity as hyper-relativity-of-simultaneity-3.

We can go on to hyper-relativity-of-simultaneity-4, hyper-relativity-of-simultaneity-5, and so-on.

We can even develop the following operator to express a limited but infinite portion of ordinated terms of hyper-relativity-of-simultaneity.

[f[Σ(j = 1; j = super-∞ ↑):(Hyper-relativity-of-simultaneity-j)]]

Here, the series limiting notation of super-∞ ↑ denotes any super-infinite value by ironically still ever greater and developing possible ordinated terms.

We may affix the above operator to formulas for gamma and related parameters to denote the above series of ordination and the underlying existential and ontological principles that enable such extreme series of possibilities.

Note that in Special Relativity, there is no such thing as absolute simultaneity. So, there is no such scenario where two events separated by space can be said to occur at the same time relative to all observer referene frames.

ESSAY 13) The Infinity Of Levels Of Light-Speed.

Now, again, we can consider our previous scenarios defined by a vector along an axis representing the super-infinite variety of speeds of light, all of which may be placed within a vanishingly small light-speed velocity spectrum.

The vector as such may serve as a gradient vector that extends from the first light-speed and undefinably beyond the first light-speed such as associated with the least infinite number.

Analogously, we can take the divergence of the hyper-relativity field **[f[Σ(j = 1; j = super-∞↑):(Hyper-relativity-of-simultaneity-j)]]** in n abstracted spatial dimensions

The divergence is:

div **[f[Σ(j = 1; j = super-∞↑):(Hyper-relativity-of-simultaneity-j)]]** = $\partial \mathbf{c_1}/\partial x_1 + \partial \mathbf{c_2}/\partial x_2 + \ldots + \partial \mathbf{c_n}/\partial x_n$

The gradient of a field is its direction of steepest change.

For more n abstracted spatial dimensions of the hyper-relativity ascension field, the gradient is:

Grad [f[Σ(j = 1; j = super-∞↑):(Hyper-relativity-of-simultaneity-j)]] = $(\partial c/\partial x_1)\mathbf{i_1} + (\partial c/\partial x_2)i_2, \ldots, + (\partial c/\partial x_n)\mathbf{i_n}$

The Laplacian is the divergence of the gradients in the hyper-relativity field and is defined abstractly in n spatial dimensions in [f[Σ(j = 1; j = super-∞↑):(Hyper-relativity-of-simultaneity-j)]]artesian coordinates as:

Dell grad [f[Σ(j = 1; j = super-∞↑):(Hyper-relativity-of-simultaneity-j)]] = $(\partial^2 c/\partial x_1^2) + (\partial^2 c/\partial x_2^2) + \ldots + (\partial^2 c/\partial x_n^2)$

Now we can consider the realm of hyper-relativity values as if an n-dimensional space in the abstract sense but not in the literal sense.

The gradient of the hyper-relativity ascension field may be infinite in magnitude and perhaps even super-infinite in magnitude.

We can also consider the divergence of the gradient of the fields of ascending hyper-relativities.

The divergence of the gradients or the associated Laplacian may plausibly have a value of less than zero but be non-negative. In other words, the Laplacian may have a modulus of less than zero but non-negative as a non-limiting option.

We can also consider analogs of the hyper-relativities ascension field for Lorentz factors, spacecraft momentum, spacecraft kinetic energy, spacecraft acceleration in the ship frame; and ironically, acceleration in the background reference frame.

In cases for which the travel upward to ever greater hyper-relativities of travel develops into an infinite modulus vector, the climb to ever greater hyper-relativities may runway not only due to a momentum-style analog, but also in a manner analogous to the acceleration of a negative refractive index light-sail for which the incident radiative power would scale as the fourth power of the temperature of a black body source of incident light used to accelerate the light-sail.

We may affix the following operator to the lengthy formulas for gamma and related parameters to denote the above hyper-relativities field concepts and the associated physics especially but not only as related to gamma.

{f{{light-speed-craft-k | 1 ≤ k ≤ ○, or k >○};{div = $\partial \mathbf{c}_1/\partial x_1 + \partial \mathbf{c}_2/\partial x_2 + \ldots + \partial \mathbf{c}_n/\partial x_n$ | mod div **[f[Σ(j = 1; j = super-∞↑):(Hyper-relativity-of-simultaneity-j)]]** ≤ 0};{Grad [f[Σ(j = 1; j = super-∞↑):(Hyper-relativity-of-simultaneity-j)]] = $(\partial c/\partial x_1)\mathbf{i}_1 + (\partial c/\partial x_2)i_2, \ldots, + (\partial c/\partial x_n)\mathbf{i}_n$ | ∞ < mod Grad [f[Σ(j = 1; j = super-∞↑):(Hyper-relativity-of-simultaneity-j)]] ≤ ○, or mod Grad [f[Σ(j = 1; j = super-∞↑):(Hyper-relativity-of-simultaneity-j)]] > ○}}}

We can take light-speed, gamma, velocity, acceleration, momentum, kinetic energy, and so on derivatives of the above operator as functionally modified to represent ship and background time, light-speed, gamma, velocity, acceleration, momentum, kinetic energy, and so on. We may affix the resulting derivative operators to formulas for gamma and related parameters to denote the operative mechanisms.

{d … {d{d {c{f{{light-speed-craft-k | 1 ≤ k ≤ ○, or k > ○};{div **[f[Σ(j = 1; j = super-∞↑):(Hyper-relativity-of-simultaneity-j)]]** = $\partial \mathbf{c}_1/\partial x_1 + \partial \mathbf{c}_2/\partial x_2 + \ldots + \partial \mathbf{c}_n/\partial x_n$ | mod div **[f[Σ(j = 1; j = super-∞↑):(Hyper-relativity-of-simultaneity-j)]]** ≤ 0};{Grad [f[Σ(j = 1; j = super-∞↑):(Hyper-relativity-of-simultaneity-j)]] = $(\partial c/\partial x_1)\mathbf{i}_1 + (\partial c/\partial x_2)i_2, \ldots, + (\partial c/\partial x_n)\mathbf{i}_n$ | ∞ < mod Grad [f[Σ(j = 1; j = super-∞↑):(Hyper-relativity-of-simultaneity-j)]] ≤ ○, or mod Grad [f[Σ(j = 1; j = super-∞↑):(Hyper-relativity-of-simultaneity-j)]] > ○}}}}/dc}/dc}/ … /d c}

{d … {d{d {c{f{{light-speed-craft-k | 1 ≤ k ≤ ○, or k > ○};{div **[f[Σ(j = 1; j = super-∞↑):(Hyper-relativity-of-simultaneity-j)]]** = $\partial \mathbf{c}_1/\partial x_1 + \partial \mathbf{c}_2/\partial x_2 + \ldots + \partial \mathbf{c}_n/\partial x_n$ | mod div **[f[Σ(j = 1; j = super-∞↑):(Hyper-relativity-of-simultaneity-j)]]** ≤ 0};{Grad [f[Σ(j = 1; j = super-∞↑):(Hyper-relativity-of-simultaneity-j)]] = $(\partial c/\partial x_1)\mathbf{i}_1 + (\partial c/\partial x_2)i_2, \ldots, + (\partial c/\partial x_n)\mathbf{i}_n$ | ∞ < mod Grad [f[Σ(j = 1; j = super-∞↑):(Hyper-relativity-of-simultaneity-j)]] ≤ ○, or mod Grad [f[Σ(j = 1; j = super-∞↑):(Hyper-relativity-of-simultaneity-j)]] > ○}}}}/dγ}/dγ}/ … /d γ}

{d … {d{d {c{f{{light-speed-craft-k | 1 ≤ k ≤ ○, or k > ○};{div **[f[Σ(j = 1; j = super-∞↑):(Hyper-relativity-of-simultaneity-j)]]** = $\partial \mathbf{c}_1/\partial x_1 + \partial \mathbf{c}_2/\partial x_2 + \ldots + \partial \mathbf{c}_n/\partial x_n$ | mod div **[f[Σ(j = 1; j = super-∞↑):(Hyper-relativity-of-simultaneity-j)]]** ≤ 0};{Grad [f[Σ(j = 1; j = super-∞↑):(Hyper-relativity-of-simultaneity-j)]] = $(\partial c/\partial x_1)\mathbf{i}_1 + (\partial c/\partial x_2)i_2, \ldots, + (\partial c/\partial x_n)\mathbf{i}_n$ | ∞ < mod Grad [f[Σ(j = 1; j = super-∞↑):(Hyper-relativity-of-simultaneity-j)]] ≤ ○, or mod Grad [f[Σ(j = 1; j = super-∞↑):(Hyper-relativity-of-simultaneity-j)]] > ○}}}}/dv}/dv}/ … /d v} {d … {d{d {c{f{{light-speed-craft-k | 1 ≤ k ≤ ◯, or k > ◯};{div **[f[Σ(j = 1; j = super-∞↑):(Hyper-relativity-of-simultaneity-j)]]** = $\partial \mathbf{c_1}/\partial x_1 + \partial \mathbf{c_2}/\partial x_2 + \ldots + \partial \mathbf{c_n}/\partial x_n$ | mod div **[f[Σ(j = 1; j = super-∞↑):(Hyper-relativity-of-simultaneity-j)]]** ≤ 0};{Grad [f[Σ(j = 1; j = super-∞↑):(Hyper-relativity-of-simultaneity-j)]] = $(\partial c/\partial x_1)\mathbf{i_1} + (\partial c/\partial x_2)i_2,\ldots, + (\partial c/\partial x_n)\mathbf{i_n}$ | ∞ < mod Grad [f[Σ(j = 1; j = super-∞↑):(Hyper-relativity-of-simultaneity-j)]] ≤ ◯, or mod Grad [f[Σ(j = 1; j = super-∞↑):(Hyper-relativity-of-simultaneity-j)]] > ◯}}}}/da}/da}/ … /d a}

{d … {d{d {c{f{{light-speed-craft-k | 1 ≤ k ≤ ◯, or k > ◯};{div **[f[Σ(j = 1; j = super-∞↑):(Hyper-relativity-of-simultaneity-j)]]** = $\partial \mathbf{c_1}/\partial x_1 + \partial \mathbf{c_2}/\partial x_2 + \ldots + \partial \mathbf{c_n}/\partial x_n$ | mod div **[f[Σ(j = 1; j = super-∞↑):(Hyper-relativity-of-simultaneity-j)]]** ≤ 0};{Grad [f[Σ(j = 1; j = super-∞↑):(Hyper-relativity-of-simultaneity-j)]] = $(\partial c/\partial x_1)\mathbf{i_1} + (\partial c/\partial x_2)i_2,\ldots, + (\partial c/\partial x_n)\mathbf{i_n}$ | ∞ < mod Grad [f[Σ(j = 1; j = super-∞↑):(Hyper-relativity-of-simultaneity-j)]] ≤ ◯, or mod Grad [f[Σ(j = 1; j = super-∞↑):(Hyper-relativity-of-simultaneity-j)]] > ◯}}}}/dP}/dP}/ … /d P}

{d … {d{d {c{f{{light-speed-craft-k | 1 ≤ k ≤ ◯, or k > ◯};{div **[f[Σ(j = 1; j = super-∞↑):(Hyper-relativity-of-simultaneity-j)]]** = $\partial \mathbf{c_1}/\partial x_1 + \partial \mathbf{c_2}/\partial x_2 + \ldots + \partial \mathbf{c_n}/\partial x_n$ | mod div **[f[Σ(j = 1; j = super-∞↑):(Hyper-relativity-of-simultaneity-j)]]** ≤ 0};{Grad [f[Σ(j = 1; j = super-∞↑):(Hyper-relativity-of-simultaneity-j)]] = $(\partial c/\partial x_1)\mathbf{i_1} + (\partial c/\partial x_2)i_2,\ldots, + (\partial c/\partial x_n)\mathbf{i_n}$ | ∞ < mod Grad [f[Σ(j = 1; j = super-∞↑):(Hyper-relativity-of-simultaneity-j)]] ≤ ◯, or mod Grad [f[Σ(j = 1; j = super-∞↑):(Hyper-relativity-of-simultaneity-j)]] > ◯}}}}/dKE}/dKE}/ … /d KE}

{d … {d{d {c{f{{light-speed-craft-k | 1 ≤ k ≤ ◯, or k > ◯};{div **[f[Σ(j = 1; j = super-∞↑):(Hyper-relativity-of-simultaneity-j)]]** = $\partial \mathbf{c_1}/\partial x_1 + \partial \mathbf{c_2}/\partial x_2 + \ldots + \partial \mathbf{c_n}/\partial x_n$ | mod div **[f[Σ(j = 1; j = super-∞↑):(Hyper-relativity-of-simultaneity-j)]]** ≤ 0};{Grad [f[Σ(j = 1; j = super-∞↑):(Hyper-relativity-of-simultaneity-j)]] = $(\partial c/\partial x_1)\mathbf{i_1} + (\partial c/\partial x_2)i_2,\ldots, + (\partial c/\partial x_n)\mathbf{i_n}$ | ∞ < mod Grad [f[Σ(j = 1; j = super-∞↑):(Hyper-relativity-of-simultaneity-j)]] ≤ ◯, or mod Grad [f[Σ(j = 1; j = super-∞↑):(Hyper-relativity-of-simultaneity-j)]] > ◯}}}}/dt}/dt}/ … /dt}

{d … {d{d {c{f{{light-speed-craft-k | 1 ≤ k ≤ ◯, or k > ◯};{div **[f[Σ(j = 1; j = super-∞↑):(Hyper-relativity-of-simultaneity-j)]]** = $\partial \mathbf{c_1}/\partial x_1 + \partial \mathbf{c_2}/\partial x_2 + \ldots + \partial \mathbf{c_n}/\partial x_n$ | mod div **[f[Σ(j = 1; j = super-∞↑):(Hyper-relativity-of-simultaneity-j)]]** ≤ 0};{Grad [f[Σ(j = 1; j = super-∞↑):(Hyper-relativity-of-simultaneity-j)]] = $(\partial c/\partial x_1)\mathbf{i_1} + (\partial c/\partial x_2)i_2,\ldots, + (\partial c/\partial x_n)\mathbf{i_n}$ | ∞ < mod Grad [f[Σ(j = 1; j = super-∞↑):(Hyper-relativity-of-simultaneity-j)]] ≤ ◯, or mod Grad [f[Σ(j = 1; j = super-∞↑):(Hyper-relativity-of-simultaneity-j)]] > ◯}}}}/dT}/dT}/ … /d T} {d … {d{d {γ{f{{light-speed-craft-k | 1 ≤ k ≤ ◯, or k > ◯};{div **[f[Σ(j = 1; j = super-∞↑):(Hyper-relativity-of-simultaneity-j)]]** = $\partial \mathbf{c_1}/\partial x_1 + \partial \mathbf{c_2}/\partial x_2 + \ldots + \partial \mathbf{c_n}/\partial x_n$ | mod div **[f[Σ(j = 1; j = super-∞↑):(Hyper-relativity-of-simultaneity-j)]]** ≤ 0};{Grad [f[Σ(j = 1; j = super-∞↑):(Hyper-relativity-of-simultaneity-j)]] = $(\partial c/\partial x_1)\mathbf{i_1} + (\partial c/\partial x_2)i_2, \ldots, + (\partial c/\partial x_n)\mathbf{i_n}$ | ∞ < mod Grad [f[Σ(j = 1; j = super-∞↑):(Hyper-relativity-of-simultaneity-j)]] ≤ ◯, or mod Grad [f[Σ(j = 1; j = super-∞↑):(Hyper-relativity-of-simultaneity-j)]] > ◯}}}}/dc}/dc}/ … /d c}

{d … {d{d {γ{f{{light-speed-craft-k | 1 ≤ k ≤ ◯, or k > ◯};{div **[f[Σ(j = 1; j = super-∞↑):(Hyper-relativity-of-simultaneity-j)]]** = $\partial \mathbf{c_1}/\partial x_1 + \partial \mathbf{c_2}/\partial x_2 + \ldots + \partial \mathbf{c_n}/\partial x_n$ | mod div **[f[Σ(j = 1; j = super-∞↑):(Hyper-relativity-of-simultaneity-j)]]** ≤ 0};{Grad [f[Σ(j = 1; j = super-∞↑):(Hyper-relativity-of-simultaneity-j)]] = $(\partial c/\partial x_1)\mathbf{i_1} + (\partial c/\partial x_2)i_2, \ldots, + (\partial c/\partial x_n)\mathbf{i_n}$ | ∞ < mod Grad [f[Σ(j = 1; j = super-∞↑):(Hyper-relativity-of-simultaneity-j)]] ≤ ◯, or mod Grad [f[Σ(j = 1; j = super-∞↑):(Hyper-relativity-of-simultaneity-j)]] > ◯}}}}/dγ}/dγ}/ … /d γ}

{d … {d{d {γ{f{{light-speed-craft-k | 1 ≤ k ≤ ◯, or k > ◯};{div **[f[Σ(j = 1; j = super-∞↑):(Hyper-relativity-of-simultaneity-j)]]** = $\partial \mathbf{c_1}/\partial x_1 + \partial \mathbf{c_2}/\partial x_2 + \ldots + \partial \mathbf{c_n}/\partial x_n$ | mod div **[f[Σ(j = 1; j = super-∞↑):(Hyper-relativity-of-simultaneity-j)]]** ≤ 0};{Grad [f[Σ(j = 1; j = super-∞↑):(Hyper-relativity-of-simultaneity-j)]] = $(\partial c/\partial x_1)\mathbf{i_1} + (\partial c/\partial x_2)i_2, \ldots, + (\partial c/\partial x_n)\mathbf{i_n}$ | ∞ < mod Grad [f[Σ(j = 1; j = super-∞↑):(Hyper-relativity-of-simultaneity-j)]] ≤ ◯, or mod Grad [f[Σ(j = 1; j = super-∞↑):(Hyper-relativity-of-simultaneity-j)]] > ◯}}}}/dv}/dv}/ … /d v}

{d … {d{d {γ{f{{light-speed-craft-k | 1 ≤ k ≤ ◯, or k > ◯};{div **[f[Σ(j = 1; j = super-∞↑):(Hyper-relativity-of-simultaneity-j)]]** = $\partial \mathbf{c_1}/\partial x_1 + \partial \mathbf{c_2}/\partial x_2 + \ldots + \partial \mathbf{c_n}/\partial x_n$ | mod div **[f[Σ(j = 1; j = super-∞↑):(Hyper-relativity-of-simultaneity-j)]]** ≤ 0};{Grad [f[Σ(j = 1; j = super-∞↑):(Hyper-relativity-of-simultaneity-j)]] = $(\partial c/\partial x_1)\mathbf{i_1} + (\partial c/\partial x_2)i_2, \ldots, + (\partial c/\partial x_n)\mathbf{i_n}$ | ∞ < mod Grad [f[Σ(j = 1; j = super-∞↑):(Hyper-relativity-of-simultaneity-j)]] ≤ ◯, or mod Grad [f[Σ(j = 1; j = super-∞↑):(Hyper-relativity-of-simultaneity-j)]] > ◯}}}}/da}/da}/ … /d a}

{d … {d{d {γ{f{{light-speed-craft-k | 1 ≤ k ≤ ◯, or k > ◯};{div **[f[Σ(j = 1; j = super-∞↑):(Hyper-relativity-of-simultaneity-j)]]** = $\partial \mathbf{c_1}/\partial x_1 + \partial \mathbf{c_2}/\partial x_2 + \ldots + \partial \mathbf{c_n}/\partial x_n$ | mod div **[f[Σ(j = 1; j = super-∞↑):(Hyper-relativity-of-simultaneity-j)]]** ≤ 0};{Grad [f[Σ(j = 1; j = super-∞↑):(Hyper-relativity-of-simultaneity-j)]] = $(\partial c/\partial x_1)\mathbf{i_1} + (\partial c/\partial x_2)i_2, \ldots, + (\partial c/\partial x_n)\mathbf{i_n}$ | ∞ < mod Grad [f[Σ(j = 1; j = super-∞↑):(Hyper-relativity-of-simultaneity-j)]] ≤ ◯, or mod Grad [f[Σ(j = 1; j = super-∞↑):(Hyper-relativity-of-simultaneity-j)]] > ◯}}}}/dP}/dP}/ … /d P}

{d … {d{d {γ{f{{light-speed-craft-k | 1 ≤ k ≤ ◯, or k > ◯};{div **[f[Σ(j = 1; j = super-∞↑):(Hyper-relativity-of-simultaneity-j)]]** = $\partial \mathbf{c_1}/\partial x_1 + \partial \mathbf{c_2}/\partial x_2 + \ldots + \partial \mathbf{c_n}/\partial x_n$ | mod div **[f[Σ(j = 1; j = super-∞↑):(Hyper-relativity-of-simultaneity-j)]]** ≤ 0};{Grad [f[Σ(j = 1; j =

super-∞↑):(Hyper-relativity-of-simultaneity-j)]] = $(\partial c/\partial x_1)\mathbf{i}_1 + (\partial c/\partial x_2)i_2,\ldots, + (\partial c/\partial x_n)\mathbf{i}_n$ | ∞ < mod Grad [f[Σ(j = 1; j = super-∞↑):(Hyper-relativity-of-simultaneity-j)]] ≤ ◯, or mod Grad [f[Σ(j = 1; j = super-∞↑):(Hyper-relativity-of-simultaneity-j)]] > ◯}}}}/dKE}/dKE}/ … /d KE}

{d … {d{d {γ{f{{light-speed-craft-k | 1 ≤ k ≤ ◯, or k > ◯};{div **[f[Σ(j = 1; j = super-∞↑):(Hyper-relativity-of-simultaneity-j)]]** = $\partial \mathbf{c}_1/\partial x_1 + \partial \mathbf{c}_2/\partial x_2 + \ldots + \partial \mathbf{c}_n/\partial x_n$ | mod div **[f[Σ(j = 1; j = super-∞↑):(Hyper-relativity-of-simultaneity-j)]]** ≤ 0};{Grad [f[Σ(j = 1; j = super-∞↑):(Hyper-relativity-of-simultaneity-j)]] = $(\partial c/\partial x_1)\mathbf{i}_1 + (\partial c/\partial x_2)i_2,\ldots, + (\partial c/\partial x_n)\mathbf{i}_n$ | ∞ < mod Grad [f[Σ(j = 1; j = super-∞↑):(Hyper-relativity-of-simultaneity-j)]] ≤ ◯, or mod Grad [f[Σ(j = 1; j = super-∞↑):(Hyper-relativity-of-simultaneity-j)]] > ◯}}}}/dt}/dt}/ … /dt}

{d … {d{d {γ{f{{light-speed-craft-k | 1 ≤ k ≤ ◯, or k > ◯};{div **[f[Σ(j = 1; j = super-∞↑):(Hyper-relativity-of-simultaneity-j)]]** = $\partial \mathbf{c}_1/\partial x_1 + \partial \mathbf{c}_2/\partial x_2 + \ldots + \partial \mathbf{c}_n/\partial x_n$ | mod div **[f[Σ(j = 1; j = super-∞↑):(Hyper-relativity-of-simultaneity-j)]]** ≤ 0};{Grad [f[Σ(j = 1; j = super-∞↑):(Hyper-relativity-of-simultaneity-j)]] = $(\partial c/\partial x_1)\mathbf{i}_1 + (\partial c/\partial x_2)i_2,\ldots, + (\partial c/\partial x_n)\mathbf{i}_n$ | ∞ < mod Grad [f[Σ(j = 1; j = super-∞↑):(Hyper-relativity-of-simultaneity-j)]] ≤ ◯, or mod Grad [f[Σ(j = 1; j = super-∞↑):(Hyper-relativity-of-simultaneity-j)]] > ◯}}}}/dT}/dT}/ … /d T}

{d … {d{d {v{f{{light-speed-craft-k | 1 ≤ k ≤ ◯, or k > ◯};{div **[f[Σ(j = 1; j = super-∞↑):(Hyper-relativity-of-simultaneity-j)]]** = $\partial \mathbf{c}_1/\partial x_1 + \partial \mathbf{c}_2/\partial x_2 + \ldots + \partial \mathbf{c}_n/\partial x_n$ | mod div **[f[Σ(j = 1; j = super-∞↑):(Hyper-relativity-of-simultaneity-j)]]** ≤ 0};{Grad [f[Σ(j = 1; j = super-∞↑):(Hyper-relativity-of-simultaneity-j)]] = $(\partial c/\partial x_1)\mathbf{i}_1 + (\partial c/\partial x_2)i_2,\ldots, + (\partial c/\partial x_n)\mathbf{i}_n$ | ∞ < mod Grad [f[Σ(j = 1; j = super-∞↑):(Hyper-relativity-of-simultaneity-j)]] ≤ ◯, or mod Grad [f[Σ(j = 1; j = super-∞↑):(Hyper-relativity-of-simultaneity-j)]] > ◯}}}}/dc}/dc}/ … /d c}

{d … {d{d {v{f{{light-speed-craft-k | 1 ≤ k ≤ ◯, or k > ◯};{div **[f[Σ(j = 1; j = super-∞↑):(Hyper-relativity-of-simultaneity-j)]]** = $\partial \mathbf{c}_1/\partial x_1 + \partial \mathbf{c}_2/\partial x_2 + \ldots + \partial \mathbf{c}_n/\partial x_n$ | mod div **[f[Σ(j = 1; j = super-∞↑):(Hyper-relativity-of-simultaneity-j)]]** ≤ 0};{Grad [f[Σ(j = 1; j = super-∞↑):(Hyper-relativity-of-simultaneity-j)]] = $(\partial c/\partial x_1)\mathbf{i}_1 + (\partial c/\partial x_2)i_2,\ldots, + (\partial c/\partial x_n)\mathbf{i}_n$ | ∞ < mod Grad [f[Σ(j = 1; j = super-∞↑):(Hyper-relativity-of-simultaneity-j)]] ≤ ◯, or mod Grad [f[Σ(j = 1; j = super-∞↑):(Hyper-relativity-of-simultaneity-j)]] > ◯}}}}/dγ}/dγ}/ … /d γ}

{d … {d{d {v{f{{light-speed-craft-k | 1 ≤ k ≤ ◯, or k > ◯};{div **[f[Σ(j = 1; j = super-∞↑):(Hyper-relativity-of-simultaneity-j)]]** = $\partial \mathbf{c}_1/\partial x_1 + \partial \mathbf{c}_2/\partial x_2 + \ldots + \partial \mathbf{c}_n/\partial x_n$ | mod div **[f[Σ(j = 1; j = super-∞↑):(Hyper-relativity-of-simultaneity-j)]]** ≤ 0};{Grad [f[Σ(j = 1; j = super-∞↑):(Hyper-relativity-of-simultaneity-j)]] = $(\partial c/\partial x_1)\mathbf{i}_1 + (\partial c/\partial x_2)i_2,\ldots, + (\partial c/\partial x_n)\mathbf{i}_n$ | ∞ < mod Grad [f[Σ(j = 1; j = super-∞↑):(Hyper-relativity-of-simultaneity-j)]] ≤ ◯, or mod Grad [f[Σ(j = 1; j = super-∞↑):(Hyper-relativity-of-simultaneity-j)]] > ◯}}}}/dv}/dv}/ … /d v} {d … {d{d {v{f{{light-speed-craft-k | 1 ≤ k ≤ ○, or k > ○};{div **[f[Σ(j = 1; j = super-∞↑):(Hyper-relativity-of-simultaneity-j)]]** = $\partial \mathbf{c_1}/\partial x_1 + \partial \mathbf{c_2}/\partial x_2 + \ldots + \partial \mathbf{c_n}/\partial x_n$ | mod div **[f[Σ(j = 1; j = super-∞↑):(Hyper-relativity-of-simultaneity-j)]]** ≤ 0};{Grad [f[Σ(j = 1; j = super-∞↑):(Hyper-relativity-of-simultaneity-j)]] = $(\partial c/\partial x_1)\mathbf{i_1}$ + $(\partial c/\partial x_2)i_2$,…, + $(\partial c/\partial x_n)\mathbf{i_n}$ | ∞ < mod Grad [f[Σ(j = 1; j = super-∞↑):(Hyper-relativity-of-simultaneity-j)]] ≤ ○, or mod Grad [f[Σ(j = 1; j = super-∞↑):(Hyper-relativity-of-simultaneity-j)]] > ○}}}}/da}/da}/ … /d a}

{d … {d{d {v{f{{light-speed-craft-k | 1 ≤ k ≤ ○, or k > ○};{div **[f[Σ(j = 1; j = super-∞↑):(Hyper-relativity-of-simultaneity-j)]]** = $\partial \mathbf{c_1}/\partial x_1 + \partial \mathbf{c_2}/\partial x_2 + \ldots + \partial \mathbf{c_n}/\partial x_n$ | mod div **[f[Σ(j = 1; j = super-∞↑):(Hyper-relativity-of-simultaneity-j)]]** ≤ 0};{Grad [f[Σ(j = 1; j = super-∞↑):(Hyper-relativity-of-simultaneity-j)]] = $(\partial c/\partial x_1)\mathbf{i_1}$ + $(\partial c/\partial x_2)i_2$,…, + $(\partial c/\partial x_n)\mathbf{i_n}$ | ∞ < mod Grad [f[Σ(j = 1; j = super-∞↑):(Hyper-relativity-of-simultaneity-j)]] ≤ ○, or mod Grad [f[Σ(j = 1; j = super-∞↑):(Hyper-relativity-of-simultaneity-j)]] > ○}}}}/dP}/dP}/ … /d P}

{d … {d{d {v{f{{light-speed-craft-k | 1 ≤ k ≤ ○, or k > ○};{div **[f[Σ(j = 1; j = super-∞↑):(Hyper-relativity-of-simultaneity-j)]]** = $\partial \mathbf{c_1}/\partial x_1 + \partial \mathbf{c_2}/\partial x_2 + \ldots + \partial \mathbf{c_n}/\partial x_n$ | mod div **[f[Σ(j = 1; j = super-∞↑):(Hyper-relativity-of-simultaneity-j)]]** ≤ 0};{Grad [f[Σ(j = 1; j = super-∞↑):(Hyper-relativity-of-simultaneity-j)]] = $(\partial c/\partial x_1)\mathbf{i_1}$ + $(\partial c/\partial x_2)i_2$,…, + $(\partial c/\partial x_n)\mathbf{i_n}$ | ∞ < mod Grad [f[Σ(j = 1; j = super-∞↑):(Hyper-relativity-of-simultaneity-j)]] ≤ ○, or mod Grad [f[Σ(j = 1; j = super-∞↑):(Hyper-relativity-of-simultaneity-j)]] > ○}}}}/dKE}/dKE}/ … /d KE}

{d … {d{d {v{f{{light-speed-craft-k | 1 ≤ k ≤ ○, or k > ○};{div **[f[Σ(j = 1; j = super-∞↑):(Hyper-relativity-of-simultaneity-j)]]** = $\partial \mathbf{c_1}/\partial x_1 + \partial \mathbf{c_2}/\partial x_2 + \ldots + \partial \mathbf{c_n}/\partial x_n$ | mod div **[f[Σ(j = 1; j = super-∞↑):(Hyper-relativity-of-simultaneity-j)]]** ≤ 0};{Grad [f[Σ(j = 1; j = super-∞↑):(Hyper-relativity-of-simultaneity-j)]] = $(\partial c/\partial x_1)\mathbf{i_1}$ + $(\partial c/\partial x_2)i_2$,…, + $(\partial c/\partial x_n)\mathbf{i_n}$ | ∞ < mod Grad [f[Σ(j = 1; j = super-∞↑):(Hyper-relativity-of-simultaneity-j)]] ≤ ○, or mod Grad [f[Σ(j = 1; j = super-∞↑):(Hyper-relativity-of-simultaneity-j)]] > ○}}}}/dt}/dt}/ … /dt}

{d … {d{d {v{f{{light-speed-craft-k | 1 ≤ k ≤ ○, or k > ○};{div **[f[Σ(j = 1; j = super-∞↑):(Hyper-relativity-of-simultaneity-j)]]** = $\partial \mathbf{c_1}/\partial x_1 + \partial \mathbf{c_2}/\partial x_2 + \ldots + \partial \mathbf{c_n}/\partial x_n$ | mod div **[f[Σ(j = 1; j = super-∞↑):(Hyper-relativity-of-simultaneity-j)]]** ≤ 0};{Grad [f[Σ(j = 1; j = super-∞↑):(Hyper-relativity-of-simultaneity-j)]] = $(\partial c/\partial x_1)\mathbf{i_1}$ + $(\partial c/\partial x_2)i_2$,…, + $(\partial c/\partial x_n)\mathbf{i_n}$ | ∞ < mod Grad [f[Σ(j = 1; j = super-∞↑):(Hyper-relativity-of-simultaneity-j)]] ≤ ○, or mod Grad [f[Σ(j = 1; j = super-∞↑):(Hyper-relativity-of-simultaneity-j)]] > ○}}}}/dT}/dT}/ … /d T}

{d … {d{d {a{f{{{light-speed-craft-k | 1 ≤ k ≤ ○, or k > ○};{div **[f[Σ(j = 1; j = super-∞↑):(Hyper-relativity-of-simultaneity-j)]]** = $\partial \mathbf{c_1}/\partial x_1 + \partial \mathbf{c_2}/\partial x_2 + \ldots + \partial \mathbf{c_n}/\partial x_n$ | mod div **[f[Σ(j = 1; j = super-∞↑):(Hyper-relativity-of-simultaneity-j)]]** ≤ 0};{Grad [f[Σ(j = 1; j =

super-∞↑):(Hyper-relativity-of-simultaneity-j)]] = $(\partial c/\partial x_1)\mathbf{i}_1 + (\partial c/\partial x_2)i_2, \ldots, + (\partial c/\partial x_n)\mathbf{i}_n$ | ∞ < mod Grad [f[Σ(j = 1; j = super-∞↑):(Hyper-relativity-of-simultaneity-j)]] ≤ ◯, or mod Grad [f[Σ(j = 1; j = super-∞↑):(Hyper-relativity-of-simultaneity-j)]] > ◯}}}}/dc}/dc}/ … /d c}

{d … {d{d {a{f{{light-speed-craft-k | 1 ≤ k ≤ ◯, or k > ◯};{div **[f[Σ(j = 1; j = super-∞↑):(Hyper-relativity-of-simultaneity-j)]]** = $\partial \mathbf{c}_1/\partial x_1 + \partial \mathbf{c}_2/\partial x_2 + \ldots + \partial \mathbf{c}_n/\partial x_n$ | mod div **[f[Σ(j = 1; j = super-∞↑):(Hyper-relativity-of-simultaneity-j)]]** ≤ 0};{Grad [f[Σ(j = 1; j = super-∞↑):(Hyper-relativity-of-simultaneity-j)]] = $(\partial c/\partial x_1)\mathbf{i}_1 + (\partial c/\partial x_2)i_2, \ldots, + (\partial c/\partial x_n)\mathbf{i}_n$ | ∞ < mod Grad [f[Σ(j = 1; j = super-∞↑):(Hyper-relativity-of-simultaneity-j)]] ≤ ◯, or mod Grad [f[Σ(j = 1; j = super-∞↑):(Hyper-relativity-of-simultaneity-j)]] > ◯}}}}/dγ}/dγ}/ … /d γ}

{d … {d{d {a{f{{light-speed-craft-k | 1 ≤ k ≤ ◯, or k > ◯};{div **[f[Σ(j = 1; j = super-∞↑):(Hyper-relativity-of-simultaneity-j)]]** = $\partial \mathbf{c}_1/\partial x_1 + \partial \mathbf{c}_2/\partial x_2 + \ldots + \partial \mathbf{c}_n/\partial x_n$ | mod div **[f[Σ(j = 1; j = super-∞↑):(Hyper-relativity-of-simultaneity-j)]]** ≤ 0};{Grad [f[Σ(j = 1; j = super-∞↑):(Hyper-relativity-of-simultaneity-j)]] = $(\partial c/\partial x_1)\mathbf{i}_1 + (\partial c/\partial x_2)i_2, \ldots, + (\partial c/\partial x_n)\mathbf{i}_n$ | ∞ < mod Grad [f[Σ(j = 1; j = super-∞↑):(Hyper-relativity-of-simultaneity-j)]] ≤ ◯, or mod Grad [f[Σ(j = 1; j = super-∞↑):(Hyper-relativity-of-simultaneity-j)]] > ◯}}}}/dv}/dv}/ … /d v}

{d … {d{d {a{f{{light-speed-craft-k | 1 ≤ k ≤ ◯, or k > ◯};{div **[f[Σ(j = 1; j = super-∞↑):(Hyper-relativity-of-simultaneity-j)]]** = $\partial \mathbf{c}_1/\partial x_1 + \partial \mathbf{c}_2/\partial x_2 + \ldots + \partial \mathbf{c}_n/\partial x_n$ | mod div **[f[Σ(j = 1; j = super-∞↑):(Hyper-relativity-of-simultaneity-j)]]** ≤ 0};{Grad [f[Σ(j = 1; j = super-∞↑):(Hyper-relativity-of-simultaneity-j)]] = $(\partial c/\partial x_1)\mathbf{i}_1 + (\partial c/\partial x_2)i_2, \ldots, + (\partial c/\partial x_n)\mathbf{i}_n$ | ∞ < mod Grad [f[Σ(j = 1; j = super-∞↑):(Hyper-relativity-of-simultaneity-j)]] ≤ ◯, or mod Grad [f[Σ(j = 1; j = super-∞↑):(Hyper-relativity-of-simultaneity-j)]] > ◯}}}}/da}/da}/ … /d a}

{d … {d{d {a{f{{light-speed-craft-k | 1 ≤ k ≤ ◯, or k > ◯};{div **[f[Σ(j = 1; j = super-∞↑):(Hyper-relativity-of-simultaneity-j)]]** = $\partial \mathbf{c}_1/\partial x_1 + \partial \mathbf{c}_2/\partial x_2 + \ldots + \partial \mathbf{c}_n/\partial x_n$ | mod div **[f[Σ(j = 1; j = super-∞↑):(Hyper-relativity-of-simultaneity-j)]]** ≤ 0};{Grad [f[Σ(j = 1; j = super-∞↑):(Hyper-relativity-of-simultaneity-j)]] = $(\partial c/\partial x_1)\mathbf{i}_1 + (\partial c/\partial x_2)i_2, \ldots, + (\partial c/\partial x_n)\mathbf{i}_n$ | ∞ < mod Grad [f[Σ(j = 1; j = super-∞↑):(Hyper-relativity-of-simultaneity-j)]] ≤ ◯, or mod Grad [f[Σ(j = 1; j = super-∞↑):(Hyper-relativity-of-simultaneity-j)]] > ◯}}}}/dP}/dP}/ … /d P}

{d … {d{d {a{f{{light-speed-craft-k | 1 ≤ k ≤ ◯, or k > ◯};{div **[f[Σ(j = 1; j = super-∞↑):(Hyper-relativity-of-simultaneity-j)]]** = $\partial \mathbf{c}_1/\partial x_1 + \partial \mathbf{c}_2/\partial x_2 + \ldots + \partial \mathbf{c}_n/\partial x_n$ | mod div **[f[Σ(j = 1; j = super-∞↑):(Hyper-relativity-of-simultaneity-j)]]** ≤ 0};{Grad [f[Σ(j = 1; j = super-∞↑):(Hyper-relativity-of-simultaneity-j)]] = $(\partial c/\partial x_1)\mathbf{i}_1 + (\partial c/\partial x_2)i_2, \ldots, + (\partial c/\partial x_n)\mathbf{i}_n$ | ∞ < mod Grad [f[Σ(j = 1; j = super-∞↑):(Hyper-relativity-of-simultaneity-j)]] ≤ ◯, or mod

Grad [f[Σ(j = 1; j = super-∞↑):(Hyper-relativity-of-simultaneity-j)]] > ◯}}}}/dKE}/dKE}/ … /d KE}

{d … {d{d {a{f{{light-speed-craft-k | 1 ≤ k ≤ ◯, or k > ◯};{div **[f[Σ(j = 1; j = super-∞↑):(Hyper-relativity-of-simultaneity-j)]]** = $\partial \mathbf{c_1}/\partial x_1 + \partial \mathbf{c_2}/\partial x_2 + \ldots + \partial \mathbf{c_n}/\partial x_n$ | mod div **[f[Σ(j = 1; j = super-∞↑):(Hyper-relativity-of-simultaneity-j)]]** ≤ 0};{Grad [f[Σ(j = 1; j = super-∞↑):(Hyper-relativity-of-simultaneity-j)]] = $(\partial c/\partial x_1)\mathbf{i_1}$ + $(\partial c/\partial x_2)i_2$,…, + $(\partial c/\partial x_n)\mathbf{i_n}$ | ∞ < mod Grad [f[Σ(j = 1; j = super-∞↑):(Hyper-relativity-of-simultaneity-j)]] ≤ ◯, or mod Grad [f[Σ(j = 1; j = super-∞↑):(Hyper-relativity-of-simultaneity-j)]] > ◯}}}}/dt}/dt}/ … /dt}

{d … {d{d {a{f{{light-speed-craft-k | 1 ≤ k ≤ ◯, or k > ◯};{div **[f[Σ(j = 1; j = super-∞↑):(Hyper-relativity-of-simultaneity-j)]]** = $\partial \mathbf{c_1}/\partial x_1 + \partial \mathbf{c_2}/\partial x_2 + \ldots + \partial \mathbf{c_n}/\partial x_n$ | mod div **[f[Σ(j = 1; j = super-∞↑):(Hyper-relativity-of-simultaneity-j)]]** ≤ 0};{Grad [f[Σ(j = 1; j = super-∞↑):(Hyper-relativity-of-simultaneity-j)]] = $(\partial c/\partial x_1)\mathbf{i_1}$ + $(\partial c/\partial x_2)i_2$,…, + $(\partial c/\partial x_n)\mathbf{i_n}$ | ∞ < mod Grad [f[Σ(j = 1; j = super-∞↑):(Hyper-relativity-of-simultaneity-j)]] ≤ ◯, or mod Grad [f[Σ(j = 1; j = super-∞↑):(Hyper-relativity-of-simultaneity-j)]] > ◯}}}}/dT}/dT}/ … /d T}

{d … {d{d {P{f{{light-speed-craft-k | 1 ≤ k ≤ ◯, or k > ◯};{div **[f[Σ(j = 1; j = super-∞↑):(Hyper-relativity-of-simultaneity-j)]]** = $\partial \mathbf{c_1}/\partial x_1 + \partial \mathbf{c_2}/\partial x_2 + \ldots + \partial \mathbf{c_n}/\partial x_n$ | mod div **[f[Σ(j = 1; j = super-∞↑):(Hyper-relativity-of-simultaneity-j)]]** ≤ 0};{Grad [f[Σ(j = 1; j = super-∞↑):(Hyper-relativity-of-simultaneity-j)]] = $(\partial c/\partial x_1)\mathbf{i_1}$ + $(\partial c/\partial x_2)i_2$,…, + $(\partial c/\partial x_n)\mathbf{i_n}$ | ∞ < mod Grad [f[Σ(j = 1; j = super-∞↑):(Hyper-relativity-of-simultaneity-j)]] ≤ ◯, or mod Grad [f[Σ(j = 1; j = super-∞↑):(Hyper-relativity-of-simultaneity-j)]] > ◯}}}}/dc}/dc}/ … /d c}

{d … {d{d {P{f{{light-speed-craft-k | 1 ≤ k ≤ ◯, or k > ◯};{div **[f[Σ(j = 1; j = super-∞↑):(Hyper-relativity-of-simultaneity-j)]]** = $\partial \mathbf{c_1}/\partial x_1 + \partial \mathbf{c_2}/\partial x_2 + \ldots + \partial \mathbf{c_n}/\partial x_n$ | mod div **[f[Σ(j = 1; j = super-∞↑):(Hyper-relativity-of-simultaneity-j)]]** ≤ 0};{Grad [f[Σ(j = 1; j = super-∞↑):(Hyper-relativity-of-simultaneity-j)]] = $(\partial c/\partial x_1)\mathbf{i_1}$ + $(\partial c/\partial x_2)i_2$,…, + $(\partial c/\partial x_n)\mathbf{i_n}$ | ∞ < mod Grad [f[Σ(j = 1; j = super-∞↑):(Hyper-relativity-of-simultaneity-j)]] ≤ ◯, or mod Grad [f[Σ(j = 1; j = super-∞↑):(Hyper-relativity-of-simultaneity-j)]] > ◯}}}}/dγ}/dγ}/ … /d γ}

{d … {d{d {P{f{{light-speed-craft-k | 1 ≤ k ≤ ◯, or k > ◯};{div **[f[Σ(j = 1; j = super-∞↑):(Hyper-relativity-of-simultaneity-j)]]** = $\partial \mathbf{c_1}/\partial x_1 + \partial \mathbf{c_2}/\partial x_2 + \ldots + \partial \mathbf{c_n}/\partial x_n$ | mod div **[f[Σ(j = 1; j = super-∞↑):(Hyper-relativity-of-simultaneity-j)]]** ≤ 0};{Grad [f[Σ(j = 1; j = super-∞↑):(Hyper-relativity-of-simultaneity-j)]] = $(\partial c/\partial x_1)\mathbf{i_1}$ + $(\partial c/\partial x_2)i_2$,…, + $(\partial c/\partial x_n)\mathbf{i_n}$ | ∞ < mod Grad [f[Σ(j = 1; j = super-∞↑):(Hyper-relativity-of-simultaneity-j)]] ≤ ◯, or mod Grad [f[Σ(j = 1; j = super-∞↑):(Hyper-relativity-of-simultaneity-j)]] > ◯}}}}/dv}/dv}/ … /d v}

{d … {d{d {P{f{{light-speed-craft-k | 1 ≤ k ≤ ◯, or k > ◯};{div **[f[Σ(j = 1; j = super-∞↑):(Hyper-relativity-of-simultaneity-j)]]** = $\partial \mathbf{c_1}/\partial x_1 + \partial \mathbf{c_2}/\partial x_2 + \ldots + \partial \mathbf{c_n}/\partial x_n$ | mod div

[f[Σ(j = 1; j = super-∞↑):(Hyper-relativity-of-simultaneity-j)]] ≤ 0};{Grad [f[Σ(j = 1; j = super-∞↑):(Hyper-relativity-of-simultaneity-j)]] = $(\partial c/\partial x_1)\mathbf{i}_1 + (\partial c/\partial x_2)i_2, \ldots, + (\partial c/\partial x_n)\mathbf{i}_n$ **|** ∞ < mod Grad [f[Σ(j = 1; j = super-∞↑):(Hyper-relativity-of-simultaneity-j)]] ≤ ○, or mod Grad [f[Σ(j = 1; j = super-∞↑):(Hyper-relativity-of-simultaneity-j)]] > ○}}}}/da}/da}/ … /d a}

{d … {d{d {P{f{{light-speed-craft-k | 1 ≤ k ≤ ○, or k > ○};{div **[f[Σ(j = 1; j = super-∞↑):(Hyper-relativity-of-simultaneity-j)]]** = $\partial \mathbf{c}_1/\partial x_1 + \partial \mathbf{c}_2/\partial x_2 + \ldots + \partial \mathbf{c}_n/\partial x_n$ | mod div **[f[Σ(j = 1; j = super-∞↑):(Hyper-relativity-of-simultaneity-j)]]** ≤ 0};{Grad [f[Σ(j = 1; j = super-∞↑):(Hyper-relativity-of-simultaneity-j)]] = $(\partial c/\partial x_1)\mathbf{i}_1 + (\partial c/\partial x_2)i_2, \ldots, + (\partial c/\partial x_n)\mathbf{i}_n$ **|** ∞ < mod Grad [f[Σ(j = 1; j = super-∞↑):(Hyper-relativity-of-simultaneity-j)]] ≤ ○, or mod Grad [f[Σ(j = 1; j = super-∞↑):(Hyper-relativity-of-simultaneity-j)]] > ○}}}}/dP}/dP}/ … /d P}

{d … {d{d {P{f{{light-speed-craft-k | 1 ≤ k ≤ ○, or k > ○};{div **[f[Σ(j = 1; j = super-∞↑):(Hyper-relativity-of-simultaneity-j)]]** = $\partial \mathbf{c}_1/\partial x_1 + \partial \mathbf{c}_2/\partial x_2 + \ldots + \partial \mathbf{c}_n/\partial x_n$ | mod div **[f[Σ(j = 1; j = super-∞↑):(Hyper-relativity-of-simultaneity-j)]]** ≤ 0};{Grad [f[Σ(j = 1; j = super-∞↑):(Hyper-relativity-of-simultaneity-j)]] = $(\partial c/\partial x_1)\mathbf{i}_1 + (\partial c/\partial x_2)i_2, \ldots, + (\partial c/\partial x_n)\mathbf{i}_n$ **|** ∞ < mod Grad [f[Σ(j = 1; j = super-∞↑):(Hyper-relativity-of-simultaneity-j)]] ≤ ○, or mod Grad [f[Σ(j = 1; j = super-∞↑):(Hyper-relativity-of-simultaneity-j)]] > ○}}}}/dKE}/dKE}/ … /d KE}

{d … {d{d {P{f{{light-speed-craft-k | 1 ≤ k ≤ ○, or k > ○};{div **[f[Σ(j = 1; j = super-∞↑):(Hyper-relativity-of-simultaneity-j)]]** = $\partial \mathbf{c}_1/\partial x_1 + \partial \mathbf{c}_2/\partial x_2 + \ldots + \partial \mathbf{c}_n/\partial x_n$ | mod div **[f[Σ(j = 1; j = super-∞↑):(Hyper-relativity-of-simultaneity-j)]]** ≤ 0};{Grad [f[Σ(j = 1; j = super-∞↑):(Hyper-relativity-of-simultaneity-j)]] = $(\partial c/\partial x_1)\mathbf{i}_1 + (\partial c/\partial x_2)i_2, \ldots, + (\partial c/\partial x_n)\mathbf{i}_n$ **|** ∞ < mod Grad [f[Σ(j = 1; j = super-∞↑):(Hyper-relativity-of-simultaneity-j)]] ≤ ○, or mod Grad [f[Σ(j = 1; j = super-∞↑):(Hyper-relativity-of-simultaneity-j)]] > ○}}}}/dt}/dt}/ … /dt}

{d … {d{d {P{f{{light-speed-craft-k | 1 ≤ k ≤ ○, or k > ○};{div **[f[Σ(j = 1; j = super-∞↑):(Hyper-relativity-of-simultaneity-j)]]** = $\partial \mathbf{c}_1/\partial x_1 + \partial \mathbf{c}_2/\partial x_2 + \ldots + \partial \mathbf{c}_n/\partial x_n$ | mod div **[f[Σ(j = 1; j = super-∞↑):(Hyper-relativity-of-simultaneity-j)]]** ≤ 0};{Grad [f[Σ(j = 1; j = super-∞↑):(Hyper-relativity-of-simultaneity-j)]] = $(\partial c/\partial x_1)\mathbf{i}_1 + (\partial c/\partial x_2)i_2, \ldots, + (\partial c/\partial x_n)\mathbf{i}_n$ **|** ∞ < mod Grad [f[Σ(j = 1; j = super-∞↑):(Hyper-relativity-of-simultaneity-j)]] ≤ ○, or mod Grad [f[Σ(j = 1; j = super-∞↑):(Hyper-relativity-of-simultaneity-j)]] > ○}}}}/dT}/dT}/ … /d T}

{d … {d{d {KE{f{{light-speed-craft-k | 1 ≤ k ≤ ○, or k > ○};{div **[f[Σ(j = 1; j = super-∞↑):(Hyper-relativity-of-simultaneity-j)]]** = $\partial \mathbf{c}_1/\partial x_1 + \partial \mathbf{c}_2/\partial x_2 + \ldots + \partial \mathbf{c}_n/\partial x_n$ | mod div **[f[Σ(j = 1; j = super-∞↑):(Hyper-relativity-of-simultaneity-j)]]** ≤ 0};{Grad [f[Σ(j = 1; j = super-∞↑):(Hyper-relativity-of-simultaneity-j)]] = $(\partial c/\partial x_1)\mathbf{i}_1 + (\partial c/\partial x_2)i_2, \ldots, + (\partial c/\partial x_n)\mathbf{i}_n$ **|**

∞ < mod Grad [f[Σ(j = 1; j = super-∞↑):(Hyper-relativity-of-simultaneity-j)]] ≤ ○, or mod Grad [f[Σ(j = 1; j = super-∞↑):(Hyper-relativity-of-simultaneity-j)]] > ○}}}}/dc}/dc}/ … /d c}

{d … {d{d {KE {f{{light-speed-craft-k | 1 ≤ k ≤ ○, or k > ○};{div **[f[Σ(j = 1; j = super-∞↑):(Hyper-relativity-of-simultaneity-j)]]** = $\partial \mathbf{c_1}/\partial x_1 + \partial \mathbf{c_2}/\partial x_2 + \ldots + \partial \mathbf{c_n}/\partial x_n$ | mod div **[f[Σ(j = 1; j = super-∞↑):(Hyper-relativity-of-simultaneity-j)]]** ≤ 0};{Grad [f[Σ(j = 1; j = super-∞↑):(Hyper-relativity-of-simultaneity-j)]] = $(\partial c/\partial x_1)\mathbf{i_1} + (\partial c/\partial x_2)i_2, \ldots, + (\partial c/\partial x_n)\mathbf{i_n}$ | ∞ < mod Grad [f[Σ(j = 1; j = super-∞↑):(Hyper-relativity-of-simultaneity-j)]] ≤ ○, or mod Grad [f[Σ(j = 1; j = super-∞↑):(Hyper-relativity-of-simultaneity-j)]] > ○}}}}/dγ}/dγ}/ … /d γ}

{d … {d{d {KE {f{{light-speed-craft-k | 1 ≤ k ≤ ○, or k > ○};{div **[f[Σ(j = 1; j = super-∞↑):(Hyper-relativity-of-simultaneity-j)]]** = $\partial \mathbf{c_1}/\partial x_1 + \partial \mathbf{c_2}/\partial x_2 + \ldots + \partial \mathbf{c_n}/\partial x_n$ | mod div **[f[Σ(j = 1; j = super-∞↑):(Hyper-relativity-of-simultaneity-j)]]** ≤ 0};{Grad [f[Σ(j = 1; j = super-∞↑):(Hyper-relativity-of-simultaneity-j)]] = $(\partial c/\partial x_1)\mathbf{i_1} + (\partial c/\partial x_2)i_2, \ldots, + (\partial c/\partial x_n)\mathbf{i_n}$ | ∞ < mod Grad [f[Σ(j = 1; j = super-∞↑):(Hyper-relativity-of-simultaneity-j)]] ≤ ○, or mod Grad [f[Σ(j = 1; j = super-∞↑):(Hyper-relativity-of-simultaneity-j)]] > ○}}}}/dv}/dv}/ … /d v}

{d … {d{d {KE {f{{light-speed-craft-k | 1 ≤ k ≤ ○, or k > ○};{div **[f[Σ(j = 1; j = super-∞↑):(Hyper-relativity-of-simultaneity-j)]]** = $\partial \mathbf{c_1}/\partial x_1 + \partial \mathbf{c_2}/\partial x_2 + \ldots + \partial \mathbf{c_n}/\partial x_n$ | mod div **[f[Σ(j = 1; j = super-∞↑):(Hyper-relativity-of-simultaneity-j)]]** ≤ 0};{Grad [f[Σ(j = 1; j = super-∞↑):(Hyper-relativity-of-simultaneity-j)]] = $(\partial c/\partial x_1)\mathbf{i_1} + (\partial c/\partial x_2)i_2, \ldots, + (\partial c/\partial x_n)\mathbf{i_n}$ | ∞ < mod Grad [f[Σ(j = 1; j = super-∞↑):(Hyper-relativity-of-simultaneity-j)]] ≤ ○, or mod Grad [f[Σ(j = 1; j = super-∞↑):(Hyper-relativity-of-simultaneity-j)]] > ○}}}}/da}/da}/ … /d a}

{d … {d{d {KE {f{{light-speed-craft-k | 1 ≤ k ≤ ○, or k > ○};{div **[f[Σ(j = 1; j = super-∞↑):(Hyper-relativity-of-simultaneity-j)]]** = $\partial \mathbf{c_1}/\partial x_1 + \partial \mathbf{c_2}/\partial x_2 + \ldots + \partial \mathbf{c_n}/\partial x_n$ | mod div **[f[Σ(j = 1; j = super-∞↑):(Hyper-relativity-of-simultaneity-j)]]** ≤ 0};{Grad [f[Σ(j = 1; j = super-∞↑):(Hyper-relativity-of-simultaneity-j)]] = $(\partial c/\partial x_1)\mathbf{i_1} + (\partial c/\partial x_2)i_2, \ldots, + (\partial c/\partial x_n)\mathbf{i_n}$ | ∞ < mod Grad [f[Σ(j = 1; j = super-∞↑):(Hyper-relativity-of-simultaneity-j)]] ≤ ○, or mod Grad [f[Σ(j = 1; j = super-∞↑):(Hyper-relativity-of-simultaneity-j)]] > ○}}}}/dP}/dP}/ … /d P}

{d … {d{d {KE {f{{light-speed-craft-k | 1 ≤ k ≤ ○, or k > ○};{div **[f[Σ(j = 1; j = super-∞↑):(Hyper-relativity-of-simultaneity-j)]]** = $\partial \mathbf{c_1}/\partial x_1 + \partial \mathbf{c_2}/\partial x_2 + \ldots + \partial \mathbf{c_n}/\partial x_n$ | mod div **[f[Σ(j = 1; j = super-∞↑):(Hyper-relativity-of-simultaneity-j)]]** ≤ 0};{Grad [f[Σ(j = 1; j = super-∞↑):(Hyper-relativity-of-simultaneity-j)]] = $(\partial c/\partial x_1)\mathbf{i_1} + (\partial c/\partial x_2)i_2, \ldots, + (\partial c/\partial x_n)\mathbf{i_n}$ | ∞ < mod Grad [f[Σ(j = 1; j = super-∞↑):(Hyper-relativity-of-simultaneity-j)]] ≤ ○, or mod Grad [f[Σ(j = 1; j = super-∞↑):(Hyper-relativity-of-simultaneity-j)]] > ○}}}}/dKE}/dKE}/ … /d KE} {d … {d{d {KE {f{{light-speed-craft-k | 1 ≤ k ≤ ◯, or k > ◯};{div **[f[Σ(j = 1; j = super-∞↑):(Hyper-relativity-of-simultaneity-j)]]** = $\partial \mathbf{c_1}/\partial x_1 + \partial \mathbf{c_2}/\partial x_2 + \ldots + \partial \mathbf{c_n}/\partial x_n$ | mod div **[f[Σ(j = 1; j = super-∞↑):(Hyper-relativity-of-simultaneity-j)]]** ≤ 0};{Grad [f[Σ(j = 1; j = super-∞↑):(Hyper-relativity-of-simultaneity-j)]] = $(\partial c/\partial x_1)\mathbf{i_1} + (\partial c/\partial x_2)i_2,\ldots, + (\partial c/\partial x_n)\mathbf{i_n}$ | ∞ < mod Grad [f[Σ(j = 1; j = super-∞↑):(Hyper-relativity-of-simultaneity-j)]] ≤ ◯, or mod Grad [f[Σ(j = 1; j = super-∞↑):(Hyper-relativity-of-simultaneity-j)]] > ◯}}}}/dt}/dt}/ … /dt}

{d … {d{d {KE {f{{light-speed-craft-k | 1 ≤ k ≤ ◯, or k > ◯};{div **[f[Σ(j = 1; j = super-∞↑):(Hyper-relativity-of-simultaneity-j)]]** = $\partial \mathbf{c_1}/\partial x_1 + \partial \mathbf{c_2}/\partial x_2 + \ldots + \partial \mathbf{c_n}/\partial x_n$ | mod div **[f[Σ(j = 1; j = super-∞↑):(Hyper-relativity-of-simultaneity-j)]]** ≤ 0};{Grad [f[Σ(j = 1; j = super-∞↑):(Hyper-relativity-of-simultaneity-j)]] = $(\partial c/\partial x_1)\mathbf{i_1} + (\partial c/\partial x_2)i_2,\ldots, + (\partial c/\partial x_n)\mathbf{i_n}$ | ∞ < mod Grad [f[Σ(j = 1; j = super-∞↑):(Hyper-relativity-of-simultaneity-j)]] ≤ ◯, or mod Grad [f[Σ(j = 1; j = super-∞↑):(Hyper-relativity-of-simultaneity-j)]] > ◯}}}}/dT}/dT}/ … /d T}

{d … {d{d {t{f{{light-speed-craft-k | 1 ≤ k ≤ ◯, or k > ◯};{div **[f[Σ(j = 1; j = super-∞↑):(Hyper-relativity-of-simultaneity-j)]]** = $\partial \mathbf{c_1}/\partial x_1 + \partial \mathbf{c_2}/\partial x_2 + \ldots + \partial \mathbf{c_n}/\partial x_n$ | mod div **[f[Σ(j = 1; j = super-∞↑):(Hyper-relativity-of-simultaneity-j)]]** ≤ 0};{Grad [f[Σ(j = 1; j = super-∞↑):(Hyper-relativity-of-simultaneity-j)]] = $(\partial c/\partial x_1)\mathbf{i_1} + (\partial c/\partial x_2)i_2,\ldots, + (\partial c/\partial x_n)\mathbf{i_n}$ | ∞ < mod Grad [f[Σ(j = 1; j = super-∞↑):(Hyper-relativity-of-simultaneity-j)]] ≤ ◯, or mod Grad [f[Σ(j = 1; j = super-∞↑):(Hyper-relativity-of-simultaneity-j)]] > ◯}}}}/dc}/dc}/ … /d c}

{d … {d{d {t{f{{light-speed-craft-k | 1 ≤ k ≤ ◯, or k > ◯};{div **[f[Σ(j = 1; j = super-∞↑):(Hyper-relativity-of-simultaneity-j)]]** = $\partial \mathbf{c_1}/\partial x_1 + \partial \mathbf{c_2}/\partial x_2 + \ldots + \partial \mathbf{c_n}/\partial x_n$ | mod div **[f[Σ(j = 1; j = super-∞↑):(Hyper-relativity-of-simultaneity-j)]]** ≤ 0};{Grad [f[Σ(j = 1; j = super-∞↑):(Hyper-relativity-of-simultaneity-j)]] = $(\partial c/\partial x_1)\mathbf{i_1} + (\partial c/\partial x_2)i_2,\ldots, + (\partial c/\partial x_n)\mathbf{i_n}$ | ∞ < mod Grad [f[Σ(j = 1; j = super-∞↑):(Hyper-relativity-of-simultaneity-j)]] ≤ ◯, or mod Grad [f[Σ(j = 1; j = super-∞↑):(Hyper-relativity-of-simultaneity-j)]] > ◯}}}}/dγ}/dγ}/ … /d γ}

{d … {d{d {t{f{{light-speed-craft-k | 1 ≤ k ≤ ◯, or k > ◯};{div **[f[Σ(j = 1; j = super-∞↑):(Hyper-relativity-of-simultaneity-j)]]** = $\partial \mathbf{c_1}/\partial x_1 + \partial \mathbf{c_2}/\partial x_2 + \ldots + \partial \mathbf{c_n}/\partial x_n$ | mod div **[f[Σ(j = 1; j = super-∞↑):(Hyper-relativity-of-simultaneity-j)]]** ≤ 0};{Grad [f[Σ(j = 1; j = super-∞↑):(Hyper-relativity-of-simultaneity-j)]] = $(\partial c/\partial x_1)\mathbf{i_1} + (\partial c/\partial x_2)i_2,\ldots, + (\partial c/\partial x_n)\mathbf{i_n}$ | ∞ < mod Grad [f[Σ(j = 1; j = super-∞↑):(Hyper-relativity-of-simultaneity-j)]] ≤ ◯, or mod Grad [f[Σ(j = 1; j = super-∞↑):(Hyper-relativity-of-simultaneity-j)]] > ◯}}}}/dv}/dv}/ … /d v}

{d … {d{d {t{f{{light-speed-craft-k | 1 ≤ k ≤ ◯, or k > ◯};{div **[f[Σ(j = 1; j = super-∞↑):(Hyper-relativity-of-simultaneity-j)]]** = $\partial \mathbf{c_1}/\partial x_1 + \partial \mathbf{c_2}/\partial x_2 + \ldots + \partial \mathbf{c_n}/\partial x_n$ | mod div **[f[Σ(j = 1; j = super-∞↑):(Hyper-relativity-of-simultaneity-j)]]** ≤ 0};{Grad [f[Σ(j = 1; j = super-∞↑):(Hyper-relativity-of-simultaneity-j)]] = $(\partial c/\partial x_1)\mathbf{i_1} + (\partial c/\partial x_2)i_2,\ldots, + (\partial c/\partial x_n)\mathbf{i_n}$ |

∞ < mod Grad [f[Σ(j = 1; j = super-∞↑):(Hyper-relativity-of-simultaneity-j)]] ≤ ◯, or mod Grad [f[Σ(j = 1; j = super-∞↑):(Hyper-relativity-of-simultaneity-j)]] > ◯}}}}/da}/da}/ … /d a}

{d … {d{d {t{f{{light-speed-craft-k | 1 ≤ k ≤ ◯, or k > ◯};{div **[f[Σ(j = 1; j = super-∞↑):(Hyper-relativity-of-simultaneity-j)]]** = $\partial \mathbf{c}_1/\partial x_1 + \partial \mathbf{c}_2/\partial x_2 + \ldots + \partial \mathbf{c}_n/\partial x_n$ | mod div **[f[Σ(j = 1; j = super-∞↑):(Hyper-relativity-of-simultaneity-j)]]** ≤ 0};{Grad [f[Σ(j = 1; j = super-∞↑):(Hyper-relativity-of-simultaneity-j)]] = $(\partial c/\partial x_1)\mathbf{i}_1 + (\partial c/\partial x_2)i_2,\ldots, + (\partial c/\partial x_n)\mathbf{i}_n$ | ∞ < mod Grad [f[Σ(j = 1; j = super-∞↑):(Hyper-relativity-of-simultaneity-j)]] ≤ ◯, or mod Grad [f[Σ(j = 1; j = super-∞↑):(Hyper-relativity-of-simultaneity-j)]] > ◯}}}}/dP}/dP}/ … /d P}

{d … {d{d {t{f{{light-speed-craft-k | 1 ≤ k ≤ ◯, or k > ◯};{div **[f[Σ(j = 1; j = super-∞↑):(Hyper-relativity-of-simultaneity-j)]]** = $\partial \mathbf{c}_1/\partial x_1 + \partial \mathbf{c}_2/\partial x_2 + \ldots + \partial \mathbf{c}_n/\partial x_n$ | mod div **[f[Σ(j = 1; j = super-∞↑):(Hyper-relativity-of-simultaneity-j)]]** ≤ 0};{Grad [f[Σ(j = 1; j = super-∞↑):(Hyper-relativity-of-simultaneity-j)]] = $(\partial c/\partial x_1)\mathbf{i}_1 + (\partial c/\partial x_2)i_2,\ldots, + (\partial c/\partial x_n)\mathbf{i}_n$ | ∞ < mod Grad [f[Σ(j = 1; j = super-∞↑):(Hyper-relativity-of-simultaneity-j)]] ≤ ◯, or mod Grad [f[Σ(j = 1; j = super-∞↑):(Hyper-relativity-of-simultaneity-j)]] > ◯}}}}/dKE}/dKE}/ … /d KE}

{d … {d{d {t{f{{light-speed-craft-k | 1 ≤ k ≤ ◯, or k > ◯};{div **[f[Σ(j = 1; j = super-∞↑):(Hyper-relativity-of-simultaneity-j)]]** = $\partial \mathbf{c}_1/\partial x_1 + \partial \mathbf{c}_2/\partial x_2 + \ldots + \partial \mathbf{c}_n/\partial x_n$ | mod div **[f[Σ(j = 1; j = super-∞↑):(Hyper-relativity-of-simultaneity-j)]]** ≤ 0};{Grad [f[Σ(j = 1; j = super-∞↑):(Hyper-relativity-of-simultaneity-j)]] = $(\partial c/\partial x_1)\mathbf{i}_1 + (\partial c/\partial x_2)i_2,\ldots, + (\partial c/\partial x_n)\mathbf{i}_n$ | ∞ < mod Grad [f[Σ(j = 1; j = super-∞↑):(Hyper-relativity-of-simultaneity-j)]] ≤ ◯, or mod Grad [f[Σ(j = 1; j = super-∞↑):(Hyper-relativity-of-simultaneity-j)]] > ◯}}}}/dt}/dt}/ … /dt}

{d … {d{d {t{f{{light-speed-craft-k | 1 ≤ k ≤ ◯, or k > ◯};{div **[f[Σ(j = 1; j = super-∞↑):(Hyper-relativity-of-simultaneity-j)]]** = $\partial \mathbf{c}_1/\partial x_1 + \partial \mathbf{c}_2/\partial x_2 + \ldots + \partial \mathbf{c}_n/\partial x_n$ | mod div **[f[Σ(j = 1; j = super-∞↑):(Hyper-relativity-of-simultaneity-j)]]** ≤ 0};{Grad [f[Σ(j = 1; j = super-∞↑):(Hyper-relativity-of-simultaneity-j)]] = $(\partial c/\partial x_1)\mathbf{i}_1 + (\partial c/\partial x_2)i_2,\ldots, + (\partial c/\partial x_n)\mathbf{i}_n$ | ∞ < mod Grad [f[Σ(j = 1; j = super-∞↑):(Hyper-relativity-of-simultaneity-j)]] ≤ ◯, or mod Grad [f[Σ(j = 1; j = super-∞↑):(Hyper-relativity-of-simultaneity-j)]] > ◯}}}}/dT}/dT}/ … /d T}

{d … {d{d {T{f{{light-speed-craft-k | 1 ≤ k ≤ ◯, or k > ◯};{div **[f[Σ(j = 1; j = super-∞↑):(Hyper-relativity-of-simultaneity-j)]]** = $\partial \mathbf{c}_1/\partial x_1 + \partial \mathbf{c}_2/\partial x_2 + \ldots + \partial \mathbf{c}_n/\partial x_n$ | mod div **[f[Σ(j = 1; j = super-∞↑):(Hyper-relativity-of-simultaneity-j)]]** ≤ 0};{Grad [f[Σ(j = 1; j = super-∞↑):(Hyper-relativity-of-simultaneity-j)]] = $(\partial c/\partial x_1)\mathbf{i}_1 + (\partial c/\partial x_2)i_2,\ldots, + (\partial c/\partial x_n)\mathbf{i}_n$ | ∞ < mod Grad [f[Σ(j = 1; j = super-∞↑):(Hyper-relativity-of-simultaneity-j)]] ≤ ◯, or mod Grad [f[Σ(j = 1; j = super-∞↑):(Hyper-relativity-of-simultaneity-j)]] > ◯}}}}/dc}/dc}/ … /d c} {d … {d{d {T{f{{light-speed-craft-k | 1 ≤ k ≤ ◯, or k > ◯};{div **[f[Σ(j = 1; j = super-∞↑):(Hyper-relativity-of-simultaneity-j)]]** = $\partial \mathbf{c}_1/\partial x_1 + \partial \mathbf{c}_2/\partial x_2 + \ldots + \partial \mathbf{c}_n/\partial x_n$ | mod div **[f[Σ(j = 1; j = super-∞↑):(Hyper-relativity-of-simultaneity-j)]]** ≤ 0};{Grad [f[Σ(j = 1; j = super-∞↑):(Hyper-relativity-of-simultaneity-j)]] = $(\partial c/\partial x_1)\mathbf{i}_1 + (\partial c/\partial x_2)i_2, \ldots, + (\partial c/\partial x_n)\mathbf{i}_n$ | ∞ < mod Grad [f[Σ(j = 1; j = super-∞↑):(Hyper-relativity-of-simultaneity-j)]] ≤ ◯, or mod Grad [f[Σ(j = 1; j = super-∞↑):(Hyper-relativity-of-simultaneity-j)]] > ◯}}}}/dγ}/dγ}/ … /d γ}

{d … {d{d {T{f{{light-speed-craft-k | 1 ≤ k ≤ ◯, or k > ◯};{div **[f[Σ(j = 1; j = super-∞↑):(Hyper-relativity-of-simultaneity-j)]]** = $\partial \mathbf{c}_1/\partial x_1 + \partial \mathbf{c}_2/\partial x_2 + \ldots + \partial \mathbf{c}_n/\partial x_n$ | mod div **[f[Σ(j = 1; j = super-∞↑):(Hyper-relativity-of-simultaneity-j)]]** ≤ 0};{Grad [f[Σ(j = 1; j = super-∞↑):(Hyper-relativity-of-simultaneity-j)]] = $(\partial c/\partial x_1)\mathbf{i}_1 + (\partial c/\partial x_2)i_2, \ldots, + (\partial c/\partial x_n)\mathbf{i}_n$ | ∞ < mod Grad [f[Σ(j = 1; j = super-∞↑):(Hyper-relativity-of-simultaneity-j)]] ≤ ◯, or mod Grad [f[Σ(j = 1; j = super-∞↑):(Hyper-relativity-of-simultaneity-j)]] > ◯}}}}/dv}/dv}/ … /d v}

{d … {d{d {T{f{{light-speed-craft-k | 1 ≤ k ≤ ◯, or k > ◯};{div **[f[Σ(j = 1; j = super-∞↑):(Hyper-relativity-of-simultaneity-j)]]** = $\partial \mathbf{c}_1/\partial x_1 + \partial \mathbf{c}_2/\partial x_2 + \ldots + \partial \mathbf{c}_n/\partial x_n$ | mod div **[f[Σ(j = 1; j = super-∞↑):(Hyper-relativity-of-simultaneity-j)]]** ≤ 0};{Grad [f[Σ(j = 1; j = super-∞↑):(Hyper-relativity-of-simultaneity-j)]] = $(\partial c/\partial x_1)\mathbf{i}_1 + (\partial c/\partial x_2)i_2, \ldots, + (\partial c/\partial x_n)\mathbf{i}_n$ | ∞ < mod Grad [f[Σ(j = 1; j = super-∞↑):(Hyper-relativity-of-simultaneity-j)]] ≤ ◯, or mod Grad [f[Σ(j = 1; j = super-∞↑):(Hyper-relativity-of-simultaneity-j)]] > ◯}}}}/da}/da}/ … /d a}

{d … {d{d {T{f{{light-speed-craft-k | 1 ≤ k ≤ ◯, or k > ◯};{div **[f[Σ(j = 1; j = super-∞↑):(Hyper-relativity-of-simultaneity-j)]]** = $\partial \mathbf{c}_1/\partial x_1 + \partial \mathbf{c}_2/\partial x_2 + \ldots + \partial \mathbf{c}_n/\partial x_n$ | mod div **[f[Σ(j = 1; j = super-∞↑):(Hyper-relativity-of-simultaneity-j)]]** ≤ 0};{Grad [f[Σ(j = 1; j = super-∞↑):(Hyper-relativity-of-simultaneity-j)]] = $(\partial c/\partial x_1)\mathbf{i}_1 + (\partial c/\partial x_2)i_2, \ldots, + (\partial c/\partial x_n)\mathbf{i}_n$ | ∞ < mod Grad [f[Σ(j = 1; j = super-∞↑):(Hyper-relativity-of-simultaneity-j)]] ≤ ◯, or mod Grad [f[Σ(j = 1; j = super-∞↑):(Hyper-relativity-of-simultaneity-j)]] > ◯}}}}/dP}/dP}/ … /d P}

{d … {d{d {T{f{{light-speed-craft-k | 1 ≤ k ≤ ◯, or k > ◯};{div **[f[Σ(j = 1; j = super-∞↑):(Hyper-relativity-of-simultaneity-j)]]** = $\partial \mathbf{c}_1/\partial x_1 + \partial \mathbf{c}_2/\partial x_2 + \ldots + \partial \mathbf{c}_n/\partial x_n$ | mod div **[f[Σ(j = 1; j = super-∞↑):(Hyper-relativity-of-simultaneity-j)]]** ≤ 0};{Grad [f[Σ(j = 1; j = super-∞↑):(Hyper-relativity-of-simultaneity-j)]] = $(\partial c/\partial x_1)\mathbf{i}_1 + (\partial c/\partial x_2)i_2, \ldots, + (\partial c/\partial x_n)\mathbf{i}_n$ | ∞ < mod Grad [f[Σ(j = 1; j = super-∞↑):(Hyper-relativity-of-simultaneity-j)]] ≤ ◯, or mod Grad [f[Σ(j = 1; j = super-∞↑):(Hyper-relativity-of-simultaneity-j)]] > ◯}}}}/dKE}/dKE}/ … /d KE}

{d … {d{d {T{f{{light-speed-craft-k | 1 ≤ k ≤ ◯, or k > ◯};{div **[f[Σ(j = 1; j = super-∞↑):(Hyper-relativity-of-simultaneity-j)]]** = $\partial \mathbf{c}_1/\partial x_1 + \partial \mathbf{c}_2/\partial x_2 + \ldots + \partial \mathbf{c}_n/\partial x_n$ | mod div

[f[Σ(j = 1; j = super-∞↑):(Hyper-relativity-of-simultaneity-j)]] ≤ 0};{Grad [f[Σ(j = 1; j = super-∞↑):(Hyper-relativity-of-simultaneity-j)]] = (∂c/∂x_1)$\mathbf{i}_1$ + (∂c/∂x_2)i_2,..., + (∂c/∂x_n)$\mathbf{i}_n$ | ∞ < mod Grad [f[Σ(j = 1; j = super-∞↑):(Hyper-relativity-of-simultaneity-j)]] ≤ ◯, or mod Grad [f[Σ(j = 1; j = super-∞↑):(Hyper-relativity-of-simultaneity-j)]] > ◯}}}}/dt}/dt}/ ... /dt}

{d ... {d{d {T{f{{light-speed-craft-k | 1 ≤ k ≤ ◯, or k > ◯};{div **[f[Σ(j = 1; j = super-∞↑):(Hyper-relativity-of-simultaneity-j)]]** = ∂$\mathbf{c}_1$/∂x_1 + ∂$\mathbf{c}_2$/∂x_2 + ... + ∂$\mathbf{c}_n$/∂x_n | mod div **[f[Σ(j = 1; j = super-∞↑):(Hyper-relativity-of-simultaneity-j)]]** ≤ 0};{Grad [f[Σ(j = 1; j = super-∞↑):(Hyper-relativity-of-simultaneity-j)]] = (∂c/∂x_1)$\mathbf{i}_1$ + (∂c/∂x_2)i_2,..., + (∂c/∂x_n)$\mathbf{i}_n$ | ∞ < mod Grad [f[Σ(j = 1; j = super-∞↑):(Hyper-relativity-of-simultaneity-j)]] ≤ ◯, or mod Grad [f[Σ(j = 1; j = super-∞↑):(Hyper-relativity-of-simultaneity-j)]] > ◯}}}}/dT}/dT}/ ... /d T}

and so on.

We can take light-speed, gamma, velocity, acceleration, momentum, kinetic energy, and so on integrals of the basic first operator as functionally modified to represent ship and background time, light-speed, gamma, velocity, acceleration, momentum, kinetic energy, and so on. We may affix the resulting integral operators to formulas for gamma and related parameters to denote the operative mechanisms without actually including these operators in the formulas.

{∫ ... {∫{∫ {c{f{{light-speed-craft-k | 1 ≤ k ≤ ◯, or k > ◯};{div **[f[Σ(j = 1; j = super-∞↑):(Hyper-relativity-of-simultaneity-j)]]** = ∂$\mathbf{c}_1$/∂x_1 + ∂$\mathbf{c}_2$/∂x_2 + ... + ∂$\mathbf{c}_n$/∂x_n | mod div **[f[Σ(j = 1; j = super-∞↑):(Hyper-relativity-of-simultaneity-j)]]** ≤ 0};{Grad [f[Σ(j = 1; j = super-∞↑):(Hyper-relativity-of-simultaneity-j)]] = (∂c/∂x_1)$\mathbf{i}_1$ + (∂c/∂x_2)i_2,..., + (∂c/∂x_n)$\mathbf{i}_n$ | ∞ < mod Grad [f[Σ(j = 1; j = super-∞↑):(Hyper-relativity-of-simultaneity-j)]] ≤ ◯, or mod Grad [f[Σ(j = 1; j = super-∞↑):(Hyper-relativity-of-simultaneity-j)]] > ◯}}}}dc}dc} ... d c}

{∫ ... {∫{∫ {c{f{{light-speed-craft-k | 1 ≤ k ≤ ◯, or k > ◯};{div **[f[Σ(j = 1; j = super-∞↑):(Hyper-relativity-of-simultaneity-j)]]** = ∂$\mathbf{c}_1$/∂x_1 + ∂$\mathbf{c}_2$/∂x_2 + ... + ∂$\mathbf{c}_n$/∂x_n | mod div **[f[Σ(j = 1; j = super-∞↑):(Hyper-relativity-of-simultaneity-j)]]** ≤ 0};{Grad [f[Σ(j = 1; j = super-∞↑):(Hyper-relativity-of-simultaneity-j)]] = (∂c/∂x_1)$\mathbf{i}_1$ + (∂c/∂x_2)i_2,..., + (∂c/∂x_n)$\mathbf{i}_n$ | ∞ < mod Grad [f[Σ(j = 1; j = super-∞↑):(Hyper-relativity-of-simultaneity-j)]] ≤ ◯, or mod Grad [f[Σ(j = 1; j = super-∞↑):(Hyper-relativity-of-simultaneity-j)]] > ◯}}}}dγ}dγ} ... d γ}

{∫ ... {∫{∫ {c{f{{light-speed-craft-k | 1 ≤ k ≤ ◯, or k > ◯};{div **[f[Σ(j = 1; j = super-∞↑):(Hyper-relativity-of-simultaneity-j)]]** = ∂$\mathbf{c}_1$/∂x_1 + ∂$\mathbf{c}_2$/∂x_2 + ... + ∂$\mathbf{c}_n$/∂x_n | mod div **[f[Σ(j = 1; j = super-∞↑):(Hyper-relativity-of-simultaneity-j)]]** ≤ 0};{Grad [f[Σ(j = 1; j = super-∞↑):(Hyper-relativity-of-simultaneity-j)]] = (∂c/∂x_1)$\mathbf{i}_1$ + (∂c/∂x_2)i_2,..., + (∂c/∂x_n)$\mathbf{i}_n$ |

∞ < mod Grad [f[Σ(j = 1; j = super-∞↑):(Hyper-relativity-of-simultaneity-j)]] ≤ ◯, or mod Grad [f[Σ(j = 1; j = super-∞↑):(Hyper-relativity-of-simultaneity-j)]] > ◯}}}}dv}dv} … d v}

{∫ … {∫{∫ {c{f{{light-speed-craft-k | 1 ≤ k ≤ ◯, or k > ◯};{div **[f[Σ(j = 1; j = super-∞↑):(Hyper-relativity-of-simultaneity-j)]]** = $\partial \mathbf{c_1}/\partial x_1 + \partial \mathbf{c_2}/\partial x_2 + \dots + \partial \mathbf{c_n}/\partial x_n$ | mod div **[f[Σ(j = 1; j = super-∞↑):(Hyper-relativity-of-simultaneity-j)]]** ≤ 0};{Grad [f[Σ(j = 1; j = super-∞↑):(Hyper-relativity-of-simultaneity-j)]] = $(\partial c/\partial x_1)\mathbf{i_1} + (\partial c/\partial x_2)i_2, \dots, + (\partial c/\partial x_n)\mathbf{i_n}$ | ∞ < mod Grad [f[Σ(j = 1; j = super-∞↑):(Hyper-relativity-of-simultaneity-j)]] ≤ ◯, or mod Grad [f[Σ(j = 1; j = super-∞↑):(Hyper-relativity-of-simultaneity-j)]] > ◯}}}}da}da} … d a}

{∫ … {∫{∫ {c{f{{light-speed-craft-k | 1 ≤ k ≤ ◯, or k > ◯};{div **[f[Σ(j = 1; j = super-∞↑):(Hyper-relativity-of-simultaneity-j)]]** = $\partial \mathbf{c_1}/\partial x_1 + \partial \mathbf{c_2}/\partial x_2 + \dots + \partial \mathbf{c_n}/\partial x_n$ | mod div **[f[Σ(j = 1; j = super-∞↑):(Hyper-relativity-of-simultaneity-j)]]** ≤ 0};{Grad [f[Σ(j = 1; j = super-∞↑):(Hyper-relativity-of-simultaneity-j)]] = $(\partial c/\partial x_1)\mathbf{i_1} + (\partial c/\partial x_2)i_2, \dots, + (\partial c/\partial x_n)\mathbf{i_n}$ | ∞ < mod Grad [f[Σ(j = 1; j = super-∞↑):(Hyper-relativity-of-simultaneity-j)]] ≤ ◯, or mod Grad [f[Σ(j = 1; j = super-∞↑):(Hyper-relativity-of-simultaneity-j)]] > ◯}}}}dP}dP} … d P}

{∫ … {∫{∫ {c{f{{light-speed-craft-k | 1 ≤ k ≤ ◯, or k > ◯};{div **[f[Σ(j = 1; j = super-∞↑):(Hyper-relativity-of-simultaneity-j)]]** = $\partial \mathbf{c_1}/\partial x_1 + \partial \mathbf{c_2}/\partial x_2 + \dots + \partial \mathbf{c_n}/\partial x_n$ | mod div **[f[Σ(j = 1; j = super-∞↑):(Hyper-relativity-of-simultaneity-j)]]** ≤ 0};{Grad [f[Σ(j = 1; j = super-∞↑):(Hyper-relativity-of-simultaneity-j)]] = $(\partial c/\partial x_1)\mathbf{i_1} + (\partial c/\partial x_2)i_2, \dots, + (\partial c/\partial x_n)\mathbf{i_n}$ | ∞ < mod Grad [f[Σ(j = 1; j = super-∞↑):(Hyper-relativity-of-simultaneity-j)]] ≤ ◯, or mod Grad [f[Σ(j = 1; j = super-∞↑):(Hyper-relativity-of-simultaneity-j)]] > ◯}}}}dKE}dKE} … d KE}

{∫ … {∫{∫ {c{f{{light-speed-craft-k | 1 ≤ k ≤ ◯, or k > ◯};{div **[f[Σ(j = 1; j = super-∞↑):(Hyper-relativity-of-simultaneity-j)]]** = $\partial \mathbf{c_1}/\partial x_1 + \partial \mathbf{c_2}/\partial x_2 + \dots + \partial \mathbf{c_n}/\partial x_n$ | mod div **[f[Σ(j = 1; j = super-∞↑):(Hyper-relativity-of-simultaneity-j)]]** ≤ 0};{Grad [f[Σ(j = 1; j = super-∞↑):(Hyper-relativity-of-simultaneity-j)]] = $(\partial c/\partial x_1)\mathbf{i_1} + (\partial c/\partial x_2)i_2, \dots, + (\partial c/\partial x_n)\mathbf{i_n}$ | ∞ < mod Grad [f[Σ(j = 1; j = super-∞↑):(Hyper-relativity-of-simultaneity-j)]] ≤ ◯, or mod Grad [f[Σ(j = 1; j = super-∞↑):(Hyper-relativity-of-simultaneity-j)]] > ◯}}}}dt}dt} ... dt}

{∫ … {∫{∫ {c{f{{light-speed-craft-k | 1 ≤ k ≤ ◯, or k > ◯};{div **[f[Σ(j = 1; j = super-∞↑):(Hyper-relativity-of-simultaneity-j)]]** = $\partial \mathbf{c_1}/\partial x_1 + \partial \mathbf{c_2}/\partial x_2 + \dots + \partial \mathbf{c_n}/\partial x_n$ | mod div **[f[Σ(j = 1; j = super-∞↑):(Hyper-relativity-of-simultaneity-j)]]** ≤ 0};{Grad [f[Σ(j = 1; j = super-∞↑):(Hyper-relativity-of-simultaneity-j)]] = $(\partial c/\partial x_1)\mathbf{i_1} + (\partial c/\partial x_2)i_2, \dots, + (\partial c/\partial x_n)\mathbf{i_n}$ | ∞ < mod Grad [f[Σ(j = 1; j = super-∞↑):(Hyper-relativity-of-simultaneity-j)]] ≤ ◯, or mod Grad [f[Σ(j = 1; j = super-∞↑):(Hyper-relativity-of-simultaneity-j)]] > ◯}}}}dT}dT} … d T}

{∫ … {∫{∫ {γ{f{{light-speed-craft-k | 1 ≤ k ≤ ◯, or k > ◯};{div **[f[Σ(j = 1; j = super-∞↑):(Hyper-relativity-of-simultaneity-j)]]** = $\partial \mathbf{c_1}/\partial x_1 + \partial \mathbf{c_2}/\partial x_2 + \dots + \partial \mathbf{c_n}/\partial x_n$ | mod div **[f[Σ(j = 1; j = super-∞↑):(Hyper-relativity-of-simultaneity-j)]]** ≤ 0};{Grad [f[Σ(j = 1; j =

super-∞↑):(Hyper-relativity-of-simultaneity-j)]] = $(\partial c/\partial x_1)\mathbf{i_1}$ + $(\partial c/\partial x_2)i_2,\ldots,$ + $(\partial c/\partial x_n)\mathbf{i_n}$ | ∞ < mod Grad [f[Σ(j = 1; j = super-∞↑):(Hyper-relativity-of-simultaneity-j)]] ≤ ○, or mod Grad [f[Σ(j = 1; j = super-∞↑):(Hyper-relativity-of-simultaneity-j)]] > ○}}}}dc}dc} … d c}

{∫ … {∫{∫ {γ{f{{light-speed-craft-k | 1 ≤ k ≤ ○, or k > ○};{div **[f[Σ(j = 1; j = super-∞↑):(Hyper-relativity-of-simultaneity-j)]]** = $\partial \mathbf{c_1}/\partial x_1 + \partial \mathbf{c_2}/\partial x_2 + \ldots + \partial \mathbf{c_n}/\partial x_n$ | mod div **[f[Σ(j = 1; j = super-∞↑):(Hyper-relativity-of-simultaneity-j)]]** ≤ 0};{Grad [f[Σ(j = 1; j = super-∞↑):(Hyper-relativity-of-simultaneity-j)]] = $(\partial c/\partial x_1)\mathbf{i_1}$ + $(\partial c/\partial x_2)i_2,\ldots,$ + $(\partial c/\partial x_n)\mathbf{i_n}$ | ∞ < mod Grad [f[Σ(j = 1; j = super-∞↑):(Hyper-relativity-of-simultaneity-j)]] ≤ ○, or mod Grad [f[Σ(j = 1; j = super-∞↑):(Hyper-relativity-of-simultaneity-j)]] > ○}}}}dγ}dγ} … d γ}

{∫ … {∫{∫ {γ{f{{light-speed-craft-k | 1 ≤ k ≤ ○, or k > ○};{div **[f[Σ(j = 1; j = super-∞↑):(Hyper-relativity-of-simultaneity-j)]]** = $\partial \mathbf{c_1}/\partial x_1 + \partial \mathbf{c_2}/\partial x_2 + \ldots + \partial \mathbf{c_n}/\partial x_n$ | mod div **[f[Σ(j = 1; j = super-∞↑):(Hyper-relativity-of-simultaneity-j)]]** ≤ 0};{Grad [f[Σ(j = 1; j = super-∞↑):(Hyper-relativity-of-simultaneity-j)]] = $(\partial c/\partial x_1)\mathbf{i_1}$ + $(\partial c/\partial x_2)i_2,\ldots,$ + $(\partial c/\partial x_n)\mathbf{i_n}$ | ∞ < mod Grad [f[Σ(j = 1; j = super-∞↑):(Hyper-relativity-of-simultaneity-j)]] ≤ ○, or mod Grad [f[Σ(j = 1; j = super-∞↑):(Hyper-relativity-of-simultaneity-j)]] > ○}}}}dv}dv} … d v}

{∫ … {∫{∫ {γ{f{{light-speed-craft-k | 1 ≤ k ≤ ○, or k > ○};{div **[f[Σ(j = 1; j = super-∞↑):(Hyper-relativity-of-simultaneity-j)]]** = $\partial \mathbf{c_1}/\partial x_1 + \partial \mathbf{c_2}/\partial x_2 + \ldots + \partial \mathbf{c_n}/\partial x_n$ | mod div **[f[Σ(j = 1; j = super-∞↑):(Hyper-relativity-of-simultaneity-j)]]** ≤ 0};{Grad [f[Σ(j = 1; j = super-∞↑):(Hyper-relativity-of-simultaneity-j)]] = $(\partial c/\partial x_1)\mathbf{i_1}$ + $(\partial c/\partial x_2)i_2,\ldots,$ + $(\partial c/\partial x_n)\mathbf{i_n}$ | ∞ < mod Grad [f[Σ(j = 1; j = super-∞↑):(Hyper-relativity-of-simultaneity-j)]] ≤ ○, or mod Grad [f[Σ(j = 1; j = super-∞↑):(Hyper-relativity-of-simultaneity-j)]] > ○}}}}da}da} … d a}

{∫ … {∫{∫ {γ{f{{light-speed-craft-k | 1 ≤ k ≤ ○, or k > ○};{div **[f[Σ(j = 1; j = super-∞↑):(Hyper-relativity-of-simultaneity-j)]]** = $\partial \mathbf{c_1}/\partial x_1 + \partial \mathbf{c_2}/\partial x_2 + \ldots + \partial \mathbf{c_n}/\partial x_n$ | mod div **[f[Σ(j = 1; j = super-∞↑):(Hyper-relativity-of-simultaneity-j)]]** ≤ 0};{Grad [f[Σ(j = 1; j = super-∞↑):(Hyper-relativity-of-simultaneity-j)]] = $(\partial c/\partial x_1)\mathbf{i_1}$ + $(\partial c/\partial x_2)i_2,\ldots,$ + $(\partial c/\partial x_n)\mathbf{i_n}$ | ∞ < mod Grad [f[Σ(j = 1; j = super-∞↑):(Hyper-relativity-of-simultaneity-j)]] ≤ ○, or mod Grad [f[Σ(j = 1; j = super-∞↑):(Hyper-relativity-of-simultaneity-j)]] > ○}}}}dP}dP} … d P}

{∫ … {∫{∫ {γ{f{{light-speed-craft-k | 1 ≤ k ≤ ○, or k > ○};{div **[f[Σ(j = 1; j = super-∞↑):(Hyper-relativity-of-simultaneity-j)]]** = $\partial \mathbf{c_1}/\partial x_1 + \partial \mathbf{c_2}/\partial x_2 + \ldots + \partial \mathbf{c_n}/\partial x_n$ | mod div **[f[Σ(j = 1; j = super-∞↑):(Hyper-relativity-of-simultaneity-j)]]** ≤ 0};{Grad [f[Σ(j = 1; j = super-∞↑):(Hyper-relativity-of-simultaneity-j)]] = $(\partial c/\partial x_1)\mathbf{i_1}$ + $(\partial c/\partial x_2)i_2,\ldots,$ + $(\partial c/\partial x_n)\mathbf{i_n}$ | ∞ < mod Grad [f[Σ(j = 1; j = super-∞↑):(Hyper-relativity-of-simultaneity-j)]] ≤ ○, or mod Grad [f[Σ(j = 1; j = super-∞↑):(Hyper-relativity-of-simultaneity-j)]] > ○}}}}dKE}dKE} … d KE}

{∫ … {∫{∫ {γ{f{{light-speed-craft-k | 1 ≤ k ≤ ◯, or k > ◯};{div **[f[Σ(j = 1; j = super-∞↑):(Hyper-relativity-of-simultaneity-j)]]** = $\partial \mathbf{c}_1/\partial x_1 + \partial \mathbf{c}_2/\partial x_2 + \ldots + \partial \mathbf{c}_n/\partial x_n$ | mod div **[f[Σ(j = 1; j = super-∞↑):(Hyper-relativity-of-simultaneity-j)]]** ≤ 0};{Grad [f[Σ(j = 1; j = super-∞↑):(Hyper-relativity-of-simultaneity-j)]] = $(\partial c/\partial x_1)\mathbf{i}_1 + (\partial c/\partial x_2)i_2, \ldots, + (\partial c/\partial x_n)\mathbf{i}_n$ | ∞ < mod Grad [f[Σ(j = 1; j = super-∞↑):(Hyper-relativity-of-simultaneity-j)]] ≤ ◯, or mod Grad [f[Σ(j = 1; j = super-∞↑):(Hyper-relativity-of-simultaneity-j)]] > ◯}}}}dt}dt} ... dt}

{∫ … {∫{∫ {γ{f{{light-speed-craft-k | 1 ≤ k ≤ ◯, or k > ◯};{div **[f[Σ(j = 1; j = super-∞↑):(Hyper-relativity-of-simultaneity-j)]]** = $\partial \mathbf{c}_1/\partial x_1 + \partial \mathbf{c}_2/\partial x_2 + \ldots + \partial \mathbf{c}_n/\partial x_n$ | mod div **[f[Σ(j = 1; j = super-∞↑):(Hyper-relativity-of-simultaneity-j)]]** ≤ 0};{Grad [f[Σ(j = 1; j = super-∞↑):(Hyper-relativity-of-simultaneity-j)]] = $(\partial c/\partial x_1)\mathbf{i}_1 + (\partial c/\partial x_2)i_2, \ldots, + (\partial c/\partial x_n)\mathbf{i}_n$ | ∞ < mod Grad [f[Σ(j = 1; j = super-∞↑):(Hyper-relativity-of-simultaneity-j)]] ≤ ◯, or mod Grad [f[Σ(j = 1; j = super-∞↑):(Hyper-relativity-of-simultaneity-j)]] > ◯}}}}dT}dT} … d T}

{∫ … {∫{∫ {v{f{{light-speed-craft-k | 1 ≤ k ≤ ◯, or k > ◯};{div **[f[Σ(j = 1; j = super-∞↑):(Hyper-relativity-of-simultaneity-j)]]** = $\partial \mathbf{c}_1/\partial x_1 + \partial \mathbf{c}_2/\partial x_2 + \ldots + \partial \mathbf{c}_n/\partial x_n$ | mod div **[f[Σ(j = 1; j = super-∞↑):(Hyper-relativity-of-simultaneity-j)]]** ≤ 0};{Grad [f[Σ(j = 1; j = super-∞↑):(Hyper-relativity-of-simultaneity-j)]] = $(\partial c/\partial x_1)\mathbf{i}_1 + (\partial c/\partial x_2)i_2, \ldots, + (\partial c/\partial x_n)\mathbf{i}_n$ | ∞ < mod Grad [f[Σ(j = 1; j = super-∞↑):(Hyper-relativity-of-simultaneity-j)]] ≤ ◯, or mod Grad [f[Σ(j = 1; j = super-∞↑):(Hyper-relativity-of-simultaneity-j)]] > ◯}}}}dc}dc} … d c}

{∫ … {∫{∫ {v{f{{light-speed-craft-k | 1 ≤ k ≤ ◯, or k > ◯};{div **[f[Σ(j = 1; j = super-∞↑):(Hyper-relativity-of-simultaneity-j)]]** = $\partial \mathbf{c}_1/\partial x_1 + \partial \mathbf{c}_2/\partial x_2 + \ldots + \partial \mathbf{c}_n/\partial x_n$ | mod div **[f[Σ(j = 1; j = super-∞↑):(Hyper-relativity-of-simultaneity-j)]]** ≤ 0};{Grad [f[Σ(j = 1; j = super-∞↑):(Hyper-relativity-of-simultaneity-j)]] = $(\partial c/\partial x_1)\mathbf{i}_1 + (\partial c/\partial x_2)i_2, \ldots, + (\partial c/\partial x_n)\mathbf{i}_n$ | ∞ < mod Grad [f[Σ(j = 1; j = super-∞↑):(Hyper-relativity-of-simultaneity-j)]] ≤ ◯, or mod Grad [f[Σ(j = 1; j = super-∞↑):(Hyper-relativity-of-simultaneity-j)]] > ◯}}}}dγ}dγ} … d γ}

{∫ … {∫{∫ {v{f{{light-speed-craft-k | 1 ≤ k ≤ ◯, or k > ◯};{div **[f[Σ(j = 1; j = super-∞↑):(Hyper-relativity-of-simultaneity-j)]]** = $\partial \mathbf{c}_1/\partial x_1 + \partial \mathbf{c}_2/\partial x_2 + \ldots + \partial \mathbf{c}_n/\partial x_n$ | mod div **[f[Σ(j = 1; j = super-∞↑):(Hyper-relativity-of-simultaneity-j)]]** ≤ 0};{Grad [f[Σ(j = 1; j = super-∞↑):(Hyper-relativity-of-simultaneity-j)]] = $(\partial c/\partial x_1)\mathbf{i}_1 + (\partial c/\partial x_2)i_2, \ldots, + (\partial c/\partial x_n)\mathbf{i}_n$ | ∞ < mod Grad [f[Σ(j = 1; j = super-∞↑):(Hyper-relativity-of-simultaneity-j)]] ≤ ◯, or mod Grad [f[Σ(j = 1; j = super-∞↑):(Hyper-relativity-of-simultaneity-j)]] > ◯}}}}dv}dv} … d v}

{∫ … {∫{∫ {v{f{{light-speed-craft-k | 1 ≤ k ≤ ◯, or k > ◯};{div **[f[Σ(j = 1; j = super-∞↑):(Hyper-relativity-of-simultaneity-j)]]** = $\partial \mathbf{c}_1/\partial x_1 + \partial \mathbf{c}_2/\partial x_2 + \ldots + \partial \mathbf{c}_n/\partial x_n$ | mod div **[f[Σ(j = 1; j = super-∞↑):(Hyper-relativity-of-simultaneity-j)]]** ≤ 0};{Grad [f[Σ(j = 1; j = super-∞↑):(Hyper-relativity-of-simultaneity-j)]] = $(\partial c/\partial x_1)\mathbf{i}_1 + (\partial c/\partial x_2)i_2, \ldots, + (\partial c/\partial x_n)\mathbf{i}_n$ |

∞ < mod Grad [f[Σ(j = 1; j = super-∞↑):(Hyper-relativity-of-simultaneity-j)]] ≤ ◯, or mod Grad [f[Σ(j = 1; j = super-∞↑):(Hyper-relativity-of-simultaneity-j)]] > ◯}}}}da}da} … d a}

{∫ … {∫∫∫ {v{f{{light-speed-craft-k | 1 ≤ k ≤ ◯, or k > ◯};{div **[f[Σ(j = 1; j = super-∞↑):(Hyper-relativity-of-simultaneity-j)]]** = $\partial \mathbf{c}_1/\partial x_1 + \partial \mathbf{c}_2/\partial x_2 + \ldots + \partial \mathbf{c}_n/\partial x_n$ | mod div **[f[Σ(j = 1; j = super-∞↑):(Hyper-relativity-of-simultaneity-j)]]** ≤ 0};{Grad [f[Σ(j = 1; j = super-∞↑):(Hyper-relativity-of-simultaneity-j)]] = $(\partial c/\partial x_1)\mathbf{i}_1 + (\partial c/\partial x_2)i_2, \ldots, + (\partial c/\partial x_n)\mathbf{i}_n$ | ∞ < mod Grad [f[Σ(j = 1; j = super-∞↑):(Hyper-relativity-of-simultaneity-j)]] ≤ ◯, or mod Grad [f[Σ(j = 1; j = super-∞↑):(Hyper-relativity-of-simultaneity-j)]] > ◯}}}}dP}dP} … d P}

{∫ … {∫∫∫ {v{f{{light-speed-craft-k | 1 ≤ k ≤ ◯, or k > ◯};{div **[f[Σ(j = 1; j = super-∞↑):(Hyper-relativity-of-simultaneity-j)]]** = $\partial \mathbf{c}_1/\partial x_1 + \partial \mathbf{c}_2/\partial x_2 + \ldots + \partial \mathbf{c}_n/\partial x_n$ | mod div **[f[Σ(j = 1; j = super-∞↑):(Hyper-relativity-of-simultaneity-j)]]** ≤ 0};{Grad [f[Σ(j = 1; j = super-∞↑):(Hyper-relativity-of-simultaneity-j)]] = $(\partial c/\partial x_1)\mathbf{i}_1 + (\partial c/\partial x_2)i_2, \ldots, + (\partial c/\partial x_n)\mathbf{i}_n$ | ∞ < mod Grad [f[Σ(j = 1; j = super-∞↑):(Hyper-relativity-of-simultaneity-j)]] ≤ ◯, or mod Grad [f[Σ(j = 1; j = super-∞↑):(Hyper-relativity-of-simultaneity-j)]] > ◯}}}}dKE}dKE} … d KE}

{∫ … {∫∫∫ {v{f{{light-speed-craft-k | 1 ≤ k ≤ ◯, or k > ◯};{div **[f[Σ(j = 1; j = super-∞↑):(Hyper-relativity-of-simultaneity-j)]]** = $\partial \mathbf{c}_1/\partial x_1 + \partial \mathbf{c}_2/\partial x_2 + \ldots + \partial \mathbf{c}_n/\partial x_n$ | mod div **[f[Σ(j = 1; j = super-∞↑):(Hyper-relativity-of-simultaneity-j)]]** ≤ 0};{Grad [f[Σ(j = 1; j = super-∞↑):(Hyper-relativity-of-simultaneity-j)]] = $(\partial c/\partial x_1)\mathbf{i}_1 + (\partial c/\partial x_2)i_2, \ldots, + (\partial c/\partial x_n)\mathbf{i}_n$ | ∞ < mod Grad [f[Σ(j = 1; j = super-∞↑):(Hyper-relativity-of-simultaneity-j)]] ≤ ◯, or mod Grad [f[Σ(j = 1; j = super-∞↑):(Hyper-relativity-of-simultaneity-j)]] > ◯}}}}dt}dt} ... dt}

{∫ … {∫∫∫ {v{f{{light-speed-craft-k | 1 ≤ k ≤ ◯, or k > ◯};{div **[f[Σ(j = 1; j = super-∞↑):(Hyper-relativity-of-simultaneity-j)]]** = $\partial \mathbf{c}_1/\partial x_1 + \partial \mathbf{c}_2/\partial x_2 + \ldots + \partial \mathbf{c}_n/\partial x_n$ | mod div **[f[Σ(j = 1; j = super-∞↑):(Hyper-relativity-of-simultaneity-j)]]** ≤ 0};{Grad [f[Σ(j = 1; j = super-∞↑):(Hyper-relativity-of-simultaneity-j)]] = $(\partial c/\partial x_1)\mathbf{i}_1 + (\partial c/\partial x_2)i_2, \ldots, + (\partial c/\partial x_n)\mathbf{i}_n$ | ∞ < mod Grad [f[Σ(j = 1; j = super-∞↑):(Hyper-relativity-of-simultaneity-j)]] ≤ ◯, or mod Grad [f[Σ(j = 1; j = super-∞↑):(Hyper-relativity-of-simultaneity-j)]] > ◯}}}}dT}dT} … d T}

{∫ … {∫∫∫ {a{f{{light-speed-craft-k | 1 ≤ k ≤ ◯, or k > ◯};{div **[f[Σ(j = 1; j = super-∞↑):(Hyper-relativity-of-simultaneity-j)]]** = $\partial \mathbf{c}_1/\partial x_1 + \partial \mathbf{c}_2/\partial x_2 + \ldots + \partial \mathbf{c}_n/\partial x_n$ | mod div **[f[Σ(j = 1; j = super-∞↑):(Hyper-relativity-of-simultaneity-j)]]** ≤ 0};{Grad [f[Σ(j = 1; j = super-∞↑):(Hyper-relativity-of-simultaneity-j)]] = $(\partial c/\partial x_1)\mathbf{i}_1 + (\partial c/\partial x_2)i_2, \ldots, + (\partial c/\partial x_n)\mathbf{i}_n$ | ∞ < mod Grad [f[Σ(j = 1; j = super-∞↑):(Hyper-relativity-of-simultaneity-j)]] ≤ ◯, or mod Grad [f[Σ(j = 1; j = super-∞↑):(Hyper-relativity-of-simultaneity-j)]] > ◯}}}}dc}dc} … d c}

{∫ … {∫∫∫ {a{f{{light-speed-craft-k | 1 ≤ k ≤ ◯, or k > ◯};{div **[f[Σ(j = 1; j = super-∞↑):(Hyper-relativity-of-simultaneity-j)]]** = $\partial \mathbf{c}_1/\partial x_1 + \partial \mathbf{c}_2/\partial x_2 + \ldots + \partial \mathbf{c}_n/\partial x_n$ | mod div

[f[Σ(j = 1; j = super-∞↑):(Hyper-relativity-of-simultaneity-j)]] ≤ 0};{Grad [f[Σ(j = 1; j = super-∞↑):(Hyper-relativity-of-simultaneity-j)]] = $(\partial c/\partial x_1)\mathbf{i_1} + (\partial c/\partial x_2)i_2, \ldots, + (\partial c/\partial x_n)\mathbf{i_n}$ | ∞ < mod Grad [f[Σ(j = 1; j = super-∞↑):(Hyper-relativity-of-simultaneity-j)]] ≤ ◯, or mod Grad [f[Σ(j = 1; j = super-∞↑):(Hyper-relativity-of-simultaneity-j)]] > ◯}}}}dγ}dγ} … d γ}

{∫ … {∫∫∫ {a{f{{light-speed-craft-k | 1 ≤ k ≤ ◯, or k > ◯};{div **[f[Σ(j = 1; j = super-∞↑):(Hyper-relativity-of-simultaneity-j)]]** = $\partial \mathbf{c_1}/\partial x_1 + \partial \mathbf{c_2}/\partial x_2 + \ldots + \partial \mathbf{c_n}/\partial x_n$ | mod div **[f[Σ(j = 1; j = super-∞↑):(Hyper-relativity-of-simultaneity-j)]]** ≤ 0};{Grad [f[Σ(j = 1; j = super-∞↑):(Hyper-relativity-of-simultaneity-j)]] = $(\partial c/\partial x_1)\mathbf{i_1} + (\partial c/\partial x_2)i_2, \ldots, + (\partial c/\partial x_n)\mathbf{i_n}$ | ∞ < mod Grad [f[Σ(j = 1; j = super-∞↑):(Hyper-relativity-of-simultaneity-j)]] ≤ ◯, or mod Grad [f[Σ(j = 1; j = super-∞↑):(Hyper-relativity-of-simultaneity-j)]] > ◯}}}}dv}dv} … d v}

{∫ … {∫∫∫ {a{f{{light-speed-craft-k | 1 ≤ k ≤ ◯, or k > ◯};{div **[f[Σ(j = 1; j = super-∞↑):(Hyper-relativity-of-simultaneity-j)]]** = $\partial \mathbf{c_1}/\partial x_1 + \partial \mathbf{c_2}/\partial x_2 + \ldots + \partial \mathbf{c_n}/\partial x_n$ | mod div **[f[Σ(j = 1; j = super-∞↑):(Hyper-relativity-of-simultaneity-j)]]** ≤ 0};{Grad [f[Σ(j = 1; j = super-∞↑):(Hyper-relativity-of-simultaneity-j)]] = $(\partial c/\partial x_1)\mathbf{i_1} + (\partial c/\partial x_2)i_2, \ldots, + (\partial c/\partial x_n)\mathbf{i_n}$ | ∞ < mod Grad [f[Σ(j = 1; j = super-∞↑):(Hyper-relativity-of-simultaneity-j)]] ≤ ◯, or mod Grad [f[Σ(j = 1; j = super-∞↑):(Hyper-relativity-of-simultaneity-j)]] > ◯}}}}da}da} … d a}

{∫ … {∫∫∫ {a{f{{light-speed-craft-k | 1 ≤ k ≤ ◯, or k > ◯};{div **[f[Σ(j = 1; j = super-∞↑):(Hyper-relativity-of-simultaneity-j)]]** = $\partial \mathbf{c_1}/\partial x_1 + \partial \mathbf{c_2}/\partial x_2 + \ldots + \partial \mathbf{c_n}/\partial x_n$ | mod div **[f[Σ(j = 1; j = super-∞↑):(Hyper-relativity-of-simultaneity-j)]]** ≤ 0};{Grad [f[Σ(j = 1; j = super-∞↑):(Hyper-relativity-of-simultaneity-j)]] = $(\partial c/\partial x_1)\mathbf{i_1} + (\partial c/\partial x_2)i_2, \ldots, + (\partial c/\partial x_n)\mathbf{i_n}$ | ∞ < mod Grad [f[Σ(j = 1; j = super-∞↑):(Hyper-relativity-of-simultaneity-j)]] ≤ ◯, or mod Grad [f[Σ(j = 1; j = super-∞↑):(Hyper-relativity-of-simultaneity-j)]] > ◯}}}}dP}dP} … d P}

{∫ … {∫∫∫ {a{f{{light-speed-craft-k | 1 ≤ k ≤ ◯, or k > ◯};{div **[f[Σ(j = 1; j = super-∞↑):(Hyper-relativity-of-simultaneity-j)]]** = $\partial \mathbf{c_1}/\partial x_1 + \partial \mathbf{c_2}/\partial x_2 + \ldots + \partial \mathbf{c_n}/\partial x_n$ | mod div **[f[Σ(j = 1; j = super-∞↑):(Hyper-relativity-of-simultaneity-j)]]** ≤ 0};{Grad [f[Σ(j = 1; j = super-∞↑):(Hyper-relativity-of-simultaneity-j)]] = $(\partial c/\partial x_1)\mathbf{i_1} + (\partial c/\partial x_2)i_2, \ldots, + (\partial c/\partial x_n)\mathbf{i_n}$ | ∞ < mod Grad [f[Σ(j = 1; j = super-∞↑):(Hyper-relativity-of-simultaneity-j)]] ≤ ◯, or mod Grad [f[Σ(j = 1; j = super-∞↑):(Hyper-relativity-of-simultaneity-j)]] > ◯}}}}dKE}dKE} … d KE}

{∫ … {∫∫∫ {a{f{{light-speed-craft-k | 1 ≤ k ≤ ◯, or k > ◯};{div **[f[Σ(j = 1; j = super-∞↑):(Hyper-relativity-of-simultaneity-j)]]** = $\partial \mathbf{c_1}/\partial x_1 + \partial \mathbf{c_2}/\partial x_2 + \ldots + \partial \mathbf{c_n}/\partial x_n$ | mod div **[f[Σ(j = 1; j = super-∞↑):(Hyper-relativity-of-simultaneity-j)]]** ≤ 0};{Grad [f[Σ(j = 1; j = super-∞↑):(Hyper-relativity-of-simultaneity-j)]] = $(\partial c/\partial x_1)\mathbf{i_1} + (\partial c/\partial x_2)i_2, \ldots, + (\partial c/\partial x_n)\mathbf{i_n}$ | ∞ < mod Grad [f[Σ(j = 1; j = super-∞↑):(Hyper-relativity-of-simultaneity-j)]] ≤ ◯, or mod Grad [f[Σ(j = 1; j = super-∞↑):(Hyper-relativity-of-simultaneity-j)]] > ◯}}}}dt}dt} ... dt}

{∫ … {∫{∫ {a{f{{light-speed-craft-k | 1 ≤ k ≤ ◯, or k > ◯};{div **[f[Σ(j = 1; j = super-∞↑):(Hyper-relativity-of-simultaneity-j)]]** = $\partial \mathbf{c_1}/\partial x_1 + \partial \mathbf{c_2}/\partial x_2 + \ldots + \partial \mathbf{c_n}/\partial x_n$ | mod div **[f[Σ(j = 1; j = super-∞↑):(Hyper-relativity-of-simultaneity-j)]]** ≤ 0};{Grad [f[Σ(j = 1; j = super-∞↑):(Hyper-relativity-of-simultaneity-j)]] = $(\partial c/\partial x_1)\mathbf{i_1} + (\partial c/\partial x_2)i_2, \ldots, + (\partial c/\partial x_n)\mathbf{i_n}$ | ∞ < mod Grad [f[Σ(j = 1; j = super-∞↑):(Hyper-relativity-of-simultaneity-j)]] ≤ ◯, or mod Grad [f[Σ(j = 1; j = super-∞↑):(Hyper-relativity-of-simultaneity-j)]] > ◯}}}}dT}dT} … d T}

{∫ … {∫{∫ {P{f{{light-speed-craft-k | 1 ≤ k ≤ ◯, or k > ◯};{div **[f[Σ(j = 1; j = super-∞↑):(Hyper-relativity-of-simultaneity-j)]]** = $\partial \mathbf{c_1}/\partial x_1 + \partial \mathbf{c_2}/\partial x_2 + \ldots + \partial \mathbf{c_n}/\partial x_n$ | mod div **[f[Σ(j = 1; j = super-∞↑):(Hyper-relativity-of-simultaneity-j)]]** ≤ 0};{Grad [f[Σ(j = 1; j = super-∞↑):(Hyper-relativity-of-simultaneity-j)]] = $(\partial c/\partial x_1)\mathbf{i_1} + (\partial c/\partial x_2)i_2, \ldots, + (\partial c/\partial x_n)\mathbf{i_n}$ | ∞ < mod Grad [f[Σ(j = 1; j = super-∞↑):(Hyper-relativity-of-simultaneity-j)]] ≤ ◯, or mod Grad [f[Σ(j = 1; j = super-∞↑):(Hyper-relativity-of-simultaneity-j)]] > ◯}}}}dc}dc} … d c}

{∫ … {∫{∫ {P{f{{light-speed-craft-k | 1 ≤ k ≤ ◯, or k > ◯};{div **[f[Σ(j = 1; j = super-∞↑):(Hyper-relativity-of-simultaneity-j)]]** = $\partial \mathbf{c_1}/\partial x_1 + \partial \mathbf{c_2}/\partial x_2 + \ldots + \partial \mathbf{c_n}/\partial x_n$ | mod div **[f[Σ(j = 1; j = super-∞↑):(Hyper-relativity-of-simultaneity-j)]]** ≤ 0};{Grad [f[Σ(j = 1; j = super-∞↑):(Hyper-relativity-of-simultaneity-j)]] = $(\partial c/\partial x_1)\mathbf{i_1} + (\partial c/\partial x_2)i_2, \ldots, + (\partial c/\partial x_n)\mathbf{i_n}$ | ∞ < mod Grad [f[Σ(j = 1; j = super-∞↑):(Hyper-relativity-of-simultaneity-j)]] ≤ ◯, or mod Grad [f[Σ(j = 1; j = super-∞↑):(Hyper-relativity-of-simultaneity-j)]] > ◯}}}}dγ}dγ} … d γ}

{∫ … {∫{∫ {P{f{{light-speed-craft-k | 1 ≤ k ≤ ◯, or k > ◯};{div **[f[Σ(j = 1; j = super-∞↑):(Hyper-relativity-of-simultaneity-j)]]** = $\partial \mathbf{c_1}/\partial x_1 + \partial \mathbf{c_2}/\partial x_2 + \ldots + \partial \mathbf{c_n}/\partial x_n$ | mod div **[f[Σ(j = 1; j = super-∞↑):(Hyper-relativity-of-simultaneity-j)]]** ≤ 0};{Grad [f[Σ(j = 1; j = super-∞↑):(Hyper-relativity-of-simultaneity-j)]] = $(\partial c/\partial x_1)\mathbf{i_1} + (\partial c/\partial x_2)i_2, \ldots, + (\partial c/\partial x_n)\mathbf{i_n}$ | ∞ < mod Grad [f[Σ(j = 1; j = super-∞↑):(Hyper-relativity-of-simultaneity-j)]] ≤ ◯, or mod Grad [f[Σ(j = 1; j = super-∞↑):(Hyper-relativity-of-simultaneity-j)]] > ◯}}}}dv}dv} … d v}

{∫ … {∫{∫ {P{f{{light-speed-craft-k | 1 ≤ k ≤ ◯, or k > ◯};{div **[f[Σ(j = 1; j = super-∞↑):(Hyper-relativity-of-simultaneity-j)]]** = $\partial \mathbf{c_1}/\partial x_1 + \partial \mathbf{c_2}/\partial x_2 + \ldots + \partial \mathbf{c_n}/\partial x_n$ | mod div **[f[Σ(j = 1; j = super-∞↑):(Hyper-relativity-of-simultaneity-j)]]** ≤ 0};{Grad [f[Σ(j = 1; j = super-∞↑):(Hyper-relativity-of-simultaneity-j)]] = $(\partial c/\partial x_1)\mathbf{i_1} + (\partial c/\partial x_2)i_2, \ldots, + (\partial c/\partial x_n)\mathbf{i_n}$ | ∞ < mod Grad [f[Σ(j = 1; j = super-∞↑):(Hyper-relativity-of-simultaneity-j)]] ≤ ◯, or mod Grad [f[Σ(j = 1; j = super-∞↑):(Hyper-relativity-of-simultaneity-j)]] > ◯}}}}da}da} … d a}

{∫ … {∫{∫ {P{f{{light-speed-craft-k | 1 ≤ k ≤ ◯, or k > ◯};{div **[f[Σ(j = 1; j = super-∞↑):(Hyper-relativity-of-simultaneity-j)]]** = $\partial \mathbf{c_1}/\partial x_1 + \partial \mathbf{c_2}/\partial x_2 + \ldots + \partial \mathbf{c_n}/\partial x_n$ | mod div **[f[Σ(j = 1; j = super-∞↑):(Hyper-relativity-of-simultaneity-j)]]** ≤ 0};{Grad [f[Σ(j = 1; j = super-∞↑):(Hyper-relativity-of-simultaneity-j)]] = $(\partial c/\partial x_1)\mathbf{i_1} + (\partial c/\partial x_2)i_2, \ldots, + (\partial c/\partial x_n)\mathbf{i_n}$ |

∞ < mod Grad [f[Σ(j = 1; j = super-∞↑):(Hyper-relativity-of-simultaneity-j)]] ≤ ◯, or mod Grad [f[Σ(j = 1; j = super-∞↑):(Hyper-relativity-of-simultaneity-j)]] > ◯}}}}dP}dP} … d P}

{∫ … {∫∫∫ {P{f{{light-speed-craft-k | 1 ≤ k ≤ ◯, or k > ◯};{div **[f[Σ(j = 1; j = super-∞↑):(Hyper-relativity-of-simultaneity-j)]]** = $\partial \mathbf{c_1}/\partial x_1 + \partial \mathbf{c_2}/\partial x_2 + \ldots + \partial \mathbf{c_n}/\partial x_n$ | mod div **[f[Σ(j = 1; j = super-∞↑):(Hyper-relativity-of-simultaneity-j)]]** ≤ 0};{Grad [f[Σ(j = 1; j = super-∞↑):(Hyper-relativity-of-simultaneity-j)]] = $(\partial c/\partial x_1)\mathbf{i_1} + (\partial c/\partial x_2)i_2, \ldots, + (\partial c/\partial x_n)\mathbf{i_n}$ | ∞ < mod Grad [f[Σ(j = 1; j = super-∞↑):(Hyper-relativity-of-simultaneity-j)]] ≤ ◯, or mod Grad [f[Σ(j = 1; j = super-∞↑):(Hyper-relativity-of-simultaneity-j)]] > ◯}}}}dKE}dKE} … d KE}

{∫ … {∫∫∫ {P{f{{light-speed-craft-k | 1 ≤ k ≤ ◯, or k > ◯};{div **[f[Σ(j = 1; j = super-∞↑):(Hyper-relativity-of-simultaneity-j)]]** = $\partial \mathbf{c_1}/\partial x_1 + \partial \mathbf{c_2}/\partial x_2 + \ldots + \partial \mathbf{c_n}/\partial x_n$ | mod div **[f[Σ(j = 1; j = super-∞↑):(Hyper-relativity-of-simultaneity-j)]]** ≤ 0};{Grad [f[Σ(j = 1; j = super-∞↑):(Hyper-relativity-of-simultaneity-j)]] = $(\partial c/\partial x_1)\mathbf{i_1} + (\partial c/\partial x_2)i_2, \ldots, + (\partial c/\partial x_n)\mathbf{i_n}$ | ∞ < mod Grad [f[Σ(j = 1; j = super-∞↑):(Hyper-relativity-of-simultaneity-j)]] ≤ ◯, or mod Grad [f[Σ(j = 1; j = super-∞↑):(Hyper-relativity-of-simultaneity-j)]] > ◯}}}}dt}dt} ... dt}

{∫ … {∫∫∫ {P{f{{light-speed-craft-k | 1 ≤ k ≤ ◯, or k > ◯};{div **[f[Σ(j = 1; j = super-∞↑):(Hyper-relativity-of-simultaneity-j)]]** = $\partial \mathbf{c_1}/\partial x_1 + \partial \mathbf{c_2}/\partial x_2 + \ldots + \partial \mathbf{c_n}/\partial x_n$ | mod div **[f[Σ(j = 1; j = super-∞↑):(Hyper-relativity-of-simultaneity-j)]]** ≤ 0};{Grad [f[Σ(j = 1; j = super-∞↑):(Hyper-relativity-of-simultaneity-j)]] = $(\partial c/\partial x_1)\mathbf{i_1} + (\partial c/\partial x_2)i_2, \ldots, + (\partial c/\partial x_n)\mathbf{i_n}$ | ∞ < mod Grad [f[Σ(j = 1; j = super-∞↑):(Hyper-relativity-of-simultaneity-j)]] ≤ ◯, or mod Grad [f[Σ(j = 1; j = super-∞↑):(Hyper-relativity-of-simultaneity-j)]] > ◯}}}}dT}dT} … d T}

{∫ … {∫∫∫ {KE{f{{light-speed-craft-k | 1 ≤ k ≤ ◯, or k > ◯};{div **[f[Σ(j = 1; j = super-∞↑):(Hyper-relativity-of-simultaneity-j)]]** = $\partial \mathbf{c_1}/\partial x_1 + \partial \mathbf{c_2}/\partial x_2 + \ldots + \partial \mathbf{c_n}/\partial x_n$ | mod div **[f[Σ(j = 1; j = super-∞↑):(Hyper-relativity-of-simultaneity-j)]]** ≤ 0};{Grad [f[Σ(j = 1; j = super-∞↑):(Hyper-relativity-of-simultaneity-j)]] = $(\partial c/\partial x_1)\mathbf{i_1} + (\partial c/\partial x_2)i_2, \ldots, + (\partial c/\partial x_n)\mathbf{i_n}$ | ∞ < mod Grad [f[Σ(j = 1; j = super-∞↑):(Hyper-relativity-of-simultaneity-j)]] ≤ ◯, or mod Grad [f[Σ(j = 1; j = super-∞↑):(Hyper-relativity-of-simultaneity-j)]] > ◯}}}}dc}dc} … d c}

{∫ … {∫∫∫ {KE {f{{light-speed-craft-k | 1 ≤ k ≤ ◯, or k > ◯};{div **[f[Σ(j = 1; j = super-∞↑):(Hyper-relativity-of-simultaneity-j)]]** = $\partial \mathbf{c_1}/\partial x_1 + \partial \mathbf{c_2}/\partial x_2 + \ldots + \partial \mathbf{c_n}/\partial x_n$ | mod div **[f[Σ(j = 1; j = super-∞↑):(Hyper-relativity-of-simultaneity-j)]]** ≤ 0};{Grad [f[Σ(j = 1; j = super-∞↑):(Hyper-relativity-of-simultaneity-j)]] = $(\partial c/\partial x_1)\mathbf{i_1} + (\partial c/\partial x_2)i_2, \ldots, + (\partial c/\partial x_n)\mathbf{i_n}$ | ∞ < mod Grad [f[Σ(j = 1; j = super-∞↑):(Hyper-relativity-of-simultaneity-j)]] ≤ ◯, or mod Grad [f[Σ(j = 1; j = super-∞↑):(Hyper-relativity-of-simultaneity-j)]] > ◯}}}}dγ}dγ} … d γ}

{∫ … {∫∫∫ {KE {f{{light-speed-craft-k | 1 ≤ k ≤ ◯, or k > ◯};{div **[f[Σ(j = 1; j = super-∞↑):(Hyper-relativity-of-simultaneity-j)]]** = $\partial \mathbf{c_1}/\partial x_1 + \partial \mathbf{c_2}/\partial x_2 + \ldots + \partial \mathbf{c_n}/\partial x_n$ | mod div

[f[Σ(j = 1; j = super-∞↑):(Hyper-relativity-of-simultaneity-j)]] ≤ 0};{Grad [f[Σ(j = 1; j = super-∞↑):(Hyper-relativity-of-simultaneity-j)]] = $(\partial c/\partial x_1)\mathbf{i}_1$ + $(\partial c/\partial x_2)i_2$,…, + $(\partial c/\partial x_n)\mathbf{i}_n$ | ∞ < mod Grad [f[Σ(j = 1; j = super-∞↑):(Hyper-relativity-of-simultaneity-j)]] ≤ ◯, or mod Grad [f[Σ(j = 1; j = super-∞↑):(Hyper-relativity-of-simultaneity-j)]] > ◯}}}}dv}dv} … d v}

{∫ … {∫{∫ {KE {f{{light-speed-craft-k | 1 ≤ k ≤ ◯, or k > ◯};{div **[f[Σ(j = 1; j = super-∞↑):(Hyper-relativity-of-simultaneity-j)]]** = $\partial \mathbf{c}_1/\partial x_1 + \partial \mathbf{c}_2/\partial x_2 + \ldots + \partial \mathbf{c}_n/\partial x_n$ | mod div **[f[Σ(j = 1; j = super-∞↑):(Hyper-relativity-of-simultaneity-j)]]** ≤ 0};{Grad [f[Σ(j = 1; j = super-∞↑):(Hyper-relativity-of-simultaneity-j)]] = $(\partial c/\partial x_1)\mathbf{i}_1$ + $(\partial c/\partial x_2)i_2$,…, + $(\partial c/\partial x_n)\mathbf{i}_n$ | ∞ < mod Grad [f[Σ(j = 1; j = super-∞↑):(Hyper-relativity-of-simultaneity-j)]] ≤ ◯, or mod Grad [f[Σ(j = 1; j = super-∞↑):(Hyper-relativity-of-simultaneity-j)]] > ◯}}}}da}da} … d a}

{∫ … {∫{∫ {KE {f{{light-speed-craft-k | 1 ≤ k ≤ ◯, or k > ◯};{div **[f[Σ(j = 1; j = super-∞↑):(Hyper-relativity-of-simultaneity-j)]]** = $\partial \mathbf{c}_1/\partial x_1 + \partial \mathbf{c}_2/\partial x_2 + \ldots + \partial \mathbf{c}_n/\partial x_n$ | mod div **[f[Σ(j = 1; j = super-∞↑):(Hyper-relativity-of-simultaneity-j)]]** ≤ 0};{Grad [f[Σ(j = 1; j = super-∞↑):(Hyper-relativity-of-simultaneity-j)]] = $(\partial c/\partial x_1)\mathbf{i}_1$ + $(\partial c/\partial x_2)i_2$,…, + $(\partial c/\partial x_n)\mathbf{i}_n$ | ∞ < mod Grad [f[Σ(j = 1; j = super-∞↑):(Hyper-relativity-of-simultaneity-j)]] ≤ ◯, or mod Grad [f[Σ(j = 1; j = super-∞↑):(Hyper-relativity-of-simultaneity-j)]] > ◯}}}}dP}dP} … d P}

{∫ … {∫{∫ {KE {f{{light-speed-craft-k | 1 ≤ k ≤ ◯, or k > ◯};{div **[f[Σ(j = 1; j = super-∞↑):(Hyper-relativity-of-simultaneity-j)]]** = $\partial \mathbf{c}_1/\partial x_1 + \partial \mathbf{c}_2/\partial x_2 + \ldots + \partial \mathbf{c}_n/\partial x_n$ | mod div **[f[Σ(j = 1; j = super-∞↑):(Hyper-relativity-of-simultaneity-j)]]** ≤ 0};{Grad [f[Σ(j = 1; j = super-∞↑):(Hyper-relativity-of-simultaneity-j)]] = $(\partial c/\partial x_1)\mathbf{i}_1$ + $(\partial c/\partial x_2)i_2$,…, + $(\partial c/\partial x_n)\mathbf{i}_n$ | ∞ < mod Grad [f[Σ(j = 1; j = super-∞↑):(Hyper-relativity-of-simultaneity-j)]] ≤ ◯, or mod Grad [f[Σ(j = 1; j = super-∞↑):(Hyper-relativity-of-simultaneity-j)]] > ◯}}}}dKE}dKE} … d KE}

{∫ … {∫{∫ {KE {f{{light-speed-craft-k | 1 ≤ k ≤ ◯, or k > ◯};{div **[f[Σ(j = 1; j = super-∞↑):(Hyper-relativity-of-simultaneity-j)]]** = $\partial \mathbf{c}_1/\partial x_1 + \partial \mathbf{c}_2/\partial x_2 + \ldots + \partial \mathbf{c}_n/\partial x_n$ | mod div **[f[Σ(j = 1; j = super-∞↑):(Hyper-relativity-of-simultaneity-j)]]** ≤ 0};{Grad [f[Σ(j = 1; j = super-∞↑):(Hyper-relativity-of-simultaneity-j)]] = $(\partial c/\partial x_1)\mathbf{i}_1$ + $(\partial c/\partial x_2)i_2$,…, + $(\partial c/\partial x_n)\mathbf{i}_n$ | ∞ < mod Grad [f[Σ(j = 1; j = super-∞↑):(Hyper-relativity-of-simultaneity-j)]] ≤ ◯, or mod Grad [f[Σ(j = 1; j = super-∞↑):(Hyper-relativity-of-simultaneity-j)]] > ◯}}}}dt}dt} ... dt}

{∫ … {∫{∫ {KE {f{{light-speed-craft-k | 1 ≤ k ≤ ◯, or k > ◯};{div **[f[Σ(j = 1; j = super-∞↑):(Hyper-relativity-of-simultaneity-j)]]** = $\partial \mathbf{c}_1/\partial x_1 + \partial \mathbf{c}_2/\partial x_2 + \ldots + \partial \mathbf{c}_n/\partial x_n$ | mod div **[f[Σ(j = 1; j = super-∞↑):(Hyper-relativity-of-simultaneity-j)]]** ≤ 0};{Grad [f[Σ(j = 1; j = super-∞↑):(Hyper-relativity-of-simultaneity-j)]] = $(\partial c/\partial x_1)\mathbf{i}_1$ + $(\partial c/\partial x_2)i_2$,…, + $(\partial c/\partial x_n)\mathbf{i}_n$ | ∞ < mod Grad [f[Σ(j = 1; j = super-∞↑):(Hyper-relativity-of-simultaneity-j)]] ≤ ◯, or mod Grad [f[Σ(j = 1; j = super-∞↑):(Hyper-relativity-of-simultaneity-j)]] > ◯}}}}dT}dT} … d T}

{∫ … {∫{∫ {t{f{{light-speed-craft-k | 1 ≤ k ≤ ◯, or k > ◯};{div **[f[Σ(j = 1; j = super-∞↑):(Hyper-relativity-of-simultaneity-j)]]** = $\partial \mathbf{c_1}/\partial x_1 + \partial \mathbf{c_2}/\partial x_2 + \ldots + \partial \mathbf{c_n}/\partial x_n$ | mod div **[f[Σ(j = 1; j = super-∞↑):(Hyper-relativity-of-simultaneity-j)]]** ≤ 0};{Grad [f[Σ(j = 1; j = super-∞↑):(Hyper-relativity-of-simultaneity-j)]] = $(\partial c/\partial x_1)\mathbf{i_1} + (\partial c/\partial x_2)i_2, \ldots, + (\partial c/\partial x_n)\mathbf{i_n}$ | ∞ < mod Grad [f[Σ(j = 1; j = super-∞↑):(Hyper-relativity-of-simultaneity-j)]] ≤ ◯, or mod Grad [f[Σ(j = 1; j = super-∞↑):(Hyper-relativity-of-simultaneity-j)]] > ◯}}}}dc}dc} … d c}

{∫ … {∫{∫ {t{f{{light-speed-craft-k | 1 ≤ k ≤ ◯, or k > ◯};{div **[f[Σ(j = 1; j = super-∞↑):(Hyper-relativity-of-simultaneity-j)]]** = $\partial \mathbf{c_1}/\partial x_1 + \partial \mathbf{c_2}/\partial x_2 + \ldots + \partial \mathbf{c_n}/\partial x_n$ | mod div **[f[Σ(j = 1; j = super-∞↑):(Hyper-relativity-of-simultaneity-j)]]** ≤ 0};{Grad [f[Σ(j = 1; j = super-∞↑):(Hyper-relativity-of-simultaneity-j)]] = $(\partial c/\partial x_1)\mathbf{i_1} + (\partial c/\partial x_2)i_2, \ldots, + (\partial c/\partial x_n)\mathbf{i_n}$ | ∞ < mod Grad [f[Σ(j = 1; j = super-∞↑):(Hyper-relativity-of-simultaneity-j)]] ≤ ◯, or mod Grad [f[Σ(j = 1; j = super-∞↑):(Hyper-relativity-of-simultaneity-j)]] > ◯}}}}dγ}dγ} … d γ}

{∫ … {∫{∫ {t{f{{light-speed-craft-k | 1 ≤ k ≤ ◯, or k > ◯};{div **[f[Σ(j = 1; j = super-∞↑):(Hyper-relativity-of-simultaneity-j)]]** = $\partial \mathbf{c_1}/\partial x_1 + \partial \mathbf{c_2}/\partial x_2 + \ldots + \partial \mathbf{c_n}/\partial x_n$ | mod div **[f[Σ(j = 1; j = super-∞↑):(Hyper-relativity-of-simultaneity-j)]]** ≤ 0};{Grad [f[Σ(j = 1; j = super-∞↑):(Hyper-relativity-of-simultaneity-j)]] = $(\partial c/\partial x_1)\mathbf{i_1} + (\partial c/\partial x_2)i_2, \ldots, + (\partial c/\partial x_n)\mathbf{i_n}$ | ∞ < mod Grad [f[Σ(j = 1; j = super-∞↑):(Hyper-relativity-of-simultaneity-j)]] ≤ ◯, or mod Grad [f[Σ(j = 1; j = super-∞↑):(Hyper-relativity-of-simultaneity-j)]] > ◯}}}}dv}dv} … d v}

{∫ … {∫{∫ {t{f{{light-speed-craft-k | 1 ≤ k ≤ ◯, or k > ◯};{div **[f[Σ(j = 1; j = super-∞↑):(Hyper-relativity-of-simultaneity-j)]]** = $\partial \mathbf{c_1}/\partial x_1 + \partial \mathbf{c_2}/\partial x_2 + \ldots + \partial \mathbf{c_n}/\partial x_n$ | mod div **[f[Σ(j = 1; j = super-∞↑):(Hyper-relativity-of-simultaneity-j)]]** ≤ 0};{Grad [f[Σ(j = 1; j = super-∞↑):(Hyper-relativity-of-simultaneity-j)]] = $(\partial c/\partial x_1)\mathbf{i_1} + (\partial c/\partial x_2)i_2, \ldots, + (\partial c/\partial x_n)\mathbf{i_n}$ | ∞ < mod Grad [f[Σ(j = 1; j = super-∞↑):(Hyper-relativity-of-simultaneity-j)]] ≤ ◯, or mod Grad [f[Σ(j = 1; j = super-∞↑):(Hyper-relativity-of-simultaneity-j)]] > ◯}}}}da}da} … d a}

{∫ … {∫{∫ {t{f{{light-speed-craft-k | 1 ≤ k ≤ ◯, or k > ◯};{div **[f[Σ(j = 1; j = super-∞↑):(Hyper-relativity-of-simultaneity-j)]]** = $\partial \mathbf{c_1}/\partial x_1 + \partial \mathbf{c_2}/\partial x_2 + \ldots + \partial \mathbf{c_n}/\partial x_n$ | mod div **[f[Σ(j = 1; j = super-∞↑):(Hyper-relativity-of-simultaneity-j)]]** ≤ 0};{Grad [f[Σ(j = 1; j = super-∞↑):(Hyper-relativity-of-simultaneity-j)]] = $(\partial c/\partial x_1)\mathbf{i_1} + (\partial c/\partial x_2)i_2, \ldots, + (\partial c/\partial x_n)\mathbf{i_n}$ | ∞ < mod Grad [f[Σ(j = 1; j = super-∞↑):(Hyper-relativity-of-simultaneity-j)]] ≤ ◯, or mod Grad [f[Σ(j = 1; j = super-∞↑):(Hyper-relativity-of-simultaneity-j)]] > ◯}}}}dP}dP} … d P}

{∫ … {∫{∫ {t{f{{light-speed-craft-k | 1 ≤ k ≤ ◯, or k > ◯};{div **[f[Σ(j = 1; j = super-∞↑):(Hyper-relativity-of-simultaneity-j)]]** = $\partial \mathbf{c_1}/\partial x_1 + \partial \mathbf{c_2}/\partial x_2 + \ldots + \partial \mathbf{c_n}/\partial x_n$ | mod div **[f[Σ(j = 1; j = super-∞↑):(Hyper-relativity-of-simultaneity-j)]]** ≤ 0};{Grad [f[Σ(j = 1; j = super-∞↑):(Hyper-relativity-of-simultaneity-j)]] = $(\partial c/\partial x_1)\mathbf{i_1} + (\partial c/\partial x_2)i_2, \ldots, + (\partial c/\partial x_n)\mathbf{i_n}$ | ∞ < mod Grad [f[Σ(j = 1; j = super-∞↑):(Hyper-relativity-of-simultaneity-j)]] ≤ ◯, or mod

Grad [f[Σ(j = 1; j = super-∞↑):(Hyper-relativity-of-simultaneity-j)]] > ◯}}}}dKE}dKE} … d KE}

{∫ … {∫∫∫ {t{f{{light-speed-craft-k | 1 ≤ k ≤ ◯, or k > ◯};{div **[f[Σ(j = 1; j = super-∞↑):(Hyper-relativity-of-simultaneity-j)]]** = $\partial \mathbf{c}_1/\partial x_1 + \partial \mathbf{c}_2/\partial x_2 + \ldots + \partial \mathbf{c}_n/\partial x_n$ | mod div **[f[Σ(j = 1; j = super-∞↑):(Hyper-relativity-of-simultaneity-j)]]** ≤ 0};{Grad [f[Σ(j = 1; j = super-∞↑):(Hyper-relativity-of-simultaneity-j)]] = $(\partial c/\partial x_1)\mathbf{i}_1 + (\partial c/\partial x_2)i_2, \ldots, + (\partial c/\partial x_n)\mathbf{i}_n$ | ∞ < mod Grad [f[Σ(j = 1; j = super-∞↑):(Hyper-relativity-of-simultaneity-j)]] ≤ ◯, or mod Grad [f[Σ(j = 1; j = super-∞↑):(Hyper-relativity-of-simultaneity-j)]] > ◯}}}}dt}dt} … dt}

{∫ … {∫∫∫ {t{f{{light-speed-craft-k | 1 ≤ k ≤ ◯, or k > ◯};{div **[f[Σ(j = 1; j = super-∞↑):(Hyper-relativity-of-simultaneity-j)]]** = $\partial \mathbf{c}_1/\partial x_1 + \partial \mathbf{c}_2/\partial x_2 + \ldots + \partial \mathbf{c}_n/\partial x_n$ | mod div **[f[Σ(j = 1; j = super-∞↑):(Hyper-relativity-of-simultaneity-j)]]** ≤ 0};{Grad [f[Σ(j = 1; j = super-∞↑):(Hyper-relativity-of-simultaneity-j)]] = $(\partial c/\partial x_1)\mathbf{i}_1 + (\partial c/\partial x_2)i_2, \ldots, + (\partial c/\partial x_n)\mathbf{i}_n$ | ∞ < mod Grad [f[Σ(j = 1; j = super-∞↑):(Hyper-relativity-of-simultaneity-j)]] ≤ ◯, or mod Grad [f[Σ(j = 1; j = super-∞↑):(Hyper-relativity-of-simultaneity-j)]] > ◯}}}}dT}dT} … d T}

{∫ … {∫∫∫ {T{f{{light-speed-craft-k | 1 ≤ k ≤ ◯, or k > ◯};{div **[f[Σ(j = 1; j = super-∞↑):(Hyper-relativity-of-simultaneity-j)]]** = $\partial \mathbf{c}_1/\partial x_1 + \partial \mathbf{c}_2/\partial x_2 + \ldots + \partial \mathbf{c}_n/\partial x_n$ | mod div **[f[Σ(j = 1; j = super-∞↑):(Hyper-relativity-of-simultaneity-j)]]** ≤ 0};{Grad [f[Σ(j = 1; j = super-∞↑):(Hyper-relativity-of-simultaneity-j)]] = $(\partial c/\partial x_1)\mathbf{i}_1 + (\partial c/\partial x_2)i_2, \ldots, + (\partial c/\partial x_n)\mathbf{i}_n$ | ∞ < mod Grad [f[Σ(j = 1; j = super-∞↑):(Hyper-relativity-of-simultaneity-j)]] ≤ ◯, or mod Grad [f[Σ(j = 1; j = super-∞↑):(Hyper-relativity-of-simultaneity-j)]] > ◯}}}}dc}dc} … d c}

{∫ … {∫∫∫ {T{f{{light-speed-craft-k | 1 ≤ k ≤ ◯, or k > ◯};{div **[f[Σ(j = 1; j = super-∞↑):(Hyper-relativity-of-simultaneity-j)]]** = $\partial \mathbf{c}_1/\partial x_1 + \partial \mathbf{c}_2/\partial x_2 + \ldots + \partial \mathbf{c}_n/\partial x_n$ | mod div **[f[Σ(j = 1; j = super-∞↑):(Hyper-relativity-of-simultaneity-j)]]** ≤ 0};{Grad [f[Σ(j = 1; j = super-∞↑):(Hyper-relativity-of-simultaneity-j)]] = $(\partial c/\partial x_1)\mathbf{i}_1 + (\partial c/\partial x_2)i_2, \ldots, + (\partial c/\partial x_n)\mathbf{i}_n$ | ∞ < mod Grad [f[Σ(j = 1; j = super-∞↑):(Hyper-relativity-of-simultaneity-j)]] ≤ ◯, or mod Grad [f[Σ(j = 1; j = super-∞↑):(Hyper-relativity-of-simultaneity-j)]] > ◯}}}}dγ}dγ} … d γ}

{∫ … {∫∫∫ {T{f{{light-speed-craft-k | 1 ≤ k ≤ ◯, or k > ◯};{div **[f[Σ(j = 1; j = super-∞↑):(Hyper-relativity-of-simultaneity-j)]]** = $\partial \mathbf{c}_1/\partial x_1 + \partial \mathbf{c}_2/\partial x_2 + \ldots + \partial \mathbf{c}_n/\partial x_n$ | mod div **[f[Σ(j = 1; j = super-∞↑):(Hyper-relativity-of-simultaneity-j)]]** ≤ 0};{Grad [f[Σ(j = 1; j = super-∞↑):(Hyper-relativity-of-simultaneity-j)]] = $(\partial c/\partial x_1)\mathbf{i}_1 + (\partial c/\partial x_2)i_2, \ldots, + (\partial c/\partial x_n)\mathbf{i}_n$ | ∞ < mod Grad [f[Σ(j = 1; j = super-∞↑):(Hyper-relativity-of-simultaneity-j)]] ≤ ◯, or mod Grad [f[Σ(j = 1; j = super-∞↑):(Hyper-relativity-of-simultaneity-j)]] > ◯}}}}dv}dv} … d v}

{∫ … {∫∫∫ {T{f{{light-speed-craft-k | 1 ≤ k ≤ ◯, or k > ◯};{div **[f[Σ(j = 1; j = super-∞↑):(Hyper-relativity-of-simultaneity-j)]]** = $\partial \mathbf{c}_1/\partial x_1 + \partial \mathbf{c}_2/\partial x_2 + \ldots + \partial \mathbf{c}_n/\partial x_n$ | mod div **[f[Σ(j = 1; j = super-∞↑):(Hyper-relativity-of-simultaneity-j)]]** ≤ 0};{Grad [f[Σ(j = 1; j =

super-∞↑):(Hyper-relativity-of-simultaneity-j)]] = $(\partial c/\partial x_1)\mathbf{i_1}$ + $(\partial c/\partial x_2)i_2,\ldots,$ + $(\partial c/\partial x_n)\mathbf{i_n}$ | ∞ < mod Grad [f[Σ(j = 1; j = super-∞↑):(Hyper-relativity-of-simultaneity-j)]] ≤ ◯, or mod Grad [f[Σ(j = 1; j = super-∞↑):(Hyper-relativity-of-simultaneity-j)]] > ◯}}}}da}da} … d a}

{∫ … {∫{∫ {T{f{{light-speed-craft-k | 1 ≤ k ≤ ◯, or k > ◯};{div **[f[Σ(j = 1; j = super-∞↑):(Hyper-relativity-of-simultaneity-j)]]** = $\partial \mathbf{c_1}/\partial x_1 + \partial \mathbf{c_2}/\partial x_2 + \ldots + \partial \mathbf{c_n}/\partial x_n$ | mod div **[f[Σ(j = 1; j = super-∞↑):(Hyper-relativity-of-simultaneity-j)]]** ≤ 0};{Grad [f[Σ(j = 1; j = super-∞↑):(Hyper-relativity-of-simultaneity-j)]] = $(\partial c/\partial x_1)\mathbf{i_1}$ + $(\partial c/\partial x_2)i_2,\ldots,$ + $(\partial c/\partial x_n)\mathbf{i_n}$ | ∞ < mod Grad [f[Σ(j = 1; j = super-∞↑):(Hyper-relativity-of-simultaneity-j)]] ≤ ◯, or mod Grad [f[Σ(j = 1; j = super-∞↑):(Hyper-relativity-of-simultaneity-j)]] > ◯}}}}dP}dP} … d P}

{∫ … {∫{∫ {T{f{{light-speed-craft-k | 1 ≤ k ≤ ◯, or k > ◯};{div **[f[Σ(j = 1; j = super-∞↑):(Hyper-relativity-of-simultaneity-j)]]** = $\partial \mathbf{c_1}/\partial x_1 + \partial \mathbf{c_2}/\partial x_2 + \ldots + \partial \mathbf{c_n}/\partial x_n$ | mod div **[f[Σ(j = 1; j = super-∞↑):(Hyper-relativity-of-simultaneity-j)]]** ≤ 0};{Grad [f[Σ(j = 1; j = super-∞↑):(Hyper-relativity-of-simultaneity-j)]] = $(\partial c/\partial x_1)\mathbf{i_1}$ + $(\partial c/\partial x_2)i_2,\ldots,$ + $(\partial c/\partial x_n)\mathbf{i_n}$ | ∞ < mod Grad [f[Σ(j = 1; j = super-∞↑):(Hyper-relativity-of-simultaneity-j)]] ≤ ◯, or mod Grad [f[Σ(j = 1; j = super-∞↑):(Hyper-relativity-of-simultaneity-j)]] > ◯}}}}dKE}dKE} … d KE}

{∫ … {∫{∫ {T{f{{light-speed-craft-k | 1 ≤ k ≤ ◯, or k > ◯};{div **[f[Σ(j = 1; j = super-∞↑):(Hyper-relativity-of-simultaneity-j)]]** = $\partial \mathbf{c_1}/\partial x_1 + \partial \mathbf{c_2}/\partial x_2 + \ldots + \partial \mathbf{c_n}/\partial x_n$ | mod div **[f[Σ(j = 1; j = super-∞↑):(Hyper-relativity-of-simultaneity-j)]]** ≤ 0};{Grad [f[Σ(j = 1; j = super-∞↑):(Hyper-relativity-of-simultaneity-j)]] = $(\partial c/\partial x_1)\mathbf{i_1}$ + $(\partial c/\partial x_2)i_2,\ldots,$ + $(\partial c/\partial x_n)\mathbf{i_n}$ | ∞ < mod Grad [f[Σ(j = 1; j = super-∞↑):(Hyper-relativity-of-simultaneity-j)]] ≤ ◯, or mod Grad [f[Σ(j = 1; j = super-∞↑):(Hyper-relativity-of-simultaneity-j)]] > ◯}}}}dt}dt} ... dt}

{∫ … {∫{∫ {T{f{{light-speed-craft-k | 1 ≤ k ≤ ◯, or k > ◯};{div **[f[Σ(j = 1; j = super-∞↑):(Hyper-relativity-of-simultaneity-j)]]** = $\partial \mathbf{c_1}/\partial x_1 + \partial \mathbf{c_2}/\partial x_2 + \ldots + \partial \mathbf{c_n}/\partial x_n$ | mod div **[f[Σ(j = 1; j = super-∞↑):(Hyper-relativity-of-simultaneity-j)]]** ≤ 0};{Grad [f[Σ(j = 1; j = super-∞↑):(Hyper-relativity-of-simultaneity-j)]] = $(\partial c/\partial x_1)\mathbf{i_1}$ + $(\partial c/\partial x_2)i_2,\ldots,$ + $(\partial c/\partial x_n)\mathbf{i_n}$ | ∞ < mod Grad [f[Σ(j = 1; j = super-∞↑):(Hyper-relativity-of-simultaneity-j)]] ≤ ◯, or mod Grad [f[Σ(j = 1; j = super-∞↑):(Hyper-relativity-of-simultaneity-j)]] > ◯}}}}dT}dT} … d T}

and so on.

We can also take multi-variable derivatives and integrals involving the following variables as non-limiting options; light-speed, gamma, velocity, acceleration, momentum, kinetic energy, and so on.

Each of the following roster elements can be the basis of expressions of differentiated functions for power, energy, velocity, momentum, and gamma and the like by differentiation in the following manners.

d f/dt,1 d(d f/dt,1)/dt,2 d(d f/dt,2)/dt,1 d[d(d f/dt,1)/dt,2]/dt,3 d[d(d f/dt,1)/dt,3]/dt,2

d[d(d f/dt,2)/dt,1]/dt,3 d[d(d f/dt,2)/dt,3]/dt,1 d[d(d f/dt,3)/dt,1]/dt,2 d[d(d f/dt,3)/dt,2]/dt,1

d f/dT,1 d(d f/dT,1)/dT,2 d(d f/dT,2)/dT,1 d[d(d f/dT,1)/dT,2]/dT,3 d[d(d f/dT,1)/dT,3]/dT,2

d[d(d f/dT,2)/dT,1]/dT,3 d[d(d f/dT,2)/dT,3]/dT,1 d[d(d f/dT,3)/dT,1]/dT,2 d[d(d f/dT,3)/dT,2]/dT,1

T in each case is a time-like blend of various sorts of multiple time dimensions that appropriately and usefully enables derivation in one iteration.

We can go further to consider.

d f/dƮ,1 d(d f/dƮ,1)/dƮ,2 d(d f/dƮ,2)/dƮ,1 d[d(d f/dƮ,1)/dƮ,2]/dƮ,3 d[d(d f/dƮ,1)/dƮ,3]/dƮ,2

d[d(d f/dƮ,2)/dƮ,1]/dƮ,3 d[d(d f/dƮ,2)/dƮ,3]/dƮ,1 d[d(d f/dƮ,3)/dƮ,1]/dƮ,2 d[d(d f/dƮ,3)/dƮ,2]/dƮ,1

Ʈ in each case is a time-like blend of time-like blends of various sorts of multiple time dimensions that appropriately and usefully enables derivation in one iteration.

By the same token, we can also integrate functions for power, energy, velocity, momentum, and gamma and the like by respect to multiple time dimensions as follows.

∫ f(dt,1) ∫∫d f(dt,1)(dt,2) ∫∫d f(dt,2)(dt,1) ∭ f(dt,1)(dt,2)(dt,3) ∭ f(dt,1)(dt,3)(dt,2)

∭ f(dt,2)(dt,1)(dt,3) ∭ f(dt,2)(dt,3)(dt,1) ∭ f(dt,3)(dt,1)(dt,2) ∭ f(dt,3)(dt,2)(dt,1)

∫ f(dT,1) ∬ f(dT,1) (dT,2) ∬ f(dT,2) ((dT,1)) ∭ f(dT,1) (dT,2) (dT,3) ∭ f(dT,1) (dT,3) (dT,2)

∭ f(dT,2) (dT,1) (dT,3) ∭ f(dT,2) (dT,3) (dT,1) ∭ f(dT,3) (dT,1) (dT,2) ∭ f(dT,3) (dT,2) (dT,1)

We can go further to consider.

∫ f(dƮ,1) ∫ ∫ f(dƮ,1)(dƮ,2) ∫ ∫ f(dƮ,2)(dƮ,1) ∫ ∫ ∫ f(dƮ,1)(dƮ,2)(dƮ,3) ∫ ∫ ∫ f(dƮ,1)(dƮ,3)(dƮ,2)

∫ ∫ ∫ f(dƮ,2)(dƮ,1)(dƮ,3) ∫ ∫ ∫ f(dƮ,2)(dƮ,3)(dƮ,1) ∫ ∫ ∫ f(dƮ,3)(dƮ,1)(dƮ,2) ∫ ∫ ∫ f(dƮ,3)(dƮ,2)(dƮ,1)

We can continue to yet higher order derivatives and integrals suffice it to say that the general patterns continue for higher numbers of dimensions.

We can also consider higher order blends of time and thus are not limited to the blends defined by T and Ʈ. We chose these two symbols for their resemblance of the letter t.

ESSAY 14) Novel Space-Time Relativistic Rocket Fuel Tank Scenarios.

Here, we consider the following novel space-time relativistic rocket fuel tank scenarios presented in clause roster form.

Space-time storage of rocket fuels where the fuel tanks are multiply space-time but not in the parallel history sense.

Space-time storage of rocket fuels where the fuel tanks are of multiplied space-time types but not in the parallel history sense.

Space-time storage of rocket fuels where the fuel tanks are multiply space-time but not in the parallel history sense and/or just simply subsist and not extended into hyperspaces, hyperspace-times, nor bulks.

Space-time storage of rocket fuels where the fuel tanks are of multiplied space-time types but not in the parallel history sense and/or just simply subsist and not extended into hyperspaces, hyperspace-times, nor bulks.

Space-time storage of rocket fuels where the fuel tanks are multiply space-time but not in the parallel history sense where the space-times are N-D-Space-M-D-Time where N is any integer equal to three or greater and M is any counting number.

Space-time storage of rocket fuels where the fuel tanks are of multiplied space-time types but not in the parallel history sense where the space-times are N-D-Space-M-D-Time where N is any integer equal to three or greater and M is any counting number.

Space-time storage of rocket fuels where the fuel tanks are multiply space-time but not in the parallel history sense and/or just simply subsist and not extended into hyperspaces, hyperspace-times, nor bulks where the space-times are N-D-Space-M-D-Time where N is any integer equal to three or greater and M is any counting number.

Space-time storage of rocket fuels where the fuel tanks are of multiplied space-time types but not in the parallel history sense and/or just simply subsist and not extended into hyperspaces, hyperspace-times, nor bulks where the space-times are N-D-Space-M-D-Time where N is any integer equal to three or greater and M is any counting number.

Space-time storage of rocket fuels where the fuel tanks are multiply space-time but not in the parallel history sense where the space-times are N-D-Space-M-D-Time where N is any integer equal to three or greater and M is any counting number and the space-times are flat, positively curved, negatively curved, positively curved and torsioned at one or more scales in arbitrary patterns including but not limited to fractals, negatively curved and torsioned at one or more scales in arbitrary patterns including but not limited to fractals, etc.

Space-time storage of rocket fuels where the fuel tanks are of multiplied space-time types but not in the parallel history sense where the space-times are N-D-Space-M-D-Time where N is any integer equal to three or greater and M is any counting number and the space-times are flat, positively curved, negatively curved, positively curved and torsioned at one or more scales in arbitrary patterns including but not limited to fractals, negatively curved and torsioned at one or more scales in arbitrary patterns including but not limited to fractals, etc.

Space-time storage of rocket fuels where the fuel tanks are multiply space-time but not in the parallel history sense where the space-times are N-D-Space-M-D-Time where N is any integer equal to three or greater and M is any counting number and the space-times are flat, positively curved, negatively curved, positively curved and torsioned at one or more scales in arbitrary patterns including but not limited to fractals, negatively curved and torsioned at one or more scales in arbitrary patterns including but not limited to fractals, etc.

Space-time storage of rocket fuels where the fuel tanks are of multiplied space-time types but not in the parallel history sense where the space-times are N-D-Space-M-D-Time where N is any integer equal to three or greater and M is any counting number and the space-times are flat, positively curved, negatively curved, positively curved and torsioned at one or more scales in arbitrary patterns including but not limited to fractals, negatively curved and torsioned at one or more scales in arbitrary patterns including but not limited to fractals, etc.

The quantity of fuel stored in the multiply space-timed tanks is as follows.

For space-times quantized at the Planck Length Scale and stored in cubical space-time, the volume of the stored fuels in numbers of cubic meters is:

{(E){(B EXP 3){ | {(A){1/{{[h/(2π)] G/(C EXP 3)} EXP (1/2)}}} | EXP (N-3)}}}

Here, A is the edge length of the cube in numbers of meters in its higher spatial dimensions, B is the edge length of the cube in its ordinary 3-D spatial dimensions in numbers of meters, and E is the number of multiplications of the interior space-time. The vertical bars represent the taking of the numerical portion of the expression within the bars thus yielding a dimensionless factor. All units are assumed to be in the MKS system.

For space-times of higher dimensionalities that are quantized at the Planck Length scales and also quantized at the Planck Length scales in their ordinary first three spatial dimensions, for which the stored fuel is nuclear fusion fuel with a 3-D density of one metric ton per 3-D cubic meter, the yield of the entire nuclear fusion fuel in Megatons of TNT is:

(~170 Megatons) {(E){(B EXP 3){ | {(A){1/{{[h/(2π)] G/(C EXP 3)} EXP (1/2)}}} | EXP (N-3)}}}

Here, A is the edge length of the cube in numbers of meters in its higher spatial dimensions, B is the edge length of the cube in its ordinary 3-D spatial dimensions in numbers of meters, and E is the number of multiplications of the interior space-time. The vertical bars represent the taking of the numerical portion of the expression within the bars thus yielding a dimensionless factor. All units are assumed to be in the MKS system.

For space-times quantized at the scale of [(1/a) Planck Length] and stored in cubical space-time where a is greater than one, the volume of the stored fuels is:

{(E){(B EXP 3){ | {(A){1/{(1/a){{[h/(2π)] G/(C EXP 3)} EXP (1/2)}}}} | EXP (N-3)}}}

Here, A is the edge length of the cube in numbers of meters in its higher spatial dimensions, B is the edge length of the cube in its ordinary 3-D spatial dimensions in numbers of meters, and E is the number of multiplications of the interior space-time. The vertical bars represent the taking of the numerical portion of the expression within the bars thus yielding a dimensionless factor. All units are assumed to be in the MKS system.

For space-times of higher dimensionalities that are quantized at the scale of [(1/a) Planck Length] scales and quantized at the Planck Length scales in their ordinary first three spatial dimensions, for which the stored fuel is nuclear fusion fuel with a 3-D density of one metric ton per 3-D cubic meter, the yield of the entire nuclear fusion fuel in Megatons of TNT is:

(~170 Megatons) {(E){(B EXP 3){ | {(A){1/{(1/a){{[h/(2π)] G/(C EXP 3)} EXP (1/2)}}}} | EXP (N-3)}}}

Here, A is the edge length of the cube in numbers of meters in its higher spatial dimensions, B is the edge length of the cube in its ordinary 3-D spatial dimensions in numbers of meters, and E is the number of multiplications of the interior space-time. The vertical bars represent the taking of the numerical portion of the expression within the bars thus yielding a dimensionless factor. All units are assumed to be in the MKS system.

For space-times quantized at the scale of [[1/[f(Aleph k)]] Planck Length] where k is any finite or infinite whole number and f denotes any functional aspects which may optionally amplify (Aleph k), the volume of the stored fuels is:

{(E){(B EXP 3){ | {(A){1/{[1/[f(Aleph k)]]{{[h/(2π)] G/(C EXP 3)} EXP (1/2)}}}} | EXP (N-3)}}}

Here, A is the edge length of the cube in numbers of meters in its higher spatial dimensions, B is the edge length of the cube in its ordinary 3-D spatial dimensions in numbers of meters, and E is the number of multiplications of the interior space-time. The vertical bars represent the taking of the numerical portion of the expression within the bars thus yielding a dimensionless factor. All units are assumed to be in the MKS system.

For space-times of higher dimensionalities that are quantized at the scale of [[1/[f(Aleph k)]] Planck Length] scales and quantized at the Planck Length scales in their ordinary first three spatial dimensions, for which the stored fuel is nuclear fusion fuel with a 3-D density of one metric ton per 3-D cubic meter, the yield of the entire nuclear fusion fuel in Megatons of TNT is:

(~170 Megatons) {(E){(B EXP 3){ | {(A){1/{[1/[f(Aleph k)]]{{[h/(2π)] G/(C EXP 3)} EXP (1/2)}}}} | EXP (N-3)}}}

Here, A is the edge length of the cube in numbers of meters in its higher spatial dimensions, B is the edge length of the cube in its ordinary 3-D spatial dimensions in numbers of meters, and E is the number of multiplications of the interior space-time. The vertical bars represent the taking of the numerical portion of the expression within the bars thus yielding a dimensionless factor. All units are assumed to be in the MKS system.

Replacing the fusion fuel with pure matter antimatter fuel with a density of one metric ton per cubic meter, we obtain the yields by replacing 170 Megatons with 22.5 Gigatons.

Where the spacecraft fuel is matter-antimatter quarkonium fuel at a density of 10 EXP 21 metric tons per cubic meter, we replace 22.5 Gigatons with [(22.5 Gigatons)(10 EXP 21)].

For cases where the spacecraft fuel is held at the Planck Mass Density, and where all of the mass of the fuel can be converted to energy, we replace [(22.5 Gigatons)(10 EXP 21)] with [(22.5 Gigatons) [5.15500 x 10^{93}]]

Note that the Planck Mass Density is equal to:

$\rho_m = m_p/l^3_p = \hbar t_p/l^5_p = c^5/[\hbar G^2]$ =

5.15500 x 10^{96} kg/m^3

Now, we are all familiar with the Cartesian Number Line, you know, that line we all learned about in basic high school algebra that extends forever in both directions and having an infinite series of positive integers and an infinite series of negative integers.

Well, we can even consider hyperextended number lines that are infinitely longer than the already infinite number line. The extent of hyperextension is arbitrary.

However, what if a light-speed spacecraft could have such an extreme Lorentz factor such that the value of the Lorentz factor would not even place on a hyperextended number line?

Accordingly, the Lorentz factor of the spacecraft would be neither an integer, rational number, irrational number, in a valid sense not even a real number.

What I am conjecturing on is a value that is also not a complex number nor a tuple such as any analog for arbitrarily high numbered dimensional space, the two dimensional example of which is commonly known as an ordered pair.

Such a spacecraft Lorentz factor would be so extreme such that it would more properly be referred to as a hyper-number.

What's more, we can consider Lorentz factors that are so extreme such that these Lorentz factors are to hyper-numbers as hyper-numbers are to real numbers. We refer to these Lorentz factors as hyper-hyper-numbers or hyper-2-numbers and hyper-numbers as hyper-1-numbers.

Even more astounding would be Lorentz factors so extreme such that these Lorentz factors would be to hyper-2-numbers as hyper-2-numbers are to hyper-1-numbers.

We can continue the conjecture accordingly to Lorentz factors of hyper-4-numbers, hyper-5-numbers, hyper-6-numbers all the way to hyper-∞-numbers and beyond forever more. The symbol ∞ can be any infinity no matter how great.

Note that Hyper4(a, n) is equal to a tetrated n or a raised to the power of itself n-1 times. The latter value is symbolically written as n subscript a. For example 3 EXP 4 = 81, but 4 subscript3 is approximately equal to 10 EXP (1,000,000,000,000).

Alternatively 4 subscript 2 = 2 EXP 2 EXP 2 EXP 2 = 2 EXP [2 EXP [2 EXP 2]] = 2 EXP (2 EXP 4) = 2 EXP 16 = 65,536.

For example, Hyper5(4, 4)is equal to 4 tetrated 4 tetrated 4 tetrated 4. This value is commonly referred to as 4 pentated 4.

Hyper 6, (4,4) is 4 pentated 4 pentated 4 pentated 4 and is also referred to as 4 hexataed 4.

Hyper 7, (4,4) is 4 hexated 4 hexated 4 hexated 4 and so-on

Aleph 0 is the infinite number of integers.

Aleph 1 according to the perhaps unprovable and thus unfalsifiable Continuum Hypotheses is the number of real numbers which is greater than Aleph 0 by a multiplicative factor of infinity.

Aleph 2 is similarly greater than Aleph 1.

Aleph 3 is similarly greater than Aleph 2.

Aleph 4 is similarly greater than Aleph 3.

and so-on

In general, Aleph n = 2 EXP [Aleph (n-1)]

The number Ω is commonly stated as the least infinite positive integer or ordinal.

Now here is a real zinger.

So we can produce the abstraction of [Hyper Aleph Ω (Aleph Ω, Aleph Ω)].

We can go to ever greater infinities.

So we can consider hyper-[Hyper Aleph Ω (Aleph Ω, Aleph Ω)]-numbers.

We consider the implied affixment of the following operator to the lengthy formulas for gamma and related parameters in Sections 1 and 2 to denote the hyper-k-number scheme for light-speed spacecraft Lorentz factors. We do not actually iterate the operator as such but merely note that it can be affix to said lengthy formulas.

[Σ(k = 1; k = [Hyper Aleph Ω (Aleph Ω, Aleph Ω)] and ↑):(hyper-k-numbers Lorentz factors)]

So do not fret any light-speed limits imposed by Mother Nature! Such limits would work out for all the better. And no one needs to stay back at home for loved ones departing on utterly extreme Lorentz factor light-speed journeys! It is conceivable that an entire cosmic light-cone, perhaps even an entire universe, multiverse, forest, biosphere and the like could come along for the journey.

For now, let us begin our interstellar travel by figuring out a mil-spec blue-printed nuclear fission reactor powered starship that can reach a Lorentz factor of a mere 1.005 or a velocity of 0.1 c. This will get us to Alpha Centauri in about 40 years and to Barnard's Star in about 59 years.

Once we have mastered nuclear fission reactor powered rockets, we can move on to nuclear fusion reactor powered rockets to attain velocities of 0.5 c for a Lorentz factor of about 1.155.

Once we learn how to make lots of antimatter, we can have positronium fueled rockets able to reach a Lorentz factor of 5 at a speed of 0.98 c. This would require a mass ratio of only about 10. For a mass ratio of 10 including purely antimatter fuel that obtains its normal matter partners from the background of space in a drag recycling manner, the Lorentz factor that accrues is about 50 with a velocity equal to 0.9998 c. It only gets better from here.

ESSAY 15) Travel Into Ontological Transcendenal Aspects Of The Cosmos.

Here, we consider travel interiorly to greater and greater extents such as into the ontological transcendental aspects of the cosmos such as its purpose, value, ontological goodness, worthiness and the like, but also deeper and deeper into the laws and physical constants of created material nature.

Here we also consider travel interiorly to greater and greater extents such as into the existential aspects of the cosmos such as its aspects distinct from its ontological transcendentals of purpose, value, ontological goodness, worthiness and the like, but also deeper and deeper into the laws and physical constants of created material nature.

Now, the physical aspects of created reality would seem to be classifiable into accidents and substances. Accidents are the visible every day discernable aspects we have observational access to. Substantial aspects are the prime matters or first created principles presumably created ex-nihilo. Then there are the ontological transcendental aspects of the physical realities such as value, worthiness, purpose, ontological goodness and then like.

However, there would also seem to be ontological-transcendental-less aspects of physical reality. As such, these aspects would be more or less of a “just is” or being kind of constructs. Note that these ontological-transcendental-less aspects are not actually worthless and the like, and thus are very, very, good.

The just is or being aspects of physical materiality are hard to intuit kinds of constructs because typically we refer only to GOD as just is or pure being. However, physical realities may be a limited class of just is or being realities. These physical aspects are as innocent as impersonal realities that exist and which are not capable of doing anything wrong. So in a limited sense, they just exist.

Now, we can consider scenarios for which spacecraft have obtained such extreme infinite Lorentz factors such that they run out of room to travel forward in time in the universe or broader realms of travel and thus over-reach these realms.

Accordingly, these spacecraft might travel into or amongst, or become existentially proximate to the ontological transcendental aspects of the universe or greater realms of travel and also into the just is or being aspects of physical reality.

These concepts are very abstract but might pose an awesome set of travel itineraries for human civilization in the eternally distance future.

We may affix the following operator to formulas for gamma and related parameters to denote light-speed or ever-so-slightly faster-than-light spacecraft travel into or amongst, or becoming existentially proximate, to the ontological transcendental aspects of the universe or greater realms of travel and also into or amongst the just is or being aspects of physical reality as well as into the laws of physics and physical constants of physical reality.

[g(Light-speed or ever-so-slightly faster-than-light spacecraft travel into or amongst, or becoming existentially proximate, to the ontological transcendental aspects of the universe or greater realms of travel and also into or amongst the just is or being aspects of physical reality as well as into the laws of physics and physical constants of physical reality)]

ESSAY 16) Travel Into Negation Or Nothing.

Now, we can intuit the concept of negation or nothing. As such, nothing in a way if it could ever fit into a reification heuristics framework would be the utter ultimate in simplicity in a manner analogous to the empty set in Set Theory.

If nothing exists outside of GOD and the created order, what exactly would such negation be or entail in theories of reified negation.

For example, does a great void exist outside of GOD and the created order? Alternatively, would such a great void be as if a vacuum devoid of space-time but analogous to completely empty abstract space-time? As another alternative, would such a great void be an existential void?

To me, the great void of negation that exists outside of GOD and the created order is part of the scope of reality.

Accordingly, I view reality as everything that previously existed, currently exists, or will exist in the future, all in terms of substantial and accidental realities, all the relations between what previously existed, currently exists, or will exist in the future, all in terms of substantial and accidental realities, all that did not previously exist, all that does not exists, all that will not exist in the future, all that was possible, all that is possible, all that will be possible, all that was impossible, all that is impossible, all that will be impossible, and so-on.

Reality thus includes not just concrete elements in a set, but the laws of reality such as laws defining what GOD can and cannot do, what GOD can and cannot be, what creatures can and cannot do, and what creature can and cannot be.

So now, you might have a sense of nothing and thus the great void or vast wasteland mentioned in the books of Genesis in the Old Testament.

Now, I have a proposal to make. Perhaps a spacecraft having attained such an infinite Lorentz factor thus having out-ran the future of a given universe or broader realm of travel can enter a great void or negation that is as nothing as it would be if all creatures were never created nor the cosmos.

Should we ever be able to enter the great void, this can be an awesome opportunity to explore nothing or negation and to study this reality which ironically would have lack of existence.

We may affix following operator to formulas for gamma and related parameters to denote the above travel into nothing and the underlying existential and ontological principles that may enable such extreme travel.

[f(Spacecraft attainment of such an infinite Lorentz factor thus having out-ran the future of a given universe or broader realm of travel to enter a great void or negation that is as nothing as it would be if all creatures were never created nor the cosmos created)]. We do not actually affix the operator in practice to the lengthy formulas for the sake of brevity.

Now, here is a zinger! Could there exist more than one nothing? If so then we can apply the following operator.

We may affix the following operator to formulas for gamma and related parameters to denote the above travel into a kth nothing and the underlying existential and ontological principles that may enable such extreme travel. We do not actually affix the operator in practice to the lengthy formulas for the sake of brevity.

[f(Spacecraft attainment of such an infinite Lorentz factor thus having out-ran the future of a given universe or broader realm of travel to enter a kth great void or negation that is as nothing as it would be if all creatures were never created nor the cosmos created where k can range from 1 to a value so infinite that the value cannot be defined)].

ESSAY 17) The Infinite Number Of Post Future Voids.

We again can consider a spacecraft traveling at the speed of light having run out of future time to travel forward in time due to sufficiently infinite Lorentz factors and commensurate infinite time dilations.

Perhaps such a spacecraft then leaves the universe or broader realm of travel to enter a not yet determined future even with respect to the spacecraft reference frame.

The post future of travel would be the future analog of negation such as preceded the natural realities according to the notion of an All-Powerful GOD.

As such, the spacecraft would in a sense become the “leader of the pack” with respect to the rest of created reality.

What’s even more profound is that there may exist untold infinite numbers of such post future voids or realities of negation.

We may affix the following operator to the lengthy formulas for gamma and related parameters to denote a suitably infinite Lorentz factor spacecraft traveling into a would be future analog of negation such as preceded the natural created realities according to the notion of an All-Powerful Catholic GOD and where the number of such post-future negations may be arbitrarily infinite.

[g(Suitably infinite Lorentz factor spacecraft traveling into a would be future analog of negation such as preceded the natural created realities according to the notion of an All-

Powerful Catholic GOD and where the number of such post-future negations may be arbitrarily infinite)]

ESSAY 18) Sufficiently Infinite Lorentz Factor Spacecraft Travel Into Non-Spatial-Temporal Realms.

We again can consider a spacecraft traveling at the speed of light having run out of future time to travel forward in time due to sufficiently infinite Lorentz factors and commensurate infinite time dilations.

Perhaps such a spacecraft then leaves the universe or broader realm of travel to enter a really exotic form of another realm.

Accordingly, it might be possible for a spacecraft to enter another continuum of extension but one that is completely distinct from space, time, space-time, hyperspace, hypertime, or hyperspace-time.

Such realms of travel may actually have a fabric of sorts as does known 4-D space-time. However, the fabric of these exotic realms of travel would be a completely different material than that of known and currently theoretical space-times.

What's more, the exotic continuums of travel may be arbitrarily infinite in number as would commensurately be the number of types of fabric materials composing such exotic continuums.

We may affix the following operator to formulas for gamma and related parameters to denote the infinite numbers of exotic continuums and associated fabrics of composition available to spacecraft having attained suitably infinite Lorentz factors.

[f(Infinite numbers of exotic continuums and associated fabrics of composition available to spacecraft having attained suitably infinite Lorentz factors)]

Regarding extremely far-out but more near term realizable scenarios, consider the following story line account.

You are about to begin a 15 billion light-year journey that will only take a few seconds in your time. Your booth operates via field effect propulsion mechanisms which are able to sweep out a very large cross-section of background magnetic energy.

You enter your travel booth which is made of 0.1 femtometer thick quarkonium sheeting.

The both when empty of fuel has invariant mass of about 10 metric tons. The booth can contain about ten billion metric tons of fuel compactly stored at a density of about 10 billion metric tons per cubic meter.

You unlock and open the entrance door to the travel both.

Then you close the door and a whimsical red light comes on which bathes you and the interior of the craft with red illumination. Red light is chosen for the fiat make and model you rented because red light will not mess with your vision should the booth have a propulsion or power failure in which case you will need to be able to see the emergency warning beacons switch and other features available to call out for assistance. The both has a nanotechnology hibernation chamber that will enable you to live billions of years in relaxed sleep until assistance comes along to tow your booth to the nearest service station and hotel accommodations.

A touch screen comes on with icons and menu selections that resemble the I-pads and I-phones in eras long-long ago.

You touch the destination menu icon and choose a star that is in a Galaxy currently about 10 billion light-years distant.

Then you touch the go icon on the screen and you hear clicks, clanks, buzzes, and other noises that sound like the photo copiers used in the first quarter of the 21st Century Earth way back in the days.

Now, your ship carries 10 billion tons of pure antimatter which it will mix with background normal matter reactants to quickly boost your craft to a Lorentz factor of 10 EXP 17.

Thanks to anti-gravatic interior acceleration cancelling technology, you feel completely normal for the 10 to 15 seconds fiat flight to your destination in your fiat reference frame.

Your craft uses reactive mechanisms with the background to rapidly slow so that onboard fuel is not needed for the deceleration phase. During the acceleration phase, the craft recycles virtually all drag energy.

In all, twenty seconds will pass fiat time as the booth accelerates and decelerates to the destination.

When you step out, you feel at first a sense of loss in knowing that your loved ones left behind died about 15 billion years ago. However, you become exhilarated by the reality that you just traveled 15 billion light-years and 15 billion years into the future in your awesome space-booth.

The Sun and Earth you left behind are gone. The Sun having swelled to red-giant phase and cooling off to become a black dwarf happened about 10 billion years ago and Earth during the process was evaporated in the outer atmosphere of the huge red giant star the Sun had evolved into.

Now that I have presented a fascinating story plot, I tease your sense of wonderlust by suggesting that the latter scenario should be possible according to current known and projected theoretical physics.

Now, within the observable universe, there are about 10 EXP 24 stars. This is about equal to the number of fine grains of table sugar that would cover the entire United States 100 meters deep. The number of planets orbiting stars in our universe seems to be about 10 times greater yet or equal to the number of fine grains of table sugar that would cover the entire United States 1,000 meters deep. The number of moons, orbiting planets, orbiting stars is estimated to be ten times greater yet or equal to the number of fine grains of table sugar that would cover the entire United States 10,000 meters deep.

Most of these stars are easily accessible should the fictional space-booths be developed within the next millennium.

OH! WHAT DREAMS WILL COME!

ESSAY 19) Huge High Intensity Magnets For Enabling Extreme Relativistic Space Flight.

Imagine creating a huge high intensity magnet, perhaps of the mass and volume of the Earth of spherical shape.

Such a magnet would enable spacecraft to travel in orbiting acceleration to relativistic velocities by the relativistic Lorentz turning force.

However, we can go a step further to consider a truss-like work supporting a magnetized spherical shell having mass equal to that of a typical star. Alternatively, the truss-like work may also support a toroidal magnet which may be permanent or super-conducting with a current continually cycling through the magnet.

We can go yet further to consider a toroidal magnet having mass of 10 EXP 18 solar masses while having a radius of 1 million light-years. The magnet may be set rotating so that it does not collapse under its own gravity.

We can go still further to consider a toroidal magnet having mass of 10 EXP 22 solar masses and radius of 10 billion light-years.

While we are at this, we can consider magnets set up having mass N times that of the total mass energy content of our observable portion of our universe while having radius of [(10)(N)] times that of our current light-cone.

Such a magnet could be made of white dwarf density material in a toroidal configuration having a linear mass density equal to about that of a stellar white dwarf.

Depending on the topology of the space-time proximate to our current cosmic light-cone boundaries, a magnet with an arcuate segment spanning the diameter of our current cosmic light-cone or future cosmic light-cone may have combined arcuate segment lengths effectively enabling a canonical ensemble or even an infinity scrapper cosmic light-cone radius units of circumferential travel distance.

Accordingly, a toroidal magnet as such might somehow be constructed starting within our current cosmic light-cone and where the boundary limits of construction extensity from the Milky Way Galaxy might only be modestly greater than one current or then future cosmic light-cone radius unit. So, spacecraft powered by such a magnet might reach speeds which are quantum mechanically indistinguishable from the speed of light. It may even be possible that the spacecraft may exceed the speed of light in the particular vacuum of our universe by a vanishingly small increment of absolute light-speed in a pure vacuum. As such, it might be possible for the craft to travel back in time.

A craft with initial Lorentz factor that is very finite, but then accelerating to a velocity faster than that of light in the particular vacuum of our universe might experience a Lorentz factor jump to infinite values along with infinite jumps in spacecraft kinetic energy. As such, the latter conjectural scenarios would put a permanent end to any future cosmic energy crises.

ESSAY 20) The Unlimited Number Of Infinite Lorentz Factors For Light-Speed Travel.

Here, we consider prospects for an unlimited number of infinite Lorentz factors associated with light-speed travel.

Accordingly, we consider that the effective velocity of light is a vanishingly small finite or literally infinitesimally small width spectrum of velocities.

Now, at the infinite Lorentz factor of a spacecraft increases, the spacecraft will presumably come ever so slightly closer and closer to the absolute ultimate speed limit. However, even if the craft never reaches the ultimate speed limit, the craft may still be traveling at c or light-speed in a vacuum.

We can intuit levels of c or the absolute speed limit that are ever so slightly progressively increasing.

One class of levels would simply be denoted as $C = c - (c/\infty)$.

Another higher class may be defined as a spectral width fraction of the ultimate speed limit having an absolute value less than zero but non-negative. We define such as:

{C > {[c – [[(c)ǁSpectral Velocity Widthǁ] | ǁSpectral Velocity Widthǁ]] is non-negative but less than zero at level 1}}

Another higher class may be defined as a spectral width fraction of the ultimate speed limit having an absolute value that is less than less than zero but non-negative. We define such as:

{C > {[c – [[(c)ǁSpectral Velocity Widthǁ] | ǁSpectral Velocity Widthǁ]] is non-negative but less than less than zero or at level 2}}

Another higher class may be defined as a spectral width fraction of the ultimate speed limit having an absolute value that is less than less than less than zero but non-negative. We define such as:

{C > {[c – [[(c)ǁSpectral Velocity Widthǁ] | ǁSpectral Velocity Widthǁ]] is non-negative but less than less than less than zero or at level 3}}

Another higher class may be defined as a spectral width fraction of the ultimate speed limit having an absolute value that is less than less than less than less than zero but non-negative. We define such as:

{C > {[c – [[(c)ǁSpectral Velocity Widthǁ] | ǁSpectral Velocity Widthǁ]] is non-negative but less than less than less than less than zero or at level 4}}

We can continue to level 5, level 6, level 7, and so-on without any upper limits to ordinality of series elements.

So we append the following operator to the lengthy formulas for gamma and related parameters in Sections 1 and 2 by implication but not in actuality to denote the spectral spread in light speed and the awesome ramifications for travel at such velocities for spacecraft and crew members.

{f{{{C > {[c – [[(c)ǁSpectral Velocity Widthǁ] | ǁSpectral Velocity Widthǁ]] is non-negative but less than zero at level 1}}, {C > {[c – [[(c)ǁSpectral Velocity Widthǁ] | ǁSpectral Velocity Widthǁ]] is non-negative but less than less than zero or at level 2}},

{C > {[c – [[(c)ǁSpectral Velocity Widthǁ] | ǁSpectral Velocity Widthǁ]] is non-negative but less than less than less than zero or at level 3}}, …↑},{Associated Lorentz factors and opportunities for ever more exotic space-time travel topologies, thermodynamics, and realms of entrance}}}

Note that the "… less than zero" terms may be either mathematically classed or so greatly classed such that the terms are progressively more extreme in qualitative ways.

So, if you think light-speed impulse travel is boring, I ask you to open your imagination and reconsider. Light-speed impulse travel is going to be awesome with never ending opportunities.

ESSAY 21) Cyclical Deep Diving Into Physical Reality To Attain Ever Greater Infinite Lorentz Factors.

We can consider traveling at light-speed at ever greater Lorentz factors to dive into reality ever deeper and to the very limits of future reality. We then may re-enter physical reality again to dive into reality ever deeper again and again to the very limits of reality.

Additionally, we can dive ever deeper and again into the very limits of future reality and then to repeat the process over and over again so that the spacecraft and crew become iteratively or multiply located at every differential element of the cosmos in time and in space. All the while, the craft and crew also travel into the new future extensions commensurate with new creation and future growth of the cosmos.

The capacity of such travel alludes to more than one temporal dimension developing for the spacecraft.

We can also intuit that as a spacecraft attains light-speed and infinite Lorentz factors, by becoming relativistic to infinite extents it becomes more and more itself or of itself in a relativistic manner with respect to the background. Accordingly, the craft becomes more and more existentially self-actualized as do the crew members although perhaps not in a conscious nor in a moral sense, instead in a more deeply ontological sense.

So we may append the following operator to formulas for gamma and related parameters to denote the above mentioned iteratively space-time presences, growth of the craft into new future created states of the cosmos, and the enhanced relativity of the craft upon obtaining ever greater Lorentz factors and the craft thus becoming more existentially transcendent and more itself as it separates from ordinary cosmic reference frames to become still more and more itself and of itself but not in a moral or ethical sense.

[f(Iteratively space-time presences of enhancing infinite Lorentz factor spacecraft, growth of the craft into new future created states of the cosmos, and the enhanced relativity of the craft upon obtaining ever greater infinite Lorentz factors and the craft thus becoming more existentially transcendent and more itself as it separates from ordinary cosmic reference frames to become still more and more itself and of itself but not in a moral or ethical sense)].

ESSAY 22) The Scaling Of The Distributions Of Realms With Realm Size In A Given Host Realm.

Now, we imagine that the number-wise distribution of universes scales inverse linearly with universal size in a given hyperspace.

Now, consider scenarios for which a light-speed craft achieving ever greater infinite Lorentz factors harnesses each universe in part or in full in its path.

The number of the smallest universes intaked scales inversely with the third power of gamma. The reason for this scaling is that the number of the frequency of smallest universes intaked from directly in front of the craft scales with gamma, and the separation of harnessed universes of smallest extent decreases with respect to both planar dimensions parallel of the perpendicular of the spacecraft velocity vector due to relativistic aberration. However, the kinetic energy of interaction with each smallest universe also scales with respect to gamma so that the energy acquisition by the spacecraft for the smallest universes scales with the fourth power of gamma. The same energy scaling occurs for each universal size already and previously acquired.

Thus, we have a factor of spacecraft time dependent energy intake proportional to the fourth power of gamma. So, we consider that the spacecraft gamma factor associated with the above scenario goes from gamma,1 = (10) gamma,1 = gamma,2.

Now we consider a scenario for which the spacecraft Lorentz factor increases by another factor of 10 to (100 gamma,1) = (10) gamma,2 = gamma,3.

The number of the smallest of the larger universes acquired at a gamma factor of gamma,3 increases in manners that scale with the third power of gamma. So the associated spacecraft time rate of increase in kinetic energy intake from these smallest of the larger universes acquired at a gamma factor of gamma,3 scales as the fourth power of gamma.

However, the invariant mass of the largest universes accessible at a Lorentz factor of gamma,3 increases in proportion to gamma. Since we assume that the number of galaxies of a given mass decreases linearly with invariant mass increase, we assume that the invariant mass specific interactive mass intake increases by another factor of gamma relative to the scenario otherwise. Thus, we obtain in total so far an additional factor of γ EXP 5 for spacecraft time dependent invariant mass increase.

However, the ship-time dependent kinetic energy per unit of invariant mass intake goes up in a way that scales with gamma. Thus, we obtain another factor of gamma for the total propulsion power available to the spacecraft.

Thus, the spacecraft propulsion power increases by a scaling factor of (γ EXP 4)(γ EXP 5)(γ EXP 1) = γ EXP 10.

Here, we assume as if the background universes are aberrated in manners similar to background resources for relativistic travel in our ordinary 4-D universe.

Note that we assume the size of an island universe is proportional to its invariant mass commensurate with the radius of a black hole being proportional to its mass. It just so happens that the quantity of mass-energy within the observable portion of our universe is about the same as that for a black hole of the same radius.

We may use the following operator to denote such γ EXP 10 scaling for light-sail craft propulsion power as the craft intakes background universes of ever increasing numbers and invariant masses as the spacecraft Lorentz factor grows.

[f(Light-Sail spacecraft propulsion power increases by a scaling factor of [(γ EXP 4)(γ EXP 5)(γ EXP 1)] = γ EXP 10 as a result of background universes mass-energy intake for a spacecraft traveling in a given hyperspace for which background universes appear to impinge on the spacecraft propulsion energy intake system(s))]

Here, we assume universal applicability of Special Relativity and associated relativistic aberration.

However, most likely, Special Relativity will not hold up under the above extreme conditions at increasing infinite Lorentz factors. So, we chose to apply the following operator instead.

[[f(Light-Sail spacecraft propulsion power increases by a scaling factor of [(γ EXP 4)(γ EXP 5)(γ EXP 1)] = γ EXP 10 as a result of background universes mass-energy intake for a spacecraft traveling in a given hyperspace for which background universes appear to impinge on the spacecraft propulsion energy intake system(s))], [g(Hyperspatial corrections to Special Relativity including changes in the; velocity of light, Lorentz factor formula, relativistic aberration, and the like)]]

The above operator may be appended to formulas for gamma and related parameters.

Now we consider scenarios for which the number of universes per invariant mass energy interval is the same across all intervals.

Now consider the average of the sum of the digits from one to ten. This average works out to 5.5.

As the number of orders of magnitude of consecutive digits summed is averaged and increases to infinity, the average is exactly equal to half the largest digit in the limit.

So, if we sum the average of all real numbers between one and ten, the average will be exactly 5.

Thus, this applies to the scenario for which the number of universes is the same for all real number masses. However, universes with a mass ten times greater than those of mass at the lowest end of the spectrum of universe intake for a change in gamma by a multiple of ten will result in intake power increases that are a multiple of ½ of ten or more generally ½ (gamma) relative to the scenario for which the number of universes of a given mass scales inversely with mass of the given universes.

Thus, the spacecraft propulsion power increases by a scaling factor of (γ EXP 4)(γ EXP 5)(γ EXP 1)(γ EXP 0.69897) = γ EXP 10.69897.

Note specifically here that ½ (gamma) = 5 = 10 EXP 0.69897.

Thus, were we to use the latter model for number-wise universal mass-energy distributions, we would replace, (γ EXP 4)(γ EXP 5)(γ EXP 1) = γ EXP 10, with (γ EXP 4)(γ EXP 5)(γ EXP 1)(γ EXP 0.69897) = γ EXP 10.69897

However, the power scaling as γ EXP 10 is more natural and follows much more closely the analogous scalings in the physical world.

Now, we can consider scenarios for which a spacecraft traveling in a bulk intakes and non-destructively reacts against a population of island multiverses each of which has a population of universes.

Accordingly, we modify the first formula presented above as:

Spacecraft propulsion power increase = [γ EXP (4 + ↑,a)][γ EXP (5 + ↑,b)][γ EXP (1+ ↑,c)] = γ EXP (10 + ↑,d)

Here, we assume as if the background multiverses are aberrated in manners similar to background resources for relativistic travel in our ordinary 4-D universe.

The upward arrow indicates irreversible growth of cosmic energy in the domains considered.

We may affix the following operator to formula for gamma for suitable light-sails to denote such γ EXP 10 scaling for said light-sail craft propulsion power as the craft intakes background multiverses of ever increasing numbers and invariant masses as the spacecraft Lorentz factor grows.

[f(Light-Sail spacecraft propulsion power increases by a scaling factor of [γ EXP (4 + ↑,a)][γ EXP (5 + ↑,b)][γ EXP (1+ ↑,c)] = γ EXP (10 + ↑,d) as a result of background multi-verses mass-energy intake for a spacecraft traveling in a given bulk for which background multi-verses appear to impinge on the spacecraft propulsion energy intake system(s))]

Here, we assume universal applicability of Special Relativity and associated relativistic aberration.

However, most likely, Special Relativity will not hold up under the above extreme conditions at increasing infinite Lorentz factors. So, we chose to apply the following operator instead.

[[f(Light-Sail spacecraft propulsion power increases by a scaling factor of [γ EXP (4 + ↑,a)][γ EXP (5 + ↑,b)][γ EXP (1+ ↑,c)] = γ EXP (10 + ↑,d) as a result of background multi-verses mass-energy intake for a spacecraft traveling in a given bulk for which background multi-verses appear to impinge on the spacecraft propulsion energy intake system(s))], [g(Bulk corrections to Special Relativity including changes in the; velocity of light, Lorentz factor formula, relativistic aberration, and the like)]]

Now we consider scenarios for which the number of multiverses per invariant mass energy interval is the same across all intervals.

Now consider the average of the sum of the digits from one to ten. This average works out to 5.5.

As the number of orders of magnitude of consecutive digits summed is averaged and increases to infinity, the average is exactly equal to half the largest digit in the limit.

So, if we sum the average of all real numbers between one and ten, the average will be exactly 5.

Thus, this applies to the scenario for which the number of multi-verses is the same for all real number masses. However, multi-verses with a mass ten times greater than those of mass at the lowest end of the spectrum of multi-verse intake for a change in gamma by a multiple of ten will result in intake power increases that are a multiple of ½ of ten or more generally ½ (gamma) relative to the scenario for which the number of multi-verses of a given mass scales inversely with mass of the given multi-verses.

Thus, the spacecraft propulsion power increases by a scaling factor of [γ EXP (4 + ↑,a)][γ EXP (5 + ↑,b)][γ EXP (1+ ↑,c)] [γ EXP (0.69897+ ↑,e)] = [γ EXP (10 + ↑,d)] [γ EXP (0.69897+ ↑,e)] = γ EXP [10.69897+ ↑,g)]

Note specifically here that ½ (gamma) = 5 = 10 EXP 0.69897.

Thus, were we to use the latter model for number-wise multi-versal mass-energy distributions, we would replace, [γ EXP (4 + ↑,a)][γ EXP (5 + ↑,b)][γ EXP (1+ ↑,c)] = γ EXP (10 + ↑,d), with [γ EXP (4 + ↑,a)][γ EXP (5 + ↑,b)][γ EXP (1+ ↑,c)] [γ EXP (0.69897+ ↑,e)] = [γ EXP (10 + ↑,d)] [γ EXP (0.69897+ ↑,e)] = γ EXP [10.69897+ ↑,g)].

Now, we can consider scenarios for which a spacecraft traveling in a bulk intakes and non-destructively reacts against a population of island forests which refer to collectives of fractal-like or tree-like multiverses each of which has a population of universes.

Accordingly, we modify the first formula presented above as:

Spacecraft propulsion power increase = [γ EXP (4 + ↑,a,forests)][γ EXP (5 + ↑,b,forests)][γ EXP (1+ ↑,c,forests)] = γ EXP (10 + ↑,d,forests)

Here, we assume as if the background forests are aberrated in manners similar to background resources for relativistic travel in our ordinary 4-D universe.

The upward arrow indicates irreversible growth of cosmic energy in the domains considered.

We may affix the following operator formula for gamma for light-sail craft of suitably novel types to denote such γ EXP 10 scaling for said light-sail craft propulsion power as the craft intakes background forests of ever increasing numbers and invariant masses as the spacecraft Lorentz factor grows.

[f(Light-Sail spacecraft propulsion power increases by a scaling factor of [γ EXP (4 + ↑,a,forests)][γ EXP (5 + ↑,b,forests)][γ EXP (1+ ↑,c,forests)] = γ EXP (10 + ↑,d,forests) as a result of background forests mass-energy intake for a spacecraft traveling in a given bulk for which background forests appear to impinge on the spacecraft propulsion energy intake system(s))]

Here, we assume universal applicability of Special Relativity and associated relativistic aberration.

However, most likely, Special Relativity will not hold up under the above extreme conditions at increasing infinite Lorentz factors. So, we chose to apply the following operator instead.

[[f(Light-Sail spacecraft propulsion power increases by a scaling factor of [γ EXP (4 + ↑,a,forests)][γ EXP (5 + ↑,b,forests)][γ EXP (1+ ↑,c,forests)] = γ EXP (10 + ↑,d,forests) as a result of background forests mass-energy intake for a spacecraft traveling in a given bulk for which background forests appear to impinge on the spacecraft propulsion energy intake system(s))], [g(Bulk corrections to Special Relativity including changes in the; velocity of light, Lorentz factor formula, relativistic aberration, and the like)]]

Now we consider scenarios for which the number of forests per invariant mass energy interval is the same across all intervals.

Now consider the average of the sum of the digits from one to ten. This average works out to 5.5.

As the number of orders of magnitude of consecutive digits summed is averaged and increases to infinity, the average is exactly equal to half the largest digit in the limit.

So, if we sum the average of all real numbers between one and ten, the average will be exactly 5.

Thus, this applies to the scenario for which the number of forests is the same for all real number masses. However, forests with a mass ten times greater than those of mass at the lowest end of the spectrum of forest intake for a change in gamma by a multiple of ten will result in intake power increases that are a multiple of ½ of ten or more generally ½ (gamma) relative to the scenario for which the number of forests of a given mass scales inversely with mass of the given forests.

Thus, the spacecraft propulsion power increases by a scaling factor of [γ EXP (4 + ↑,a,forests)][γ EXP (5 + ↑,b,forests)][γ EXP (1+ ↑,c,forests)] [γ EXP (0.69897+ ↑,e,forests)] = [γ EXP (10 + ↑,d,forests)] [γ EXP (0.69897+ ↑,e,forests)] = γ EXP [10.69897+ ↑,g,forests)]

Note specifically here that ½ (gamma) = 5 = 10 EXP 0.69897.

Thus, were we to use the latter model for number-wise forest mass-energy distributions, we would replace, [γ EXP (4 + ↑,a,forests)][γ EXP (5 + ↑,b,forests)][γ EXP (1+ ↑,c,forests)] = γ EXP (10 + ↑,d,forests), with [γ EXP (4 + ↑,a,forests)][γ EXP (5 + ↑,b,forests)][γ EXP (1+ ↑,c,forests)] [γ EXP (0.69897+ ↑,e,forests)] = [γ EXP (10 + ↑,d)] [γ EXP (0.69897+ ↑,e,forests)] = γ EXP [10.69897+ ↑,g,forests)].

Now, we can consider scenarios for which a spacecraft traveling in a bulk intakes and non-destructively reacts against a population of island biospheres which refer to collectives of fractal-like or tree-like forests each of which has a population of multi-verses.

Accordingly, we modify the first formula presented above as:

Spacecraft propulsion power increase = [γ EXP (4 + ↑,a, biospheres)][γ EXP (5 + ↑,b, biospheres)][γ EXP (1+ ↑,c, biospheres)] = γ EXP (10 + ↑,d, biospheres)

Here, we assume as if the background biospheres are aberrated in `manners similar to background resources for relativistic travel in our ordinary 4-D universe.

The upward arrow indicates irreversible growth of cosmic energy in the domains considered.

We may affix the following operator to formulas for gamma for sufficiently novel light-sails to denote such γ EXP 10 scaling for said light-sail craft propulsion power as the craft intakes background biospheres of ever increasing numbers and invariant masses as the spacecraft Lorentz factor grows.

[f(Light-Sail spacecraft propulsion power increases by a scaling factor of [γ EXP (4 + ↑,a, biospheres)][γ EXP (5 + ↑,b, biospheres)][γ EXP (1+ ↑,c, biospheres)] = γ EXP (10 + ↑,d, biospheres) as a result of background biospheres mass-energy intake for a spacecraft traveling in a given bulk for which background biospheres appear to impinge on the spacecraft propulsion energy intake system(s))]

Here, we assume universal applicability of Special Relativity and associated relativistic aberration.

However, most likely, Special Relativity will not hold up under the above extreme conditions at increasing infinite Lorentz factors. So, we chose to apply the following operator instead.

[[f(Light-Sail spacecraft propulsion power increases by a scaling factor of [γ EXP (4 + ↑,a, biospheres)][γ EXP (5 + ↑,b, biospheres)][γ EXP (1+ ↑,c, biospheres)] = γ EXP (10 + ↑,d, biospheres) as a result of background biospheres mass-energy intake for a spacecraft traveling in a given bulk for which background biospheres appear to impinge on the spacecraft propulsion energy intake system(s))], [g(Bulk corrections to Special Relativity including changes in the; velocity of light, Lorentz factor formula, relativistic aberration, and the like)]]

Now we consider scenarios for which the number of biospheres per invariant mass energy interval is the same across all intervals.

Now consider the average of the sum of the digits from one to ten. This average works out to 5.5.

As the number of orders of magnitude of consecutive digits summed is averaged and increases to infinity, the average is exactly equal to half the largest digit in the limit.

So, if we sum the average of all real numbers between one and ten, the average will be exactly 5.

Thus, this applies to the scenario for which the number of biospheres is the same for all real number masses. However, biospheres with a mass ten times greater than those of mass at the lowest end of the spectrum of biosphere intake for a change in gamma by a multiple of ten will result in intake power increases that are a multiple of ½ of ten or more generally ½ (gamma) relative to the scenario for which the number of biospheres of a given mass scales inversely with mass of the given biospheres.

Thus, the spacecraft propulsion power increases by a scaling factor of [γ EXP (4 + ↑,a, biospheres)][γ EXP (5 + ↑,b, biospheres)][γ EXP (1+ ↑,c, biospheres)] [γ EXP (0.69897+ ↑,e, biospheres)] = [γ EXP (10 + ↑,d, biospheres)] [γ EXP (0.69897+ ↑,e, biospheres)] = γ EXP [10.69897+ ↑,g, biospheres)]

Note specifically here that ½ (gamma) = 5 = 10 EXP 0.69897.

Thus, were we to use the latter model for number-wise biospheres mass-energy distributions, we would replace, [γ EXP (4 + ↑,a, biospheres)][γ EXP (5 + ↑,b, biospheres)][γ EXP (1+ ↑,c,biospheres)] = γ EXP (10 + ↑,d, biospheres), with [γ EXP (4 + ↑,a, biospheres)][γ EXP (5 + ↑,b, biospheres][γ EXP (1+ ↑,c, biospheres)] [γ EXP (0.69897+ ↑,e, biospheres)] = [γ EXP (10 + ↑,d)] [γ EXP (0.69897+ ↑,e, biospheres)] = γ EXP [10.69897+ ↑,g, biospheres)].

Now, we can consider scenarios for which a spacecraft traveling in a block intakes and non-destructively reacts against a population of island bulks.

A bulk is a general realm that hosts hyperspaces, universes, multiverses, forests, biospheres, and the like.

A block is a general realm that hosts bulks.

Accordingly, we modify the first formula presented above as:

Spacecraft propulsion power increase = [γ EXP (4 + ↑,a, bulks)][γ EXP (5 + ↑,b, bulks)][γ EXP (1+ ↑,c, bulks)] = γ EXP (10 + ↑,d, bulks)

Here, we assume as if the background bulks are aberrated in manners similar to background resources for relativistic travel in our ordinary 4-D universe.

The upward arrow indicates irreversible growth of cosmic energy in the domains considered.

We may append the following operator to formulas for gamma for sufficiently novel and suitable light-sail craft to denote such γ EXP 10 scaling for said light-sail craft propulsion power as the craft intakes background bulks of ever increasing numbers and invariant masses as the spacecraft Lorentz factor grows.

[f(Light-Sail spacecraft propulsion power increases by a scaling factor of [γ EXP (4 + ↑,a, bulks)][γ EXP (5 + ↑,b, bulks)][γ EXP (1+ ↑,c, bulks)] = γ EXP (10 + ↑,d, bulks) as a result of background bulks mass-energy intake for a spacecraft traveling in a given block for which background bulks appear to impinge on the spacecraft propulsion energy intake system(s))]

Here, we assume universal applicability of Special Relativity and associated relativistic aberration.

However, most likely, Special Relativity will not hold up under the above extreme conditions at increasing infinite Lorentz factors. So, we chose to apply the following operator instead.

[[f(Light-Sail spacecraft propulsion power increases by a scaling factor of [γ EXP (4 + ↑,a, bulks)][γ EXP (5 + ↑,b, bulks)][γ EXP (1+ ↑,c, bulks)] = γ EXP (10 + ↑,d, bulks) as a result of background bulks mass-energy intake for a spacecraft traveling in a given block for which background bulks appear to impinge on the spacecraft propulsion energy intake system(s))], [g(Block corrections to Special Relativity including changes in the; velocity of light, Lorentz factor formula, relativistic aberration, and the like)]]

Now we consider scenarios for which the number of bulks per invariant mass energy interval is the same across all intervals.

Now consider the average of the sum of the digits from one to ten. This average works out to 5.5.

As the number of orders of magnitude of consecutive digits summed is averaged and increases to infinity, the average is exactly equal to half the largest digit in the limit.

So, if we sum the average of all real numbers between one and ten, the average will be exactly 5.

Thus, this applies to the scenario for which the number of bulks is the same for all real number masses. However, bulks with a mass ten times greater than those of mass at the lowest end of the spectrum of bulk intake for a change in gamma by a multiple of ten will result in intake power increases that are a multiple of ½ of ten or more generally ½ (gamma) relative to the scenario for which the number of bulks of a given mass scales inversely with mass of the given bulks.

Thus, the spacecraft propulsion power increases by a scaling factor of [γ EXP (4 + ↑,a, bulks)][γ EXP (5 + ↑,b, bulks)][γ EXP (1+ ↑,c, bulks)] [γ EXP (0.69897+ ↑,e, bulks)] = [γ EXP (10 + ↑,d, bulks)] [γ EXP (0.69897+ ↑,e, bulks)] = γ EXP [10.69897+ ↑,g, bulks)]

Note specifically here that ½ (gamma) = 5 = 10 EXP 0.69897.

Thus, were we to use the latter model for number-wise bulks mass-energy distributions, we would replace, [γ EXP (4 + ↑,a, bulks)][γ EXP (5 + ↑,b, bulks)][γ EXP (1+ ↑,c,bulks)] = γ EXP (10 + ↑,d, bulks), with [γ EXP (4 + ↑,a, bulks)][γ EXP (5 + ↑,b, bulks][γ EXP (1+ ↑,c, bulks)] [γ EXP (0.69897+ ↑,e, bulks)] = [γ EXP (10 + ↑,d)] [γ EXP (0.69897+ ↑,e, bulks)] = γ EXP [10.69897+ ↑,g, bulks)].

ESSAY 23) Levels Of Hyper-Electromagnetic Radiation And Infinite Lorentz Factor Spacecraft.

Now, we can consider fields of energy that are to electromagnetic radiation as electromagnetic radiation is to ordinary periodic table elemental matter.

Accordingly, there may be hyper-electromagnetic fields or hyper-1-electromagnetic fields which are to electromagnetic fields as electromagnetic fields are to ordinary periodic table elemental matter.

Now let us consider hyper-hyper-electromagnetic fields or hyper-2-electromagnetic fields which are to hyper-1-electromagnetic fields as hyper-1-electromagnetic fields are to electromagnetic fields as electromagnetic fields are to ordinary periodic table elemental matter.

We will go further to consider hyper-hyper-hyper-electromagnetic fields or hyper-3-electromagnetic fields which are to hyper-2-electromagnetic fields as hyper-2-electromagnetic fields are to hyper-1-electromagnetic fields as hyper-1-electromagnetic fields are to electromagnetic fields as electromagnetic fields are to ordinary periodic table elemental matter.

Perhaps there is a never ending hierarchy of ascension.

Thus, we consider scenarios of hyper-k-electromagnetic fields where k can be any finite or infinite positive integer, even integers on a hyper-extended number line.

Perhaps one way of generating such fields includes beaming laser, maser, phased array microwave, phased array radiowave, nuclear reaction gamma rays, and the like ahead of light-speed or infinite Lorentz factor spacecraft.

Accordingly, the new hyper-k-electromagnetic fields would naturally result from velocity compositions of light-speed reference frames. As such, each consecutive level of hyper-k-electromagnetic fields starting with k = 1, would be generated by a spacecraft having increasingly infinite Lorentz factors.

Now Aleph 0 is the infinite number of positive integers.

Aleph 1 is according to the perhaps unprovable and unfalsifiable Continuum Hypothesis equal to the number of real numbers.

Either way, Aleph 1 = 2 EXP (Aleph 0).

Aleph 2 = 2 EXP (Aleph 1).

Aleph 3 = 2 EXP (Aleph 2).

Aleph 4 = 2 EXP (Aleph 3).

and so-on.

In general, [Aleph (N +1)] = 2 EXP (Aleph N) where N is any whole number.

Note that any radiation forms of hyper-k-electromagnetic fields may be generated and/or naturally source for propelling spacecraft such as relativistic rockets, and/or beam sail or other beam powered spacecraft.

We may append the operator to formulas for gamma and related parameters.

[f(Radiation forms of hyper-k-electromagnetic fields generated and/or naturally sourced for propelling spacecraft such as relativistic rockets, and/or beam sail or other beam powered spacecraft)].

The prefaced letter, f, denotes appropriate functionality which is context dependent and for with the functions can have differing values in various locations within the formulas or portions thereof to which it is affixed. The value of f can also vary in ship-time.

ESSAY 24) Ever Growing Future Time Dimensions For Infinite Lorentz Factor Spacecraft.

Now, we consider scenarios for travel to the future cosmic limits of time but where the future of time dimensions keeps growing for ever deeper future time travel for infinite Lorentz factor spacecraft.

As such, the spacecraft and crew can participate in eternally distant future technological development and work raw physical matter and energy into evolved forms and perhaps even into new forms of life.

In a sense, human persons in that eternally distant future era can work matter-energy-space-time into higher levels of existence which in turn can provide for enhancements in the bodies especially the brains of these persons. The feedback between body and soul can enable the accidental aspects of the soul to be enhanced and perhaps also the semi-substantial aspects of the human soul as well.

ESSAY 25) Various Levels Of Hyper-Mass And Infinite Lorentz Factor Spacecraft Travel.

Here, we consider a few broad constructs.

The first construct to consider is the distinction between mass and something that is more mass than mass or a form of hyper-mass or hyper-1-mass.

We can consider hyper-hyper-mass or hyper-2-mass which is to hyper-1-mass as hyper-1-mass is to mass.

Likewise, we consider hyper-3-mass, hyper-4-mass, and so-on in a never ending ascending series.

The second construct to consider is the distinction between matter and something that is more matter than matter or a form of hyper-matter or hyper-1-matter.

Here, there is a distinction made with mass in that matter can be any form of physical corporality whereas mass can refer to negative, positive, or imaginary mass and then like.

We can consider hyper-hyper-matter or hyper-2-matter which is to hyper-1- matter as hyper-1-matter is to matter.

Likewise, we consider hyper-3-matter, hyper-4-matter, and so-on in a never ending ascending series.

The third construct to consider is the distinction between material and something that is more material than material or a form of hyper- material or hyper-1- material.

Here, there is a distinction made with mass and matter in that material can be any form of materiality, physical or non-physical, whereas mass can refer to negative, positive, or imaginary mass and then like, and matter generally refers to physical corporality.

We can consider hyper-hyper-material or hyper-2-material which is to hyper-1- material as hyper-1-material is to matter.

Likewise, we consider hyper-3- material, hyper-4-material, and so-on in a never ending ascending series.

The above three classes of compositions may find use as energy or power sources for spacecraft able to attain infinite Lorentz factors in light-speed travel.

For example, these compositions can serve as carried aboard spacecraft fuel, pellet runway fuels, beamed fuels, and the like.

The latent energy density of these fuels in the spacecraft reference frame may be E >> ɱ[f(c EXP 2)] for hyper-masses, E >> ʃɧ[f(c EXP 2)] for hyper-matters, and E >> Ꮑ[f(c EXP 2)] for hyper-materials.

The following operator may be affixed to formulas for gamma of novel light-sail craft to denote the mechanisms for the three compositions considered above as power sources for the light-sail spacecraft.

{f[[[E >> ɱ,k,[f(c EXP 2)]], [E >> ʃɧ,l,[f(c EXP 2)]], [E >> ղ,m,[f(c EXP 2)]]] for sail mode, mass-ratio rocket mode, interstellar ramjet mode, fuel pellet runway mode, beam ship mode, etc.]}

Here we assume super-relativistic energy yields for the proposed fuels.

Here, the indices, k, l, and m, range from 1 to any infinite value on a hyper-extended number line for which the degree of hyper-extension can be quantitative, and even more fantastically be qualitative as well.

ESSAY 26) Serial Light-Speed Scout Craft Launched From A Light-Speed Mother Ship.

Now, a spacecraft traveling at the velocity of light would under Special Relativity have an infinite Lorentz factor.

A light-speed spacecraft might be able to launch a scout craft which may develop a relative light-speed and infinite Lorentz factor relative to the mother-ship.

The relative light-speed scout craft may in turn launch its own scout craft which might be able to attain a relative light-speed and infinite Lorentz factor relative to the first scout craft. The second scout craft may launch its own relative light-speed scout craft or third scout craft and the series can likewise continue from here.

Perhaps a mothership traveling at the speed of light with respect to the background would have relatively infinite kinetic energy. However, the light-speed mothership would need to be perfectly cloaked from the universe of travel so as not to be annihilated nor annihilate the universe of travel. Such cloaking may be inherently applied as no spacecraft could reach light-speed in inertial travel anyhow if the craft was not already suitably cloaked.

Regardless of whether a light-speed spacecraft was infinitely cloaked but also having infinite kinetic energy or perhaps having a finite energy but having its invariant mass reduced to zero in magnitude, such craft might essentially become pure light.

The first scout craft would become pure light relative to the mothership which the second scout craft would become pure light relative to the primary scout craft and so-on.

So, the so-called “lightness” of the mothership would be toped by the so-called “hyper-1-lightness” of the first scout craft while the“hyper-1-lightness” of the first scout craft would be toped by the so-called “hyper-2-lightness” of the second scout craft and so-on.

Thus, you can see that as the series of relative light-speed craft have increased infinite Lorentz factors, the craft in ascending order will become more and more refined types of light.

Perhaps, given an eternity of technological development, these mothership and scout craft systems will become a created form and state of Archetypal Light. Thus, they may grow more and more to resemble the Archetypal Divine Light while never obviously become as refined nor equal to the Archetypal Divine Light of GOD.

Now, my conjecture may seem a little but intellectually immature and an over-exuberant expression for a desire to reify abstract principles. However, I believe GOD has willed that humanity develop technology and the wonders it enables humanity to develop.

We may affix the following operator to formulas for novel light-sails as appropriate to denote the above progression toward created archetypal light.

[f(Light to light to light and so-on to become a created form and state of Archetypal Light)]

As such, the above operator implies advancing in importance, purpose, and nature of physical reality.

ESSAY 27) Complete Halting Of Time Passage For Sufficiently Infinite Lorentz Factored, Light-Speed, Spacecraft.

A fascinating construct regarding reaching the speed of light in astoundingly great infinite Lorentz factors might be the complete halting of time passage.

Accordingly, the brain activity of exuberantly light-speed spacecraft would completely be frozen in place as would currently known psychological processes.

As such, we are considering a certain succession of all conscious processes.

One consequence of such psychic static states is that the human psyche might take on new modes of operation or development but where the subjective sense of normal time passage is absent. Such new modes of operation and development we will refer to as New-Psychic-States-1.

Should a spacecraft acquire still greater infinite Lorentz factors, perhaps the craft crew members will ascend to still newer and higher yet modes of operation and development. We refer to these modes of operation as New-Psychic-States-2.

Should a spacecraft acquire still yet greater infinite Lorentz factors, perhaps the craft crew members will ascend to still newer and higher yet modes of operation and development. We refer to these modes of operation as New-Psychic-States-3.

We can continue the series perhaps forever.

We may affix the following operator to formulas for gamma and related parameters to denote the consciousness elevating aspects of exuberantly expressed light-speed travel.

[Σ(k = 1; k = ∞↑):(New-Psychic-States-k)], [d(...(d [Σ(k = 1; k = ∞↑):(Rate Of Development Of New-Psychic-States-k)])/dγ)/...)/dγ], [∫...∫[Σ(k = 1; k = ∞↑):(Rate Of Development Of New-Psychic-States-k)]dγ)...dγ)]

The magnitude of the above conjectured derivatives and integrals can be infinite and perhaps even super-infinite.

We can also consider the onset of super-static mind/brain processes for increasing infinite.

Accordingly, a level-1 super-static-mind/brane would involve ordinary conscious processes being more changeless than processes with zero first time derivatives.

Accordingly, the level-1 processes would be defined as follows.

|[d(Rate of change of level-1 super-static-mind/brane processes)/dγ]| << zero

A level-2 super-static-mind/brane would involve ordinary conscious processes being more changeless than processes with zero first time derivatives level-1 super-static-mind/brane processes.

Accordingly, the level-2 processes would be defined as follows.

|[d(Rate of change of level-2 super-static-mind/brane processes)/dγ]| << zero/{f[|[d(Rate of change of level-1 super-static-mind/brane processes)/dγ]|}

A level-3 super-static-mind/brane would involve ordinary conscious processes being more changeless than processes with zero first time derivatives level-2 super-static-mind/brane processes.

Accordingly, the level-3 processes would be defined as follows.

|[d(Rate of change of level-3 super-static-mind/brane processes)/dγ]| << zero/{f[|[d(Rate of change of level-2 super-static-mind/brane processes)/dγ]|}

A level-4 super-static-mind/brane would involve ordinary conscious processes being more changeless than processes with zero first time derivatives level-3 super-static-mind/brane processes.

Accordingly, the level-4 processes would be defined as follows.

|[d(Rate of change of level-4 super-static-mind/brane processes)/dγ]| << zero/{f[|[d(Rate of change of level-3 super-static-mind/brane processes)/dγ]|}

In general, a level-(n+1) super-static-mind/brane would involve ordinary conscious processes being more changeless than processes with zero first time derivatives level-n super-static-mind/brane processes.

Accordingly, the level-(n+1) processes would be defined as follows.

{|[d(Rate of change of level-(n+1) super-static-mind/brane processes)/dγ]| << zero/{f[|[d(Rate of change of level-n super-static-mind/brane processes)/dγ]|}}

We may affix the following operator to formulas for gamma and related parameters to denote the mathematics of the super-static-mind/brane processes.

{f{|[d(Rate of change of level-(n+1) super-static-mind/brane processes)/dγ]| << zero/{f[|[d(Rate of change of level-n super-static-mind/brane processes)/dγ]|}}

An even more fascinating construct is the establishing of a meta-static analog of static states of the spacecraft and also in the psychic processes of the crew members in a light-speed spacecraft that is made to undergo an unspecified kinetic or topological mode of operation.

Accordingly, we refer to a first level of a first species of meta-static analog of stationary states as meta-static-analog-1,1.

We can also refer to a second level of a first species of meta-static analog of stationary states as meta-static-analog-1,2. This level may be obtained as a result of suitably modulating the kinematics and topology of a spacecraft having obtained sufficiently infinite Lorentz factors.

Likewise we can consider still more extreme cases of a first species such as meta-static-analog-1,3, meta-static-analog-1,4, meta-static-analog-1,5 and so-on all the way to meta-static-analog-1-∞'s and beyond. We represent the first species of meta-static-analogs as meta-static-analog-1-k

We can go on to consider other species meta-static-analogs such as meta-static-analog-2-k, meta-static-analog-3-k, meta-static-analog-4-k, and all the way up to and beyond meta-static-analog-∞-k.

In general, we define the set of species of meta-static-analogs by the following set builder notation.

{Meta-static-analog-h-k| h and k range from integers of one to ∞ and beyond and may but need not be equal}

We can consider gamma derivatives of arbitrary order in the rate of ordinal increase in acquisition of elements of the set {Meta-static-analog-h-k| h = 1, 2, 3, …, ꝏ, … and k = 1, 2, 3, …, ꝏ, …} by a spacecraft as follows.

d{…d{{Meta-static-analog-h-k| h = 1, 2, 3, …, ꝏ, … and k = 1, 2, 3, …, ꝏ, …}/γ}/… }/dγ.

Additionally, we may define a new set of distance of travel constructs by the following integral.

∫{…∫{{Meta-static-analog-h-k| h = 1, 2, 3, …, ꝏ, … and k = 1, 2, 3, …, ꝏ, …}dγ}… }dγ.

Accordingly, we consider novel constructs of what it means to travel.

We may affix the following operator to formulas for gamma and related parameter to denote the above conjectured phenomenon.

{Rate of charge and growth in Meta-static-analog-h-k| h & h = 1, 2, 3, …, ꝏ, … }:{d{…d{{Rate of charge and growth in Meta-static-analog-h-k| h & h = 1, 2, 3, …, ꝏ, … }/γ}/… }/dγ}:{∫{…∫{{Rate of charge and growth in Meta-static-analog-h-k| h & h = 1, 2, 3, …, ꝏ, … }dγ}… }dγ}

ESSAY 28) Creation-Stars As Prime Matter Cosmic Existential Underpinnings.

Here we consider Creation-Stars as the prime matter underpinnings of the existence and development of the cosmos and all creatures that are so coupled causally, thermodynamically, and topologically including human persons, human persons' souls, angels good and bad, ETI's and UTI's.

I seems as if the current cosmic order supported by one or more, but perhaps only one Creation-Star.

Perhaps it may be that GOD is drawing into being in the future, or whatever the heck, uncountable infinities of Creation-Stars.

Along with GOD's drawing into being in the future, or whatever the heck, uncountable infinities of Creation-Stars, perhaps these prime matters will include uncountable infinities of species of Creation-Stars and Creation-Stars of increasing immensity, sustaining power, extent of degrees as first principles, and so-on.

GOD may enhance the glory in natural and super-natural aspects of our current Creation-Star and all other Creation-Stars existing and/or yet to be created.

Now, perhaps spacecraft having obtain sufficiently infinite Lorentz factors and employing proper modulation of Lorentz varying effects, the craft may travel out of the ontological bulk of coupled creation and toward or into one or more Creation-Stars.

We may affix the following operator to formulas for gamma to denote mechanisms of travel toward and perhaps into the highly conjectural Creation-Stars.

Note that Creation-Stars are not the heavenly bodies we commonly refer to as stars, but instead represent fundamental ontological principles.

[f(Mechanisms of travel toward and perhaps into the highly conjectural Creation-Stars)].

According to the Special Theory Of Relativity, the speed of light is an inviolable limit.

Additionally, the velocity of light is a constant such that everything else physical and perhaps created spiritual that moves has thermodynamic states relative to the speed of light.

The constant, c, shows up virtually ubiquitously in physics and in electrical engineering.

In a sense, the velocity of light is so special that it is as if a created analog of the divine.

Achieving travel at the speed of light provides for plausible mystical ontological conditions of spacecraft and crew members. Moreover, for a spacecraft traveling at the speed of light in infinite Lorentz factors, the speed of light might become ever more ingrained or imprinted into the craft almost has if such a scenario would be simply Mother Nature developing a habit and a permanent trait.

As a light-speed craft excelled to ever greater infinite Lorentz factors, the speed of light may become more and more imprinted in the craft with the craft taking on more and more features related to the speed of light.

The ramifications of craft attaining the speed of light, and then in ever greater infinite Lorentz factors, in my opinion, will allow the crew members to develop more and more mystical aspects in body and in soul, in accidents and in substance of the soul.

The number of mystical aspects or features identifiable as distinct may well grow to infinities and then so to ever greater infinities.

We affix the following operator to the lengthy formulas for gamma and related parameters that follow to denote the number of and extent of the development of mystical aspects of light-speed traveling crew members, in body, soul, and of accidents and substantial aspects of the soul.

{r{Development of k mystical aspects of light-speed traveling crew members, in body, soul, and of accidents and substantial aspects of the soul considering increasing infinite Lorentz factors | k = 1, 2, ...}}

When we see a bright star at night, we all generally have a sense of the mysterious. Looking out into the midnight sky on a clear Winter night provides for a profound sense of mystique.

Those who claim near death out of body experiences often report seeing a wonderful bright warm light emitting and radiating love. Many folks perceive the light not as white but as whiter than white, as if a clear transparent light.

Perhaps intrepid future human travelers attaining the speed of light in their spacecraft will experience analogs of clear transparent light, whether these analogs manifest in their conscious identity or as profound light of intellectual realization and abstract wisdom gained during the process of climbing to ever greater infinite Lorentz factors.

Truly, the speed of light is a fact of nature and can guide us onward to do research and development of craft that can travel ever more close to the speed of light in ever greater finite Lorentz factors with the cosmically distant goal of craft being designed, crewed, and flown at the speed of light in ever greater infinite Lorentz factors.

One thing is for sure! The speed of light is both real and concrete as well as ephemeral and mystical.

ESSAY 29) Derivatives And Integrals Of Development Rates Of Mystical Aspects Of Light-Speed Of Crew Members Of Light-Speed Spacecraft

Now, we can take the ship-time, gamma, kinetic energy, and acceleration derivatives of the rate of development of k mystical aspects of light-speed traveling crew members, in body, soul, and of accidents and substantial aspects of the soul considering increasing infinite Lorentz factors as follows.

d{…d{r{Rate of development of k mystical aspects of light-speed traveling crew members | k = 1, 2, …}}/d(tship or γ or a or KE)}/…}/ d(tship or γ or a or KE)

We may affix the following operator to formulas for gamma and related parameters to denote the ship-time, gamma, kinetic energy, and acceleration derivatives of the rate of development of k mystical aspects of light-speed traveling crew members, in body, soul, and of accidents and substantial aspects of the soul considering increasing infinite Lorentz factors.

{s{d{…d{r{Rate of development of k mystical aspects of light-speed traveling crew members | k = 1, 2, …}}/d(tship or γ or a or KE)}/…}/ d(tship or γ or a or KE)}}

Now we can consider travel as a form of progression in the rate of development in the above conjectural mystical aspects.

Accordingly we can integrate the rate of development in the above conjectural mystical aspects with respect to ship-time, gamma, kinetic energy, and acceleration to yield different meanings of such travel distance.

∫{…{∫{r{Rate of development of k mystical aspects of light-speed traveling crew members | k = 1, 2, …}}d(tship or γ or a or KE)}…} d(tship or γ or a or KE)

We may affix the following operator to formulas for gamma and related parameters to indicate the operation of these exotic meanings of travel in the context of the formulas for gamma and related parameters.

{s{∫{…{∫{r{Rate of development of k mystical aspects of light-speed traveling crew members | k = 1, 2, …}}d(tship or γ or a or KE)}…} d(tship or γ or a or KE)}}

ESSAY 30) Levels Of Meta-Nows And Light-Speed Travel Of Sufficiently Infinite Lorentz Factors.

Here, we consider the essence of the nature of now to ever more extreme levels.

For example, an essence which is more now than now or Meta-now-essence-1 is a profound construct and loosely translates as more eternal now than the eternal now. We refer to more eternal now than the eternal now as More-eternal-now-than-now-1.

An essence which is more now than Meta-now-essence-1 or Meta-now-essence-2 is a profound construct and loosely translates as more eternal now than More-eternal-now-than-now-1. We refer to more eternal now than More-eternal-now-than-now-1 as More-eternal-now-than-now-2.

An essence which is more now than Meta-now-essence-2 or Meta-now-essence-3 is an even more profound construct and loosely translates as more eternal now than More-eternal-now-than-now-2. We refer to more eternal now than More-eternal-now-than-now-2 as More-eternal-now-than-now-3.

An essence which is more now than Meta-now-essence-3 or Meta-now-essence-4 is an even more profound construct and loosely translates as more eternal now than More-eternal-now-than-now-3. We refer to more eternal now than More-eternal-now-than-now-3 as More-eternal-now-than-now-4.

An essence which is more now than Meta-now-essence-4 or Meta-now-essence-5 is an even more profound construct and loosely translates as more eternal now than More-eternal-now-than-now-4. We refer to more eternal now than More-eternal-now-than-now-4 as More-eternal-now-than-now-5.

An essence which is more now than Meta-now-essence-5 or Meta-now-essence-6 is a yet still more profound construct and loosely translates as more eternal now than More-eternal-now-than-now-5. We refer to more eternal now than More-eternal-now-than-now-5 as More-eternal-now-than-now-6.

In general, we can consider Meta-now-essence-r, and More-eternal-now-than-now-r where r is any positive integer, finite or infinite, and may be so large such that its values can only be expressed as remote locations of the right-side of a hyper-extended number line.

Now, we can consider scenarios for which a spacecraft having attained the speed of light in the least infinite Lorentz factor continues to increase its Lorentz factor. Eventually, the spacecraft may run out of future eternity room to travel forward in time end.

Having run out of temporal room to travel forward in time, an extremely infinite Lorentz factor craft it would at first seem conceptually as if the craft would travel instantaneously as a bridging mechanism to the conceptualized backward time travel of faster-than-light impulse travel systems.

However, I propose that a sufficiently infinite Lorentz factor craft would instead in a sense "pop" out of time and enter an eternal now state. The extremity and number of eternal now states available to the spacecraft may be described by the set Meta-now-essence-r, and More-eternal-now-than-now-r where r is any positive integer, finite or infinite, and may be so large such that its values can only be expressed as remote locations on the right-side of a hyper-extended number line.

We can also take the spacecraft time, gamma, kinetic energy, and acceleration derivatives of the spacecraft as referenced to the travel in the various levels of nows and meta-nows. Taking the integrals instead can give us a measure of abstract alternative meanings of travel distance.

We may affix the following operator to formulas for gamma and related parameters to indicate the above conjectured mechanisms of sufficiently infinite Lorentz factor spacecraft entering eternal now states.

{Σ(r = 1; r = ∞↑):→Meta-now-essence-r;More-eternal-now-than-now-r}:[d(…(d→Meta-now-essence-r;More-eternal-now-than-now-r)/dγ,ts,a,k)/…)/d dγ,ts,a,k]:[∫(…(∫→Meta-now-essence-r;More-eternal-now-than-now-r) dγ,ts,a,k)…)dγ,ts,a,k]

We go on to consider analogs of now and of Meta-now-essence-r and More-eternal-now-than-now-r.

Such analogs are denoted by appending the above two terms with a hyphen and the lower-case letter, b.

These analogs would be distinct from now, Meta-now-essence-r, and More-eternal-now-than-now-r.

We may affix the following operator to formulas for gamma and related parameters to indicate the above conjectured mechanisms of sufficiently infinite Lorentz factor spacecraft entering analog eternal now states.

{Σ(r & b = 1; r & b = ∞↑):→Meta-now-essence-r-b;More-eternal-now-than-now-r-b}:[d(…(d→Meta-now-essence-r-b;More-eternal-now-than-now-r-b)/dγ,ts,a,k)/…)/d dγ,ts,a,k]:[∫(…(∫→Meta-now-essence-r-b;More-eternal-now-than-now-r-b) dγ,ts,a,k)…)dγ,ts,a,k]

The terminology of γ,ts,a,k stands for gamma, or ship-time, or acceleration, or kinetic energy.

We can consider scenarios for which the velocity of a spacecraft traveling into a Meta-now-essence-r or More-eternal-now-than-now-r may be greater than that of light or translate loosely as equivalent to greater than that of light.

Perhaps the spacecraft velocity in a Meta-now-essence-r or More-eternal-now-than-now-r may existentially translate loosely into infinite multiples of c. Such infinities may be arbitrarily great.

Aleph 0 is the number of positive integers.

Aleph 1 = 2 EXP (Aleph 0).

Aleph 2 = 2 EXP (Aleph 1).

Aleph 3 = 2 EXP (Aleph 2).

In general,

Aleph (N + 1) = 2 EXP (Aleph N)

Consider the hyper-operator notation that was designed to express huge values not otherwise expressible.

For example, Note that Hyper4(a, n) is equal to a tetrated n or a raised to the power of itself n-1 times. The latter value is symbolically written as n subscript a. For example 3 EXP 4 = 81, but 4 subscript3 is approximately equal to 10 EXP (1,000,000,000,000).

Alternatively 4 subscript 2 = 2 EXP 2 EXP 2 EXP 2 = 2 EXP [2 EXP [2 EXP 2]] = 2 EXP (2 EXP 4) = 2 EXP 16 = 65,536.

For example, Hyper5(4, 4)is equal to 4 tetrated 4 tetrated 4 tetrated 4. This value is commonly referred to as 4 pentated 4.

Hyper 6, (4,4) is 4 pentated 4 pentated 4 pentated 4 and is also referred to as 4 hexataed 4.

Hyper 7, (4,4) is 4 hexated 4 hexated 4 hexated 4 and so-on

Again, Aleph 0 is the infinite number of positive integers.

Aleph 1 according to the perhaps unprovable and thus unfalsifiable Continuum Hypotheses is the number of real numbers which is greater than Aleph 0 by a multiplicative factor of infinity.

The number Ω is commonly stated as the least infinite positive integer or ordinal.

Now here is a real zinger.

So we can produce the abstraction of Hyper Aleph Ω (Aleph Ω, Aleph Ω).

Now uncountable infinities are mainly viewed as qualitatively distinct from countably infinities as well as qualitatively larger.

So, can you imagine the abstraction of a value such as,

Hyper An Extreme Uncountable Infinity (An Extreme Uncountable Infinity, An Extreme Uncountable Infinity).

Now, how about a velocity that existentially translates loosely into Hyper An Extreme Uncountable Infinity (An Extreme Uncountable Infinity, An Extreme Uncountable Infinity) multiples of c.

We may affix the following operator to formulas for gamma and related parameters to denote the above uncountably infinite multiples of light speed in a Meta-now-essence-r or More-eternal-now-than-now-r

[[Hyper An Extreme Uncountable Infinity ↑ (An Extreme Uncountable Infinity ↑, An Extreme Uncountable Infinity ↑)] multiples of c in a Meta-now-essence-r or More-eternal-now-than-now-r]

The upward pointing arrows indicate unbounded increases of uncountable infinities as possible options.

Again, we consider analogs of now and of Meta-now-essence-r and More-eternal-now-than-now-r.

Again, such analogs are denoted by appending the above two terms with a hyphen and the lower-case letter, b.

Still again, these analogs would be distinct from now, Meta-now-essence-r, and More-eternal-now-than-now-r.

We affix the following operator to the lengthy formulas for gamma and related parameters to denote the above uncountably infinite multiples of light speed in a Meta-now-essence-r-b or More-eternal-now-than-now-r-b.

[[Hyper An Extreme Uncountable Infinity ↑ (An Extreme Uncountable Infinity ↑, An Extreme Uncountable Infinity ↑)] multiples of c in a Meta-now-essence-r-b or More-eternal-now-than-now-r-b]

The notion of the above infinite numbers of indexed terms and associated infinite Lorentz factors in my writings about light-speed travel need not and does not take away from GOD's pre-ordinance. In the following digression, we explain why.

Now again, consider that Aleph 0 is the number of positive integers.

Aleph 1 = 2 EXP (Aleph 0).

Aleph 2 = 2 EXP (Aleph 1).

Aleph 3 = 2 EXP (Aleph 2).

In general,

Aleph N = 2 EXP [Aleph (N-1)]

Consider again the hyper-operator notation that was designed to express huge values not otherwise expressible.

For example, Note that Hyper4(a, n) is equal to a tetrated n or a raised to the power of itself n-1 times. The latter value is symbolically written as n subscript a. For example 3 EXP 4 = 81, but 4 subscript3 is approximately equal to 10 EXP (1,000,000,000,000).

Alternatively 4 subscript 2 = 2 EXP 2 EXP 2 EXP 2 = 2 EXP [2 EXP [2 EXP 2]] = 2 EXP (2 EXP 4) = 2 EXP 16 = 65,536.

For example, Hyper5(4, 4)is equal to 4 tetrated 4 tetrated 4 tetrated 4. This value is commonly referred to as 4 pentated 4.

Hyper 6, (4,4) is 4 pentated 4 pentated 4 pentated 4 and is also referred to as 4 hexataed 4.

Hyper 7, (4,4) is 4 hexated 4 hexated 4 hexated 4 and so-on

Aleph 0 is the infinite number of integers.

Aleph 1 according to the perhaps unprovable and thus unfalsifiable Continuum Hypotheses is the number of real numbers which is greater than Aleph 0 by a multiplicative factor of infinity.

In general, Aleph n = 2 EXP [Aleph (n-1)] where n is any counting number.

The number Ω is commonly stated as the least infinite positive integer or ordinal.

Now here is a real zinger.

So we can produce the abstraction of Hyper Aleph Ω (Aleph Ω, Aleph Ω).

Now uncountable infinities are mainly viewed as qualitatively distinct from countably infinities as well as qualitatively larger.

So, can you imagine the abstraction of a value such as,

Hyper An Extreme Uncountable Infinity (An Extreme Uncountable Infinity, An Extreme Uncountable Infinity).

So, here we have covered countable infinities and uncountable infinities.

However, we can also consider unknown infinities above uncountable infinities.

Additionally, we can conceptualize infinities that are so great they cannot be abstractly defined except to say that they are so great that they cannot be abstractly defined.

Still further, we can consider infinities that are so great that we simply cannot talk about them, neither from a rational state of mine nor an irrational state.

Now we consider ontological transcendentals which are qualities such goodness, value, purpose, and the like.

Ontological transcendentals are properties that transcend or are above properties such as holiness, love, power, strength, immortality, incorruptibility, intelligence, wisdom, and the like.

We can likely state that GOD has so infinitely many ontological transcendentals such that the infinity of GOD's ontological transcendentals cannot be abstractly defined except to say that they are so great that they cannot be abstractly defined.

We will go one step further to state that GOD has so infinitely many ontological transcendentals such that the infinity of GOD's ontological transcendentals cannot be mentioned, neither from a rational state of mine nor an irrational state.

GOD is so absolutely real that GOD does not need to exist to be real. Such a construct is bizarrely far-out but does not denigrate GOD's existence because GOD's existence is incomprehensibly and unstatably absolute.

Now here is a real big zinger! We can imagine infinities that are so great that only GOD can understand and behold them. Accordingly, not even Jesus Human Nature can understand and behold them. Neither can Saint Michael The Archangel, nor the Blessed Virgin Mary.

Spacecraft Lorentz factors are limited to countable infinities for the most part because the light-speed Lorentz factors can be imprinted on the spacecraft as concrete existential physical conditions. Thus, since these infinite Lorentz factors can be existentially and concretely imbued within a spacecraft state, they can also be counted.

Uncountable infinite values of spacecraft Lorentz factors likely are not possible in the current cosmic era but may serve as implicit possibilities to work towards obtaining in the depths of future eternity. Uncountable infinite values of Lorentz factors might loosely manifest in lower order terms for spacecraft in the depths of eternities from now. However, no matter how hard we will ever try, we will never be able to attain Lorentz factors greater than the lowest of uncountable infinities. However, we can aim for ever greater countable infinities as values of Lorentz factors.

The reference to extreme uncountably infinities in the context of the starship parameters thusly described is a relative concept. For every uncountably infinity, there is always an uncountably infinite number of greater uncountable infinities. I use the word extreme more as a "flashy term" to imply the hugeness of any uncountable infinities, even the least ones.

ESSAY 31) The Infinite Number Of Levels Of Not-Time-Like Related Light-Speed Spacecraft.

Herein, we consider travel to a location of cosmic distance away from a location on Earth or any other location in our cosmic light-cone but in ways and of aspects not related to time, neither past, present, nor future.

Accordingly, a craft at light-speed in an infinite Lorentz factor would run out of future temporal room to travel forward in time in. Upon obtaining sufficiently infinite Lorentz factors, the craft would enter a state that is interstitial between further forward time travel and backward time travel while still not violating Special Relativity thus not traveling faster than light.

The light-speed craft would thus experience location said cosmic distance away from its point of origin without temporal reference to the point of origin nor with respect to the intervening space between the origin and destination of the craft.

We refer to this travel in ways and of aspects not related to time, neither past, present, nor future in a first level as [s(Travel-not-time-related-1)]

Still more extreme non-temporal relatedness travel as a more extreme analog to [s(Travel-not-time-related-1)] we refer to as [s(Travel-not-time-related-2)]

Still more extreme non-temporal relatedness travel as a more extreme analog to [s(Travel-not-time-related-2)] we refer to as [s(Travel-not-time-related-3)]

Still more extreme non-temporal relatedness travel as a more extreme analog to [s(Travel-not-time-related-3)] we refer to as [s(Travel-not-time-related-4)]

Still more extreme non-temporal relatedness travel as a more extreme analog to [s(Travel-not-time-related-4)] we refer to as [s(Travel-not-time-related-5)]

We can continue the progression to [s(Travel-not-time-related-∞)].

Here, the infinity can be as great as we choose.

We denote the above travel mechanisms and the relevant physics by the set notion of:

{[s(Travel-not-time-related-k)]| k goes from 1 to arbitrarily great ∞'s}

Here, by progressively more extreme non-temporal relatedness travel indices, we are referring [s(Travel-not-time-related-g)] is to [s(Travel-not-time-related-(g-1))] as [s(Travel-not-time-related-(g-1))] is to [s(Travel-not-time-related-(g-2))] as [s(Travel-not-time-related-(g-2))] is to [s(Travel-not-time-related-(g-3))] and so-on all the way down to [s(Travel-not-time-related-1)]

Additionally, the distance or separation between the space of origin and the space of destination can be as great as arbitrarily infinite distances. Moreover, the number of spatial dimensions of the space traveled in is arbitrary but likely would range from 3 dimensions to arbitrarily infinite numbers of dimensions. Still further, the travel can be in our universe, another universe, our multiverse, another multiverse, our forest, another forest, and so-on and even into so-called bulks in which hyperspaces are theoretically manifest or existentially underwritten.

We may affix the following operator to formulas for gamma and related parameters to denote the above travel possibilities and related parameters and physics.

{f{{{r{[s(Travel-not-time-related-k)]| k goes from 1 to arbitrarily great ꝏ's}} in j spatial dimensionality of the realm of travel}| j is any real positive number but most likely and usefully from 3 to arbitrarily great ꝏ's}}

ESSAY 32) Light-Speed Spacecraft "Pop Out" Of Realm Of Origin.

We can intuit a craft having reached the speed of light and attaining infinite Lorentz factors as running out of future room to travel forward in time in.

Thus, a light-speed craft may in a sense simply "pop out" of a realm of origin.

One Planck Time Unit in a light-speed craft of increasing infinite Lorentz factors and kinetic energies might be said to per due forever in the background of spacecraft origin.

As the Lorentz factor of the spacecraft becomes ever more infinite, the craft might embark in a journey into hyper-forever which we will label hyper-forever-1.

Hyper-forever-1 would be to forever as we typically conceptualize it and perhaps even forever in the Biblical or Catholic theological sense as the latter concept of forever is to temporary.

Should a spacecraft traveling at light-speed achieve ever greater infinite Lorentz factors, the spacecraft might journey into hyper-hyper-forever which we will label as hyper-forever-2.

Hyper-forever-2 would be to hyper-forever-1 as hyper-forever-1 is to standard Scriptural concepts of forever as standard Scriptural concepts of forever are to the realm of the temporary.

Should a spacecraft traveling at light-speed achieve ever greater infinite Lorentz factors, the spacecraft might journey into hyper-hyper-hyper-forever which we will label as hyper-forever-3.

Hyper-forever-3 would be to hyper-forever-2 as hyper-forever-2 would be to hyper-forever-1 as hyper-forever-1 is to standard Scriptural concepts of forever as standard Scriptural concepts of forever are to the realm of the temporary.

By now you obtain the picture.

We can continue to hyper-forever-4, hyper-forever-5 and so-on up to and beyond analogs labeled as hyper-forever-ꝏ.

We denote the set of such analogs by the following general notation.

{hyper-forever-k| k = 1, 2, …}

We can take the ship time derivatives of growth in the rate of hyper-forever-k achievement as,

d{…{d{Rate of growth in {hyper-forever-k| k = 1, 2, …}}/dts}/…}/dts.

We can take the gamma derivatives with respect of the rate of hyper-forever-k achievement as,

d{…{d{Rate of growth in {hyper-forever-k| k = 1, 2, …}}/dγ}/…}/dγ.

We can take the acceleration derivatives with respect of the rate of hyper-forever-k achievement as,

d{…{d{Rate of growth in {hyper-forever-k| k = 1, 2, …}}/da}/…}/da.

We can take the kinetic energy derivatives with respect of the rate of hyper-forever-k achievement as,

d{…{d{Rate of growth in {hyper-forever-k| k = 1, 2,…}}/dKE}/…}/dKE.

In order to include the above possibilities in an operator affixed to formulas for gamma and related parameters, the subject operator is stated as,

d{…{d{Rate of growth in {hyper-forever-k| k = 1, 2, …}}/dts,γ,a,KE}/…}/ dts,γ,a,KE.

As measures of distances traveled by various interpretations of the meaning of distance, we affix the following operator to the lengthy formulas for gamma and related parameters in Section 1. Here instead of differentiating, we integrate.

∫{…{∫{Rate of growth in {hyper-forever-k| k = 1, 2, …}}d(ts,γ,a,KE)}…}d(ts,γ,a,KE).

We bundle the above three main operators as follows and affix them to the lengthy formulas for gamma and related parameters.

{hyper-forever-k| k = 1, 2, …}: {d{…{d{Rate of growth in {hyper-forever-k| k = 1, 2, …}}/dts,γ,a,KE}/…}/ dts,γ,a,KE}: {∫{…{∫{Rate of growth in {hyper-forever-k| k = 1, 2, …}}d(ts,γ,a,KE)}…}d(ts,γ,a,KE)}

Assuming there are analogs of forever that are above and beyond forever, perhaps travelers at light-speed will arrive at a realm in which something awesomely good has happened for the souls in Hell and the condemned angels. I will speak no further on this matter suffice it to say that this conjecture gives me grounding and hope. The idea of any soul going to Hell no matter how badly they acted in life is as if a part of me goes to Hell. I am definitely an empath and do not enjoy any creature suffering.

Now, before we can design, build, and fly starships capable of infinite Lorentz factors in light-speed travel, more than likely we will need to develop technology for a period of an infinite number of human familial generations. However, this is not a bad thing because we will have loads of whimsical joy overcoming each next Lorentz factor hurdle. It will almost as if be like achieving light-speed in infinite Lorentz factors will be like working toward an infinitely distance goal as if a subtly dim beacon of mystery that is infinitely distance.

Now, most stars are red dwarfs. However, when we first fly starships capable of a Lorentz factor of 2, the red dwarfs will appear blue with more stars in view as you look further toward the front of the spacecraft heading relative to perpendicular to the spacecraft heading. The red-dwarf stars in back will look brownish red.

Eventually, we will be able to reach a Lorentz factor of 10 which will place any stars within a 500 light-year radius from Earth in range of the starship in one average working life time of a human.

Our next major goal will be obtaining a Lorentz factor of about 1,500. As such, the outer hull of the spacecraft would be made of tantalum-hafnium-carbide which melts at about 7,610 degrees F. Accordingly, the cosmic microwave background radiation will appear to be about 7,000 degrees F so a really refractory compound will be required to prevent the ship from being destroyed by the heat. Additionally, the outer hull will need to be polished to near perfect reflectiveness as well as being highly elongated and pointed much like a sowing needle.

Now, can you imagine a time, perhaps 200 years into the future from now, booking a flight on Galactic Starlines.

When everyone has boarded, space traffic control will provide flight clearance and the pilot will radio in setting course for the Pleiades at G-1,500. The term G-1,500 will mean a value of gamma or Lorentz factor equal to 1,500.

Eventually, we will develop even better refractories to enable all the greater spacecraft Lorentz factors.

At some point, I believe we will figure out how to make bulk neutronium, then quarkonium, then higgsinium, the mono-higgsinium, then monopolium, and onward from there.

ESSAY 33) Light-Speed Extreme Infinite Lorentz Factors Spacecraft Out Run Of Future Time.

Here, we consider travel at such extreme infinite Lorentz factors in light-speed travel that a spacecraft runs out of future time to travel forward in time in a given universe, multiverse, forest, biosphere, and the like of travel.

A spacecraft having thus popped out of a realm of origin might end up leaving materiality all together as a realm and enter mathematical objects of abstract eternal forms.

As such, the number of mathematical objects that can be entered is huge, as large as the number of real numbers, imaginary numbers, complex number, tuples, functions, transforms, relations, operators, integrals, derivatives, and differentials, as well as the absolutely huge number of expressions and equations.

These concepts are really wackily far-out but perhaps makes sense as creation overall forever continues to evolve according to Divine Providence. Hopefully you get a kick out of the neologism I used in the first sentence of this paragraph. After all, a little humor is sometimes called for.

After having entered into mathematical objects, the craft may then enter logical objects or the abstract eternal forms of logic equations, proofs, objects, and operators.

We then consider the spacecraft time, gamma, acceleration, and kinetic energy derivatives and integrals of the rate of progression of entrance into mathematical and logical objects. The integrals in a way serve to denote novel aspects or definitions of metaphorical distances of travel.

We may affix the following operator to formulas for gamma and related parameters to denote these rates of entrance into mathematical objects and derivatives and integrals thereof.

{[Σ(Rate of entrance into math & logic objects)]:{d[…[d(Rate of entrance into math & logic objects)/dts,γ,a,KE]/…,/ dts,γ,a,KE]}:{∫…,[∫(Rate of entrance into math & logic objects)dts,γ,a,KE]…, dts,γ,a,KE]}}

We can also consider amalgams of entrance into mathematical and logical objects and/or into the realm of mathematical and logical objects and created material realities. These material realities may be substantial and/accidental.

We then consider the spacecraft time, gamma, acceleration, and kinetic energy derivatives and integrals of the rate of progression of entrance into amalgams of mathematical and logical objects and physical realities. The integrals in a way serve to denote novel aspects or definitions of metaphorical distances of travel.

We may affix the following operator to formulas for gamma and related parameters to denote these rates of entrance into amalgams of mathematical and logical objects and created material realities and derivatives and integrals thereof.

{[Σ(Δ entrance into amalgams of m & l objects & MR)]:{d[…[d(Δ entrance into amalgams of m & l objects & MR)/dts,γ,a,KE]/…,/ dts,γ,a,KE]}:{∫…,[∫(Δ entrance into amalgams of m & l objects & MR)dts,γ,a,KE]…, dts,γ,a,KE]}}

We may further consider entering the realms of mathematical and logical objects and amalgams of mathematical and logical objects and physical realities.

Accordingly, such novel realms would serve as a meta-physical sea or location where mathematical and logical objects and amalgams of mathematical and logical objects and physical realities would be “located” as a background which serves as a mechanism by which mathematical and logical objects are related to each other. As such, the meta-physical sea would be immeasurably infinite in scale and contain an undefinable uncountable infinity of objects, an uncountable infinity which may forever grow in numbers.

We may affix the following operator to formulas for gamma and related parameters to denote the entrance into and unbounded opportunities for travel within the meta-physical sea or location of mathematical and logical objects and amalgams of mathematical and logical objects and physical realities.

[r(Unbounded opportunities for travel within the meta-physical sea or location of mathematical and logical objects and amalgams of mathematical and logical objects and physical realities)]

ESSAY 34) Light-Speed Travel Into Wave-Functions Of Objects Ranging From Particles To Hyper-Bulks, Etc..

Here, we consider travel facilitated at light-speed and suitably infinite Lorentz factors into wave-functions of particles, coherent collections of particles, universal light-cones, universes, multiverses, forests, biospheres and the like and hyperspaces, hyperspace-times, bulks, hyper-bulks, and the like.

As such, we consider travel more quantum-mechanically than classically and instead of being present classically in the normally considered ways to said particles, coherent collections of particles, universal light-cones, universes, multiverses, forests, biospheres and the like, and hyperspaces, hyperspace-times, bulks, hyper-bulks, and the like, we become immersed in these objects quantum-mechanical wave-functions.

We can also consider travel into the individual components of the overall wave functions of the above subject objects.

We then consider the spacecraft time, gamma, acceleration, and kinetic energy derivatives and integrals of the rate of progression of entrance into the objects wave-functions. The integrals in a way serve to denote novel aspects or definitions of metaphorical distances of travel.

We may affix the following operator to formulas for gamma and related parameters to denote the rate of the above conjectured travel mechanisms and derivatives and integrals thereof.

{(Δ trav into indiv comp of and overall wave-f):{d[…[d(Δ trav into indiv comp of and overall wave-f)/dts,γ,a,KE]/…,/ dts,γ,a,KE]}:{∫…,[∫(Δ trav into indiv comp of and overall wave-f)dts,γ,a,KE]…, dts,γ,a,KE]}}

ESSAY 35) Travel Into Meta-Mathematical Objects.

We have delved into craft travel into mathematical objects where the craft may then travel into logical objects or the abstract eternal forms of logic equations, proofs, objects, and operators.

Now we consider travel into meta-mathematical objects and meta-logical objects.

We now consider the spacecraft time, gamma, acceleration, and kinetic energy derivatives and integrals of the rate of progression of entrance into meta-mathematical and meta-logical objects. The integrals in a way serve to denote novel aspects or definitions of metaphorical distances of travel.

We may affix the following operator to formulas for gamma and related parameters to denote these rates of entrance into meta-mathematical and meta-logical objects and derivatives and integrals thereof.

{[Σ(Rate of entrance into mmath & mlogic objects)]:{d[…[d(Rate of entrance into mmath & mlogic objects)/dts,γ,a,KE]/…,/ dts,γ,a,KE]}:{∫…,[∫(Rate of entrance into mmath & mlogic objects)dts,γ,a,KE]…, dts,γ,a,KE]}}

We have delved into craft travel into mathematical objects where the craft may then travel into logical objects or the abstract eternal forms of logic equations, proofs, objects, and operators. We have also considered craft travel into amalgams of mathematical and logical objects and physical realities.

Now we consider travel into amalgams meta-mathematical objects and meta-logical objects and physical realities.

We now consider the spacecraft time, gamma, acceleration, and kinetic energy derivatives and integrals of the rate of progression of entrance into amalgams of meta-mathematical and meta-logical objects and physical reality. The integrals in a way serve to denote novel aspects or definitions of metaphorical distances of travel.

We affix the following operator to formulas for gamma and related parameters to denote these rates of entrance into amalgams of meta-mathematical and meta-logical objects and physical realities and derivatives and integrals thereof.

{[Σ(Δ entrance into amal of mm & ml objects & MR)]:{d[…[d(Δ entrance into amal of mm & ml objects & MR)/dts,γ,a,KE]/…,/ dts,γ,a,KE]}:{∫…,[∫(Δ entrance into amal of mm & ml objects & MR)dts,γ,a,KE]…, dts,γ,a,KE]}}

We may further consider entering the realms of meta-mathematical and meta-logical objects and amalgams of meta-mathematical and meta-logical objects and physical realities.

Accordingly, such novel realms would serve as a meta-physical sea or location where meta-mathematical and meta-logical objects and amalgams of meta-mathematical and meta-logical objects and physical realities would be "located" as a background which serves as a mechanism by which meta-mathematical and meta-logical objects are related to each other. As such, the meta-physical sea would be immeasurably infinite in scale and contain an undefinable uncountable infinity of objects, an uncountable infinity which may forever grow in numbers.

We may affix the following operator to formulas for gamma and related parameters to denote the entrance into and unbounded opportunities for travel within the meta-physical sea or location of meta-mathematical and meta-logical objects and amalgams of meta-mathematical and meta-logical objects and physical realities.

[s(Unbounded opportunities for travel within the meta-physical sea or location of meta-mathematical and meta-logical objects and amalgams of meta-mathematical and meta-logical objects and physical realities)]

ESSAY 36) Light-Speed Travel Into Any Existent Hidden Variables Of Wave-Functions Of Objects Ranging From Particles To Hyper-Bulks, Etc..

Here, we consider travel facilitated at light-speed and suitably infinite Lorentz factors into any existent hidden variables of wave-functions of particles, coherent collections of particles, universal light-cones, universes, multiverses, forests, biospheres and the like and hyperspaces, hyperspace-times, bulks, hyper-bulks, and the like.

As such, we consider travel more quantum-mechanically than classically and instead of being present classically in the normally considered ways to said particles, coherent collections of particles, universal light-cones, universes, multiverses, forests, biospheres and the like, and hyperspaces, hyperspace-times, bulks, hyper-bulks, and the like, we become immersed in these objects any existent hidden variables of quantum-mechanical wave-functions.

We can also consider travel into the individual components of the overall any existent hidden variables of wave functions of the above subject objects.

We then consider the spacecraft time, gamma, acceleration, and kinetic energy derivatives and integrals of the rate of progression of entrance into the objects any existent hidden variables of wave-functions. The integrals in a way serve to denote novel aspects or definitions of metaphorical distances of travel.

We may affix the following operator to formulas for gamma and related parameters to denote the rate of the above conjectured travel mechanisms and derivatives and integrals thereof.

{(Δ trav into indiv comp of and overall hv-wave-f):{d[…[d(Δ trav into indiv comp of and overall hv-wave-f)/dts,γ,a,KE]/…,/ dts,γ,a,KE]}:{∫…,[∫(Δ trav into indiv comp of and overall hv-wave-f)dts,γ,a,KE]…, dts,γ,a,KE]}}

Just as we consider two dimensions of a 2-D plane having y-axis analogs of the x-axis of the Cartesian coordinate plane, we can also in principle have as many axes as there are positive integers which are infinite in number. In fact, we can have as many axes as there are positive real numbers, a huge set indeed.

Now we consider analogs of; mathematical objects, logical objects, meta-mathematical objects, meta-logical objects, seas or realms containing mathematical objects, logical objects, meta-mathematical objects, meta-logical objects, quantum wave-functions, components of quantum wave functions, quantum hidden variables, and components of quantum hidden variables. We also consider analogs of amalgams of mathematical objects, logical objects, meta-mathematical objects, meta-logical objects, and physical realities.

We consider travel in meta-seas or meta-realms for which the number of analogs of each of the above parameters is equal to the number of positive real numbers.

So, we consider that a light-speed sufficiently infinite Lorentz factor spacecraft having run out of room in a physical realm of travel to travel forward in time in may have all of the above analog options open for them in terms of travel.

We denote this enormously infinite range of options by the following operator which we affix to the lengthy formulas for gamma and related parameters presented in Section 1 of this book.

(R-Tuple analogs of: mo, lo, mmo, mlo; physicalized amalgams of mo, lo, mmo, mlo; seas of mo, lo, mmo, mlo; qwf, cqwf, qhv, cqhv).

ESSAY 37) The Numerous Methods Of Light-Speed Travel.

Now there are several methods of light-speed travel. These are; 1) impulse travel, 2) light-speed classical teleportation, 3) light-speed quantum teleportation, 4) light-speed wormhole travel, 5) classically beamed travel, and 6) light-speed establishment of multiple space-time connectivity.

The number of subsets including those of combinations of two or more of the above travel methods is (2 EXP 6) -1 or 63.

The power series set of the number of travel combinations is (2 EXP 63) – 1.

The number of subsets of the power series set of the number of travel combinations methods is 2 EXP [(2 EXP 63) – 1] which is a canonical ensemble. We can continue from here suffice it so say, the number of back up transport mechanisms is huge and can include more than one method.

However, there may be other forms of light-speed travel that we have never thought of previously.

The number of such forms of light-speed travel may be finite or infinite and as such, the prospects for light-speed travel superabound with awesome potential.

We may affix the following operator to formulas for gamma and related parameters to denote the operations of wide open ended varieties of light-speed transport methods.

(# trav meth = 6 + n);{# subsets trav meth = {[2 EXP (6 + n)] -1}};{# sets in power series set of trav meth = {{2 EXP {[2 EXP (6 + n)] -1}} - 1}};{# subsets of the power series set of trav meth = {2 EXP {{2 EXP {[2 EXP (6 + n)] -1}} - 1} -1}}

Assuming that each method, or set of methods, can be represented by an ontological wave-function or set of ontological wave-functions, any combinations in the above sets or extended sets has transport merit.

Light-speed travel over cosmic distances, using multiple propulsion methods in series and/or parallel including any of the above implied sets can act as a "rope braid" to reinforce the integrity of the propulsion events and spacecraft and crew integrity.

ESSAY 38) Apparent Superluminal Events And Spacecraft Entanglement With The Wave-Functions Of Such Events.

We understand apparent superluminal events or conditions such as a shadow moving across a huge spherical screen for which an observer would be in the center and watch a shadow revolve around on the inside of the screen several times per second. Here, the projection screen would have radius of one A.U. or perhaps even one light-year. In theory, the screen may have radius of one cosmic light-cone unit.

We also understand the apparent correlation between the outcome of one measure entangled photon and the measurement of the second entangled photon of an entangled pair. Here, the measured outcome is random and a distant beamed to observer receives information about the first photon via a classical feedback signal so that no violation of Special Relativity manifests.

The immediacy of correlation of entangled photons upon measurement no matter how far apart they are from each other may or may not be superluminal depending on whether hidden quantum variables exist. I am a “GOD does not place dice” kind of guy in the spirit of Einstein. Either way, the requirement of a classical feedback loop assures that no signal travels faster than light.

Other forms of apparent faster than light travel may exist but these cases are also not real.

Now, what if the above methods and correlations might be validly stated as methods of pseudo travel at superluminal speeds whereby a spacecraft would be conveyed in shadow or quantum entanglement state and actually transported superluminally but where the spacecraft could not experience its new surroundings at a destination of arrival until a classical feedback signal was transmitted between the two nodes of travel. Accordingly, the spacecraft would be completely cloaked and invisible at its destination from outside and the outside environment of the spacecraft would be invisible to the spacecraft until the classical feedback loop arrived to unlock the presence of the spacecraft.

An analogous mechanism might also work for superluminal shadow casting mechanisms for transporting a spacecraft.

The beautiful thing about these proposed methods is that the spacecraft may bypass dangerous intermediate conditions along the otherwise straightest flight path the spacecraft would take.

There are likely several if not many forms of such muted travel.

The above travel mechanisms may conceivably be possible in ordinary 4-D space-time, in N-D-Space-M-D-Time scales where N is any integer greater than two and M is any counting number, where N is any integer greater that two and M is any positive rational number, where N is any integer greater that two and M is any positive irrational number, where N is any integer greater that two and M is any positive real number, where N is any positive rational number and M is any counting number, where N is any positive rational number and M is any positive rational number, where N is any positive rational number and M is any positive irrational number, where N any positive rational number and M is any positive real number, where N is any positive irrational number and M is any counting number, where N is any positive irrational number and M is any positive rational number, where N is any positive irrational number and M is any positive irrational number, where N is any positive irrational number and M is any positive real number, where N is any positive real number and M is any counting number, where N is any positive real number and M where N is any positive real number, where N is any positive real number and M is any positive irrational number, where N is any positive real number and M is any positive real number: for which the space-time is flat, positively curved, negatively curved, positively curved and torsioned at one or more scales in arbitrary patterns including but not limited to fractals, negatively curved and torsioned at one of more scales in arbitrary patterns including but not limited to fractals, positively super-curved, negatively super-curved, positively super-curved and torsioned at one or more scales in arbitrary patterns including but not limited to fractals, negatively super-curved and torsioned at one of more scales in arbitrary patterns including but not limited to fractals, positively super-super-curved, negatively super-super-curved, positively super-super-curved and torsioned at one or more scales in arbitrary patterns including but not limited to fractals, negatively super-super-curved and torsioned at one of more scales in arbitrary patterns including but not limited to fractals, positively super-super-super-curved, negatively super-super-super-curved, positively super-super-super-curved and torsioned at one or more scales in arbitrary patterns including but not limited to fractals, negatively super-super-super-curved and torsioned at one of more scales in arbitrary patterns including but not limited to fractals, positively super-super- ... -super-curved, negatively super-super- ... -super-curved, positively super-super- ... -super-curved and torsioned at one or more scales in arbitrary patterns including but not limited to fractals, negatively super-super- ... -super-curved and torsioned at one of more scales in arbitrary patterns including but not limited to fractals.

We affix the following operator to the lengthy formulas for gamma and related parameters to indicate the set of such delayed manifestation superluminal travel methods.

(# > c meth = 2 + n);{# ss > c meth = {[2 EXP (2 + n)] -1}};{# s pow ser s of > c = {{2 EXP {[2 EXP (2 + n)] -1}} - 1}};{# ss pow ser s of trav > c = {2 EXP {{2 EXP {[2 EXP (2 + n)] -1}} - 1} -1}}

Here we assume the two types of travel considered above added to any additional delayed manifestation travel methods.

ESSAY 39) The Numerous Classes Of Space Travel.

Now, there are several classes of space travel.

Accordingly, there is inertial or impulse travel, warp-drive, wormhole transport, quantum teleportation, and classical beaming.

Inertial or impulse travel consists in principle of a canonical ensemble of different methods if multimode travel scenarios are considered.

Inertial or impulse travel seems to get into the real meaning of travel in a visceral if not more primitive manner.

Now here is a fascinating concept.

We can consider hyper-inertial or hyper-impulse travel that is to inertial or impulse travel as inertial or impulse travel are to warp-drive, wormhole transport, quantum teleportation, and classical beaming.

Now, hyper-inertial or hyper-impulse travel may or may not have superluminal velocity modes. However, at least light-speed modes should be available. It is conceivable that hyper-inertial or hyper-impulse travel can manifest in a light-speed that is ever-so-slightly greater than the light-speed of known inertial or impulse travel.

Additionally, although hyper-inertial or hyper-impulse travel may manifest Lorentz factors for a given velocity less than that which is standard for inertial or impulse travel, the available Lorentz factors for hyper-inertial or hyper-impulse travel at the speed of light associated with hyper-inertial or hyper-impulse travel may have greater to much greater infinite boundary limits than the infinite Lorentz factors available to standard impulse and inertial light-speed travel.

We can also consider hyper-hyper-inertial or hyper-hyper-impulse travel. We refer to hyper-inertial travel or hyper-impulse travel as hyper-inertial-1 or hyper-impulse-1. Hyper-hyper-inertial travel or hyper-hyper-impulse travel are referred to as hyper-inertial-2 or hyper-impulse-2.

Hyper-inertial-2 or hyper-impulse-2 travel would be to hyper-inertial-1 or hyper-impulse-1 as hyper-inertial-1 or hyper-impulse-1 is to inertial or impulse travel.

Additionally, although hyper-inertial-2 or hyper-impulse-2 travel may manifest Lorentz factors for a given velocity less than that which is standard for hyper-inertial-1 or hyper-impulse-1, the available Lorentz factors for hyper-inertial-2 or hyper-impulse travel-2 at the speed of light associated with hyper-inertial-2 or hyper-impulse-2 travel may have greater to much greater infinite boundary limits than the infinite Lorentz factors available to hyper-inertial-1 or hyper-impulse-1.

We can also consider hyper-hyper-hyper-inertial or hyper-hyper-hyper-impulse travel. We refer to hyper-hyper-hyper-inertial travel or hyper-hyper-hyper-impulse travel as hyper-inertial-3 or hyper-impulse-3.

Hyper-inertial-3 or hyper-impulse-3 travel would be to hyper-inertial-2 or hyper-impulse-2 as hyper-inertial-2 or hyper-impulse-2 is to hyper-inertial-1 or hyper-impulse-1.

Additionally, although hyper-inertial-3 or hyper-impulse-3 travel may manifest Lorentz factors for a given velocity less than that which is standard for hyper-inertial-2 or hyper-impulse-2, the available Lorentz factors for hyper-inertial-3 or hyper-impulse travel-3 at the speed of light associated with hyper-inertial-3 or hyper-impulse-3 travel may have greater to much greater infinite boundary limits than the infinite Lorentz factors available to hyper-inertial-2 or hyper-impulse-2.

We can also consider hyper-hyper-hyper-hyper-inertial or hyper-hyper-hyper-hyper-impulse travel. We refer to hyper-hyper-hyper-hyper-inertial travel or hyper-hyper-hyper-hyper-impulse travel as hyper-inertial-4 or hyper-impulse-4.

Hyper-inertial-4 or hyper-impulse-4 travel would be to hyper-inertial-3 or hyper-impulse-3 as hyper-inertial-3 or hyper-impulse-3 is to hyper-inertial-2 or hyper-impulse-2.

Additionally, although hyper-inertial-4 or hyper-impulse-4 travel may manifest Lorentz factors for a given velocity less than that which is standard for hyper-inertial-3 or hyper-impulse-3, the available Lorentz factors for hyper-inertial-4 or hyper-impulse travel-4 at the speed of light associated with hyper-inertial-4 or hyper-impulse-4 travel may have greater to much greater infinite boundary limits than the infinite Lorentz factors available to hyper-inertial-3 or hyper-impulse-3.

We can continue the general numerically indexed pattern perhaps to infinitely ordinated terms.

We denote the range of possible iteratively “hyper” indexed inertial or impulse travel methods by the following set builder notation.

{hyper-inertial-k or hyper-impulse-k| k = 1, 2, 3, …}.

It is even plausible that k can be as great as super-infinite integers.

We denote the range of scenarios and operative physics of hyper indexed inertial or impulse travel methods by the following operator which we may affix to formulas for gamma and related parameters.

{r{{hyper-inertial-k or hyper-impulse-k| k = 1, 2, 3, …}}}

The velocities of light and the maximum infinite Lorentz factors associated with each advancing level are assumed to be greater than that of the previous levels.

ESSAY 40) Hyper-Operator Functions Of Infinite Arguments As Spacecraft Lorentz Factors And Levels Of Something Else Light-Speed.

Now, consider the hyper-operator notation that was designed to express huge values not otherwise expressible.

For example, Note that Hyper4(a, n) is equal to a tetrated n or a raised to the power of itself n-1 times. The latter value is symbolically written as n subscripta. For example 3 EXP 4 = 81, but 4 subscript3 is approximately equal to 10 EXP (1,000,000,000,000).

Alternatively 4 subscript 2 = 2 EXP 2 EXP 2 EXP 2 = 2 EXP [2 EXP [2 EXP 2]] = 2 EXP (2 EXP 4) = 2 EXP 16 = 65,536.

For example, Hyper5(4, 4)is equal to 4 tetrated 4 tetrated 4 tetrated 4. This value is commonly referred to as 4 pentated 4.

Hyper 6, (4,4) is 4 pentated 4 pentated 4 pentated 4 and is also referred to as 4 hexataed 4.

Hyper 7, (4,4) is 4 hexated 4 hexated 4 hexated 4 and so-on

Aleph 0 is the number of positive integers and is thus infinite.

Aleph 1 is the number of real numbers according to the perhaps unprovable or unfalsifiable Continuum Hypothesis but either way is equal to 2 EXP (Aleph 0).

Aleph 2 is like-wise equal to 2 EXP (Aleph 1).

Aleph 3 is like-wise equal to 2 EXP (Aleph 3).

In general, Aleph N is like-wise equal to 2 EXP [Aleph(N-1)] where N can be any finite or any infinite positive integer.

Aleph numbers go to at least Aleph Ω where Ω is the least transfinite ordinal but can be as big as any one chooses.

So, imagine the number Hyper Aleph Ω, (Aleph Ω, Aleph Ω)

Now, there are countable and uncountable infinities. Uncountable infinities are generally greater than countable infinities.

Just as there can be no greatest countable infinity, there can be no greatest uncountable infinity. You can always add one element to a set even if only a duplicate of an element already included in a set no matter how infinitely many elements are in the set.

Now, what about the next level up from infinities?

Well, we can consider hyper-extended number lines for which there exist super-infinite integers. Accordingly, the number line would be so extended that the super-infinite integers would be larger than any infinite integers. We will denote the least super-infinite ordinal by the notation, ◯.

So, can you imagine how big of a value the following expression would define!

Hyper ◯, (◯, ◯)

Next, we move on to values that are so large they are not numbers and thus are beyond infinities countable and uncountable and beyond super-infinities.

We define these non-number values by the symbol,֍.

Now, can you imagine the following value!

Hyper ֍, (֍,֍)

Next we move on up to yet another qualitative level to expressions that represent somethings that are so immeasurably immense that these realities or abstract objects are non-quantitative and thus are not along the metric that ranges from less than, equal to, nor greater than.

These abstract objects are distinct from qualitative infinities and thus are much more sublime. We use the symbol, ☼, to denote objects in this class.

Well, given sufficient numbers and levels of future eternities to behold, it is plausible that a light-speed spacecraft would morph into something else speed and take on at least the smallest transfinite ordinal commonly referred to as the least infinite number. With all of the constructs considered in the above several page digression, there is likely no concrete limit to how infinite or beyond infinite a spacecraft Lorentz factor can become.

If we assume that here and now light-speed is an inviolable limit and in a good sense will remain so, we can consider that the light-speed of a light-speed spacecraft can morph into what we will call, something-else-light-speed.

As a first level of such, we will refer to the something-else-light-speed as something-else-light-speed-1.

For a something-else-light-speed of the next or 2nd level, we use the reference something-else-light-speed-2.

For a something-else-light-speed of the next or 3rd level, we use the reference something-else-light-speed-3.

For a something-else-light-speed of the next or 4th level, we use the reference something-else-light-speed-4 and continue from there

We go all the way to something-else-light-speed-[Hyper Aleph Ω, (Aleph Ω, Aleph Ω) and onward from there. Here, Ω, can be as large of a transfinite integer as we desire. Thus, even using this hyper-operator expression, there is no concrete limit to the levels of something-else-light-speed.

In order to convey the ramifications, we affix the following operator to the lengthy formulas for gamma and related parameters in Section 1 of this book.

[z(Light-speed → Something-else-light-speed-[Hyper Aleph Ω, (Aleph Ω, Aleph Ω)], and forever beyond and infinite, super-infinite, and so-on Lorentz factors)]

So, if you have ever been off-put at any light-speed limits, do not fret at all. This actually works out better and no one needs to be left behind on an associated travel mission because future technologies will enable us to build ever larger craft.

Even at light-speed and a mere Lorentz factor of Aleph 0, we can travel about Aleph 0 light-years through space in one year ship time and Aleph 0 years into the future in one year ship time. Not bad at all!

ESSAY 41) Nuclear Explosives That Produce Light-Speed Waves Of Infinite Energy Densities.

Here, we consider nuclear explosives that produce a light-speed wave of infinite energy density that travels forward and backward in time from the point of detonation in ordinary 4-D spacetime

As for hyperspatial applications, we consider nuclear explosives that produce a light-speed wave of infinite energy density that travels forward and backward in time from the point of detonation in hyperspaces.

We conjecture that sufficiently concentrated shaped charge nuclear explosive blasts can produce such high temperatures such that a local portion of a cosmic realm of detonation location is classically blasted out of its potential energy well or is induced to tunnel through the well walls to another energy state that is lower.

Such hyperspaces can be N-Space-M-Time where N is an integer greater than or equal to 3 and M is any counting number.

Alternatively, N can be any rational number equal to three or greater and M is any rational number greater than or equal to one.

As another set of scenarios, N can be any irrational number equal to three or greater and M is any irrational number greater than or equal to one.

Alternatively, N can be any rational number greater than zero and M is any rational number greater than zero.

As another set of scenarios, N can be any irrational number greater than zero and M is any irrational number greater zero.

The N-Space-M-Time may optionally be flat, positively curved, negatively curved, positively curved and torsioned at one or more scales in arbitrary patterns including but not limited to fractals, negatively curved and torsioned at one or more scales in arbitrary patterns including but not limited to fractals, positively super-curved, negatively super-curved, positively super-curved and torsioned at one or more scales in arbitrary patterns including but not limited to fractals, negatively super-curved and torsioned at one or more scales in arbitrary patterns including but not limited to fractals, positively super-curved and super-torsioned at one or more scales in arbitrary patterns including but not limited to fractals, negatively super-curved and super-torsioned at one or more scales in arbitrary patterns including but not limited fractals, positively curved and positively torsioned at one or more scales in arbitrary patterns including but not limited to fractals, negatively curved and positively torsioned at one or more scales in arbitrary patterns including but not limited to fractals, positively super-curved, negatively super-curved, positively super-curved and positively torsioned at one or more scales in arbitrary patterns including but not limited to fractals, negatively super-curved and positively torsioned at one or more scales in arbitrary patterns including but not limited to fractals, positively super-curved and positively super-torsioned at one or more scales in arbitrary patterns including but not limited to fractals, negatively super-curved and positively super-torsioned at one or more scales in arbitrary patterns including but not limited fractals, positively curved and negatively torsioned at one or more scales in arbitrary patterns including but not limited to fractals, negatively curved and negatively torsioned at one or more scales in arbitrary patterns including but not limited to fractals, positively super-curved, negatively super-curved, positively super-curved and negatively torsioned at one or more scales in arbitrary patterns including but not limited to fractals, negatively super-curved and negatively torsioned at one or more scales in arbitrary patterns including but not limited to fractals, positively super-curved and negatively super-torsioned at one or more scales in arbitrary patterns including but not limited to fractals, negatively super-curved and negatively super-torsioned at one or more scales in arbitrary patterns including but not limited fractals.

We consider that such explosive energy waves have a radial coordinate for which a spherical wedge of the explosion travels so directly back in time such that the backward temporal progression is straighter or more direct than conventionally direct backward time travel.

For example, consider an angle with two rays pointing directly backward in time and with an angle measure of zero degrees.

Well, what I am considering is the equivalent of an angle with two backward pointing rays but with a modulus measure of less than zero or measure = 0/u where u ranges from just greater than one to arbitrary uncountable infinities. We denote such a condition by $\cos \Theta > 1$.

Likewise, the explosive wave front can have a temporal spherical sector traveling forward in time such that the backward temporal progression is straighter or more direct than conventionally direct backward time travel.

Again, consider an angle with two rays pointing directly backward in time and with an angle measure of zero degrees.

Here we are considering the equivalent of an angle with two backward pointing rays but with a modulus measure of less than zero or measure = 0/u where u ranges from just greater than one to arbitrary uncountable infinities. We denote such a condition by $\cos \Theta > 1$.

Now, it is conceivable that a spacecraft could couple to be located in front of the wave but remain outside the wave front thus being thrust forward to slightly superluminal velocities into the future.

Analogs for more than one temporal dimension are plausible.

Such considerations may be relatable to the notion of a temporal volume future-wise being filled up by the explosion. When the temporal cup is filled, the explosion would boil over the rim of the cup to create brand new; space-time-mass-energy, universal light-cones, universal domains, universes, multiverses, forests, biospheres, etc, hyperspaces, hyperspace-times, hyper-times, bulks, hyper-bulks, and the like.

We may affix the following operator to formulas for gamma and related parameters to denote the above superluminal M-D-time travel constructs based in infinite energy density propulsive wave-fronts caused by exotic nuclear explosions.

[z(Superluminal M-D-time travel constructs based in infinite energy density propulsive wave-fronts caused by exotic nuclear explosions with direct time travel wedges of angular measure less than 0 radians and cos ϴ >1)]

ESSAY 42) Spacecraft Travel At Such Great Infinite And Super-Infinite Lorentz Factors Such That These Lorentz Factors Are Not Defined.

Here, we consider travel at infinite and super-infinite Lorentz factors such that the spacecraft travel Lorentz factor is not definite and is thus not definitely defined commensurate with infinite values being non-concrete and not specifically ordinated and not locally numerically well defined.

Such a mechanism may hyper-cohere a spacecraft wave-function to make it more spread out and available for targeted decoherence for exotic spacecraft propulsion methods. The hyper-coherent spacecraft may thus have a nebulous location in space and in time thus perhaps going beyond the physics of special relativity and the speed of light.

The spacecraft hyper-coherence mechanism can occur in ordinary 4-D space-time, or in N-D-Space-M-D-Time scales where N is any integer greater than two and M is any counting number, where N is any integer greater that two and M is any positive rational number, where N is any integer greater that two and M is any positive irrational number, where N is any integer greater that two and M is any positive real number, where N is any positive rational number and M is any counting number, where N is any positive rational number and M is any positive rational number, where N is any positive rational number and M is any positive irrational number, where N any positive rational number and M is any positive real number, where N is any positive irrational number and M is any counting number, where N is any positive irrational number and M is any positive rational number, where N is any positive irrational number and M is any positive irrational number, where N is any positive irrational number and M is any positive real number, where N is any positive real number and M is any counting number, where N is any positive real number and M where N is any positive real number, where N is any positive real number and M is any positive irrational number, where N is any positive real number and M is any positive real number: for which the space-time is flat, positively curved, negatively curved, positively curved and torsioned at one or more scales in arbitrary patterns including but not limited to fractals, negatively curved and torsioned at one of more scales in arbitrary patterns including but not limited to fractals, positively super-curved, negatively super-curved, positively super-curved and torsioned at one or more scales in arbitrary patterns including but not limited to fractals, negatively super-curved and torsioned at one of more scales in arbitrary patterns including but not limited to fractals, positively super-super-curved, negatively super-super-curved, positively super-super-curved and torsioned at one or more scales in arbitrary patterns including but not limited to fractals, negatively super-super-curved and torsioned at one of more scales in arbitrary patterns including but not limited to fractals, positively super-super-super-curved, negatively super-super-super-curved, positively super-super-super-curved and torsioned at one or more scales in arbitrary patterns including but not limited to fractals, negatively super-super-super-curved and torsioned at one of more scales in arbitrary patterns including but not limited to fractals, positively super-super- ... -super-curved, negatively super-super- ... -super-curved, positively super-super- ... -super-curved and torsioned at one or more scales in arbitrary patterns including but not limited to fractals, negatively super-super- ... -super-curved and torsioned at one of more scales in arbitrary patterns including but not limited to fractals.

We denote the spacecraft hyper-coherence as a function of blurred Lorentz factors, derivatives and integrals of functions of gamma, velocity, kinetic energy, acceleration, and ship-time by the following operator which we affix to formulas for gamma and related parameters.

{(SC h-c):[(γ), (v), (KE), (a), (ts)]:{d{...{d{(SC h-c):[(γ), (v), (KE), (a), (tship)]}/dγ,v,KE,a, or ts}/.../dγ,v, KE,a,ts}}:{∫{...{∫{(SC h-c):[(γ), (v), (KE), (a), (tship)]}dγ,v,KE,a, or ts}... dγ,v,KE,a, or ts}}}

ESSAY 43) Light-Speed Spacecraft At Sufficient Infinite And Super-Infinite Lorentz Factors Travel Ahead Of Itself.

We now consider travel at infinite and super-infinite Lorentz factors of more than one value per differential volumetric element of a light-speed spacecraft so as to enable the spacecraft to travel ahead of itself, its average velocity, its average Lorentz factor, and perhaps travel ahead of the great bulk of itself by forward skewing of its expectation values of its Lorentz factor, velocity, position, acceleration, and kinetic energy wave-functions.

We also consider travel at light-speed at qualitatively and super-qualitatively infinite Lorentz factors. Accordingly, a spacecraft Lorentz factor would be qualitative and then so infinite.

Such travel can occur not only in ordinary 4-D space-time, but also in N-D-Space-M-D-Time scales where N is any integer greater than two and M is any counting number, where N is any integer greater that two and M is any positive rational number, where N is any integer greater that two and M is any positive irrational number, where N is any integer greater that two and M is any positive real number, where N is any positive rational number and M is any counting number, where N is any positive rational number and M is any positive rational number, where N is any positive rational number and M is any positive irrational number, where N any positive rational number and M is any positive real number, where N is any positive irrational number and M is any counting number, where N is any positive irrational number and M is any positive rational number, where N is any positive irrational number and M is any positive irrational number, where N is any positive irrational number and M is any positive real number, where N is any positive real number and M is any counting number, where N is any positive real number and M where N is any positive real number, where N is any positive real number and M is any positive irrational number, where N is any positive real number and M is any positive real number: for which the space-time is flat, positively curved, negatively curved, positively curved and torsioned at one or more scales in arbitrary patterns including but not limited to fractals, negatively curved and torsioned at one of more scales in arbitrary patterns including but not limited to fractals, positively super-curved, negatively super-curved, positively super-curved and torsioned at one or more scales in arbitrary patterns including but not limited to fractals, negatively super-curved and torsioned at one of more scales in arbitrary patterns including but not limited to fractals, positively super-super-curved, negatively super-super-curved, positively super-super-curved and torsioned at one or more scales in arbitrary patterns including but not limited to fractals, negatively super-super-curved and torsioned at one of more scales in arbitrary patterns including but not limited to fractals, positively super-super-super-curved, negatively super-super-super-curved, positively super-super-super-curved and torsioned at one or more scales in arbitrary patterns including but not limited to fractals, negatively super-super-super-curved and torsioned at one of more scales in arbitrary patterns including but not limited to fractals, positively super-super- … -super-curved, negatively super-super- … -super-curved, positively super-super- … -super-curved and torsioned at one or more scales in arbitrary patterns including but not limited to fractals, negatively super-super- … -super-curved and torsioned at one of more scales in arbitrary patterns including but not limited to fractals.

We denote the qualitative values of gamma for the spacecraft and the spacecraft expectation values for gamma, velocity, position, acceleration, and kinetic energy and derivatives and integrals thereof for functions of gamma, velocity, kinetic energy, acceleration, and ship-time by the following operator which we may affix to formulas for gamma and related parameters.

{Qualitative ∞ γ{[z(Skewing of SC Ex for γ, v, p, a, and KE)]: {d{…{d{(SC Ex):[(γ), (v), (KE), (a), (p)]}/dγ,v,KE,a, or ts}/…/dγ,v, KE,a,ts}}:{∫{…{∫{(SC Ex):[(γ), (v), (KE), (a), (p)]}dγ,v,KE,a, or ts}… dγ,v,KE,a, or ts}}}}

ESSAY 44) Travel In Conventional Meanings Of The Word.

Herein, we consider travel in unconventional meanings of the word. Accordingly, we consider travel distinct from that in space and/or forward in time. The following roster of unconventional travel scenarios is briefly and clearly self-defined in terms of lexicography. Note that these travel scenarios are physical but exotic indeed.

Traveling into electric charge, magnetic charge, blue or green or red color charge, upness, down-ness, charmed-ness, strange-ness, topness, bottom-ness, weak force charges, gravitational charge, higgs-boson charge, electromagnetism, weak force, strong force, gravitational force, electro-weak, electro-strong, electrogravatic, electro-weak-strong, electro-weak-gravatic, electro-strong-gravatic, weak-strong, weak-gravatic, weak-strong-gravatic, strong-gravatic, electro-weak-strong-gravatic, higgs-electromagnetism, higgs-weak force, higgs-strong force, higgs-gravitational force, higgs-electro-weak, higgs-electro-strong, higgs-electrogravatic, higgs-electro-weak-strong, higgs-electro-weak-gravatic, higgs-electro-strong-gravatic, higgs-weak-strong, higgs-weak-gravatic, higgs-weak-strong-gravatic, higgs-strong-gravatic, higgs-electro-weak-strong-gravatic, SUSY-electromagnetism, SUSY-weak force, SUSY-strong force, SUSY-gravitational force, SUSY-electro-weak, SUSY-electro-strong, SUSY-electrogravatic, SUSY-electro-weak-strong, SUSY-electro-weak-gravatic, SUSY-electro-strong-gravatic, SUSY-weak-strong, SUSY-weak-gravatic, SUSY-weak-strong-gravatic, SUSY-strong-gravatic, SUSY-electro-weak-strong-gravatic, SUSY-higgs-electromagnetism, SUSY-higgs-weak force, SUSY-higgs-strong force, SUSY-higgs-gravitational force, SUSY-higgs-electro-weak, SUSY-higgs-electro-strong, SUSY-higgs-electrogravatic, SUSY-higgs-electro-weak-strong, SUSY-higgs-electro-weak-gravatic, SUSY-higgs-electro-strong-gravatic, SUSY-higgs-weak-strong, SUSY-higgs-weak-gravatic, SUSY-higgs-weak-strong-gravatic, SUSY-higgs-strong-gravatic, SUSY-higgs-electro-weak-strong-gravatic.

Traveling into wave-functions of; electric charge, magnetic charge, blue or green or red color charge, upness, down-ness, charmed-ness, strange-ness, topness, bottom-ness, weak force charges, gravitational charge, higgs-boson charge, electromagnetism, weak force, strong force, gravitational force, electro-weak, electro-strong, electrogravatic, electro-weak-strong, electro-weak-gravatic, electro-strong-gravatic, weak-strong, weak-gravatic, weak-strong-gravatic, strong-gravatic, electro-weak-strong-gravatic, higgs-electromagnetism, higgs-weak force, higgs-strong force, higgs-gravitational force, higgs-electro-weak, higgs-electro-strong, higgs-electrogravatic, higgs-electro-weak-strong, higgs-electro-weak-gravatic, higgs-electro-strong-gravatic, higgs-weak-strong, higgs-weak-gravatic, higgs-weak-strong-gravatic, higgs-strong-gravatic, higgs-electro-weak-strong-gravatic, SUSY-electromagnetism, SUSY-weak force, SUSY-strong force, SUSY-gravitational force, SUSY-electro-weak, SUSY-electro-strong, SUSY-electrogravatic, SUSY-electro-weak-strong, SUSY-electro-weak-gravatic, SUSY-electro-strong-gravatic, SUSY-weak-strong, SUSY-weak-gravatic, SUSY-weak-strong-gravatic, SUSY-strong-gravatic, SUSY-electro-weak-strong-gravatic, SUSY-higgs-electromagnetism, SUSY-higgs-weak force, SUSY-higgs-strong force, SUSY-higgs-gravitational force, SUSY-higgs-electro-weak, SUSY-higgs-electro-strong, SUSY-higgs-electrogravatic, SUSY-higgs-electro-weak-strong, SUSY-higgs-electro-weak-gravatic, SUSY-higgs-electro-strong-gravatic, SUSY-higgs-weak-strong, SUSY-higgs-weak-gravatic, SUSY-higgs-weak-strong-gravatic, SUSY-higgs-strong-gravatic, SUSY-higgs-electro-weak-strong-gravatic.

Traveling into components of wave-functions of; electric charge, magnetic charge, blue or green or red color charge, upness, down-ness, charmed-ness, strange-ness, topness, bottom-ness, weak force charges, gravitational charge, higgs-boson charge, electromagnetism, weak force, strong force, gravitational force, electro-weak, electro-strong, electrogravatic, electro-weak-strong, electro-weak-gravatic, electro-strong-gravatic, weak-strong, weak-gravatic, weak-strong-gravatic, strong-gravatic, electro-weak-strong-gravatic, higgs-electromagnetism, higgs-weak force, higgs-strong force, higgs-gravitational force, higgs-electro-weak, higgs-electro-strong, higgs-electrogravatic, higgs-electro-weak-strong, higgs-electro-weak-gravatic, higgs-electro-strong-gravatic, higgs-weak-strong, higgs-weak-gravatic, higgs-weak-strong-gravatic, higgs-strong-gravatic, higgs-electro-weak-strong-gravatic, SUSY-electromagnetism, SUSY-weak force, SUSY-strong force, SUSY-gravitational force, SUSY-electro-weak, SUSY-electro-strong, SUSY-electrogravatic, SUSY-electro-weak-strong, SUSY-electro-weak-gravatic, SUSY-electro-strong-gravatic, SUSY-weak-strong, SUSY-weak-gravatic, SUSY-weak-strong-gravatic, SUSY-strong-gravatic, SUSY-electro-weak-strong-gravatic, SUSY-higgs-electromagnetism, SUSY-higgs-weak force, SUSY-higgs-strong force, SUSY-higgs-gravitational force, SUSY-higgs-electro-weak, SUSY-higgs-electro-strong, SUSY-higgs-electrogravatic, SUSY-higgs-electro-weak-strong, SUSY-higgs-electro-weak-gravatic, SUSY-higgs-electro-strong-gravatic, SUSY-higgs-weak-strong, SUSY-higgs-weak-gravatic, SUSY-higgs-weak-strong-gravatic, SUSY-higgs-strong-gravatic, SUSY-higgs-electro-weak-strong-gravatic.

Traveling into quantum hidden variables of; electric charge, magnetic charge, blue or green or red color charge, upness, down-ness, charmed-ness, strange-ness, topness, bottom-ness, weak force charges, gravitational charge, higgs-boson charge, electromagnetism, weak force, strong force, gravitational force, electro-weak, electro-strong, electrogravatic, electro-weak-strong, electro-weak-gravatic, electro-strong-gravatic, weak-strong, weak-gravatic, weak-strong-gravatic, strong-gravatic, electro-weak-strong-gravatic, higgs-electromagnetism, higgs-weak force, higgs-strong force, higgs-gravitational force, higgs-electro-weak, higgs-electro-strong, higgs-electrogravatic, higgs-electro-weak-strong, higgs-electro-weak-gravatic, higgs-electro-strong-gravatic, higgs-weak-strong, higgs-weak-gravatic, higgs-weak-strong-gravatic, higgs-strong-gravatic, higgs-electro-weak-strong-gravatic, SUSY-electromagnetism,

SUSY-weak force, SUSY-strong force, SUSY-gravitational force, SUSY-electro-weak, SUSY-electro-strong, SUSY-electrogravatic, SUSY-electro-weak-strong, SUSY-electro-weak-gravatic, SUSY-electro-strong-gravatic, SUSY-weak-strong, SUSY-weak-gravatic, SUSY-weak-strong-gravatic, SUSY-strong-gravatic, SUSY-electro-weak-strong-gravatic, SUSY-higgs-electromagnetism, SUSY-higgs-weak force, SUSY-higgs-strong force, SUSY-higgs-gravitational force, SUSY-higgs-electro-weak, SUSY-higgs-electro-strong, SUSY-higgs-electrogravatic, SUSY-higgs-electro-weak-strong, SUSY-higgs-electro-weak-gravatic, SUSY-higgs-electro-strong-gravatic, SUSY-higgs-weak-strong, SUSY-higgs-weak-gravatic, SUSY-higgs-weak-strong-gravatic, SUSY-higgs-strong-gravatic, SUSY-higgs-electro-weak-strong-gravatic.

Traveling into pseudo-spatial-temporal realities in the sense of travel into electric charge, magnetic charge, blue or green or red color charge, upness, down-ness, charmed-ness, strange-ness, topness, bottom-ness, weak force charges, gravitational charge, higgs-boson charge, electromagnetism, weak force, strong force, gravitational force, electro-weak, electro-strong, electrogravatic, electro-weak-strong, electro-weak-gravatic, electro-strong-gravatic, weak-strong, weak-gravatic, weak-strong-gravatic, strong-gravatic, electro-weak-strong-gravatic, higgs-electromagnetism, higgs-weak force, higgs-strong force, higgs-gravitational force, higgs-electro-weak, higgs-electro-strong, higgs-electrogravatic, higgs-electro-weak-strong, higgs-electro-weak-gravatic, higgs-electro-strong-gravatic, higgs-weak-strong, higgs-weak-gravatic, higgs-weak-strong-gravatic, higgs-strong-gravatic, higgs-electro-weak-strong-gravatic, SUSY-electromagnetism, SUSY-weak force, SUSY-strong force, SUSY-gravitational force, SUSY-electro-weak, SUSY-electro-strong, SUSY-electrogravatic, SUSY-electro-weak-strong, SUSY-electro-weak-gravatic, SUSY-electro-strong-gravatic, SUSY-weak-strong, SUSY-weak-gravatic, SUSY-weak-strong-gravatic, SUSY-strong-gravatic, SUSY-electro-weak-strong-gravatic, SUSY-higgs-electromagnetism, SUSY-higgs-weak force, SUSY-higgs-strong force, SUSY-higgs-gravitational force, SUSY-higgs-electro-weak, SUSY-higgs-electro-strong, SUSY-higgs-electrogravatic, SUSY-higgs-electro-weak-strong, SUSY-higgs-electro-weak-gravatic, SUSY-higgs-electro-strong-gravatic, SUSY-higgs-weak-strong, SUSY-higgs-weak-gravatic, SUSY-higgs-weak-strong-gravatic, SUSY-higgs-strong-gravatic, SUSY-higgs-electro-weak-strong-gravatic.

Traveling into pseudo-spatial-temporal realities in the sense of travel into wave-functions of; electric charge, magnetic charge, blue or green or red color charge, upness, down-ness, charmed-ness, strange-ness, topness, bottom-ness, weak force charges, gravitational charge, higgs-boson charge, electromagnetism, weak force, strong force, gravitational force, electro-weak, electro-strong, electrogravatic, electro-weak-strong, electro-weak-gravatic, electro-strong-gravatic, weak-strong, weak-gravatic, weak-strong-gravatic, strong-gravatic, electro-weak-strong-gravatic, higgs-electromagnetism, higgs-weak force, higgs-strong force, higgs-gravitational force, higgs-electro-weak, higgs-electro-strong, higgs-electrogravatic, higgs-electro-weak-strong, higgs-electro-weak-gravatic, higgs-electro-strong-gravatic, higgs-weak-strong, higgs-weak-gravatic, higgs-weak-strong-gravatic, higgs-strong-gravatic, higgs-electro-weak-strong-gravatic, SUSY-electromagnetism, SUSY-weak force, SUSY-strong force, SUSY-gravitational force, SUSY-electro-weak, SUSY-electro-strong, SUSY-electrogravatic, SUSY-electro-weak-strong, SUSY-electro-weak-gravatic, SUSY-electro-strong-gravatic, SUSY-weak-strong, SUSY-weak-gravatic, SUSY-weak-strong-gravatic, SUSY-strong-gravatic, SUSY-electro-weak-strong-gravatic, SUSY-higgs-electromagnetism, SUSY-higgs-weak force, SUSY-higgs-strong force, SUSY-higgs-gravitational force, SUSY-higgs-electro-weak, SUSY-higgs-electro-strong, SUSY-higgs-electrogravatic, SUSY-higgs-electro-weak-strong, SUSY-higgs-electro-weak-gravatic, SUSY-higgs-electro-strong-gravatic, SUSY-higgs-weak-strong, SUSY-higgs-weak-gravatic, SUSY-higgs-weak-strong-gravatic, SUSY-higgs-strong-gravatic, SUSY-higgs-electro-weak-strong-gravatic.

Traveling into pseudo-spatial-temporal realities in the sense of travel into components of wave-functions of; electric charge, magnetic charge, blue or green or red color charge, upness, down-ness, charmed-ness, strange-ness, topness, bottom-ness, weak force charges, gravitational charge, higgs-boson charge, electromagnetism, weak force, strong force, gravitational force, electro-weak, electro-strong, electrogravatic, electro-weak-strong, electro-weak-gravatic, electro-strong-gravatic, weak-strong, weak-gravatic, weak-strong-gravatic, strong-gravatic, electro-weak-strong-gravatic, higgs-electromagnetism, higgs-weak force, higgs-strong force, higgs-gravitational force, higgs-electro-weak, higgs-electro-strong, higgs-electrogravatic, higgs-electro-weak-strong, higgs-electro-weak-gravatic, higgs-electro-strong-gravatic, higgs-weak-strong, higgs-weak-gravatic, higgs-weak-strong-gravatic, higgs-strong-gravatic, higgs-electro-weak-strong-gravatic, SUSY-electromagnetism, SUSY-weak force, SUSY-strong force, SUSY-gravitational force, SUSY-electro-weak, SUSY-electro-strong, SUSY-electrogravatic, SUSY-electro-weak-strong, SUSY-electro-weak-gravatic, SUSY-electro-strong-gravatic, SUSY-weak-strong, SUSY-weak-gravatic, SUSY-weak-strong-gravatic, SUSY-strong-gravatic, SUSY-electro-weak-strong-gravatic, SUSY-higgs-electromagnetism, SUSY-higgs-weak force, SUSY-higgs-strong force, SUSY-higgs-gravitational force, SUSY-higgs-electro-weak, SUSY-higgs-electro-strong, SUSY-higgs-electrogravatic, SUSY-higgs-electro-weak-strong, SUSY-higgs-electro-weak-gravatic, SUSY-higgs-electro-strong-gravatic, SUSY-higgs-weak-strong, SUSY-higgs-weak-gravatic, SUSY-higgs-weak-strong-gravatic, SUSY-higgs-strong-gravatic, SUSY-higgs-electro-weak-strong-gravatic.

Traveling into pseudo-spatial-temporal realities in the sense of travel into quantum hidden variables of; electric charge, magnetic charge, blue or green or red color charge, upness, down-ness, charmed-ness, strange-ness, topness, bottom-ness, weak force charges, gravitational charge, higgs-boson charge, electromagnetism, weak force, strong force, gravitational force, electro-weak, electro-strong, electrogravatic, electro-weak-strong, electro-weak-gravatic, electro-strong-gravatic, weak-strong, weak-gravatic, weak-strong-gravatic, strong-gravatic, electro-weak-strong-gravatic, higgs-electromagnetism, higgs-weak force, higgs-strong force, higgs-gravitational force, higgs-electro-weak, higgs-electro-strong, higgs-electrogravatic, higgs-electro-weak-strong, higgs-electro-weak-gravatic, higgs-electro-strong-gravatic, higgs-weak-strong, higgs-weak-gravatic, higgs-weak-strong-gravatic, higgs-strong-gravatic, higgs-electro-weak-strong-gravatic, SUSY-electromagnetism, SUSY-weak force, SUSY-strong force, SUSY-gravitational force, SUSY-electro-weak, SUSY-electro-strong, SUSY-electrogravatic, SUSY-electro-weak-strong, SUSY-electro-weak-gravatic, SUSY-electro-strong-gravatic, SUSY-weak-strong, SUSY-weak-gravatic, SUSY-weak-strong-gravatic, SUSY-strong-gravatic, SUSY-electro-weak-strong-gravatic, SUSY-higgs-electromagnetism, SUSY-higgs-weak force, SUSY-higgs-strong force, SUSY-higgs-gravitational force, SUSY-higgs-electro-weak, SUSY-higgs-electro-strong, SUSY-higgs-electrogravatic, SUSY-higgs-electro-weak-strong, SUSY-higgs-electro-weak-gravatic, SUSY-higgs-electro-strong-gravatic, SUSY-higgs-weak-strong, SUSY-higgs-weak-gravatic, SUSY-higgs-weak-strong-gravatic, SUSY-higgs-strong-gravatic, SUSY-higgs-electro-weak-strong-gravatic.

We denote the above travel scenarios by the very abstract operator, [Σ(z = 1; z = sub-canonical ensemble ↑):(Ƶ,z)], which we may affix to formulas for gamma and related parameters.

The upward pointing arrow in the series limit implies open ended boundary values for the number of travel possibilities beyond sub-canonical ensembles. Accordingly, any multiple of individual travel methods conjectured can be included to enable hybrid or compound travel methods.

ESSAY 45) Analogs Of Black-Hole Surface Boundary Conditions.

Here we consider a reasonably full range of concepts akin to black-hole surface boundary conditions and central singularities such as hyper-electromagnetism at various electromagnetic levels. These levels have relevance for travel and high energy density materials. In essence, we consider various levels of hyper indices for all of the following.

electric charge, magnetic charge, blue or green or red color charge, upness, down-ness, charmed-ness, strange-ness, topness, bottom-ness, weak force charges, gravitational charge, higgs-boson charge, electromagnetism, weak force, strong force, gravitational force, electro-weak, electro-strong, electrogravatic, electro-weak-strong, electro-weak-gravatic, electro-strong-gravatic, weak-strong, weak-gravatic, weak-strong-gravatic, strong-gravatic, electro-weak-strong-gravatic, higgs-electromagnetism, higgs-weak force, higgs-strong force, higgs-gravitational force, higgs-electro-weak, higgs-electro-strong, higgs-electrogravatic, higgs-electro-weak-strong, higgs-electro-weak-gravatic, higgs-electro-strong-gravatic, higgs-weak-strong, higgs-weak-gravatic, higgs-weak-strong-gravatic, higgs-strong-gravatic, higgs-electro-weak-strong-gravatic, SUSY-electromagnetism, SUSY-weak force, SUSY-strong force, SUSY-gravitational force, SUSY-electro-weak, SUSY-electro-strong, SUSY-electrogravatic, SUSY-electro-weak-strong, SUSY-electro-weak-gravatic, SUSY-electro-strong-gravatic, SUSY-weak-strong, SUSY-weak-gravatic, SUSY-weak-strong-gravatic, SUSY-strong-gravatic, SUSY-electro-weak-strong-gravatic, SUSY-higgs-electromagnetism, SUSY-higgs-weak force, SUSY-higgs-strong force, SUSY-higgs-gravitational force, SUSY-higgs-electro-weak, SUSY-higgs-electro-strong, SUSY-higgs-electrogravatic, SUSY-higgs-electro-weak-strong, SUSY-higgs-electro-weak-gravatic, SUSY-higgs-electro-strong-gravatic, SUSY-higgs-weak-strong, SUSY-higgs-weak-gravatic, SUSY-higgs-weak-strong-gravatic, SUSY-higgs-strong-gravatic, SUSY-higgs-electro-weak-strong-gravatic.

Traveling into wave-functions of various levels of hyper indices for all of the following.

electric charge, magnetic charge, blue or green or red color charge, upness, down-ness, charmed-ness, strange-ness, topness, bottom-ness, weak force charges, gravitational charge, higgs-boson charge, electromagnetism, weak force, strong force, gravitational force, electro-weak, electro-strong, electrogravatic, electro-weak-strong, electro-weak-gravatic, electro-strong-gravatic, weak-strong, weak-gravatic, weak-strong-gravatic, strong-gravatic, electro-weak-strong-gravatic, higgs-electromagnetism, higgs-weak force, higgs-strong force, higgs-gravitational force, higgs-electro-weak, higgs-electro-strong, higgs-electrogravatic, higgs-electro-weak-strong, higgs-electro-weak-gravatic, higgs-electro-strong-gravatic, higgs-weak-strong, higgs-weak-gravatic, higgs-weak-strong-gravatic, higgs-strong-gravatic, higgs-electro-weak-strong-gravatic, SUSY-electromagnetism, SUSY-weak force, SUSY-strong force, SUSY-gravitational force, SUSY-electro-weak, SUSY-electro-strong, SUSY-electrogravatic, SUSY-electro-weak-strong, SUSY-electro-weak-gravatic, SUSY-electro-strong-gravatic, SUSY-weak-strong, SUSY-weak-gravatic, SUSY-weak-strong-gravatic, SUSY-strong-gravatic, SUSY-electro-weak-strong-gravatic, SUSY-higgs-electromagnetism, SUSY-higgs-weak force, SUSY-higgs-strong force, SUSY-higgs-gravitational force, SUSY-higgs-electro-weak, SUSY-higgs-electro-strong, SUSY-higgs-electrogravatic, SUSY-higgs-electro-weak-strong, SUSY-higgs-electro-weak-gravatic, SUSY-higgs-electro-strong-gravatic,

SUSY-higgs-weak-strong, SUSY-higgs-weak-gravatic, SUSY-higgs-weak-strong-gravatic, SUSY-higgs-strong-gravatic, SUSY-higgs-electro-weak-strong-gravatic.

Traveling into components of wave-functions of various levels of hyper indices for all of the following

electric charge, magnetic charge, blue or green or red color charge, upness, down-ness, charmed-ness, strange-ness, topness, bottom-ness, weak force charges, gravitational charge, higgs-boson charge, electromagnetism, weak force, strong force, gravitational force, electro-weak, electro-strong, electrogravatic, electro-weak-strong, electro-weak-gravatic, electro-strong-gravatic, weak-strong, weak-gravatic, weak-strong-gravatic, strong-gravatic, electro-weak-strong-gravatic, higgs-electromagnetism, higgs-weak force, higgs-strong force, higgs-gravitational force, higgs-electro-weak, higgs-electro-strong, higgs-electrogravatic, higgs-electro-weak-strong, higgs-electro-weak-gravatic, higgs-electro-strong-gravatic, higgs-weak-strong, higgs-weak-gravatic, higgs-weak-strong-gravatic, higgs-strong-gravatic, higgs-electro-weak-strong-gravatic, SUSY-electromagnetism, SUSY-weak force, SUSY-strong force, SUSY-gravitational force, SUSY-electro-weak, SUSY-electro-strong, SUSY-electrogravatic, SUSY-electro-weak-strong, SUSY-electro-weak-gravatic, SUSY-electro-strong-gravatic, SUSY-weak-strong, SUSY-weak-gravatic, SUSY-weak-strong-gravatic, SUSY-strong-gravatic, SUSY-electro-weak-strong-gravatic, SUSY-higgs-electromagnetism, SUSY-higgs-weak force, SUSY-higgs-strong force, SUSY-higgs-gravitational force, SUSY-higgs-electro-weak, SUSY-higgs-electro-strong, SUSY-higgs-electrogravatic, SUSY-higgs-electro-weak-strong, SUSY-higgs-electro-weak-gravatic, SUSY-higgs-electro-strong-gravatic, SUSY-higgs-weak-strong, SUSY-higgs-weak-gravatic, SUSY-higgs-weak-strong-gravatic, SUSY-higgs-strong-gravatic, SUSY-higgs-electro-weak-strong-gravatic.

Traveling into quantum hidden variables of various levels of hyper indices for all of the following.

electric charge, magnetic charge, blue or green or red color charge, upness, down-ness, charmed-ness, strange-ness, topness, bottom-ness, weak force charges, gravitational charge, higgs-boson charge, electromagnetism, weak force, strong force, gravitational force, electro-weak, electro-strong, electrogravatic, electro-weak-strong, electro-weak-gravatic, electro-strong-gravatic, weak-strong, weak-gravatic, weak-strong-gravatic, strong-gravatic, electro-weak-strong-gravatic, higgs-electromagnetism, higgs-weak force, higgs-strong force, higgs-gravitational force, higgs-electro-weak, higgs-electro-strong, higgs-electrogravatic, higgs-electro-weak-strong, higgs-electro-weak-gravatic, higgs-electro-strong-gravatic, higgs-weak-strong, higgs-weak-gravatic, higgs-weak-strong-gravatic, higgs-strong-gravatic, higgs-electro-weak-strong-gravatic, SUSY-electromagnetism, SUSY-weak force, SUSY-strong force, SUSY-gravitational force, SUSY-electro-weak, SUSY-electro-strong, SUSY-electrogravatic, SUSY-electro-weak-strong, SUSY-electro-weak-gravatic, SUSY-electro-strong-gravatic, SUSY-weak-strong, SUSY-weak-gravatic, SUSY-weak-strong-gravatic, SUSY-strong-gravatic, SUSY-electro-weak-strong-gravatic, SUSY-higgs-electromagnetism, SUSY-higgs-weak force, SUSY-higgs-strong force, SUSY-higgs-gravitational force, SUSY-higgs-electro-weak, SUSY-higgs-electro-strong, SUSY-higgs-electrogravatic, SUSY-higgs-electro-weak-strong, SUSY-higgs-electro-weak-gravatic, SUSY-higgs-electro-strong-gravatic, SUSY-higgs-weak-strong, SUSY-higgs-weak-gravatic, SUSY-higgs-weak-strong-gravatic, SUSY-higgs-strong-gravatic, SUSY-higgs-electro-weak-strong-gravatic.

Traveling into pseudo-spatial-temporal realities in the sense of travel into various levels of hyper indices for all of the following.

electric charge, magnetic charge, blue or green or red color charge, upness, down-ness, charmed-ness, strange-ness, topness, bottom-ness, weak force charges, gravitational charge, higgs-boson charge, electromagnetism, weak force, strong force, gravitational force, electro-weak, electro-strong, electrogravatic, electro-weak-strong, electro-weak-gravatic, electro-strong-gravatic, weak-strong, weak-gravatic, weak-strong-gravatic, strong-gravatic, electro-weak-strong-gravatic, higgs-electromagnetism, higgs-weak force, higgs-strong force, higgs-gravitational force, higgs-electro-weak, higgs-electro-strong, higgs-electrogravatic, higgs-electro-weak-strong, higgs-electro-weak-gravatic, higgs-electro-strong-gravatic, higgs-weak-strong, higgs-weak-gravatic, higgs-weak-strong-gravatic, higgs-strong-gravatic, higgs-electro-weak-strong-gravatic, SUSY-electromagnetism, SUSY-weak force, SUSY-strong force, SUSY-gravitational force, SUSY-electro-weak, SUSY-electro-strong, SUSY-electrogravatic, SUSY-electro-weak-strong, SUSY-electro-weak-gravatic, SUSY-electro-strong-gravatic, SUSY-weak-strong, SUSY-weak-gravatic, SUSY-weak-strong-gravatic, SUSY-strong-gravatic, SUSY-electro-weak-strong-gravatic, SUSY-higgs-electromagnetism, SUSY-higgs-weak force, SUSY-higgs-strong force, SUSY-higgs-gravitational force, SUSY-higgs-electro-weak, SUSY-higgs-electro-strong, SUSY-higgs-electrogravatic, SUSY-higgs-electro-weak-strong, SUSY-higgs-electro-weak-gravatic, SUSY-higgs-electro-strong-gravatic, SUSY-higgs-weak-strong, SUSY-higgs-weak-gravatic, SUSY-higgs-weak-strong-gravatic, SUSY-higgs-strong-gravatic, SUSY-higgs-electro-weak-strong-gravatic.

Traveling into pseudo-spatial-temporal realities in the sense of travel into wave-functions of various levels of hyper indices for all of the following.

electric charge, magnetic charge, blue or green or red color charge, upness, down-ness, charmed-ness, strange-ness, topness, bottom-ness, weak force charges, gravitational charge, higgs-boson charge, electromagnetism, weak force, strong force, gravitational force, electro-weak, electro-strong, electrogravatic, electro-weak-strong, electro-weak-gravatic, electro-strong-gravatic, weak-strong, weak-gravatic, weak-strong-gravatic, strong-gravatic, electro-weak-strong-gravatic, higgs-electromagnetism, higgs-weak force, higgs-strong force, higgs-gravitational force, higgs-electro-weak, higgs-electro-strong, higgs-electrogravatic, higgs-electro-weak-strong, higgs-electro-weak-gravatic, higgs-electro-strong-gravatic, higgs-weak-strong, higgs-weak-gravatic, higgs-weak-strong-gravatic, higgs-strong-gravatic, higgs-electro-weak-strong-gravatic, SUSY-electromagnetism, SUSY-weak force, SUSY-strong force, SUSY-gravitational force, SUSY-electro-weak, SUSY-electro-strong, SUSY-electrogravatic, SUSY-electro-weak-strong, SUSY-electro-weak-gravatic, SUSY-electro-strong-gravatic, SUSY-weak-strong, SUSY-weak-gravatic, SUSY-weak-strong-gravatic, SUSY-strong-gravatic, SUSY-electro-weak-strong-gravatic, SUSY-higgs-electromagnetism, SUSY-higgs-weak force, SUSY-higgs-strong force, SUSY-higgs-gravitational force, SUSY-higgs-electro-weak, SUSY-higgs-electro-strong, SUSY-higgs-electrogravatic, SUSY-higgs-electro-weak-strong, SUSY-higgs-electro-weak-gravatic, SUSY-higgs-electro-strong-gravatic, SUSY-higgs-weak-strong, SUSY-higgs-weak-gravatic, SUSY-higgs-weak-strong-gravatic, SUSY-higgs-strong-gravatic, SUSY-higgs-electro-weak-strong-gravatic.

Traveling into pseudo-spatial-temporal realities in the sense of travel into components of wave-functions of; electric charge, magnetic charge, blue or green or red color charge, upness, down-ness, charmed-ness, strange-ness, topness, bottom-ness, weak force charges, gravitational charge, higgs-boson charge, electromagnetism, weak force, strong force, gravitational force, electro-weak, electro-strong, electrogravatic, electro-weak-strong, electro-weak-gravatic, electro-strong-gravatic, weak-strong, weak-gravatic, weak-strong-gravatic, strong-gravatic, electro-weak-strong-gravatic, higgs-electromagnetism, higgs-weak force, higgs-strong force, higgs-gravitational force, higgs-electro-weak, higgs-electro-strong, higgs-electrogravatic, higgs-electro-weak-strong, higgs-electro-weak-gravatic, higgs-electro-strong-gravatic, higgs-weak-strong, higgs-weak-gravatic, higgs-weak-strong-gravatic, higgs-strong-gravatic, higgs-electro-weak-strong-gravatic, SUSY-electromagnetism, SUSY-weak force, SUSY-strong force, SUSY-gravitational force, SUSY-electro-weak, SUSY-electro-strong, SUSY-electrogravatic, SUSY-electro-weak-strong, SUSY-electro-weak-gravatic, SUSY-electro-strong-gravatic, SUSY-weak-strong, SUSY-weak-gravatic, SUSY-weak-strong-gravatic, SUSY-strong-gravatic, SUSY-electro-weak-strong-gravatic, SUSY-higgs-electromagnetism, SUSY-higgs-weak force, SUSY-higgs-strong force, SUSY-higgs-gravitational force, SUSY-higgs-electro-weak, SUSY-higgs-electro-strong, SUSY-higgs-electrogravatic, SUSY-higgs-electro-weak-strong, SUSY-higgs-electro-weak-gravatic, SUSY-higgs-electro-strong-gravatic, SUSY-higgs-weak-strong, SUSY-higgs-weak-gravatic, SUSY-higgs-weak-strong-gravatic, SUSY-higgs-strong-gravatic, SUSY-higgs-electro-weak-strong-gravatic.

Traveling into pseudo-spatial-temporal realities in the sense of travel into quantum hidden variables of various levels of hyper indices for all of the following.

electric charge, magnetic charge, blue or green or red color charge, upness, down-ness, charmed-ness, strange-ness, topness, bottom-ness, weak force charges, gravitational charge, higgs-boson charge, electromagnetism, weak force, strong force, gravitational force, electro-weak, electro-strong, electrogravatic, electro-weak-strong, electro-weak-gravatic, electro-strong-gravatic, weak-strong, weak-gravatic, weak-strong-gravatic, strong-gravatic, electro-weak-strong-gravatic, higgs-electromagnetism, higgs-weak force, higgs-strong force, higgs-gravitational force, higgs-electro-weak, higgs-electro-strong, higgs-electrogravatic, higgs-electro-weak-strong, higgs-electro-weak-gravatic, higgs-electro-strong-gravatic, higgs-weak-strong, higgs-weak-gravatic, higgs-weak-strong-gravatic, higgs-strong-gravatic, higgs-electro-weak-strong-gravatic, SUSY-electromagnetism, SUSY-weak force, SUSY-strong force, SUSY-gravitational force, SUSY-electro-weak, SUSY-electro-strong, SUSY-electrogravatic, SUSY-electro-weak-strong, SUSY-electro-weak-gravatic, SUSY-electro-strong-gravatic, SUSY-weak-strong, SUSY-weak-gravatic, SUSY-weak-strong-gravatic, SUSY-strong-gravatic, SUSY-electro-weak-strong-gravatic, SUSY-higgs-electromagnetism, SUSY-higgs-weak force, SUSY-higgs-strong force, SUSY-higgs-gravitational force, SUSY-higgs-electro-weak, SUSY-higgs-electro-strong, SUSY-higgs-electrogravatic, SUSY-higgs-electro-weak-strong, SUSY-higgs-electro-weak-gravatic, SUSY-higgs-electro-strong-gravatic, SUSY-higgs-weak-strong, SUSY-higgs-weak-gravatic, SUSY-higgs-weak-strong-gravatic, SUSY-higgs-strong-gravatic, SUSY-higgs-electro-weak-strong-gravatic.

We denote the above travel scenarios by the very abstract operator, [[Σ (A few thru variously infinite levels of hyper-indexed):[Σ(z = 1; z = sub-canonical ensemble ↑):(Ƶ,z)]] as such pertains to travel and high energy density materials], which we may affix to formulas for gamma and related parameters.

The upward pointing arrow in the series limit implies open ended boundary values for the number of travel possibilities beyond sub-canonical ensembles. Accordingly, any multiple of individual travel methods conjectured can be included to enable hybrid or compound travel methods.

ESSAY 46) Travel At Ascending Light-Speeds Into Higher Planes Of Physical Reality.

Now we consider some whimsically far-out light-speed travel notions.

Consider the following clauses anticipating extreme manifestations of light-speed travel.

Travel at a first light-speed sufficiently great finite or infinite Lorentz factor as manifesting entering another plane of physical existence and perhaps a first higher plane of physical existence.

Travel at a second light-speed sufficiently great finite or infinite Lorentz factor as manifesting entering a second alternative plane of existence or a second higher plane of existence relative to the one we live in but a first higher plane of existence beyond the first higher plane of existence relative to the one we live in.

Travel at a third light-speed sufficiently great finite or infinite Lorentz factor as manifesting entering a third alternative plane of existence or a third higher plane of existence relative to the one we live in but a first higher plane of existence beyond the second higher plane of existence relative to the one we live in and a second higher plane of existence relative to the first higher plane of existence relative to the one we live in.

Travel at a fourth light-speed sufficiently great finite or infinite Lorentz factor as manifesting entering a fourth alternative plane of existence or a fourth higher plane of existence relative to the one we live in but a first higher plane of existence beyond the third higher plane of existence relative to the one we live in and a second higher plane of existence relative to the second higher plane of existence relative to the one we live in and a third higher plane of existence relative to the first higher plane of existence relative to the one we live in.

The number of levels as such should be infinite and perhaps potentially of an uncountable infinity number of levels.

The notion of higher planes of existence available for spacecraft, or more properly fiats, is derived from the reality that a light-speed spacecraft technically and traditionally would have an infinite Lorentz factor. Thus, the spacecraft would be eternalized with respect to the background of origin and in a sense made relatively immutable with respect to the background of origin.

Even in cases for which light-speed would manifest in finite but obviously huge Lorentz factors, massive objects traveling at light-speed may simply pass through and thus not be present to the background objects in its path. The reason for this is that the material aspects of the craft would not have enough time to interact with the matter they pass through. In a sense, the space craft would arrive at and pass material objects before the background matter and objects would have time to react with the spacecraft. Thus, a finite Lorentz factor light-speed craft would become ephemeral with respect to the background of origin.

We affix the following operator to the length formulas for gamma and related parameters in Section 62 of this book to denote the multi-level planes available to the spacecraft in its travel at light speed.

We can also take the ship-time, gamma, kinetic energy, and acceleration derivatives and integrals of the rate of progression of a spacecraft through elevated physical planes of existences.

{Σ(N = 1; N = ∞↑):{Nth Higher Plane Of Presence And Entrance Of A Light-Speed Spacecraft & Progression Among Such Levels And 1st & Higher Order Differentiation And Integration Thereof With Respect To γ, a, KE, tship}}

The infinity symbol followed by the upwardly pointing arrow indicates open ended infinite boundary conditions and thus implies no limits to the number of levels of higher planes.

ESSAY 47) Spacecraft Quantum Mechanical Entanglement With Light-Speed.

Here, we consider a spacecraft becoming quantum mechanically entangled with the speed of light, the many speeds of light, analogs of the speed of light, analogs of the many speeds of light, speeds of light in alternative first derivatives of location with respect to other temporal or time-like dimensions.

Accordingly, a spacecraft can be so entangled and then teleported quantum mechanically to the speed of light, the many speeds of light, analogs of the speed of light, analogs of the many speeds of light, speeds of light in alternative first derivatives of location with respect to other temporal or time-like dimensions.

The spacecraft may also be entangled with associated specific vectors of the above light-speeds and related constructs thereof.

Additionally, a spacecraft may plausibly be quantum mechanically entangled with and teleported to infinite values of; acceleration, gamma, and kinetic energy.

We may affix the following operator to formulas for gamma and related parameters to denote the above teleportation schemes.

[w(Fiat q-m-e w/ & of spec vectors thereof: c, the many c's, analogs of c, analogs of the many c's, c's in alternative 1st deriv of location w/ respect to other temporal or time-like dimensions & ∞; a's, γ's, and KE's)].

ESSAY 48) Need For New Math Of Infinity To Describe Light-Speed Spacecraft Travel.

Consider the hyper-operator notation that was designed to express huge values not otherwise expressible.

For example, Note that Hyper4(a, n) is equal to a tetrated n or a raised to the power of itself n-1 times. The latter value is symbolically written as n subscript a. For example 3 EXP 4 = 81, but 4 subscript3 is approximately equal to 10 EXP (1,000,000,000,000).

Alternatively 4 subscript 2 = 2 EXP 2 EXP 2 EXP 2 = 2 EXP [2 EXP [2 EXP 2]] = 2 EXP (2 EXP 4) = 2 EXP 16 = 65,536.

For example, Hyper5(4, 4)is equal to 4 tetrated 4 tetrated 4 tetrated 4. This value is commonly referred to as 4 pentated 4.

Hyper 6, (4,4) is 4 pentated 4 pentated 4 pentated 4 and is also referred to as 4 hexataed 4.

Hyper 7, (4,4) is 4 hexated 4 hexated 4 hexated 4 and so-on

Aleph 0 is the infinite number of integers.

Aleph 1 according to the perhaps unprovable and thus unfalsifiable Continuum Hypotheses is the number of real numbers which is greater than Aleph 0 by a multiplicative factor of infinity.

Aleph 2 is similarly greater than Aleph 1.

Aleph 3 is similarly greater than Aleph 2.

Aleph 4 is similarly greater than Aleph 3.

and so-on

In general, Aleph n = 2 EXP [Aleph (n-1)]

The number Ω is commonly stated as the least infinite positive integer or ordinal.

Now here is a real zinger.

So we can produce the abstraction of Hyper Aleph Ω (Aleph Ω, Aleph Ω).

Uncountable infinities are generally larger infinities than countable infinities.

So, you can imagine

Hyper Uncountable Infinity (Aleph Uncountable Infinity, Aleph Uncountable Infinity).

However, we can go a little circular here to formulate,

So we can produce the abstraction of Hyper Aleph Uncountable Infinity (Aleph Uncountable Infinity, Aleph Uncountable Infinity).

We can even delve into notions of super-infinities which loosely can be interpreted as merely infinities albeit on proverbial steroids.

Now, we talk about the finite and the infinite.

A fascinating construct is to conjure eternal forms which are to infinities as infinities are to finite values. Here, we are not simply considering quantitative differences, but instead entire differing constructs at a philosophical level.

It is often said that a need for new math results in new mathematical constructs.

So, in my wide variety of free wheeling speculations on light-speed travel and infinite Lorentz factors, what I have been doing is developing applied infinity mathematics.

Traveling at infinite Lorentz factors will require detailed and mil-spec application of mathematics of the infinite.

Regarding travel at infinite Lorentz factors, time for the spacecraft would be infinitely dilated.

Due to the reality that time on the Planck scale of [5.39 x (10 EXP – 44)] second is nebulous, the Planck Time Unit being the smallest time unit in our universe that seems to be defined, a spacecraft having infinite Lorentz factors may have a nebulous location due to the uncertainty of the spacecraft at the Planck Time scales and the associated uncertainty in spacecraft Lorentz factor.

Accordingly, the spacecraft might enter a state for which it is located ahead of itself and is thus in a valid sense, traveling faster than itself and perhaps ever-so-slightly faster than light. Likewise, the spacecraft may be located slightly ahead of itself in forward time travel.

Additionally, because the spacecraft would have a wave-function, the leading edge of the spacecraft wave-function may extend in a forward manner from the root-mean-square or whatever statistical consideration is appropriate for defining the averaged location of the spacecraft.

We may affix the following operator to formulas for gamma and related parameters to denote the above mentioned applicability of math of the infinite to infinite Lorentz factor spacecraft, the temporally-based nebulous location of the spacecraft in space and time, and the slightly super-luminal travel aspects of the spacecraft due to its continuous wave-function distribution.

[z[(Need for new ∞ math to describe light-speed spacecraft),(Nebulous location of light-speed spacecraft in time and space),(Spatial-temporal spacecraft travel ahead of itself per ∞ γ values)]]

ESSAY 49) The Uncountably Infinite Number Of Ever Greater Infinite Spacecraft Lorentz Factors.

Attaining the speed of light might otherwise seem like the end of a long game or vacation resulting forever afterward in boredom.

However, at light-speed, there are perhaps an uncountable infinite number of ever greater infinite Lorentz factor to attain.

First consider again the following.

For example, that Hyper4(a, n) is equal to a tetrated n or a raised to the power of itself n-1 times. The latter value is symbolically written as n subscript a. For example 3 EXP 4 = 81, but 4 subscript3 is approximately equal to 10 EXP (1,000,000,000,000).

Alternatively 4 subscript 2 = 2 EXP 2 EXP 2 EXP 2 = 2 EXP [2 EXP [2 EXP 2]] = 2 EXP (2 EXP 4) = 2 EXP 16 = 65,536.

For example, Hyper5(4, 4)is equal to 4 tetrated 4 tetrated 4 tetrated 4. This value is commonly referred to as 4 pentated 4.

Hyper 6, (4,4) is 4 pentated 4 pentated 4 pentated 4 and is also referred to as 4 hexataed 4.

Hyper 7, (4,4) is 4 hexated 4 hexated 4 hexated 4 and so-on

Aleph 0 is the infinite number of integers.

Aleph 1 according to the perhaps unprovable and thus unfalsifiable Continuum Hypotheses is the number of real numbers which is greater than Aleph 0 by a multiplicative factor of infinity.

Aleph 2 is similarly greater than Aleph 1.

Aleph 3 is similarly greater than Aleph 2.

Aleph 4 is similarly greater than Aleph 3.

and so-on

In general, Aleph n = 2 EXP [Aleph (n-1)]

The number Ω is commonly stated as the least infinite positive integer or ordinal.

Now here is a real zinger.

So we can produce the abstraction of Hyper Aleph Ω (Aleph Ω, Aleph Ω).

Uncountable infinities are generally larger infinities than countable infinities.

So, you can imagine

Hyper Uncountable Infinity (Aleph Uncountable Infinity, Aleph Uncountable Infinity).

However, we can go a little circular here to formulate,

So we can produce the abstraction of Hyper Aleph Uncountable Infinity (Aleph Uncountable Infinity, Aleph Uncountable Infinity).

We can even delve into notions of super-infinities which loosely can be interpreted as merely infinities albeit on proverbial steroids.

Now, we have talked about the finite and the infinite.

A fascinating construct is to conjure eternal forms which are to infinities as infinities are to finite values. Here, we are not simply considering quantitative differences, but instead entire differing constructs at a philosophical level.

Now, upon reaching the speed of light, we may then travel from one infinite Lorentz factor to another. In a sense, we will have an infinite abstract region to travel in as we progress along the proverbial road to ever greater Lorentz factors.

The rate of increase in Lorentz factors may in itself be viewed as a form of velocity. The ship-time integration of such velocity may be viewed as a form of distance traveled.

There can be no highest countable infinity and even no highest uncountable infinity as we can always add another object to any infinite set even if a duplicate object. So, there can be no limit in theory to how many Lorentz factors are available and perhaps no limit to the velocity of a spacecraft presented as ship-time derivatives of spacecraft Lorentz factors and distance traveled as ship-time integrals of derivatives of gamma with respect to ship time.

So, we have entire new schemata for defining travel velocity and distance.

Moving on to ever greater Lorentz factors may be viewed as embodiments in a set of geometric lines or line segments.

Accordingly, we can in our light-speed increasing Lorentz factor travel draw beautiful landscapes, crystalline, and even fractal patterns.

As another way of consideration, we can in our light-speed increasing Lorentz factor travel produce wonderful amorphous, crystalline, and/or fractal patterns of any abstract dimensionality from just greater than zero to arbitrary uncountable infinities of dimensions.

The same can be said for the rate of increase in spacecraft time derivatives not only of gamma, but also of ship kinetic energy and acceleration. Thus, these rates of increase can be considered as lines or line segments, planes or plane sectors, volumes, curved lines or line segments, curved planes or plane segments, and warped volumes, and the like.

We can also consider travel analogs based on the attainment and rate thereof of novel phenomenon and mathematical representations thereof associated with increasingly infinite Lorentz factors.

Truly, the stories of light-speed travel will never be complete. There will always be ever grander vistas to behold and ever bolder and more beautiful trips and sceneries to behold.

So fasten your seat belts, we are in for a never ending eternal trip.

We may affix the following operator to formulas for gamma and related parameters to denote the above novel travel schemes.

[w(Amorph, crystal, and fractal trips of > 0 thru (uncount ∞)-D-abstract space and as functions of γ, ship-time, KE, a, and novel phenomenon onset, and of derivatives and integrals of these abstract defect growths)]

ESSAY 50) The Infinite Number Of Infra-Light-Speed Boundary Conditions.

We now consider scenarios for which the speed of light and each distinguishable ever so slightly superluminal velocity have intra-boundary step-functions.

For example, upon attaining the velocity of light, we arrive at an infinite Lorentz factor.

Upon pushing the Lorentz factor to greater infinite values, we conjecture that we arrive at another intra-light-speed-boundary-condition.

Upon pushing the Lorentz factor to still greater infinite values, we conjecture that we arrive at another intra-light-speed-boundary-condition.

We conjecture that the number of intra-boundary conditions for light speed and each successive ever so slightly super-luminal boundary condition is large and perhaps boundary condition specific infinite.

We affix the following operator to the lengthy formulas for gamma and related parameters to denote the intra-boundary-condition step-functions which can include merely quantitative conditions or qualitative manifestations.

{ΣG[[(h or k = 1; h or k = ∞↑)]:[hth intra-light-speed-boundary-condition or hth intra-boundary-condition for kth ever so slightly superluminal boundary condition as applicable]]}

ESSAY 51) The Infinite Modes Of And Types Of Created Entities And Created Existence.

Now, it is plausible that there may be different modes of created existence and perhaps even different degrees and types of existence in the set of created entities.

For example, imagine a spacecraft attaining the velocity of light in a least infinite Lorentz factor and traveling for one Planck Time Unit ship frame.

Accordingly, the spacecraft would be eternalized with respect to the background sub-light-speed realms and experience no time dependent interior change relative to the background reference frame.

Now, suppose a spacecraft traveling at the speed of light launches a scout craft that travels at the speed of light in a least infinite Lorentz factor ahead and away from the mothership light-speed craft. The first scout craft would be eternalized relative to the mothership and experience no time dependent interior change relative to the mothership.

Now suppose the first scout ship traveling at the speed of light relative to the mothership launches its own or a second scout craft which then travels at the speed of light in a least infinite Lorentz factor relative to the first scout craft and ahead of the first scout craft. The second scout craft would be eternalized relative to the first scout craft and twice eternalized with respect to the mothership and experience no time dependent interior change relative to the first scout craft and the mothership as well as with respect to the background.

Now suppose the second scout ship traveling at the speed of light relative to the first scout ship launches its own or a third scout craft which then travels at the speed of light in a least infinite Lorentz factor relative to the second scout craft and ahead of the second scout craft. The third scout craft would be eternalized relative to the second scout craft, twice eternalized with respect to the first scout ship and thrice eternalized with respect to the mothership and experience no time dependent interior change relative to the first and second scout craft and the mothership as well as with respect to the background.

Now suppose the third scout ship traveling at the speed of light relative to the second scout ship launches its own or a fourth scout craft which then travels at the speed of light in a least infinite Lorentz factor relative to the third scout craft and ahead of the third scout craft. The fourth scout craft would be eternalized relative to the third scout craft, twice eternalized with respect to the second scout ship and thrice eternalized with respect to the first scout ship, and four-fold eternalized with respect to the mothership and experience no time dependent interior change relative to the first, second, and third scout craft and the mothership as well as with respect to the background.

We can continue the digression forever.

So, in a sense, the number of degrees of existence is unlimited commensurate with an unlimited potential of the numbers of levels of eternalization as more and more scout ships are launched. The scout ships in principle might be fabricated out of dark energy, background Standard Model mass and energy, reified zero-point field energy, energy intaked from hyperspace, and then like.

Since light-speed may be the ultimate speed limit, we humans and any existent extra-terrestrial and ultra-terrestrial friend species we may have now and/or in the future may need to settle into goals of excelling in other ways, the most natural of which would be enabling travel at ever greater infinite Lorentz factors.

Since we humans and the ETIs and UTIs could work together to enable ever greater Lorentz factor capable craft and associated infrastructure and rationally and free-willed informed portions of the cosmos, it is almost as if our progression and drives to obtain ever greater Lorentz factors is also manifest in an intra-cosmic drive of the cosmos to develop itself, grow, and thrive to an ever greater extent.

Some theologians have posited that since life seems to evolve efficiently anywhere it can, raw space-time-mass-energy may validly said to be the urge to live.

So, as we humans, the ETIs and UTIs strive to develop spacecraft that can obtain ever greater finite and then ever greater infinite Lorentz factors, we in a sense are collectively living out our call to become evermore eternalized and to thus achieve higher levels of created existence.

Accordingly, it may be written into the nature of we humans and our ETI and UTI friends to advance ourselves, each other, and the cosmic order in general, in terms of levels of existence and evolutionary development by developing systems that can enable ever greater Lorentz factors.

In a sense, the cosmos as an organism has at least one aspect of reason, purpose, and free-will informed primal drives to make itself evermore temporally connected and extreme.

As we progress toward an ever more eternally distant goal of perhaps ultimate evolution, we will have an entire huge eternity of delights along the way.

We may affix the following operator to formulas for gamma and related parameters to denote the above conjectured cosmic development mechanisms via stacked light-speed reference frame craft.

{w{{(k = 1; k = ∞↑): serially kth light-speed craft rel to pred craft}| craft manif increasing levels of existence through assoc levels of eternalization implying cosmic evolution and growth in a primal urge to develop and live}}

ESSAY 52) Shortcuts In Space-Time.

Now, consider the following lengthy list of clauses that are self-explanatory in terms of introducing the related ideas.

Taking a path in 3 dimensions that enables a shorter route to a destination than does a straight line in Newtonian flat space.

Taking a path in 3 dimensions that enables a shorter route to a destination than does the shortest path in 4-D Einsteinian space-time.

Taking a path in 3 dimensions that enables a shorter route to a destination than does a straight line in Newtonian flat space where the 3-D space is the same space through which said straight line is located.

Taking a path in 3 dimensions that enables a shorter route to a destination than does the shortest path in 4-D Einsteinian space-time where said 3 dimensions are in the same realm as the straightest path in the 4-D Einsteinian space-time.

Taking a path in 3 dimensions that enables a shorter route to a destination than does a straight line in Newtonian flat space but where warped space or space-time such as wormholes, and the like are not applied.

Taking a path in 3 dimensions that enables a shorter route to a destination than does the shortest path in 4-D Einsteinian space-time but where warped space or space-time such as wormholes, and the like are not applied.

Taking a path in 3 dimensions that enables a shorter route to a destination than does a straight line in Newtonian flat space where the 3-D space is the same space through which said straight line is located but where warped space or space-time such as wormholes, and the like are not applied.

Taking a path in 3 dimensions that enables a shorter route to a destination than does the shortest path in 4-D Einsteinian space-time where said 3 dimensions are in the same realm as the straightest path in the 4-D Einsteinian space-time but where warped space or space-time such as wormholes, and the like are not applied.

Taking a path in 3 dimensions that enables a shorter route to a destination than does a straight line in Newtonian flat space where said taken path involves microscopic adjustments of Planck Length scale or smaller scales along a substantially linear said shorter route.

Taking a path in 3 dimensions that enables a shorter route to a destination than does the shortest path in 4-D Einsteinian space-time where said taken path involves microscopic adjustments of Planck Length scale or smaller scales along a substantially linear said shorter route.

Taking a path in 3 dimensions that enables a shorter route to a destination than does a straight line in Newtonian flat space where the 3-D space is the same space through which said straight line is located where said taken path involves microscopic adjustments of Planck Length scale or smaller scales along a substantially linear said shorter route.

Taking a path in 3 dimensions that enables a shorter route to a destination than does the shortest path in 4-D Einsteinian space-time where said 3 dimensions are in the same realm as the straightest path in the 4-D Einsteinian space-time where said taken path involves microscopic adjustments of Planck Length scale or smaller scales along a substantially linear said shorter route.

Taking a path in 3 dimensions that enables a shorter route to a destination than does a straight line in Newtonian flat space but where warped space or space-time such as wormholes, and the like are not applied where said taken path involves microscopic adjustments of Planck Length scale or smaller scales along a substantially linear said shorter route.

Taking a path in 3 dimensions that enables a shorter route to a destination than does the shortest path in 4-D Einsteinian space-time but where warped space or space-time such as wormholes, and the like are not applied where said taken path involves microscopic adjustments of Planck Length scale or smaller scales along a substantially linear said shorter route.

Taking a path in 3 dimensions that enables a shorter route to a destination than does a straight line in Newtonian flat space where the 3-D space is the same space through which said straight line is located but where warped space or space-time such as wormholes, and the like are not applied where said taken path involves microscopic adjustments of Planck Length scale or smaller scales along a substantially linear said shorter route.

Taking a path in 3 dimensions that enables a shorter route to a destination than does the shortest path in 4-D Einsteinian space-time where said 3 dimensions are in the same realm as the straightest path in the 4-D Einsteinian space-time but where warped space or space-time such as wormholes, and the like are not applied where said taken path involves microscopic adjustments of Planck Length scale or smaller scales along a substantially linear said shorter route.

Taking a path in 3 dimensions that enables a shorter route to a destination than does a straight line in Newtonian flat space where said taken path involves microscopic adjustments of Planck Length scale or smaller scales along a substantially non-linear said shorter route.

Taking a path in 3 dimensions that enables a shorter route to a destination than does the shortest path in 4-D Einsteinian space-time where said taken path involves microscopic adjustments of Planck Length scale or smaller scales along a substantially non-linear said shorter route.

Taking a path in 3 dimensions that enables a shorter route to a destination than does a straight line in Newtonian flat space where the 3-D space is the same space through which said straight line is located where said taken path involves microscopic adjustments of Planck Length scale or smaller scales along a substantially non-linear said shorter route.

Taking a path in 3 dimensions that enables a shorter route to a destination than does the shortest path in 4-D Einsteinian space-time where said 3 dimensions are in the same realm as the straightest path in the 4-D Einsteinian space-time where said taken path involves microscopic adjustments of Planck Length scale or smaller scales along a substantially non-linear said shorter route.

Taking a path in 3 dimensions that enables a shorter route to a destination than does a straight line in Newtonian flat space but where warped space or space-time such as wormholes, and the like are not applied where said taken path involves microscopic adjustments of Planck Length scale or smaller scales along a substantially non-linear said shorter route.

Taking a path in 3 dimensions that enables a shorter route to a destination than does the shortest path in 4-D Einsteinian space-time but where warped space or space-time such as wormholes, and the like are not applied where said taken path involves microscopic adjustments of Planck Length scale or smaller scales along a substantially non-linear said shorter route.

Taking a path in 3 dimensions that enables a shorter route to a destination than does a straight line in Newtonian flat space where the 3-D space is the same space through which said straight line is located but where warped space or space-time such as wormholes, and the like are not applied where said taken path involves microscopic adjustments of Planck Length scale or smaller scales along a substantially non-linear said shorter route.

Taking a path in 3 dimensions that enables a shorter route to a destination than does the shortest path in 4-D Einsteinian space-time where said 3 dimensions are in the same realm as the straightest path in the 4-D Einsteinian space-time but where warped space or space-time such as wormholes, and the like are not applied where said taken path involves microscopic adjustments of Planck Length scale or smaller scales along a substantially non-linear said shorter route.

Taking a path in 3 dimensions that enables a shorter route to a destination than does a straight line in Newtonian flat space where said shorter route is of one or more than one very strange shapes.

Taking a path in 3 dimensions that enables a shorter route to a destination than does the shortest path in 4-D Einsteinian space-time where said shorter route is of one or more than one very strange shapes.

Taking a path in 3 dimensions that enables a shorter route to a destination than does a straight line in Newtonian flat space where the 3-D space is the same space through which said straight line is located where said shorter route is of one or more than one very strange shapes.

Taking a path in 3 dimensions that enables a shorter route to a destination than does the shortest path in 4-D Einsteinian space-time where said 3 dimensions are in the same realm as the straightest path in the 4-D Einsteinian space-time where said shorter route is of one or more than one very strange shapes.

Taking a path in 3 dimensions that enables a shorter route to a destination than does a straight line in Newtonian flat space but where warped space or space-time such as wormholes, and the like are not applied where said shorter route is of one or more than one very strange shapes.

Taking a path in 3 dimensions that enables a shorter route to a destination than does the shortest path in 4-D Einsteinian space-time but where warped space or space-time such as wormholes, and the like are not applied where said shorter route is of one or more than one very strange shapes.

Taking a path in 3 dimensions that enables a shorter route to a destination than does a straight line in Newtonian flat space where the 3-D space is the same space through which said straight line is located but where warped space or space-time such as wormholes, and the like are not applied where said shorter route is of one or more than one very strange shapes.

Taking a path in 3 dimensions that enables a shorter route to a destination than does the shortest path in 4-D Einsteinian space-time where said 3 dimensions are in the same realm as the straightest path in the 4-D Einsteinian space-time but where warped space or space-time such as wormholes, and the like are not applied where said shorter route is of one or more than one very strange shapes.

Taking a path in 4 dimensions that enables a shorter route to a destination than does a straight line in Newtonian flat space.

Taking a path in 4 dimensions that enables a shorter route to a destination than does the shortest path in 4-D-Space-1-D-Time.

Taking a path in 4 dimensions that enables a shorter route to a destination than does a straight line in Newtonian flat space where the 4-D space is the same space through which said straight line is located.

Taking a path in 4 dimensions that enables a shorter route to a destination than does the shortest path in 4-D-Space-1-D-Time where said 4 dimensions are in the same realm as the straightest path in the 4-D-Space-1-D-Time.

Taking a path in 4 dimensions that enables a shorter route to a destination than does a straight line in Newtonian flat space but where warped space or space-time such as wormholes, and the like are not applied.

Taking a path in 4 dimensions that enables a shorter route to a destination than does the shortest path in 4-D-Space-1-D-Time but where warped space or space-time such as wormholes, and the like are not applied.

Taking a path in 4 dimensions that enables a shorter route to a destination than does a straight line in Newtonian flat space where the 4-D space is the same space through which said straight line is located but where warped space or space-time such as wormholes, and the like are not applied.

Taking a path in 4 dimensions that enables a shorter route to a destination than does the shortest path in 4-D-Space-1-D-Time where said 4 dimensions are in the same realm as the straightest path in the 4-D-Space-1-D-Time but where warped space or space-time such as wormholes, and the like are not applied.

Taking a path in 4 dimensions that enables a shorter route to a destination than does a straight line in Newtonian flat space where said taken path involves microscopic adjustments of Planck Length scale or smaller scales along a substantially linear said shorter route.

Taking a path in 4 dimensions that enables a shorter route to a destination than does the shortest path in 4-D-Space-1-D-Time where said taken path involves microscopic adjustments of Planck Length scale or smaller scales along a substantially linear said shorter route.

Taking a path in 4 dimensions that enables a shorter route to a destination than does a straight line in Newtonian flat space where the 4-D space is the same space through which said straight line is located where said taken path involves microscopic adjustments of Planck Length scale or smaller scales along a substantially linear said shorter route.

Taking a path in 4 dimensions that enables a shorter route to a destination than does the shortest path in 4-D-Space-1-D-Time where said 4 dimensions are in the same realm as the straightest path in the 4-D-Space-1-D-Time where said taken path involves microscopic adjustments of Planck Length scale or smaller scales along a substantially linear said shorter route.

Taking a path in 4 dimensions that enables a shorter route to a destination than does a straight line in Newtonian flat space but where warped space or space-time such as wormholes, and the like are not applied where said taken path involves microscopic adjustments of Planck Length scale or smaller scales along a substantially linear said shorter route.

Taking a path in 4 dimensions that enables a shorter route to a destination than does the shortest path in 4-D-Space-1-D-Time but where warped space or space-time such as wormholes, and the like are not applied where said taken path involves microscopic adjustments of Planck Length scale or smaller scales along a substantially linear said shorter route.

Taking a path in 4 dimensions that enables a shorter route to a destination than does a straight line in Newtonian flat space where the 4-D space is the same space through which said straight line is located but where warped space or space-time such as wormholes, and the like are not applied where said taken path involves microscopic adjustments of Planck Length scale or smaller scales along a substantially linear said shorter route.

Taking a path in 4 dimensions that enables a shorter route to a destination than does the shortest path in 4-D-Space-1-D-Time where said 4 dimensions are in the same realm as the straightest path in the 4-D-Space-1-D-Time but where warped space or space-time such as wormholes, and the like are not applied where said taken path involves microscopic adjustments of Planck Length scale or smaller scales along a substantially linear said shorter route.

Taking a path in 4 dimensions that enables a shorter route to a destination than does a straight line in Newtonian flat space where said taken path involves microscopic adjustments of Planck Length scale or smaller scales along a substantially non-linear said shorter route.

Taking a path in 4 dimensions that enables a shorter route to a destination than does the shortest path in 4-D-Space-1-D-Time where said taken path involves microscopic adjustments of Planck Length scale or smaller scales along a substantially non-linear said shorter route.

Taking a path in 4 dimensions that enables a shorter route to a destination than does a straight line in Newtonian flat space where the 4-D space is the same space through which said straight line is located where said taken path involves microscopic adjustments of Planck Length scale or smaller scales along a substantially non-linear said shorter route.

Taking a path in 4 dimensions that enables a shorter route to a destination than does the shortest path in 4-D-Space-1-D-Time where said 4 dimensions are in the same realm as the straightest path in the 4-D-Space-1-D-Time where said taken path involves microscopic adjustments of Planck Length scale or smaller scales along a substantially non-linear said shorter route.

Taking a path in 4 dimensions that enables a shorter route to a destination than does a straight line in Newtonian flat space but where warped space or space-time such as wormholes, and the like are not applied where said taken path involves microscopic adjustments of Planck Length scale or smaller scales along a substantially non-linear said shorter route.

Taking a path in 4 dimensions that enables a shorter route to a destination than does the shortest path in 4-D-Space-1-D-Time but where warped space or space-time such as wormholes, and the like are not applied where said taken path involves microscopic adjustments of Planck Length scale or smaller scales along a substantially non-linear said shorter route.

Taking a path in 4 dimensions that enables a shorter route to a destination than does a straight line in Newtonian flat space where the 4-D space is the same space through which said straight line is located but where warped space or space-time such as wormholes, and the like are not applied where said taken path involves microscopic adjustments of Planck Length scale or smaller scales along a substantially non-linear said shorter route.

Taking a path in 4 dimensions that enables a shorter route to a destination than does the shortest path in 4-D-Space-1-D-Time where said 4 dimensions are in the same realm as the straightest path in the 4-D-Space-1-D-Time but where warped space or space-time such as wormholes, and the like are not applied where said taken path involves microscopic adjustments of Planck Length scale or smaller scales along a substantially non-linear said shorter route.

Taking a path in 4 dimensions that enables a shorter route to a destination than does a straight line in Newtonian flat space where said shorter route is of one or more than one very strange shapes.

Taking a path in 4 dimensions that enables a shorter route to a destination than does the shortest path in 4-D-Space-1-D-Time where said shorter route is of one or more than one very strange shapes.

Taking a path in 4 dimensions that enables a shorter route to a destination than does a straight line in Newtonian flat space where the 4-D space is the same space through which said straight line is located where said shorter route is of one or more than one very strange shapes.

Taking a path in 4 dimensions that enables a shorter route to a destination than does the shortest path in 4-D-Space-1-D-Time where said 4 dimensions are in the same realm as the straightest path in the 4-D-Space-1-D-Time where said shorter route is of one or more than one very strange shapes.

Taking a path in 4 dimensions that enables a shorter route to a destination than does a straight line in Newtonian flat space but where warped space or space-time such as wormholes, and the like are not applied where said shorter route is of one or more than one very strange shapes.

Taking a path in 4 dimensions that enables a shorter route to a destination than does the shortest path in 4-D-Space-1-D-Time but where warped space or space-time such as wormholes, and the like are not applied where said shorter route is of one or more than one very strange shapes.

Taking a path in 4 dimensions that enables a shorter route to a destination than does a straight line in Newtonian flat space where the 4-D space is the same space through which said straight line is located but where warped space or space-time such as wormholes, and the like are not applied where said shorter route is of one or more than one very strange shapes.

Taking a path in 4 dimensions that enables a shorter route to a destination than does the shortest path in 4-D-Space-1-D-Time where said 4 dimensions are in the same realm as the straightest path in the 4-D-Space-1-D-Time but where warped space or space-time such as wormholes, and the like are not applied where said shorter route is of one or more than one very strange shapes.

Taking a path in 5 dimensions that enables a shorter route to a destination than does a straight line in Newtonian flat space.

Taking a path in 5 dimensions that enables a shorter route to a destination than does the shortest path in 5-D-Space-1-D-Time.

Taking a path in 5 dimensions that enables a shorter route to a destination than does a straight line in Newtonian flat space where the 5-D space is the same space through which said straight line is located.

Taking a path in 5 dimensions that enables a shorter route to a destination than does the shortest path in 5-D-Space-1-D-Time where said 5 dimensions are in the same realm as the straightest path in the 5-D-Space-1-D-Time.

Taking a path in 5 dimensions that enables a shorter route to a destination than does a straight line in Newtonian flat space but where warped space or space-time such as wormholes, and the like are not applied.

Taking a path in 5 dimensions that enables a shorter route to a destination than does the shortest path in 5-D-Space-1-D-Time but where warped space or space-time such as wormholes, and the like are not applied.

Taking a path in 5 dimensions that enables a shorter route to a destination than does a straight line in Newtonian flat space where the 5-D space is the same space through which said straight line is located but where warped space or space-time such as wormholes, and the like are not applied.

Taking a path in 5 dimensions that enables a shorter route to a destination than does the shortest path in 5-D-Space-1-D-Time where said 5 dimensions are in the same realm as the straightest path in the 5-D-Space-1-D-Time but where warped space or space-time such as wormholes, and the like are not applied.

Taking a path in 5 dimensions that enables a shorter route to a destination than does a straight line in Newtonian flat space where said taken path involves microscopic adjustments of Planck Length scale or smaller scales along a substantially linear said shorter route.

Taking a path in 5 dimensions that enables a shorter route to a destination than does the shortest path in 5-D-Space-1-D-Time where said taken path involves microscopic adjustments of Planck Length scale or smaller scales along a substantially linear said shorter route.

Taking a path in 5 dimensions that enables a shorter route to a destination than does a straight line in Newtonian flat space where the 5-D space is the same space through which said straight line is located where said taken path involves microscopic adjustments of Planck Length scale or smaller scales along a substantially linear said shorter route.

Taking a path in 5 dimensions that enables a shorter route to a destination than does the shortest path in 5-D-Space-1-D-Time where said 5 dimensions are in the same realm as the straightest path in the 5-D-Space-1-D-Time where said taken path involves microscopic adjustments of Planck Length scale or smaller scales along a substantially linear said shorter route.

Taking a path in 5 dimensions that enables a shorter route to a destination than does a straight line in Newtonian flat space but where warped space or space-time such as wormholes, and the like are not applied where said taken path involves microscopic adjustments of Planck Length scale or smaller scales along a substantially linear said shorter route.

Taking a path in 5 dimensions that enables a shorter route to a destination than does the shortest path in 5-D-Space-1-D-Time but where warped space or space-time such as wormholes, and the like are not applied where said taken path involves microscopic adjustments of Planck Length scale or smaller scales along a substantially linear said shorter route.

Taking a path in 5 dimensions that enables a shorter route to a destination than does a straight line in Newtonian flat space where the 5-D space is the same space through which said straight line is located but where warped space or space-time such as wormholes, and the like are not applied where said taken path involves microscopic adjustments of Planck Length scale or smaller scales along a substantially linear said shorter route.

Taking a path in 5 dimensions that enables a shorter route to a destination than does the shortest path in 5-D-Space-1-D-Time where said 5 dimensions are in the same realm as the straightest path in the 5-D-Space-1-D-Time but where warped space or space-time such as wormholes, and the like are not applied where said taken path involves microscopic adjustments of Planck Length scale or smaller scales along a substantially linear said shorter route.

Taking a path in 5 dimensions that enables a shorter route to a destination than does a straight line in Newtonian flat space where said taken path involves microscopic adjustments of Planck Length scale or smaller scales along a substantially non-linear said shorter route.

Taking a path in 5 dimensions that enables a shorter route to a destination than does the shortest path in 5-D-Space-1-D-Time where said taken path involves microscopic adjustments of Planck Length scale or smaller scales along a substantially non-linear said shorter route.

Taking a path in 5 dimensions that enables a shorter route to a destination than does a straight line in Newtonian flat space where the 5-D space is the same space through which said straight line is located where said taken path involves microscopic adjustments of Planck Length scale or smaller scales along a substantially non-linear said shorter route.

Taking a path in 5 dimensions that enables a shorter route to a destination than does the shortest path in 5-D-Space-1-D-Time where said 5 dimensions are in the same realm as the straightest path in the 5-D-Space-1-D-Time where said taken path involves microscopic adjustments of Planck Length scale or smaller scales along a substantially non-linear said shorter route.

Taking a path in 5 dimensions that enables a shorter route to a destination than does a straight line in Newtonian flat space but where warped space or space-time such as wormholes, and the like are not applied where said taken path involves microscopic adjustments of Planck Length scale or smaller scales along a substantially non-linear said shorter route.

Taking a path in 5 dimensions that enables a shorter route to a destination than does the shortest path in 5-D-Space-1-D-Time but where warped space or space-time such as wormholes, and the like are not applied where said taken path involves microscopic adjustments of Planck Length scale or smaller scales along a substantially non-linear said shorter route.

Taking a path in 5 dimensions that enables a shorter route to a destination than does a straight line in Newtonian flat space where the 5-D space is the same space through which said straight line is located but where warped space or space-time such as wormholes, and the like are not applied where said taken path involves microscopic adjustments of Planck Length scale or smaller scales along a substantially non-linear said shorter route.

Taking a path in 5 dimensions that enables a shorter route to a destination than does the shortest path in 5-D-Space-1-D-Time where said 5 dimensions are in the same realm as the straightest path in the 5-D-Space-1-D-Time but where warped space or space-time such as wormholes, and the like are not applied where said taken path involves microscopic adjustments of Planck Length scale or smaller scales along a substantially non-linear said shorter route.

Taking a path in 5 dimensions that enables a shorter route to a destination than does a straight line in Newtonian flat space where said shorter route is of one or more than one very strange shapes.

Taking a path in 5 dimensions that enables a shorter route to a destination than does the shortest path in 5-D-Space-1-D-Time where said shorter route is of one or more than one very strange shapes.

Taking a path in 5 dimensions that enables a shorter route to a destination than does a straight line in Newtonian flat space where the 5-D space is the same space through which said straight line is located where said shorter route is of one or more than one very strange shapes.

Taking a path in 5 dimensions that enables a shorter route to a destination than does the shortest path in 5-D-Space-1-D-Time where said 5 dimensions are in the same realm as the straightest path in the 5-D-Space-1-D-Time where said shorter route is of one or more than one very strange shapes.

Taking a path in 5 dimensions that enables a shorter route to a destination than does a straight line in Newtonian flat space but where warped space or space-time such as wormholes, and the like are not applied where said shorter route is of one or more than one very strange shapes.

Taking a path in 5 dimensions that enables a shorter route to a destination than does the shortest path in 5-D-Space-1-D-Time but where warped space or space-time such as wormholes, and the like are not applied where said shorter route is of one or more than one very strange shapes.

Taking a path in 5 dimensions that enables a shorter route to a destination than does a straight line in Newtonian flat space where the 5-D space is the same space through which said straight line is located but where warped space or space-time such as wormholes, and the like are not applied where said shorter route is of one or more than one very strange shapes.

Taking a path in 5 dimensions that enables a shorter route to a destination than does the shortest path in 5-D-Space-1-D-Time where said 5 dimensions are in the same realm as the straightest path in the 5-D-Space-1-D-Time but where warped space or space-time such as wormholes, and the like are not applied where said shorter route is of one or more than one very strange shapes.

Taking a path in 6 dimensions that enables a shorter route to a destination than does a straight line in Newtonian flat space.

Taking a path in 6 dimensions that enables a shorter route to a destination than does the shortest path in 6-D-Space-1-D-Time.

Taking a path in 6 dimensions that enables a shorter route to a destination than does a straight line in Newtonian flat space where the 6-D space is the same space through which said straight line is located.

Taking a path in 6 dimensions that enables a shorter route to a destination than does the shortest path in 6-D-Space-1-D-Time where said 6 dimensions are in the same realm as the straightest path in the 6-D-Space-1-D-Time.

Taking a path in 6 dimensions that enables a shorter route to a destination than does a straight line in Newtonian flat space but where warped space or space-time such as wormholes, and the like are not applied.

Taking a path in 6 dimensions that enables a shorter route to a destination than does the shortest path in 6-D-Space-1-D-Time but where warped space or space-time such as wormholes, and the like are not applied.

Taking a path in 6 dimensions that enables a shorter route to a destination than does a straight line in Newtonian flat space where the 6-D space is the same space through which said straight line is located but where warped space or space-time such as wormholes, and the like are not applied.

Taking a path in 6 dimensions that enables a shorter route to a destination than does the shortest path in 6-D-Space-1-D-Time where said 6 dimensions are in the same realm as the straightest path in the 6-D-Space-1-D-Time but where warped space or space-time such as wormholes, and the like are not applied.

Taking a path in 6 dimensions that enables a shorter route to a destination than does a straight line in Newtonian flat space where said taken path involves microscopic adjustments of Planck Length scale or smaller scales along a substantially linear said shorter route.

Taking a path in 6 dimensions that enables a shorter route to a destination than does the shortest path in 6-D-Space-1-D-Time where said taken path involves microscopic adjustments of Planck Length scale or smaller scales along a substantially linear said shorter route.

Taking a path in 6 dimensions that enables a shorter route to a destination than does a straight line in Newtonian flat space where the 6-D space is the same space through which said straight line is located where said taken path involves microscopic adjustments of Planck Length scale or smaller scales along a substantially linear said shorter route.

Taking a path in 6 dimensions that enables a shorter route to a destination than does the shortest path in 6-D-Space-1-D-Time where said 6 dimensions are in the same realm as the straightest path in the 6-D-Space-1-D-Time where said taken path involves microscopic adjustments of Planck Length scale or smaller scales along a substantially linear said shorter route.

Taking a path in 6 dimensions that enables a shorter route to a destination than does a straight line in Newtonian flat space but where warped space or space-time such as wormholes, and the like are not applied where said taken path involves microscopic adjustments of Planck Length scale or smaller scales along a substantially linear said shorter route.

Taking a path in 6 dimensions that enables a shorter route to a destination than does the shortest path in 6-D-Space-1-D-Time but where warped space or space-time such as wormholes, and the like are not applied where said taken path involves microscopic adjustments of Planck Length scale or smaller scales along a substantially linear said shorter route.

Taking a path in 6 dimensions that enables a shorter route to a destination than does a straight line in Newtonian flat space where the 6-D space is the same space through which said straight line is located but where warped space or space-time such as wormholes, and the like are not applied where said taken path involves microscopic adjustments of Planck Length scale or smaller scales along a substantially linear said shorter route.

Taking a path in 6 dimensions that enables a shorter route to a destination than does the shortest path in 6-D-Space-1-D-Time where said 6 dimensions are in the same realm as the straightest path in the 6-D-Space-1-D-Time but where warped space or space-time such as wormholes, and the like are not applied where said taken path involves microscopic adjustments of Planck Length scale or smaller scales along a substantially linear said shorter route.

Taking a path in 6 dimensions that enables a shorter route to a destination than does a straight line in Newtonian flat space where said taken path involves microscopic adjustments of Planck Length scale or smaller scales along a substantially non-linear said shorter route.

Taking a path in 6 dimensions that enables a shorter route to a destination than does the shortest path in 6-D-Space-1-D-Time where said taken path involves microscopic adjustments of Planck Length scale or smaller scales along a substantially non-linear said shorter route.

Taking a path in 6 dimensions that enables a shorter route to a destination than does a straight line in Newtonian flat space where the 6-D space is the same space through which said straight line is located where said taken path involves microscopic adjustments of Planck Length scale or smaller scales along a substantially non-linear said shorter route.

Taking a path in 6 dimensions that enables a shorter route to a destination than does the shortest path in 6-D-Space-1-D-Time where said 6 dimensions are in the same realm as the straightest path in the 6-D-Space-1-D-Time where said taken path involves microscopic adjustments of Planck Length scale or smaller scales along a substantially non-linear said shorter route.

Taking a path in 6 dimensions that enables a shorter route to a destination than does a straight line in Newtonian flat space but where warped space or space-time such as wormholes, and the like are not applied where said taken path involves microscopic adjustments of Planck Length scale or smaller scales along a substantially non-linear said shorter route.

Taking a path in 6 dimensions that enables a shorter route to a destination than does the shortest path in 6-D-Space-1-D-Time but where warped space or space-time such as wormholes, and the like are not applied where said taken path involves microscopic adjustments of Planck Length scale or smaller scales along a substantially non-linear said shorter route.

Taking a path in 6 dimensions that enables a shorter route to a destination than does a straight line in Newtonian flat space where the 6-D space is the same space through which said straight line is located but where warped space or space-time such as wormholes, and the like are not applied where said taken path involves microscopic adjustments of Planck Length scale or smaller scales along a substantially non-linear said shorter route.

Taking a path in 6 dimensions that enables a shorter route to a destination than does the shortest path in 6-D-Space-1-D-Time where said 6 dimensions are in the same realm as the straightest path in the 6-D-Space-1-D-Time but where warped space or space-time such as wormholes, and the like are not applied where said taken path involves microscopic adjustments of Planck Length scale or smaller scales along a substantially non-linear said shorter route.

Taking a path in 6 dimensions that enables a shorter route to a destination than does a straight line in Newtonian flat space where said shorter route is of one or more than one very strange shapes.

Taking a path in 6 dimensions that enables a shorter route to a destination than does the shortest path in 6-D-Space-1-D-Time where said shorter route is of one or more than one very strange shapes.

Taking a path in 6 dimensions that enables a shorter route to a destination than does a straight line in Newtonian flat space where the 6-D space is the same space through which said straight line is located where said shorter route is of one or more than one very strange shapes.

Taking a path in 6 dimensions that enables a shorter route to a destination than does the shortest path in 6-D-Space-1-D-Time where said 6 dimensions are in the same realm as the straightest path in the 6-D-Space-1-D-Time where said shorter route is of one or more than one very strange shapes.

Taking a path in 6 dimensions that enables a shorter route to a destination than does a straight line in Newtonian flat space but where warped space or space-time such as wormholes, and the like are not applied where said shorter route is of one or more than one very strange shapes.

Taking a path in 6 dimensions that enables a shorter route to a destination than does the shortest path in 6-D-Space-1-D-Time but where warped space or space-time such as wormholes, and the like are not applied where said shorter route is of one or more than one very strange shapes.

Taking a path in 6 dimensions that enables a shorter route to a destination than does a straight line in Newtonian flat space where the 6-D space is the same space through which said straight line is located but where warped space or space-time such as wormholes, and the like are not applied where said shorter route is of one or more than one very strange shapes.

Taking a path in 6 dimensions that enables a shorter route to a destination than does the shortest path in 6-D-Space-1-D-Time where said 6 dimensions are in the same realm as the straightest path in the 6-D-Space-1-D-Time but where warped space or space-time such as wormholes, and the like are not applied where said shorter route is of one or more than one very strange shapes.

Taking a path in N dimensions that enables a shorter route to a destination than does a straight line in Newtonian flat space.

Taking a path in N dimensions that enables a shorter route to a destination than does the shortest path in N-D-Space-1-D-Time.

Taking a path in N dimensions that enables a shorter route to a destination than does a straight line in Newtonian flat space where the N-D space is the same space through which said straight line is located.

Taking a path in N dimensions that enables a shorter route to a destination than does the shortest path in N-D-Space-1-D-Time where said N dimensions are in the same realm as the straightest path in the N-D-Space-1-D-Time.

Taking a path in N dimensions that enables a shorter route to a destination than does a straight line in Newtonian flat space but where warped space or space-time such as wormholes, and the like are not applied.

Taking a path in N dimensions that enables a shorter route to a destination than does the shortest path in N-D-Space-1-D-Time but where warped space or space-time such as wormholes, and the like are not applied.

Taking a path in N dimensions that enables a shorter route to a destination than does a straight line in Newtonian flat space where the N-D space is the same space through which said straight line is located but where warped space or space-time such as wormholes, and the like are not applied.

Taking a path in N dimensions that enables a shorter route to a destination than does the shortest path in N-D-Space-1-D-Time where said N dimensions are in the same realm as the straightest path in the N-D-Space-1-D-Time but where warped space or space-time such as wormholes, and the like are not applied.

Taking a path in N dimensions that enables a shorter route to a destination than does a straight line in Newtonian flat space where said taken path involves microscopic adjustments of Planck Length scale or smaller scales along a substantially linear said shorter route.

Taking a path in N dimensions that enables a shorter route to a destination than does the shortest path in N-D-Space-1-D-Time where said taken path involves microscopic adjustments of Planck Length scale or smaller scales along a substantially linear said shorter route.

Taking a path in N dimensions that enables a shorter route to a destination than does a straight line in Newtonian flat space where the N-D space is the same space through which said straight line is located where said taken path involves microscopic adjustments of Planck Length scale or smaller scales along a substantially linear said shorter route.

Taking a path in N dimensions that enables a shorter route to a destination than does the shortest path in N-D-Space-1-D-Time where said N dimensions are in the same realm as the straightest path in the N-D-Space-1-D-Time where said taken path involves microscopic adjustments of Planck Length scale or smaller scales along a substantially linear said shorter route.

Taking a path in N dimensions that enables a shorter route to a destination than does a straight line in Newtonian flat space but where warped space or space-time such as wormholes, and the like are not applied where said taken path involves microscopic adjustments of Planck Length scale or smaller scales along a substantially linear said shorter route.

Taking a path in N dimensions that enables a shorter route to a destination than does the shortest path in N-D-Space-1-D-Time but where warped space or space-time such as wormholes, and the like are not applied where said taken path involves microscopic adjustments of Planck Length scale or smaller scales along a substantially linear said shorter route.

Taking a path in N dimensions that enables a shorter route to a destination than does a straight line in Newtonian flat space where the N-D space is the same space through which said straight line is located but where warped space or space-time such as wormholes, and the like are not applied where said taken path involves microscopic adjustments of Planck Length scale or smaller scales along a substantially linear said shorter route.

Taking a path in N dimensions that enables a shorter route to a destination than does the shortest path in N-D-Space-1-D-Time where said N dimensions are in the same realm as the straightest path in the N-D-Space-1-D-Time but where warped space or space-time such as wormholes, and the like are not applied where said taken path involves microscopic adjustments of Planck Length scale or smaller scales along a substantially linear said shorter route.

Taking a path in N dimensions that enables a shorter route to a destination than does a straight line in Newtonian flat space where said taken path involves microscopic adjustments of Planck Length scale or smaller scales along a substantially non-linear said shorter route.

Taking a path in N dimensions that enables a shorter route to a destination than does the shortest path in N-D-Space-1-D-Time where said taken path involves microscopic adjustments of Planck Length scale or smaller scales along a substantially non-linear said shorter route.

Taking a path in N dimensions that enables a shorter route to a destination than does a straight line in Newtonian flat space where the N-D space is the same space through which said straight line is located where said taken path involves microscopic adjustments of Planck Length scale or smaller scales along a substantially non-linear said shorter route.

Taking a path in N dimensions that enables a shorter route to a destination than does the shortest path in N-D-Space-1-D-Time where said N dimensions are in the same realm as the straightest path in the N-D-Space-1-D-Time where said taken path involves microscopic adjustments of Planck Length scale or smaller scales along a substantially non-linear said shorter route.

Taking a path in N dimensions that enables a shorter route to a destination than does a straight line in Newtonian flat space but where warped space or space-time such as wormholes, and the like are not applied where said taken path involves microscopic adjustments of Planck Length scale or smaller scales along a substantially non-linear said shorter route.

Taking a path in N dimensions that enables a shorter route to a destination than does the shortest path in N-D-Space-1-D-Time but where warped space or space-time such as wormholes, and the like are not applied where said taken path involves microscopic adjustments of Planck Length scale or smaller scales along a substantially non-linear said shorter route.

Taking a path in N dimensions that enables a shorter route to a destination than does a straight line in Newtonian flat space where the N-D space is the same space through which said straight line is located but where warped space or space-time such as wormholes, and the like are not applied where said taken path involves microscopic adjustments of Planck Length scale or smaller scales along a substantially non-linear said shorter route.

Taking a path in N dimensions that enables a shorter route to a destination than does the shortest path in N-D-Space-1-D-Time where said N dimensions are in the same realm as the straightest path in the N-D-Space-1-D-Time but where warped space or space-time such as wormholes, and the like are not applied where said taken path involves microscopic adjustments of Planck Length scale or smaller scales along a substantially non-linear said shorter route.

Taking a path in N dimensions that enables a shorter route to a destination than does a straight line in Newtonian flat space where said shorter route is of one or more than one very strange shapes.

Taking a path in N dimensions that enables a shorter route to a destination than does the shortest path in N-D-Space-1-D-Time where said shorter route is of one or more than one very strange shapes.

Taking a path in N dimensions that enables a shorter route to a destination than does a straight line in Newtonian flat space where the N-D space is the same space through which said straight line is located where said shorter route is of one or more than one very strange shapes.

Taking a path in N dimensions that enables a shorter route to a destination than does the shortest path in N-D-Space-1-D-Time where said N dimensions are in the same realm as the straightest path in the N-D-Space-1-D-Time where said shorter route is of one or more than one very strange shapes.

Taking a path in N dimensions that enables a shorter route to a destination than does a straight line in Newtonian flat space but where warped space or space-time such as wormholes, and the like are not applied where said shorter route is of one or more than one very strange shapes.

Taking a path in N dimensions that enables a shorter route to a destination than does the shortest path in N-D-Space-1-D-Time but where warped space or space-time such as wormholes, and the like are not applied where said shorter route is of one or more than one very strange shapes.

Taking a path in N dimensions that enables a shorter route to a destination than does a straight line in Newtonian flat space where the N-D space is the same space through which said straight line is located but where warped space or space-time such as wormholes, and the like are not applied where said shorter route is of one or more than one very strange shapes.

Taking a path in N dimensions that enables a shorter route to a destination than does the shortest path in N-D-Space-1-D-Time where said N dimensions are in the same realm as the straightest path in the N-D-Space-1-D-Time but where warped space or space-time such as wormholes, and the like are not applied where said shorter route is of one or more than one very strange shapes.

Taking a path in N dimensions that enables a shorter route to a destination than does a straight line in Newtonian flat space.

Taking a path in N dimensions that enables a shorter route to a destination than does the shortest path in N-D-Space-M-D-Time.

Taking a path in N dimensions that enables a shorter route to a destination than does a straight line in Newtonian flat space where the N-D space is the same space through which said straight line is located.

Taking a path in N dimensions that enables a shorter route to a destination than does the shortest path in N-D-Space-M-D-Time where said N dimensions are in the same realm as the straightest path in the N-D-Space-M-D-Time.

Taking a path in N dimensions that enables a shorter route to a destination than does a straight line in Newtonian flat space but where warped space or space-time such as wormholes, and the like are not applied.

Taking a path in N dimensions that enables a shorter route to a destination than does the shortest path in N-D-Space-M-D-Time but where warped space or space-time such as wormholes, and the like are not applied.

Taking a path in N dimensions that enables a shorter route to a destination than does a straight line in Newtonian flat space where the N-D space is the same space through which said straight line is located but where warped space or space-time such as wormholes, and the like are not applied.

Taking a path in N dimensions that enables a shorter route to a destination than does the shortest path in N-D-Space-M-D-Time where said N dimensions are in the same realm as the straightest path in the N-D-Space-M-D-Time but where warped space or space-time such as wormholes, and the like are not applied.

Taking a path in N dimensions that enables a shorter route to a destination than does a straight line in Newtonian flat space where said taken path involves microscopic adjustments of Planck Length scale or smaller scales along a substantially linear said shorter route.

Taking a path in N dimensions that enables a shorter route to a destination than does the shortest path in N-D-Space-M-D-Time where said taken path involves microscopic adjustments of Planck Length scale or smaller scales along a substantially linear said shorter route.

Taking a path in N dimensions that enables a shorter route to a destination than does a straight line in Newtonian flat space where the N-D space is the same space through which said straight line is located where said taken path involves microscopic adjustments of Planck Length scale or smaller scales along a substantially linear said shorter route.

Taking a path in N dimensions that enables a shorter route to a destination than does the shortest path in N-D-Space-M-D-Time where said N dimensions are in the same realm as the straightest path in the N-D-Space-M-D-Time where said taken path involves microscopic adjustments of Planck Length scale or smaller scales along a substantially linear said shorter route.

Taking a path in N dimensions that enables a shorter route to a destination than does a straight line in Newtonian flat space but where warped space or space-time such as wormholes, and the like are not applied where said taken path involves microscopic adjustments of Planck Length scale or smaller scales along a substantially linear said shorter route.

Taking a path in N dimensions that enables a shorter route to a destination than does the shortest path in N-D-Space-M-D-Time but where warped space or space-time such as wormholes, and the like are not applied where said taken path involves microscopic adjustments of Planck Length scale or smaller scales along a substantially linear said shorter route.

Taking a path in N dimensions that enables a shorter route to a destination than does a straight line in Newtonian flat space where the N-D space is the same space through which said straight line is located but where warped space or space-time such as wormholes, and the like are not applied where said taken path involves microscopic adjustments of Planck Length scale or smaller scales along a substantially linear said shorter route.

Taking a path in N dimensions that enables a shorter route to a destination than does the shortest path in N-D-Space-M-D-Time where said N dimensions are in the same realm as the straightest path in the N-D-Space-M-D-Time but where warped space or space-time such as wormholes, and the like are not applied where said taken path involves microscopic adjustments of Planck Length scale or smaller scales along a substantially linear said shorter route.

Taking a path in N dimensions that enables a shorter route to a destination than does a straight line in Newtonian flat space where said taken path involves microscopic adjustments of Planck Length scale or smaller scales along a substantially non-linear said shorter route.

Taking a path in N dimensions that enables a shorter route to a destination than does the shortest path in N-D-Space-M-D-Time where said taken path involves microscopic adjustments of Planck Length scale or smaller scales along a substantially non-linear said shorter route.

Taking a path in N dimensions that enables a shorter route to a destination than does a straight line in Newtonian flat space where the N-D space is the same space through which said straight line is located where said taken path involves microscopic adjustments of Planck Length scale or smaller scales along a substantially non-linear said shorter route.

Taking a path in N dimensions that enables a shorter route to a destination than does the shortest path in N-D-Space-M-D-Time where said N dimensions are in the same realm as the straightest path in the N-D-Space-M-D-Time where said taken path involves microscopic adjustments of Planck Length scale or smaller scales along a substantially non-linear said shorter route.

Taking a path in N dimensions that enables a shorter route to a destination than does a straight line in Newtonian flat space but where warped space or space-time such as wormholes, and the like are not applied where said taken path involves microscopic adjustments of Planck Length scale or smaller scales along a substantially non-linear said shorter route.

Taking a path in N dimensions that enables a shorter route to a destination than does the shortest path in N-D-Space-M-D-Time but where warped space or space-time such as wormholes, and the like are not applied where said taken path involves microscopic adjustments of Planck Length scale or smaller scales along a substantially non-linear said shorter route.

Taking a path in N dimensions that enables a shorter route to a destination than does a straight line in Newtonian flat space where the N-D space is the same space through which said straight line is located but where warped space or space-time such as wormholes, and the like are not applied where said taken path involves microscopic adjustments of Planck Length scale or smaller scales along a substantially non-linear said shorter route.

Taking a path in N dimensions that enables a shorter route to a destination than does the shortest path in N-D-Space-M-D-Time where said N dimensions are in the same realm as the straightest path in the N-D-Space-M-D-Time but where warped space or space-time such as wormholes, and the like are not applied where said taken path involves microscopic adjustments of Planck Length scale or smaller scales along a substantially non-linear said shorter route.

Taking a path in N dimensions that enables a shorter route to a destination than does a straight line in Newtonian flat space where said shorter route is of one or more than one very strange shapes.

Taking a path in N dimensions that enables a shorter route to a destination than does the shortest path in N-D-Space-M-D-Time where said shorter route is of one or more than one very strange shapes.

Taking a path in N dimensions that enables a shorter route to a destination than does a straight line in Newtonian flat space where the N-D space is the same space through which said straight line is located where said shorter route is of one or more than one very strange shapes.

Taking a path in N dimensions that enables a shorter route to a destination than does the shortest path in N-D-Space-M-D-Time where said N dimensions are in the same realm as the straightest path in the N-D-Space-M-D-Time where said shorter route is of one or more than one very strange shapes.

Taking a path in N dimensions that enables a shorter route to a destination than does a straight line in Newtonian flat space but where warped space or space-time such as wormholes, and the like are not applied where said shorter route is of one or more than one very strange shapes.

Taking a path in N dimensions that enables a shorter route to a destination than does the shortest path in N-D-Space-M-D-Time but where warped space or space-time such as wormholes, and the like are not applied where said shorter route is of one or more than one very strange shapes.

Taking a path in N dimensions that enables a shorter route to a destination than does a straight line in Newtonian flat space where the N-D space is the same space through which said straight line is located but where warped space or space-time such as wormholes, and the like are not applied where said shorter route is of one or more than one very strange shapes.

Taking a path in N dimensions that enables a shorter route to a destination than does the shortest path in N-D-Space-M-D-Time where said N dimensions are in the same realm as the straightest path in the N-D-Space-M-D-Time but where warped space or space-time such as wormholes, and the like are not applied where said shorter route is of one or more than one very strange shapes.

ESSAY 53) Travel Into Matter And/Or Energy And/Or Space-Time To Small Sub-Composition Level.

Here, we digress on travel into matter and/or energy and/or space-time to vanishingly small sub-composition levels.

Travel into matter and/or energy and/or space-time to such a small sub-composition level such that the distance between the outside region of origin and the point of entrance in to an associated quanta is infinite to increasing extents as the travel progresses into smaller and smaller portions of a point.

Travel into matter and/or energy and/or space-time to such a small sub-composition level such that the distance between the outside region of origin and the fraction of a geometric point of entrance into an associated quanta is infinite to increasing extents as the travel progresses into smaller and smaller portions of a sub-point.

Travel into matter and/or energy and/or space-time to such a small sub-composition level such that the distance between the outside region of origin and a fraction of a point of a geometric point of entrance into an associated quanta is infinite to increasing extents as the travel progresses into smaller and smaller portions of a sub-point within a point.

Travel into matter and/or energy and/or space-time to such a small sub-composition level such that the distance between the outside region of origin and a fraction of a point of a point of a geometric point of entrance into an associated quanta is infinite to increasing extents as the travel progresses into smaller and smaller portions of a sub-point of a point within a point.

Travel into matter and/or energy and/or space-time to such a small sub-composition level such that the distance between the outside region of origin and a fraction of a point of a point of a point of a geometric point of entrance into an associated quanta is infinite to increasing extents as the travel progresses into smaller and smaller portions of a sub-point of a point within a point within a point.

We can continue this digression forever to yield ever greater distances of travel.

Now, we consider a geometric object such as a point within a point. Here, a point is zero dimensional and a point within a point is (zero)(zero) dimensional.

Now, we consider a geometric object such as a point within a point within a point. Here, a point is zero dimensional and a point within a point within a point is (zero)(zero)(zero) dimensional.

Now, we consider a geometric object such as a point within a point within a point within a point. Here, a point is zero dimensional and a point within a point within a point is (zero)(zero)(zero)(zero) dimensional.

We can take the series to (zero EXP ∞'s) where the infinities can be arbitrarily large.

We can consider intra-topology of point that is three fold. Accordingly: analogs of a; cube, solid sphere, block, solid pyramid, solid tetrahedron, and the like are zero dimensional.

We can again consider intra-topology of a point that is three fold. Accordingly: analogs of a; circle, plane, square, rectangle, triangle, parallelogram, trapezoid, ellipse, hollow sphere, holly block, hollow pyramid, hollow tetrahedron and the like are (2/3)(zero) dimensional.

We can again consider intra-topology of a point that is three fold. Accordingly: analogs of a; straight-line, and the like are (1/3)(zero) dimensional.

We can consider intra-topology of point that is four fold. Accordingly: analogs of a; 4-D cube, 4-D solid sphere, 4-D block, 4-D solid pyramid, 4-D solid tetrahedron, and the like are zero dimensional.

We can consider intra-topology of point that is four fold. Accordingly: analogs of a; 3-D cube, 3-D solid sphere, 3-D block, 3-D solid pyramid, 3-D solid tetrahedron, and the like are [(3/4)(zero)] dimensional.

We can again consider intra-topology of a point that is four fold.

Accordingly: analogs of a; circle, plane, square, rectangle, triangle, parallelogram, trapezoid, ellipse, hollow sphere, holly block, hollow pyramid, hollow tetrahedron and the like are (2/4)(zero) dimensional.

We can again consider intra-topology of a point that is four fold. Accordingly: analogs of a; straight-line, and the like are (1/4)(zero) dimensional.

We can consider intra-topology of point that is five fold. Accordingly: analogs of a; 5-D cube, 5-D solid sphere, 5-D block, 5-D solid pyramid, 5-D solid tetrahedron, and the like are zero dimensional.

We can consider intra-topology of point that is five fold. Accordingly: analogs of a; 4-D cube, 4-D solid sphere, 4-D block, 4-D solid pyramid, 4-D solid tetrahedron, and the like are [(4/5)(zero)] dimensional.

We can consider intra-topology of point that is five fold. Accordingly: analogs of a; 3-D cube, 3-D solid sphere, 3-D block, 3-D solid pyramid, 3-D solid tetrahedron, and the like are [(3/5)(zero)] dimensional.

We can again consider intra-topology of a point that is five fold.

Accordingly: analogs of a; circle, plane, square, rectangle, triangle, parallelogram, trapezoid, ellipse, hollow sphere, holly block, hollow pyramid, hollow tetrahedron and the like are (2/5)(zero) dimensional.

We can again consider intra-topology of a point that is five fold. Accordingly: analogs of a; straight-line, and the like are (1/5)(zero) dimensional.

We can consider intra-topology of point that is six fold. Accordingly: analogs of a; 6-D cube, 6-D solid sphere, 6-D block, 6-D solid pyramid, 6-D solid tetrahedron, and the like are zero dimensional.

We can consider intra-topology of point that is six fold. Accordingly: analogs of a; 5-D cube, 5-D solid sphere, 5-D block, 5-D solid pyramid, 5-D solid tetrahedron, and the like are (5/6)(zero) dimensional.

We can consider intra-topology of point that is six fold. Accordingly: analogs of a; 4-D cube, 4-D solid sphere, 4-D block, 4-D solid pyramid, 4-D solid tetrahedron, and the like are [(4/6)(zero)] dimensional.

We can consider intra-topology of point that is six fold.

Accordingly: analogs of a; 3-D cube, 3-D solid sphere, 3-D block, 3-D solid pyramid, 3-D solid tetrahedron, and the like are [(3/6)(zero)] dimensional.

We can again consider intra-topology of a point that is six fold.

Accordingly: analogs of a; circle, plane, square, rectangle, triangle, parallelogram, trapezoid, ellipse, hollow sphere, holly block, hollow pyramid, hollow tetrahedron and the like are (2/6)(zero) dimensional.

We can again consider intra-topology of a point that is six fold. Accordingly: analogs of a; straight-line, and the like are (1/6)(zero) dimensional.

By now you can see the pattern.

We can again consider intra-topology of a point that is a given infinity fold. Accordingly: analogs of a; (said given infinity fold)-D cube, (said given infinity fold)-D solid sphere, (said given infinity fold)-D block, (said given infinity fold)-D solid pyramid, (said given infinity fold)-D solid tetrahedron, and the like are zero dimensional.

We can consider intra-topology of a point that is a given infinity fold. Accordingly: analogs of a; 5-D cube, 5-D solid sphere, 5-D block, 5-D solid pyramid, 5-D solid tetrahedron, and the like are [5/(said given infinity)](zero) dimensional.

We can consider intra-topology of a point that is a given infinity fold. Accordingly: analogs of a; 4-D cube, 4-D solid sphere, 4-D block, 4-D solid pyramid, 4-D solid tetrahedron, and the like are [4/(said given infinity)] dimensional.

We can consider intra-topology of a point that is a given infinity fold. Accordingly: analogs of a; 3-D cube, 3-D solid sphere, 3-D block, 3-D solid pyramid, 3-D solid tetrahedron, and the like are [3/(said given infinity)] dimensional.

We can consider intra-topology of a point that is a given infinity fold. Accordingly: analogs of a; circle, plane, square, rectangle, triangle, parallelogram, trapezoid, ellipse, hollow sphere, holly block, hollow pyramid, hollow tetrahedron and the like are [2/(said given infinity)] dimensional.

We can consider intra-topology of a point that is a given infinity fold. Accordingly: analogs of a; straight-line, and the like are [1/(said given infinity)] dimensional.

We affix the following optional operator to the lengthy formulas for gamma and related parameters to denote the extreme infinite travel distances by traveling into ever smaller locations with proximate ports of entry.

{Π(m = 1; m = ∞↑):[[n/(said given infinity)][(zero) EXP m]-D] travel in infinite distances | n ranges from zero to as great as any arbitrary applicable infinity}

ESSAY 54) Compounded Light-Speed Velocity States.

We can intuit a spacecraft attaining the maximum velocity of light which is that in a pure vacuum. Accordingly, the Lorentz factor of the spacecraft would be a limiting infinite value which we will refer to as Ω,1.

However, there is no reason to believe that the kinetic energy of the spacecraft could be limited as we can always add on joule to the spacecraft kinetic energy while holding its invariant mass the same.

So, we now consider scenarios for which a spacecraft traveling inertially would have an increase in kinetic energy Ω,1 to a next suitable infinite boundary condition relative to its previous reference frame velocity state of Ω,1.

Since the velocity state, Ω,1, cannot be added to in its own velocity continuum, we will assume that the velocity of the spacecraft having reached its next available infinite boundary condition of kinetic energy will manifest as a velocity vector of, Ω,1 + Ω,2.

The latter compound velocity is simply the addition of two vectors head to tail where the vectors are orthogonal.

So the magnitude of the vector resulting from the addition of the two vectors will be {[(Ω,1) EXP 2] + [(Ω,2) EXP 2]} EXP (1/2).

Thus, the craft in a sense will slip out of the potential velocity vector space associated with ordinary 4-D space-time and into another potential velocity vector space. Such a space might manifest as a scalar field in a roiling sea of randomly oriented virtual velocity vectors for which the average net velocity associated with a region within the sea is zero in magnitude.

Since the velocity state, Ω,1 + Ω,2, cannot be added to in its own 2-fold velocity continuum, we will assume that the velocity of the spacecraft having reached its next available infinite boundary condition of kinetic energy will manifest as a velocity vector of, Ω,1 + Ω,2 + Ω,3.

The latter compound velocity is simply the addition of three vectors head to tail where the vectors are orthogonal.

So the magnitude of the vector resulting from the addition of the three vectors will be {[(Ω,1) EXP 2] + [(Ω,2) EXP 2] + [(Ω,3) EXP 2]} EXP (1/2).

Thus, the craft in a sense will have slipped out of its own 2-fold velocity continuum and into another potential velocity vector space characterized as a continuum of [f[Ω,1, Ω,2, Ω,3]]. Such a space might manifest as a scalar field in a roiling sea of randomly oriented virtual velocity vectors for which the average net velocity associated with a region within the sea is zero in magnitude.

Since the velocity state, Ω,1 + Ω,2 + Ω,3, cannot be added to in its own 3-fold velocity continuum, we will assume that the velocity of the spacecraft having reached its next available infinite boundary condition of kinetic energy will manifest as a velocity vector of, Ω,1 + Ω,2 + Ω,3 + Ω,4.

The latter compound velocity is simply the addition of four vectors head to tail where the vectors are orthogonal.

So the magnitude of the vector resulting from the addition of the four vectors will be {[(Ω,1) EXP 2] + [(Ω,2) EXP 2] + [(Ω,3) EXP 2] + [(Ω,4) EXP 2]} EXP (1/2).

Thus, the craft in a sense will have slipped out of its own 3-fold velocity continuum and into another potential velocity vector space characterized as a continuum of [f[Ω,1, Ω,2, Ω,3, Ω,4]]. Such a space might manifest as a scalar field in a roiling sea of randomly oriented virtual velocity vectors for which the average net velocity associated with a region within the sea is zero in magnitude.

We can consider arbitrary numbers of iterations to

[f[Ω,1, Ω,2, Ω,3, …, Ω,N]]

to yield the velocity vector,

{[(Ω,1) EXP 2] + [(Ω,2) EXP 2] + [(Ω,3) EXP 2] + … + [(Ω,N) EXP 2]} EXP (1/2)

We can consider hyperspatial analogs as well as curved space-times and velocity realm analogs. The curvature can be actually or abstractly in ordinary 4-D space-time, in N-D-Space-M-D-Time scales where N is any integer greater than two and M is any counting number, where N is any integer greater that two and M is any positive rational number, where N is any integer greater that two and M is any positive irrational number, where N is any integer greater that two and M is any positive real number, where N is any positive rational number and M is any counting number, where N is any positive rational number and M is any positive rational number, where N is any positive rational number and M is any positive irrational number, where N any positive rational number and M is any positive real number, where N is any positive irrational number and M is any counting number, where N is any positive irrational number and M is any positive rational number, where N is any positive irrational number and M is any positive irrational number, where N is any positive irrational number and M is any positive real number, where N is any positive real number and M is any counting number, where N is any positive real number and M where N is any positive real number, where N is any positive real number and M is any positive irrational number, where N is any positive real number and M is any positive real number: for which the space-time is flat, positively curved, negatively curved, positively curved and torsioned at one or more scales in arbitrary patterns including but not limited to fractals, negatively curved and torsioned at one of more scales in arbitrary patterns including but not limited to fractals, positively super-curved, negatively super-curved, positively super-curved and torsioned at one or more scales in arbitrary patterns including but not limited to fractals, negatively super-curved and torsioned at one of more scales in arbitrary patterns including but not limited to fractals, positively super-super-curved, negatively super-super-curved, positively super-super-curved and torsioned at one or more scales in arbitrary patterns including but not limited to fractals, negatively super-super-curved and torsioned at one of more scales in arbitrary patterns including but not limited to fractals, positively super-super-super-curved, negatively super-super-super-curved, positively super-super-super-curved and torsioned at one or more scales in arbitrary patterns including but not limited to fractals, negatively super-super-super-curved and torsioned at one of more scales in arbitrary patterns including but not limited to fractals, positively super-super- … -super-curved, negatively super-super- … -super-curved, positively super-super- … -super-curved and torsioned at one or more scales in arbitrary patterns including but not limited to fractals, negatively super-super- … -super-curved and torsioned at one of more scales in arbitrary patterns including but not limited to fractals.

So, we affix the following operator to the lengthy formulas for gamma and related parameters to denote the above mechanisms of velocity vector scalar seas.

{(Context Specific Curvature Function){f{{[(Ω,1) EXP 2] + [(Ω,2) EXP 2] + [(Ω,3) EXP 2] + … + [(Ω,N) EXP 2]} EXP (1/2)}}}

Note that each iteration produces half a step-wise loop thus indicating one-way temporal progression even if multiple time dimensions are implies. The return portion of the loop would signify backward time travel.

We can take the first and higher order derivatives and integrals of the above light-speed velocities. These derivatives can be with respect to ship-time, gamma, ship-frame acceleration, spacecraft kinetic energy and the like.

We denote possibilities by the following operator which we may affix to formulas for gamma and related parameters.

{{(Context Spec Curv Funct){f{(1st der or 1st ∫), …,(nth der or mth ∫) {{[(Ω,1) EXP 2] + [(Ω,2) EXP 2] + [(Ω,3) EXP 2] + … + [(Ω,N) EXP 2]} EXP (1/2)}}}} where the ders and/or ∫s are w/ respect to arb pars}

ESSAY 55) Light-Speed Travel In Levels Outside Of Eternity.

Now, we have all heard of the notion of “outside of time and space” while referring to an all-knowing GOD who knows the past as well as the present and future in complete detail.

Accordingly, we will state that GOD is thus eternal.

Now, we may consider that a light-speed spacecraft having achieved a great enough infinite Lorentz factor may run out of temporal room to travel forward in time in. Thus, we have pondered previously in this book that such a spacecraft will force itself out of time and into a first level of eternity.

We have also previously considered spacecraft progressing into higher and higher levels of eternity as the spacecraft Lorentz factors become all the more infinite.

Now, here is a new concept.

What if a spacecraft Lorentz factor became so infinite that the spacecraft traveled outside of eternities? Of course, we are talking about created cosmic eternities and not about transcendent GOD here.

Such travel in a first level we will call outside-of-eternity-1.

We can like-wise consider even more extreme infinite Lorentz factors such that a spacecraft would travel into the next level of outside eternity or outside-of-eternity-2.

We can like-wise consider even more extreme infinite Lorentz factors such that a spacecraft would travel into the next level of outside eternity or outside-of-eternity-3.

We can continue the progression all the way to outside-of-eternity-∞ where the infinity can be arbitrarily great.

We affix the following operator to the lengthy formulas for gamma and related parameters to denote the above extra-eternities travel.

{Travel outside-of-eternity-k| k ranges from 1 into arbitrary infinite integers}

We can also consider scenarios for which the passage of time ceases in a Lorentz variant way for craft attaining suitably infinite Lorentz factors.

Accordingly, the craft may needed sudden boosting to such Lorentz factors since otherwise, the craft would experience ship-time dependent increases or decreases in gamma, kinetic energy, Lorentz contraction, and modifications of outside reference frames relativistic aberration.

We refer to a first level of time-less-ness as time-less-ness-1.

Boosting the craft suddenly to a more extreme infinite Lorentz factor may result in the craft attaining a second level of time-less-ness which we refer to as time-less-ness-2.

Boosting the craft suddenly to a yet more extreme infinite Lorentz factor may result in the craft attaining a third level of time-less-ness which we refer to as time-less-ness-3.

We can continue this progression forever perhaps without limit.

We affix the following operator to the lengthy formulas for gamma and related parameters to denote the above time-less-ness options.

{Time-less-ness-k| k ranges from 1 into arbitrary infinite integers}

ESSAY 57) Instantaneous Jumps To Light-Speed.

Now, we who love science fiction while also enjoying speculations about wormholes, warp-drive, and the like often will bemoan any light-speed limits nature may impose on us. However, this is not a bad option imposed by Mother Nature. After all, light-speed travel in variously infinite Lorentz factors enables us to travel arbitrarily infinite numbers of light-years through space and infinite years into the future even in one Planck Time Unit ship-frame. The Planck Time Unit is very small at about [5.39 x (10 EXP -44)]. The value is expressed as tp = {[h/(2π)](G)/(c EXP 5)} EXP (1/2).

There is another interesting metric with which to fathom light-speed travel. This metric is the notion that a spacecraft may jump to light-speed in an infinitesimal time period. To enable such, the spacecraft would need to have its acceleration pre-calibrated such that the craft accelerates all at once instead of just the leading edge or the trailing edge reaching light-speed before the rest of the craft does so.

Accordingly, the effective acceleration vector may be arbitrarily infinite. In some cases, the infinite acceleration will result in infinite space time dilation with respect to the background.

The above conjectures on infinite acceleration provide another measure of the exotic nature of light-speed prospects.

We can go additional steps to consider super-instantaneous acceleration to the speed of light. The degree of super-instantaneous acceleration is arbitrary and thus whimsically farout to contemplate.

We may affix the following operator to formulas for gamma and related parameters that follow to denote the prospects of instantaneous acceleration to the speed of light as well as super-instantaneous acceleration to the speed of light.

[f(Prospects of instantaneous or super-instantaneous acceleration to the speed of light)].

Now, the instantaneous and super-instantaneous accelerations can be in the ship-frame or the background reference frame. As a first order approximation, the ship-frame acceleration will equal the background rate of acceleration multiplied by the instantaneous value of gamma for the ship. Since the value of gamma increases asymptotically as the spacecraft instantaneously or super-instantaneously approaches light-speed, the ship frame acceleration instantaneously supercedes background acceleration by various context specific values of infinity. Thus, the ship frame acceleration is thus non-linear but asymptotic.

ESSAY 57) Cloning The Creation Precursor Conditions For Powering Light-Speed Spacecraft.

Herein we consider extremely speculative classes of propulsion systems.

First, we consider scenarios for which the precursor conditions associated with creation exnihilo by GOD are grafted or cloned for purposes of powering light-speed spacecraft. Thus, the very first conditions distinct from GOD for the process of creation ex-nihilo are harnessed and may be propagated or otherwise cloned for powering spacecraft to unheard of infinitely great Lorentz factors.

As another less extreme analogous style of propulsion, we consider the natural process of the formation of big bangs, multiverses, forests, biospheres, and even the birth of hyperspaces of arbitrary dimensionalies as mechanisms for spacecraft propulsion.

We may affix the following operator to formulas for gamma and related parameters that follow to denote the scenarios for which the precursor conditions associated with creation ex-nihilo by GOD are grafted or cloned for purposes for powering spacecraft as well as the same for big bangs, multiverses, forests, biospheres, and even the birth of hyperspaces.

[w(Propulsion by precursor conditions of creation ex-nihilo by GOD and grafting or cloning thereof as well as by big bangs, multiverses, forests, biospheres, and birth of k-dimensional hyperspaces)]

Here k can be any positive real number, finite or infinite.

The craft can be propelled by ordinary 4-D space-time, in N-D-Space-M-D-Time scales where N is any integer greater than two and M is any counting number, where N is any integer greater that two and M is any positive rational number, where N is any integer greater that two and M is any positive irrational number, where N is any integer greater that two and M is any positive real number, where N is any positive rational number and M is any counting number, where N is any positive rational number and M is any positive rational number, where N is any positive rational number and M is any positive irrational number, where N any positive rational number and M is any positive real number, where N is any positive irrational number and M is any counting number, where N is any positive irrational number and M is any positive rational number, where N is any positive irrational number and M is any positive irrational number, where N is any positive irrational number and M is any positive real number, where N is any positive real number and M is any counting number, where N is any positive real number and M where N is any positive real number, where N is any positive real number and M is any positive irrational number, where N is any positive real number and M is any positive real number: for which the space-time is flat, positively curved, negatively curved, positively curved and torsioned at one or more scales in arbitrary patterns including but not limited to fractals, negatively curved and torsioned at one of more scales in arbitrary patterns including but not limited to fractals, positively super-curved, negatively super-curved, positively super-curved and torsioned at one or more scales in arbitrary patterns including but not limited to fractals, negatively super-curved and torsioned at one of more scales in arbitrary patterns including but not limited to fractals, positively super-super-curved, negatively super-super-curved, positively super-super-curved and torsioned at one or more scales in arbitrary patterns including but not limited to fractals, negatively super-super-curved and torsioned at one of more scales in arbitrary patterns including but not limited to fractals, positively super-super-super-curved, negatively super-super-super-curved, positively super-super-super-curved and torsioned at one or more scales in arbitrary patterns including but not limited to fractals, negatively super-super-super-curved and torsioned at one of more scales in arbitrary patterns including but not limited to fractals, positively super-super- ... -super-curved, negatively super-super- ... -super-curved, positively super-super- ... -super-curved and torsioned at one or more scales in arbitrary patterns including but not limited to fractals, negatively super-super- ... -super-curved and torsioned at one of more scales in arbitrary patterns including but not limited to fractals.

ESSAY 58) Acquisition Of Propulsion Energy From Hyperspatial Big Bang Analogs.

Next, we consider prospects for the acquisition of propulsion energy from natural and/or artificial hyperspatial big bang analogs.

Accordingly, hyperspaces exploding into existence would provide near infinite to variously infinite quantities energy supplies to a spacecraft.

We may also consider next levels up of hyper-hyper-spaces, hyper-hyper-hyper-spaces and the like which we denote by the lexicography of hyper-2-spaces, hyper-3-spaces, and generically hyper-k-spaces where k ranges from 1 thru infinity. Ordinary hyperspaces are simply labeled hyperspaces or hyper-1-spaces.

We may also take the ship-time, gamma, acceleration, and kinetic energy derivatives and integrals of arbitrary orders of spacecraft travel distances and/or number of hyper-k-spaces utilized to provide a more complete picture of the range of options for exploring and defining these hyperspatial propulsion mechanisms.

Also, note that the hyperspatial explosive energies may be directly quantum-mechanically imprinted on the spacecraft wave-function via teleportation and tunneling mechanisms. Thus, the spacecraft may suddenly jump to light-speed with no perceived acceleration-based forces.

We may affix the following operator to formulas for gamma and related parameters that follow to denote the scenarios for which hyper-k-spaces are harnessed for propulsive energy as well as differentiation and integration of space travel distance and also that of number of hyper-k-spaces utilized for energy with respect to ship-time, gamma, acceleration, and spacecraft kinetic energy.

[q(hyper-k-spaces harnessed and nth order diff & int of space travel distance & of number of hyper-k-spaces utilized for energy with respect to ship-time, γ, a, and spacecraft K.E.)]

ESSAY 59) Acquisition Of Propulsion Energy From Bulkial And Hyper-Bulkial Big Bang Analogs.

Next, we consider prospects for the acquisition of propulsion energy from natural and/or artificial bulkial and hyperbulkial big bang analogs.

Accordingly, bulks and hyperbulks exploding into existence would provide near infinite to variously infinite quantities energy supplies to a spacecraft.

We may also consider next levels up of hyper-hyper-bulks, hyper-hyper-hyper-bulks and the like which we denote by the lexicography of hyper-2-bulks, hyper-3-bulks, and generically hyper-k-bulks where k ranges from 1 thru infinity. Ordinary hyperbulks are simply labeled hyperbulks or hyper-1-bulks.

We may also take the ship-time, gamma, acceleration, and kinetic energy derivatives and integrals of arbitrary orders of spacecraft travel distances and/or number of hyper-k-bulks utilized to provide a more complete picture of the range of options for exploring and defining these hyperbulkial propulsion mechanisms.

Also, note that the hyperbulkial explosive energies may be directly quantum-mechanically imprinted on the spacecraft wave-function via teleportation and tunneling mechanisms. Thus, the spacecraft may suddenly jump to light-speed with no perceived acceleration-based forces.

We may affix the following operator to formulas for gamma and related parameters that follow to denote the scenarios for which hyper-k-bulks are harnessed for propulsive energy as well as differentiation and integration of space travel distance and also that of number of hyper-k-bulks utilized for energy with respect to ship-time, gamma, acceleration, and spacecraft kinetic energy.

Note that for the sake of brevity, bulks without a prefix, hyper, are denoted as hyper-0-bulks.

[q(hyper-k-bulks harnessed and nth order diff & int of space travel distance & of number of hyper-k-spaces utilized for energy with respect to ship-time, γ, a, and spacecraft K.E.)]

ESSAY 60) Lorentz Factors Of Even Greater Infinities.

Now Aleph 0 is the infinite number of positive integers.

Aleph 1 is according to the perhaps unprovable and unfalsifiable Continuum Hypothesis equal to the number of real numbers.

Either way, Aleph 1 = 2 EXP (Aleph 0).

Aleph 2 = 2 EXP (Aleph 1).

Aleph 3 = 2 EXP (Aleph 2).

Aleph 4 = 2 EXP (Aleph 3).

and so-on.

In general, [Aleph (N +1)] = 2 EXP (Aleph N) where N is any whole number.

Now, there are countable and uncountable infinities.

But before we go further, consider the hyper-operator notation that was designed to express huge values not otherwise expressible.

For example, note that Hyper4(a, n) is equal to a tetrated n or a raised to the power of itself n-1 times. The latter value is symbolically written as n subscript a. For example 3 EXP 4 = 81, but 4 subscript3 is approximately equal to 10 EXP (1,000,000,000,000).

Alternatively 4 subscript 2 = 2 EXP 2 EXP 2 EXP 2 = 2 EXP [2 EXP [2 EXP 2]] = 2 EXP (2 EXP 4) = 2 EXP 16 = 65,536.

For example, Hyper5(4, 4)is equal to 4 tetrated 4 tetrated 4 tetrated 4. This value is commonly referred to as 4 pentated 4.

Hyper 6, (4,4) is 4 pentated 4 pentated 4 pentated 4 and is also referred to as 4 hexataed 4.

Hyper 7, (4,4) is 4 hexated 4 hexated 4 hexated 4 and so-on.

So we can produce the abstraction of Hyper Aleph Ω (Aleph Ω, Aleph Ω).

We can abstractly construct infinities in other ways. A particularly fascinating way to to state something like, zero is to Aleph 0 as Aleph 0 is to 1st level analog infinity.

We can likewise make a statement that zero is to Aleph 0 as Aleph 0 is to 1st level analog infinity as 1st level analog infinity is to 2nd level analog infinity.

We can likewise make a statement that zero is to Aleph 0 as Aleph 0 is to 1st level analog infinity as 1st level analog infinity is to 2nd level analog infinity as 2nd level analog infinity is to 3rd level analog infinity.

We can progress to the kth level analog infinity where k is any finite or infinite positive integer.

We can go additional steps further to develop extended levels of analog infinities.

Accordingly, 0 is to 1st level analog infinity as 1st level analog infinity is to 1st extended-1 level analog infinity as 1st extended-1 level analog infinity is to 2nd extended-1 level analog infinity as 2nd extended-1 level analog infinity is to 3rd extended-1 level analog infinity and so-on all the way to kth extended-1 level of analog infinity.

We can go still further to develop the following concepts.

Accordingly, 0 is to 1st extended-1 level analog infinity as 1st extended-1 level analog infinity is to 1st extended-2 level analog infinity as 1st extended-2 level analog infinity is to 2nd extended-2 level analog infinity as 2nd extended-2 level analog infinity is to 3rd extended-2 level analog infinity as 3rd extended-2 level analog infinity is to 4th extended-2 level analog infinity and so-on all the way to kth extended-2 level of analog infinity.

We can go still further to develop the following concepts.

Accordingly, 0 is to 1st extended-2 level analog infinity as 1st extended-2 level analog infinity is to 1st extended-3 level analog infinity as 1st extended-3 level analog infinity is to 2nd extended-3 level analog infinity as 2nd extended-3 level analog infinity is to 3rd extended-3 level analog infinity as 3rd extended-3 level analog infinity is to 4th extended-3 level analog infinity and so-on all the way to kth extended-3 level of analog infinity.

We can go still further to develop the following concepts.

Accordingly, 0 is to 1st extended-3 level analog infinity as 1st extended-3 level analog infinity is to 1st extended-4 level analog infinity as 1st extended-4 level analog infinity is to 2nd extended-4 level analog infinity as 2nd extended-4 level analog infinity is to 3rd extended-4 level analog infinity as 3rd extended-4 level analog infinity is to 4th extended-4 level analog infinity and so-on all the way to kth extended-4 level of analog infinity.

In general, we can go all the way up to and beyond kth extended-j level of analog infinity where k and j can be arbitrary countable or uncountable infinities.

All of these new constructs can be said to be infinities as they are abstractly derived as functions of infinities.

Now, imagine a spacecraft attaining Lorentz factors of kth extended-j level of analog infinity where k and j are arbitrary countable or uncountable infinities.

ESSAY 61) Light-Speed Options Via Exotic QCD Fuels.

Nuclear energy comes in many forms.

Nuclear fission involves splitting of atomic nuclei motivated mainly by neutron flux. Uranium-235 is a good nuclear fission fuel. Complete nuclear fission of a portion of pure nuclear fission fuel converts about 0.1 percent of its mass to energy.

Nuclear fusion involves fusing atomic nuclei together which can start with simple hydrogen or protons and provides exothermic yields until the most low energy isotope of iron is produced. Completed fusion processes convert about 1 percent of the mass of the reactants to energy.

Radioactive decay by beta or alpha processes respectively involve the emission of electrons or helium nuclei from various unstable isotopes. The fraction of mass converted to energy is usually roughly in the range of 0.001 percent.

Nuclear isomers provide about the same mass specific energy yield as radio-active decay and involve stored energy in exited atomic nuclei of various but not all isotopes.

And there is a new kid on the block that is actively being studied. The energy forms involve hyper-nuclei which are atomic nuclei of atoms that have lumps of neutrons and/or protons in one or more groups separated but proximate to the primary interior portion of the nuclei.

Another less considered form of nuclear energy is more properly referred to as quantum-chromo-dynamics energy or QCD energy. Accordingly, QCD energy may enable a greater portion of the reactants being converted to energy.

QCD energy as anticipated above would involve the production of stable strangelets, charmlets, bottom-lets and/or top-lets which could catalyze ordinary baryonic matter into strange, charmed, bottomed, and/or top matter. In some cases, such a reaction might result in the production of stable exotic quarkonium and may liberate most of the mass of the reactants as energy including heat, gamma radiation, neutrinos, beta and alpha rays. The residual quarkonium may be very dangerously catalytic. If the quark matter produced is stable, or long lived, a run-away chain reaction may insue converting every thing in its path to exotic quark matter in a near light-speed reaction wave front that gradually consumes an entire cosmic light cone and any universel locations accessible by black holes within the light-cone.

QCD reactions as such might not be as powerful pure matter-antimatter reactions, but the former would be more transformative by far. We will refer to such transformative processes and forms of matter as “acerbic”.

Acerbic forms of matter when of suitably limited half-lives in decaying into non-acerbic species may provide a fantastic resource for powering relativistic rockets, pellet stream drives, fuel beam drives, and interstellar ramjets to velocities limited only by the speed of light.

At the very least, I believe nuclear acerbic fuels can enable spacecraft to attain velocities which are quantum-mechanically indistinguishable from the speed of light.

We may affix the following operator to formulas for gamma and related parameters in this section and the next to denote the effectively light-speed options enabled by acerbic QCD fuels.

[q(Effectively light-speed spacecraft options enabled by acerbic QCD fuels)].

ESSAY 62) Spacecraft Acquisition Of Fuel Pellets Located Beyond Our Cosmic Light-Cone.

Now, we consider scenarios for which a spacecraft would access a series of pellets that would ultimately extend beyond our current light-cone in conventional non-spatially increased length partitions.

Herein, we consider utrarelativistic pellets fired out of a gun, perhaps a staged electron gas compression gun, for which the pellets uptake background matter to build themselves up while propelling themselves to overcome drag and undergo positive acceleration. These pellet when fully assembled will fire their own seed pellets at ultra-relativistic velocities with respect to themselves for which these secondary pellets uptake background matter to build themselves up. These secondary pellets while propelling themselves to overcome drag and undergoing positive acceleration then fire tertiary pellets at ultra-relativistic velocities with respect to themselves. These ternary pellets uptake background matter to build themselves up while propelling themselves to overcome drag and undergo positive acceleration for which these tertiary pellets then fire 4th series pellets and so-on for an eternity.

Here, we assume that the pellets are made of sub-white-dwarf density matter, white dwarf density matter, neutronium, quarkonium, higgsinium, mono-higgsinium, monopolium, raw mass-energy-space-time, wormhole wall fabric and the like.

Now, consider scenarios for which the very first pellets attain a Lorentz factor of 5,000 upon reaching a distance of about 13.8 billion light-years. This is the approximately equal to about one cosmic light-cone radius.

Upon reaching this value of distance traveled and going immediately further, the pellet recessional velocity will become equal to the speed of light plus the speed of the spacecraft relative to its then local background. So the spacecraft recessional velocity with respect to the location of departure will be [(1 + 0.99999998) c].

The ratio of the distance of travel by the super-light-cone traveling spacecraft to the distance viewable from the Milky Way Galaxy assuming that the cosmic light-cone radius remains the same due to constant universal expansion rate per unit of distance for the above scenario is:

{{[(1 + 0.99999998) c](Background Travel Time)}/(Cosmic Light-Cone Radius)}

For constant ship frame Lorentz factors and ship-frame time, the following formula applies:

{{{[(1 + 0.99999998) c](Ship-Frame Travel Time)}/(Cosmic Light-Cone Radius)}(γ)}

The following formula provides elapsed time for constant acceleration ship-frame and starting velocity, V_o

$$T_o = \{(c/g) \ln \{\{[[(C^2) + (V_o^2)]^{1/2}] - [V_o/[[1 - [(V_o/C)^2]]^{1/2}]]\} \{[[C^2] + [[(g)(t) + [V_o/[1 - [(V_o/C)^2]]^{1/2}]^2]]^{1/2}] + [(g)(t)] + [V_o/[[1 - [(V_o/C)^2]]^{1/2}]]\} / [C^2]\}\}$$

For constant ship frame acceleration, the formula becomes:

$$\{\{\{[(1 + 0.99999998)\, c](\text{Ship-Frame Travel Time})\}/(\text{Cosmic Light-Cone Radius})\} |\{(c/g) \ln \{\{[[(C^2) + (V_o^2)]^{1/2}] - [V_o/[[1 - [(V_o/C)^2]]^{1/2}]]\} \{[[C^2] + [[(g)(t) + [V_o/[1 - [(V_o/C)^2]]^{1/2}]^2]]^{1/2}] + [(g)(t)] + [V_o/[[1 - [(V_o/C)^2]]^{1/2}]]\} / [C^2]\}\}|\}$$

Note the vertical lines bounding the expression for elapsed time. This denotes taking the absolute value and is thus dimensionless. However, the results are accurate if the same unit systems are used throughout the formulas so modified, for example, the MKS system.

The following formula provides elapsed time for varying acceleration ship-frame, g_{0i} for each background time step, t_i, and starting velocity, V_{oi}, for each ship-time step, t_i.

T_o = {Σ (I =1, I =n) {(C/g_i) ln {{[[(C^2) + ($V_{0i}{}^2$)] $^{1/2}$] – [V_{0i}/[[1 – [(V_{0i}/C) ^{2}]] $^{1/2}$]]} { [(C ^{2}) + [[(g_i)(t_i) + [V_{0i} /[1 – [(V_{0i}/C) ^{2}]] $^{1/2}$] ^{2}]] $^{1/2}$] + [(g_i)(t_i)] + [V_{0i}/[[1 – [(V_{0i}/C) ^{2}]] $^{1/2}$]]} / (C ^{2})}}

So for varying ship-frame acceleration, the formula becomes:

{{{[(1 + 0.99999998) c](Ship-Frame Travel Time)}/(Cosmic Light-Cone Radius)} |{Σ (I =1, I =n) {(C/g_i) ln {{[[(C^2) + ($V_{0i}{}^2$)] $^{1/2}$] – [V_{0i}/[[1 – [(V_{0i}/C) ^{2}]] $^{1/2}$]]} { [(C ^{2}) + [[(g_i)(t_i) + [V_{0i} /[1 – [(V_{0i}/C) ^{2}]] $^{1/2}$] ^{2}]] $^{1/2}$] + [(g_i)(t_i)] + [V_{0i}/[[1 – [(V_{0i}/C) ^{2}]] $^{1/2}$]]} / (C ^{2})}}|}

Now, the starting recessional velocity of the spacecraft from its point of origin for the next cosmic time-step will be [(1 + 0.99999998) c]. We assume that this first pellet craft builds up and fires a second pellet which builds up to a second spacecraft which travels for about 13.8 billion light-years into this new toy model cosmic light-cone, we will assume that it attains a Lorentz factor of 50,000,000 with respect to the background. So, the velocity of the second pellet spacecraft after traveling one cosmic light-cone radius beyond the second location of toy model origin with respect to said second toy model point of origin will be [(1 + 0.9999999999999998) c] while the recessional velocity of the spacecraft from the first point of origin will be {[(1 + 0.99999998) c] + (1c)} + [(1 + 0.9999999999999998) c].

The ratio of the distance of travel by the super-light-cone traveling spacecraft to the distance viewable from the Milky Way Galaxy assuming that the cosmic light-cone radius remains the same due to constant universal expansion rate per unit of distance for the above scenario is:

{{{[(1 + 0.99999998) c] + (1c)} + [(1 + 0.9999999999999998) c] (Background Travel Time)}/(Cosmic Light-Cone Radius)}

For constant ship frame Lorentz factors and ship-frame time, the following formula applies:

{{{{[(1 + 0.99999998) c] + (1c)} + [(1 + 0.9999999999999998) c] (Spacecraft Reference Frame Travel Time)}/(Cosmic Light-Cone Radius)}(γ)}

For constant ship frame acceleration, the formula becomes:

{{{{[(1 + 0.99999998) c] + (1c)} + [(1 + 0.9999999999999998) c] (Spacecraft Reference Frame Travel Time)}/(Cosmic Light-Cone Radius)} |{(c/g) ln {{[[(C ^{2}) + ($V_o{}^2$)]$^{1/2}$] - [V_o/[[1 - [(V_o/C) ^{2}]]$^{1/2}$]]} {[[C^2] + [[(g)(t) + [V_o /[1 - [(V_o/C) ^{2}]] $^{1/2}$] ^{2}]] $^{1/2}$] + [(g)(t)] + [V_o/[[1 - [(V_o/C) ^{2}]] $^{1/2}$]]} / [C ^{2}]}}|}

So for varying ship-frame acceleration, the formula becomes:

{{{{[(1 + 0.99999998) c] + (1c)} + [(1 + 0.9999999999999998) c] (Spacecraft Reference Frame Travel Time)}/(Cosmic Light-Cone Radius)} |{Σ (I =1, I =n) {(C/g_i) ln {{[[(C^2) + (V_{0i}

^{2})] $^{1/2}$] – [V_{0i}/[[1 – [(V_{0i}/C) ^{2}]] $^{1/2}$]]} { [(C ^{2}) + [[(g_i)(t_i) + [V_{0i} /[1 – [(V_{0i}/C) ^{2}]] $^{1/2}$] ^{2}]] $^{1/2}$] + [(g_i)(t_i)] + [V_{0i}/[[1 – [(V_{0i}/C) ^{2}]] $^{1/2}$]]} / (C ^{2})}}|}

Now, we consider spacecraft continuance into a 3rd cosmic light-cone journey. We will consider that a third pellet derived spacecraft attains a Lorentz factor of γ_3 = [5 x (10 EXP 12)] after traveling across this third toy model cosmic light-cone to obtain a velocity of {c – [(2)[c EXP [-[(2)[log (2γ_3)]]]]]} with respect to its then proximate space-time background. The third pellet derived spacecraft recessional velocity with respect to the first pellet spacecraft's very first point of departure will be {[(1 + 0.99999998) c] + (1c) + (2c)} + {[(1 + 0.9999999999999998) c] + (1c)} + {1c + {c – [(2)(c)[10 EXP [-[(2)[log (2γ_3)]]]]]}}

The ratio of the distance of travel by the super-light-cone traveling spacecraft to the distance viewable from the Milky Way Galaxy assuming that the cosmic light-cone radius remains the same due to constant universal expansion rate per unit of distance for the above scenario is:

{{{[(1 + 0.99999998) c] + (1c) + (2c)} + {[(1 + 0.9999999999999998) c] + (1c)} + {1c + {c – [(2)(c)[10 EXP [-[(2)[log (2γ_3)]]]]]}} (Background Travel Time)}/(Cosmic Light-Cone Radius)}

For constant ship frame Lorentz factors and ship-frame time, the following formula applies:

{{{{[(1 + 0.99999998) c] + (1c) + (2c)} + {[(1 + 0.9999999999999998) c] + (1c)} + {1c + {c – [(2)(c)[10 EXP [-[(2)[log (2γ_3)]]]]]}} (Spacecraft Reference Frame Travel Time)}/(Cosmic Light-Cone Radius)}(γ)}

For constant ship frame acceleration, the formula becomes:

{{{{[(1 + 0.99999998) c] + (1c) + (2c)} + {[(1 + 0.9999999999999998) c] + (1c)} + {1c + {c – [(2)(c)[10 EXP [-[(2)[log (2γ_3)]]]]]}} (Spacecraft Reference Frame Travel Time)}/(Cosmic Light-Cone Radius)} |{(c/g) ln {{[[(C ^{2}) + (V_o ^{2})]$^{1/2}$] - [V_o/[[1 - [(V_o/C) ^{2}]]$^{1/2}$]]} {[[C^{2}] + [[(g)(t) + [V_o /[1 - [(V_o/C) ^{2}]] $^{1/2}$] ^{2}]] $^{1/2}$] + [(g)(t)] + [V_o/[[1 - [(V_o/C) ^{2}]] $^{1/2}$]]} / [C ^{2}]}}|}

So for varying ship-frame acceleration, the formula becomes:

{{{{[(1 + 0.99999998) c] + (1c) + (2c)} + {[(1 + 0.9999999999999998) c] + (1c)} + {1c + {c – [(2)(c)[10 EXP [-[(2)[log (2γ_3)]]]]]}} (Spacecraft Reference Frame Travel Time)}/(Cosmic Light-Cone Radius)} |{Σ (I =1, I =n) {(C/g_i) ln {{[[(C^{2}) + (V_{0i} ^{2})] $^{1/2}$] – [V_{0i}/[[1 – [(V_{0i}/C) ^{2}]] $^{1/2}$]]} { [(C ^{2}) + [[(g_i)(t_i) + [V_{0i} /[1 – [(V_{0i}/C) ^{2}]] $^{1/2}$] ^{2}]] $^{1/2}$] + [(g_i)(t_i)] + [V_{0i}/[[1 – [(V_{0i}/C) ^{2}]] $^{1/2}$]]} / (C ^{2})}}|}

Now, we consider spacecraft continuance into a 4th cosmic light-cone journey. We will consider that the fourth pellet derived spacecraft attains a Lorentz factor of γ_4 = [5 x (10 EXP 16)] after traveling across this fourth toy model cosmic light-cone to obtain a velocity of {c – [(2)(c)[10 EXP [-[(2)[log (2γ_4)]]]]]} with respect to its then proximate space-time background. The spacecraft recessional velocity with respect to the first pellet derived spacecraft's very first point of departure will be

{[(1 + 0.99999998) c] + (1c) + (2c) + (4c)} + {[(1 + 0.9999999999999998) c] + (1c) + (2c)} + {1c + {c – [(2)(c)[10 EXP [-[(2)[log (2γ_3)]]]]]} + (1c)} + {1c + {c – [(2)(c)[10 EXP [-[(2)[log (2γ_4)]]]]]}}

The ratio of the distance of travel by the super-light-cone traveling spacecraft to the distance viewable from the Milky Way Galaxy assuming that the cosmic light-cone radius remains the same due to constant universal expansion rate per unit of distance for the above scenario is:

{{{[(1 + 0.99999998) c] + (1c) + (2c) + (4c)} + {[(1 + 0.9999999999999998) c] + (1c) + (2c)} + {1c + {c – [(2)(c)[10 EXP [-[(2)[log (2γ_3)]]]]]} + (1c)} + {1c + {c – [(2)(c)[10 EXP [-[(2)[log (2γ_4)]]]]]}} (Background Travel Time)}/(Cosmic Light-Cone Radius)}

For constant ship frame Lorentz factors and ship-frame time, the following formula applies:

{{{{[(1 + 0.99999998) c] + (1c) + (2c) + (4c)} + {[(1 + 0.9999999999999998) c] + (1c) + (2c)} + {1c + {c – [(2)(c)[10 EXP [-[(2)[log (2γ_3)]]]]]} + (1c)} + {1c + {c – [(2)(c)[10 EXP [-[(2)[log (2γ_4)]]]]]}} (Spacecraft Reference Frame Travel Time)}/(Cosmic Light-Cone Radius)}(γ)}

For constant ship frame acceleration, the formula becomes:

{{{{[(1 + 0.99999998) c] + (1c) + (2c) + (4c)} + {[(1 + 0.9999999999999998) c] + (1c) + (2c)} + {1c + {c – [(2)(c)[10 EXP [-[(2)[log (2γ_3)]]]]]} + (1c)} + {1c + {c – [(2)(c)[10 EXP [-[(2)[log (2γ_4)]]]]]}} (Spacecraft Reference Frame Travel Time)}/(Cosmic Light-Cone Radius)} |{(c/g) ln {{[[(C^2) + (V_o^2)]$^{1/2}$] - [V_o/[[1 - [(V_o/C)^2]]$^{1/2}$]]} {[[C^2] + [[(g)(t) + [V_o /[1 - [(V_o/C)^2]]$^{1/2}$]^2]]$^{1/2}$] + [(g)(t)] + [V_o/[[1 - [(V_o/C)^2]]$^{1/2}$]]} / [C^2]}}|}

So for varying ship-frame acceleration, the formula becomes:

{{{{[(1 + 0.99999998) c] + (1c) + (2c) + (4c)} + {[(1 + 0.9999999999999998) c] + (1c) + (2c)} + {1c + {c – [(2)(c)[10 EXP [-[(2)[log (2γ_3)]]]]]} + (1c)} + {1c + {c – [(2)(c)[10 EXP [-[(2)[log (2γ_4)]]]]]}} (Spacecraft Reference Frame Travel Time)}/(Cosmic Light-Cone Radius)} |{Σ (I =1, I =n) {(C/g_i) ln {{[[(C^2) + (V_{0i}^2)]$^{1/2}$] – [V_{0i}/[[1 – [(V_{0i}/C)^2]]$^{1/2}$]]} { [(C

^{2}) + [[(g_i)(t_i) + [V_{0i} /[1 – [(V_{0i}/C) ^{2}]] $^{1/2}$] ^{2}]] $^{1/2}$] + [(g_i)(t_i)] + [V_{0i}/[[1 – [(V_{0i}/C) ^{2}]] $^{1/2}$]]} / (C ^{2})}}|}

Now, we consider spacecraft continuance into a 5th cosmic light-cone journey. We will consider that the fifth pellet derived spacecraft attains a Lorentz factor of γ_5 = [5 x (10 EXP 20)] after traveling across this fifth toy model cosmic light-cone to obtain a velocity of {c – [(2)(c)[10 EXP [-[(2)[log (2γ_5)]]]]]} with respect to its then proximate space-time background. The spacecraft recessional velocity with respect to the first pellet derived spacecraft's very first point of departure will be

{[(1 + 0.99999998) c] + (1c) + (2c) + (4c) + (8c)} + {[(1 + 0.9999999999999998) c] + (1c) + (2c) + (4c)} + {1c + {c – [(2)(c)[10 EXP [-[(2)[log (2γ_3)]]]]]} + (1c) + (2c)} + {1c + {c – [(2)(c)[10 EXP [-[(2)[log (2γ_4)]]]]]} + (1c)} + {1 + {c – [(2)(c)[10 EXP [-[(2)[log (2γ_5)]]]]]}}

The ratio of the distance of travel by the super-light-cone traveling spacecraft to the distance viewable from the Milky Way Galaxy assuming that the cosmic light-cone radius remains the same due to constant universal expansion rate per unit of distance for the above scenario is:

{{{[(1 + 0.99999998) c] + (1c) + (2c) + (4c) + (8c)} + {[(1 + 0.9999999999999998) c] + (1c) + (2c) + (4c)} + {1c + {c – [(2)(c)[10 EXP [-[(2)[log (2γ_3)]]]]]} + (1c) + (2c)} + {1c + {c – [(2)(c)[10 EXP [-[(2)[log (2γ_4)]]]]]} + (1c)} + {1 + {c – [(2)(c)[10 EXP [-[(2)[log (2γ_5)]]]]]}} (Background Travel Time)}/(Cosmic Light-Cone Radius)}

For constant ship frame Lorentz factors and ship-frame time, the following formula applies:

{{{{[(1 + 0.99999998) c] + (1c) + (2c) + (4c) + (8c)} + {[(1 + 0.9999999999999998) c] + (1c) + (2c) + (4c)} + {1c + {c – [(2)(c)[10 EXP [-[(2)[log (2γ_3)]]]]]} + (1c) + (2c)} + {1c + {c – [(2)(c)[10 EXP [-[(2)[log (2γ_4)]]]]]} + (1c)} + {1 + {c – [(2)(c)[10 EXP [-[(2)[log (2γ_5)]]]]]}} (Spacecraft Reference Frame Travel Time)}/(Cosmic Light-Cone Radius)}(γ)}

For constant ship frame acceleration, the formula becomes:

{{{{[(1 + 0.99999998) c] + (1c) + (2c) + (4c) + (8c)} + {[(1 + 0.9999999999999998) c] + (1c) + (2c) + (4c)} + {1c + {c – [(2)(c)[10 EXP [-[(2)[log (2γ_3)]]]]]} + (1c) + (2c)} + {1c + {c – [(2)(c)[10 EXP [-[(2)[log (2γ_4)]]]]]} + (1c)} + {1 + {c – [(2)(c)[10 EXP [-[(2)[log (2γ_5)]]]]]}} (Spacecraft Reference Frame Travel Time)}/(Cosmic Light-Cone Radius)} |{(c/g) ln {{[[(C ^{2}) + (V_o ^{2})]$^{1/2}$] - [V_o/[[1 - [(V_o/C) ^{2}]]$^{1/2}$]]} {[[C^{2}] + [[(g)(t) + [V_o /[1 - [(V_o/C) ^{2}]] $^{1/2}$] ^{2}]] $^{1/2}$] + [(g)(t)] + [V_o/[[1 - [(V_o/C) ^{2}]] $^{1/2}$]]} / [C ^{2}]}}|}

So for varying ship-frame acceleration, the formula becomes:

{{{{[(1 + 0.99999998) c] + (1c) + (2c) + (4c) + (8c)} + {[(1 + 0.9999999999999998) c] + (1c) + (2c) + (4c)} + {1c + {c – [(2)(c)[10 EXP [-[(2)[log (2γ_3)]]]]]} + (1c) + (2c)} + {1c + {c – [(2)(c)[10 EXP [-[(2)[log (2γ_4)]]]]]} + (1c)} + {1 + {c – [(2)(c)[10 EXP [-[(2)[log (2γ_5)]]]]]}} (Spacecraft Reference Frame Travel Time)}/(Cosmic Light-Cone Radius)} |{Σ (I =1, I =n) {(C/g_i) ln {{[[(C^2) + (V_{0i}^2)] $^{1/2}$] – [V_{0i}/[[1 – [(V_{0i}/C) ^{2}]] $^{1/2}$]]} { [(C^2) + [[(g_i)(t_i) + [V_{0i} /[1 – [(V_{0i}/C) ^{2}]] $^{1/2}$] ^{2}]] $^{1/2}$] + [(g_i)(t_i)] + [V_{0i}/[[1 – [(V_{0i}/C) ^{2}]] $^{1/2}$]]} / (C^2)}}|}

Now, we consider spacecraft continuance into a 6th cosmic light-cone journey. We will consider that the sixth pellet derived spacecraft attains a Lorentz factor of γ_6 = [5 x (10 EXP 24)] after traveling across this 6th toy model cosmic light-cone to obtain a velocity of {c – [(2)(c)[10 EXP [-[(2)[log (2γ_6)]]]]]} with respect to its then proximate space-time background. The spacecraft recessional velocity with respect to the first pellet derived spacecraft's very first point of departure will be

{[(1 + 0.99999998) c] + (1c) + (2c) + (4c) + (8c) + (16c)} + {[(1 + 0.9999999999999998) c] + (1c) + (2c) + (4c) +(8c)} + {1c + {c – [(2)(c)[10 EXP [-[(2)[log (2γ_3)]]]]]} + (1c) + (2c) + (4c)} + {1c + {c – [(2)(c)[10 EXP [-[(2)[log (2γ_4)]]]]]} + (1c) + (2c)} + {1c + {c – [(2)(c)[10 EXP [-[(2)[log (2γ_5)]]]]]} + (1c)} + {1c + {c – [(2)(c)[10 EXP [-[(2)[log (2γ_6)]]]]]}}

The ratio of the distance of travel by the super-light-cone traveling spacecraft to the distance viewable from the Milky Way Galaxy assuming that the cosmic light-cone radius remains the same due to constant universal expansion rate per unit of distance for the above scenario is:

{{{[(1 + 0.99999998) c] + (1c) + (2c) + (4c) + (8c) + (16c)} + {[(1 + 0.9999999999999998) c] + (1c) + (2c) + (4c) +(8c)} + {1c + {c – [(2)(c)[10 EXP [-[(2)[log (2γ_3)]]]]]} + (1c) + (2c) + (4c)} + {1c + {c – [(2)(c)[10 EXP [-[(2)[log (2γ_4)]]]]]} + (1c) + (2c)} + {1c + {c – [(2)(c)[10 EXP [-[(2)[log (2γ_5)]]]]]} + (1c)} + {1c + {c – [(2)(c)[10 EXP [-[(2)[log (2γ_6)]]]]]}} (Background Travel Time)}/(Cosmic Light-Cone Radius)}

For constant ship frame Lorentz factors and ship-frame time, the following formula applies:

{{{{[(1 + 0.99999998) c] + (1c) + (2c) + (4c) + (8c) + (16c)} + {[(1 + 0.9999999999999998) c] + (1c) + (2c) + (4c) +(8c)} + {1c + {c – [(2)(c)[10 EXP [-[(2)[log (2γ_3)]]]]]} + (1c) + (2c) + (4c)} + {1c + {c – [(2)(c)[10 EXP [-[(2)[log (2γ_4)]]]]]} + (1c) + (2c)} + {1c + {c – [(2)(c)[10 EXP [-[(2)[log (2γ_5)]]]]]} + (1c)} + {1c + {c – [(2)(c)[10 EXP [-[(2)[log (2γ_6)]]]]]}} (Spacecraft Reference Frame Travel Time)}/(Cosmic Light-Cone Radius)}(γ)}

For constant ship frame acceleration, the formula becomes:

{{{{[(1 + 0.99999998) c] + (1c) + (2c) + (4c) + (8c) + (16c)} + {[(1 + 0.9999999999999998) c] + (1c) + (2c) + (4c) +(8c)} + {1c + {c – [(2)(c)[10 EXP [-[(2)[log (2γ_3)]]]]]} + (1c) + (2c) + (4c)} + {1c + {c – [(2)(c)[10 EXP [-[(2)[log (2γ_4)]]]]]]} + (1c) + (2c)} + {1c + {c – [(2)(c)[10 EXP [-[(2)[log (2γ_5)]]]]]]} + (1c)} + {1c + {c – [(2)(c)[10 EXP [-[(2)[log (2γ_6)]]]]]]}} (Spacecraft Reference Frame Travel Time)}/(Cosmic Light-Cone Radius)} |{(c/g) ln {{[[(C ^{2}) + (V_o ^{2})]$^{1/2}$] - [V_o/[[1 - [(V_o/C) ^{2}]]$^{1/2}$]]} {[[C^2] + [[(g)(t) + [V_o /[1 - [(V_o/C) ^{2}]] $^{1/2}$] ^{2}]] $^{1/2}$] + [(g)(t)] + [V_o/[[1 - [(V_o/C) ^{2}]] $^{1/2}$]]} / [C ^{2}]}}|}

So for varying ship-frame acceleration, the formula becomes:

{{{{[(1 + 0.99999998) c] + (1c) + (2c) + (4c) + (8c) + (16c)} + {[(1 + 0.9999999999999998) c] + (1c) + (2c) + (4c) +(8c)} + {1c + {c – [(2)(c)[10 EXP [-[(2)[log (2γ_3)]]]]]} + (1c) + (2c) + (4c)} + {1c + {c – [(2)(c)[10 EXP [-[(2)[log (2γ_4)]]]]]]} + (1c) + (2c)} + {1c + {c – [(2)(c)[10 EXP [-[(2)[log (2γ_5)]]]]]]} + (1c)} + {1c + {c – [(2)(c)[10 EXP [-[(2)[log (2γ_6)]]]]]]}} (Spacecraft Reference Frame Travel Time)}/(Cosmic Light-Cone Radius)} |{Σ (I =1, I =n) {(C/g_i) ln {{[[(C^2) + (V_{0i} ^{2})] $^{1/2}$] – [V_{0i}/[[1 – [(V_{0i}/C) ^{2}]] $^{1/2}$]]} { [(C ^{2}) + [[(g_i)(t_i) + [V_{0i} /[1 – [(V_{0i}/C) ^{2}]] $^{1/2}$] ^{2}]] $^{1/2}$] + [(g_i)(t_i)] + [V_{0i}/[[1 – [(V_{0i}/C) ^{2}]] $^{1/2}$]]} / (C ^{2})}}}|}

Now, we consider spacecraft continuance into a 7th cosmic light-cone journey. We will consider that the 7th pellet derived spacecraft attains a Lorentz factor of γ_7 = [5 x (10 EXP 28)] after traveling across this 7th toy model cosmic light-cone to obtain a velocity of {c – [(2)(c)[10 EXP [-[(2)[log (2γ_7)]]]]]} with respect to its then proximate space-time background. The spacecraft recessional velocity with respect to the first pellet derived spacecraft's very first point of departure will be

{{[(1 + 0.99999998) c] + (1c)} + (2c) + (4c) + (8c) + (16c) + (32c)} + {[(1 + 0.9999999999999998) c] + (1c) + (2c) + (4c) +(8c) + (16c)} + {1c + {c – [(2)(c)[10 EXP [-[(2)[log (2γ_3)]]]]]} + (1c) + (2c) + (4c) + (8c)} + {1 + {c – [(2)(c)[10 EXP [-[(2)[log (2γ_4)]]]]]} + (1c) + (2c) + (4)} + {1c + {c – [(2)(c)[10 EXP [-[(2)[log (2γ_5)]]]]]]} + (1c) + (2c)} + {1c + {c – [(2)(c)[10 EXP [-[(2)[log (2γ_6)]]]]]} + (1c)} + {1c + {c – [(2)(c)[10 EXP [-[(2)[log (2γ_7)]]]]]]}}

The ratio of the distance of travel by the super-light-cone traveling spacecraft to the distance viewable from the Milky Way Galaxy assuming that the cosmic light-cone radius remains the same due to constant universal expansion rate per unit of distance for the above scenario is:

{{{{[(1 + 0.99999998) c] + (1c)} + (2c) + (4c) + (8c) + (16c) + (32c)} + {[(1 + 0.9999999999999998) c] + (1c) + (2c) + (4c) +(8c) + (16c)} + {1c + {c – [(2)(c)[10 EXP [-[(2)[log (2γ_3)]]]]]} + (1c) + (2c) + (4c) + (8c)} + {1 + {c – [(2)(c)[10 EXP [-[(2)[log (2γ_4)]]]]]} + (1c) + (2c) + (4)} + {1c + {c – [(2)(c)[10 EXP [-[(2)[log (2γ_5)]]]]]]} + (1c) + (2c)}

+ {1c + {c – [(2)(c)[10 EXP [-[(2)[log (2γ_6)]]]]]} + (1c)} + {1c + {c – [(2)(c)[10 EXP [-[(2)[log (2γ_7)]]]]]}} (Background Travel Time)}/(Cosmic Light-Cone Radius)}

For constant ship frame Lorentz factors and ship-frame time, the following formula applies:

{{{{{[(1 + 0.99999998) c] + (1c)} + (2c) + (4c) + (8c) + (16c) + (32c)} + {[(1 + 0.9999999999999998) c] + (1c) + (2c) + (4c) +(8c) + (16c)} + {1c + {c – [(2)(c)[10 EXP [-[(2)[log (2γ_3)]]]]]} + (1c) + (2c) + (4c) + (8c)} + {1 + {c – [(2)(c)[10 EXP [-[(2)[log (2γ_4)]]]]]} + (1c) + (2c) + (4)} + {1c + {c – [(2)(c)[10 EXP [-[(2)[log (2γ_5)]]]]]} + (1c) + (2c)} + {1c + {c – [(2)(c)[10 EXP [-[(2)[log (2γ_6)]]]]]} + (1c)} + {1c + {c – [(2)(c)[10 EXP [-[(2)[log (2γ_7)]]]]]}} (Spacecraft Reference Frame Travel Time)}/(Cosmic Light-Cone Radius)}(γ)}

For constant ship frame acceleration, the formula becomes:

{{{{{[(1 + 0.99999998) c] + (1c)} + (2c) + (4c) + (8c) + (16c) + (32c)} + {[(1 + 0.9999999999999998) c] + (1c) + (2c) + (4c) +(8c) + (16c)} + {1c + {c – [(2)(c)[10 EXP [-[(2)[log (2γ_3)]]]]]} + (1c) + (2c) + (4c) + (8c)} + {1 + {c – [(2)(c)[10 EXP [-[(2)[log (2γ_4)]]]]]} + (1c) + (2c) + (4)} + {1c + {c – [(2)(c)[10 EXP [-[(2)[log (2γ_5)]]]]]} + (1c) + (2c)} + {1c + {c – [(2)(c)[10 EXP [-[(2)[log (2γ_6)]]]]]} + (1c)} + {1c + {c – [(2)(c)[10 EXP [-[(2)[log (2γ_7)]]]]]}} (Spacecraft Reference Frame Travel Time)}/(Cosmic Light-Cone Radius)} |{(c/g) ln {{[[(C ^{2}) + (V_o ^{2})]$^{1/2}$] - [V_o/[[1 - [(V_o/C) ^{2}]]$^{1/2}$]]} {[[C^2] + [[(g)(t) + [V_o /[1 - [(V_o/C) ^{2}]] $^{1/2}$] ^{2}]] $^{1/2}$] + [(g)(t)] + [V_o/[[1 - [(V_o/C) ^{2}]] $^{1/2}$]]} / [C ^{2}]}}|}

So for varying ship-frame acceleration, the formula becomes:

{{{{{[(1 + 0.99999998) c] + (1c)} + (2c) + (4c) + (8c) + (16c) + (32c)} + {[(1 + 0.9999999999999998) c] + (1c) + (2c) + (4c) +(8c) + (16c)} + {1c + {c – [(2)(c)[10 EXP [-[(2)[log (2γ_3)]]]]]} + (1c) + (2c) + (4c) + (8c)} + {1 + {c – [(2)(c)[10 EXP [-[(2)[log (2γ_4)]]]]]} + (1c) + (2c) + (4)} + {1c + {c – [(2)(c)[10 EXP [-[(2)[log (2γ_5)]]]]]} + (1c) + (2c)} + {1c + {c – [(2)(c)[10 EXP [-[(2)[log (2γ_6)]]]]]} + (1c)} + {1c + {c – [(2)(c)[10 EXP [-[(2)[log (2γ_7)]]]]]}} (Spacecraft Reference Frame Travel Time)}/(Cosmic Light-Cone Radius)} |{Σ (I =1, I =n) {(C/g_i) ln {{[[(C^2) + (V_{0i} ^{2})] $^{1/2}$] – [V_{0i}/[[1 – [(V_{0i}/C) ^{2}]] $^{1/2}$]]} { [(C ^{2}) + [[(g_i)(t_i) + [V_{0i} /[1 – [(V_{0i}/C) ^{2}]] $^{1/2}$] ^{2}]] $^{1/2}$] + [(g_i)(t_i)] + [V_{0i}/[[1 – [(V_{0i}/C) ^{2}]] $^{1/2}$]]} / (C ^{2})}}|}

Now, we consider spacecraft continuance into an 8th cosmic light-cone journey. We will consider that the 8th pellet derived spacecraft attains a Lorentz factor of γ_8 = [5 x (10 EXP 32)] after traveling across this 8th toy model cosmic light-cone to obtain a velocity of {c – [(2)(c)[10 EXP [-[(2)[log (2γ_8)]]]]]} with respect to its then proximate space-time background. The spacecraft recessional velocity with respect to the first pellet derived spacecraft's very first point of departure will be

{[(1 + 0.99999998) c] + (1c) + (2c) + (4c) + (8c) + (16c) + (32c) + (64c)} + {[(1 + 0.9999999999999998) c] + (1c) + (2c) + (4c) +(8c) + (16c) + (32c)} + {1c + {c – [(2)(c)[10 EXP [-[(2)[log (2γ_3)]]]]]} + (1c) + (2c) + (4c) + (8c) + (16c)} + {1c + {c – [(2)(c)[10 EXP [-[(2)[log (2γ_4)]]]]]} + (1c) + (2c) + (4c) + (8c)} + {1c + {c – [(2)(c)[10 EXP [-[(2)[log (2γ_5)]]]]]} + (1c) + (2c) + (4c)} + {1c + {c – [(2)(c)[10 EXP [-[(2)[log (2γ_6)]]]]]} + (1c) + (2c)} + {1c + {c– [(2)(c)[10 EXP [-[(2)[log (2γ_7)]]]]]} + (1c)} + {1c + {c – [(2)(c)[10 EXP [-[(2)[log (2γ_8)]]]]]}}

The ratio of the distance of travel by the super-light-cone traveling spacecraft to the distance viewable from the Milky Way Galaxy assuming that the cosmic light-cone radius remains the same due to constant universal expansion rate per unit of distance for the above scenario is:

{{{[(1 + 0.99999998) c] + (1c) + (2c) + (4c) + (8c) + (16c) + (32c) + (64c)} + {[(1 + 0.9999999999999998) c] + (1c) + (2c) + (4c) +(8c) + (16c) + (32c)} + {1c + {c – [(2)(c)[10 EXP [-[(2)[log (2γ_3)]]]]]} + (1c) + (2c) + (4c) + (8c) + (16c)} + {1c + {c – [(2)(c)[10 EXP [-[(2)[log (2γ_4)]]]]]} + (1c) + (2c) + (4c) + (8c)} + {1c + {c – [(2)(c)[10 EXP [-[(2)[log (2γ_5)]]]]]} + (1c) + (2c) + (4c)} + {1c + {c – [(2)(c)[10 EXP [-[(2)[log (2γ_6)]]]]]} + (1c) + (2c)} + {1c + {c– [(2)(c)[10 EXP [-[(2)[log (2γ_7)]]]]]} + (1c)} + {1c + {c – [(2)(c)[10 EXP [-[(2)[log (2γ_8)]]]]]}} (Background Travel Time)}/(Cosmic Light-Cone Radius)}

For constant ship frame Lorentz factors and ship-frame time, the following formula applies:

{{{{[(1 + 0.99999998) c] + (1c) + (2c) + (4c) + (8c) + (16c) + (32c) + (64c)} + {[(1 + 0.9999999999999998) c] + (1c) + (2c) + (4c) +(8c) + (16c) + (32c)} + {1c + {c – [(2)(c)[10 EXP [-[(2)[log (2γ_3)]]]]]} + (1c) + (2c) + (4c) + (8c) + (16c)} + {1c + {c – [(2)(c)[10 EXP [-[(2)[log (2γ_4)]]]]]} + (1c) + (2c) + (4c) + (8c)} + {1c + {c – [(2)(c)[10 EXP [-[(2)[log (2γ_5)]]]]]} + (1c) + (2c) + (4c)} + {1c + {c – [(2)(c)[10 EXP [-[(2)[log (2γ_6)]]]]]} + (1c) + (2c)} + {1c + {c– [(2)(c)[10 EXP [-[(2)[log (2γ_7)]]]]]} + (1c)} + {1c + {c – [(2)(c)[10 EXP [-[(2)[log (2γ_8)]]]]]}} (Spacecraft Reference Frame Travel Time)}/(Cosmic Light-Cone Radius)}(γ)}

For constant ship frame acceleration, the formula becomes:

{{{{[(1 + 0.99999998) c] + (1c) + (2c) + (4c) + (8c) + (16c) + (32c) + (64c)} + {[(1 + 0.9999999999999998) c] + (1c) + (2c) + (4c) +(8c) + (16c) + (32c)} + {1c + {c – [(2)(c)[10 EXP [-[(2)[log (2γ_3)]]]]]} + (1c) + (2c) + (4c) + (8c) + (16c)} + {1c + {c – [(2)(c)[10 EXP [-[(2)[log (2γ_4)]]]]]} + (1c) + (2c) + (4c) + (8c)} + {1c + {c – [(2)(c)[10 EXP [-[(2)[log (2γ_5)]]]]]} + (1c) + (2c) + (4c)} + {1c + {c – [(2)(c)[10 EXP [-[(2)[log (2γ_6)]]]]]} + (1c) + (2c)} + {1c + {c– [(2)(c)[10 EXP [-[(2)[log (2γ_7)]]]]]} + (1c)} + {1c + {c – [(2)(c)[10

EXP [-[(2)[log $(2\gamma_8)$]]]]]}} (Spacecraft Reference Frame Travel Time)}/(Cosmic Light-Cone Radius)} |{(c/g) ln {{[[(C ^{2}) + (V_o ^{2})]$^{1/2}$] - [V_o/[[1 - [(V_o/C) ^{2}]]$^{1/2}$]]} {[[C^2] + [[(g)(t) + [V_o /[1 - [(V_o/C) ^{2}]] $^{1/2}$] ^{2}]] $^{1/2}$] + [(g)(t)] + [V_o/[[1 - [(V_o/C) ^{2}]] $^{1/2}$]]} / [C ^{2}]}}|}

So for varying ship-frame acceleration, the formula becomes:

{{{{[(1 + 0.99999998) c] + (1c) + (2c) + (4c) + (8c) + (16c) + (32c) + (64c)} + {[(1 + 0.9999999999999998) c] + (1c) + (2c) + (4c) +(8c) + (16c) + (32c)} + {1c + {c – [(2)(c)[10 EXP [-[(2)[log $(2\gamma_3)$]]]]]} + (1c) + (2c) + (4c) + (8c) + (16c)} + {1c + {c – [(2)(c)[10 EXP [-[(2)[log $(2\gamma_4)$]]]]]} + (1c) + (2c) + (4c) + (8c)} + {1c + {c – [(2)(c)[10 EXP [-[(2)[log $(2\gamma_5)$]]]]]} + (1c) + (2c) + (4c)} + {1c + {c – [(2)(c)[10 EXP [-[(2)[log $(2\gamma_6)$]]]]]} + (1c) + (2c)} + {1c + {c– [(2)(c)[10 EXP [-[(2)[log $(2\gamma_7)$]]]]]} + (1c)} + {1c + {c – [(2)(c)[10 EXP [-[(2)[log $(2\gamma_8)$]]]]]}} (Spacecraft Reference Frame Travel Time)}/(Cosmic Light-Cone Radius)} |{Σ (I =1, I =n) {(C/g_i) ln {{[[(C^2) + (V_{0i} ^{2})] $^{1/2}$] – [V_{0i}/[[1 – [(V_{0i}/C) ^{2}]] $^{1/2}$]]} { [(C ^{2}) + [[(g_i)(t_i) + [V_{0i} /[1 – [(V_{0i}/C) ^{2}]] $^{1/2}$] ^{2}]] $^{1/2}$] + [(g_i)(t_i)] + [V_{0i}/[[1 – [(V_{0i}/C) ^{2}]] $^{1/2}$]]} / (C ^{2})}}}|}

Now, we consider spacecraft continuance into a 9th cosmic light-cone journey. We will consider that the 9th pellet derived spacecraft attains a Lorentz factor of γ_9 = [5 x (10 EXP 36)] after traveling across this 9th toy model cosmic light-cone to obtain a velocity of {c –[(2)(c)[10 EXP [-[(2)[log $(2\gamma_9)$]]]]]} with respect to its then proximate space-time background. The spacecraft recessional velocity with respect to the first pellet derived spacecraft's very first point of departure will be

{{[(1 + 0.99999998) c] + (1c)} + (2c) + (4c) + (8c) + (16c) + (32c) + (64c) + (128c)} + {[(1 + 0.9999999999999998) c] + (1c) + (2c) + (4c) +(8c) + (16c) + (32c) + (64c)} + {1c + {c – [(2)(c)[10 EXP [-[(2)[log $(2\gamma_3)$]]]]]} + (1c) + (2c) + (4c) + (8c) + (16c) + (32c)} + {1c + {c – [(2)(c)[10 EXP [-[(2)[log $(2\gamma_4)$]]]]] } + (1c) + (2c) + (4c) + (8c) + (16c)} + {1c + {c – [(2)(c)[10 EXP [-[(2)[log $(2\gamma_5)$]]]]]} + (1c) + (2c) + (4c) +(8c)} + {1c + {c – [(2)(c)[10 EXP [-[(2)[log $(2\gamma_6)$]]]]]} + (1c) + (2c) + (4c)} + {1c + {c – [(2)(c)[10 EXP [-[(2)[log $(2\gamma_7)$]]]]]} + (1c) + (2)} + {1c + {c – [(2)(c)[10 EXP [-[(2)[log $(2\gamma_8)$]]]]]} + (1c)} + {1c + {c – [(2)(c)[10 EXP [-[(2)[log $(2\gamma_9)$]]]]]}}

The ratio of the distance of travel by the super-light-cone traveling spacecraft to the distance viewable from the Milky Way Galaxy assuming that the cosmic light-cone radius remains the same due to constant universal expansion rate per unit of distance for the above scenario is:

{{{{[(1 + 0.99999998) c] + (1c)} + (2c) + (4c) + (8c) + (16c) + (32c) + (64c) + (128c)} + {[(1 + 0.9999999999999998) c] + (1c) + (2c) + (4c) +(8c) + (16c) + (32c) + (64c)} + {1c + {c – [(2)(c)[10 EXP [-[(2)[log $(2\gamma_3)$]]]]]} + (1c) + (2c) + (4c) + (8c) + (16c) + (32c)} + {1c + {c – [(2)(c)[10 EXP [-[(2)[log $(2\gamma_4)$]]]]] } + (1c) + (2c) + (4c) + (8c) + (16c)} + {1c + {c – [(2)(c)[10 EXP [-[(2)[log $(2\gamma_5)$]]]]]} + (1c) + (2c) + (4c) +(8c)} + {1c + {c – [(2)(c)[10 EXP [-

[(2)[log (2γ_6)]]]]]} + (1c) + (2c) + (4c)} + {1c + {c – [(2)(c)[10 EXP [-[(2)[log (2γ_7)]]]]]} + (1c) + (2)} + {1c + {c – [(2)(c)[10 EXP [-[(2)[log (2γ_8)]]]]]} + (1c)} + {1c + {c – [(2)(c)[10 EXP [-[(2)[log (2γ_9)]]]]]]}} (Background Travel Time)}/(Cosmic Light-Cone Radius)}

For constant ship frame Lorentz factors and ship-frame time, the following formula applies:

{{{{{[(1 + 0.99999998) c] + (1c)} + (2c) + (4c) + (8c) + (16c) + (32c) + (64c) + (128c)} + {[(1 + 0.9999999999999998) c] + (1c) + (2c) + (4c) +(8c) + (16c) + (32c) + (64c)} + {1c + {c – [(2)(c)[10 EXP [-[(2)[log (2γ_3)]]]]]]} + (1c) + (2c) + (4c) + (8c) + (16c) + (32c)} + {1c + {c – [(2)(c)[10 EXP [-[(2)[log (2γ_4)]]]]] } + (1c) + (2c) + (4c) + (8c) + (16c)} + {1c + {c – [(2)(c)[10 EXP [-[(2)[log (2γ_5)]]]]]} + (1c) + (2c) + (4c) +(8c)} + {1c + {c – [(2)(c)[10 EXP [-[(2)[log (2γ_6)]]]]]} + (1c) + (2c) + (4c)} + {1c + {c – [(2)(c)[10 EXP [-[(2)[log (2γ_7)]]]]]} + (1c) + (2)} + {1c + {c – [(2)(c)[10 EXP [-[(2)[log (2γ_8)]]]]]} + (1c)} + {1c + {c – [(2)(c)[10 EXP [-[(2)[log (2γ_9)]]]]]]}} (Spacecraft Reference Frame Travel Time)}/(Cosmic Light-Cone Radius)}(γ)}

For constant ship frame acceleration, the formula becomes:

{{{{{[(1 + 0.99999998) c] + (1c)} + (2c) + (4c) + (8c) + (16c) + (32c) + (64c) + (128c)} + {[(1 + 0.9999999999999998) c] + (1c) + (2c) + (4c) +(8c) + (16c) + (32c) + (64c)} + {1c + {c – [(2)(c)[10 EXP [-[(2)[log (2γ_3)]]]]]]} + (1c) + (2c) + (4c) + (8c) + (16c) + (32c)} + {1c + {c – [(2)(c)[10 EXP [-[(2)[log (2γ_4)]]]]] } + (1c) + (2c) + (4c) + (8c) + (16c)} + {1c + {c – [(2)(c)[10 EXP [-[(2)[log (2γ_5)]]]]]} + (1c) + (2c) + (4c) +(8c)} + {1c + {c – [(2)(c)[10 EXP [-[(2)[log (2γ_6)]]]]]} + (1c) + (2c) + (4c)} + {1c + {c – [(2)(c)[10 EXP [-[(2)[log (2γ_7)]]]]]} + (1c) + (2)} + {1c + {c – [(2)(c)[10 EXP [-[(2)[log (2γ_8)]]]]]} + (1c)} + {1c + {c – [(2)(c)[10 EXP [-[(2)[log (2γ_9)]]]]]]}} (Spacecraft Reference Frame Travel Time)}/(Cosmic Light-Cone Radius)} |{(c/g) ln {{[[(C^2) + ($V_o{}^2$)]$^{1/2}$] - [V_o/[[1 - [(V_o/C) ^{2}]]$^{1/2}$]]} {[[C^2] + [[(g)(t) + [V_o /[1 - [(V_o/C) ^{2}]] $^{1/2}$] ^{2}]] $^{1/2}$] + [(g)(t)] + [V_o/[[1 - [(V_o/C) ^{2}]] $^{1/2}$]]} / [C ^{2}]}}|}

So for varying ship-frame acceleration, the formula becomes:

{{{{{[(1 + 0.99999998) c] + (1c)} + (2c) + (4c) + (8c) + (16c) + (32c) + (64c) + (128c)} + {[(1 + 0.9999999999999998) c] + (1c) + (2c) + (4c) +(8c) + (16c) + (32c) + (64c)} + {1c + {c – [(2)(c)[10 EXP [-[(2)[log (2γ_3)]]]]]]} + (1c) + (2c) + (4c) + (8c) + (16c) + (32c)} + {1c + {c – [(2)(c)[10 EXP [-[(2)[log (2γ_4)]]]]] } + (1c) + (2c) + (4c) + (8c) + (16c)} + {1c + {c – [(2)(c)[10 EXP [-[(2)[log (2γ_5)]]]]]} + (1c) + (2c) + (4c) +(8c)} + {1c + {c – [(2)(c)[10 EXP [-[(2)[log (2γ_6)]]]]]} + (1c) + (2c) + (4c)} + {1c + {c – [(2)(c)[10 EXP [-[(2)[log (2γ_7)]]]]]} + (1c) + (2)} + {1c + {c – [(2)(c)[10 EXP [-[(2)[log (2γ_8)]]]]]} + (1c)} + {1c + {c – [(2)(c)[10 EXP [-[(2)[log (2γ_9)]]]]]]}} (Spacecraft Reference Frame Travel Time)}/(Cosmic Light-Cone Radius)} |{Σ (I =1, I =n) {(C/g_i) ln {{[[(C^2) + ($V_{0i}{}^2$)] $^{1/2}$] – [V_{0i}/[[1 – [(V_{0i}/C) ^{2}]] $^{1/2}$]]} { [(C^2) + [[(g_i)(t_i) + [V_{0i} /[1 – [(V_{0i}/C) ^{2}]] $^{1/2}$] ^{2}]] $^{1/2}$] + [(g_i)(t_i)] + [V_{0i}/[[1 – [(V_{0i}/C) ^{2}]] $^{1/2}$]]} / (C^2)}}|}

Now, we consider spacecraft continuance into a 10th cosmic light-cone journey. We will consider that the 10th pellet derived spacecraft attains a Lorentz factor of γ_{10} = [5 x (10 EXP 40)] after traveling across this 10th toy model cosmic light-cone to obtain a velocity of {c – [(2)(c)[10 EXP [-[(2)[log (2γ_{10})]]]]]} with respect to its then proximate space-time background. The spacecraft recessional velocity with respect to the first pellet derived spacecraft's very first point of departure will be

{{[(1 + 0.99999998) c] + (1c)} + (2c) + (4c) + (8c) + (16c) + (32c) + (64c) + (128c) + (256c)} + {[(1 + 0.9999999999999998) c] + (1c) + (2c) + (4c) +(8c) + (16c) + (32c) + (64c) + (128c)} + {1c + {c – [(2)(c)[10 EXP [-[(2)[log (2γ_3)]]]]]} + (1c) + (2c) + (4c) + (8c) + (16c) + (32c) + (64c)} + {1c + {c – [(2)(c)[10 EXP [-[(2)[log (2γ_4)]]]]]} + (1c) + (2c) + (4c) + (8c) + (16c) + (32c)} + {1c + {c – [(2)(c)[10 EXP [-[(2)[log (2γ_5)]]]]]} + (1c) + (2c) + (4c) +(8c) + (16c)} + {1c + {1c – [(2)(c)[10 EXP [-[(2)[log (2γ_6)]]]]]} + (1c) + (2c) + (4c) + (8c)} + {1c + {c – [(2)(c)[10 EXP [-[(2)[log (2γ_7)]]]]]} + (1c) + (2c) + (4c)} + {1c + {c – [(2)(c)[10 EXP [-[(2)[log (2γ_8)]]]]]} + (1c) + (2c)} + {1c + {c – [(2)(c)[10 EXP [-[(2)[log (2γ_9)]]]]]} + (1c)} + {c - [(2)(c)[10 EXP [-[(2)[log (2γ_{10})]]]]]}

The ratio of the distance of travel by the super-light-cone traveling spacecraft to the distance viewable from the Milky Way Galaxy assuming that the cosmic light-cone radius remains the same due to constant universal expansion rate per unit of distance for the above scenario is:

{{{{[(1 + 0.99999998) c] + (1c)} + (2c) + (4c) + (8c) + (16c) + (32c) + (64c) + (128c) + (256c)} + {[(1 + 0.9999999999999998) c] + (1c) + (2c) + (4c) +(8c) + (16c) + (32c) + (64c) + (128c)} + {1c + {c – [(2)(c)[10 EXP [-[(2)[log (2γ_3)]]]]]} + (1c) + (2c) + (4c) + (8c) + (16c) + (32c) + (64c)} + {1c + {c – [(2)(c)[10 EXP [-[(2)[log (2γ_4)]]]]]} + (1c) + (2c) + (4c) + (8c) + (16c) + (32c)} + {1c + {c – [(2)(c)[10 EXP [-[(2)[log (2γ_5)]]]]]} + (1c) + (2c) + (4c) +(8c) + (16c)} + {1c + {1c – [(2)(c)[10 EXP [-[(2)[log (2γ_6)]]]]]} + (1c) + (2c) + (4c) + (8c)} + {1c + {c – [(2)(c)[10 EXP [-[(2)[log (2γ_7)]]]]]} + (1c) + (2c) + (4c)} + {1c + {c – [(2)(c)[10 EXP [-[(2)[log (2γ_8)]]]]]} + (1c) + (2c)} + {1c + {c – [(2)(c)[10 EXP [-[(2)[log (2γ_9)]]]]]} + (1c)} + {c - [(2)(c)[10 EXP [-[(2)[log (2γ_{10})]]]]]} (Background Travel Time)}/(Cosmic Light-Cone Radius)}

For constant ship frame Lorentz factors and ship-frame time, the following formula applies:

{{{{{[(1 + 0.99999998) c] + (1c)} + (2c) + (4c) + (8c) + (16c) + (32c) + (64c) + (128c) + (256c)} + {[(1 + 0.9999999999999998) c] + (1c) + (2c) + (4c) +(8c) + (16c) + (32c) + (64c) + (128c)} + {1c + {c – [(2)(c)[10 EXP [-[(2)[log (2γ_3)]]]]]} + (1c) + (2c) + (4c) + (8c) + (16c) + (32c) + (64c)} + {1c + {c – [(2)(c)[10 EXP [-[(2)[log (2γ_4)]]]]]} + (1c) + (2c) + (4c) + (8c) + (16c) + (32c)} + {1c + {c – [(2)(c)[10 EXP [-[(2)[log (2γ_5)]]]]]} + (1c) + (2c) + (4c) +(8c) + (16c)} + {1c + {1c – [(2)(c)[10 EXP [-[(2)[log (2γ_6)]]]]]} + (1c) + (2c) + (4c) + (8c)}

+ {1c + {c – [(2)(c)[10 EXP [-[(2)[log (2γ_7)]]]]]} + (1c) + (2c) + (4c)} + {1c + {c – [(2)(c)[10 EXP [-[(2)[log (2γ_8)]]]]]} + (1c) + (2c)} + {1c + {c – [(2)(c)[10 EXP [-[(2)[log (2γ_9)]]]]]} + (1c)} + {c - [(2)(c)[10 EXP [-[(2)[log (2γ_{10})]]]]]} (Spacecraft Reference Frame Travel Time)}/(Cosmic Light-Cone Radius)}(γ)}

For constant ship frame acceleration, the formula becomes:

{{{{{[(1 + 0.99999998) c] + (1c)} + (2c) + (4c) + (8c) + (16c) + (32c) + (64c) + (128c) + (256c)} + {[(1 + 0.9999999999999998) c] + (1c) + (2c) + (4c) +(8c) + (16c) + (32c) + (64c) + (128c)} + {1c + {c – [(2)(c)[10 EXP [-[(2)[log (2γ_3)]]]]]} + (1c) + (2c) + (4c) + (8c) + (16c) + (32c) + (64c)} + {1c + {c – [(2)(c)[10 EXP [-[(2)[log (2γ_4)]]]]]} + (1c) + (2c) + (4c) + (8c) + (16c) + (32c)} + {1c + {c – [(2)(c)[10 EXP [-[(2)[log (2γ_5)]]]]]} + (1c) + (2c) + (4c) +(8c) + (16c)} + {1c + {1c – [(2)(c)[10 EXP [-[(2)[log (2γ_6)]]]]]} + (1c) + (2c) + (4c) + (8c)} + {1c + {c – [(2)(c)[10 EXP [-[(2)[log (2γ_7)]]]]]} + (1c) + (2c) + (4c)} + {1c + {c – [(2)(c)[10 EXP [-[(2)[log (2γ_8)]]]]]} + (1c) + (2c)} + {1c + {c – [(2)(c)[10 EXP [-[(2)[log (2γ_9)]]]]]} + (1c)} + {c - [(2)(c)[10 EXP [-[(2)[log (2γ_{10})]]]]]} (Spacecraft Reference Frame Travel Time)}/(Cosmic Light-Cone Radius)} |{(c/g) ln {{[[(C^2) + (V_o^2)]$^{1/2}$] - [V_o/[[1 - [(V_o/C)^2]]$^{1/2}$]]} {[[C^2] + [[(g)(t) + [V_o /[1 - [(V_o/C)^2]]$^{1/2}$]^2]]$^{1/2}$] + [(g)(t)] + [V_o/[[1 - [(V_o/C)^2]]$^{1/2}$]]} / [C^2]}}|}

So for varying ship-frame acceleration, the formula becomes:

{{{{{[(1 + 0.99999998) c] + (1c)} + (2c) + (4c) + (8c) + (16c) + (32c) + (64c) + (128c) + (256c)} + {[(1 + 0.9999999999999998) c] + (1c) + (2c) + (4c) +(8c) + (16c) + (32c) + (64c) + (128c)} + {1c + {c – [(2)(c)[10 EXP [-[(2)[log (2γ_3)]]]]]} + (1c) + (2c) + (4c) + (8c) + (16c) + (32c) + (64c)} + {1c + {c – [(2)(c)[10 EXP [-[(2)[log (2γ_4)]]]]]} + (1c) + (2c) + (4c) + (8c) + (16c) + (32c)} + {1c + {c – [(2)(c)[10 EXP [-[(2)[log (2γ_5)]]]]]} + (1c) + (2c) + (4c) +(8c) + (16c)} + {1c + {1c – [(2)(c)[10 EXP [-[(2)[log (2γ_6)]]]]]} + (1c) + (2c) + (4c) + (8c)} + {1c + {c – [(2)(c)[10 EXP [-[(2)[log (2γ_7)]]]]]} + (1c) + (2c) + (4c)} + {1c + {c – [(2)(c)[10 EXP [-[(2)[log (2γ_8)]]]]]} + (1c) + (2c)} + {1c + {c – [(2)(c)[10 EXP [-[(2)[log (2γ_9)]]]]]} + (1c)} + {c - [(2)(c)[10 EXP [-[(2)[log (2γ_{10})]]]]]} (Spacecraft Reference Frame Travel Time)}/(Cosmic Light-Cone Radius)} |{Σ (I =1, I =n) {(C/g_i) ln {{[[(C^2) + (V_{0i}^2)] $^{1/2}$] – [V_{0i}/[[1 – [(V_{0i}/C)^2]] $^{1/2}$]]} { [(C^2) + [[(g_i)(t_i) + [V_{0i} /[1 – [(V_{0i}/C)^2]] $^{1/2}$]^2]] $^{1/2}$] + [(g_i)(t_i)] + [V_{0i}/[[1 – [(V_{0i}/C)^2]] $^{1/2}$]]} / (C^2)}}|}

We could have chosen any value for gamma in terms of terminal pellet ship velocity for each pellet at the time said pellet fires the next pellet.

So the above formula may be expressed more generally as:

{{1c + {c – [(2)(c)[10 EXP [-[(2)[log (2γ_1)]]]]]} + (1c)} + (2c) + (4c) + (8c) + (16c) + (32c) + (64c) + (128c) + (256c)} + {1c + {c – [(2)(c)[10 EXP [-[(2)[log (2γ_2)]]]]]} + (1c) + (2c) + (4c) +(8c) + (16c) + (32c) + (64c) + (128c)} + {1c + {c – [(2)(c)[10 EXP [-[(2)[log (2γ_3)]]]]]

} + (1c) + (2c) + (4c) + (8c) + (16c) + (32c) + (64c)} + {1c + {c – [(2)(c)[10 EXP [-[(2)[log (2γ_{4})]]]]] } + (1c) + (2c) + (4c) + (8c) + (16c) + (32c)} + {1c + {c – [(2)(c)[10 EXP [-[(2)[log (2γ_{5})]]]]]} + (1c) + (2c) + (4c) +(8c) + (16c)} + {1c + {c – [(2)(c)[10 EXP [-[(2)[log (2γ_{6})]]]]]} + (1c) + (2c) + (4c) + (8c)} + {1c + {c – [(2)(c)[10 EXP [-[(2)[log (2γ_{7})]]]]]} + (1c) + (2c) + (4c)} + {1c + {c – [(2)(c)[10 EXP [-[(2)[log (2γ_{8})]]]]]} + (1c) + (2c)} + {1c + {c – [(2)(c)[10 EXP [-[(2)[log (2γ_{9})]]]]]} + (1c)} + {c - [(2)(c)[10 EXP [-[(2)[log (2γ_{10})]]]]]}

The ratio of the distance of travel by the super-light-cone traveling spacecraft to the distance viewable from the Milky Way Galaxy assuming that the cosmic light-cone radius remains the same due to constant universal expansion rate per unit of distance for the above scenario is:

{{{{1c + {c – [(2)(c)[10 EXP [-[(2)[log (2γ_{1})]]]]]} + (1c)} + (2c) + (4c) + (8c) + (16c) + (32c) + (64c) + (128c) + (256c)} + {1c + {c – [(2)(c)[10 EXP [-[(2)[log (2γ_{2})]]]]]} + (1c) + (2c) + (4c) +(8c) + (16c) + (32c) + (64c) + (128c)} + {1c + {c – [(2)(c)[10 EXP [-[(2)[log (2γ_{3})]]]]] } + (1c) + (2c) + (4c) + (8c) + (16c) + (32c) + (64c)} + {1c + {c – [(2)(c)[10 EXP [-[(2)[log (2γ_{4})]]]]] } + (1c) + (2c) + (4c) + (8c) + (16c) + (32c)} + {1c + {c – [(2)(c)[10 EXP [-[(2)[log (2γ_{5})]]]]]} + (1c) + (2c) + (4c) +(8c) + (16c)} + {1c + {c – [(2)(c)[10 EXP [-[(2)[log (2γ_{6})]]]]]} + (1c) + (2c) + (4c) + (8c)} + {1c + {c – [(2)(c)[10 EXP [-[(2)[log (2γ_{7})]]]]]} + (1c) + (2c) + (4c)} + {1c + {c – [(2)(c)[10 EXP [-[(2)[log (2γ_{8})]]]]]} + (1c) + (2c)} + {1c + {c – [(2)(c)[10 EXP [-[(2)[log (2γ_{9})]]]]]} + (1c)} + {c - [(2)(c)[10 EXP [-[(2)[log (2γ_{10})]]]]]} (Background Travel Time)}/(Cosmic Light-Cone Radius)}

For constant ship frame Lorentz factors and ship-frame time, the following formula applies:

{{{{1c + {c – [(2)(c)[10 EXP [-[(2)[log (2γ_{1})]]]]]} + (1c)} + (2c) + (4c) + (8c) + (16c) + (32c) + (64c) + (128c) + (256c)} + {1c + {c – [(2)(c)[10 EXP [-[(2)[log (2γ_{2})]]]]]} + (1c) + (2c) + (4c) +(8c) + (16c) + (32c) + (64c) + (128c)} + {1c + {c – [(2)(c)[10 EXP [-[(2)[log (2γ_{3})]]]]] } + (1c) + (2c) + (4c) + (8c) + (16c) + (32c) + (64c)} + {1c + {c – [(2)(c)[10 EXP [-[(2)[log (2γ_{4})]]]]] } + (1c) + (2c) + (4c) + (8c) + (16c) + (32c)} + {1c + {c – [(2)(c)[10 EXP [-[(2)[log (2γ_{5})]]]]]} + (1c) + (2c) + (4c) +(8c) + (16c)} + {1c + {c – [(2)(c)[10 EXP [-[(2)[log (2γ_{6})]]]]]} + (1c) + (2c) + (4c) + (8c)} + {1c + {c – [(2)(c)[10 EXP [-[(2)[log (2γ_{7})]]]]]} + (1c) + (2c) + (4c)} + {1c + {c – [(2)(c)[10 EXP [-[(2)[log (2γ_{8})]]]]]} + (1c) + (2c)} + {1c + {c – [(2)(c)[10 EXP [-[(2)[log (2γ_{9})]]]]]} + (1c)} + {c - [(2)(c)[10 EXP [-[(2)[log (2γ_{10})]]]]]} (Spacecraft Reference Frame Travel Time)}/(Cosmic Light-Cone Radius)}(γ)}

For constant ship frame acceleration, the formula becomes:

{{{{1c + {c – [(2)(c)[10 EXP [-[(2)[log (2γ_{1})]]]]]} + (1c)} + (2c) + (4c) + (8c) + (16c) + (32c) + (64c) + (128c) + (256c)} + {1c + {c – [(2)(c)[10 EXP [-[(2)[log (2γ_{2})]]]]]} + (1c) + (2c) + (4c) +(8c) + (16c) + (32c) + (64c) + (128c)} + {1c + {c – [(2)(c)[10 EXP [-[(2)[log (2γ_{3})]]]]] } + (1c) + (2c) + (4c) + (8c) + (16c) + (32c) + (64c)} + {1c + {c – [(2)(c)[10 EXP [-[(2)[log

(2γ_4)]]]]] } + (1c) + (2c) + (4c) + (8c) + (16c) + (32c)} + {1c + {c – [(2)(c)[10 EXP [-[(2)[log (2γ_5)]]]]]} + (1c) + (2c) + (4c) +(8c) + (16c)} + {1c + {c – [(2)(c)[10 EXP [-[(2)[log (2γ_6)]]]]]} + (1c) + (2c) + (4c) + (8c)} + {1c + {c – [(2)(c)[10 EXP [-[(2)[log (2γ_7)]]]]]} + (1c) + (2c) + (4c)} + {1c + {c – [(2)(c)[10 EXP [-[(2)[log (2γ_8)]]]]]} + (1c) + (2c)} + {1c + {c – [(2)(c)[10 EXP [-[(2)[log (2γ_9)]]]]]} + (1c)} + {c - [(2)(c)[10 EXP [-[(2)[log (2γ_{10})]]]]]} (Spacecraft Reference Frame Travel Time)}/(Cosmic Light-Cone Radius)} |{(c/g) ln {{[[(C ^{2}) + (V_o ^{2})]$^{1/2}$] - [V_o/[[1 - [(V_o/C) ^{2}]]$^{1/2}$]]} {[[C^{2}] + [[(g)(t) + [V_o /[1 - [(V_o/C) ^{2}]] $^{1/2}$] ^{2}]] $^{1/2}$] + [(g)(t)] + [V_o/[[1 - [(V_o/C) ^{2}]] $^{1/2}$]]} / [C ^{2}]}}|}

So for varying ship-frame acceleration, the formula becomes:

{{{{1c + {c – [(2)(c)[10 EXP [-[(2)[log (2γ_1)]]]]]} + (1c)} + (2c) + (4c) + (8c) + (16c) + (32c) + (64c) + (128c) + (256c)} + {1c + {c – [(2)(c)[10 EXP [-[(2)[log (2γ_2)]]]]]} + (1c) + (2c) + (4c) +(8c) + (16c) + (32c) + (64c) + (128c)} + {1c + {c – [(2)(c)[10 EXP [-[(2)[log (2γ_3)]]]]] } + (1c) + (2c) + (4c) + (8c) + (16c) + (32c) + (64c)} + {1c + {c – [(2)(c)[10 EXP [-[(2)[log (2γ_4)]]]]] } + (1c) + (2c) + (4c) + (8c) + (16c) + (32c)} + {1c + {c – [(2)(c)[10 EXP [-[(2)[log (2γ_5)]]]]]} + (1c) + (2c) + (4c) +(8c) + (16c)} + {1c + {c – [(2)(c)[10 EXP [-[(2)[log (2γ_6)]]]]]} + (1c) + (2c) + (4c) + (8c)} + {1c + {c – [(2)(c)[10 EXP [-[(2)[log (2γ_7)]]]]]} + (1c) + (2c) + (4c)} + {1c + {c – [(2)(c)[10 EXP [-[(2)[log (2γ_8)]]]]]} + (1c) + (2c)} + {1c + {c – [(2)(c)[10 EXP [-[(2)[log (2γ_9)]]]]]} + (1c)} + {c - [(2)(c)[10 EXP [-[(2)[log (2γ_{10})]]]]]} (Spacecraft Reference Frame Travel Time)}/(Cosmic Light-Cone Radius)} |{Σ (I =1, I =n) {(C/g_i) ln {{[[(C^{2}) + (V_{0i} ^{2})] $^{1/2}$] – [V_{0i}/[[1 – [(V_{0i}/C) ^{2}]] $^{1/2}$]]} { [(C ^{2}) + [[(g_i)(t_i) + [V_{0i} /[1 – [(V_{0i}/C) ^{2}]] $^{1/2}$] ^{2}]] $^{1/2}$] + [(g_i)(t_i)] + [V_{0i}/[[1 – [(V_{0i}/C) ^{2}]] $^{1/2}$]]} / (C ^{2})}}|}

We can develop analogous formulas for which the rate of expansion per unit space-time length varies with time and/or the translational gamma through a given light-cone varies arbitrarily. So, the gamma factor can but need not increase by constant multiples from one cosmic light-cone unit modeled to the next one modeled.

In a sense, the above digression implies truly superluminal travel, albeit, in the sense of recessionary velocities between two reference frames separated by a super-cosmic light-cone distance.

ESSAY 63) Enourmously Infinite Multiples Of Light-Speed In Spacecraft Recessional Velocities From The Remains Of The Milky Way Galaxy Or Other Origin Of Location.

For a universe such as ours that roughly doubles in size every 13.75 billion years background reference frame time, a spacecraft having achieved a velocity close to light-speed, say 98 percent of the speed of light in a reasonable Lorentz factor of about 5 and then having slowed to a stationary state a distance of 13.75 billion light-years from its origin of the Milky Way Galaxy, would be receding from the Milky Way at the speed of light.

Given another 13.75 billion years background reference frame time, the spacecraft will be receding from the Milky Way at 2 c.

Given another 13.75 billion years background reference frame time, the spacecraft will be receding from the Milky Way at 4 c.

Given another 13.75 billion years background reference frame time, the spacecraft will be receding from the Milky Way at 8 c.

Given about 10 EXP 25 current universal age units, the spacecraft will be receding from the Milk Way then composed of a central ultra-massive black hole at a velocity of about [2 EXP (10 EXP 25)] times the speed of light.

A fascinating prospect involves a civilization traveling at close to light speed then slowing down at a location 13.75 billion light-years from Earth to build a habitat in a large ice giant planet, perhaps one with a layer of hydrogen ice tens of thousands of kilometers thick.

Accordingly, the ice can act as a thermal insulator all the while neutrinos passing through the planet would provide the required warmth to support an interior civilization, perhaps 2/3 of the radius of the planet down into the depths of the planet. Computations suggest that temperate conditions could be maintained within the hab potentially for 10 EXP 35 years.

Due to the reduction of background temperature as the universe ages, the layer of ice will experience increasingly smaller thermal radiation power emissions. A steady state may be reached for which the temperature of the background reaches a lower limit thereby enabling the space-hab to acquire heat enough to enable the hab to function for infinite number of years.

Thus, for a universe that doubles size every 13.75 years or every current universal age unit, over an Aleph 0 number of current universal age units, the spacecraft will be receding from its initial location at [2 EXP (Aleph 0)] times the speed of light which will be equal to Aleph 1 times the speed of light.

Aleph 0 is the number of positive integers.

Aleph 1 = 2 EXP (Aleph 0).

Aleph 2 = 2 EXP (Aleph 1).

Aleph 3 = 2 EXP (Aleph 2).

Aleph N = 2 EXP [Aleph (N-1)]

For a universe that doubles size every 13.75 years or every current universal age unit, over an Aleph Ω number of current universal age units, the spacecraft will be receding from its initial location as [2 EXP (Aleph Ω)] times the speed of light. Omega is the least transfinite ordinal or least infinite integer.

Note that Hyper4(a, n) is equal to a tetrated n or a raised to the power of itself n-1 times. The latter value is symbolically written as n subscript a. For example 3 EXP 4 = 81, but 4 subscript3 is approximately equal to 10 EXP (1,000,000,000,000).

Alternatively 4 subscript 2 = 2 EXP 2 EXP 2 EXP 2 = 2 EXP [2 EXP [2 EXP 2]] = 2 EXP (2 EXP 4) = 2 EXP 16 = 65,536.

For example, Hyper5(4, 4)is equal to 4 tetrated 4 tetrated 4 tetrated 4. This value is commonly referred to as 4 pentated 4.

Hyper 6, (4,4) is 4 pentated 4 pentated 4 pentated 4 and is also referred to as 4 hexataed 4.

Hyper 7, (4,4) is 4 hexated 4 hexated 4 hexated 4 and so-on

Again, Aleph 0 is the infinite number of integers.

Aleph 1 according to the perhaps unprovable and thus unfalsifiable Continuum Hypotheses is the number of real numbers which is greater than Aleph 0 by a multiplicative factor of infinity.

Aleph 2 is similarly greater than Aleph 1.

Aleph 3 is similarly greater than Aleph 2.

Aleph 4 is similarly greater than Aleph 3.

and so-on

In general, Aleph n = 2 EXP [Aleph (n-1)]

The number Ω is commonly stated as the least infinite positive integer or ordinal.

Now here is a real zinger.

So we can produce the abstraction of Hyper Aleph Ω (Aleph Ω, Aleph Ω).

For a universe that doubles size every 13.75 billion years or every current universal age unit, over an Hyper Aleph Ω (Aleph Ω, Aleph Ω) number of current universal age units, the spacecraft will be receding from its initial location as [2 EXP [Hyper Aleph Ω (Aleph Ω, Aleph Ω)]] times the speed of light.

We may affix the following operator to formulas for gamma and related parameters that follow to denote the extremity of the recessional velocity of an icy hab world from the Milky Way or other region of origin.

[f(Recessional velocity of icy world from Milky Way or other location of causative origin after: [Hyper Aleph Ω (Aleph Ω, Aleph Ω)] current universal age units of time passage of [Hyper Aleph Ω (Aleph Ω, Aleph Ω)] c; {[Hyper Aleph Ω (Aleph Ω, Aleph Ω)], {[Hyper Aleph Ω (Aleph Ω, Aleph Ω)], [Hyper Aleph Ω (Aleph Ω, Aleph Ω)]}} current universal age units of time passage of {[Hyper Aleph Ω (Aleph Ω, Aleph Ω)], {[Hyper Aleph Ω (Aleph Ω, Aleph Ω)], [Hyper Aleph Ω (Aleph Ω, Aleph Ω)]}} c; {{[Hyper Aleph Ω (Aleph Ω, Aleph Ω)], {[Hyper Aleph Ω (Aleph Ω, Aleph Ω)], [Hyper Aleph Ω (Aleph Ω, Aleph Ω)]}} {{[Hyper Aleph Ω (Aleph Ω, Aleph Ω)], {[Hyper Aleph Ω (Aleph Ω, Aleph Ω)], [Hyper Aleph Ω (Aleph Ω, Aleph Ω)]}}, {[Hyper Aleph Ω (Aleph Ω, Aleph Ω)], {[Hyper Aleph Ω (Aleph Ω, Aleph Ω)], [Hyper Aleph Ω (Aleph Ω, Aleph Ω)]}}}} current universal age units of time passage of {[Hyper Aleph Ω (Aleph Ω, Aleph Ω)], {[Hyper Aleph Ω (Aleph Ω, Aleph Ω)], [Hyper Aleph Ω (Aleph Ω, Aleph Ω)]}} c; {{[Hyper Aleph Ω (Aleph Ω, Aleph Ω)], {[Hyper Aleph Ω (Aleph Ω, Aleph Ω)], [Hyper Aleph Ω (Aleph Ω, Aleph Ω)]}} {{[Hyper Aleph Ω (Aleph Ω, Aleph Ω)], {[Hyper Aleph Ω (Aleph Ω, Aleph Ω)], [Hyper Aleph Ω (Aleph Ω, Aleph Ω)]}}, {[Hyper Aleph Ω (Aleph Ω, Aleph Ω)], {[Hyper Aleph Ω (Aleph Ω, Aleph Ω)], [Hyper Aleph Ω (Aleph Ω, Aleph Ω)]}}}} c; {{{[Hyper Aleph Ω (Aleph Ω, Aleph Ω)], {[Hyper Aleph Ω (Aleph Ω, Aleph Ω)], [Hyper Aleph Ω (Aleph Ω, Aleph Ω)]}} {{[Hyper Aleph Ω (Aleph Ω, Aleph Ω)], {[Hyper Aleph Ω (Aleph Ω, Aleph Ω)], [Hyper Aleph Ω (Aleph Ω, Aleph Ω)]}}, {[Hyper Aleph Ω (Aleph Ω, Aleph Ω)], {[Hyper Aleph Ω (Aleph Ω, Aleph Ω)], [Hyper Aleph Ω (Aleph Ω, Aleph Ω)]}}}} {{{[Hyper Aleph Ω (Aleph Ω, Aleph Ω)], {[Hyper Aleph Ω (Aleph Ω, Aleph Ω)], [Hyper Aleph Ω (Aleph Ω, Aleph Ω)]}} {{[Hyper Aleph Ω (Aleph Ω, Aleph Ω)], {[Hyper Aleph Ω (Aleph Ω, Aleph Ω)], [Hyper Aleph Ω (Aleph Ω, Aleph Ω)]}}, {[Hyper Aleph Ω (Aleph Ω, Aleph Ω)], {[Hyper Aleph Ω (Aleph Ω, Aleph Ω)], [Hyper Aleph Ω (Aleph Ω, Aleph Ω)]}}}}}, {{[Hyper Aleph Ω (Aleph Ω, Aleph Ω)], {[Hyper Aleph Ω (Aleph Ω, Aleph Ω)], [Hyper Aleph Ω (Aleph Ω, Aleph Ω)]}} {{[Hyper Aleph Ω (Aleph Ω, Aleph Ω)], {[Hyper Aleph Ω (Aleph Ω, Aleph Ω)], [Hyper Aleph Ω (Aleph Ω, Aleph Ω)]}}, {[Hyper Aleph Ω (Aleph Ω, Aleph Ω)], {[Hyper Aleph Ω (Aleph Ω, Aleph Ω)], [Hyper Aleph Ω (Aleph Ω, Aleph Ω)]}}}}}} current universal age units of time passage of {[Hyper Aleph Ω (Aleph Ω, Aleph Ω)], {[Hyper Aleph Ω (Aleph Ω, Aleph Ω)], [Hyper Aleph Ω (Aleph Ω, Aleph Ω)]}} c; {{[Hyper Aleph Ω (Aleph Ω, Aleph Ω)], {[Hyper Aleph Ω (Aleph Ω, Aleph Ω)], [Hyper Aleph Ω (Aleph Ω, Aleph Ω)]}} {{[Hyper Aleph Ω (Aleph Ω, Aleph Ω)], {[Hyper Aleph Ω (Aleph Ω, Aleph Ω)],

[Hyper Aleph Ω (Aleph Ω, Aleph Ω)]}}, {[Hyper Aleph Ω (Aleph Ω, Aleph Ω)], {[Hyper Aleph Ω (Aleph Ω, Aleph Ω)], [Hyper Aleph Ω (Aleph Ω, Aleph Ω)]}}}} c; {{{[Hyper Aleph Ω (Aleph Ω, Aleph Ω)], {[Hyper Aleph Ω (Aleph Ω, Aleph Ω)], [Hyper Aleph Ω (Aleph Ω, Aleph Ω)]}} {{[Hyper Aleph Ω (Aleph Ω, Aleph Ω)], {[Hyper Aleph Ω (Aleph Ω, Aleph Ω)], [Hyper Aleph Ω (Aleph Ω, Aleph Ω)]}}, {[Hyper Aleph Ω (Aleph Ω, Aleph Ω)], {[Hyper Aleph Ω (Aleph Ω, Aleph Ω)], [Hyper Aleph Ω (Aleph Ω, Aleph Ω)]}}}} {{{[Hyper Aleph Ω (Aleph Ω, Aleph Ω)], {[Hyper Aleph Ω (Aleph Ω, Aleph Ω)], [Hyper Aleph Ω (Aleph Ω, Aleph Ω)]}} {{[Hyper Aleph Ω (Aleph Ω, Aleph Ω)], {[Hyper Aleph Ω (Aleph Ω, Aleph Ω)], [Hyper Aleph Ω (Aleph Ω, Aleph Ω)]}}, {[Hyper Aleph Ω (Aleph Ω, Aleph Ω)], {[Hyper Aleph Ω (Aleph Ω, Aleph Ω)], [Hyper Aleph Ω (Aleph Ω, Aleph Ω)]}}}}, {{[Hyper Aleph Ω (Aleph Ω, Aleph Ω)], {[Hyper Aleph Ω (Aleph Ω, Aleph Ω)], [Hyper Aleph Ω (Aleph Ω, Aleph Ω)]}} {{[Hyper Aleph Ω (Aleph Ω, Aleph Ω)], {[Hyper Aleph Ω (Aleph Ω, Aleph Ω)], [Hyper Aleph Ω (Aleph Ω, Aleph Ω)]}}, {[Hyper Aleph Ω (Aleph Ω, Aleph Ω)], {[Hyper Aleph Ω (Aleph Ω, Aleph Ω)], [Hyper Aleph Ω (Aleph Ω, Aleph Ω)]}}}}}} c and so-on)].

ESSAY 64) Light-Speed Travel And Travel Into Levels Of Voids Preordinate To Space-Time.

Often times, folks new to contemporary cosmology will ponder what did the universe expand into after the big bang.

Then as the inquisitive folks become more mature in their study of cosmology, they will come to see that our universe was the very start of the space-time within our universe.

Then, eventually, they will accept the plausibility that there may be countless infinities of universes which may dot a hyperspace as if countless islands in the sea.

Having come this far, they have advanced in descriptive cosmological principles that are as advanced as almost all current cosmologists can accept as plausible.

However, more advanced thinkers in cosmology will posit bulks as media that contain any number of hyperspaces which can have up to infinite numbers of dimensions.

Still other cosmologist hold that scalar fields exist.

Accordingly, the scalar field that would have given our universe its birth in the big bang is as if a roiling sea of randomly or semi-randomly fluctuating field vectors on an ultra-microscopic scale. Every so often, a chance alignment of the vectors in a local portion of the scalar field results in the development of huge potential energies which then result in a big bang and birth of a new universe. Thus, the field is referred to as a scalar field because on macroscopic levels, the otherwise vector fields do not exist and the microscopic vectors cancel each other out most of the time in a given location within the scalar field.

According to the above chaotic inflationary scalar field theories, big bangs are happening all of the time almost everywhere in the scalar field and also in our own universe where the baby universes that branch off from our universe will themselves often give birth to their own lineage universes and so the process continues. Such considerations are common in the chaotic inflationary fractalverse theories.

Accordingly, a multiverse is a fractal tree of universes and descendants thereof.

A forest is a related group of multiverses.

A biosphere is a related group of forests and so the pattern continues on eternally.

However, there are additional concepts that may have realities at play here.

For example, we may posit a great existential void that is preordinate to the creation of the very first big bang in our given fractalverse. This void would in a sense be the space of our space-time or a form of a void-preordinate-to-space-time.

A first level of such a void we refer to as void-1-preordinate-to-space-time.

A second level is referred to as void-2-preordinate-to-space-time which would be to void-1-preordinate-to-space-time as void-1-preordinate-to-space-time is to ordinary space-time.

A third level is referred to as void-3-preordinate-to-space-time which would be to void-2-preordinate-to-space-time as void-2-preordinate-to-space-time would be to void-1-preordinate-to-space-time as void-1-preordinate-to-space-time is to ordinary space-time.

The number of hierarchies as such may be arbitrarily infinite.

Travel into these voids may not even be considerable to Special and General Relativity because such voids are neither space nor time. So to assign a Lorentz factor to craft plying these voids does not make immediate sense but perhaps could if void traveling craft could pop back into ordinary 4-D space-time at locations in time and place of cosmic scale distances from the origin of the craft in the ordinary 4-D space-time of mission commencement.

ESSAY 65) Serial Entrance Into Ever Larger Realms Of Light-Speed Craft With Increasing Infinite Lorentz Factors.

A fascinating construct is the notion that a spacecraft having achieved the speed of light in a most true sense thus having an infinite Lorentz factor would run out of future extension to travel forward in time in within a given universe, multiverse, forest, biosphere, and the like realm of travel.

Thus, such a spacecraft might leave its realm of initial travel completely to enter another larger more eternal realm. Thus, at least one class of novel travel constructs manefests.

Likewise, a spacecraft somehow becoming imbued with yet greater Lorentz factors may once again remove itself from said second class of travel realms to enter a third class of even more eternal realms than the second class of eternal realms.

The step-by-step process of leaving realms to enter evermore eternal realms may continue forever.

Thus, the number of classes of travel at light speed and more broadly, at ever greater infinite Lorentz factors would seem to be an ever growing infinite value.

We may affix the following operator to formulas for gamma and related parameters to denote the growing infinite number of classes of travel enabled by increasingly infinite Lorentz factor spacecraft.

[f(Number of classes of travel at light speed, and more broadly, at ever greater infinite Lorentz factors equal to ↑∞)]

Now as the number and extremity of classes of travel grows, perhaps other superseding kinematic processes occur which we will refer to as outside-of-travel-kinematics.

As a first level of outside-of-travel-kinematics is attained, the spacecraft or more properly the fiat, can go on to a second level of outside-of-travel-kinematics.

As a second level of outside-of-travel-kinematics is attained, the spacecraft or more properly the fiat, can go on to a third level of outside-of-travel-kinematics.

We label these kinematics as; outside-1-of-travel-kinematics, outside-2-of-travel-kinematics, outside-3-of-travel-kinematics.

We can go on to more extremes all the way up to and beyond the lowest infinite levels.

We use the following set builder notation to denote the range of the outside-of-travel-kinematics and as an operator indicating these mechanisms in formulas for gamma and related parameters.

{f{Outside-k-of-travel-kinematics: k = 1 to ↑∞}}

Eventually, our Lorentz factors may become so infinite such that the itinerary of the spacecraft transcends or supersedes the scope and meaning of kinematics altogether. We may affix the following operator to formulas for gamma and related parameters to denote the range of extra-kinematic-iteneraries.

{f{Extra-k-of-outside-of-kinematics: k = 1 to ↑∞}}

We bundle these three operators as follows into a compound operator and then produce the following operator which takes into account possibilities for differentiation and integration of the three classes of mechanisms with respect to ship-time, acceleration, ship kinetic energy, and gamma.

{[f(Number of classes of travel at light speed, and more broadly, at ever greater infinite Lorentz factors equal to ↑∞)], {f{Outside-k-of-travel-kinematics: k = 1 to ↑∞}},{f{Extra-k-of-outside-of-kinematics: k = 1 to ↑∞}}} : {Arbitrary order ship-time, ship-frame acceleration, ship kinetic energy, and gamma derivatives and integrals of the mechanisms expressed in the following compound operator{[f(Number of classes of travel at light speed, and more broadly, at ever greater infinite Lorentz factors equal to ↑∞)], {f{Outside-k-of-travel-kinematics: k = 1 to ↑∞}},{f{Extra-k-of-outside-of-kinematics: k = 1 to ↑∞}}}}

ESSAY 66) Time Embedded In Time.

We can consider a kind of time embedded in time.

Accordingly, time may have hidden or otherwise manifestable additional time dimensions for which known 1 dimensional time would act as an analog of hyperspace as hyperspace over-reaches or encompasses ordinary 4-D space-time.

As such, the hidden time dimensions time travelers may have a form of security and be able to travel to do surveillance. Another benefit to traveling in hidden time dimensions is the ability to access one or more hidden realms. Such realms may be open for exploration and habitation.

The number of hidden or manifestable time dimensions may range from vanishingly small finite sub-unitary values to arbitrarily infinite values.

We can imagine a spacecraft traveling in zero dimensional time or occupying a realm for which time is zero dimensional. Accordingly, within the 0th time dimension, there may exist numbers of hidden or manifestable time dimensions ranging from vanishingly small finite sub-unitary values to arbitrarily infinite values.

The time dimensions of travel may also be hidden within the overall external or over-arching time continuums even in cases where the latter continuums have more than one dimension of time.

For example, the hidden time dimension of travel might not be a sub-set of just one temporal dimension of an over-arching time continuum but instead may be manifest as a subset of the over-arching time continuum, for example, having placement holistically within an entire 3 time dimensional continuum and more generally within an entire N-dimensional time continuum.

An interesting prospect is entering a sub-set of hidden time within a one or more temporal dimensional over-arching time continuum for which the sub-set of hidden time is of less than zero dimensions but still non-negative in number. Accordingly, the sub-time of travel may include dimensionalities of [(Zero)/n] where n is positive and finite or positive and arbitrarily infinite.

In a way, a spacecraft traveling or persons existing in hidden sub-time of dimensionalities of [(Zero)/n] where n is positive and finite or positive and arbitrarily infinite may be extremely secure and be located in a state of extreme ontological simplicity that is an existential portal to a new class of continuum which includes eternities but where the hidden sub-times are associated with more incorruptibility than objects or created persons that are eternal. The proposed analogs are different than eternities just as the color green is distinct from the color blue but where the entire spectrum of colors is suffuse with infinite numbers of specific colors. Here, eternities no matter how many would be represented by just one color of blue where the number of non-temporal analogs would be infinite and be represented by all of the other colors.

The prospect of more non-eternity analogs of colors invented or created by GOD or yet to be manifested by GOD may extend beyond the visible spectrum as if by analogs of electromagnetic radiation beyond visible light such as ultra-violet, x-ray, gamma-ray, and also infrared, microwave, radio-frequency and the like radiations. The greater than visible light frequencies may represent enhanced existentiality, whereas the lower than visible light frequencies may represent enhanced simplicity and existential stability and quietness.

We may affix the following operator to the lenthy formulas for gamma and related parameters to denote the extremities of sub-sets of time dimensional travel and habitation.

{[[f(Travel into kinds of time embedded in time of dimensionality ranging from (Zero/n) where n is greater than one and may be arbitrarily infinite to embedded time dimensionalities as great as [1/(Zero/n)]], [f(Travel or habitation within kinds of embedded time for which the time travelers or residents are defined with reference to other planes or modes of ontology of which eternities are as a set are as if one color among the infinite variety of possible colors with the other colors representing said other planes or modes of ontology)],[f(Travel or habitation in above said other planes or modes of ontology that are yet to be realized but which GOD can create as representable by non-visible frequencies of light with higher frequencies representing all the more extreme existentialities and the lower frequencies representing all the more calm and quiet existentialities)]}

ESSAY 67) Infinite Lorentz Factor Craft Attaining New Forms Of Velocity.

Consider the following clauses that are for the most part self-descriptive and knowable from first principles as axiomatic truths provided the usual caveats stated throughout this book for light-speed inertial travel are met.

A spacecraft traveling at the velocity of light and then acquiring even greater infinite Lorentz factors traveling ahead of itself and faster than itself.

A spacecraft having attained suitably great finite or infinite Lorentz factors running out of space-time to travel forward in due to the contraction of the universe and space in front of the spacecraft and due to time dilation taking on a new type of velocity to fill the lack of natural opportunity for the spacecraft.

A spacecraft having attained suitably great finite or infinite Lorentz factors running out of space-time to travel forward in due to the contraction of the universe and space in front of the spacecraft and due to time dilation taking on a new type of velocity analog to fill the lack of natural opportunity for the spacecraft.

A spacecraft having attained suitably great finite or infinite Lorentz factors running out of space-time to travel forward in due to the contraction of the universe and space in front of the spacecraft and due to time dilation taking on a new type of velocity-based parameter of [f(Velocity)] where the functional parameter [f(Velocity)] can be exponential, trigonometric, logarithmic, transcendental, linear, complex, and the like and which can have magnitude ranging from that of the velocity of light to various arbitrary infinite values to fill the lack of natural opportunity for the spacecraft.

A spacecraft with acceleration so extreme such that the acceleration increases exponentially in the background reference frame even while the spacecraft approaches or obtains the velocity of light in the background reference frame.

A spacecraft having attained suitably great finite or infinite Lorentz factors running out of space-time to travel forward in due to the contraction of the universe and space in front of the spacecraft and due to time dilation taking on a new type of kinetic energy to fill the lack of natural opportunity for the spacecraft.

A spacecraft having attained suitably great finite or infinite Lorentz factors running out of space-time to travel forward in due to the contraction of the universe and space in front of the spacecraft and due to time dilation taking on a new type of kinetic energy analog to fill the lack of natural opportunity for the spacecraft.

A spacecraft having attained suitably great finite or infinite Lorentz factors running out of space-time to travel forward in due to the contraction of the universe and space in front of the spacecraft and due to time dilation taking on a new type of kinetic-energy-based parameter of [f(Kinetic Energy)] where the functional parameter [f(Kinetic Energy)] can be exponential, trigonometric, logarithmic, transcendental, linear, complex, and the like and which can have magnitude near infinite values to various arbitrary infinite values to fill the lack of natural opportunity for the spacecraft.

A spacecraft having attained suitably great finite or infinite Lorentz factors running out of space-time to travel forward in due to the contraction of the universe and space in front of the spacecraft and due to time dilation taking on a new type of Lorentz factor to fill the lack of natural opportunity for the spacecraft.

A spacecraft having attained suitably great finite or infinite Lorentz factors running out of space-time to travel forward in due to the contraction of the universe and space in front of the spacecraft and due to time dilation taking on a new type of Lorentz Factor analog to fill the lack of natural opportunity for the spacecraft.

A spacecraft having attained suitably great finite or infinite Lorentz factors running out of space-time to travel forward in due to the contraction of the universe and space in front of the spacecraft and due to time dilation taking on a new type of Lorentz-factor-based parameter of [f(Lorentz Factor)] where the functional parameter [f(Lorentz Factor)] can be exponential, trigonometric, logarithmic, transcendental, linear, complex, and the like and which can have magnitude near infinite values to various arbitrary infinite values to fill the lack of natural opportunity for the spacecraft.

We may affix the following operator to formulas for gamma and related parameters to denote the operations of the above travel mechanisms.

The open ended infinite series limit for the sigma notation indicates possibly unlimited associated thermodynamic, kinematic, and topological states for the proposed travel mechanisms.

{Σ(k = 1; k = ∞↑):[[f(Spacecraft traveling at the velocity of light and then acquiring even greater infinite Lorentz factors traveling ahead of itself and faster than itself) k],[f(Spacecraft having attained suitably great finite or infinite Lorentz factors running out of space-time to travel forward in due to the contraction of the universe and space in front of the spacecraft and due to time dilation taking on a new type of velocity to fill the lack of natural opportunity for the spacecraft) k],[f(Spacecraft having attained suitably great finite or infinite Lorentz factors running out of space-time to travel forward in due to the contraction of the universe and space in front of the spacecraft and due to time dilation taking on a new type of velocity analog to fill the lack of natural opportunity for the spacecraft) k], [f(Spacecraft having attained suitably great finite or infinite Lorentz factors running out of space-time to travel forward in due to the contraction of the universe and space in front of the spacecraft and due to time dilation taking on a new type of velocity-based parameter of [f(Velocity)] where the functional parameter [f(Velocity)] can be exponential, trigonometric, logarithmic, transcendental, linear, complex, and the like and which can have magnitude ranging from that of the velocity of light to various arbitrary infinite values to fill the lack of natural opportunity for the spacecraft) k], [f(Spacecraft with acceleration so extreme such that the acceleration increases exponentially in the background reference frame even while the spacecraft approaches or obtains the velocity of light in the background reference frame) k], [f(Spacecraft having attained suitably great finite or infinite Lorentz factors running out of space-time to travel forward in due to the contraction of the universe and space in front of the spacecraft and due to time dilation taking on a new type of kinetic energy to fill the lack of natural opportunity for the spacecraft) k], [f(Spacecraft having attained suitably great finite or infinite Lorentz factors running out of space-time to travel forward in due to the contraction of the universe and space in front of the spacecraft and due to time dilation taking on a new type of kinetic energy analog to fill the lack of natural opportunity for the spacecraft) k], [f(Spacecraft having attained suitably great finite or infinite Lorentz factors running out of space-time to travel forward in due to the contraction of the universe and space in front of the spacecraft and due to time dilation taking on a new type of kinetic-energy-based parameter of [f(Kinetic Energy)] where the functional parameter [f(Kinetic Energy)] can be exponential, trigonometric, logarithmic, transcendental, linear, complex, and the like and which can have magnitude near infinite values to various arbitrary infinite values to fill the lack of natural opportunity for the spacecraft) k], [f(Spacecraft having attained suitably great finite or infinite Lorentz factors running out of space-time to travel forward in due to the contraction of the universe and space in front of the spacecraft and due to time dilation taking on a new type of Lorentz factor to fill the lack of natural opportunity for the spacecraft) k],

[f(A spacecraft having attained suitably great finite or infinite Lorentz factors running out of space-time to travel forward in due to the contraction of the universe and space in front of the spacecraft and due to time dilation taking on a new type of Lorentz Factor analog to fill the lack of natural opportunity for the spacecraft) k], [f(Spacecraft having attained suitably great finite or infinite Lorentz factors running out of space-time to travel forward in due to the contraction of the universe and space in front of the spacecraft and due to time dilation taking on a new type of Lorentz-factor-based parameter of [f(Lorentz Factor)] where the functional parameter [f(Lorentz Factor)] can be exponential, trigonometric, logarithmic, transcendental, linear, complex, and the like and which can have magnitude near infinite values to various arbitrary infinite values to fill the lack of natural opportunity for the spacecraft) k]]}

ESSAY 68) Serial Flight Of Light-Speed Scout Ships From A Light-Speed Mother Ship And Stacked Light-Speed Velocities.

Now, we may intuit that spacecraft able to reach the speed of light in infinite Lorentz factors may effectively leave the universe or broader realm of travel as the spacecraft would run out of future temporal eternity in said realm of travel. Accordingly, the space craft would only sense realities of objects moving at Newtonian or every day velocities with respect to itself. Objects stationary with respect of the spacecraft will be assumed moving at the same Lorentz factor as the spacecraft.

So, we can further consider spacecraft that would be launched from the former light-speed spacecraft after being assembled on the former light-speed spacecraft or carried along from the start of a mission of said former light-speed spacecraft. This first scout craft would travel at headings parallel to that of the mothership in the mother-ships reference frame. Thus, the first scout craft may accelerate to the speed of light relative to the mother-ship and run out of room in the new light-speed acquired realm of the mothership. Thus, the scout ship may become eternalized with respect to the mothership to enter its own new realm.

The speed of light of the first scout ship we will refer to as doubly-stacked-light-speed or stacked-2-light-speed.

Now, we can further consider spacecraft that would be launched from the first scout ship after being assembled on the first scout ship or carried along from the start of a mission of the first scout ship. This second scout craft would travel at headings parallel to that of the first scout ship in the first scout ship's reference frame. Thus, the second scout craft may accelerate to the speed of light relative to the first scout ship and run out of room in the new light-speed acquired realm of the first scout ship. Thus, the second scout ship may become eternalized with respect to the first scout ship to enter its own new realm.

The speed of light of the second scout ship we will refer to as triple-stacked-light-speed or stacked-3-light-speed.

Now, we can further consider spacecraft that would be launched from the second scout ship after being assembled on the second scout ship or carried along from the start of a mission of the second scout ship. This third scout craft would travel at headings parallel to that of the second scout ship in the second scout ship's reference frame. Thus, the third scout craft may accelerate to the speed of light relative to the second scout ship and run out of room in the new light-speed acquired realm of the second scout ship. Thus, the third scout ship may become eternalized with respect to the second scout ship to enter its own new realm.

The speed of light of the third scout ship we will refer to as quadruple-stacked-light-speed or stacked-4-light-speed.

We can consider additional scout ships without limit to infinity and then to ever greater infinities.

We denote the range of possibilities and distances traveled from the origin of the mother-ship by the single time integration in the following operator which we may affix formulas for gamma and related parameters. Note that the operator also includes notation for arbitrary time, gamma, acceleration, and kinetic energy derivatives of the process of light-speed stacking.

{kth-stacked-light-speed or stacked-k-light-speed of (k-1)th scouts ship| k ranges from 2 to arbitrary infinities}:{First ship-time derivative of {kth-stacked-light-speed or stacked-k-light-speed of (k-1)th scouts ship| k ranges from 2 to arbitrary infinities}}:{{Nth order ship-time derivative of {kth-stacked-light-speed or stacked-k-light-speed of (k-1)th scouts ship| k ranges from 2 to arbitrary infinities}}| N ranges from 2 to open ended arbitrary infinities}:{First gamma derivative of {kth-stacked-light-speed or stacked-k-light-speed of (k-1)th scouts ship| k ranges from 2 to arbitrary infinities}}:{{Nth order gamma derivative of {kth-stacked-light-speed or stacked-k-light-speed of (k-1)th scouts ship| k ranges from 2 to arbitrary infinities}}| N ranges from 2 to open ended arbitrary infinities}:{First ship-frame acceleration derivative of {kth-stacked-light-speed or stacked-k-light-speed of (k-1)th scouts ship| k ranges from 2 to arbitrary infinities}}:{{Nth order ship-frame acceleration derivative of {kth-stacked-light-speed or stacked-k-light-speed of (k-1)th scouts ship| k ranges from 2 to arbitrary infinities}}| N ranges from 2 to open ended arbitrary infinities}:{First spacecraft kinetic energy derivative of {kth-stacked-light-speed or stacked-k-light-speed of (k-1)th scouts ship| k ranges from 2 to arbitrary infinities}}:{{Nth order spacecraft kinetic energy derivative of {kth-stacked-light-speed or stacked-k-light-speed of (k-1)th scouts ship| k ranges from 2 to arbitrary infinities}}| N ranges from 2 to open ended arbitrary infinities}:{(Ship time integral of mothership light-speed)(mothership Lorentz factor)} + {(Ship time integral of first scout ship light-speed)(mothership Lorentz factor)(first scout ship Lorentz factor with respect to mothership)} + {(Ship time integral of second scout ship light-speed)(mothership Lorentz factor)(first scout ship Lorentz factor with respect to mothership)(second scout ship Lorentz factor with respect to first scout ship)} + {(Ship time integral of third scout ship light-speed)(mothership Lorentz factor)(first scout ship Lorentz factor with respect to mothership)(second scout ship Lorentz factor with respect to first scout ship) (third scout ship Lorentz factor with respect to second scout ship)} + … + {(Ship time integral of kth scout ship light-speed)(mothership Lorentz factor)(first scout ship Lorentz factor with respect to mothership)(second scout ship Lorentz factor with respect to first scout ship) … (kth scout ship Lorentz factor with respect to (k-1)th scout ship)}:{(Ship time integral of mothership light-speed)(mothership Lorentz factor)(Context dependent factor for non-constant velocity)} + {(Ship time integral of first scout ship light-speed)(mothership Lorentz factor)(first scout ship Lorentz factor with respect to mothership) (Context dependent factor for non-constant velocities)} + {(Ship time integral of second scout ship light-speed)(mothership Lorentz factor)(first scout ship Lorentz factor with respect to mothership)(second scout ship Lorentz factor with respect to first scout ship) (Context dependent factor for non-constant velocities)} + {(Ship time integral of third scout ship light-speed)(mothership Lorentz factor)(first scout ship Lorentz factor with respect to mothership)(second scout ship Lorentz factor with respect to first scout ship) (third scout ship Lorentz factor with respect to second scout ship) (Context dependent factor for non-constant velocities)} + … + {(Ship time integral of kth scout ship light-speed)(mothership Lorentz factor)(first scout ship Lorentz factor with respect to mothership)(second scout ship Lorentz factor with respect to first scout ship) … (kth scout ship Lorentz factor with respect to (k-1)th scout ship) (Context dependent factor for non-constant velocities)}

For cases where the velocities of the mothership and the scout ship vary as they usually would, we include distance modification factors to yield more logical formulas. These factors are generally denoted by the expression (Context dependent factor for non-constant velocity) or (Context dependent factor for non-constant velocities).

ESSAY 69) Light-Speed Impulse Travel And Unveiling Levels Of Pure Physical Being.

Now, we can contemplate probing extreme conditions by attaining extreme infinite Lorentz factors in light-speed impulse travel to unveil one or more levels of pure physical being.

The idea here goes deeper than unveiling prime matter. Instead, we consider diving deeper into the physical created analog of the divine notion of GOD as pure being.

So, in a sense exploring physical reality beyond prime matter is akin to searching for the essence of the substantial principles of prime matter.

Now, I conjecture that the created natural order is as old as GOD and that the age of the physical natural order is infinite. It does not make much sense that the three Divine Persons of GOD would wait around for all eternity then decide on a whim to start the process of creation, whether of spiritual or physical creatures.

Regarding pursuits of study all the way down into natural physical being, doing so may take a future eternity. By the time we accomplish study of pure physical being, we likely will have a never ending series of analogs to study on the same class as the spiritual and the physical belong to.

It may even be the case that the two classes of created reality, the physical and the spiritual are just the beginning of GOD's creative work. Depending on how we humans, and any extraterrestrials and ultraterrestrials behave, GOD may decide if HE has not already, to create an undefinably infinite number of additional analogs that are as different from each other as they are different from the physical and the spiritual as the physical and spiritual are different from each other.

So let us all behave in a kinder gentler manner so that GOD will not be pissed off at us and will not wonder why HE even created us.

We have a lot of fun opportunity for exploring physical reality right here in our own cosmic light-cone of radius about 13.75 billion light-years.

Therein is approximately two trillion galaxies and 10 EXP 24 stars or about the number of fine grains of table sugar that would cover the entire United States 100 meters deep. The number of planets orbiting these stars is estimated to be about ten times greater yet or equal to the number of fine grains of table sugar that would cover the entire United States 1,000 meters deep. The number of moons orbiting these planets is estimated to be about ten times greater yet or equal to the number of fine grains of table sugar that would cover the entire United States 10,000 meters deep. Our cosmic light-cone is likely a miniscule portion of our universe which may be just one universe amongst infinitely many which may be placed within one hyperspace amongst infinitely many hyperspaces where the dimensionality of hyperspaces can range from 4 spatial dimensions to untold infinite numbers of dimensions.

We affix the first operator below to the lengthy formulas for gamma and related parameters in this section to provisionally denote the access to one more principles of pure physical being as we develop ever more infinite Lorentz factor capable spacecraft.

[w(Access to one more principles of pure physical being as we develop ever more infinite Lorentz factor capable spacecraft)]

We like-wise affix the second operator below to the lengthy formulas for gamma and related parameters in this section to provisionally denote the access to one more principles of analogs of pure physical being and pure created spiritual being as we develop ever more infinite Lorentz factor capable spacecraft.

[w(Access to one more principles of pure physical being as we develop ever more infinite Lorentz factor capable spacecraft)]

[z(Access to one more principles of analogs of pure physical being and pure created spiritual being as we develop ever more infinite Lorentz factor capable spacecraft where the number of analogs may superabundantly grow to ever greater infinite values)]

Study of pure physical being to elucidate knowledge understanding of the spiritual being of the human soul at the natural level.

ESSAY 70) Some Really Huge Infinite Lorentz Factors And Travel Not In Space And Not In Time.

Regarding how big infinities can be consider the following digression which is repeated again in this book.

Now, there are countable and uncountable infinities.

But before we go further, consider the hyper-operator notation that was designed to express huge values not otherwise expressible.

For example, Note that Hyper4(a, n) is equal to a tetrated n or a raised to the power of itself n-1 times. The latter value is symbolically written as n subscript a. For example 3 EXP 4 = 81, but 4 subscript3 is approximately equal to 10 EXP (1,000,000,000,000).

Alternatively 4 subscript 2 = 2 EXP 2 EXP 2 EXP 2 = 2 EXP [2 EXP [2 EXP 2]] = 2 EXP (2 EXP 4) = 2 EXP 16 = 65,536.

For example, Hyper5(4, 4)is equal to 4 tetrated 4 tetrated 4 tetrated 4. This value is commonly referred to as 4 pentated 4.

Hyper 6, (4,4) is 4 pentated 4 pentated 4 pentated 4 and is also referred to as 4 hexataed 4.

Hyper 7, (4,4) is 4 hexated 4 hexated 4 hexated 4 and so-on

Aleph 0 is the infinite number of integers.

Aleph 1 according to the perhaps unprovable and thus unfalsifiable Continuum Hypotheses is the number of real numbers which is greater than Aleph 0 by a multiplicative factor of infinity.

Aleph 2 is similarly greater than Aleph 1.

Aleph 3 is similarly greater than Aleph 2.

Aleph 4 is similarly greater than Aleph 3.

and so-on

In general, Aleph n = 2 EXP [Aleph (n-1)] where n can be any finite or infinite positive integer.

The number Ω is commonly stated as the least infinite positive integer or ordinal.

Now here is a real zinger.

So we can produce the abstraction of Hyper Aleph Ω (Aleph Ω, Aleph Ω).

W e can also produce the abstraction of Hyper [Hyper Aleph Ω (Aleph Ω, Aleph Ω)] ((Hyper Aleph Ω (Aleph Ω, Aleph Ω)), Hyper Aleph Ω (Aleph Ω, Aleph Ω)).

W e can also produce the abstraction of Hyper [Hyper Aleph Ω (Aleph Ω, Aleph Ω)] ((Hyper Aleph Ω (Aleph Ω, Aleph Ω)), Hyper Aleph Ω (Aleph Ω, Aleph Ω)) ([Hyper Aleph Ω (Aleph Ω, Aleph Ω)] ((Hyper Aleph Ω (Aleph Ω, Aleph Ω)), Hyper Aleph Ω (Aleph Ω, Aleph Ω)), [Hyper Aleph Ω (Aleph Ω, Aleph Ω)] ((Hyper Aleph Ω (Aleph Ω, Aleph Ω)), Hyper Aleph Ω (Aleph Ω, Aleph Ω)))

We can go one to produce an infinity using the above nesting style of convention with all potential computer storage media currently on Earth.

We can go one to produce an infinity using the above nesting style of convention with all potential computer storage media producible by the mineral resources within the observable portion of our universe.

We can go one to produce an infinity using the above nesting style of convention with all potential computer storage media producible by the mineral resources within our entire universe.

Now here is a real zinger.

So we can produce the abstraction of Hyper Aleph (Large Uncountable Infinities) (Aleph (Large Uncountable Infinities), Aleph (Large Uncountable Infinities)).

We can also produce the abstraction of Hyper [Hyper Aleph (Large Uncountable Infinities) (Aleph (Large Uncountable Infinities), Aleph (Large Uncountable Infinities))] ((Hyper Aleph (Large Uncountable Infinities) (Aleph (Large Uncountable Infinities), Aleph (Large Uncountable Infinities))), Hyper Aleph (Large Uncountable Infinities) (Aleph (Large Uncountable Infinities), Aleph (Large Uncountable Infinities))).

We can also produce the abstraction of Hyper [Hyper Aleph (Large Uncountable Infinities) (Aleph (Large Uncountable Infinities), Aleph (Large Uncountable Infinities))]

((Hyper Aleph (Large Uncountable Infinities) (Aleph (Large Uncountable Infinities), Aleph (Large Uncountable Infinities))), Hyper Aleph (Large Uncountable Infinities) (Aleph (Large Uncountable Infinities), Aleph (Large Uncountable Infinities))) ([Hyper Aleph (Large Uncountable Infinities) (Aleph (Large Uncountable Infinities), Aleph (Large Uncountable Infinities))] ((Hyper Aleph (Large Uncountable Infinities) (Aleph (Large Uncountable Infinities), Aleph (Large Uncountable Infinities))), Hyper Aleph (Large Uncountable Infinities) (Aleph (Large Uncountable Infinities), Aleph (Large Uncountable Infinities))), [Hyper Aleph (Large Uncountable Infinities) (Aleph (Large Uncountable Infinities), Aleph (Large Uncountable Infinities))] ((Hyper Aleph (Large Uncountable Infinities) (Aleph (Large Uncountable Infinities), Aleph (Large Uncountable Infinities))), Hyper Aleph (Large Uncountable Infinities) (Aleph (Large Uncountable Infinities), Aleph (Large Uncountable Infinities))))

We can go one to produce an infinity using the above nesting style of convention with all potential computer storage media currently on Earth.

We can go one to produce an infinity using the above nesting style of convention with all potential computer storage media producible by the mineral resources within the observable portion of our universe.

We can go one to produce an infinity using the above nesting style of convention with all potential computer storage media producible by the mineral resources within our entire universe.

So, you can see how large of infinities we could abstractly denote using all of the mineral resources in our universe to construct the storage media.

However, our universe is likely just one of untold infinite numbers of universes.

Moreover, we have a potential untold infinite numbers of hyperspaces of 4 thru untold infinite numbers of dimensions, at least some of which may have some forms of mineral material extractible and convertible to storage media to define ever larger infinities.

So, we can arrive at an abstract understanding of how great a light-speed spacecraft's infinite Lorentz factor can become and the associated infinite number of light-years of travel in one year ship-frame and associated infinite number of years of travel into the future on one year ship-frame.

Since the more we colonize space or the broader cosmos in the future, the more human persons can be procreated, you can obtain a sense of how many human persons can be conceived, raised to happy adulthood, followed by a unlimitedly joy, peace, love, and eternal progress filled afterlife.

Now we consider scenarios that are utterly exotic even amidst the prospects that the speed of light may be an ultimate travel limit.

Accordingly, we can consider travel methods that are not defined by velocity or travel time.

For example, we are not considering instantaneous transport nor space-time shortcuts here. Instead, we are contemplating entering realms after suitable acquisition of infinite Lorentz factors in light-speed travel such that the new realm of travel implies travel without velocity nor travel time.

We can go further to consider entering realms where we travel without speed.

We can modify these concepts even more to consider travel in the material cosmos without velocity nor travel time as well as travel without any form of speed.

The first such realm of entrance and associated aspect of velocity-less travel we respectively refer to as realm-of-entrance-1 and velocity-less-travel-1.

Upon going to additional extremes, we may be able to enter a second realm and a yet greater ordinality of velocity-less travel which we respectively refer to as realm-of-entrance-2 and velocity-less-travel-2.

Upon going to additional extremes, we may be able to enter a third realm and a yet greater ordinality of velocity-less travel which we respectively refer to as realm-of-entrance-3 and velocity-less-travel-3.

Upon going to still additional extremes, we may be able to enter a fourth realm and a yet greater ordinality of velocity-less travel which we respectively refer to as realm-of-entrance-4 and velocity-less-travel-4.

Regarding how big infinities can be consider the following digression which is repeated again in this book.

Now, there are countable and uncountable infinities.

But before we go further, consider the hyper-operator notation that was designed to express huge values not otherwise expressible.

For example, Note that Hyper4(a, n) is equal to a tetrated n or a raised to the power of itself n-1 times. The latter value is symbolically written as n subscript a. For example 3 EXP 4 = 81, but 4 subscript3 is approximately equal to 10 EXP (1,000,000,000,000).

Alternatively 4 subscript 2 = 2 EXP 2 EXP 2 EXP 2 = 2 EXP [2 EXP [2 EXP 2]] = 2 EXP (2 EXP 4) = 2 EXP 16 = 65,536.

For example, Hyper5(4, 4)is equal to 4 tetrated 4 tetrated 4 tetrated 4. This value is commonly referred to as 4 pentated 4.

Hyper 6, (4,4) is 4 pentated 4 pentated 4 pentated 4 and is also referred to as 4 hexataed 4.

Hyper 7, (4,4) is 4 hexated 4 hexated 4 hexated 4 and so-on

Aleph 0 is the infinite number of integers.

Aleph 1 according to the perhaps unprovable and thus unfalsifiable Continuum Hypotheses is the number of real numbers which is greater than Aleph 0 by a multiplicative factor of infinity.

Aleph 2 is similarly greater than Aleph 1.

Aleph 3 is similarly greater than Aleph 2.

Aleph 4 is similarly greater than Aleph 3.

and so-on

In general, Aleph n = 2 EXP [Aleph (n-1)] where n can be any finite or infinite positive integer.

The number Ω is commonly stated as the least infinite positive integer or ordinal.

Now here is a real zinger.

So we can produce the abstraction of Hyper Aleph Ω (Aleph Ω, Aleph Ω).

W e can also produce the abstraction of Hyper [Hyper Aleph Ω (Aleph Ω, Aleph Ω)] ((Hyper Aleph Ω (Aleph Ω, Aleph Ω)), Hyper Aleph Ω (Aleph Ω, Aleph Ω)).

W e can also produce the abstraction of Hyper [Hyper Aleph Ω (Aleph Ω, Aleph Ω)] ((Hyper Aleph Ω (Aleph Ω, Aleph Ω)), Hyper Aleph Ω (Aleph Ω, Aleph Ω)) ([Hyper Aleph Ω (Aleph Ω, Aleph Ω)] ((Hyper Aleph Ω (Aleph Ω, Aleph Ω)), Hyper Aleph Ω (Aleph Ω, Aleph Ω)), [Hyper Aleph Ω (Aleph Ω, Aleph Ω)] ((Hyper Aleph Ω (Aleph Ω, Aleph Ω)), Hyper Aleph Ω (Aleph Ω, Aleph Ω)))

Upon going to exotic additional extremes, we may be eventually able to enter a fourth realm and a yet greater ordinality of velocity-less travel which we respectively refer to as realm-of-entrance- Hyper [Hyper Aleph Ω (Aleph Ω, Aleph Ω)] ((Hyper Aleph Ω (Aleph Ω, Aleph Ω)), Hyper Aleph Ω (Aleph Ω, Aleph Ω)) ([Hyper Aleph Ω (Aleph Ω, Aleph Ω)] ((Hyper Aleph Ω (Aleph Ω, Aleph Ω)), Hyper Aleph Ω (Aleph Ω, Aleph Ω)), [Hyper Aleph Ω (Aleph Ω, Aleph Ω)] ((Hyper Aleph Ω (Aleph Ω, Aleph Ω)), Hyper Aleph Ω (Aleph Ω,

Aleph Ω))) and velocity-less-travel- Hyper [Hyper Aleph Ω (Aleph Ω, Aleph Ω)] ((Hyper Aleph Ω (Aleph Ω, Aleph Ω)), Hyper Aleph Ω (Aleph Ω, Aleph Ω)) ([Hyper Aleph Ω (Aleph Ω, Aleph Ω)] ((Hyper Aleph Ω (Aleph Ω, Aleph Ω)), Hyper Aleph Ω (Aleph Ω, Aleph Ω)), [Hyper Aleph Ω (Aleph Ω, Aleph Ω)] ((Hyper Aleph Ω (Aleph Ω, Aleph Ω)), Hyper Aleph Ω (Aleph Ω, Aleph Ω))).

We may replace Ω with what we will call a high-end mathematically recognized uncountable infinity to develop the following construct. We will refer to this infinity as Ω'

Upon going to exotic additional extremes, we may be eventually able to enter a fourth realm and a yet greater ordinality of velocity-less travel which we respectively refer to as realm-of-entrance- Hyper [Hyper Aleph Ω' (Aleph Ω', Aleph Ω')] ((Hyper Aleph Ω' (Aleph Ω', Aleph Ω')), Hyper Aleph Ω' (Aleph Ω', Aleph Ω')) ([Hyper Aleph Ω' (Aleph Ω', Aleph Ω')] ((Hyper Aleph Ω' (Aleph Ω', Aleph Ω')), Hyper Aleph Ω' (Aleph Ω', Aleph Ω')), [Hyper Aleph Ω' (Aleph Ω', Aleph Ω')] ((Hyper Aleph Ω' (Aleph Ω', Aleph Ω')), Hyper Aleph Ω' (Aleph Ω', Aleph Ω'))) and velocity-less-travel- Hyper [Hyper Aleph Ω' (Aleph Ω', Aleph Ω')] ((Hyper Aleph Ω' (Aleph Ω', Aleph Ω')), Hyper Aleph Ω' (Aleph Ω', Aleph Ω')) ([Hyper Aleph Ω' (Aleph Ω', Aleph Ω')] ((Hyper Aleph Ω' (Aleph Ω', Aleph Ω')), Hyper Aleph Ω' (Aleph Ω', Aleph Ω')), [Hyper Aleph Ω' (Aleph Ω', Aleph Ω')] ((Hyper Aleph Ω' (Aleph Ω', Aleph Ω')), Hyper Aleph Ω' (Aleph Ω', Aleph Ω'))).

We may affix the following operator to formulas for gamma and related parameters in this book to denote the above range of possibilities.

{f{From realm-of-entrance-1 and velocity-less-travel-1 to realm-of-entrance-Hyper [Hyper Aleph Ω' (Aleph Ω', Aleph Ω')] ((Hyper Aleph Ω' (Aleph Ω', Aleph Ω')), Hyper Aleph Ω' (Aleph Ω', Aleph Ω')) ([Hyper Aleph Ω' (Aleph Ω', Aleph Ω')] ((Hyper Aleph Ω' (Aleph Ω', Aleph Ω')), Hyper Aleph Ω' (Aleph Ω', Aleph Ω')), [Hyper Aleph Ω' (Aleph Ω', Aleph Ω')] ((Hyper Aleph Ω' (Aleph Ω', Aleph Ω')), Hyper Aleph Ω' (Aleph Ω', Aleph Ω'))) and so-on without symbolic limit and velocity-less-travel-Hyper [Hyper Aleph Ω' (Aleph Ω', Aleph Ω')] ((Hyper Aleph Ω' (Aleph Ω', Aleph Ω')), Hyper Aleph Ω' (Aleph Ω', Aleph Ω')) ([Hyper Aleph Ω' (Aleph Ω', Aleph Ω')] ((Hyper Aleph Ω' (Aleph Ω', Aleph Ω')), Hyper Aleph Ω' (Aleph Ω', Aleph Ω')), [Hyper Aleph Ω' (Aleph Ω', Aleph Ω')] ((Hyper Aleph Ω' (Aleph Ω', Aleph Ω')), Hyper Aleph Ω' (Aleph Ω', Aleph Ω'))) and so-on without symbolic limit}}.

We can take arbitrary orders of derivatives and integrals of the above expression with respect to spacecraft; ship-time, acceleration, analogs of acceleration, gamma, kinetic energy and the like. Thus, we produce and affix the following operator to the lengthy formulas for gamma and related parameters in this section to denote these prospects.

Note that the integrals in a sense offer values akin to distances travel. So, built into the mathematical conventions provided below are mechanisms that are not defined in conventional terms for distance but instead of forms of travel that are not defined by distance but instead some other suitable parameters than can be arbitrarily infinite. Thus, just as we can have infinite spatial and/or temporal distances of travel, the alternatives to conventional distances of travel may also be infinite.

{w{{d{…d{{f{From realm-of-entrance-1 and velocity-less-travel-1 to realm-of-entrance-Hyper [Hyper Aleph Ω' (Aleph Ω', Aleph Ω')] ((Hyper Aleph Ω' (Aleph Ω', Aleph Ω')), Hyper Aleph Ω' (Aleph Ω', Aleph Ω')) ([Hyper Aleph Ω' (Aleph Ω', Aleph Ω')] ((Hyper Aleph Ω' (Aleph Ω', Aleph Ω')), Hyper Aleph Ω' (Aleph Ω', Aleph Ω')), [Hyper Aleph Ω' (Aleph Ω', Aleph Ω')] ((Hyper Aleph Ω' (Aleph Ω', Aleph Ω')), Hyper Aleph Ω' (Aleph Ω', Aleph Ω'))) and so-on without symbolic limit and velocity-less-travel-Hyper [Hyper Aleph Ω' (Aleph Ω', Aleph Ω')] ((Hyper Aleph Ω' (Aleph Ω', Aleph Ω')), Hyper Aleph Ω' (Aleph Ω', Aleph Ω')) ([Hyper Aleph Ω' (Aleph Ω', Aleph Ω')] ((Hyper Aleph Ω' (Aleph Ω', Aleph Ω')), Hyper Aleph Ω' (Aleph Ω', Aleph Ω')), [Hyper Aleph Ω' (Aleph Ω', Aleph Ω')] ((Hyper Aleph Ω' (Aleph Ω', Aleph Ω')), Hyper Aleph Ω' (Aleph Ω', Aleph Ω'))) and so-on without symbolic limit}}/d ship-time, or acceleration, or analogs of acceleration, or gamma, or kinetic energy and the like}/ …}/d ship-time, or acceleration, or analogs of acceleration, or gamma, or kinetic energy and the like}:{∫{…∫{{f{From realm-of-entrance-1 and velocity-less-travel-1 to realm-of-entrance-Hyper [Hyper Aleph Ω' (Aleph Ω', Aleph Ω')] ((Hyper Aleph Ω' (Aleph Ω', Aleph Ω')), Hyper Aleph Ω' (Aleph Ω', Aleph Ω')) ([Hyper Aleph Ω' (Aleph Ω', Aleph Ω')] ((Hyper Aleph Ω' (Aleph Ω', Aleph Ω')), Hyper Aleph Ω' (Aleph Ω', Aleph Ω')), [Hyper Aleph Ω' (Aleph Ω', Aleph Ω')] ((Hyper Aleph Ω' (Aleph Ω', Aleph Ω')), Hyper Aleph Ω' (Aleph Ω', Aleph Ω'))) and so-on without symbolic limit and velocity-less-travel-Hyper [Hyper Aleph Ω' (Aleph Ω', Aleph Ω')] ((Hyper Aleph Ω' (Aleph Ω', Aleph Ω')), Hyper Aleph Ω' (Aleph Ω', Aleph Ω')) ([Hyper Aleph Ω' (Aleph Ω', Aleph Ω')] ((Hyper Aleph Ω' (Aleph Ω', Aleph Ω')), Hyper Aleph Ω' (Aleph Ω', Aleph Ω')), [Hyper Aleph Ω' (Aleph Ω', Aleph Ω')] ((Hyper Aleph Ω' (Aleph Ω', Aleph Ω')), Hyper Aleph Ω' (Aleph Ω', Aleph Ω'))) and so-on without symbolic limit}}d ship-time, or acceleration, or analogs of acceleration, or gamma, or kinetic energy and the like} …}d ship-time, or acceleration, or analogs of acceleration, or gamma, or kinetic energy and the like}}}.

We can thus consider forms of travel that are not defined by distance but instead some other suitable parameters than can be arbitrarily infinite.

ESSAY 71) Pre-Creational Light Sailing At Infinite Lorentz Factors.

Now, assuming that GOD is indeed the source of all creation, it behooves us to consider that there exists some forms of natural existential boundaries between GOD as creator and each creature, but more broadly, between GOD and the created organism of the Cosmos.

Since all things related have boundary conditions between themselves and the rest of real elements of existence, it would seem as though between boundary conditions of the cosmos and GOD, there is some interstitial medium that is neither GOD nor the substantial aspects of the cosmos but which is generated by GOD.

There may even be more than one interstitial medium such as between GOD and each substantial creation. If the latter speculation is correct, then the number of interstitial mediums is likely an increasing uncountable infinity.

There may even be more than one cosmos, perhaps an uncountable infinite number of cosmoses. This infinity likely would keep growing.

So, this notion of such interstitial mediums alludes to the mediums as being forms of precreation light.

A suitably technologically advanced civilization may by Divine Providence be able to harness these precreational lights for some quite literal and concrete spacecraft propulsion mechanism.

Accordingly, we refer to these systems principally as precreational light-sails which are literal analogs of electromagnetic light-sails.

The vast energies available to harness in precreational forms of light is utterly infinitely fantastic.

We may affix the following operator to formulas for gamma and related parameters to denote the mechanisms of precreational forms of light for light-sailing craft.

{[Σ (k = 1; k = ꝏ↑):[[w(Mechanisms of precreational forms of light of the same level for light-sailing craft)],k]]: [Σ (j = 1; j = ꝏ↑):[w(Mechanisms of j levels of precreational forms of light for light-sailing craft)]]}

The upper series limits indicate open ended infinities with opportunities for indefinite growth.

ESSAY 72) Considering Infinitely Many Futures As Boundary Conditions For Light-Speed Travel.

We can consider the future or futures as boundary conditions that every creature, universe, multiverse, forest, biosphere and the like, every hyperspace or hyperspace-time and so-on runs toward via the laws of entropy and thermodynamics.

We can also consider a future after a future after a future and so-on that every creature, universe, multiverse, forest, biosphere and the like, every hyperspace or hyperspace-time and so-on runs toward via the laws of entropy and thermodynamics.

Furthermore, we can consider a future embedded within a future embedded within a future and so-on that every creature, universe, multiverse, forest, biosphere and the like, every hyperspace or hyperspace-time and so-on runs toward via the laws of entropy and thermodynamics.

The idea here is to maximize net disorder while promoting the development of pockets of organization and order.

Light-speed travel at infinite Lorentz factors may enable a spacecraft so disposed to running out of future room to travel forward in time in for a given universe or broader realm of travel. Thus, a light-speed spacecraft may excel at running forward to ever grander future boundary conditions.

We may affix the following operator to formulas for gamma and related parameters of the book to denote the mechanisms presented the previous five paragraphs.

[u(Light-speed spacecraft at ever greater infinite Lorentz factors excelling at running forward to ever grander future boundary conditions such as considered in a future or futures, infinite series of futures, and/or infinite levels of futures embedded within futures embedded within futures and so-on)]

We can take the ship-time, gamma, acceleration, kinetic energy derivatives and so-on as useful or necessary as well as the same for integrals of the above operator. In cases where integrations are done, the result is a variety of different classes of distances traveled or analogs thereof. Thus, we develop the following two operators, one for derivatives and one for integrals of the previous operator provided above.

{d{…{d[u(Light-speed spacecraft at ever greater infinite Lorentz factors excelling at running forward to ever grander future boundary conditions such as considered in a future or futures, infinite series of futures, and/or infinite levels of futures embedded within futures embedded within futures and so-on)]/d ship-time or gamma or acceleration or kinetic energy derivatives or so-on}/…}/ d ship-time or gamma or acceleration or kinetic energy derivatives or so-on}

{∫{…{∫[u(Light-speed spacecraft at ever greater infinite Lorentz factors excelling at running forward to ever grander future boundary conditions such as considered in a future or futures, infinite series of futures, and/or infinite levels of futures embedded within futures embedded within futures and so-on)]d ship-time or gamma or acceleration or kinetic energy derivatives or so-on}…}d ship-time or gamma or acceleration or kinetic energy derivatives or so-on}.

We combine all three subject operators as a compound operator listed as follows to denote the associated mechanisms and may affix the compound operator to formulas for gamma and related parameters.

{[u(Light-speed spacecraft at ever greater infinite Lorentz factors excelling at running forward to ever grander future boundary conditions such as considered in a future or futures, infinite series of futures, and/or infinite levels of futures embedded within futures embedded within futures and so-on)]: {d{…{d[u(Light-speed spacecraft at ever greater infinite Lorentz factors excelling at running forward to ever grander future boundary conditions such as considered in a future or futures, infinite series of futures, and/or infinite levels of futures embedded within futures embedded within futures and so-on)]/d ship-time or gamma or acceleration or kinetic energy derivatives or so-on}/…}/ d ship-time or gamma or acceleration or kinetic energy derivatives or so-on}:{∫{…{∫[u(Light-speed spacecraft at ever greater infinite Lorentz factors excelling at running forward to ever grander future boundary conditions such as considered in a future or futures, infinite series of futures, and/or infinite levels of futures embedded within futures embedded within futures and so-on)]d ship-time or gamma or acceleration or kinetic energy derivatives or so-on}…}d ship-time or gamma or acceleration or kinetic energy derivatives or so-on}}.

So, we have never ending glories upon glories to look forward to, and when we have progressed an infinite number of years, we've no less time than when we first begun.

ESSAY 73) Special Light-Sail Craft Of Special Shapes For Perpetual Acceleration At The Speed Of Light.

A fascinating class of spacecraft propulsion methods involves a craft taking on either permanently or temporarily a special shape that enables the craft to perpetually accelerate.

For example, a light-sail may perpetually accelerate as long as one side stays electromagnetically active while the other side is always completely transmissive. The action may be reflective, or negative electromagnetic index of refraction pulling. Such a sail would approach light-speed ever more closely given time and sufficient erosion resistance and refractory characteristics.

However, a spacecraft including but not limited to a light-sail craft may have a huge amount of possible geometric shapes and then so shapes that are conducive to perpetual acceleration.

For example, a spacecraft inscribed within a 3-D spatial cubic volume in our universe may have 2 EXP {{(Cube edge length)/{{[h/(2π)] G/(c EXP 3)} EXP (1/2)}} EXP 3} possible shapes as a first order estimate.

For a spacecraft inscribed within a hyperspatial cube of 4 dimensions for which the higher dimension or the fourth dimension is quantized at the level of (1/a){{[h/(2π)] G/(c EXP 3)} EXP (1/2)} where a is any greater than one finite or infinite positive real number, the number of possible shapes is in first order equal to {2 EXP { {(Cube edge length)/{{[h/(2π)] G/(c EXP 3)} EXP (1/2)} } EXP 3}} EXP {(Cube edge length)/ {(1/a){{[h/(2π)] G/(c EXP 3)} EXP (1/2)}}}

For a spacecraft inscribed within a hyperspatial cube of 5 dimensions for which the two higher dimensions or the fourth and fifth dimensions are each quantized at the level of (1/a){{[h/(2π)] G/(c EXP 3)} EXP (1/2)} where a is any greater than one finite or infinite positive real number, the number of possible shapes is in first order equal to {2 EXP { {(Cube edge length)/{{[h/(2π)] G/(c EXP 3)} EXP (1/2)} } EXP 3}} EXP {{(Cube edge length)/ {(1/a){{[h/(2π)] G/(c EXP 3)} EXP (1/2)}}} EXP 2}

For a spacecraft inscribed within a hyperspatial cube of 6 dimensions for which the three higher dimensions or the fourth, fifth, and sixth dimensions are each quantized at the level of (1/a){{[h/(2π)] G/(c EXP 3)} EXP (1/2)} where a is any greater than one finite or infinite positive real number, the number of possible shapes is in first order equal to {2 EXP { {(Cube edge length)/{{[h/(2π)] G/(c EXP 3)} EXP (1/2)} } EXP 3}} EXP {{(Cube edge length)/ {(1/a){{[h/(2π)] G/(c EXP 3)} EXP (1/2)}}} EXP 3}

For a spacecraft inscribed within a hyperspatial cube of (3 + n) dimensions for which the n higher dimensions are each quantized at the level of (1/a){{[h/(2π)] G/(c EXP 3)} EXP (1/2)} where a is any greater than one finite or infinite positive real number, the number of possible shapes is in first order equal to {2 EXP { {(Cube edge length)/{{[h/(2π)] G/(c EXP 3)} EXP (1/2)} } EXP 3}} EXP {{(Cube edge length)/ {(1/a){{[h/(2π)] G/(c EXP 3)} EXP (1/2)}}} EXP n}

In the cases where a spacecraft inscribed within a 3-D spatial cubic volume in our universe may can have g material compositions, the number of possible spacecraft forms is equal to {2 EXP {{(Cube edge length)/{{[h/(2π)] G/(c EXP 3)} EXP (1/2)}} EXP 3}}(g)

In the cases where a spacecraft inscribed within a hyperspatial cube of 4 dimensions for which the higher dimension or the fourth dimensions is quantized at the level of (1/a){{[h/(2π)] G/(c EXP 3)} EXP (1/2)} where a is any greater than one finite or infinite positive real number, and can have g material compositions, the number of possible spacecraft forms is in first order equal to {{2 EXP {{(Cube edge length)/{{[h/(2π)] G/(c EXP 3)} EXP (1/2)}} EXP 3}} EXP {(Cube edge length)/ {(1/a){{[h/(2π)] G/(c EXP 3)} EXP (1/2)}}}}(g)

In the cases where a spacecraft inscribed within a hyperspatial cube of 5 dimensions for which the two higher dimensions or the fourth and fifth dimensions are each quantized at the level of (1/a){{[h/(2π)] G/(c EXP 3)} EXP (1/2)} where a is any greater than one finite or infinite positive real number, and can have g material compositions, the number of possible spacecraft forms is in first order equal to {{2 EXP {{(Cube edge length)/{{[h/(2π)] G/(c EXP 3)} EXP (1/2)}} EXP 3}} EXP {{(Cube edge length)/ {(1/a){{[h/(2π)] G/(c EXP 3)} EXP (1/2)}}} EXP 2}}(g)

In the cases where a spacecraft inscribed within a hyperspatial cube of 6 dimensions for which the three higher dimensions or the fourth, fifth, and sixth dimensions are each quantized at the level of (1/a){{[h/(2π)] G/(c EXP 3)} EXP (1/2)} where a is any greater than one finite or infinite positive real number, and can have g material compositions, the number of possible spacecraft forms is in first order equal to {{2 EXP {{(Cube edge length)/{{[h/(2π)] G/(c EXP 3)} EXP (1/2)}} EXP 3}} EXP {{(Cube edge length)/ {(1/a){{[h/(2π)] G/(c EXP 3)} EXP (1/2)}}} EXP 3}}(g)

In the cases where a spacecraft inscribed within a hyperspatial cube of (3 + n) dimensions for which the n higher dimensions are each quantized at the level of (1/a){{[h/(2π)] G/(c EXP 3)} EXP (1/2)} where a is any greater than one finite or infinite positive real number, and can have g material compositions, the number of possible spacecraft forms is in first order equal to {{2 EXP {{(Cube edge length)/{{[h/(2π)] G/(c EXP 3)} EXP (1/2)}} EXP 3}} EXP {{(Cube edge length)/ {(1/a){{[h/(2π)] G/(c EXP 3)} EXP (1/2)}}} EXP n}}(g)

Note that in this digression, we assume a spacecraft as non-limiting options can be embodied in only one or a few particles which may or may not be connected. This is possible in cases where the particles serve as the carrier medium of the spacecraft via classical beaming of Star Trek TV series style or by quantum-mechanical entanglement based tunneling.

We affix the following operator to the lengthy formulas for gamma and related parameters in this part of this section of the book to denote the above maximized range of spacecraft forms.

{u{# of possible spacecraft forms in n space in 1st order of {{2 EXP {{(Cube edge length)/{{[h/(2π)] G/(c EXP 3)} EXP (1/2)}} EXP 3}} EXP {{(Cube edge length)/ {(1/a){{[h/(2π)] G/(c EXP 3)} EXP (1/2)}}} EXP n}}(g)}}

Additionally, we can consider that the spacecraft can be imbued with secondary motions of relativistic velocities in the spacecraft reference frame thus enabling various relativistic contractions even to sizes of a point or a line segment or analogs thereof in finitely quantized space. However, this is a more difficult problem but is addressed in the next topic below.

ESSAY 74) Special Light-Sail Craft Of Special Shapes For Perpetual Acceleration At The Speed Of Light And Where The Craft Has Multiple Light-Speed Degrees Of Oscillation.

Now, we can consider that a spacecraft inscribed within a cubic volume, in a given space may have rotational or other oscillatory motion-based energies that modify its geometric shape.

The number of possible kinetic-energy kinematic states is equal to the number of possible rotational and other oscillatory cycle paths raised to the power of the total number of available kinetic energy levels associated with said paths.

For example, a spacecraft inscribed within a 3-D spatial cubic volume in our universe may have 2 EXP {{(Cube edge length)/{{[h/(2π)] G/(c EXP 3)} EXP (1/2)}} EXP 3} possible shapes as a first order estimate and total number of thermodynamic states equal to {2 EXP {{(Cube edge length)/{{[h/(2π)] G/(c EXP 3)} EXP (1/2)}} EXP 3}} EXP [(number of possible rotational and other oscillatory cycle paths) EXP (total number of available kinetic energy levels associated with said paths)].

For a spacecraft inscribed within a hyperspatial cube of 4 dimensions for which the higher dimension or the fourth dimension is quantized at the level of (1/a){{[h/(2π)] G/(c EXP 3)} EXP (1/2)} where a is any greater than one finite or infinite positive real number, the number of possible shapes is in first order equal to {2 EXP { {(Cube edge length)/{{[h/(2π)] G/(c EXP 3)} EXP (1/2)} } EXP 3}} EXP {(Cube edge length)/ {(1/a){{[h/(2π)] G/(c EXP 3)} EXP (1/2)}}} and total number of thermodynamic states equal to {{2 EXP { {(Cube edge length)/{{[h/(2π)] G/(c EXP 3)} EXP (1/2)} } EXP 3}} EXP {(Cube edge length)/ {(1/a){{[h/(2π)] G/(c EXP 3)} EXP (1/2)}}}} EXP [(number of possible rotational and other oscillatory cycle paths) EXP (total number of available kinetic energy levels associated with said paths)].

For a spacecraft inscribed within a hyperspatial cube of 5 dimensions for which the two higher dimensions or the fourth and fifth dimensions are each quantized at the level of (1/a){{[h/(2π)] G/(c EXP 3)} EXP (1/2)} where a is any greater than one finite or infinite positive real number, the number of possible shapes is in first order equal to {2 EXP { {(Cube edge length)/{{[h/(2π)] G/(c EXP 3)} EXP (1/2)} } EXP 3}} EXP {{(Cube edge length)/ {(1/a){{[h/(2π)] G/(c EXP 3)} EXP (1/2)}}} EXP 2} and total number of thermodynamic states equal to {{2 EXP { {(Cube edge length)/{{[h/(2π)] G/(c EXP 3)} EXP (1/2)} } EXP 3}} EXP {{(Cube edge length)/ {(1/a){{[h/(2π)] G/(c EXP 3)} EXP (1/2)}}} EXP 2}} EXP [(number of possible rotational and other oscillatory cycle paths) EXP (total number of available kinetic energy levels associated with said paths)].

For a spacecraft inscribed within a hyperspatial cube of 6 dimensions for which the three higher dimensions or the fourth, fifth, and sixth dimensions are each quantized at the level of (1/a){{[h/(2π)] G/(c EXP 3)} EXP (1/2)} where a is any greater than one finite or infinite positive real number, the number of possible shapes is in first order equal to {2 EXP { {(Cube edge length)/{{[h/(2π)] G/(c EXP 3)} EXP (1/2)} } EXP 3}} EXP {{(Cube edge length)/ {(1/a){{[h/(2π)] G/(c EXP 3)} EXP (1/2)}}} EXP 3} and total number of thermodynamic states equal to {{2 EXP { {(Cube edge length)/{{[h/(2π)] G/(c EXP 3)} EXP (1/2)} } EXP 3}} EXP {{(Cube edge length)/ {(1/a){{[h/(2π)] G/(c EXP 3)} EXP (1/2)}}} EXP 3}} EXP [(number of possible rotational and other oscillatory cycle paths) EXP (total number of available kinetic energy levels associated with said paths)].

For a spacecraft inscribed within a hyperspatial cube of (3 + n) dimensions for which the n higher dimensions are each quantized at the level of (1/a){{[h/(2π)] G/(c EXP 3)} EXP (1/2)} where a is any greater than one finite or infinite positive real number, the number of possible shapes is in first order equal to {2 EXP { {(Cube edge length)/{{[h/(2π)] G/(c EXP 3)} EXP (1/2)} } EXP 3}} EXP {{(Cube edge length)/ {(1/a){{[h/(2π)] G/(c EXP 3)} EXP (1/2)}}} EXP n} and total number of thermodynamic states equal to {{2 EXP { {(Cube edge length)/{{[h/(2π)] G/(c EXP 3)} EXP (1/2)} } EXP 3}} EXP {{(Cube edge length)/ {(1/a){{[h/(2π)] G/(c EXP 3)} EXP (1/2)}}} EXP n}} EXP [(number of possible rotational and other oscillatory cycle paths) EXP (total number of available kinetic energy levels associated with said paths)].

In the cases where a spacecraft inscribed within a 3-D spatial cubic volume in our universe may can have g material compositions, the number of possible spacecraft forms is equal to {2 EXP {{(Cube edge length)/{{[h/(2π)] G/(c EXP 3)} EXP (1/2)}} EXP 3}}(g) and total number of thermodynamic states equal to {{2 EXP {{(Cube edge length)/{{[h/(2π)] G/(c EXP 3)} EXP (1/2)}} EXP 3}}(g)} EXP [(number of possible rotational and other oscillatory cycle paths) EXP (total number of available kinetic energy levels associated with said paths)].

In the cases where a spacecraft inscribed within a hyperspatial cube of 4 dimensions for which the higher dimension or the fourth dimensions is quantized at the level of (1/a){{[h/(2π)] G/(c EXP 3)} EXP (1/2)} where a is any greater than one finite or infinite positive real number, and can have g material compositions, the number of possible spacecraft forms is in first order equal to {{2 EXP {{(Cube edge length)/{{[h/(2π)] G/(c EXP 3)} EXP (1/2)}} EXP 3}} EXP {(Cube edge length)/ {(1/a){{[h/(2π)] G/(c EXP 3)} EXP (1/2)}}}}(g) and total number of thermodynamic states equal to {{{2 EXP {{(Cube edge length)/{{[h/(2π)] G/(c EXP 3)} EXP (1/2)}} EXP 3}} EXP {(Cube edge length)/ {(1/a){{[h/(2π)] G/(c EXP 3)} EXP (1/2)}}}}(g)} EXP [(number of possible rotational and other oscillatory cycle paths) EXP (total number of available kinetic energy levels associated with said paths)].

In the cases where a spacecraft inscribed within a hyperspatial cube of 5 dimensions for which the two higher dimensions or the fourth and fifth dimensions are each quantized at the level of (1/a){{[h/(2π)] G/(c EXP 3)} EXP (1/2)} where a is any greater than one finite or infinite positive real number, and can have g material compositions, the number of possible spacecraft forms is in first order equal to {{2 EXP {{(Cube edge length)/{{[h/(2π)] G/(c EXP 3)} EXP (1/2)}} EXP 3}} EXP {{(Cube edge length)/ {(1/a){{[h/(2π)] G/(c EXP 3)} EXP (1/2)}}} EXP 2}}(g) and total number of thermodynamic states equal to {{{2 EXP {{(Cube edge length)/{{[h/(2π)] G/(c EXP 3)} EXP (1/2)}} EXP 3}} EXP {{(Cube edge length)/ {(1/a){{[h/(2π)] G/(c EXP 3)} EXP (1/2)}}} EXP 2}}(g)} EXP [(number of possible rotational and other oscillatory cycle paths) EXP (total number of available kinetic energy levels associated with said paths)].

In the cases where a spacecraft inscribed within a hyperspatial cube of 6 dimensions for which the three higher dimensions or the fourth, fifth, and sixth dimensions are each quantized at the level of (1/a){{[h/(2π)] G/(c EXP 3)} EXP (1/2)} where a is any greater than one finite or infinite positive real number, and can have g material compositions, the number of possible spacecraft forms is in first order equal to {{2 EXP {{(Cube edge length)/{{[h/(2π)] G/(c EXP 3)} EXP (1/2)}} EXP 3}} EXP {{(Cube edge length)/ {(1/a){{[h/(2π)] G/(c EXP 3)} EXP (1/2)}}} EXP 3}}(g) and total number of thermodynamic states equal to {{{2 EXP {{(Cube edge length)/{{[h/(2π)] G/(c EXP 3)} EXP (1/2)}} EXP 3}} EXP {{(Cube edge length)/ {(1/a){{[h/(2π)] G/(c EXP 3)} EXP (1/2)}}} EXP 3}}(g)} EXP [(number of possible rotational and other oscillatory cycle paths) EXP (total number of available kinetic energy levels associated with said paths)].

In the cases where a spacecraft inscribed within a hyperspatial cube of (3 + n) dimensions for which the n higher dimensions are each quantized at the level of (1/a){{[h/(2π)] G/(c EXP 3)} EXP (1/2)} where a is any greater than one finite or infinite positive real number, and can have g material compositions, the number of possible spacecraft forms is in first order equal to {{2 EXP {{(Cube edge length)/{{[h/(2π)] G/(c EXP 3)} EXP (1/2)}} EXP 3}} EXP {{(Cube edge length)/ {(1/a){{[h/(2π)] G/(c EXP 3)} EXP (1/2)}}} EXP n}}(g) and total number of thermodynamic states equal to {{{2 EXP {{(Cube edge length)/{{[h/(2π)] G/(c EXP 3)} EXP (1/2)}} EXP 3}} EXP {{(Cube edge length)/ {(1/a){{[h/(2π)] G/(c EXP 3)} EXP (1/2)}}} EXP n}}(g)} EXP [(number of possible rotational and other oscillatory cycle paths) EXP (total number of available kinetic energy levels associated with said paths)].

Note, again, that in this digression, we assume a spacecraft as non-limiting options can be embodied in only one or a few particles which may or may not be connected. This is possible in cases where the particles serve as the carrier medium of the spacecraft via classical beaming of Star Trek TV series style or by quantum-mechanical entanglement based tunneling.

We affix the following operator to the lengthy formulas for gamma and related parameters in this part of this section of the book to denote the above maximized range of thermodynamic spacecraft forms.

{u{# of thermodynamic spacecraft forms in n space in 1st order of {{{{2 EXP {{(Cube edge length)/{{[h/(2π)] G/(c EXP 3)} EXP (1/2)}} EXP 3}} EXP {{(Cube edge length)/{(1/a){{[h/(2π)] G/(c EXP 3)} EXP (1/2)}}} EXP n}}(g)} EXP [(number of possible rotational and other oscillatory cycle paths) EXP (total number of available kinetic energy levels associated with said paths)]}}}

ESSAY 75) Big Bang(s) Powered Starships.

Now, we of Catholic faith, protestant faiths, evangelical faiths, Judaism, Islam, Hindu, Buddhist, and so-on, even of UFO new age belief systems are well familiar with the notion of free will and the faculty of will.

We in the West tend to down-play the importance of our emotions except psychologists, councilors, therapists, psychiatrists and the like.

For those of us familiar with psychodynamic theories, we understand that attitudes are as if a cross between emotions, the will, and thinking to various degrees in various ways that are context and individual specific.

However, Sacred Scripture most notably the Holy Bible does not fault folks so much for their actions as it faults folks for their attitudes.

Thus, it seems as though attitudes have great sway over our free wills and perhaps allude to a faculty higher than the free will but related to our attitudes.

We will refer to this conjectured faculty as hyper-will, not in the hyper-activity behavioral sense, but in the sense of supersedence. An analogy is the notion of hypersonic travel in aircraft that can fly several times the speed of sound.

So, this faculty of the hyperwill is very powerful and transcendent and is perhaps the mechanism by which we ultimately choose to follow the path of good or follow the path of evil.

We affix the following operator to the lengthy formulas for the existential size and significance of the soul to denote the transcendence and sublimity of the conjectured human hyperwill.

[u(Transcendence and sublimity of the conjectured human hyperwill)]

Big Bangs as the origins of the levels of Heaven.

Big Bangs as the origins of astral planes and levels thereof.

Big Bangs as the origins of the laws of physics.

Big Bangs as the origins of the laws of mathematics.

Big Bangs as the originis of the laws of logic.

Divine Big Bangs as the origins of the three Divine Persons.

Big Bangs to produce realms with different laws of math.

Big Bangs to produe realms with different laws of logic.

Big Bangs to produce realms without quantitative measures.

Big Bangs to produce realms without physical quantities.

Big Bangs as the origins of the levels of Heaven to power spacecraft.

Big Bangs as the origins of astral planes and levels thereof to power spacecraft.

Big Bangs as the origins of the laws of physics to power spacecraft.

Big Bangs as the origins of the laws of mathematics to power spacecraft.

Big Bangs as the originis of the laws of logic to power spacecraft.

Divine Big Bangs as the origins of the three Divine Persons to power spacecraft.

Big Bangs to produce realms with different laws of math to power spacecraft.

Big Bangs to produe realms with different laws of logic to power spacecraft.

Big Bangs to produce realms without quantitative measures to power spacecraft.

Big Bangs to produce realms without physical quantities to power spacecraft

Big Bangs as the origins of the levels of Heaven to power spacecraft where the spacecraft travel back in time, meta-time, or eternity to capture the Big Bangs' Energies, ride the Big Bangs' waves or wave-functions, buy backward classical travel, quantum teleportation, backward quantum tunneling, backward wormhole travel and/or the like.

Big Bangs as the origins of astral planes and levels thereof to power spacecraft where the spacecraft travel back in time, meta-time, or eternity to capture the Big Bangs' Energies, ride the Big Bangs' waves or wave-functions, buy backward classical travel, quantum teleportation, backward quantum tunneling, backward wormhole travel and/or the like.

Big Bangs as the origins of the laws of physics to power spacecraft where the spacecraft travel back in time, meta-time, or eternity to capture the Big Bangs' Energies, ride the Big Bangs' waves or wave-functions, buy backward classical travel, quantum teleportation, backward quantum tunneling, backward wormhole travel and/or the like.

Big Bangs as the origins of the laws of mathematics to power spacecraft where the spacecraft travel back in time, meta-time, or eternity to capture the Big Bangs' Energies, ride the Big Bangs' waves or wave-functions, buy backward classical travel, quantum teleportation, backward quantum tunneling, backward wormhole travel and/or the like.

Big Bangs as the originis of the laws of logic to power spacecraft where the spacecraft travel back in time, meta-time, or eternity to capture the Big Bangs' Energies, ride the Big Bangs' waves or wave-functions, buy backward classical travel, quantum teleportation, backward quantum tunneling, backward wormhole travel and/or the like.

Divine Big Bangs as the origins of the three Divine Persons to power spacecraft where the spacecraft travel back in time, meta-time, or eternity to capture the Big Bangs' Energies, ride the Big Bangs' waves or wave-functions, buy backward classical travel, quantum teleportation, backward quantum tunneling, backward wormhole travel and/or the like.

Big Bangs to produce realms with different laws of math to power spacecraft where the spacecraft travel back in time, meta-time, or eternity to capture the Big Bangs' Energies, ride the Big Bangs' waves or wave-functions, buy backward classical travel, quantum teleportation, backward quantum tunneling, backward wormhole travel and/or the like.

Big Bangs to produe realms with different laws of logic to power spacecraft where the spacecraft travel back in time, meta-time, or eternity to capture the Big Bangs' Energies, ride the Big Bangs' waves or wave-functions, buy backward classical travel, quantum teleportation, backward quantum tunneling, backward wormhole travel and/or the like where the spacecraft travel back in time, meta-time, or eternity to capture the Big Bangs' Energies, ride the Big Bangs' waves or wave-functions, buy backward classical travel, quantum teleportation, backward quantum tunneling, backward wormhole travel and/or the like.

Big Bangs to produce realms without quantitative measures to power spacecraft where the spacecraft travel back in time, meta-time, or eternity to capture the Big Bangs' Energies, ride the Big Bangs' waves or wave-functions, buy backward classical travel, quantum teleportation, backward quantum tunneling, backward wormhole travel and/or the like.

Big Bangs to produce realms without physical quantities to power spacecraft to power spacecraft where the spacecraft travel back in time, meta-time, or eternity to capture the Big Bangs' Energies, ride the Big Bangs' waves or wave-functions, buy backward classical travel, quantum teleportation, backward quantum tunneling, backward wormhole travel and/or the like.

ESSAY 76) Effective Instantaneous Travel.

We can consider effective instantaneous travel with respect to our ordinary 4-D space-time as a result of traveling along a completely orthogonal temporal coordinate or multiple orthogonal temporal coordinates relative to the known single temporal coordinates where all or some additional temporal coordinates are parallel to each other or some or all additional temporal coordinates are orthogonal to each other.

We can consider effective faster than instantaneous travel with respect to our ordinary 4-D space-time as a result of traveling along a completely super-orthogonal temporal coordinate or multiple super-orthogonal temporal coordinates relative to the known single temporal coordinates where all or some additional temporal coordinates are super-parallel to each other or some or all additional temporal coordinates are super-orthogonal to each other.

We affix the following operator to the lengthy formulas for gamma and related parameters to denote the above orthogonally, or super-orthogonally temporal travel scenarios.

[q(Faster than light, instantaneous, or super-instantaneous travel with respect to ordinary 4-D space-time by travel along one or more orthogonal or super-orthogonal temporal coordinates)].

Here, Sin θ > 1 where θ is the angle between the one or more super-orthogonal temporal coordinates and the known temporal coordinate.

Now, we can consider scenarios of light-speed craft having infinite Lorentz factors.

Accordingly, a spacecraft may attain an alignment of its velocity vector with its acceleration and thrust vectors becoming super-parallel.

This may enable the spacecraft to travel translationally more efficiently and to travel ever so slightly faster than light.

Accordingly, the spacecraft may slip out of forwardly located space and into other dimensions or realms.

We may affix the following operator to formulas for gamma and related parameters in this part to denote the super-parallel alignments mentioned above.

[r(Super-parallel V, A, T for which V•A, V•T, A•T > ||V|| ||A|| cos 0, ||V|| ||T|| cos 0, ||A|| ||T|| cos 0, respectively to yield slightly super-luminal velocities)]

Note that the acceleration, velocity, and temporal travel can be super-parallel to the direction of otherwise spacecraft momentum or the precursor spacecraft momentum immediately before activation of the super-parallel mechanisms.

We can also consider super-relativistically aberrated incident light such as in cases where a spacecraft Lorentz factor goes to various infinite values or perhaps above a certain finite or infinite value threshold.

Such craft when operating as a light-sail of negative electromagnetic refraction types would experience a geometry of light incidence on a flat or other form of sail for which (sin ϴ) > 1 to (sin ϴ) = ∞↑. Here, the upwardly pointing arrow indicates unbounded infinite growth. The angle, ϴ, is the angle of incidence of the light on the light-sail spacecraft with respect to the plane orthogonal to the spacecraft velocity vector. The light would actually be impinging super-antiparallel to the spacecraft velocity.

The above light may be CMBR, star and/or quasar light or other light sources. Analogs for incident gravitational radiation are also plausible.

Other related mechanisms may include rocket exhaust that is directed in super-anti-parallel angles with respect to a relativistic rocket vehicle.

Such craft would experience an exhaust geometry of photons, gravitational waves, or massons with respect to the plane orthogonal to the spacecraft heading for which (sin ϴ) > 1 to (sin ϴ) = ∞↑. Here, the upwardly pointing arrow indicates unbounded infinite growth. The angle, ϴ, is the angle of direction of the exhaust with respect to the plane orthogonal to the spacecraft heading. The exhaust would actually be traveling super-antiparallel to the spacecraft velocity vector.

The light in these cases may as non-limiting options be intaked from hyperspace(s) of higher and/or alternate dimensionalities. The exhaust for the rocket concepts may as non-limiting options be directed into hyperspace(s) and/or alternate dimensionalities.

Note that as stated before, the acceleration, velocity, and temporal travel can be super-parallel to the direction of otherwise spacecraft momentum or the precursor spacecraft momentum immediately before activation of the super-parallel mechanisms.

We may affix the following operator to formulas for gamma and related parameters to denote the super-trigonometric vectors for superluminal and super-relativistic velocities and effective spacecraft propulsion energy levels including but not limited to those of relativistic rocket fuels.

[r(Super-parallel V, A, T, Sailing Radiation Direction (SSD), Exhaust Direction (ED), for which V•A, V•T, A•T, V•SSD, V•ED, A•SSD, A•ED, T•SSD, T•ED, > llVll llAll cos 0, llVll llTll cos 0, llAll llTll cos 0, llVll llSSDll cos 0, llVll llEDll cos 0, llAll llSSDll cos 0, llEDll llAll cos 0, llSSDll llTll cos 0, llEDll llTll cos 0, respectively to yield slightly to infinite super-luminal velocities and for which V•A, V•T, A•T, V•SSD, V•ED, A•SSD, A•ED, T•SSD, T•ED, can range from just greater than one to arbitrary infinite values)]

ESSAY 77) Infra-Spiritual High Energy Density Materials.

A fascinating energy source for spacecraft propulsion is class of materials which I refer to as infra-spiritual high energy density materials.

Accordingly, these materials may have yields of E > > M(c EXP 2).

Alternatively, these materials may be mass-less to a complete extent thus having no inertia.

Accordingly, infra-spiritual rocket fuels can be loaded into a relativistic rocket that would carry massive fuels but where the infra-spiritual fuel would be used to heat the massive reaction mass to provide specific impulses greater than or equal to 1 c all the way to sub-canonical ensemble multiples of light-speed. A finite limit on specific impulse makes sense for carried aboard reaction mass because upon heating the reaction mass to greatly will result in a black hole state being formed by the mass imbued with sufficient kinetic energy.

Another fascinating class of spacecraft energized by infra-spiritual fuels would include ones for which the infra-spiritual fuel would engage non-rocket propulsion systems.

Accordingly, the following propulsion systems could be energized by the infra-spiritual fuel.

1) Electro-hydrodynamic-plasma-drives.
2) Magneto-hydrodynamic-plasma-drives.
3) Electro-Magneto-hydrodynamic-plasma-drives.

4) Electromagnetic-hydrodynamic-plasma-drives.
5) Magnetic plasma sails.
6) Magnetic induction plasma drives
7) Magnetic induction field effect propulsion systems
8) Magnetic field effect propulsion drives
9) Combinations of two or more of the above and/or other suitable systems

There may be more than one species of infra-spiritual materials. In possibility, there could be infinite numbers of species of infra-spiritual materials some of which may be convertible to physical energy.

Additionally, some infra-spiritual materials may be convertible to pure spiritual energy in full or in part.

Any of the above spiritual materials may plausibly be non-inertial while having relativistic spacecraft reference frame Lorentz invariant yields.

Now there are the usual modes of high gamma factor transport such as any one of the following mechanisms which also might be employed in N-D space where N is any integer greater than or equal to 3: Relativistic craft can include any and all of the following: 1) Fusion Rockets, 2) Fission Rockets, 3) Fission Fragment Drives, 4) Fusion powered ion, electron, photon, and/or neutrino rockets, 5) Fission powered ion, electron, photon, and/or neutrino rockets, 6) matter antimatter rockets that carry both components of fuel on board from the start of the mission, 7) matter antimatter rockets that carry only their antimatter fuel component(s) along from the start of the mission, 8] matter antimatter reactor powered ion, electron, photon, and/or neutrino rockets, 9) fusion fuel pellet linear runway powered craft, 10) fission fuel pellet linear runway powered craft, 11) fission-fusion fuel pellet linear runway powered craft, 12) nuclear isomer fuel pellet linear runway powered craft, 13) matter antimatter fuel pellet linear runway powered craft, 14) antimatter fuel pellet linear runway powered craft, 15) fusion fuel pellet circulinear runway powered craft, 16) fission fuel pellet circulinear runway powered craft, 17) fission-fusion fuel pellet circulinear runway powered craft, 18] nuclear isomer fuel pellet circulinear runway powered craft, 19) matter antimatter fuel pellet circulinear runway powered craft, 20) antimatter fuel pellet circulinear runway powered craft, 21) nuclear fission powered electro-hydrodynamic-plasma drive craft, 22) nuclear fusion powered electro-hydrodynamic-plasma drive craft, 23) matter antimatter reaction powered electro-hydrodynamic-plasma drive craft, 24) nuclear fission powered magneto-hydrodynamic-plasma drive craft, 25) nuclear fusion powered magneto-hydrodynamic-plasma drive craft, 26) matter antimatter reaction powered electro-hydrodynamic-plasma drive craft, 27) nuclear fission powered electro-magneto-hydrodynamic-plasma drive craft, 28] nuclear fusion powered electro-magneto-hydrodynamic-plasma drive craft, 29) matter-antimatter powered electro-magneto-hydrodynamic-plasma drive craft, 30) fusion powered magnetic field effect drive, 31) fission powered magnetic field effect drive, 32) matter antimatter reaction powered field effect drive, 33) single pass solar dive and fry sail driven craft, 34) single pass stellar dive and fry sail driven craft, 35) single pass quasar dive and fry sail driven craft, 36) multi-pass stellar cycler solar dive and fry sail driven craft, 37) multi-pass stellar cycler dive and fry sail driven craft, 38] multi-pass cycler quasar dive and fry sail driven craft, 38] laser beam driven relativistic sail craft, 40) microwave beam driven relativistic sail craft, 41) radio-frequency beam driven relativistic sail craft, 42) massive neutral particle beam driven sail craft, 43) massive charged particle beam driven sail craft, 44) massive particle beam fission fuel powered craft, 45) massive particle beam fusion fuel powered craft, 46) massive particle beam matter antimatter beam fuel powered craft, 47) antimatter beam fuel powered craft, 48] nuclear bomb pulse driven propulsion of the original Project Orion forms, 49) pure nuclear fusion bomb pulse driven propulsion analogous to the original Project Orion forms, 50) matter antimatter bomb driven propulsion analogous to the original Project Orion forms, 51) one side reflective cosmic microwave background radiation sails, 52) multiple beam bounce propulsion methods, 53) any improved interstellar ramjet craft, 54) fusion rocket or fusion powered electron, ion, photon, or neutrino rockets utilizing single body or serially multiple body powered gravitational assists, 55) fission rocket or fission powered electron, ion, photon, or neutrino rockets utilizing single body or serially multiple body powered gravitational assists, 56) matter antimatter rocket or matter antimatter reaction electron, ion, photon, or neutrino rockets utilizing single body or serially multiple body powered gravitational assists and others. If the propulsion system is multi-modal, then even if only one stage is used for each mode, the number of combinations and thus the number of possible propulsion systems is at least equal to (2 EXP 56) – 1 = 7.205 x 10 EXP 16 = 72.05 Quadrillion = 72,050 trillion!

In each of these 56 categories, several sub-methods have been proposed and so the number of possible multi-mode/multi-stage propulsion systems is many, many orders of magnitude greater yet. If say, each category permits 10 sub-categories, which I can reasonably assure you is likely a conservative estimate, then the total number of possible propulsion systems, all else being the same is equal to about (2 EXP 560) - 1 which is approximately equal to 10 EXP 160. This is roughly 10 EXP 70 times the number of atoms, electrons, photons, and neutrinos in the observable universe. Most of these methods permit very high gamma factors, in many cases, virtually unlimited relativistic gamma factors given the virtually if not actually unlimited future time periods available to sequester ever greater resources of fuel via an ever expanding human space based resource collection infrastructure.

We may affix the following operator to formulas for gamma and related parameters to denote the mechanisms of the conjectured infra-spiritual high energy density materials.

[Σ(k = 1; k = ∞↑):[[Espiritual,k > > mM(c EXP 2) where m ranges from 1 to arbitrary infinite real numbers]:[kth Infra-spiritual fuel having zero mass and zero inertia and storable in infinite quantities in finite volumes]: [kth Infra-spiritual fuel equivalent to kth quantity of infra-spiritual energy]:[Σ(h = 1; h = canonical ensemble):hth propulsion system powerable by kth infra-spiritual fuel]]]

ESSAY 78) Sufficiently Infinite Lorentz Factors And Kinematic And Topological Processes Distinct From The Meaning Of Travel.

Herein, we consider travel at the speed of light in sufficiently infinite Lorentz factors such that a spacecraft maxes out on travel options and begins to under another kinematic and topological process distinct from travel and the conventional meaning of travel.

We can consider travel as if a ray that extends from the origin of a Cartesian plane and along the positive x-axis. The y-axis would represent the kinematic and topological process distinct from travel and the conventional meaning of travel. Rays extending into the positive quadrant of the x-y plane would be some forms of hybrid between conventional travel and the kinematic and topological process distinct from travel and the conventional meaning of travel.

We can generalize the above process with a second kinematic and topological process distinct from travel and the conventional meaning of travel and distinct from the first kinematic and topological process distinct from travel and the conventional meaning of travel indicated by the positive y-axis.

Thus second kinematic and topological process distinct from travel and the conventional meaning of travel would be represented by the z-axis along a Cartesian 3-D coordinate system. Rays extending from the origin of the coordinate system and into the positive octant of the 3-D coordinate system would be hybrids of the mechanisms defined by the positive x and y axis and/or hybrids of either the x and y axis or both the x and y axis.

Naturally, we can go on to consider a flat coordinate system having at least Aleph 0 dimensions but optionally Aleph K dimensions where K can be any finite or infinite number. The usual axial representation of a pure non-travel state and the hybrid options are available to any infinite dimensional coordinate system.

We affix the following operator to the lengthy formulas for gamma and related parameters in this portion of this section to denote the range of pure and hybrid kinematic and topological process distinct from travel and the conventional meaning of travel. We can also consider arbitrarily curved and/or torsioned coordinate systems and even super-curved and super-torsioned coordinate systems where the sign of the curvature can be arbitrary as can be the degree of curvature. The curved and/or torsion coordinate systems allow for biased numbers of said hybrids kinematic and topological processes. Additionally, we can take ship-frame time derivatives and integrals of the rate of progression or change of kinematic and topological processes and/or hybrids thereof.

The options summarized in the preceding paragraph are thus denoted by the following compound operator.

[w(Arbitrary infinite numbers non-travel states and the hybrid options thereof available for representation in any finite or infinite dimensional coordinate system where the coordinate system can be flat or arbitrarily curved and/or torsioned)]: [w(Arbitrary orders of ship-time differentiation or integration of the acquisition rate of the arbitrary infinite numbers non-travel states and the hybrid options thereof available for representation in any finite or infinite dimensional coordinate system where the coordinate system can be flat or arbitrarily curved and/or torsioned)]

We also consider mechanisms according to the following detailed list of clauses in the next several dozen pages.

High energy density materials of periodic table composition for which the atoms comprising the materials have one suppressed spatial dimension of location thus resulting in a macroscopic portion of the material being a substantially flat or curved planar section or two dimension manifold.

High energy density materials of periodic table composition for which the atomic nuclei and electrons comprising the materials have one suppressed spatial dimension of location thus resulting in a macroscopic portion of the material being a substantially flat or curved planar section or two dimension manifold.

High energy density materials of neutronium composition for which the neutrons comprising the materials have one suppressed spatial dimension of location thus resulting in a macroscopic portion of the material being a substantially flat or curved planar section or two dimension manifold.

High energy density materials of proto-neutronium composition for which the protons and neutrons comprising the materials have one suppressed spatial dimension of location thus resulting in a macroscopic portion of the material being a substantially flat or curved planar section or two dimension manifold and where almost all or all coulombic expansive tension is screened.

High energy density materials of quarkonium composition for which the quarks comprising the materials have one suppressed spatial dimension of location thus resulting in a macroscopic portion of the material being a substantially flat or curved planar section or two dimension manifold and where any of the 6 flavors of quarks and/or antiquarks compose the materials.

High energy density materials of quarkonium composition for which the quarks comprising the materials have one suppressed spatial dimension of location thus resulting in a macroscopic portion of the material being a substantially flat or curved planar section or two dimension manifold and where any of the 6 flavors of quarks and/or antiquarks compose the materials and where most of the quarks have the same sign electric charge and where most or all of the expansive coulombic tension is screened.

High energy density materials of higgsinium composition for which the higgsinos comprising the materials have one suppressed spatial dimension of location thus resulting in a macroscopic portion of the material being a substantially flat or curved planar section or two dimension manifold.

High energy density materials of higgsinium composition for which the higgsinos comprising the materials have one suppressed spatial dimension of location thus resulting in a macroscopic portion of the material being a substantially flat or curved planar section or two dimension manifold and where all or most of the higgsinos have the same sign electric charge and where most or all of the expansive coulombic tension is screened.

High energy density materials of quarkonium and higgsinium composition for which the quarks higgsinos comprising the materials have one suppressed spatial dimension of location thus resulting in a macroscopic portion of the material being a substantially flat or curved planar section or two dimension manifold and where any of the 6 flavors of quarks and/or of the antiquarks can be included in the materials.

High energy density materials of quarkonium and higgsinium composition for which the quarks higgsinos comprising the materials have one suppressed spatial dimension of location thus resulting in a macroscopic portion of the material being a substantially flat or curved planar section or two dimension manifold and where any of the 6 flavors of quarks and/or of the antiquarks can be included in the materials and where all or most of the quarks and higgsinos have the same sign electric charge and where most or all of the expansive coulombic tension is screened.

High energy density materials of monopolium, quarkonium, and higgsinium composition for which the monopoles, quarks, and higgsinos comprising the materials have one suppressed spatial dimension of location thus resulting in a macroscopic portion of the material being a substantially flat or curved planar section or two dimension manifold and where any of the 6 flavors of quarks and/or of the antiquarks can be included in the materials.

High energy density materials of monopolium, quarkonium, and higgsinium composition for which the monopoles, quarks, and higgsinos comprising the materials have one suppressed spatial dimension of location thus resulting in a macroscopic portion of the material being a substantially flat or curved planar section or two dimension manifold and where any of the 6 flavors of quarks and/or of the antiquarks can be included in the materials and where all or most of the quarks and higgsinos have the same sign electric charge and most of the monopoles have the same sign magnetic charge and where most or all of the expansive coulombic and magnetic tension is screened.

High energy density materials of monopolium, and quarkonium composition for which the monopoles and quarks comprising the materials have one suppressed spatial dimension of location thus resulting in a macroscopic portion of the material being a substantially flat or curved planar section or two dimension manifold and where any of the 6 flavors of quarks and/or of the antiquarks can be included in the materials.

High energy density materials of monopolium and quarkonium composition for which the monopoles and quarks comprising the materials have one suppressed spatial dimension of location thus resulting in a macroscopic portion of the material being a substantially flat or curved planar section or two dimension manifold and where any of the 6 flavors of quarks and/or of the antiquarks can be included in the materials and where all or most of the quarks have the same sign electric charge and most of the monopoles have the same sign magnetic charge and where most or all of the expansive coulombic and magnetic tension is screened.

High energy density materials of monopolium and higgsinium composition for which the monopoles and higgsinos comprising the materials have one suppressed spatial dimension of location thus resulting in a macroscopic portion of the material being a substantially flat or curved planar section or two dimension manifold.

High energy density materials of monopolium and higgsinium composition for which the monopoles and higgsinos comprising the materials have one suppressed spatial dimension of location thus resulting in a macroscopic portion of the material being a substantially flat or curved planar section or two dimension manifold and where all or most of the higgsinos have the same sign electric charge and most of the monopoles have the same sign magnetic charge and where most or all of the expansive coulombic and magnetic tension is screened.

High energy density materials of monopolium for which the monopoles comprising the materials have one suppressed spatial dimension of location thus resulting in a macroscopic portion of the material being a substantially flat or curved planar section or two dimension manifold.

High energy density materials of monopolium composition for which the monopoles comprising the materials have one suppressed spatial dimension of location thus resulting in a macroscopic portion of the material being a substantially flat or curved planar section or two dimension manifold and where most of the monopoles have the same sign magnetic charge and where most or all of the magnetic tension is screened.

High energy density materials of periodic table composition for which the atoms comprising the materials have two suppressed spatial dimension of location thus resulting in a macroscopic portion of the material being a substantially straight or curved line.

High energy density materials of periodic table composition for which the atomic nuclei and electrons comprising the materials have two suppressed spatial dimensions of location thus resulting in a macroscopic portion of the material being a substantially straight or curved line.

High energy density materials of neutronium composition for which the neutrons comprising the materials have two suppressed spatial dimensions of location thus resulting in a macroscopic portion of the material being a substantially straight or curved line.

High energy density materials of proto-neutronium composition for which the protons and neutrons comprising the materials have two suppressed spatial dimensions of location thus resulting in a macroscopic portion of the material being a substantially straight or curved line and where almost all or all expansive coulombic expansive tension is screened.

High energy density materials of quarkonium composition for which the quarks comprising the materials have two suppressed spatial dimensions of location thus resulting in a macroscopic portion of the material being a substantially straight or curved line and where any of the 6 flavors of quarks and/or antiquarks compose the materials.

High energy density materials of quarkonium composition for which the quarks comprising the materials have two suppressed spatial dimensions of location thus resulting in a macroscopic portion of the material being a substantially straight or curved line and where any of the 6 flavors of quarks and/or antiquarks compose the materials and where most of the quarks have the same sign electric charge and all or most of the expansive coulombic expansive tension is screened.

High energy density materials of higgsinium composition for which the higgsinos comprising the materials have two suppressed spatial dimensions of location thus resulting in a macroscopic portion of the material being a substantially straight or curved line.

High energy density materials of higgsinium composition for which the higgsinos comprising the materials have two suppressed spatial dimensions of location thus resulting in a macroscopic portion of the material being a substantially straight or curved line and where all or most of the higgsinos have the same sign electric charge and where most or all of the expansive coulombic tension is screened.

High energy density materials of quarkonium and higgsinium composition for which the quarks higgsinos comprising the materials have two suppressed spatial dimensions of location thus resulting in a macroscopic portion of the material being a substantially straight or curved line and where any of the 6 flavors of quarks and/or of the antiquarks can be included in the materials.

High energy density materials of quarkonium and higgsinium composition for which the quarks higgsinos comprising the materials have two suppressed spatial dimensions of location thus resulting in a macroscopic portion of the material being a substantially straight or curved line and where any of the 6 flavors of quarks and/or of the antiquarks can be included in the materials and where all or most of the quarks and higgsinos have the same sign electric charge and where almost all or all of the expansive coulombic tension is screened.

High energy density materials of monopolium, quarkonium, and higgsinium composition for which the monopoles, quarks, and higgsinos comprising the materials have two suppressed spatial dimensions of location thus resulting in a macroscopic portion of the material being a substantially straight or curved line and where any of the 6 flavors of quarks and/or of the antiquarks can be included in the materials.

High energy density materials of monopolium, quarkonium, and higgsinium composition for which the monopoles, quarks, and higgsinos comprising the materials have two suppressed spatial dimensions of location thus resulting in a macroscopic portion of the material being a substantially straight or curved line and where any of the 6 flavors of quarks and/or of the antiquarks can be included in the materials and where all or most of the quarks and higgsinos have the same sign electric charge and most of the monopoles have the same sign magnetic charge and where most or all of the expansive coulombic and magnetic tension is screened.

High energy density materials of monopolium, and quarkonium composition for which the monopoles and quarks comprising the materials have two suppressed spatial dimensions of location thus resulting in a macroscopic portion of the material being a substantially straight or curved line and where any of the 6 flavors of quarks and/or of the antiquarks can be included in the materials.

High energy density materials of monopolium and quarkonium composition for which the monopoles and quarks comprising the materials have two suppressed spatial dimensions of location thus resulting in a macroscopic portion of the material being a substantially straight or curved line and where any of the 6 flavors of quarks and/or of the antiquarks can be included in the materials and where all or most of the quarks have the same sign electric charge and most of the monopoles have the same sign magnetic charge and where most of the or all of the expansive coulombic and magnetic tension is screened.

High energy density materials of monopolium and higgsinium composition for which the monopoles and higgsinos comprising the materials have two suppressed spatial dimensions of location thus resulting in a macroscopic portion of the material being a substantially straight or curved line.

High energy density materials of monopolium and higgsinium composition for which the monopoles and higgsinos comprising the materials have two suppressed spatial dimensions of location thus resulting in a macroscopic portion of the material being a substantially straight or curved line and where all or most of the higgsinos have the same sign electric charge and most of the monopoles have the same sign magnetic charge and where most or all of the expansive coulombic and magnetic tension is screened.

High energy density materials of monopolium for which the monopoles comprising the materials have two suppressed spatial dimensions of location thus resulting in a macroscopic portion of the material being a substantially straight or curved line.

High energy density materials of monopolium composition for which the monopoles comprising the materials have two suppressed spatial dimensions of location thus resulting in a macroscopic portion of the material being a substantially straight or curved line and where most of the monopoles have the same sign magnetic charge and most or all of the expansive magnetic tension is screened.

The mechanisms of dimensional suppression may include a continual back and forth quantum-mechanical entanglement based teleportation of each particle and its electric and/or magnetic field, and the same for the particles' strong and weak force fields as applicable, as well as with the particles' gravitational fields.

Another mechanism of dimensional suppression may include a continual back and forth quantum-mechanical tunneling of each particle and its electric and/or magnetic field, and the same for the particles' strong and weak force fields as applicable, as well as with the particles' gravitational fields.

High energy density materials of periodic table composition for which the atoms comprising the materials have one suppressed spatial dimension of location thus resulting in a macroscopic portion of the material being a substantially flat or curved planar section or two dimension manifold and/or as a non-limiting option, screening of most or all of the gravitational implosive tension amongst the particles and thus of the associated bulk materials.

High energy density materials of periodic table composition for which the atomic nuclei and electrons comprising the materials have one suppressed spatial dimension of location thus resulting in a macroscopic portion of the material being a substantially flat or curved planar section or two dimension manifold and/or as a non-limiting option, screening of most or all of the gravitational implosive tension amongst the particles and thus of the associated bulk materials.

High energy density materials of neutronium composition for which the neutrons comprising the materials have one suppressed spatial dimension of location thus resulting in a macroscopic portion of the material being a substantially flat or curved planar section or two dimension manifold and/or as a non-limiting option, screening of most or all of the gravitational implosive tension amongst the particles and thus of the associated bulk materials.

High energy density materials of proto-neutronium composition for which the protons and neutrons comprising the materials have one suppressed spatial dimension of location thus resulting in a macroscopic portion of the material being a substantially flat or curved planar section or two dimension manifold and where almost all or all coulombic expansive tension is screened and/or as a non-limiting option, screening of most or all of the gravitational implosive tension amongst the particles and thus of the associated bulk materials.

High energy density materials of quarkonium composition for which the quarks comprising the materials have one suppressed spatial dimension of location thus resulting in a macroscopic portion of the material being a substantially flat or curved planar section or two dimension manifold and where any of the 6 flavors of quarks and/or antiquarks compose the materials and/or as a non-limiting option, screening of most or all of the gravitational implosive tension amongst the particles and thus of the associated bulk materials.

High energy density materials of quarkonium composition for which the quarks comprising the materials have one suppressed spatial dimension of location thus resulting in a macroscopic portion of the material being a substantially flat or curved planar section or two dimension manifold and where any of the 6 flavors of quarks and/or antiquarks compose the materials and where most of the quarks have the same sign electric charge and where most or all of the expansive coulombic tension is screened and/or as a non-limiting option, screening of most or all of the gravitational implosive tension amongst the particles and thus of the associated bulk materials.

High energy density materials of higgsinium composition for which the higgsinos comprising the materials have one suppressed spatial dimension of location thus resulting in a macroscopic portion of the material being a substantially flat or curved planar section or two dimension manifold and/or as a non-limiting option, screening of most or all of the gravitational implosive tension amongst the particles and thus of the associated bulk materials.

High energy density materials of higgsinium composition for which the higgsinos comprising the materials have one suppressed spatial dimension of location thus resulting in a macroscopic portion of the material being a substantially flat or curved planar section or two dimension manifold and where all or most of the higgsinos have the same sign electric charge and where most or all of the expansive coulombic tension is screened and/or as a non-limiting option, screening of most or all of the gravitational implosive tension amongst the particles and thus of the associated bulk materials.

High energy density materials of quarkonium and higgsinium composition for which the quarks higgsinos comprising the materials have one suppressed spatial dimension of location thus resulting in a macroscopic portion of the material being a substantially flat or curved planar section or two dimension manifold and where any of the 6 flavors of quarks and/or of the antiquarks can be included in the materials and/or as a non-limiting option, screening of most or all of the gravitational implosive tension amongst the particles and thus of the associated bulk materials.

High energy density materials of quarkonium and higgsinium composition for which the quarks higgsinos comprising the materials have one suppressed spatial dimension of location thus resulting in a macroscopic portion of the material being a substantially flat or curved planar section or two dimension manifold and where any of the 6 flavors of quarks and/or of the antiquarks can be included in the materials and where all or most of the quarks and higgsinos have the same sign electric charge and where most or all of the expansive coulombic tension is screened and/or as a non-limiting option, screening of most or all of the gravitational implosive tension amongst the particles and thus of the associated bulk materials.

High energy density materials of monopolium, quarkonium, and higgsinium composition for which the monopoles, quarks, and higgsinos comprising the materials have one suppressed spatial dimension of location thus resulting in a macroscopic portion of the material being a substantially flat or curved planar section or two dimension manifold and where any of the 6 flavors of quarks and/or of the antiquarks can be included in the materials and/or as a non-limiting option, screening of most or all of the gravitational implosive tension amongst the particles and thus of the associated bulk materials.

High energy density materials of monopolium, quarkonium, and higgsinium composition for which the monopoles, quarks, and higgsinos comprising the materials have one suppressed spatial dimension of location thus resulting in a macroscopic portion of the material being a substantially flat or curved planar section or two dimension manifold and where any of the 6 flavors of quarks and/or of the antiquarks can be included in the materials and where all or most of the quarks and higgsinos have the same sign electric charge and most of the monopoles have the same sign magnetic charge and where most or all of the expansive coulombic and magnetic tension is screened and/or as a non-limiting option, screening of most or all of the gravitational implosive tension amongst the particles and thus of the associated bulk materials.

High energy density materials of monopolium, and quarkonium composition for which the monopoles and quarks comprising the materials have one suppressed spatial dimension of location thus resulting in a macroscopic portion of the material being a substantially flat or curved planar section or two dimension manifold and where any of the 6 flavors of quarks and/or of the antiquarks can be included in the materials and/or as a non-limiting option, screening of most or all of the gravitational implosive tension amongst the particles and thus of the associated bulk materials.

High energy density materials of monopolium and quarkonium composition for which the monopoles and quarks comprising the materials have one suppressed spatial dimension of location thus resulting in a macroscopic portion of the material being a substantially flat or curved planar section or two dimension manifold and where any of the 6 flavors of quarks and/or of the antiquarks can be included in the materials and where all or most of the quarks have the same sign electric charge and most of the monopoles have the same sign magnetic charge and where most or all of the expansive coulombic and magnetic tension is screened and/or as a non-limiting option, screening of most or all of the gravitational implosive tension amongst the particles and thus of the associated bulk materials.

High energy density materials of monopolium and higgsinium composition for which the monopoles and higgsinos comprising the materials have one suppressed spatial dimension of location thus resulting in a macroscopic portion of the material being a substantially flat or curved planar section or two dimension manifold and/or as a non-limiting option, screening of most or all of the gravitational implosive tension amongst the particles and thus of the associated bulk materials.

High energy density materials of monopolium and higgsinium composition for which the monopoles and higgsinos comprising the materials have one suppressed spatial dimension of location thus resulting in a macroscopic portion of the material being a substantially flat or curved planar section or two dimension manifold and where all or most of the higgsinos have the same sign electric charge and most of the monopoles have the same sign magnetic charge and where most or all of the expansive coulombic and magnetic tension is screened and/or as a non-limiting option, screening of most or all of the gravitational implosive tension amongst the particles and thus of the associated bulk materials.

High energy density materials of monopolium for which the monopoles comprising the materials have one suppressed spatial dimension of location thus resulting in a macroscopic portion of the material being a substantially flat or curved planar section or two dimension manifold and/or as a non-limiting option, screening of most or all of the gravitational implosive tension amongst the particles and thus of the associated bulk materials.

High energy density materials of monopolium composition for which the monopoles comprising the materials have one suppressed spatial dimension of location thus resulting in a macroscopic portion of the material being a substantially flat or curved planar section or two dimension manifold and where most of the monopoles have the same sign magnetic charge and where most or all of the magnetic tension is screened and/or as a non-limiting option, screening of most or all of the gravitational implosive tension amongst the particles and thus of the associated bulk materials.

High energy density materials of periodic table composition for which the atoms comprising the materials have two suppressed spatial dimension of location thus resulting in a macroscopic portion of the material being a substantially straight or curved line and/or as a non-limiting option, screening of most or all of the gravitational implosive tension amongst the particles and thus of the associated bulk materials.

High energy density materials of periodic table composition for which thc atomic nuclei and electrons comprising the materials have two suppressed spatial dimensions of location thus resulting in a macroscopic portion of the material being a substantially straight or curved line and/or as a non-limiting option, screening of most or all of the gravitational implosive tension amongst the particles and thus of the associated bulk materials.

High energy density materials of neutronium composition for which the neutrons comprising the materials have two suppressed spatial dimensions of location thus resulting in a macroscopic portion of the material being a substantially straight or curved line and/or as a non-limiting option, screening of most or all of the gravitational implosive tension amongst the particles and thus of the associated bulk materials.

High energy density materials of proto-neutronium composition for which the protons and neutrons comprising the materials have two suppressed spatial dimensions of location thus resulting in a macroscopic portion of the material being a substantially straight or curved line and where almost all or all expansive coulombic expansive tension is screened and/or as a non-limiting option, screening of most or all of the gravitational implosive tension amongst the particles and thus of the associated bulk materials.

High energy density materials of quarkonium composition for which the quarks comprising the materials have two suppressed spatial dimensions of location thus resulting in a macroscopic portion of the material being a substantially straight or curved line and where any of the 6 flavors of quarks and/or antiquarks compose the materials and/or as a non-limiting option, screening of most or all of the gravitational implosive tension amongst the particles and thus of the associated bulk materials.

High energy density materials of quarkonium composition for which the quarks comprising the materials have two suppressed spatial dimensions of location thus resulting in a macroscopic portion of the material being a substantially straight or curved line and where any of the 6 flavors of quarks and/or antiquarks compose the materials and where most of the quarks have the same sign electric charge and all or most of the expansive coulombic expansive tension is screened and/or as a non-limiting option, screening of most or all of the gravitational implosive tension amongst the particles and thus of the associated bulk materials.

High energy density materials of higgsinium composition for which the higgsinos comprising the materials have two suppressed spatial dimensions of location thus resulting in a macroscopic portion of the material being a substantially straight or curved line and/or as a non-limiting option, screening of most or all of the gravitational implosive tension amongst the particles and thus of the associated bulk materials.

High energy density materials of higgsinium composition for which the higgsinos comprising the materials have two suppressed spatial dimensions of location thus resulting in a macroscopic portion of the material being a substantially straight or curved line and where all or most of the higgsinos have the same sign electric charge and where most or all of the expansive coulombic tension is screened and/or as a non-limiting option, screening of most or all of the gravitational implosive tension amongst the particles and thus of the associated bulk materials.

High energy density materials of quarkonium and higgsinium composition for which the quarks higgsinos comprising the materials have two suppressed spatial dimensions of location thus resulting in a macroscopic portion of the material being a substantially straight or curved line and where any of the 6 flavors of quarks and/or of the antiquarks can be included in the materials and/or as a non-limiting option, screening of most or all of the gravitational implosive tension amongst the particles and thus of the associated bulk materials.

High energy density materials of quarkonium and higgsinium composition for which the quarks higgsinos comprising the materials have two suppressed spatial dimensions of location thus resulting in a macroscopic portion of the material being a substantially straight or curved line and where any of the 6 flavors of quarks and/or of the antiquarks can be included in the materials and where all or most of the quarks and higgsinos have the same sign electric charge and where almost all or all of the expansive coulombic tension is screened and/or as a non-limiting option, screening of most or all of the gravitational implosive tension amongst the particles and thus of the associated bulk materials.

High energy density materials of monopolium, quarkonium, and higgsinium composition for which the monopoles, quarks, and higgsinos comprising the materials have two suppressed spatial dimensions of location thus resulting in a macroscopic portion of the material being a substantially straight or curved line and where any of the 6 flavors of quarks and/or of the antiquarks can be included in the materials and/or as a non-limiting option, screening of most or all of the gravitational implosive tension amongst the particles and thus of the associated bulk materials.

High energy density materials of monopolium, quarkonium, and higgsinium composition for which the monopoles, quarks, and higgsinos comprising the materials have two suppressed spatial dimensions of location thus resulting in a macroscopic portion of the material being a substantially straight or curved line and where any of the 6 flavors of quarks and/or of the antiquarks can be included in the materials and where all or most of the quarks and higgsinos have the same sign electric charge and most of the monopoles have the same sign magnetic charge and where most or all of the expansive coulombic and magnetic tension is screened and/or as a non-limiting option, screening of most or all of the gravitational implosive tension amongst the particles and thus of the associated bulk materials.

High energy density materials of monopolium, and quarkonium composition for which the monopoles and quarks comprising the materials have two suppressed spatial dimensions of location thus resulting in a macroscopic portion of the material being a substantially straight or curved line and where any of the 6 flavors of quarks and/or of the antiquarks can be included in the materials and/or as a non-limiting option, screening of most or all of the gravitational implosive tension amongst the particles and thus of the associated bulk materials.

High energy density materials of monopolium and quarkonium composition for which the monopoles and quarks comprising the materials have two suppressed spatial dimensions of location thus resulting in a macroscopic portion of the material being a substantially straight or curved line and where any of the 6 flavors of quarks and/or of the antiquarks can be included in the materials and where all or most of the quarks have the same sign electric charge and most of the monopoles have the same sign magnetic charge and where most of the or all of the expansive coulombic and magnetic tension is screened and/or as a non-limiting option, screening of most or all of the gravitational implosive tension amongst the particles and thus of the associated bulk materials.

High energy density materials of monopolium and higgsinium composition for which the monopoles and higgsinos comprising the materials have two suppressed spatial dimensions of location thus resulting in a macroscopic portion of the material being a substantially straight or curved line and/or as a non-limiting option, screening of most or all of the gravitational implosive tension amongst the particles and thus of the associated bulk materials.

High energy density materials of monopolium and higgsinium composition for which the monopoles and higgsinos comprising the materials have two suppressed spatial dimensions of location thus resulting in a macroscopic portion of the material being a substantially straight or curved line and where all or most of the higgsinos have the same sign electric charge and most of the monopoles have the same sign magnetic charge and where most or all of the expansive coulombic and magnetic tension is screened and/or as a non-limiting option, screening of most or all of the gravitational implosive tension amongst the particles and thus of the associated bulk materials.

High energy density materials of monopolium for which the monopoles comprising the materials have two suppressed spatial dimensions of location thus resulting in a macroscopic portion of the material being a substantially straight or curved line and/or as a non-limiting option, screening of most or all of the gravitational implosive tension amongst the particles and thus of the associated bulk materials.

High energy density materials of monopolium composition for which the monopoles comprising the materials have two suppressed spatial dimensions of location thus resulting in a macroscopic portion of the material being a substantially straight or curved line and where most of the monopoles have the same sign magnetic charge and most or all of the expansive magnetic tension is screened and/or as a non-limiting option, screening of most or all of the gravitational implosive tension amongst the particles and thus of the associated bulk materials.

We can also consider fractional and irrational numbers of suppressed dimensions real number values ranging from just greater than zero to three. Accordingly, commensurately great portions of fuels can be carried along on a spacecraft as non-limiting examples to enable truly extreme Lorentz factors attainable.

So we include the following formulas which take into account the dimension suppressing mechanisms and associated efficiency of spacecraft fuel storage and spacecraft dimensional contraction.

Herein, we consider a theory beyond the Standard Model for which there exist 6 magnetically positively charged monopolar quarks and 6 magnetically negatively charged monopolar quarks, and twelve corresponding antimatter versions.

Likewise consider three magnetic monopoles and three antimatter magnetic monopoles to mirror the three charged leptons and antimatter versions thereof.

We likewise consider three magneto-weak interacting analogs of neutrinos and three antimatter analogs so as to mirror the three flavors of neutrinos and their antimatter counterparts.

We also consider an analog of the higgs boson(s) as mechanisms for the magneto-weak symmetry breaking mechanism.

We further consider three analogs of the weak force bosons, two of which are magnetically monopolar charges and one of which is magnetically neutral.

Thus we consider 36 additional fermions and 4 additional bosons.

We consider super-symmetric partners to these magnetically monopolar charged particles to yield 36 additional bosonic particles and 4 additional fermions.

Regarding supersymmetry, we have 24 bosons corresponding to the 12 Standard Model species of fermions and their 12 antimatter versions, and 13 species of fermions corresponding to the total of 8 species of gluons, three species of weak force bosons, and the photon and the one known higgs boson. When considering antimatter versions, we have a total 50 particles of supersymmetric partnership of the Standard Model particles. Some of these particles may be their own anti-particles.

ESSAY 79) Light-Speed Sufficiently Infinite Lorentz Factor Craft Ascending Levels Of Physical Realms.

Here, we consider light-speed sufficiently infinite Lorentz factor space travel for which a never ending ascending series of levels of a universe, multiverse, forest, biosphere, hyperspace, hyper-space-time, bulk, hyper-bulk, scalar field and the like, of travel are available and progressively serially entered in on of multiple instantiations. Each such level is more elevated and evolved then then previous level in ways such that natural and/or divine mechanisms assist in the growth and development of each said ascending level.

The above conjecture makes sense as the craft will travel an entire future eternity for every Planck Time Unit that is realized in the spacecraft reference frame.

Another way of considering these concepts is that a light-speed infinite Lorentz factor craft cannot help but progress to one ascending level to another.

For example, a craft having attained a sufficiently huge finite or infinite Lorentz factor will still be around to traveling in our universe when the next symmetry breaking event or phase change occurs. How this will affect the craft and crew is uncertain but adverse effects are likely strongly mitigated or avoiding altogether if the craft is suitably thermodynamically cloaked from the background medium of space-time but still topologically present and related.

As another example, a craft that would attain a huge finite or even infinite Lorentz factor may still be in flight when if traveling in a cyclical big bang universe, the universe re-collapses and rebounds in another expansion cycle. Thus, the crew will be able to witness a brand new universe being born.

As yet another example, a universe of travel may attain a state where disorder has reached its maximum possible level. Such a universe may thus need divine assistance for any organized behavior to occur or develop. Accordingly, perhaps GOD will take the universe as it is and bring the universe to another higher existential level for which accidental aspects of the universe have restocked global order as a set thus enabling additional modes and portions of the universe to become more organized even as the overall level of disorder increases.

As still another example, a universe instantiation in its near eternal or eternal death rows may be divinely rescued and directly and deliberately elevated by GOD.

Regardless, a spacecraft traveling at an infinite Lorentz factor will by definition be able to witness eternity after eternity go by perhaps somehow even during a given Planck Time Unit before the next Planck Time Unit begins in the spacecraft reference frame observer time.

Such a spacecraft as conjectured above cannot help but witness one cosmic era after another one. If such a program of travel can be realized for a spacecraft and her crew, then, light-speed travel has additional profound implications, so much so, that those who would fret any light-speed limits imposed by Mother Nature can instead become and should become alive with joy at these prospects of infinite Lorentz factor travel.

We may affix the following operator to formulas for gamma and related parameters to denote the above conjectured mechanisms.

[w(Light-speed sufficiently infinite Lorentz factor space travel for which a never ended ascending series of levels of a universe, multiverse, forest, biosphere, hyperspace, hyper-space-time, bulk, hyper-bulk, scalar field, and the like, of travel are available and progressively serially entered where each such level can be more elevated and evolved then then previous level in ways such that natural and/or divine mechanisms assist in the growth and development of each said ascending level)].

ESSAY 80) A Whimsical Tour Of Infinities, Light-Speed, Ever So Lightly Faster Than Light Speed, And Effects On Ontology.

Now, we are all well aware of the concept of existence. We ascribe the notion of existence to ourselves and our loved ones and colleagues. Most especially, we Catholics refer to GOD as pure existence.

We can also intuit notions such as hyper-existence at one or more levels just as we can intuit notions of hyperspace and then like.

Hyper-existence, hyper-hyper-existence, hyper-hyper-hyper-existence all seem as belonging on the same ontological continuum as these constructs are more or less ever more extreme levels of the existence spectrum.

However, we can spin another set of notions such as else-than-existence.

For example, else-than-existence would be another parameter that is completely distinct from existence. Were we to consider just this one item, we could refer to it as else-than-existence-1.

However, there may be additional items that are just as distinct as existence and else-than-existence-1 as these latter two parameters are from each other. We refer to these additional items as else-than-existence-2, else-than-existence-3, else-than-existence-4 and so-on.

Altogether, we can intuit the series [Σ(k = 1; k = ∞↑):(else-than-existence-k)]

Presumably, GOD excels to ultimate and infinite extents in being each one of the items in the series [Σ(k = 1; k = ∞↑):(else-than-existence-k)]. For GOD, HIS being each one of these items does not diminish HIM being other items in the set nor does it diminish HIS existence. For GOD is pure existence itself. Note that the series limit of “k = ∞↑” implies open ended immensity of infinite values.

So, reality it appears is fantastically huge.

We do not have to look to eternity to obtain a sense of the immensity of reality.

For example, we can ponder our local light-cone which has radius of about 13.75 billion light-years. In this region, there exist about 10 EXP 24 stars. This is equivalent to the number of fine grains of table sugar that would cover the entire United States 100 meters deep. The number of planets orbiting stars in our cosmic light-cone is estimated to be ten times greater yet or about equal to the number of fine grains of table sugar that would cover the entire United States 1,000 meters deep. The number of moons orbiting planets orbiting stars is estimated to be about ten times greater yet or about equal to the number of fine grains of table sugar that would cover the entire United States 10,000 meters deep.

Our cosmic light-cone is generally believed by contemporary cosmologists to be just a tiny fraction of our universe, which may be infinite in size and be just one of an infinite number of universes that manifest as "island universes" that dot a larger hyperspace. What's more is that there may be infinitely many hyperspaces, some of which have infinite numbers of dimensions.

To travel about in our cosmic light-cone, all we need is nuclear fusion fuel. For moderately relativistic colony ships, say those that travel between 10 percent to 90 percent of the speed of light as purely relativistic rockets, the nuclear fuel can be carried aboard the craft from the start of a mission.

To travel to Lorentz factors ranging from about 3 to one trillion, interstellar ramjets will suffice as will nuclear fuel pellet streams. Once a craft would attain a Lorentz factor of about one trillion, the craft could travel a distance equal to about the radius of the currently observable portion of our universe which is simply our cosmic light-cone in about one week ship time. Materials of spacecraft construction such as neutronium, quarkonium, higgsinnium, mono-higgsinium, monopolium, wormhole wall material, and the like may in theory enable such rapid ship-frame travel since the collision and/or diversion of photons and chargons in the background of spacetime would interact with the craft in almost purely the laboratory frame. We may even come up with more exotic materials. Regardless, such ultra-relativistic craft may be constructed like long and thin and extremely sharp needles to reduce drag and effective cross-sectional area.

Perhaps there is no finite upper boundary limit to spacecraft Lorentz factor. It may even be the case that variously infinite Lorentz factor can be obtained, once again, perhaps without limits.

Part of the enjoyment of a drive to a great vacation get-away or to a new residential location surrounded by beautiful pine forests and mountains is the road trip itself. The many city and rural sights one may travel through on a trip across the continental United States are enough to delight almost any travel enthusiast.

For me, the notion of attaining ever greater spacecraft Lorentz factors in relativistic impulse travel represents a journey, perhaps infinite journey to realizing infinite Lorentz factors. I hope to influence public, private, government, and military organizations to take on the same mantra and to dream the dream of what commonly is viewed as impossible and work toward making the dreams reality.

As humanity ventures out into cis-lunar space and beyond, we can all take joy in knowing that these baby steps are on the first few stepping stones into eternity.

As a Catholic man and member of the Knights Of Columbus, I make no apologies for expressing my faith in open literature. I believe that this life itself is but a stepping stone into eternity. How travel will be lived out in the next life and just about anyone's guess. Such travel will be the subject of additional series of books I will write.

However, we do not need to wait for the next life to go boldly where no human has gone before.

As of the time of this writing, I am anticipating the launch of Artemis 1, a mission that will take a large unmanned crew module to lunar orbit and return to Earth. Artemis 2 planned for next year anticipates carrying astronauts to lunar orbit and bringing them back safely. Artemis 3 to launch sometime around 2025 will land astronauts on the Moon. From there, practice for manned missions to Mars and beyond will be accomplished.

In a sense, here and now and over the rest of this decade, we are privileged to be a part of making all of this happen. So fasten your seat-belts and look forward to the ride into eternity and enjoy the scenery along the way.

You are about to begin a 15 billion light-year journey that will only take a few seconds in your time. Your booth operates via field effect propulsion mechanisms which are able to sweep out a very large cross-section of background magnetic energy.

You enter your travel booth which is made of 0.1 femtometer thick quarkonium sheeting.

The booth when empty of fuel has invariant mass of about 10 metric tons. The booth can contain about ten billion metric tons of fuel compactly stored at a density of about 10 billion metric tons per cubic meter.

You unlock and open the entrance door to the travel both.

Then you close the door and a whimsical red light comes on which bathes you and the interior of the craft with red illumination. Red light is chosen for the fiat make and model you rented because red light will not mess with your vision should the booth have a propulsion or power failure in which case you will need to be able to see the emergency warning beacons switch and other features available to call out for assistance. The booth has a nanotechnology hibernation chamber that will enable you to live billions of years in relaxed sleep until assistance comes along to tow your booth to the nearest service station and hotel accommodations.

A touch screen comes on with icons and menu selections that resemble the I-pads and I-phones in eras long-long ago.

You touch the destination menu icon and choose a star that is in a Galaxy currently about 10 billion light-years distant.

Then you touch the go icon on the screen and you hear clicks, clanks, buzzes, and other noises that sound like the photo copiers used in the first quarter of the 21st Century Earth way back in the days.

Now, your ship carries 10 billion tons of pure antimatter which it will mix with background normal matter reactants to quickly boost your craft to a Lorentz factor of 10 EXP 17.

Thanks to anti-gravatic interior acceleration cancelling technology, you feel completely normal for the 10 to 15 seconds fiat flight to your destination in your fiat reference frame.

Your craft uses reactive mechanisms with the background to rapidly slow so that onboard fuel is not needed for the deceleration phase. During the acceleration phase, the craft recycles virtually all drag energy.

In all, twenty seconds will pass fiat time as the booth accelerates and decelerates to the destination.

When you step out, you feel at first a sense of loss in knowing that your loved ones left behind died about 15 billion years ago. However, you become exhilarated by the reality that you just traveled 15 billion light-years and 15 billion years into the future in your awesome space-booth.

The Sun and Earth you left behind are gone. The Sun having swelled to red-giant phase and cooling off to become a black dwarf happened about 10 billion years ago and Earth during the process was evaporated in the outer atmosphere of the huge red giant star the Sun had evolved into.

Now that I have presented a fascinating story plot, I tease your sense of wonderlust by suggesting that the latter scenario should be possible according to current known and projected theoretical physics.

Now, within the observable universe, there are about 10 EXP 24 stars. This is about equal to the number of fine grains of table sugar that would cover the entire United States 100 meters deep. The number of planets orbiting stars in our universe seems to be about 10 times greater yet or equal to the number of fine grains of table sugar that would cover the entire United States 1,000 meters deep. The number of moons, orbiting planets, orbiting stars is estimated to be ten times greater yet or equal to the number of fine grains of table sugar that would cover the entire United States 10,000 meters deep.

Most of these stars are easily accessible should the fictional space-booths be developed within the next millennium.

OH! WHAT DREAMS WILL COME!

A Future Star-farers Story Yet To Be Told

It is sometime in the early 23rd Century, Earth time. You have entered orbit around an exo-gas-giant-planet within the habitable zone of a star about 50 light-years from Earth. As you glance at the beautiful blue, green, and white ribbons in a cacophony of shades, and tones, making up the planet's atmosphere, you also catch a glimpse of a moon that resembles Earth in tone and color but not in land mass.

After traveling about 10 years ship time having reached a gamma factor of about 8, you muze that you and your crew members as mere human persons have arrived as the ETs visiting this planetary system which has all of the biochemical signatures of infusement with life. Your beam sail and very high mass ratio fusion rocket, dual mode starship, has made this all possible.

You also notice the star studded background and realize that humanity now collectively has in theory the capacity to send explorers and colonists beyond an Earth centered cosmic light cone. You muse at the technical hurdles needed to be overcome to bridge the gaps between galaxies, and even more so, between galaxy clusters. However, you realize that self-contained world ships with large crews in theory are up to the task even if manned star travel is limited to roughly the gamma factor achieved by the mothership from which you now gaze upon this lovely planetary system.

Being an explorer at heart and never content to stay still, you begin to day-dream with a sense of exuberance at the ensemble of similar planetary systems that have long been demonstrated to exist within the observable universe, perhaps upwards of 10 EXP 20 such systems in total just within your cosmic light-cone.

As you look out a carbonaceous super-material spacecraft window, you start to contemplate the number of similar possible scenes perceivable by a typical human. First, you consider a typical human vision of angular resolution roughly equal to 10,000 by 10,000 pixels. Then you recall a childhood present handed down through your familial lineage from your many times great uncle, back in the days when he worked along with numerous other space-heads on the basic thermodynamics and kinematics of the type of ship you now crew. You then recall reading in this gift of a copy of the Guinness Book Of World Records dating back to the 20th Century about the ability of humans to sense about 10 million different colors if placed side by side.

Now you make a mental estimate of the number of possible planetary sights that can in theory be viewed and sensed by a human person as (10 EXP 4)(10 EXP 4) EXP 10,000,000 or 10 EXP 80,000,000. Next you assume the capacity to detect roughly 10,000 tones for each color to derive a number of possible sights equal to (10 EXP 4)(10 EXP 4) EXP [(10,000,000)(10,000)] or 10 EXP 800,000,000,000. You now become satisfied from a purely artistic standpoint of the huge rough estimate of the number of possible vistas.

As the nature of human emotions, attitudes, feelings, and thoughts in terms of the still unconfirmed continuity or super-continuity of such subjective experiences is brought into the estimate, a sense of exhilaration begins to flow within your head and body along with a mild general relaxed euphoria of the perhaps infinite numbers of individual subjective experiences which are possible in consideration of the coupling of these psychodynamic variables and the mere perceptual field of visual field pixilations.

Now that my brief fictional account is complete, day dreams similar to the above story help keep me grounded in my life's calling to help pave the theoretical path to the stars. This is not simply a Jim show, but is motivated for the benefit of all those yet to be born, in a great hope of future communion in an ever growing cosmic civilization of love for which Humanity will live in fraternal love and in lasting peace amongst the perhaps untold numbers of any existent extra-terrestrial and ultra-terrestrial peoples currently unknown to humanity.

Such extreme gamma factors permit extreme real time travel into the future. This future is not merely some extrapolation of the present nor is it the kind of holographic future according to conspiracy theorists that can be traveled into to find out what folks are going to do so that dangerous actions can be stopped before they occur. Such an apparent future would not really be the true future but would instead be a mere latent or potential projection from the present. Extreme gamma factor space travel at below the velocity of light permits crew members to travel into the cosmically distant future as it will actually unfold in an unalterable manner. Now, according to Stephen Hawking, the future of our universe if closed may be indistinguishable from its birth. Perhaps traveling at suitably extreme sub-light speeds would enable future distant descendants or emissaries from our cosmic era to travel into the past of our universe as it actually unfolded or perhaps into the future of another Big Bang incarnation of our current universe. The caveat, is well, the ability to pass through the transitional energy state of our collapsing universe into our past and/or the future incarnation of our universe upon its eventual collapse.

A suitably cloaked space craft may in some poorly understood, but valid sense, travel within a margin of 4-D space-time or within a boundary seam of 4-D space-time. Such travel may thus occur near or at the boundary transition between 4-D space-time and not 4-D space-time. It is thus conceivable that such travel may be properly viewed as more exotic than travel in higher dimensional space, hyperspace, other universes, or parallel space-times of the Many Worlds Interpretation of Quantum Theory forms. It may even be reasonable to suggest that such travel marginally occurs in space-time in an absolute sense regardless of the space-time considered. Can you imagine traveling but not so in space nor in space-time of any form!

With the passing near the time of this writing of the 67th anniversary of the destruction of Hiroshima and Nagasaki, I have been contemplating the terrible power of the atomic nucleus, which if unchecked in its applications for warfare, can certainly lead to our destruction. However, this cosmic energy source can be applied for great and profound good, in its enablement of the unlimited extension of the human presence into the eternal blackness of space-time. Atoms for Peace must replace Atoms for War. Atoms for peace can be an ultimate charitable mantra by promoting untold future living space for humanity. Atoms for War can certainly lead to the death of Humanity. The choice of paths to follow is profoundly ours, here and now, in this generation and the next one or two to follow. Choose wisely. With Peace, all things are possible including eternal progress.

I am especially interested in nuclear fusion because this is the dominant real energy source within our universe. Since the mean lifetime of the proton has been verified to be greater than about 10 EXP 35 years and may in fact be eternally stable. Nuclear fusion fuel will be available for commensurately long time periods.

Nuclear fusion fuels can enable achievement of virtually unlimited special relativistic Lorentz transformation factors for cases where the fuel is sequestered by space craft in situ as it travels in interstellar or intergalactic space such as may be possible with the latest interstellar ramjet concepts. Alternatively, the nuclear fuels can be embodied in fusion fuel runways thus permitting extreme gamma factors with mass ratios of essentially one.

Antimatter may well be producible in large bulk quantities and safely stored as rocket fuel at some future time. However, most of the Standard Model baryonic matter within our universe is hydrogen and helium isotopes and thus may be more effectively harnessed for its direct latent energy as opposed to using nuclear fusion power sources to produce antimatter fuels.

As someone who is a nuclear energy freak, I see great potential for Humanity to harness nuclear energy in the decades, centuries, and millennia that lie ahead.

Nuclear fusion will always be available to turn the light on, whether for night-time illumination, or the light of nuclear energy for energizing our near term and distant future starships.

In a sense, nuclear energy is the most natural form of energy in our present cosmic era. For nuclear energy runs the stars, and heats the cores of Earth-like planets to thus assist with the provision of a thermodynamic gradient and planetary chemistry necessary for the development and evolution of any extraterrestrial life forms.

However, protonic and neutronic nuclear energy may not be the only useful form of nuclear energy. Science fiction is familiar with the concept of a quark bomb where protons would be induced to fission thus yielding energy virtually identical in quantity to the reaction of an equal mass consisting of half matter and half antimatter of the same species. Strangelets, charmedlets, and bottomlets when produced in any stable or metastable varieties in a responsible manner may be useful for catalyzing exotic quarkonium transformative reactions within rocket fuel or within captured interstellar Standard Model baryonic matter. Perhaps additional nuclear forces will be discovered that provide a higher mass specific yield in reactants than the best known nuclear fusion fuels.

This book is a glance at some plausible and possible outcomes that can evolve from the mantra, “Atoms For Peace”. This book is just one glimpse of a possible future full of excitement, pioneering exploration, and the expansion of human civilization, and stories yet to be told. This is the story of The Galactic Explorer!

Regarding how big infinities can be consider the following digression which is repeated again in this book.

Now, there are countable and uncountable infinities.

But before we go further, consider the hyper-operator notation that was designed to express huge values not otherwise expressible.

For example, Note that Hyper4(a, n) is equal to a tetrated n or a raised to the power of itself n-1 times. The latter value is symbolically written as n subscript a. For example 3 EXP 4 = 81, but 4 subscript3 is approximately equal to 10 EXP (1,000,000,000,000).

Alternatively 4 subscript 2 = 2 EXP 2 EXP 2 EXP 2 = 2 EXP [2 EXP [2 EXP 2]] = 2 EXP (2 EXP 4) = 2 EXP 16 = 65,536.

For example, Hyper5(4, 4)is equal to 4 tetrated 4 tetrated 4 tetrated 4. This value is commonly referred to as 4 pentated 4.

Hyper 6, (4,4) is 4 pentated 4 pentated 4 pentated 4 and is also referred to as 4 hexataed 4.

Hyper 7, (4,4) is 4 hexated 4 hexated 4 hexated 4 and so-on

Aleph 0 is the infinite number of integers.

Aleph 1 according to the perhaps unprovable and thus unfalsifiable Continuum Hypotheses is the number of real numbers which is greater than Aleph 0 by a multiplicative factor of infinity.

Aleph 2 is similarly greater than Aleph 1.

Aleph 3 is similarly greater than Aleph 2.

Aleph 4 is similarly greater than Aleph 3.

and so-on

In general, Aleph n = 2 EXP [Aleph (n-1)]

The number Ω is commonly stated as the least infinite positive integer or ordinal.

Now here is a real zinger.

So we can produce the abstraction of Hyper Aleph Ω (Aleph Ω, Aleph Ω).

Now, how about a spacecraft attaining a Lorentz factor of Hyper Aleph Ω (Aleph Ω, Aleph Ω)!

Such a spacecraft could travel about Hyper Aleph Ω (Aleph Ω, Aleph Ω) light-years thru space in one year ship time and Hyper Aleph Ω (Aleph Ω, Aleph Ω) years into the future in one year ship time.

We can contemplate far, far, greater infinities than Hyper Aleph Ω (Aleph Ω, Aleph Ω).

Now, here is another zinger!

How about a spacecraft attaining so great of a Lorentz factor such that the Lorentz factor is beyond any infinity.

For example, the Lorentz factor would not even be a real number.

In fact, the Lorentz factor of our craft would not be any kind of number nor any kind of quantity.

Accordingly, the number of distinct Lorentz factors as such itself would not be a real number nor any kind of quantity.

Imagine the possible itineraries for such a craft.

Here is another way of looking at the immensity of infinite Lorentz factors.

Now Aleph 0 is the infinite number of positive integers.

Aleph 1 is according to the perhaps unprovable and unfalsifiable Continuum Hypothesis equal to the number of real numbers.

Either way, Aleph 1 = 2 EXP (Aleph 0).

Aleph 2 = 2 EXP (Aleph 1).

Aleph 3 = 2 EXP (Aleph 2).

Aleph 4 = 2 EXP (Aleph 3).

and so-on.

In general, [Aleph (N +1)] = 2 EXP (Aleph N) where N is any whole number. Now, there are countable and uncountable infinities.

But before we go further, consider the hyper-operator notation that was designed to express huge values not otherwise expressible.

For example, note that Hyper4(a, n) is equal to a tetrated n or a raised to the power of itself n-1 times. The latter value is symbolically written as n subscript a. For example 3 EXP 4 = 81, but 4 subscript3 is approximately equal to 10 EXP (1,000,000,000,000).

Alternatively 4 subscript 2 = 2 EXP 2 EXP 2 EXP 2 = 2 EXP [2 EXP [2 EXP 2]] = 2 EXP (2 EXP 4) = 2 EXP 16 = 65,536.

For example, Hyper5(4, 4)is equal to 4 tetrated 4 tetrated 4 tetrated 4. This value is commonly referred to as 4 pentated 4.

Hyper 6, (4,4) is 4 pentated 4 pentated 4 pentated 4 and is also referred to as 4 hexataed 4.

Hyper 7, (4,4) is 4 hexated 4 hexated 4 hexated 4 and so-on.

So we can produce the abstraction of Hyper Aleph Ω (Aleph Ω, Aleph Ω).

We can abstractly construct infinities in other ways. A particularly fascinating way to to state something like, zero is to Aleph 0 as Aleph 0 is to 1st level analog infinity.

We can likewise make a statement that zero is to Aleph 0 as Aleph 0 is to 1st level analog infinity as 1st level analog infinity is to 2nd level analog infinity.

We can likewise make a statement that zero is to Aleph 0 as Aleph 0 is to 1st level analog infinity as 1st level analog infinity is to 2nd level analog infinity as 2nd level analog infinity is to 3rd level analog infinity.

We can progress to the kth level analog infinity where k is any finite or infinite positive integer.

We can go additional steps further to develop extended levels of analog infinities.

Accordingly, 0 is to 1st level analog infinity as 1st level analog infinity is to 1st extended-1 level analog infinity as 1st extended-1 level analog infinity is to 2nd extended-1 level analog infinity as 2nd extended-1 level analog infinity is to 3rd extended-1 level analog infinity and so-on all the way to kth extended-1 level of analog infinity.

We can go still further to develop the following concepts.

Accordingly, 0 is to 1st extended-1 level analog infinity as 1st extended-1 level analog infinity is to 1st extended-2 level analog infinity as 1st extended-2 level analog infinity is to 2nd extended-2 level analog infinity as 2nd extended-2 level analog infinity is to 3rd extended-2 level analog infinity as 3rd extended-2 level analog infinity is to 4th extended-2 level analog infinity and so-on all the way to kth extended-2 level of analog infinity.

We can go still further to develop the following concepts.

Accordingly, 0 is to 1st extended-2 level analog infinity as 1st extended-2 level analog infinity is to 1st extended-3 level analog infinity as 1st extended-3 level analog infinity is to 2nd extended-3 level analog infinity as 2nd extended-3 level analog infinity is to 3rd extended-3 level analog infinity as 3rd extended-3 level analog infinity is to 4th extended-3 level analog infinity and so-on all the way to kth extended-3 level of analog infinity.

We can go still further to develop the following concepts.

Accordingly, 0 is to 1st extended-3 level analog infinity as 1st extended-3 level analog infinity is to 1st extended-4 level analog infinity as 1st extended-4 level analog infinity is to 2nd extended-4 level analog infinity as 2nd extended-4 level analog infinity is to 3rd extended-4 level analog infinity as 3rd extended-4 level analog infinity is to 4th extended-4 level analog infinity and so-on all the way to kth extended-4 level of analog infinity.

In general, we can go all the way up to and beyond kth extended-j level of analog infinity where k and j can be arbitrary countable or uncountable infinities.

All of these new constructs can be said to be infinities as they are abstractly derived as functions of infinities.

Now, imagine a spacecraft attaining Lorentz factors of kth extended-j level of analog infinity where k and j are arbitrary countable or uncountable infinities.

Regarding how big infinities can be consider the following digression which is repeated again in this book.

Now, there are countable and uncountable infinities.

W e can also produce the abstraction of Hyper [Hyper Aleph Ω (Aleph Ω, Aleph Ω)] ((Hyper Aleph Ω (Aleph Ω, Aleph Ω)), Hyper Aleph Ω (Aleph Ω, Aleph Ω)).

W e can also produce the abstraction of Hyper [Hyper Aleph Ω (Aleph Ω, Aleph Ω)] ((Hyper Aleph Ω (Aleph Ω, Aleph Ω)), Hyper Aleph Ω (Aleph Ω, Aleph Ω)) ([Hyper Aleph Ω (Aleph Ω, Aleph Ω)] ((Hyper Aleph Ω (Aleph Ω, Aleph Ω)), Hyper Aleph Ω (Aleph Ω, Aleph Ω)), [Hyper Aleph Ω (Aleph Ω, Aleph Ω)] ((Hyper Aleph Ω (Aleph Ω, Aleph Ω)), Hyper Aleph Ω (Aleph Ω, Aleph Ω)))

We can go one to produce an infinity using the above nesting style of convention with all potential computer storage media currently on Earth.

We can go one to produce an infinity using the above nesting style of convention with all potential computer storage media producible by the mineral resources within the observable portion of our universe.

We can go one to produce an infinity using the above nesting style of convention with all potential computer storage media producible by the mineral resources within our entire universe.

Now here is a real zinger.

So we can produce the abstraction of Hyper Aleph (Large Uncountable Infinities) (Aleph (Large Uncountable Infinities), Aleph (Large Uncountable Infinities)).

We can also produce the abstraction of Hyper [Hyper Aleph (Large Uncountable Infinities) (Aleph (Large Uncountable Infinities), Aleph (Large Uncountable Infinities))] ((Hyper Aleph (Large Uncountable Infinities) (Aleph (Large Uncountable Infinities), Aleph (Large Uncountable Infinities))), Hyper Aleph (Large Uncountable Infinities) (Aleph (Large Uncountable Infinities), Aleph (Large Uncountable Infinities))).

We can also produce the abstraction of Hyper [Hyper Aleph (Large Uncountable Infinities) (Aleph (Large Uncountable Infinities), Aleph (Large Uncountable Infinities))] ((Hyper Aleph (Large Uncountable Infinities) (Aleph (Large Uncountable Infinities), Aleph (Large Uncountable Infinities))), Hyper Aleph (Large Uncountable Infinities) (Aleph (Large Uncountable Infinities), Aleph (Large Uncountable Infinities))) ([Hyper Aleph (Large Uncountable Infinities) (Aleph (Large Uncountable Infinities), Aleph (Large Uncountable Infinities))] ((Hyper Aleph (Large Uncountable Infinities) (Aleph (Large Uncountable Infinities), Aleph (Large Uncountable Infinities))), Hyper Aleph (Large Uncountable Infinities) (Aleph (Large Uncountable Infinities), Aleph (Large Uncountable Infinities))), [Hyper Aleph (Large Uncountable Infinities) (Aleph (Large Uncountable Infinities), Aleph (Large Uncountable Infinities))] ((Hyper Aleph (Large Uncountable Infinities) (Aleph (Large Uncountable Infinities), Aleph (Large Uncountable Infinities))), Hyper Aleph (Large Uncountable Infinities) (Aleph (Large Uncountable Infinities), Aleph (Large Uncountable Infinities))))

We can go one to produce an infinity using the above nesting style of convention with all potential computer storage media currently on Earth.

We can go one to produce an infinity using the above nesting style of convention with all potential computer storage media producible by the mineral resources within the observable portion of our universe.

We can go one to produce an infinity using the above nesting style of convention with all potential computer storage media producible by the mineral resources within our entire universe.

So, you can see how large of infinities we could abstractly denote using all of the mineral resources in our universe to construct the storage media.

However, our universe is likely just one of untold infinite numbers of universes.

Moreover, we have a potential untold infinite numbers of hyperspaces of 4 thru untold infinite numbers of dimensions, at least some of which may have some forms of mineral material extractible and convertible to storage media to define ever larger infinities.

So, we can arrive at an abstract understanding of how great a light-speed spacecraft's infinite Lorentz factor can become and the associated infinite number of light-years of travel in one year ship-frame and associated infinite number of years of travel into the future on one year ship-frame.

Since the more we colonize space or the broader cosmos in the future, the more human persons can be procreated, you can obtain a sense of how many human persons can be conceived, raised to happy adulthood, followed by a unlimitedly joy, peace, love, and eternal progress filled afterlife.

Here is another way of looking at reality and the dreams to come.

Note that Aleph 0 is the infinite number of positive integers.

Aleph 1 = 2 EXP (Aleph 0) and is according to the perhaps unprovable and thus not disprovable Continuum Hypotheses equal to the number of real numbers.

Aleph 2 = 2 EXP (Aleph 1)

Aleph 3 = 2 EXP (Aleph 2)

Aleph 4 = 2 EXP (Aleph 3)

and so-on.

In general, Aleph n = 2 EXP Aleph (n - 1) where n is any finite or infinite positive integer.

But before we go further, consider the hyper-operator notation that was designed to express huge values not otherwise expressible.

For example, Note that Hyper4(a, n) is equal to a tetrated n or a raised to the power of itself n-1 times. The latter value is symbolically written as n subscript a. For example 3 EXP 4 = 81, but 4 subscript3 is approximately equal to 10 EXP (1,000,000,000,000).

Alternatively 4 subscript 2 = 2 EXP 2 EXP 2 EXP 2 = 2 EXP [2 EXP [2 EXP 2]] = 2 EXP (2 EXP 4) = 2 EXP 16 = 65,536.

For example, Hyper5(4, 4)is equal to 4 tetrated 4 tetrated 4 tetrated 4. This value is commonly referred to as 4 pentated 4.

Hyper 6, (4,4) is 4 pentated 4 pentated 4 pentated 4 and is also referred to as 4 hexataed 4.

Hyper 7, (4,4) is 4 hexated 4 hexated 4 hexated 4 and so-on

Aleph 0 is the infinite number of integers.

Aleph 1 according to the perhaps unprovable and thus unfalsifiable Continuum Hypotheses is the number of real numbers which is greater than Aleph 0 by a multiplicative factor of infinity.

The number Ω is commonly stated as the least infinite positive integer or ordinal.

Now here is a real zinger.

So we can produce the abstraction of Hyper Aleph Ω (Aleph Ω, Aleph Ω).

Infinities can be as big as we would like. For example, consider the following infinity.

Hyper Hyper Aleph Ω (Aleph Ω, Aleph Ω)[Hyper Aleph Ω (Aleph Ω, Aleph Ω), Hyper Aleph Ω (Aleph Ω, Aleph Ω)]

We can go still larger to develop the following infinity.

Hyper Hyper Hyper Aleph Ω (Aleph Ω, Aleph Ω)[Hyper Aleph Ω (Aleph Ω, Aleph Ω), Hyper Aleph Ω (Aleph Ω, Aleph Ω)] {Hyper Hyper Aleph Ω (Aleph Ω, Aleph Ω)[Hyper Aleph Ω (Aleph Ω, Aleph Ω), Hyper Aleph Ω (Aleph Ω, Aleph Ω)], Hyper Hyper Aleph Ω (Aleph Ω, Aleph Ω)[Hyper Aleph Ω (Aleph Ω, Aleph Ω), Hyper Aleph Ω (Aleph Ω, Aleph Ω)]}

We can go still larger to develop the following infinity.

Hyper Hyper Hyper Hyper Aleph Ω (Aleph Ω, Aleph Ω)[Hyper Aleph Ω (Aleph Ω, Aleph Ω), Hyper Aleph Ω (Aleph Ω, Aleph Ω)] {Hyper Hyper Aleph Ω (Aleph Ω, Aleph Ω)[Hyper Aleph Ω (Aleph Ω, Aleph Ω), Hyper Aleph Ω (Aleph Ω, Aleph Ω)], Hyper Hyper Aleph Ω (Aleph Ω, Aleph Ω)[Hyper Aleph Ω (Aleph Ω, Aleph Ω), Hyper Aleph Ω (Aleph Ω, Aleph Ω)]} { Hyper Hyper Hyper Aleph Ω (Aleph Ω, Aleph Ω)[Hyper Aleph Ω (Aleph Ω, Aleph Ω), Hyper Aleph Ω (Aleph Ω, Aleph Ω)] {Hyper Hyper Aleph Ω (Aleph Ω, Aleph Ω)[Hyper Aleph Ω (Aleph Ω, Aleph Ω), Hyper Aleph Ω (Aleph Ω, Aleph Ω)], Hyper Hyper Aleph Ω (Aleph Ω, Aleph Ω)[Hyper Aleph Ω (Aleph Ω, Aleph Ω), Hyper Aleph Ω (Aleph Ω, Aleph Ω)]}, Hyper Hyper Hyper Aleph Ω (Aleph Ω, Aleph Ω)[Hyper Aleph Ω (Aleph Ω, Aleph Ω), Hyper Aleph Ω (Aleph Ω, Aleph Ω)] {Hyper Hyper Aleph Ω (Aleph Ω, Aleph Ω)[Hyper Aleph Ω (Aleph Ω, Aleph Ω), Hyper Aleph Ω (Aleph Ω, Aleph Ω)], Hyper Hyper Aleph Ω (Aleph Ω, Aleph Ω)[Hyper Aleph Ω (Aleph Ω, Aleph Ω), Hyper Aleph Ω (Aleph Ω, Aleph Ω)]}}

We can go still larger to develop the following infinity.

Hyper Hyper Hyper Hyper Hyper Aleph Ω (Aleph Ω, Aleph Ω)[Hyper Aleph Ω (Aleph Ω, Aleph Ω), Hyper Aleph Ω (Aleph Ω, Aleph Ω)] {Hyper Hyper Aleph Ω (Aleph Ω, Aleph Ω)[Hyper Aleph Ω (Aleph Ω, Aleph Ω), Hyper Aleph Ω (Aleph Ω, Aleph Ω)], Hyper Hyper Aleph Ω (Aleph Ω, Aleph Ω)[Hyper Aleph Ω (Aleph Ω, Aleph Ω), Hyper Aleph Ω (Aleph Ω, Aleph Ω)]} { Hyper Hyper Hyper Aleph Ω (Aleph Ω, Aleph Ω)[Hyper Aleph Ω (Aleph Ω, Aleph Ω), Hyper Aleph Ω (Aleph Ω, Aleph Ω)] {Hyper Hyper Aleph Ω (Aleph Ω, Aleph Ω)[Hyper Aleph Ω (Aleph Ω, Aleph Ω), Hyper Aleph Ω (Aleph Ω, Aleph Ω)], Hyper Hyper Aleph Ω (Aleph Ω, Aleph Ω)[Hyper Aleph Ω (Aleph Ω, Aleph Ω), Hyper Aleph Ω (Aleph Ω, Aleph Ω)]}, Hyper Hyper Hyper Aleph Ω (Aleph Ω, Aleph Ω)[Hyper Aleph Ω (Aleph Ω, Aleph Ω), Hyper Aleph Ω (Aleph Ω, Aleph Ω)] {Hyper Hyper Aleph Ω (Aleph Ω, Aleph Ω)[Hyper Aleph Ω (Aleph Ω, Aleph Ω), Hyper Aleph Ω (Aleph Ω, Aleph Ω)], Hyper Hyper Aleph Ω (Aleph Ω, Aleph Ω)[Hyper Aleph Ω (Aleph Ω, Aleph Ω), Hyper Aleph Ω (Aleph Ω, Aleph Ω)]}}{ Hyper Hyper Hyper Hyper Aleph Ω (Aleph Ω, Aleph Ω)[Hyper Aleph Ω (Aleph Ω, Aleph Ω), Hyper Aleph Ω (Aleph Ω, Aleph Ω)] {Hyper Hyper Aleph Ω (Aleph Ω, Aleph Ω)[Hyper Aleph Ω (Aleph Ω, Aleph Ω), Hyper Aleph Ω (Aleph Ω, Aleph Ω)], Hyper Hyper Aleph Ω (Aleph Ω, Aleph Ω)[Hyper Aleph Ω (Aleph Ω, Aleph Ω), Hyper Aleph Ω (Aleph Ω, Aleph Ω)]} { Hyper Hyper Hyper Aleph Ω (Aleph Ω, Aleph Ω)[Hyper Aleph Ω (Aleph Ω, Aleph Ω), Hyper Aleph Ω (Aleph Ω, Aleph Ω)] {Hyper Hyper Aleph Ω (Aleph Ω, Aleph Ω)[Hyper Aleph Ω (Aleph Ω, Aleph Ω), Hyper Aleph Ω (Aleph Ω, Aleph Ω)], Hyper Hyper Aleph Ω (Aleph Ω, Aleph Ω)[Hyper Aleph Ω (Aleph Ω, Aleph Ω), Hyper Aleph Ω (Aleph Ω, Aleph Ω)]}, Hyper Hyper Hyper Aleph Ω (Aleph Ω, Aleph Ω)[Hyper Aleph Ω (Aleph Ω, Aleph Ω), Hyper Aleph Ω (Aleph Ω, Aleph Ω)] {Hyper Hyper Aleph Ω (Aleph Ω, Aleph Ω)[Hyper Aleph Ω (Aleph Ω, Aleph Ω), Hyper Aleph Ω (Aleph Ω, Aleph Ω)], Hyper Hyper Aleph Ω (Aleph Ω, Aleph Ω)[Hyper Aleph Ω (Aleph Ω, Aleph Ω), Hyper Aleph Ω (Aleph Ω, Aleph Ω)]}}, Hyper Hyper Hyper Hyper Aleph Ω (Aleph Ω, Aleph Ω)[Hyper Aleph Ω (Aleph Ω, Aleph Ω), Hyper Aleph Ω (Aleph Ω, Aleph Ω)] {Hyper Hyper Aleph Ω (Aleph Ω, Aleph Ω)[Hyper Aleph Ω (Aleph Ω, Aleph Ω), Hyper Aleph Ω (Aleph Ω, Aleph Ω)], Hyper Hyper Aleph Ω (Aleph Ω, Aleph Ω)[Hyper Aleph Ω (Aleph Ω, Aleph Ω), Hyper Aleph Ω (Aleph Ω, Aleph Ω)]} { Hyper Hyper Hyper Aleph Ω (Aleph Ω, Aleph Ω)[Hyper Aleph Ω (Aleph Ω, Aleph Ω), Hyper Aleph Ω (Aleph Ω, Aleph Ω)] {Hyper Hyper Aleph Ω (Aleph Ω, Aleph Ω)[Hyper Aleph Ω (Aleph Ω, Aleph Ω), Hyper Aleph Ω (Aleph Ω, Aleph Ω)], Hyper Hyper Aleph Ω (Aleph Ω, Aleph Ω)[Hyper Aleph Ω (Aleph Ω, Aleph Ω), Hyper Aleph Ω (Aleph Ω, Aleph Ω)]}, Hyper Hyper Hyper Aleph Ω (Aleph Ω, Aleph Ω)[Hyper Aleph Ω (Aleph Ω, Aleph Ω), Hyper Aleph Ω (Aleph Ω, Aleph Ω)] {Hyper Hyper Aleph Ω (Aleph Ω, Aleph Ω)[Hyper Aleph Ω (Aleph Ω, Aleph Ω), Hyper Aleph Ω (Aleph Ω, Aleph Ω)], Hyper Hyper Aleph Ω (Aleph Ω, Aleph Ω)[Hyper Aleph Ω (Aleph Ω, Aleph Ω), Hyper Aleph Ω (Aleph Ω, Aleph Ω)]}}}

We can go still larger to develop the following infinity.

Hyper Hyper Hyper Hyper Hyper Hyper Aleph Ω (Aleph Ω, Aleph Ω)[Hyper Aleph Ω (Aleph Ω, Aleph Ω), Hyper Aleph Ω (Aleph Ω, Aleph Ω)] {Hyper Hyper Aleph Ω (Aleph Ω, Aleph Ω)[Hyper Aleph Ω (Aleph Ω, Aleph Ω), Hyper Aleph Ω (Aleph Ω, Aleph Ω)], Hyper Hyper Aleph Ω (Aleph Ω, Aleph Ω)[Hyper Aleph Ω (Aleph Ω, Aleph Ω), Hyper Aleph Ω (Aleph Ω, Aleph Ω)]} { Hyper Hyper Hyper Aleph Ω (Aleph Ω, Aleph Ω)[Hyper Aleph Ω (Aleph Ω, Aleph Ω), Hyper Aleph Ω (Aleph Ω, Aleph Ω)] {Hyper Hyper Aleph Ω (Aleph Ω, Aleph Ω)[Hyper Aleph Ω (Aleph Ω, Aleph Ω), Hyper Aleph Ω (Aleph Ω, Aleph Ω)], Hyper Hyper Aleph Ω (Aleph Ω, Aleph Ω)[Hyper Aleph Ω (Aleph Ω, Aleph Ω), Hyper Aleph Ω (Aleph Ω, Aleph Ω)]}, Hyper Hyper Hyper Aleph Ω (Aleph Ω, Aleph Ω)[Hyper Aleph Ω (Aleph Ω, Aleph Ω), Hyper Aleph Ω (Aleph Ω, Aleph Ω)] {Hyper Hyper Aleph Ω (Aleph Ω, Aleph Ω)[Hyper Aleph Ω (Aleph Ω, Aleph Ω), Hyper Aleph Ω (Aleph Ω, Aleph Ω)], Hyper Hyper Aleph Ω (Aleph Ω, Aleph Ω)[Hyper Aleph Ω (Aleph Ω, Aleph Ω), Hyper Aleph Ω (Aleph Ω, Aleph Ω)]}}{ Hyper Hyper Hyper Hyper Aleph Ω (Aleph Ω, Aleph Ω)[Hyper Aleph Ω (Aleph Ω, Aleph Ω), Hyper Aleph Ω (Aleph Ω, Aleph Ω)] {Hyper Hyper Aleph Ω (Aleph Ω, Aleph Ω)[Hyper Aleph Ω (Aleph Ω, Aleph Ω), Hyper Aleph Ω (Aleph Ω, Aleph Ω)], Hyper Hyper Aleph Ω (Aleph Ω, Aleph Ω)[Hyper Aleph Ω (Aleph Ω, Aleph Ω), Hyper Aleph Ω (Aleph Ω, Aleph Ω)]} { Hyper Hyper Hyper Aleph Ω (Aleph Ω, Aleph Ω)[Hyper Aleph Ω (Aleph Ω, Aleph Ω), Hyper Aleph Ω (Aleph Ω, Aleph Ω)] {Hyper Hyper Aleph Ω (Aleph Ω, Aleph Ω)[Hyper Aleph Ω (Aleph Ω, Aleph Ω), Hyper Aleph Ω (Aleph Ω, Aleph Ω)], Hyper Hyper Aleph Ω (Aleph Ω, Aleph Ω)[Hyper Aleph Ω (Aleph Ω, Aleph Ω), Hyper Aleph Ω (Aleph Ω, Aleph Ω)]}, Hyper Hyper Hyper Aleph Ω (Aleph Ω, Aleph Ω)[Hyper Aleph Ω (Aleph Ω, Aleph Ω), Hyper Aleph Ω (Aleph Ω, Aleph Ω)] {Hyper Hyper Aleph Ω (Aleph Ω, Aleph Ω)[Hyper Aleph Ω (Aleph Ω, Aleph Ω), Hyper Aleph Ω (Aleph Ω, Aleph Ω)], Hyper Hyper Aleph Ω (Aleph Ω, Aleph Ω)[Hyper Aleph Ω (Aleph Ω, Aleph Ω), Hyper Aleph Ω (Aleph Ω, Aleph Ω)]}}, Hyper Hyper Hyper Hyper Aleph Ω (Aleph Ω, Aleph Ω)[Hyper Aleph Ω (Aleph Ω, Aleph Ω), Hyper Aleph Ω (Aleph Ω, Aleph Ω)] {Hyper Hyper Aleph Ω (Aleph Ω, Aleph Ω)[Hyper Aleph Ω (Aleph Ω, Aleph Ω), Hyper Aleph Ω (Aleph Ω, Aleph Ω)], Hyper Hyper Aleph Ω (Aleph Ω, Aleph Ω)[Hyper Aleph Ω (Aleph Ω, Aleph Ω), Hyper Aleph Ω (Aleph Ω, Aleph Ω)]} { Hyper Hyper Hyper Aleph Ω (Aleph Ω, Aleph Ω)[Hyper Aleph Ω (Aleph Ω, Aleph Ω), Hyper Aleph Ω (Aleph Ω, Aleph Ω)] {Hyper Hyper Aleph Ω (Aleph Ω, Aleph Ω)[Hyper Aleph Ω (Aleph Ω, Aleph Ω), Hyper Aleph Ω (Aleph Ω, Aleph Ω)], Hyper Hyper Aleph Ω (Aleph Ω, Aleph Ω)[Hyper Aleph Ω (Aleph Ω, Aleph Ω), Hyper Aleph Ω (Aleph Ω, Aleph Ω)]}, Hyper Hyper Hyper Aleph Ω (Aleph Ω, Aleph Ω)[Hyper Aleph Ω (Aleph Ω, Aleph Ω), Hyper Aleph Ω (Aleph Ω, Aleph Ω)] {Hyper Hyper Aleph Ω (Aleph Ω, Aleph Ω)[Hyper Aleph Ω (Aleph Ω, Aleph Ω), Hyper Aleph Ω (Aleph Ω, Aleph Ω)], Hyper Hyper Aleph Ω (Aleph Ω, Aleph Ω)[Hyper Aleph Ω (Aleph Ω, Aleph Ω), Hyper Aleph Ω (Aleph Ω, Aleph Ω)]}}}{ Hyper Hyper Hyper Hyper Hyper Aleph Ω (Aleph Ω, Aleph Ω)[Hyper Aleph Ω (Aleph Ω, Aleph Ω), Hyper Aleph Ω (Aleph Ω, Aleph Ω)] {Hyper Hyper Aleph Ω (Aleph Ω, Aleph Ω)[Hyper Aleph Ω (Aleph Ω, Aleph Ω), Hyper Aleph Ω (Aleph Ω, Aleph Ω)], Hyper Hyper Aleph Ω (Aleph Ω, Aleph Ω)[Hyper

Aleph Ω (Aleph Ω, Aleph Ω), Hyper Aleph Ω (Aleph Ω, Aleph Ω)]} { Hyper Hyper Hyper Aleph Ω (Aleph Ω, Aleph Ω)[Hyper Aleph Ω (Aleph Ω, Aleph Ω), Hyper Aleph Ω (Aleph Ω, Aleph Ω)] {Hyper Hyper Aleph Ω (Aleph Ω, Aleph Ω)[Hyper Aleph Ω (Aleph Ω, Aleph Ω), Hyper Aleph Ω (Aleph Ω, Aleph Ω)], Hyper Hyper Aleph Ω (Aleph Ω, Aleph Ω)[Hyper Aleph Ω (Aleph Ω, Aleph Ω), Hyper Aleph Ω (Aleph Ω, Aleph Ω)]}, Hyper Hyper Hyper Aleph Ω (Aleph Ω, Aleph Ω)[Hyper Aleph Ω (Aleph Ω, Aleph Ω), Hyper Aleph Ω (Aleph Ω, Aleph Ω)] {Hyper Hyper Aleph Ω (Aleph Ω, Aleph Ω)[Hyper Aleph Ω (Aleph Ω, Aleph Ω), Hyper Aleph Ω (Aleph Ω, Aleph Ω)], Hyper Hyper Aleph Ω (Aleph Ω, Aleph Ω)[Hyper Aleph Ω (Aleph Ω, Aleph Ω), Hyper Aleph Ω (Aleph Ω, Aleph Ω)]}}{ Hyper Hyper Hyper Hyper Aleph Ω (Aleph Ω, Aleph Ω)[Hyper Aleph Ω (Aleph Ω, Aleph Ω), Hyper Aleph Ω (Aleph Ω, Aleph Ω)] {Hyper Hyper Aleph Ω (Aleph Ω, Aleph Ω)[Hyper Aleph Ω (Aleph Ω, Aleph Ω), Hyper Aleph Ω (Aleph Ω, Aleph Ω)], Hyper Hyper Aleph Ω (Aleph Ω, Aleph Ω)[Hyper Aleph Ω (Aleph Ω, Aleph Ω), Hyper Aleph Ω (Aleph Ω, Aleph Ω)]} { Hyper Hyper Hyper Aleph Ω (Aleph Ω, Aleph Ω)[Hyper Aleph Ω (Aleph Ω, Aleph Ω), Hyper Aleph Ω (Aleph Ω, Aleph Ω)] {Hyper Hyper Aleph Ω (Aleph Ω, Aleph Ω)[Hyper Aleph Ω (Aleph Ω, Aleph Ω), Hyper Aleph Ω (Aleph Ω, Aleph Ω)], Hyper Hyper Aleph Ω (Aleph Ω, Aleph Ω)[Hyper Aleph Ω (Aleph Ω, Aleph Ω), Hyper Aleph Ω (Aleph Ω, Aleph Ω)]}, Hyper Hyper Hyper Aleph Ω (Aleph Ω, Aleph Ω)[Hyper Aleph Ω (Aleph Ω, Aleph Ω), Hyper Aleph Ω (Aleph Ω, Aleph Ω)] {Hyper Hyper Aleph Ω (Aleph Ω, Aleph Ω)[Hyper Aleph Ω (Aleph Ω, Aleph Ω), Hyper Aleph Ω (Aleph Ω, Aleph Ω)], Hyper Hyper Aleph Ω (Aleph Ω, Aleph Ω)[Hyper Aleph Ω (Aleph Ω, Aleph Ω), Hyper Aleph Ω (Aleph Ω, Aleph Ω)]}}, Hyper Hyper Hyper Hyper Aleph Ω (Aleph Ω, Aleph Ω)[Hyper Aleph Ω (Aleph Ω, Aleph Ω), Hyper Aleph Ω (Aleph Ω, Aleph Ω)] {Hyper Hyper Aleph Ω (Aleph Ω, Aleph Ω)[Hyper Aleph Ω (Aleph Ω, Aleph Ω), Hyper Aleph Ω (Aleph Ω, Aleph Ω)], Hyper Hyper Aleph Ω (Aleph Ω, Aleph Ω)[Hyper Aleph Ω (Aleph Ω, Aleph Ω), Hyper Aleph Ω (Aleph Ω, Aleph Ω)]} { Hyper Hyper Hyper Aleph Ω (Aleph Ω, Aleph Ω)[Hyper Aleph Ω (Aleph Ω, Aleph Ω), Hyper Aleph Ω (Aleph Ω, Aleph Ω)] {Hyper Hyper Aleph Ω (Aleph Ω, Aleph Ω)[Hyper Aleph Ω (Aleph Ω, Aleph Ω), Hyper Aleph Ω (Aleph Ω, Aleph Ω)], Hyper Hyper Aleph Ω (Aleph Ω, Aleph Ω)[Hyper Aleph Ω (Aleph Ω, Aleph Ω), Hyper Aleph Ω (Aleph Ω, Aleph Ω)]}, Hyper Hyper Hyper Aleph Ω (Aleph Ω, Aleph Ω)[Hyper Aleph Ω (Aleph Ω, Aleph Ω), Hyper Aleph Ω (Aleph Ω, Aleph Ω)] {Hyper Hyper Aleph Ω (Aleph Ω, Aleph Ω)[Hyper Aleph Ω (Aleph Ω, Aleph Ω), Hyper Aleph Ω (Aleph Ω, Aleph Ω)], Hyper Hyper Aleph Ω (Aleph Ω, Aleph Ω)[Hyper Aleph Ω (Aleph Ω, Aleph Ω), Hyper Aleph Ω (Aleph Ω, Aleph Ω)]}}}, Hyper Hyper Hyper Hyper Hyper Aleph Ω (Aleph Ω, Aleph Ω)[Hyper Aleph Ω (Aleph Ω, Aleph Ω), Hyper Aleph Ω (Aleph Ω, Aleph Ω)] {Hyper Hyper Aleph Ω (Aleph Ω, Aleph Ω)[Hyper Aleph Ω (Aleph Ω, Aleph Ω), Hyper Aleph Ω (Aleph Ω, Aleph Ω)], Hyper Hyper Aleph Ω (Aleph Ω, Aleph Ω)[Hyper Aleph Ω (Aleph Ω, Aleph Ω), Hyper Aleph Ω (Aleph Ω, Aleph Ω)]} { Hyper Hyper Hyper Aleph Ω (Aleph Ω, Aleph Ω)[Hyper Aleph Ω (Aleph Ω, Aleph Ω), Hyper Aleph Ω (Aleph Ω, Aleph Ω)] {Hyper Hyper Aleph Ω (Aleph Ω, Aleph Ω)[Hyper Aleph Ω (Aleph Ω, Aleph Ω), Hyper Aleph Ω (Aleph Ω, Aleph

Ω)], Hyper Hyper Aleph Ω (Aleph Ω, Aleph Ω)[Hyper Aleph Ω (Aleph Ω, Aleph Ω), Hyper Aleph Ω (Aleph Ω, Aleph Ω)]}, Hyper Hyper Hyper Aleph Ω (Aleph Ω, Aleph Ω)[Hyper Aleph Ω (Aleph Ω, Aleph Ω), Hyper Aleph Ω (Aleph Ω, Aleph Ω)] {Hyper Hyper Aleph Ω (Aleph Ω, Aleph Ω)[Hyper Aleph Ω (Aleph Ω, Aleph Ω), Hyper Aleph Ω (Aleph Ω, Aleph Ω)], Hyper Hyper Aleph Ω (Aleph Ω, Aleph Ω)[Hyper Aleph Ω (Aleph Ω, Aleph Ω), Hyper Aleph Ω (Aleph Ω, Aleph Ω)]}}{ Hyper Hyper Hyper Hyper Aleph Ω (Aleph Ω, Aleph Ω)[Hyper Aleph Ω (Aleph Ω, Aleph Ω), Hyper Aleph Ω (Aleph Ω, Aleph Ω)] {Hyper Hyper Aleph Ω (Aleph Ω, Aleph Ω)[Hyper Aleph Ω (Aleph Ω, Aleph Ω), Hyper Aleph Ω (Aleph Ω, Aleph Ω)], Hyper Hyper Aleph Ω (Aleph Ω, Aleph Ω)[Hyper Aleph Ω (Aleph Ω, Aleph Ω), Hyper Aleph Ω (Aleph Ω, Aleph Ω)]} { Hyper Hyper Hyper Aleph Ω (Aleph Ω, Aleph Ω)[Hyper Aleph Ω (Aleph Ω, Aleph Ω), Hyper Aleph Ω (Aleph Ω, Aleph Ω)] {Hyper Hyper Aleph Ω (Aleph Ω, Aleph Ω)[Hyper Aleph Ω (Aleph Ω, Aleph Ω), Hyper Aleph Ω (Aleph Ω, Aleph Ω)], Hyper Hyper Aleph Ω (Aleph Ω, Aleph Ω)[Hyper Aleph Ω (Aleph Ω, Aleph Ω), Hyper Aleph Ω (Aleph Ω, Aleph Ω)]}, Hyper Hyper Hyper Aleph Ω (Aleph Ω, Aleph Ω)[Hyper Aleph Ω (Aleph Ω, Aleph Ω), Hyper Aleph Ω (Aleph Ω, Aleph Ω)] {Hyper Hyper Aleph Ω (Aleph Ω, Aleph Ω)[Hyper Aleph Ω (Aleph Ω, Aleph Ω), Hyper Aleph Ω (Aleph Ω, Aleph Ω)], Hyper Hyper Aleph Ω (Aleph Ω, Aleph Ω)[Hyper Aleph Ω (Aleph Ω, Aleph Ω), Hyper Aleph Ω (Aleph Ω, Aleph Ω)]}}, Hyper Hyper Hyper Hyper Aleph Ω (Aleph Ω, Aleph Ω)[Hyper Aleph Ω (Aleph Ω, Aleph Ω), Hyper Aleph Ω (Aleph Ω, Aleph Ω)] {Hyper Hyper Aleph Ω (Aleph Ω, Aleph Ω)[Hyper Aleph Ω (Aleph Ω, Aleph Ω), Hyper Aleph Ω (Aleph Ω, Aleph Ω)], Hyper Hyper Aleph Ω (Aleph Ω, Aleph Ω)[Hyper Aleph Ω (Aleph Ω, Aleph Ω), Hyper Aleph Ω (Aleph Ω, Aleph Ω)]} { Hyper Hyper Hyper Aleph Ω (Aleph Ω, Aleph Ω)[Hyper Aleph Ω (Aleph Ω, Aleph Ω), Hyper Aleph Ω (Aleph Ω, Aleph Ω)] {Hyper Hyper Aleph Ω (Aleph Ω, Aleph Ω)[Hyper Aleph Ω (Aleph Ω, Aleph Ω), Hyper Aleph Ω (Aleph Ω, Aleph Ω)], Hyper Hyper Aleph Ω (Aleph Ω, Aleph Ω)[Hyper Aleph Ω (Aleph Ω, Aleph Ω), Hyper Aleph Ω (Aleph Ω, Aleph Ω)]}, Hyper Hyper Hyper Aleph Ω (Aleph Ω, Aleph Ω)[Hyper Aleph Ω (Aleph Ω, Aleph Ω), Hyper Aleph Ω (Aleph Ω, Aleph Ω)] {Hyper Hyper Aleph Ω (Aleph Ω, Aleph Ω)[Hyper Aleph Ω (Aleph Ω, Aleph Ω), Hyper Aleph Ω (Aleph Ω, Aleph Ω)], Hyper Hyper Aleph Ω (Aleph Ω, Aleph Ω)[Hyper Aleph Ω (Aleph Ω, Aleph Ω), Hyper Aleph Ω (Aleph Ω, Aleph Ω)]}}}}

We can go still yet further to develop the following infinity.

Hyper Hyper Hyper Hyper Hyper Hyper Hyper Aleph Ω (Aleph Ω, Aleph Ω)[Hyper Aleph Ω (Aleph Ω, Aleph Ω), Hyper Aleph Ω (Aleph Ω, Aleph Ω)] {Hyper Hyper Aleph Ω (Aleph Ω, Aleph Ω)[Hyper Aleph Ω (Aleph Ω, Aleph Ω), Hyper Aleph Ω (Aleph Ω, Aleph Ω)], Hyper Hyper Aleph Ω (Aleph Ω, Aleph Ω)[Hyper Aleph Ω (Aleph Ω, Aleph Ω), Hyper Aleph Ω (Aleph Ω, Aleph Ω)]} { Hyper Hyper Hyper Aleph Ω (Aleph Ω, Aleph Ω)[Hyper Aleph Ω (Aleph Ω, Aleph Ω), Hyper Aleph Ω (Aleph Ω, Aleph Ω)] {Hyper Hyper Aleph Ω (Aleph Ω, Aleph Ω)[Hyper Aleph Ω (Aleph Ω, Aleph Ω), Hyper Aleph Ω (Aleph Ω, Aleph Ω)], Hyper Hyper Aleph Ω (Aleph Ω, Aleph Ω)[Hyper Aleph Ω (Aleph Ω, Aleph Ω), Hyper

Aleph Ω (Aleph Ω, Aleph Ω)]}, Hyper Hyper Hyper Aleph Ω (Aleph Ω, Aleph Ω)[Hyper Aleph Ω (Aleph Ω, Aleph Ω), Hyper Aleph Ω (Aleph Ω, Aleph Ω)] {Hyper Hyper Aleph Ω (Aleph Ω, Aleph Ω)[Hyper Aleph Ω (Aleph Ω, Aleph Ω), Hyper Aleph Ω (Aleph Ω, Aleph Ω)], Hyper Hyper Aleph Ω (Aleph Ω, Aleph Ω)[Hyper Aleph Ω (Aleph Ω, Aleph Ω), Hyper Aleph Ω (Aleph Ω, Aleph Ω)]}}{ Hyper Hyper Hyper Hyper Aleph Ω (Aleph Ω, Aleph Ω)[Hyper Aleph Ω (Aleph Ω, Aleph Ω), Hyper Aleph Ω (Aleph Ω, Aleph Ω)] {Hyper Hyper Aleph Ω (Aleph Ω, Aleph Ω)[Hyper Aleph Ω (Aleph Ω, Aleph Ω), Hyper Aleph Ω (Aleph Ω, Aleph Ω)], Hyper Hyper Aleph Ω (Aleph Ω, Aleph Ω)[Hyper Aleph Ω (Aleph Ω, Aleph Ω), Hyper Aleph Ω (Aleph Ω, Aleph Ω)]} { Hyper Hyper Hyper Aleph Ω (Aleph Ω, Aleph Ω)[Hyper Aleph Ω (Aleph Ω, Aleph Ω), Hyper Aleph Ω (Aleph Ω, Aleph Ω)] {Hyper Hyper Aleph Ω (Aleph Ω, Aleph Ω)[Hyper Aleph Ω (Aleph Ω, Aleph Ω), Hyper Aleph Ω (Aleph Ω, Aleph Ω)], Hyper Hyper Aleph Ω (Aleph Ω, Aleph Ω)[Hyper Aleph Ω (Aleph Ω, Aleph Ω), Hyper Aleph Ω (Aleph Ω, Aleph Ω)]}, Hyper Hyper Hyper Aleph Ω (Aleph Ω, Aleph Ω)[Hyper Aleph Ω (Aleph Ω, Aleph Ω), Hyper Aleph Ω (Aleph Ω, Aleph Ω)] {Hyper Hyper Aleph Ω (Aleph Ω, Aleph Ω)[Hyper Aleph Ω (Aleph Ω, Aleph Ω), Hyper Aleph Ω (Aleph Ω, Aleph Ω)], Hyper Hyper Aleph Ω (Aleph Ω, Aleph Ω)[Hyper Aleph Ω (Aleph Ω, Aleph Ω), Hyper Aleph Ω (Aleph Ω, Aleph Ω)]}}, Hyper Hyper Hyper Hyper Aleph Ω (Aleph Ω, Aleph Ω)[Hyper Aleph Ω (Aleph Ω, Aleph Ω), Hyper Aleph Ω (Aleph Ω, Aleph Ω)] {Hyper Hyper Aleph Ω (Aleph Ω, Aleph Ω)[Hyper Aleph Ω (Aleph Ω, Aleph Ω), Hyper Aleph Ω (Aleph Ω, Aleph Ω)], Hyper Hyper Aleph Ω (Aleph Ω, Aleph Ω)[Hyper Aleph Ω (Aleph Ω, Aleph Ω), Hyper Aleph Ω (Aleph Ω, Aleph Ω)]} { Hyper Hyper Hyper Aleph Ω (Aleph Ω, Aleph Ω)[Hyper Aleph Ω (Aleph Ω, Aleph Ω), Hyper Aleph Ω (Aleph Ω, Aleph Ω)] {Hyper Hyper Aleph Ω (Aleph Ω, Aleph Ω)[Hyper Aleph Ω (Aleph Ω, Aleph Ω), Hyper Aleph Ω (Aleph Ω, Aleph Ω)], Hyper Hyper Aleph Ω (Aleph Ω, Aleph Ω)[Hyper Aleph Ω (Aleph Ω, Aleph Ω), Hyper Aleph Ω (Aleph Ω, Aleph Ω)]}, Hyper Hyper Hyper Aleph Ω (Aleph Ω, Aleph Ω)[Hyper Aleph Ω (Aleph Ω, Aleph Ω), Hyper Aleph Ω (Aleph Ω, Aleph Ω)] {Hyper Hyper Aleph Ω (Aleph Ω, Aleph Ω)[Hyper Aleph Ω (Aleph Ω, Aleph Ω), Hyper Aleph Ω (Aleph Ω, Aleph Ω)], Hyper Hyper Aleph Ω (Aleph Ω, Aleph Ω)[Hyper Aleph Ω (Aleph Ω, Aleph Ω), Hyper Aleph Ω (Aleph Ω, Aleph Ω)]}}}{ Hyper Hyper Hyper Hyper Hyper Aleph Ω (Aleph Ω, Aleph Ω)[Hyper Aleph Ω (Aleph Ω, Aleph Ω), Hyper Aleph Ω (Aleph Ω, Aleph Ω)] {Hyper Hyper Aleph Ω (Aleph Ω, Aleph Ω)[Hyper Aleph Ω (Aleph Ω, Aleph Ω), Hyper Aleph Ω (Aleph Ω, Aleph Ω)], Hyper Hyper Aleph Ω (Aleph Ω, Aleph Ω)[Hyper Aleph Ω (Aleph Ω, Aleph Ω), Hyper Aleph Ω (Aleph Ω, Aleph Ω)]} { Hyper Hyper Hyper Aleph Ω (Aleph Ω, Aleph Ω)[Hyper Aleph Ω (Aleph Ω, Aleph Ω), Hyper Aleph Ω (Aleph Ω, Aleph Ω)] {Hyper Hyper Aleph Ω (Aleph Ω, Aleph Ω)[Hyper Aleph Ω (Aleph Ω, Aleph Ω), Hyper Aleph Ω (Aleph Ω, Aleph Ω)], Hyper Hyper Aleph Ω (Aleph Ω, Aleph Ω)[Hyper Aleph Ω (Aleph Ω, Aleph Ω), Hyper Aleph Ω (Aleph Ω, Aleph Ω)]}, Hyper Hyper Hyper Aleph Ω (Aleph Ω, Aleph Ω)[Hyper Aleph Ω (Aleph Ω, Aleph Ω), Hyper Aleph Ω (Aleph Ω, Aleph Ω)] {Hyper Hyper Aleph Ω (Aleph Ω, Aleph Ω)[Hyper Aleph Ω (Aleph Ω, Aleph Ω), Hyper Aleph Ω (Aleph Ω, Aleph Ω)], Hyper Hyper Aleph Ω (Aleph Ω, Aleph Ω)[Hyper

Aleph Ω (Aleph Ω, Aleph Ω), Hyper Aleph Ω (Aleph Ω, Aleph Ω)]}}{ Hyper Hyper Hyper Hyper Aleph Ω (Aleph Ω, Aleph Ω)[Hyper Aleph Ω (Aleph Ω, Aleph Ω), Hyper Aleph Ω (Aleph Ω, Aleph Ω)] {Hyper Hyper Aleph Ω (Aleph Ω, Aleph Ω)[Hyper Aleph Ω (Aleph Ω, Aleph Ω), Hyper Aleph Ω (Aleph Ω, Aleph Ω)], Hyper Hyper Aleph Ω (Aleph Ω, Aleph Ω)[Hyper Aleph Ω (Aleph Ω, Aleph Ω), Hyper Aleph Ω (Aleph Ω, Aleph Ω)]} { Hyper Hyper Hyper Aleph Ω (Aleph Ω, Aleph Ω)[Hyper Aleph Ω (Aleph Ω, Aleph Ω), Hyper Aleph Ω (Aleph Ω, Aleph Ω)] {Hyper Hyper Aleph Ω (Aleph Ω, Aleph Ω)[Hyper Aleph Ω (Aleph Ω, Aleph Ω), Hyper Aleph Ω (Aleph Ω, Aleph Ω)], Hyper Hyper Aleph Ω (Aleph Ω, Aleph Ω)[Hyper Aleph Ω (Aleph Ω, Aleph Ω), Hyper Aleph Ω (Aleph Ω, Aleph Ω)]}, Hyper Hyper Hyper Aleph Ω (Aleph Ω, Aleph Ω)[Hyper Aleph Ω (Aleph Ω, Aleph Ω), Hyper Aleph Ω (Aleph Ω, Aleph Ω)] {Hyper Hyper Aleph Ω (Aleph Ω, Aleph Ω)[Hyper Aleph Ω (Aleph Ω, Aleph Ω), Hyper Aleph Ω (Aleph Ω, Aleph Ω)], Hyper Hyper Aleph Ω (Aleph Ω, Aleph Ω)[Hyper Aleph Ω (Aleph Ω, Aleph Ω), Hyper Aleph Ω (Aleph Ω, Aleph Ω)]}}, Hyper Hyper Hyper Hyper Aleph Ω (Aleph Ω, Aleph Ω)[Hyper Aleph Ω (Aleph Ω, Aleph Ω), Hyper Aleph Ω (Aleph Ω, Aleph Ω)] {Hyper Hyper Aleph Ω (Aleph Ω, Aleph Ω)[Hyper Aleph Ω (Aleph Ω, Aleph Ω), Hyper Aleph Ω (Aleph Ω, Aleph Ω)], Hyper Hyper Aleph Ω (Aleph Ω, Aleph Ω)[Hyper Aleph Ω (Aleph Ω, Aleph Ω), Hyper Aleph Ω (Aleph Ω, Aleph Ω)]} { Hyper Hyper Hyper Aleph Ω (Aleph Ω, Aleph Ω)[Hyper Aleph Ω (Aleph Ω, Aleph Ω), Hyper Aleph Ω (Aleph Ω, Aleph Ω)] {Hyper Hyper Aleph Ω (Aleph Ω, Aleph Ω)[Hyper Aleph Ω (Aleph Ω, Aleph Ω), Hyper Aleph Ω (Aleph Ω, Aleph Ω)], Hyper Hyper Aleph Ω (Aleph Ω, Aleph Ω)[Hyper Aleph Ω (Aleph Ω, Aleph Ω), Hyper Aleph Ω (Aleph Ω, Aleph Ω)]}, Hyper Hyper Hyper Aleph Ω (Aleph Ω, Aleph Ω)[Hyper Aleph Ω (Aleph Ω, Aleph Ω), Hyper Aleph Ω (Aleph Ω, Aleph Ω)] {Hyper Hyper Aleph Ω (Aleph Ω, Aleph Ω)[Hyper Aleph Ω (Aleph Ω, Aleph Ω), Hyper Aleph Ω (Aleph Ω, Aleph Ω)], Hyper Hyper Aleph Ω (Aleph Ω, Aleph Ω)[Hyper Aleph Ω (Aleph Ω, Aleph Ω), Hyper Aleph Ω (Aleph Ω, Aleph Ω)]}}}, Hyper Hyper Hyper Hyper Hyper Aleph Ω (Aleph Ω, Aleph Ω)[Hyper Aleph Ω (Aleph Ω, Aleph Ω), Hyper Aleph Ω (Aleph Ω, Aleph Ω)] {Hyper Hyper Aleph Ω (Aleph Ω, Aleph Ω)[Hyper Aleph Ω (Aleph Ω, Aleph Ω), Hyper Aleph Ω (Aleph Ω, Aleph Ω)], Hyper Hyper Aleph Ω (Aleph Ω, Aleph Ω)[Hyper Aleph Ω (Aleph Ω, Aleph Ω), Hyper Aleph Ω (Aleph Ω, Aleph Ω)]} { Hyper Hyper Hyper Aleph Ω (Aleph Ω, Aleph Ω)[Hyper Aleph Ω (Aleph Ω, Aleph Ω), Hyper Aleph Ω (Aleph Ω, Aleph Ω)] {Hyper Hyper Aleph Ω (Aleph Ω, Aleph Ω)[Hyper Aleph Ω (Aleph Ω, Aleph Ω), Hyper Aleph Ω (Aleph Ω, Aleph Ω)], Hyper Hyper Aleph Ω (Aleph Ω, Aleph Ω)[Hyper Aleph Ω (Aleph Ω, Aleph Ω), Hyper Aleph Ω (Aleph Ω, Aleph Ω)]}, Hyper Hyper Hyper Aleph Ω (Aleph Ω, Aleph Ω)[Hyper Aleph Ω (Aleph Ω, Aleph Ω), Hyper Aleph Ω (Aleph Ω, Aleph Ω)] {Hyper Hyper Aleph Ω (Aleph Ω, Aleph Ω)[Hyper Aleph Ω (Aleph Ω, Aleph Ω), Hyper Aleph Ω (Aleph Ω, Aleph Ω)], Hyper Hyper Aleph Ω (Aleph Ω, Aleph Ω)[Hyper Aleph Ω (Aleph Ω, Aleph Ω), Hyper Aleph Ω (Aleph Ω, Aleph Ω)]}}{ Hyper Hyper Hyper Hyper Aleph Ω (Aleph Ω, Aleph Ω)[Hyper Aleph Ω (Aleph Ω, Aleph Ω), Hyper Aleph Ω (Aleph Ω, Aleph Ω)] {Hyper Hyper Aleph Ω (Aleph Ω, Aleph Ω)[Hyper Aleph Ω (Aleph Ω, Aleph Ω), Hyper Aleph Ω (Aleph

Ω, Aleph Ω)], Hyper Hyper Aleph Ω (Aleph Ω, Aleph Ω)[Hyper Aleph Ω (Aleph Ω, Aleph Ω), Hyper Aleph Ω (Aleph Ω, Aleph Ω)]} { Hyper Hyper Hyper Aleph Ω (Aleph Ω, Aleph Ω)[Hyper Aleph Ω (Aleph Ω, Aleph Ω), Hyper Aleph Ω (Aleph Ω, Aleph Ω)] {Hyper Hyper Aleph Ω (Aleph Ω, Aleph Ω)[Hyper Aleph Ω (Aleph Ω, Aleph Ω), Hyper Aleph Ω (Aleph Ω, Aleph Ω)], Hyper Hyper Aleph Ω (Aleph Ω, Aleph Ω)[Hyper Aleph Ω (Aleph Ω, Aleph Ω), Hyper Aleph Ω (Aleph Ω, Aleph Ω)]}, Hyper Hyper Hyper Aleph Ω (Aleph Ω, Aleph Ω)[Hyper Aleph Ω (Aleph Ω, Aleph Ω), Hyper Aleph Ω (Aleph Ω, Aleph Ω)] {Hyper Hyper Aleph Ω (Aleph Ω, Aleph Ω)[Hyper Aleph Ω (Aleph Ω, Aleph Ω), Hyper Aleph Ω (Aleph Ω, Aleph Ω)], Hyper Hyper Aleph Ω (Aleph Ω, Aleph Ω)[Hyper Aleph Ω (Aleph Ω, Aleph Ω), Hyper Aleph Ω (Aleph Ω, Aleph Ω)]}}, Hyper Hyper Hyper Hyper Aleph Ω (Aleph Ω, Aleph Ω)[Hyper Aleph Ω (Aleph Ω, Aleph Ω), Hyper Aleph Ω (Aleph Ω, Aleph Ω)] {Hyper Hyper Aleph Ω (Aleph Ω, Aleph Ω)[Hyper Aleph Ω (Aleph Ω, Aleph Ω), Hyper Aleph Ω (Aleph Ω, Aleph Ω)], Hyper Hyper Aleph Ω (Aleph Ω, Aleph Ω)[Hyper Aleph Ω (Aleph Ω, Aleph Ω), Hyper Aleph Ω (Aleph Ω, Aleph Ω)]} { Hyper Hyper Hyper Aleph Ω (Aleph Ω, Aleph Ω)[Hyper Aleph Ω (Aleph Ω, Aleph Ω), Hyper Aleph Ω (Aleph Ω, Aleph Ω)] {Hyper Hyper Aleph Ω (Aleph Ω, Aleph Ω)[Hyper Aleph Ω (Aleph Ω, Aleph Ω), Hyper Aleph Ω (Aleph Ω, Aleph Ω)], Hyper Hyper Aleph Ω (Aleph Ω, Aleph Ω)[Hyper Aleph Ω (Aleph Ω, Aleph Ω), Hyper Aleph Ω (Aleph Ω, Aleph Ω)]}, Hyper Hyper Hyper Aleph Ω (Aleph Ω, Aleph Ω)[Hyper Aleph Ω (Aleph Ω, Aleph Ω), Hyper Aleph Ω (Aleph Ω, Aleph Ω)] {Hyper Hyper Aleph Ω (Aleph Ω, Aleph Ω)[Hyper Aleph Ω (Aleph Ω, Aleph Ω), Hyper Aleph Ω (Aleph Ω, Aleph Ω)], Hyper Hyper Aleph Ω (Aleph Ω, Aleph Ω)[Hyper Aleph Ω (Aleph Ω, Aleph Ω), Hyper Aleph Ω (Aleph Ω, Aleph Ω)]}}}}{ Hyper Hyper Hyper Hyper Hyper Hyper Aleph Ω (Aleph Ω, Aleph Ω)[Hyper Aleph Ω (Aleph Ω, Aleph Ω), Hyper Aleph Ω (Aleph Ω, Aleph Ω)] {Hyper Hyper Aleph Ω (Aleph Ω, Aleph Ω)[Hyper Aleph Ω (Aleph Ω, Aleph Ω), Hyper Aleph Ω (Aleph Ω, Aleph Ω)], Hyper Hyper Aleph Ω (Aleph Ω, Aleph Ω)[Hyper Aleph Ω (Aleph Ω, Aleph Ω), Hyper Aleph Ω (Aleph Ω, Aleph Ω)]} { Hyper Hyper Hyper Aleph Ω (Aleph Ω, Aleph Ω)[Hyper Aleph Ω (Aleph Ω, Aleph Ω), Hyper Aleph Ω (Aleph Ω, Aleph Ω)] {Hyper Hyper Aleph Ω (Aleph Ω, Aleph Ω)[Hyper Aleph Ω (Aleph Ω, Aleph Ω), Hyper Aleph Ω (Aleph Ω, Aleph Ω)], Hyper Hyper Aleph Ω (Aleph Ω, Aleph Ω)[Hyper Aleph Ω (Aleph Ω, Aleph Ω), Hyper Aleph Ω (Aleph Ω, Aleph Ω)]}, Hyper Hyper Hyper Aleph Ω (Aleph Ω, Aleph Ω)[Hyper Aleph Ω (Aleph Ω, Aleph Ω), Hyper Aleph Ω (Aleph Ω, Aleph Ω)] {Hyper Hyper Aleph Ω (Aleph Ω, Aleph Ω)[Hyper Aleph Ω (Aleph Ω, Aleph Ω), Hyper Aleph Ω (Aleph Ω, Aleph Ω)], Hyper Hyper Aleph Ω (Aleph Ω, Aleph Ω)[Hyper Aleph Ω (Aleph Ω, Aleph Ω), Hyper Aleph Ω (Aleph Ω, Aleph Ω)]}}{ Hyper Hyper Hyper Hyper Aleph Ω (Aleph Ω, Aleph Ω)[Hyper Aleph Ω (Aleph Ω, Aleph Ω), Hyper Aleph Ω (Aleph Ω, Aleph Ω)] {Hyper Hyper Aleph Ω (Aleph Ω, Aleph Ω)[Hyper Aleph Ω (Aleph Ω, Aleph Ω), Hyper Aleph Ω (Aleph Ω, Aleph Ω)], Hyper Hyper Aleph Ω (Aleph Ω, Aleph Ω)[Hyper Aleph Ω (Aleph Ω, Aleph Ω), Hyper Aleph Ω (Aleph Ω, Aleph Ω)]} { Hyper Hyper Hyper Aleph Ω (Aleph Ω, Aleph Ω)[Hyper Aleph Ω (Aleph Ω, Aleph Ω), Hyper Aleph Ω (Aleph Ω, Aleph Ω)] {Hyper Hyper Aleph Ω (Aleph Ω, Aleph

Ω)[Hyper Aleph Ω (Aleph Ω, Aleph Ω), Hyper Aleph Ω (Aleph Ω, Aleph Ω)], Hyper Hyper Aleph Ω (Aleph Ω, Aleph Ω)[Hyper Aleph Ω (Aleph Ω, Aleph Ω), Hyper Aleph Ω (Aleph Ω, Aleph Ω)]}, Hyper Hyper Hyper Aleph Ω (Aleph Ω, Aleph Ω)[Hyper Aleph Ω (Aleph Ω, Aleph Ω), Hyper Aleph Ω (Aleph Ω, Aleph Ω)] {Hyper Hyper Aleph Ω (Aleph Ω, Aleph Ω)[Hyper Aleph Ω (Aleph Ω, Aleph Ω), Hyper Aleph Ω (Aleph Ω, Aleph Ω)], Hyper Hyper Aleph Ω (Aleph Ω, Aleph Ω)[Hyper Aleph Ω (Aleph Ω, Aleph Ω), Hyper Aleph Ω (Aleph Ω, Aleph Ω)]}}, Hyper Hyper Hyper Hyper Aleph Ω (Aleph Ω, Aleph Ω)[Hyper Aleph Ω (Aleph Ω, Aleph Ω), Hyper Aleph Ω (Aleph Ω, Aleph Ω)] {Hyper Hyper Aleph Ω (Aleph Ω, Aleph Ω)[Hyper Aleph Ω (Aleph Ω, Aleph Ω), Hyper Aleph Ω (Aleph Ω, Aleph Ω)], Hyper Hyper Aleph Ω (Aleph Ω, Aleph Ω)[Hyper Aleph Ω (Aleph Ω, Aleph Ω), Hyper Aleph Ω (Aleph Ω, Aleph Ω)]} { Hyper Hyper Hyper Aleph Ω (Aleph Ω, Aleph Ω)[Hyper Aleph Ω (Aleph Ω, Aleph Ω), Hyper Aleph Ω (Aleph Ω, Aleph Ω)] {Hyper Hyper Aleph Ω (Aleph Ω, Aleph Ω)[Hyper Aleph Ω (Aleph Ω, Aleph Ω), Hyper Aleph Ω (Aleph Ω, Aleph Ω)], Hyper Hyper Aleph Ω (Aleph Ω, Aleph Ω)[Hyper Aleph Ω (Aleph Ω, Aleph Ω), Hyper Aleph Ω (Aleph Ω, Aleph Ω)]}, Hyper Hyper Hyper Aleph Ω (Aleph Ω, Aleph Ω)[Hyper Aleph Ω (Aleph Ω, Aleph Ω), Hyper Aleph Ω (Aleph Ω, Aleph Ω)] {Hyper Hyper Aleph Ω (Aleph Ω, Aleph Ω)[Hyper Aleph Ω (Aleph Ω, Aleph Ω), Hyper Aleph Ω (Aleph Ω, Aleph Ω)], Hyper Hyper Aleph Ω (Aleph Ω, Aleph Ω)[Hyper Aleph Ω (Aleph Ω, Aleph Ω), Hyper Aleph Ω (Aleph Ω, Aleph Ω)]}}}{ Hyper Hyper Hyper Hyper Hyper Aleph Ω (Aleph Ω, Aleph Ω)[Hyper Aleph Ω (Aleph Ω, Aleph Ω), Hyper Aleph Ω (Aleph Ω, Aleph Ω)] {Hyper Hyper Aleph Ω (Aleph Ω, Aleph Ω)[Hyper Aleph Ω (Aleph Ω, Aleph Ω), Hyper Aleph Ω (Aleph Ω, Aleph Ω)], Hyper Hyper Aleph Ω (Aleph Ω, Aleph Ω)[Hyper Aleph Ω (Aleph Ω, Aleph Ω), Hyper Aleph Ω (Aleph Ω, Aleph Ω)]} { Hyper Hyper Hyper Aleph Ω (Aleph Ω, Aleph Ω)[Hyper Aleph Ω (Aleph Ω, Aleph Ω), Hyper Aleph Ω (Aleph Ω, Aleph Ω)] {Hyper Hyper Aleph Ω (Aleph Ω, Aleph Ω)[Hyper Aleph Ω (Aleph Ω, Aleph Ω), Hyper Aleph Ω (Aleph Ω, Aleph Ω)], Hyper Hyper Aleph Ω (Aleph Ω, Aleph Ω)[Hyper Aleph Ω (Aleph Ω, Aleph Ω), Hyper Aleph Ω (Aleph Ω, Aleph Ω)]}, Hyper Hyper Hyper Aleph Ω (Aleph Ω, Aleph Ω)[Hyper Aleph Ω (Aleph Ω, Aleph Ω), Hyper Aleph Ω (Aleph Ω, Aleph Ω)] {Hyper Hyper Aleph Ω (Aleph Ω, Aleph Ω)[Hyper Aleph Ω (Aleph Ω, Aleph Ω), Hyper Aleph Ω (Aleph Ω, Aleph Ω)], Hyper Hyper Aleph Ω (Aleph Ω, Aleph Ω)[Hyper Aleph Ω (Aleph Ω, Aleph Ω), Hyper Aleph Ω (Aleph Ω, Aleph Ω)]}}{ Hyper Hyper Hyper Hyper Aleph Ω (Aleph Ω, Aleph Ω)[Hyper Aleph Ω (Aleph Ω, Aleph Ω), Hyper Aleph Ω (Aleph Ω, Aleph Ω)] {Hyper Hyper Aleph Ω (Aleph Ω, Aleph Ω)[Hyper Aleph Ω (Aleph Ω, Aleph Ω), Hyper Aleph Ω (Aleph Ω, Aleph Ω)], Hyper Hyper Aleph Ω (Aleph Ω, Aleph Ω)[Hyper Aleph Ω (Aleph Ω, Aleph Ω), Hyper Aleph Ω (Aleph Ω, Aleph Ω)]} { Hyper Hyper Hyper Aleph Ω (Aleph Ω, Aleph Ω)[Hyper Aleph Ω (Aleph Ω, Aleph Ω), Hyper Aleph Ω (Aleph Ω, Aleph Ω)] {Hyper Hyper Aleph Ω (Aleph Ω, Aleph Ω)[Hyper Aleph Ω (Aleph Ω, Aleph Ω), Hyper Aleph Ω (Aleph Ω, Aleph Ω)], Hyper Hyper Aleph Ω (Aleph Ω, Aleph Ω)[Hyper Aleph Ω (Aleph Ω, Aleph Ω), Hyper Aleph Ω (Aleph Ω, Aleph Ω)]}, Hyper Hyper Hyper Aleph Ω (Aleph Ω, Aleph Ω)[Hyper Aleph Ω (Aleph Ω, Aleph Ω), Hyper Aleph Ω (Aleph Ω, Aleph

Ω)] {Hyper Hyper Aleph Ω (Aleph Ω, Aleph Ω)[Hyper Aleph Ω (Aleph Ω, Aleph Ω), Hyper Aleph Ω (Aleph Ω, Aleph Ω)], Hyper Hyper Aleph Ω (Aleph Ω, Aleph Ω)[Hyper Aleph Ω (Aleph Ω, Aleph Ω), Hyper Aleph Ω (Aleph Ω, Aleph Ω)]}}, Hyper Hyper Hyper Hyper Aleph Ω (Aleph Ω, Aleph Ω)[Hyper Aleph Ω (Aleph Ω, Aleph Ω), Hyper Aleph Ω (Aleph Ω, Aleph Ω)] {Hyper Hyper Aleph Ω (Aleph Ω, Aleph Ω)[Hyper Aleph Ω (Aleph Ω, Aleph Ω), Hyper Aleph Ω (Aleph Ω, Aleph Ω)], Hyper Hyper Aleph Ω (Aleph Ω, Aleph Ω)[Hyper Aleph Ω (Aleph Ω, Aleph Ω), Hyper Aleph Ω (Aleph Ω, Aleph Ω)]} { Hyper Hyper Hyper Aleph Ω (Aleph Ω, Aleph Ω)[Hyper Aleph Ω (Aleph Ω, Aleph Ω), Hyper Aleph Ω (Aleph Ω, Aleph Ω)] {Hyper Hyper Aleph Ω (Aleph Ω, Aleph Ω)[Hyper Aleph Ω (Aleph Ω, Aleph Ω), Hyper Aleph Ω (Aleph Ω, Aleph Ω)], Hyper Hyper Aleph Ω (Aleph Ω, Aleph Ω)[Hyper Aleph Ω (Aleph Ω, Aleph Ω), Hyper Aleph Ω (Aleph Ω, Aleph Ω)]}, Hyper Hyper Hyper Aleph Ω (Aleph Ω, Aleph Ω)[Hyper Aleph Ω (Aleph Ω, Aleph Ω), Hyper Aleph Ω (Aleph Ω, Aleph Ω)] {Hyper Hyper Aleph Ω (Aleph Ω, Aleph Ω)[Hyper Aleph Ω (Aleph Ω, Aleph Ω), Hyper Aleph Ω (Aleph Ω, Aleph Ω)], Hyper Hyper Aleph Ω (Aleph Ω, Aleph Ω)[Hyper Aleph Ω (Aleph Ω, Aleph Ω), Hyper Aleph Ω (Aleph Ω, Aleph Ω)]}}}, Hyper Hyper Hyper Hyper Hyper Aleph Ω (Aleph Ω, Aleph Ω)[Hyper Aleph Ω (Aleph Ω, Aleph Ω), Hyper Aleph Ω (Aleph Ω, Aleph Ω)] {Hyper Hyper Aleph Ω (Aleph Ω, Aleph Ω)[Hyper Aleph Ω (Aleph Ω, Aleph Ω), Hyper Aleph Ω (Aleph Ω, Aleph Ω)], Hyper Hyper Aleph Ω (Aleph Ω, Aleph Ω)[Hyper Aleph Ω (Aleph Ω, Aleph Ω), Hyper Aleph Ω (Aleph Ω, Aleph Ω)]} { Hyper Hyper Hyper Aleph Ω (Aleph Ω, Aleph Ω)[Hyper Aleph Ω (Aleph Ω, Aleph Ω), Hyper Aleph Ω (Aleph Ω, Aleph Ω)] {Hyper Hyper Aleph Ω (Aleph Ω, Aleph Ω)[Hyper Aleph Ω (Aleph Ω, Aleph Ω), Hyper Aleph Ω (Aleph Ω, Aleph Ω)], Hyper Hyper Aleph Ω (Aleph Ω, Aleph Ω)[Hyper Aleph Ω (Aleph Ω, Aleph Ω), Hyper Aleph Ω (Aleph Ω, Aleph Ω)]}, Hyper Hyper Hyper Aleph Ω (Aleph Ω, Aleph Ω)[Hyper Aleph Ω (Aleph Ω, Aleph Ω), Hyper Aleph Ω (Aleph Ω, Aleph Ω)] {Hyper Hyper Aleph Ω (Aleph Ω, Aleph Ω)[Hyper Aleph Ω (Aleph Ω, Aleph Ω), Hyper Aleph Ω (Aleph Ω, Aleph Ω)], Hyper Hyper Aleph Ω (Aleph Ω, Aleph Ω)[Hyper Aleph Ω (Aleph Ω, Aleph Ω), Hyper Aleph Ω (Aleph Ω, Aleph Ω)]}}{ Hyper Hyper Hyper Hyper Aleph Ω (Aleph Ω, Aleph Ω)[Hyper Aleph Ω (Aleph Ω, Aleph Ω), Hyper Aleph Ω (Aleph Ω, Aleph Ω)] {Hyper Hyper Aleph Ω (Aleph Ω, Aleph Ω)[Hyper Aleph Ω (Aleph Ω, Aleph Ω), Hyper Aleph Ω (Aleph Ω, Aleph Ω)], Hyper Hyper Aleph Ω (Aleph Ω, Aleph Ω)[Hyper Aleph Ω (Aleph Ω, Aleph Ω), Hyper Aleph Ω (Aleph Ω, Aleph Ω)]} { Hyper Hyper Hyper Aleph Ω (Aleph Ω, Aleph Ω)[Hyper Aleph Ω (Aleph Ω, Aleph Ω), Hyper Aleph Ω (Aleph Ω, Aleph Ω)] {Hyper Hyper Aleph Ω (Aleph Ω, Aleph Ω)[Hyper Aleph Ω (Aleph Ω, Aleph Ω), Hyper Aleph Ω (Aleph Ω, Aleph Ω)], Hyper Hyper Aleph Ω (Aleph Ω, Aleph Ω)[Hyper Aleph Ω (Aleph Ω, Aleph Ω), Hyper Aleph Ω (Aleph Ω, Aleph Ω)]}, Hyper Hyper Hyper Aleph Ω (Aleph Ω, Aleph Ω)[Hyper Aleph Ω (Aleph Ω, Aleph Ω), Hyper Aleph Ω (Aleph Ω, Aleph Ω)] {Hyper Hyper Aleph Ω (Aleph Ω, Aleph Ω)[Hyper Aleph Ω (Aleph Ω, Aleph Ω), Hyper Aleph Ω (Aleph Ω, Aleph Ω)], Hyper Hyper Aleph Ω (Aleph Ω, Aleph Ω)[Hyper Aleph Ω (Aleph Ω, Aleph Ω), Hyper Aleph Ω (Aleph Ω, Aleph Ω)]}}, Hyper Hyper Hyper Hyper Aleph Ω (Aleph Ω, Aleph Ω)[Hyper Aleph Ω

(Aleph Ω, Aleph Ω), Hyper Aleph Ω (Aleph Ω, Aleph Ω)] {Hyper Hyper Aleph Ω (Aleph Ω, Aleph Ω)[Hyper Aleph Ω (Aleph Ω, Aleph Ω), Hyper Aleph Ω (Aleph Ω, Aleph Ω)], Hyper Hyper Aleph Ω (Aleph Ω, Aleph Ω)[Hyper Aleph Ω (Aleph Ω, Aleph Ω), Hyper Aleph Ω (Aleph Ω, Aleph Ω)]} { Hyper Hyper Hyper Aleph Ω (Aleph Ω, Aleph Ω)[Hyper Aleph Ω (Aleph Ω, Aleph Ω), Hyper Aleph Ω (Aleph Ω, Aleph Ω)] {Hyper Hyper Aleph Ω (Aleph Ω, Aleph Ω)[Hyper Aleph Ω (Aleph Ω, Aleph Ω), Hyper Aleph Ω (Aleph Ω, Aleph Ω)], Hyper Hyper Aleph Ω (Aleph Ω, Aleph Ω)[Hyper Aleph Ω (Aleph Ω, Aleph Ω), Hyper Aleph Ω (Aleph Ω, Aleph Ω)]}, Hyper Hyper Hyper Aleph Ω (Aleph Ω, Aleph Ω)[Hyper Aleph Ω (Aleph Ω, Aleph Ω), Hyper Aleph Ω (Aleph Ω, Aleph Ω)] {Hyper Hyper Aleph Ω (Aleph Ω, Aleph Ω)[Hyper Aleph Ω (Aleph Ω, Aleph Ω), Hyper Aleph Ω (Aleph Ω, Aleph Ω)], Hyper Hyper Aleph Ω (Aleph Ω, Aleph Ω)[Hyper Aleph Ω (Aleph Ω, Aleph Ω), Hyper Aleph Ω (Aleph Ω, Aleph Ω)]}}}}, Hyper Hyper Hyper Hyper Hyper Hyper Aleph Ω (Aleph Ω, Aleph Ω)[Hyper Aleph Ω (Aleph Ω, Aleph Ω), Hyper Aleph Ω (Aleph Ω, Aleph Ω)] {Hyper Hyper Aleph Ω (Aleph Ω, Aleph Ω)[Hyper Aleph Ω (Aleph Ω, Aleph Ω), Hyper Aleph Ω (Aleph Ω, Aleph Ω)], Hyper Hyper Aleph Ω (Aleph Ω, Aleph Ω)[Hyper Aleph Ω (Aleph Ω, Aleph Ω), Hyper Aleph Ω (Aleph Ω, Aleph Ω)]} { Hyper Hyper Hyper Aleph Ω (Aleph Ω, Aleph Ω)[Hyper Aleph Ω (Aleph Ω, Aleph Ω), Hyper Aleph Ω (Aleph Ω, Aleph Ω)] {Hyper Hyper Aleph Ω (Aleph Ω, Aleph Ω)[Hyper Aleph Ω (Aleph Ω, Aleph Ω), Hyper Aleph Ω (Aleph Ω, Aleph Ω)], Hyper Hyper Aleph Ω (Aleph Ω, Aleph Ω)[Hyper Aleph Ω (Aleph Ω, Aleph Ω), Hyper Aleph Ω (Aleph Ω, Aleph Ω)]}, Hyper Hyper Hyper Aleph Ω (Aleph Ω, Aleph Ω)[Hyper Aleph Ω (Aleph Ω, Aleph Ω), Hyper Aleph Ω (Aleph Ω, Aleph Ω)] {Hyper Hyper Aleph Ω (Aleph Ω, Aleph Ω)[Hyper Aleph Ω (Aleph Ω, Aleph Ω), Hyper Aleph Ω (Aleph Ω, Aleph Ω)], Hyper Hyper Aleph Ω (Aleph Ω, Aleph Ω)[Hyper Aleph Ω (Aleph Ω, Aleph Ω), Hyper Aleph Ω (Aleph Ω, Aleph Ω)]}}{ Hyper Hyper Hyper Hyper Aleph Ω (Aleph Ω, Aleph Ω)[Hyper Aleph Ω (Aleph Ω, Aleph Ω), Hyper Aleph Ω (Aleph Ω, Aleph Ω)] {Hyper Hyper Aleph Ω (Aleph Ω, Aleph Ω)[Hyper Aleph Ω (Aleph Ω, Aleph Ω), Hyper Aleph Ω (Aleph Ω, Aleph Ω)], Hyper Hyper Aleph Ω (Aleph Ω, Aleph Ω)[Hyper Aleph Ω (Aleph Ω, Aleph Ω), Hyper Aleph Ω (Aleph Ω, Aleph Ω)]} { Hyper Hyper Hyper Aleph Ω (Aleph Ω, Aleph Ω)[Hyper Aleph Ω (Aleph Ω, Aleph Ω), Hyper Aleph Ω (Aleph Ω, Aleph Ω)] {Hyper Hyper Aleph Ω (Aleph Ω, Aleph Ω)[Hyper Aleph Ω (Aleph Ω, Aleph Ω), Hyper Aleph Ω (Aleph Ω, Aleph Ω)], Hyper Hyper Aleph Ω (Aleph Ω, Aleph Ω)[Hyper Aleph Ω (Aleph Ω, Aleph Ω), Hyper Aleph Ω (Aleph Ω, Aleph Ω)]}, Hyper Hyper Hyper Aleph Ω (Aleph Ω, Aleph Ω)[Hyper Aleph Ω (Aleph Ω, Aleph Ω), Hyper Aleph Ω (Aleph Ω, Aleph Ω)] {Hyper Hyper Aleph Ω (Aleph Ω, Aleph Ω)[Hyper Aleph Ω (Aleph Ω, Aleph Ω), Hyper Aleph Ω (Aleph Ω, Aleph Ω)], Hyper Hyper Aleph Ω (Aleph Ω, Aleph Ω)[Hyper Aleph Ω (Aleph Ω, Aleph Ω), Hyper Aleph Ω (Aleph Ω, Aleph Ω)]}}, Hyper Hyper Hyper Hyper Aleph Ω (Aleph Ω, Aleph Ω)[Hyper Aleph Ω (Aleph Ω, Aleph Ω), Hyper Aleph Ω (Aleph Ω, Aleph Ω)] {Hyper Hyper Aleph Ω (Aleph Ω, Aleph Ω)[Hyper Aleph Ω (Aleph Ω, Aleph Ω), Hyper Aleph Ω (Aleph Ω, Aleph Ω)], Hyper Hyper Aleph Ω (Aleph Ω, Aleph Ω)[Hyper Aleph Ω (Aleph Ω, Aleph Ω), Hyper Aleph Ω (Aleph Ω, Aleph Ω)]} { Hyper

Hyper Hyper Aleph Ω (Aleph Ω, Aleph Ω)[Hyper Aleph Ω (Aleph Ω, Aleph Ω), Hyper Aleph Ω (Aleph Ω, Aleph Ω)] {Hyper Hyper Aleph Ω (Aleph Ω, Aleph Ω)[Hyper Aleph Ω (Aleph Ω, Aleph Ω), Hyper Aleph Ω (Aleph Ω, Aleph Ω)], Hyper Hyper Aleph Ω (Aleph Ω, Aleph Ω)[Hyper Aleph Ω (Aleph Ω, Aleph Ω), Hyper Aleph Ω (Aleph Ω, Aleph Ω)]}, Hyper Hyper Hyper Aleph Ω (Aleph Ω, Aleph Ω)[Hyper Aleph Ω (Aleph Ω, Aleph Ω), Hyper Aleph Ω (Aleph Ω, Aleph Ω)] {Hyper Hyper Aleph Ω (Aleph Ω, Aleph Ω)[Hyper Aleph Ω (Aleph Ω, Aleph Ω), Hyper Aleph Ω (Aleph Ω, Aleph Ω)], Hyper Hyper Aleph Ω (Aleph Ω, Aleph Ω)[Hyper Aleph Ω (Aleph Ω, Aleph Ω), Hyper Aleph Ω (Aleph Ω, Aleph Ω)]}}}{ Hyper Hyper Hyper Hyper Hyper Aleph Ω (Aleph Ω, Aleph Ω)[Hyper Aleph Ω (Aleph Ω, Aleph Ω), Hyper Aleph Ω (Aleph Ω, Aleph Ω)] {Hyper Hyper Aleph Ω (Aleph Ω, Aleph Ω)[Hyper Aleph Ω (Aleph Ω, Aleph Ω), Hyper Aleph Ω (Aleph Ω, Aleph Ω)], Hyper Hyper Aleph Ω (Aleph Ω, Aleph Ω)[Hyper Aleph Ω (Aleph Ω, Aleph Ω), Hyper Aleph Ω (Aleph Ω, Aleph Ω)]} { Hyper Hyper Hyper Aleph Ω (Aleph Ω, Aleph Ω)[Hyper Aleph Ω (Aleph Ω, Aleph Ω), Hyper Aleph Ω (Aleph Ω, Aleph Ω)] {Hyper Hyper Aleph Ω (Aleph Ω, Aleph Ω)[Hyper Aleph Ω (Aleph Ω, Aleph Ω), Hyper Aleph Ω (Aleph Ω, Aleph Ω)], Hyper Hyper Aleph Ω (Aleph Ω, Aleph Ω)[Hyper Aleph Ω (Aleph Ω, Aleph Ω), Hyper Aleph Ω (Aleph Ω, Aleph Ω)]}, Hyper Hyper Hyper Aleph Ω (Aleph Ω, Aleph Ω)[Hyper Aleph Ω (Aleph Ω, Aleph Ω), Hyper Aleph Ω (Aleph Ω, Aleph Ω)] {Hyper Hyper Aleph Ω (Aleph Ω, Aleph Ω)[Hyper Aleph Ω (Aleph Ω, Aleph Ω), Hyper Aleph Ω (Aleph Ω, Aleph Ω)], Hyper Hyper Aleph Ω (Aleph Ω, Aleph Ω)[Hyper Aleph Ω (Aleph Ω, Aleph Ω), Hyper Aleph Ω (Aleph Ω, Aleph Ω)]}}{ Hyper Hyper Hyper Hyper Aleph Ω (Aleph Ω, Aleph Ω)[Hyper Aleph Ω (Aleph Ω, Aleph Ω), Hyper Aleph Ω (Aleph Ω, Aleph Ω)] {Hyper Hyper Aleph Ω (Aleph Ω, Aleph Ω)[Hyper Aleph Ω (Aleph Ω, Aleph Ω), Hyper Aleph Ω (Aleph Ω, Aleph Ω)], Hyper Hyper Aleph Ω (Aleph Ω, Aleph Ω)[Hyper Aleph Ω (Aleph Ω, Aleph Ω), Hyper Aleph Ω (Aleph Ω, Aleph Ω)]} { Hyper Hyper Hyper Aleph Ω (Aleph Ω, Aleph Ω)[Hyper Aleph Ω (Aleph Ω, Aleph Ω), Hyper Aleph Ω (Aleph Ω, Aleph Ω)] {Hyper Hyper Aleph Ω (Aleph Ω, Aleph Ω)[Hyper Aleph Ω (Aleph Ω, Aleph Ω), Hyper Aleph Ω (Aleph Ω, Aleph Ω)], Hyper Hyper Aleph Ω (Aleph Ω, Aleph Ω)[Hyper Aleph Ω (Aleph Ω, Aleph Ω), Hyper Aleph Ω (Aleph Ω, Aleph Ω)]}, Hyper Hyper Hyper Aleph Ω (Aleph Ω, Aleph Ω)[Hyper Aleph Ω (Aleph Ω, Aleph Ω), Hyper Aleph Ω (Aleph Ω, Aleph Ω)] {Hyper Hyper Aleph Ω (Aleph Ω, Aleph Ω)[Hyper Aleph Ω (Aleph Ω, Aleph Ω), Hyper Aleph Ω (Aleph Ω, Aleph Ω)], Hyper Hyper Aleph Ω (Aleph Ω, Aleph Ω)[Hyper Aleph Ω (Aleph Ω, Aleph Ω), Hyper Aleph Ω (Aleph Ω, Aleph Ω)]}}, Hyper Hyper Hyper Hyper Aleph Ω (Aleph Ω, Aleph Ω)[Hyper Aleph Ω (Aleph Ω, Aleph Ω), Hyper Aleph Ω (Aleph Ω, Aleph Ω)] {Hyper Hyper Aleph Ω (Aleph Ω, Aleph Ω)[Hyper Aleph Ω (Aleph Ω, Aleph Ω), Hyper Aleph Ω (Aleph Ω, Aleph Ω)], Hyper Hyper Aleph Ω (Aleph Ω, Aleph Ω)[Hyper Aleph Ω (Aleph Ω, Aleph Ω), Hyper Aleph Ω (Aleph Ω, Aleph Ω)]} { Hyper Hyper Hyper Aleph Ω (Aleph Ω, Aleph Ω)[Hyper Aleph Ω (Aleph Ω, Aleph Ω), Hyper Aleph Ω (Aleph Ω, Aleph Ω)] {Hyper Hyper Aleph Ω (Aleph Ω, Aleph Ω)[Hyper Aleph Ω (Aleph Ω, Aleph Ω), Hyper Aleph Ω (Aleph Ω, Aleph Ω)], Hyper Hyper Aleph Ω (Aleph Ω, Aleph Ω)[Hyper Aleph Ω (Aleph Ω,

Aleph Ω), Hyper Aleph Ω (Aleph Ω, Aleph Ω)]}, Hyper Hyper Hyper Aleph Ω (Aleph Ω, Aleph Ω)[Hyper Aleph Ω (Aleph Ω, Aleph Ω), Hyper Aleph Ω (Aleph Ω, Aleph Ω)] {Hyper Hyper Aleph Ω (Aleph Ω, Aleph Ω)[Hyper Aleph Ω (Aleph Ω, Aleph Ω), Hyper Aleph Ω (Aleph Ω, Aleph Ω)], Hyper Hyper Aleph Ω (Aleph Ω, Aleph Ω)[Hyper Aleph Ω (Aleph Ω, Aleph Ω), Hyper Aleph Ω (Aleph Ω, Aleph Ω)]}}}, Hyper Hyper Hyper Hyper Hyper Aleph Ω (Aleph Ω, Aleph Ω)[Hyper Aleph Ω (Aleph Ω, Aleph Ω), Hyper Aleph Ω (Aleph Ω, Aleph Ω)] {Hyper Hyper Aleph Ω (Aleph Ω, Aleph Ω)[Hyper Aleph Ω (Aleph Ω, Aleph Ω), Hyper Aleph Ω (Aleph Ω, Aleph Ω)], Hyper Hyper Aleph Ω (Aleph Ω, Aleph Ω)[Hyper Aleph Ω (Aleph Ω, Aleph Ω), Hyper Aleph Ω (Aleph Ω, Aleph Ω)]} { Hyper Hyper Hyper Aleph Ω (Aleph Ω, Aleph Ω)[Hyper Aleph Ω (Aleph Ω, Aleph Ω), Hyper Aleph Ω (Aleph Ω, Aleph Ω)] {Hyper Hyper Aleph Ω (Aleph Ω, Aleph Ω)[Hyper Aleph Ω (Aleph Ω, Aleph Ω), Hyper Aleph Ω (Aleph Ω, Aleph Ω)], Hyper Hyper Aleph Ω (Aleph Ω, Aleph Ω)[Hyper Aleph Ω (Aleph Ω, Aleph Ω), Hyper Aleph Ω (Aleph Ω, Aleph Ω)]}, Hyper Hyper Hyper Aleph Ω (Aleph Ω, Aleph Ω)[Hyper Aleph Ω (Aleph Ω, Aleph Ω), Hyper Aleph Ω (Aleph Ω, Aleph Ω)] {Hyper Hyper Aleph Ω (Aleph Ω, Aleph Ω)[Hyper Aleph Ω (Aleph Ω, Aleph Ω), Hyper Aleph Ω (Aleph Ω, Aleph Ω)], Hyper Hyper Aleph Ω (Aleph Ω, Aleph Ω)[Hyper Aleph Ω (Aleph Ω, Aleph Ω), Hyper Aleph Ω (Aleph Ω, Aleph Ω)]}}{ Hyper Hyper Hyper Hyper Aleph Ω (Aleph Ω, Aleph Ω)[Hyper Aleph Ω (Aleph Ω, Aleph Ω), Hyper Aleph Ω (Aleph Ω, Aleph Ω)] {Hyper Hyper Aleph Ω (Aleph Ω, Aleph Ω)[Hyper Aleph Ω (Aleph Ω, Aleph Ω), Hyper Aleph Ω (Aleph Ω, Aleph Ω)], Hyper Hyper Aleph Ω (Aleph Ω, Aleph Ω)[Hyper Aleph Ω (Aleph Ω, Aleph Ω), Hyper Aleph Ω (Aleph Ω, Aleph Ω)]} { Hyper Hyper Hyper Aleph Ω (Aleph Ω, Aleph Ω)[Hyper Aleph Ω (Aleph Ω, Aleph Ω), Hyper Aleph Ω (Aleph Ω, Aleph Ω)] {Hyper Hyper Aleph Ω (Aleph Ω, Aleph Ω)[Hyper Aleph Ω (Aleph Ω, Aleph Ω), Hyper Aleph Ω (Aleph Ω, Aleph Ω)], Hyper Hyper Aleph Ω (Aleph Ω, Aleph Ω)[Hyper Aleph Ω (Aleph Ω, Aleph Ω), Hyper Aleph Ω (Aleph Ω, Aleph Ω)]}, Hyper Hyper Hyper Aleph Ω (Aleph Ω, Aleph Ω)[Hyper Aleph Ω (Aleph Ω, Aleph Ω), Hyper Aleph Ω (Aleph Ω, Aleph Ω)] {Hyper Hyper Aleph Ω (Aleph Ω, Aleph Ω)[Hyper Aleph Ω (Aleph Ω, Aleph Ω), Hyper Aleph Ω (Aleph Ω, Aleph Ω)], Hyper Hyper Aleph Ω (Aleph Ω, Aleph Ω)[Hyper Aleph Ω (Aleph Ω, Aleph Ω), Hyper Aleph Ω (Aleph Ω, Aleph Ω)]}}, Hyper Hyper Hyper Hyper Aleph Ω (Aleph Ω, Aleph Ω)[Hyper Aleph Ω (Aleph Ω, Aleph Ω), Hyper Aleph Ω (Aleph Ω, Aleph Ω)] {Hyper Hyper Aleph Ω (Aleph Ω, Aleph Ω)[Hyper Aleph Ω (Aleph Ω, Aleph Ω), Hyper Aleph Ω (Aleph Ω, Aleph Ω)], Hyper Hyper Aleph Ω (Aleph Ω, Aleph Ω)[Hyper Aleph Ω (Aleph Ω, Aleph Ω), Hyper Aleph Ω (Aleph Ω, Aleph Ω)]} { Hyper Hyper Hyper Aleph Ω (Aleph Ω, Aleph Ω)[Hyper Aleph Ω (Aleph Ω, Aleph Ω), Hyper Aleph Ω (Aleph Ω, Aleph Ω)] {Hyper Hyper Aleph Ω (Aleph Ω, Aleph Ω)[Hyper Aleph Ω (Aleph Ω, Aleph Ω), Hyper Aleph Ω (Aleph Ω, Aleph Ω)], Hyper Hyper Aleph Ω (Aleph Ω, Aleph Ω)[Hyper Aleph Ω (Aleph Ω, Aleph Ω), Hyper Aleph Ω (Aleph Ω, Aleph Ω)]}, Hyper Hyper Hyper Aleph Ω (Aleph Ω, Aleph Ω)[Hyper Aleph Ω (Aleph Ω, Aleph Ω), Hyper Aleph Ω (Aleph Ω, Aleph Ω)] {Hyper Hyper Aleph Ω (Aleph Ω, Aleph Ω)[Hyper Aleph Ω (Aleph Ω, Aleph Ω), Hyper Aleph Ω (Aleph Ω, Aleph Ω)], Hyper Hyper

Aleph Ω (Aleph Ω, Aleph Ω)[Hyper Aleph Ω (Aleph Ω, Aleph Ω), Hyper Aleph Ω (Aleph Ω, Aleph Ω)]}}}}}

In reality, we will not simply say that we can use all of the manufacturable computer storage media from materials in our cosmic light-cone to store the largest expression analog of the above expression that can be developed in stored on said storage media, but instead we can consider building out the above style of expression to ones that are so big that they would take up all possible computer storage media manufacturable from the entire mineral content of our universe, of which only a tiny portion is visible or within our cosmic light-cone.

Accordingly, such a huge infinity gives you a hint at how big infinities can be.

Such a huge infinity can give you a glimpse about how many ontological transcendental GOD has in HIS ESSENCE. I believe GOD has at least as many ontological transcendentals as the above explicitly formulated infinity but also even more, So much more that it would take all eternity to develop an analogous expression to define the infinite number of ontological transcendentals GOD has,

Ontological transcendentals are properties such as ontological goodness, value, purpose, reason for existing.

In human persons, the above four ontological transcendentals are about all we can think of but perhaps we could throw in a few more if we held a think tank meeting to come up with them.

Note that the symbol Ω can denote an infinity as large as we would like.

Now again, within the observable universe, there are about 10 EXP 24 stars. This is about equal to the number of fine grains of table sugar that would cover the entire United States 100 meters deep. The number of planets orbiting stars in our universe seems to be about 10 times greater yet or equal to the number of fine grains of table sugar that would cover the entire United States 1,000 meters deep. The number of moons, orbiting planets, orbiting stars is estimated to be ten times greater yet or equal to the number of fine grains of table sugar that would cover the entire United States 10,000 meters deep.

So let us go boldly into the future to explore the awesome realm GOD has made. The next 13.78 billion years will keep us very busy as we travel distances from Earth equal to one current cosmic light-cone radius unit. However, our universe extends far beyond the observable portion.

Now, we can consider analogs of the above infinite expressions but by replacing Ω with a high-end uncountable infinity which is must greater than Ω by far.

Now here is a real zinger.

So we can produce the abstraction of Hyper Aleph (high-end uncountable infinity) (Aleph (high-end uncountable infinity), Aleph (high-end uncountable infinity)).

Infinities can be as big as we would like. For example, consider the following infinity.

Hyper Hyper Aleph (high-end uncountable infinity) (Aleph (high-end uncountable infinity), Aleph (high-end uncountable infinity))[Hyper Aleph (high-end uncountable infinity) (Aleph (high-end uncountable infinity), Aleph (high-end uncountable infinity)), Hyper Aleph (high-end uncountable infinity) (Aleph (high-end uncountable infinity), Aleph (high-end uncountable infinity))]

We can go still larger to develop the following infinity.

Hyper Hyper Hyper Aleph (high-end uncountable infinity) (Aleph (high-end uncountable infinity), Aleph (high-end uncountable infinity))[Hyper Aleph (high-end uncountable infinity) (Aleph (high-end uncountable infinity), Aleph (high-end uncountable infinity)), Hyper Aleph (high-end uncountable infinity) (Aleph (high-end uncountable infinity), Aleph (high-end uncountable infinity))] {Hyper Hyper Aleph (high-end uncountable infinity) (Aleph (high-end uncountable infinity), Aleph (high-end uncountable infinity))[Hyper Aleph (high-end uncountable infinity) (Aleph (high-end uncountable infinity), Aleph (high-end uncountable infinity)), Hyper Aleph (high-end uncountable infinity) (Aleph (high-end uncountable infinity), Aleph (high-end uncountable infinity))], Hyper Hyper Aleph (high-end uncountable infinity) (Aleph (high-end uncountable infinity), Aleph (high-end uncountable infinity))[Hyper Aleph (high-end uncountable infinity) (Aleph (high-end uncountable infinity), Aleph (high-end uncountable infinity)), Hyper Aleph (high-end uncountable infinity) (Aleph (high-end uncountable infinity), Aleph (high-end uncountable infinity))]}

We can go still larger to develop the following infinity.

Hyper Hyper Hyper Hyper Aleph (high-end uncountable infinity) (Aleph (high-end uncountable infinity), Aleph (high-end uncountable infinity))[Hyper Aleph (high-end uncountable infinity) (Aleph (high-end uncountable infinity), Aleph (high-end uncountable infinity)), Hyper Aleph (high-end uncountable infinity) (Aleph (high-end uncountable infinity), Aleph (high-end uncountable infinity))] {Hyper Hyper Aleph (high-end uncountable infinity) (Aleph (high-end uncountable infinity), Aleph (high-end uncountable infinity))[Hyper Aleph (high-end uncountable infinity) (Aleph (high-end uncountable infinity), Aleph (high-end uncountable infinity)), Hyper Aleph (high-end uncountable infinity) (Aleph (high-end uncountable infinity), Aleph (high-end uncountable infinity))], Hyper Hyper Aleph (high-end uncountable infinity) (Aleph (high-end uncountable infinity), Aleph (high-end uncountable infinity))[Hyper Aleph (high-end uncountable infinity) (Aleph (high-end uncountable infinity), Aleph (high-end uncountable infinity)), Hyper Aleph (high-end uncountable infinity) (Aleph (high-end uncountable infinity), Aleph (high-end uncountable infinity))]} { Hyper Hyper Hyper Aleph (high-end uncountable infinity) (Aleph (high-end uncountable infinity), Aleph (high-end uncountable infinity))[Hyper Aleph (high-end uncountable infinity) (Aleph (high-end uncountable infinity), Aleph (high-end uncountable infinity)), Hyper Aleph (high-end uncountable infinity) (Aleph (high-end uncountable infinity), Aleph (high-end uncountable infinity))] {Hyper Hyper Aleph (high-end uncountable infinity) (Aleph (high-end uncountable infinity), Aleph (high-end uncountable infinity))[Hyper Aleph (high-end uncountable infinity) (Aleph (high-end uncountable infinity), Aleph (high-end uncountable infinity)), Hyper Aleph (high-end uncountable infinity) (Aleph (high-end uncountable infinity), Aleph (high-end uncountable infinity))], Hyper Hyper Aleph (high-end uncountable infinity) (Aleph (high-end uncountable infinity), Aleph (high-end uncountable infinity))[Hyper Aleph (high-end uncountable infinity) (Aleph (high-end uncountable infinity), Aleph (high-end uncountable infinity)), Hyper Aleph (high-end uncountable infinity) (Aleph (high-end uncountable infinity), Aleph (high-end uncountable infinity))]}, Hyper Hyper Hyper Aleph (high-end uncountable infinity) (Aleph (high-end uncountable infinity), Aleph (high-end uncountable infinity))[Hyper Aleph (high-end uncountable infinity) (Aleph (high-end uncountable infinity), Aleph (high-end uncountable infinity)), Hyper Aleph (high-end uncountable infinity) (Aleph (high-end uncountable infinity), Aleph (high-end uncountable infinity))] {Hyper Hyper Aleph (high-end uncountable infinity) (Aleph (high-end uncountable infinity), Aleph (high-end uncountable infinity))[Hyper Aleph (high-end uncountable infinity) (Aleph (high-end uncountable infinity), Aleph (high-end uncountable infinity)), Hyper Aleph (high-end uncountable infinity) (Aleph (high-end uncountable infinity), Aleph (high-end uncountable infinity))], Hyper Hyper Aleph (high-end uncountable infinity) (Aleph (high-end uncountable infinity), Aleph (high-end uncountable infinity))[Hyper Aleph (high-end uncountable infinity) (Aleph (high-end uncountable infinity), Aleph (high-end uncountable infinity)), Hyper Aleph (high-end uncountable infinity) (Aleph (high-end uncountable infinity), Aleph (high-end uncountable infinity))]}}

We can go still larger to develop the following infinity.

Hyper Hyper Hyper Hyper Hyper Aleph (high-end uncountable infinity) (Aleph (high-end uncountable infinity), Aleph (high-end uncountable infinity))[Hyper Aleph (high-end uncountable infinity) (Aleph (high-end uncountable infinity), Aleph (high-end uncountable infinity)), Hyper Aleph (high-end uncountable infinity) (Aleph (high-end uncountable infinity), Aleph (high-end uncountable infinity))] {Hyper Hyper Aleph (high-end uncountable infinity) (Aleph (high-end uncountable infinity), Aleph (high-end uncountable infinity))[Hyper Aleph (high-end uncountable infinity) (Aleph (high-end uncountable infinity), Aleph (high-end uncountable infinity)), Hyper Aleph (high-end uncountable infinity) (Aleph (high-end uncountable infinity), Aleph (high-end uncountable infinity))], Hyper Hyper Aleph (high-end uncountable infinity) (Aleph (high-end uncountable infinity), Aleph (high-end uncountable infinity))[Hyper Aleph (high-end uncountable infinity) (Aleph (high-end uncountable infinity), Aleph (high-end uncountable infinity)), Hyper Aleph (high-end uncountable infinity) (Aleph (high-end uncountable infinity), Aleph (high-end uncountable infinity))]} { Hyper Hyper Hyper Aleph (high-end uncountable infinity) (Aleph (high-end uncountable infinity), Aleph (high-end uncountable infinity))[Hyper Aleph (high-end uncountable infinity) (Aleph (high-end uncountable infinity), Aleph (high-end uncountable infinity)), Hyper Aleph (high-end uncountable infinity) (Aleph (high-end uncountable infinity), Aleph (high-end uncountable infinity))] {Hyper Hyper Aleph (high-end uncountable infinity) (Aleph (high-end uncountable infinity), Aleph (high-end uncountable infinity))[Hyper Aleph (high-end uncountable infinity) (Aleph (high-end uncountable infinity), Aleph (high-end uncountable infinity)), Hyper Aleph (high-end uncountable infinity) (Aleph (high-end uncountable infinity), Aleph (high-end uncountable infinity))], Hyper Hyper Aleph (high-end uncountable infinity) (Aleph (high-end uncountable infinity), Aleph (high-end uncountable infinity))[Hyper Aleph (high-end uncountable infinity) (Aleph (high-end uncountable infinity), Aleph (high-end uncountable infinity)), Hyper Aleph (high-end uncountable infinity) (Aleph (high-end uncountable infinity), Aleph (high-end uncountable infinity))]}, Hyper Hyper Hyper Aleph (high-end uncountable infinity) (Aleph (high-end uncountable infinity), Aleph (high-end uncountable infinity))[Hyper Aleph (high-end uncountable infinity) (Aleph (high-end uncountable infinity), Aleph (high-end uncountable infinity)), Hyper Aleph (high-end uncountable infinity) (Aleph (high-end uncountable infinity), Aleph (high-end uncountable infinity))] {Hyper Hyper Aleph (high-end uncountable infinity) (Aleph (high-end uncountable infinity), Aleph (high-end uncountable infinity))[Hyper Aleph (high-end uncountable infinity) (Aleph (high-end uncountable infinity), Aleph (high-end uncountable infinity)), Hyper Aleph (high-end uncountable infinity) (Aleph (high-end uncountable infinity), Aleph (high-end uncountable infinity))], Hyper Hyper Aleph (high-end uncountable infinity) (Aleph (high-end uncountable infinity), Aleph (high-end uncountable infinity))[Hyper Aleph (high-end uncountable infinity) (Aleph (high-end uncountable infinity), Aleph (high-end uncountable infinity)), Hyper Aleph (high-end uncountable infinity) (Aleph (high-end uncountable infinity), Aleph (high-end uncountable infinity))]}}{ Hyper Hyper Hyper Hyper Aleph (high-end uncountable infinity) (Aleph (high-end uncountable infinity), Aleph (high-end uncountable infinity))[Hyper Aleph (high-end uncountable infinity) (Aleph (high-end uncountable infinity), Aleph (high-end uncountable infinity)), Hyper Aleph (high-end uncountable infinity) (Aleph (high-end uncountable infinity), Aleph (high-end uncountable infinity))] {Hyper Hyper Aleph (high-end uncountable infinity) (Aleph (high-end uncountable infinity), Aleph (high-end uncountable infinity))[Hyper Aleph (high-end uncountable infinity) (Aleph (high-end uncountable infinity), Aleph (high-end uncountable infinity)), Hyper Aleph (high-end uncountable infinity) (Aleph (high-end uncountable infinity), Aleph (high-end uncountable infinity))], Hyper Hyper Aleph (high-end uncountable infinity) (Aleph (high-end uncountable infinity), Aleph (high-end uncountable infinity))[Hyper Aleph (high-end uncountable infinity) (Aleph (high-end uncountable infinity), Aleph (high-end uncountable infinity)), Hyper Aleph (high-end uncountable infinity) (Aleph (high-end uncountable infinity), Aleph (high-end uncountable infinity))]} { Hyper Hyper Hyper Aleph (high-end uncountable infinity) (Aleph (high-end uncountable infinity), Aleph (high-end uncountable infinity))[Hyper Aleph (high-end uncountable infinity) (Aleph (high-end uncountable infinity), Aleph (high-end uncountable infinity)), Hyper Aleph (high-end uncountable infinity) (Aleph (high-end uncountable infinity), Aleph (high-end uncountable infinity))] {Hyper Hyper Aleph (high-end uncountable infinity) (Aleph (high-end uncountable infinity), Aleph (high-end uncountable infinity))[Hyper Aleph (high-end uncountable infinity) (Aleph (high-end uncountable infinity), Aleph (high-end uncountable infinity)), Hyper Aleph (high-end uncountable infinity) (Aleph (high-end uncountable infinity), Aleph (high-end uncountable infinity))], Hyper Hyper Aleph (high-end uncountable infinity) (Aleph (high-end uncountable infinity), Aleph (high-end uncountable infinity))[Hyper Aleph (high-end uncountable infinity) (Aleph (high-end uncountable infinity), Aleph (high-end uncountable infinity)), Hyper Aleph (high-end uncountable infinity) (Aleph (high-end uncountable infinity), Aleph (high-end uncountable infinity))]}, Hyper Hyper Hyper Aleph (high-end uncountable infinity) (Aleph (high-end uncountable infinity), Aleph (high-end uncountable infinity))[Hyper Aleph (high-end uncountable infinity) (Aleph (high-end uncountable infinity), Aleph (high-end uncountable infinity)), Hyper Aleph (high-end uncountable infinity) (Aleph (high-end uncountable infinity), Aleph (high-end uncountable infinity))] {Hyper Hyper Aleph (high-end uncountable infinity) (Aleph (high-end uncountable infinity), Aleph (high-end uncountable infinity))[Hyper Aleph (high-end uncountable infinity) (Aleph (high-end uncountable infinity), Aleph (high-end uncountable infinity)), Hyper Aleph (high-end uncountable infinity) (Aleph (high-end uncountable infinity), Aleph (high-end uncountable infinity))], Hyper Hyper Aleph (high-end uncountable infinity) (Aleph (high-end uncountable infinity), Aleph (high-end uncountable infinity))[Hyper Aleph (high-end uncountable infinity) (Aleph (high-end uncountable infinity), Aleph (high-end uncountable infinity)), Hyper Aleph (high-end uncountable infinity) (Aleph (high-end uncountable infinity), Aleph (high-end uncountable infinity))]}}, Hyper Hyper Hyper Hyper Aleph (high-end uncountable infinity) (Aleph (high-end uncountable infinity), Aleph (high-end uncountable infinity))[Hyper Aleph (high-end uncountable infinity) (Aleph (high-end uncountable infinity), Aleph (high-end uncountable infinity)), Hyper Aleph (high-end uncountable infinity) (Aleph (high-end uncountable infinity), Aleph (high-end uncountable infinity))] {Hyper Hyper Aleph (high-end uncountable infinity) (Aleph (high-end uncountable infinity), Aleph (high-end uncountable infinity))[Hyper Aleph (high-end uncountable infinity) (Aleph (high-end uncountable infinity), Aleph (high-end uncountable infinity)), Hyper Aleph (high-end uncountable infinity) (Aleph (high-end uncountable infinity), Aleph (high-end uncountable infinity))], Hyper Hyper Aleph (high-end uncountable infinity) (Aleph (high-end uncountable infinity), Aleph (high-end uncountable infinity))[Hyper Aleph (high-end uncountable infinity) (Aleph (high-end uncountable infinity), Aleph (high-end uncountable infinity)), Hyper Aleph (high-end uncountable infinity) (Aleph (high-end uncountable infinity), Aleph (high-end uncountable infinity))]} { Hyper Hyper Hyper Aleph (high-end uncountable infinity) (Aleph (high-end uncountable infinity), Aleph (high-end uncountable infinity))[Hyper Aleph (high-end uncountable infinity) (Aleph (high-end uncountable infinity), Aleph (high-end uncountable infinity)), Hyper Aleph (high-end uncountable infinity) (Aleph (high-end uncountable infinity), Aleph (high-end uncountable infinity))] {Hyper Hyper Aleph (high-end uncountable infinity) (Aleph (high-end uncountable infinity), Aleph (high-end uncountable infinity))[Hyper Aleph (high-end uncountable infinity) (Aleph (high-end uncountable infinity), Aleph (high-end uncountable infinity)), Hyper Aleph (high-end uncountable infinity) (Aleph (high-end uncountable infinity), Aleph (high-end uncountable infinity))], Hyper Hyper Aleph (high-end uncountable infinity) (Aleph (high-end uncountable infinity), Aleph (high-end uncountable infinity))[Hyper Aleph (high-end uncountable infinity) (Aleph (high-end uncountable infinity), Aleph (high-end uncountable infinity)), Hyper Aleph (high-end uncountable infinity) (Aleph (high-end uncountable infinity), Aleph (high-end uncountable infinity))]}, Hyper Hyper Hyper Aleph (high-end uncountable infinity) (Aleph (high-end uncountable infinity), Aleph (high-end uncountable infinity))[Hyper Aleph (high-end uncountable infinity) (Aleph (high-end uncountable infinity), Aleph (high-end uncountable infinity)), Hyper Aleph (high-end uncountable infinity) (Aleph (high-end uncountable infinity), Aleph (high-end uncountable infinity))] {Hyper Hyper Aleph (high-end uncountable infinity) (Aleph (high-end uncountable infinity), Aleph (high-end uncountable infinity))[Hyper Aleph (high-end uncountable infinity) (Aleph (high-end uncountable infinity), Aleph (high-end uncountable infinity)), Hyper Aleph (high-end uncountable infinity) (Aleph (high-end uncountable infinity), Aleph (high-end uncountable infinity))], Hyper Hyper Aleph (high-end uncountable infinity) (Aleph (high-end uncountable infinity), Aleph (high-end uncountable infinity))[Hyper Aleph (high-end uncountable infinity) (Aleph (high-end uncountable infinity), Aleph (high-end uncountable infinity)), Hyper Aleph (high-end uncountable infinity) (Aleph (high-end uncountable infinity), Aleph (high-end uncountable infinity))]}}}

We can go still larger to develop the following infinity.

Hyper Hyper Hyper Hyper Hyper Hyper Aleph (high-end uncountable infinity) (Aleph (high-end uncountable infinity), Aleph (high-end uncountable infinity))[Hyper Aleph (high-end uncountable infinity) (Aleph (high-end uncountable infinity), Aleph (high-end uncountable infinity)), Hyper Aleph (high-end uncountable infinity) (Aleph (high-end uncountable infinity), Aleph (high-end uncountable infinity))] {Hyper Hyper Aleph (high-end uncountable infinity) (Aleph (high-end uncountable infinity), Aleph (high-end uncountable infinity))[Hyper Aleph (high-end uncountable infinity) (Aleph (high-end uncountable infinity), Aleph (high-end uncountable infinity)), Hyper Aleph (high-end uncountable infinity) (Aleph (high-end uncountable infinity), Aleph (high-end uncountable infinity))], Hyper Hyper Aleph (high-end uncountable infinity) (Aleph (high-end uncountable infinity), Aleph (high-end uncountable infinity))[Hyper Aleph (high-end uncountable infinity) (Aleph (high-end uncountable infinity), Aleph (high-end uncountable infinity)), Hyper Aleph (high-end uncountable infinity) (Aleph (high-end uncountable infinity), Aleph (high-end uncountable infinity))]} { Hyper Hyper Hyper Aleph (high-end uncountable infinity) (Aleph (high-end uncountable infinity), Aleph (high-end uncountable infinity))[Hyper Aleph (high-end uncountable infinity) (Aleph (high-end uncountable infinity), Aleph (high-end uncountable infinity)), Hyper Aleph (high-end uncountable infinity) (Aleph (high-end uncountable infinity), Aleph (high-end uncountable infinity))] {Hyper Hyper Aleph (high-end uncountable infinity) (Aleph (high-end uncountable infinity), Aleph (high-end uncountable infinity))[Hyper Aleph (high-end uncountable infinity) (Aleph (high-end uncountable infinity), Aleph (high-end uncountable infinity)), Hyper Aleph (high-end uncountable infinity) (Aleph (high-end uncountable infinity), Aleph (high-end uncountable infinity))], Hyper Hyper Aleph (high-end uncountable infinity) (Aleph (high-end uncountable infinity), Aleph (high-end uncountable infinity))[Hyper Aleph (high-end uncountable infinity) (Aleph (high-end uncountable infinity), Aleph (high-end uncountable infinity)), Hyper Aleph (high-end uncountable infinity) (Aleph (high-end uncountable infinity), Aleph (high-end uncountable infinity))]}, Hyper Hyper Hyper Aleph (high-end uncountable infinity) (Aleph (high-end uncountable infinity), Aleph (high-end uncountable infinity))[Hyper Aleph (high-end uncountable infinity) (Aleph (high-end uncountable infinity), Aleph (high-end uncountable infinity)), Hyper Aleph (high-end uncountable infinity) (Aleph (high-end uncountable infinity), Aleph (high-end uncountable infinity))] {Hyper Hyper Aleph (high-end uncountable infinity) (Aleph (high-end uncountable infinity), Aleph (high-end uncountable infinity))[Hyper Aleph (high-end uncountable infinity) (Aleph (high-end uncountable infinity), Aleph (high-end uncountable infinity)), Hyper Aleph (high-end uncountable infinity) (Aleph (high-end uncountable infinity), Aleph (high-end uncountable infinity))], Hyper Hyper Aleph (high-end uncountable infinity) (Aleph (high-end uncountable infinity), Aleph (high-end uncountable infinity))[Hyper Aleph (high-end uncountable infinity) (Aleph (high-end uncountable infinity), Aleph (high-end uncountable infinity)), Hyper Aleph (high-end uncountable infinity) (Aleph (high-end uncountable infinity), Aleph (high-end uncountable infinity))]}}{ Hyper Hyper Hyper Hyper Aleph (high-end uncountable infinity) (Aleph (high-end uncountable infinity), Aleph (high-end uncountable infinity))[Hyper Aleph (high-end uncountable infinity) (Aleph (high-end uncountable infinity), Aleph (high-end uncountable infinity)), Hyper Aleph (high-end uncountable infinity) (Aleph (high-end uncountable infinity), Aleph (high-end uncountable infinity))] {Hyper Hyper Aleph (high-end uncountable infinity) (Aleph (high-end uncountable infinity), Aleph (high-end uncountable infinity))[Hyper Aleph (high-end uncountable infinity) (Aleph (high-end uncountable infinity), Aleph (high-end uncountable infinity)), Hyper Aleph (high-end uncountable infinity) (Aleph (high-end uncountable infinity), Aleph (high-end uncountable infinity))], Hyper Hyper Aleph (high-end uncountable infinity) (Aleph (high-end uncountable infinity), Aleph (high-end uncountable infinity))[Hyper Aleph (high-end uncountable infinity) (Aleph (high-end uncountable infinity), Aleph (high-end uncountable infinity)), Hyper Aleph (high-end uncountable infinity) (Aleph (high-end uncountable infinity), Aleph (high-end uncountable infinity))]} { Hyper Hyper Hyper Aleph (high-end uncountable infinity) (Aleph (high-end uncountable infinity), Aleph (high-end uncountable infinity))[Hyper Aleph (high-end uncountable infinity) (Aleph (high-end uncountable infinity), Aleph (high-end uncountable infinity)), Hyper Aleph (high-end uncountable infinity) (Aleph (high-end uncountable infinity), Aleph (high-end uncountable infinity))] {Hyper Hyper Aleph (high-end uncountable infinity) (Aleph (high-end uncountable infinity), Aleph (high-end uncountable infinity))[Hyper Aleph (high-end uncountable infinity) (Aleph (high-end uncountable infinity), Aleph (high-end uncountable infinity)), Hyper Aleph (high-end uncountable infinity) (Aleph (high-end uncountable infinity), Aleph (high-end uncountable infinity))], Hyper Hyper Aleph (high-end uncountable infinity) (Aleph (high-end uncountable infinity), Aleph (high-end uncountable infinity))[Hyper Aleph (high-end uncountable infinity) (Aleph (high-end uncountable infinity), Aleph (high-end uncountable infinity)), Hyper Aleph (high-end uncountable infinity) (Aleph (high-end uncountable infinity), Aleph (high-end uncountable infinity))]}, Hyper Hyper Hyper Aleph (high-end uncountable infinity) (Aleph (high-end uncountable infinity), Aleph (high-end uncountable infinity))[Hyper Aleph (high-end uncountable infinity) (Aleph (high-end uncountable infinity), Aleph (high-end uncountable infinity)), Hyper Aleph (high-end uncountable infinity) (Aleph (high-end uncountable infinity), Aleph (high-end uncountable infinity))] {Hyper Hyper Aleph (high-end uncountable infinity) (Aleph (high-end uncountable infinity), Aleph (high-end uncountable infinity))[Hyper Aleph (high-end uncountable infinity) (Aleph (high-end uncountable infinity), Aleph (high-end uncountable infinity)), Hyper Aleph (high-end uncountable infinity) (Aleph (high-end uncountable infinity), Aleph (high-end uncountable infinity))], Hyper Hyper Aleph (high-end uncountable infinity) (Aleph (high-end uncountable infinity), Aleph (high-end uncountable infinity))[Hyper Aleph (high-end uncountable infinity) (Aleph (high-end uncountable infinity), Aleph (high-end uncountable infinity)), Hyper Aleph (high-end uncountable infinity) (Aleph (high-end uncountable infinity), Aleph (high-end uncountable infinity))]}}, Hyper Hyper Hyper Hyper Aleph (high-end uncountable infinity) (Aleph (high-end uncountable infinity), Aleph (high-end uncountable infinity))[Hyper Aleph (high-end uncountable infinity) (Aleph (high-end uncountable infinity), Aleph (high-end uncountable infinity)), Hyper Aleph (high-end uncountable infinity) (Aleph (high-end uncountable infinity), Aleph (high-end uncountable infinity))] {Hyper Hyper Aleph (high-end uncountable infinity) (Aleph (high-end uncountable infinity), Aleph (high-end uncountable infinity))[Hyper Aleph (high-end uncountable infinity) (Aleph (high-end uncountable infinity), Aleph (high-end uncountable infinity)), Hyper Aleph (high-end uncountable infinity) (Aleph (high-end uncountable infinity), Aleph (high-end uncountable infinity))], Hyper Hyper Aleph (high-end uncountable infinity) (Aleph (high-end uncountable infinity), Aleph (high-end uncountable infinity))[Hyper Aleph (high-end uncountable infinity) (Aleph (high-end uncountable infinity), Aleph (high-end uncountable infinity)), Hyper Aleph (high-end uncountable infinity) (Aleph (high-end uncountable infinity), Aleph (high-end uncountable infinity))]} { Hyper Hyper Hyper Aleph (high-end uncountable infinity) (Aleph (high-end uncountable infinity), Aleph (high-end uncountable infinity))[Hyper Aleph (high-end uncountable infinity) (Aleph (high-end uncountable infinity), Aleph (high-end uncountable infinity)), Hyper Aleph (high-end uncountable infinity) (Aleph (high-end uncountable infinity), Aleph (high-end uncountable infinity))] {Hyper Hyper Aleph (high-end uncountable infinity) (Aleph (high-end uncountable infinity), Aleph (high-end uncountable infinity))[Hyper Aleph (high-end uncountable infinity) (Aleph (high-end uncountable infinity), Aleph (high-end uncountable infinity)), Hyper Aleph (high-end uncountable infinity) (Aleph (high-end uncountable infinity), Aleph (high-end uncountable infinity))], Hyper Hyper Aleph (high-end uncountable infinity) (Aleph (high-end uncountable infinity), Aleph (high-end uncountable infinity))[Hyper Aleph (high-end uncountable infinity) (Aleph (high-end uncountable infinity), Aleph (high-end uncountable infinity)), Hyper Aleph (high-end uncountable infinity) (Aleph (high-end uncountable infinity), Aleph (high-end uncountable infinity))]}, Hyper Hyper Hyper Aleph (high-end uncountable infinity) (Aleph (high-end uncountable infinity), Aleph (high-end uncountable infinity))[Hyper Aleph (high-end uncountable infinity) (Aleph (high-end uncountable infinity), Aleph (high-end uncountable infinity)), Hyper Aleph (high-end uncountable infinity) (Aleph (high-end uncountable infinity), Aleph (high-end uncountable infinity))] {Hyper Hyper Aleph (high-end uncountable infinity) (Aleph (high-end uncountable infinity), Aleph (high-end uncountable infinity))[Hyper Aleph (high-end uncountable infinity) (Aleph (high-end uncountable infinity), Aleph (high-end uncountable infinity)), Hyper Aleph (high-end uncountable infinity) (Aleph (high-end uncountable infinity), Aleph (high-end uncountable infinity))], Hyper Hyper Aleph (high-end uncountable infinity) (Aleph (high-end uncountable infinity), Aleph (high-end uncountable infinity))[Hyper Aleph (high-end uncountable infinity) (Aleph (high-end uncountable infinity), Aleph (high-end uncountable infinity)), Hyper Aleph (high-end uncountable infinity) (Aleph (high-end uncountable infinity), Aleph (high-end uncountable infinity))]}}}{ Hyper Hyper Hyper Hyper Hyper Aleph (high-end uncountable infinity) (Aleph (high-end uncountable infinity), Aleph (high-end uncountable infinity))[Hyper Aleph (high-end uncountable infinity) (Aleph (high-end uncountable infinity), Aleph (high-end uncountable infinity)), Hyper Aleph (high-end uncountable infinity) (Aleph (high-end uncountable infinity), Aleph (high-end uncountable infinity))] {Hyper Hyper Aleph (high-end uncountable infinity) (Aleph (high-end uncountable infinity), Aleph (high-end uncountable infinity))[Hyper Aleph (high-end uncountable infinity) (Aleph (high-end uncountable infinity), Aleph (high-end uncountable infinity)), Hyper Aleph (high-end uncountable infinity) (Aleph (high-end uncountable infinity), Aleph (high-end uncountable infinity))], Hyper Hyper Aleph (high-end uncountable infinity) (Aleph (high-end uncountable infinity), Aleph (high-end uncountable infinity))[Hyper Aleph (high-end uncountable infinity) (Aleph (high-end uncountable infinity), Aleph (high-end uncountable infinity)), Hyper Aleph (high-end uncountable infinity) (Aleph (high-end uncountable infinity), Aleph (high-end uncountable infinity))]} { Hyper Hyper Hyper Aleph (high-end uncountable infinity) (Aleph (high-end uncountable infinity), Aleph (high-end uncountable infinity))[Hyper Aleph (high-end uncountable infinity) (Aleph (high-end uncountable infinity), Aleph (high-end uncountable infinity)), Hyper Aleph (high-end uncountable infinity) (Aleph (high-end uncountable infinity), Aleph (high-end uncountable infinity))] {Hyper Hyper Aleph (high-end uncountable infinity) (Aleph (high-end uncountable infinity), Aleph (high-end uncountable infinity))[Hyper Aleph (high-end uncountable infinity) (Aleph (high-end uncountable infinity), Aleph (high-end uncountable infinity)), Hyper Aleph (high-end uncountable infinity) (Aleph (high-end uncountable infinity), Aleph (high-end uncountable infinity))], Hyper Hyper Aleph (high-end uncountable infinity) (Aleph (high-end uncountable infinity), Aleph (high-end uncountable infinity))[Hyper Aleph (high-end uncountable infinity) (Aleph (high-end uncountable infinity), Aleph (high-end uncountable infinity)), Hyper Aleph (high-end uncountable infinity) (Aleph (high-end uncountable infinity), Aleph (high-end uncountable infinity))]}, Hyper Hyper Hyper Aleph (high-end uncountable infinity) (Aleph (high-end uncountable infinity), Aleph (high-end uncountable infinity))[Hyper Aleph (high-end uncountable infinity) (Aleph (high-end uncountable infinity), Aleph (high-end uncountable infinity)), Hyper Aleph (high-end uncountable infinity) (Aleph (high-end uncountable infinity), Aleph (high-end uncountable infinity))] {Hyper Hyper Aleph (high-end uncountable infinity) (Aleph (high-end uncountable infinity), Aleph (high-end uncountable infinity))[Hyper Aleph (high-end uncountable infinity) (Aleph (high-end uncountable infinity), Aleph (high-end uncountable infinity)), Hyper Aleph (high-end uncountable infinity) (Aleph (high-end uncountable infinity), Aleph (high-end uncountable infinity))], Hyper Hyper Aleph (high-end uncountable infinity) (Aleph (high-end uncountable infinity), Aleph (high-end uncountable infinity))[Hyper Aleph (high-end uncountable infinity) (Aleph (high-end uncountable infinity), Aleph (high-end uncountable infinity)), Hyper Aleph (high-end uncountable infinity) (Aleph (high-end uncountable infinity), Aleph (high-end uncountable infinity))]}}{ Hyper Hyper Hyper Hyper Aleph (high-end uncountable infinity) (Aleph (high-end uncountable infinity), Aleph (high-end uncountable infinity))[Hyper Aleph (high-end uncountable infinity) (Aleph (high-end uncountable infinity), Aleph (high-end uncountable infinity)), Hyper Aleph (high-end uncountable infinity) (Aleph (high-end uncountable infinity), Aleph (high-end uncountable infinity))] {Hyper Hyper Aleph (high-end uncountable infinity) (Aleph (high-end uncountable infinity), Aleph (high-end uncountable infinity))[Hyper Aleph (high-end uncountable infinity) (Aleph (high-end uncountable infinity), Aleph (high-end uncountable infinity)), Hyper Aleph (high-end uncountable infinity) (Aleph (high-end uncountable infinity), Aleph (high-end uncountable infinity))], Hyper Hyper Aleph (high-end uncountable infinity) (Aleph (high-end uncountable infinity), Aleph (high-end uncountable infinity))[Hyper Aleph (high-end uncountable infinity) (Aleph (high-end uncountable infinity), Aleph (high-end uncountable infinity)), Hyper Aleph (high-end uncountable infinity) (Aleph (high-end uncountable infinity), Aleph (high-end uncountable infinity))]} { Hyper Hyper Hyper Aleph (high-end uncountable infinity) (Aleph (high-end uncountable infinity), Aleph (high-end uncountable infinity))[Hyper Aleph (high-end uncountable infinity) (Aleph (high-end uncountable infinity), Aleph (high-end uncountable infinity)), Hyper Aleph (high-end uncountable infinity) (Aleph (high-end uncountable infinity), Aleph (high-end uncountable infinity))] {Hyper Hyper Aleph (high-end uncountable infinity) (Aleph (high-end uncountable infinity), Aleph (high-end uncountable infinity))[Hyper Aleph (high-end uncountable infinity) (Aleph (high-end uncountable infinity), Aleph (high-end uncountable infinity)), Hyper Aleph (high-end uncountable infinity) (Aleph (high-end uncountable infinity), Aleph (high-end uncountable infinity))], Hyper Hyper Aleph (high-end uncountable infinity) (Aleph (high-end uncountable infinity), Aleph (high-end uncountable infinity))[Hyper Aleph (high-end uncountable infinity) (Aleph (high-end uncountable infinity), Aleph (high-end uncountable infinity)), Hyper Aleph (high-end uncountable infinity) (Aleph (high-end uncountable infinity), Aleph (high-end uncountable infinity))]}, Hyper Hyper Hyper Aleph (high-end uncountable infinity) (Aleph (high-end uncountable infinity), Aleph (high-end uncountable infinity))[Hyper Aleph (high-end uncountable infinity) (Aleph (high-end uncountable infinity), Aleph (high-end uncountable infinity)), Hyper Aleph (high-end uncountable infinity) (Aleph (high-end uncountable infinity), Aleph (high-end uncountable infinity))] {Hyper Hyper Aleph (high-end uncountable infinity) (Aleph (high-end uncountable infinity), Aleph (high-end uncountable infinity))[Hyper Aleph (high-end uncountable infinity) (Aleph (high-end uncountable infinity), Aleph (high-end uncountable infinity)), Hyper Aleph (high-end uncountable infinity) (Aleph (high-end uncountable infinity), Aleph (high-end uncountable infinity))], Hyper Hyper Aleph (high-end uncountable infinity) (Aleph (high-end uncountable infinity), Aleph (high-end uncountable infinity))[Hyper Aleph (high-end uncountable infinity) (Aleph (high-end uncountable infinity), Aleph (high-end uncountable infinity)), Hyper Aleph (high-end uncountable infinity) (Aleph (high-end uncountable infinity), Aleph (high-end uncountable infinity))]}}, Hyper Hyper Hyper Hyper Aleph (high-end uncountable infinity) (Aleph (high-end uncountable infinity), Aleph (high-end uncountable infinity))[Hyper Aleph (high-end uncountable infinity) (Aleph (high-end uncountable infinity), Aleph (high-end uncountable infinity)), Hyper Aleph (high-end uncountable infinity) (Aleph (high-end uncountable infinity), Aleph (high-end uncountable infinity))] {Hyper Hyper Aleph (high-end uncountable infinity) (Aleph (high-end uncountable infinity), Aleph (high-end uncountable infinity))[Hyper Aleph (high-end uncountable infinity) (Aleph (high-end uncountable infinity), Aleph (high-end uncountable infinity)), Hyper Aleph (high-end uncountable infinity) (Aleph (high-end uncountable infinity), Aleph (high-end uncountable infinity))], Hyper Hyper Aleph (high-end uncountable infinity) (Aleph (high-end uncountable infinity), Aleph (high-end uncountable infinity))[Hyper Aleph (high-end uncountable infinity) (Aleph (high-end uncountable infinity), Aleph (high-end uncountable infinity)), Hyper Aleph (high-end uncountable infinity) (Aleph (high-end uncountable infinity), Aleph (high-end uncountable infinity))]} { Hyper Hyper Hyper Aleph (high-end uncountable infinity) (Aleph (high-end uncountable infinity), Aleph (high-end uncountable infinity))[Hyper Aleph (high-end uncountable infinity) (Aleph (high-end uncountable infinity), Aleph (high-end uncountable infinity)), Hyper Aleph (high-end uncountable infinity) (Aleph (high-end uncountable infinity), Aleph (high-end uncountable infinity))] {Hyper Hyper Aleph (high-end uncountable infinity) (Aleph (high-end uncountable infinity), Aleph (high-end uncountable infinity))[Hyper Aleph (high-end uncountable infinity) (Aleph (high-end uncountable infinity), Aleph (high-end uncountable infinity)), Hyper Aleph (high-end uncountable infinity) (Aleph (high-end uncountable infinity), Aleph (high-end uncountable infinity))], Hyper Hyper Aleph (high-end uncountable infinity) (Aleph (high-end uncountable infinity), Aleph (high-end uncountable infinity))[Hyper Aleph (high-end uncountable infinity) (Aleph (high-end uncountable infinity), Aleph (high-end uncountable infinity)), Hyper Aleph (high-end uncountable infinity) (Aleph (high-end uncountable infinity), Aleph (high-end uncountable infinity))]}, Hyper Hyper Hyper Aleph (high-end uncountable infinity) (Aleph (high-end uncountable infinity), Aleph (high-end uncountable infinity))[Hyper Aleph (high-end uncountable infinity) (Aleph (high-end uncountable infinity), Aleph (high-end uncountable infinity)), Hyper Aleph (high-end uncountable infinity) (Aleph (high-end uncountable infinity), Aleph (high-end uncountable infinity))] {Hyper Hyper Aleph (high-end uncountable infinity) (Aleph (high-end uncountable infinity), Aleph (high-end uncountable infinity))[Hyper Aleph (high-end uncountable infinity) (Aleph (high-end uncountable infinity), Aleph (high-end uncountable infinity)), Hyper Aleph (high-end uncountable infinity) (Aleph (high-end uncountable infinity), Aleph (high-end uncountable infinity))], Hyper Hyper Aleph (high-end uncountable infinity) (Aleph (high-end uncountable infinity), Aleph (high-end uncountable infinity))[Hyper Aleph (high-end uncountable infinity) (Aleph (high-end uncountable infinity), Aleph (high-end uncountable infinity)), Hyper Aleph (high-end uncountable infinity) (Aleph (high-end uncountable infinity), Aleph (high-end uncountable infinity))]}}}, Hyper Hyper Hyper Hyper Hyper Aleph (high-end uncountable infinity) (Aleph (high-end uncountable infinity), Aleph (high-end uncountable infinity))[Hyper Aleph (high-end uncountable infinity) (Aleph (high-end uncountable infinity), Aleph (high-end uncountable infinity)), Hyper Aleph (high-end uncountable infinity) (Aleph (high-end uncountable infinity), Aleph (high-end uncountable infinity))] {Hyper Hyper Aleph (high-end uncountable infinity) (Aleph (high-end uncountable infinity), Aleph (high-end uncountable infinity))[Hyper Aleph (high-end uncountable infinity) (Aleph (high-end uncountable infinity), Aleph (high-end uncountable infinity)), Hyper Aleph (high-end uncountable infinity) (Aleph (high-end uncountable infinity), Aleph (high-end uncountable infinity))], Hyper Hyper Aleph (high-end uncountable infinity) (Aleph (high-end uncountable infinity), Aleph (high-end uncountable infinity))[Hyper Aleph (high-end uncountable infinity) (Aleph (high-end uncountable infinity), Aleph (high-end uncountable infinity)), Hyper Aleph (high-end uncountable infinity) (Aleph (high-end uncountable infinity), Aleph (high-end uncountable infinity))]} { Hyper Hyper Hyper Aleph (high-end uncountable infinity) (Aleph (high-end uncountable infinity), Aleph (high-end uncountable infinity))[Hyper Aleph (high-end uncountable infinity) (Aleph (high-end uncountable infinity), Aleph (high-end uncountable infinity)), Hyper Aleph (high-end uncountable infinity) (Aleph (high-end uncountable infinity), Aleph (high-end uncountable infinity))] {Hyper Hyper Aleph (high-end uncountable infinity) (Aleph (high-end uncountable infinity), Aleph (high-end uncountable infinity))[Hyper Aleph (high-end uncountable infinity) (Aleph (high-end uncountable infinity), Aleph (high-end uncountable infinity)), Hyper Aleph (high-end uncountable infinity) (Aleph (high-end uncountable infinity), Aleph (high-end uncountable infinity))], Hyper Hyper Aleph (high-end uncountable infinity) (Aleph (high-end uncountable infinity), Aleph (high-end uncountable infinity))[Hyper Aleph (high-end uncountable infinity) (Aleph (high-end uncountable infinity), Aleph (high-end uncountable infinity)), Hyper Aleph (high-end uncountable infinity) (Aleph (high-end uncountable infinity), Aleph (high-end uncountable infinity))]}, Hyper Hyper Hyper Aleph (high-end uncountable infinity) (Aleph (high-end uncountable infinity), Aleph (high-end uncountable infinity))[Hyper Aleph (high-end uncountable infinity) (Aleph (high-end uncountable infinity), Aleph (high-end uncountable infinity)), Hyper Aleph (high-end uncountable infinity) (Aleph (high-end uncountable infinity), Aleph (high-end uncountable infinity))] {Hyper Hyper Aleph (high-end uncountable infinity) (Aleph (high-end uncountable infinity), Aleph (high-end uncountable infinity))[Hyper Aleph (high-end uncountable infinity) (Aleph (high-end uncountable infinity), Aleph (high-end uncountable infinity)), Hyper Aleph (high-end uncountable infinity) (Aleph (high-end uncountable infinity), Aleph (high-end uncountable infinity))], Hyper Hyper Aleph (high-end uncountable infinity) (Aleph (high-end uncountable infinity), Aleph (high-end uncountable infinity))[Hyper Aleph (high-end uncountable infinity) (Aleph (high-end uncountable infinity), Aleph (high-end uncountable infinity)), Hyper Aleph (high-end uncountable infinity) (Aleph (high-end uncountable infinity), Aleph (high-end uncountable infinity))]}}{ Hyper Hyper Hyper Hyper Aleph (high-end uncountable infinity) (Aleph (high-end uncountable infinity), Aleph (high-end uncountable infinity))[Hyper Aleph (high-end uncountable infinity) (Aleph (high-end uncountable infinity), Aleph (high-end uncountable infinity)), Hyper Aleph (high-end uncountable infinity) (Aleph (high-end uncountable infinity), Aleph (high-end uncountable infinity))] {Hyper Hyper Aleph (high-end uncountable infinity) (Aleph (high-end uncountable infinity), Aleph (high-end uncountable infinity))[Hyper Aleph (high-end uncountable infinity) (Aleph (high-end uncountable infinity), Aleph (high-end uncountable infinity)), Hyper Aleph (high-end uncountable infinity) (Aleph (high-end uncountable infinity), Aleph (high-end uncountable infinity))], Hyper Hyper Aleph (high-end uncountable infinity) (Aleph (high-end uncountable infinity), Aleph (high-end uncountable infinity))[Hyper Aleph (high-end uncountable infinity) (Aleph (high-end uncountable infinity), Aleph (high-end uncountable infinity)), Hyper Aleph (high-end uncountable infinity) (Aleph (high-end uncountable infinity), Aleph (high-end uncountable infinity))]} { Hyper Hyper Hyper Aleph (high-end uncountable infinity) (Aleph (high-end uncountable infinity), Aleph (high-end uncountable infinity))[Hyper Aleph (high-end uncountable infinity) (Aleph (high-end uncountable infinity), Aleph (high-end uncountable infinity)), Hyper Aleph (high-end uncountable infinity) (Aleph (high-end uncountable infinity), Aleph (high-end uncountable infinity))] {Hyper Hyper Aleph (high-end uncountable infinity) (Aleph (high-end uncountable infinity), Aleph (high-end uncountable infinity))[Hyper Aleph (high-end uncountable infinity) (Aleph (high-end uncountable infinity), Aleph (high-end uncountable infinity)), Hyper Aleph (high-end uncountable infinity) (Aleph (high-end uncountable infinity), Aleph (high-end uncountable infinity))], Hyper Hyper Aleph (high-end uncountable infinity) (Aleph (high-end uncountable infinity), Aleph (high-end uncountable infinity))[Hyper Aleph (high-end uncountable infinity) (Aleph (high-end uncountable infinity), Aleph (high-end uncountable infinity)), Hyper Aleph (high-end uncountable infinity) (Aleph (high-end uncountable infinity), Aleph (high-end uncountable infinity))]}, Hyper Hyper Hyper Aleph (high-end uncountable infinity) (Aleph (high-end uncountable infinity), Aleph (high-end uncountable infinity))[Hyper Aleph (high-end uncountable infinity) (Aleph (high-end uncountable infinity), Aleph (high-end uncountable infinity)), Hyper Aleph (high-end uncountable infinity) (Aleph (high-end uncountable infinity), Aleph (high-end uncountable infinity))] {Hyper Hyper Aleph (high-end uncountable infinity) (Aleph (high-end uncountable infinity), Aleph (high-end uncountable infinity))[Hyper Aleph (high-end uncountable infinity) (Aleph (high-end uncountable infinity), Aleph (high-end uncountable infinity)), Hyper Aleph (high-end uncountable infinity) (Aleph (high-end uncountable infinity), Aleph (high-end uncountable infinity))], Hyper Hyper Aleph (high-end uncountable infinity) (Aleph (high-end uncountable infinity), Aleph (high-end uncountable infinity))[Hyper Aleph (high-end uncountable infinity) (Aleph (high-end uncountable infinity), Aleph (high-end uncountable infinity)), Hyper Aleph (high-end uncountable infinity) (Aleph (high-end uncountable infinity), Aleph (high-end uncountable infinity))]}}, Hyper Hyper Hyper Hyper Aleph (high-end uncountable infinity) (Aleph (high-end uncountable infinity), Aleph (high-end uncountable infinity))[Hyper Aleph (high-end uncountable infinity) (Aleph (high-end uncountable infinity), Aleph (high-end uncountable infinity)), Hyper Aleph (high-end uncountable infinity) (Aleph (high-end uncountable infinity), Aleph (high-end uncountable infinity))] {Hyper Hyper Aleph (high-end uncountable infinity) (Aleph (high-end uncountable infinity), Aleph (high-end uncountable infinity))[Hyper Aleph (high-end uncountable infinity) (Aleph (high-end uncountable infinity), Aleph (high-end uncountable infinity)), Hyper Aleph (high-end uncountable infinity) (Aleph (high-end uncountable infinity), Aleph (high-end uncountable infinity))], Hyper Hyper Aleph (high-end uncountable infinity) (Aleph (high-end uncountable infinity), Aleph (high-end uncountable infinity))[Hyper Aleph (high-end uncountable infinity) (Aleph (high-end uncountable infinity), Aleph (high-end uncountable infinity)), Hyper Aleph (high-end uncountable infinity) (Aleph (high-end uncountable infinity), Aleph (high-end uncountable infinity))]} { Hyper Hyper Hyper Aleph (high-end uncountable infinity) (Aleph (high-end uncountable infinity), Aleph (high-end uncountable infinity))[Hyper Aleph (high-end uncountable infinity) (Aleph (high-end uncountable infinity), Aleph (high-end uncountable infinity)), Hyper Aleph (high-end uncountable infinity) (Aleph (high-end uncountable infinity), Aleph (high-end uncountable infinity))] {Hyper Hyper Aleph (high-end uncountable infinity) (Aleph (high-end uncountable infinity), Aleph (high-end uncountable infinity))[Hyper Aleph (high-end uncountable infinity) (Aleph (high-end uncountable infinity), Aleph (high-end uncountable infinity)), Hyper Aleph (high-end uncountable infinity) (Aleph (high-end uncountable infinity), Aleph (high-end uncountable infinity))], Hyper Hyper Aleph (high-end uncountable infinity) (Aleph (high-end uncountable infinity), Aleph (high-end uncountable infinity))[Hyper Aleph (high-end uncountable infinity) (Aleph (high-end uncountable infinity), Aleph (high-end uncountable infinity)), Hyper Aleph (high-end uncountable infinity) (Aleph (high-end uncountable infinity), Aleph (high-end uncountable infinity))]}, Hyper Hyper Hyper Aleph (high-end uncountable infinity) (Aleph (high-end uncountable infinity), Aleph (high-end uncountable infinity))[Hyper Aleph (high-end uncountable infinity) (Aleph (high-end uncountable infinity), Aleph (high-end uncountable infinity)), Hyper Aleph (high-end uncountable infinity) (Aleph (high-end uncountable infinity), Aleph (high-end uncountable infinity))] {Hyper Hyper Aleph (high-end uncountable infinity) (Aleph (high-end uncountable infinity), Aleph (high-end uncountable infinity))[Hyper Aleph (high-end uncountable infinity) (Aleph (high-end uncountable infinity), Aleph (high-end uncountable infinity)), Hyper Aleph (high-end uncountable infinity) (Aleph (high-end uncountable infinity), Aleph (high-end uncountable infinity))], Hyper Hyper Aleph (high-end uncountable infinity) (Aleph (high-end uncountable infinity), Aleph (high-end uncountable infinity))[Hyper Aleph (high-end uncountable infinity) (Aleph (high-end uncountable infinity), Aleph (high-end uncountable infinity)), Hyper Aleph (high-end uncountable infinity) (Aleph (high-end uncountable infinity), Aleph (high-end uncountable infinity))]}}}}

We can go still yet further to develop the following infinity.

Hyper Hyper Hyper Hyper Hyper Hyper Hyper Aleph (high-end uncountable infinity) (Aleph (high-end uncountable infinity), Aleph (high-end uncountable infinity))[Hyper Aleph (high-end uncountable infinity) (Aleph (high-end uncountable infinity), Aleph (high-end uncountable infinity)), Hyper Aleph (high-end uncountable infinity) (Aleph (high-end uncountable infinity), Aleph (high-end uncountable infinity))] {Hyper Hyper Aleph (high-end uncountable infinity) (Aleph (high-end uncountable infinity), Aleph (high-end uncountable infinity))[Hyper Aleph (high-end uncountable infinity) (Aleph (high-end uncountable infinity), Aleph (high-end uncountable infinity)), Hyper Aleph (high-end uncountable infinity) (Aleph (high-end uncountable infinity), Aleph (high-end uncountable infinity))], Hyper Hyper Aleph (high-end uncountable infinity) (Aleph (high-end uncountable infinity), Aleph (high-end uncountable infinity))[Hyper Aleph (high-end uncountable infinity) (Aleph (high-end uncountable infinity), Aleph (high-end uncountable infinity)), Hyper Aleph (high-end uncountable infinity) (Aleph (high-end uncountable infinity), Aleph (high-end uncountable infinity))]} { Hyper Hyper Hyper Aleph (high-end uncountable infinity) (Aleph (high-end uncountable infinity), Aleph (high-end uncountable infinity))[Hyper Aleph (high-end uncountable infinity) (Aleph (high-end uncountable infinity), Aleph (high-end uncountable infinity)), Hyper Aleph (high-end uncountable infinity) (Aleph (high-end uncountable infinity), Aleph (high-end uncountable infinity))] {Hyper Hyper Aleph (high-end uncountable infinity) (Aleph (high-end uncountable infinity), Aleph (high-end uncountable infinity))[Hyper Aleph (high-end uncountable infinity) (Aleph (high-end uncountable infinity), Aleph (high-end uncountable infinity)), Hyper Aleph (high-end uncountable infinity) (Aleph (high-end uncountable infinity), Aleph (high-end uncountable infinity))], Hyper Hyper Aleph (high-end uncountable infinity) (Aleph (high-end uncountable infinity), Aleph (high-end uncountable infinity))[Hyper Aleph (high-end uncountable infinity) (Aleph (high-end uncountable infinity), Aleph (high-end uncountable infinity)), Hyper Aleph (high-end uncountable infinity) (Aleph (high-end uncountable infinity), Aleph (high-end uncountable infinity))]}, Hyper Hyper Hyper Aleph (high-end uncountable infinity) (Aleph (high-end uncountable infinity), Aleph (high-end uncountable infinity))[Hyper Aleph (high-end uncountable infinity) (Aleph (high-end uncountable infinity), Aleph (high-end uncountable infinity)), Hyper Aleph (high-end uncountable infinity) (Aleph (high-end uncountable infinity), Aleph (high-end uncountable infinity))] {Hyper Hyper Aleph (high-end uncountable infinity) (Aleph (high-end uncountable infinity), Aleph (high-end uncountable infinity))[Hyper Aleph (high-end uncountable infinity) (Aleph (high-end uncountable infinity), Aleph (high-end uncountable infinity)), Hyper Aleph (high-end uncountable infinity) (Aleph (high-end uncountable infinity), Aleph (high-end uncountable infinity))], Hyper Hyper Aleph (high-end uncountable infinity) (Aleph (high-end uncountable infinity), Aleph (high-end uncountable infinity))[Hyper Aleph (high-end uncountable infinity) (Aleph (high-end uncountable infinity), Aleph (high-end uncountable infinity)), Hyper Aleph (high-end uncountable infinity) (Aleph (high-end uncountable infinity), Aleph (high-end uncountable infinity))]}}{ Hyper Hyper Hyper Hyper Aleph (high-end uncountable infinity) (Aleph (high-end uncountable infinity), Aleph (high-end uncountable infinity))[Hyper Aleph (high-end uncountable infinity) (Aleph (high-end uncountable infinity), Aleph (high-end uncountable infinity)), Hyper Aleph (high-end uncountable infinity) (Aleph (high-end uncountable infinity), Aleph (high-end uncountable infinity))] {Hyper Hyper Aleph (high-end uncountable infinity) (Aleph (high-end uncountable infinity), Aleph (high-end uncountable infinity))[Hyper Aleph (high-end uncountable infinity) (Aleph (high-end uncountable infinity), Aleph (high-end uncountable infinity)), Hyper Aleph (high-end uncountable infinity) (Aleph (high-end uncountable infinity), Aleph (high-end uncountable infinity))], Hyper Hyper Aleph (high-end uncountable infinity) (Aleph (high-end uncountable infinity), Aleph (high-end uncountable infinity))[Hyper Aleph (high-end uncountable infinity) (Aleph (high-end uncountable infinity), Aleph (high-end uncountable infinity)), Hyper Aleph (high-end uncountable infinity) (Aleph (high-end uncountable infinity), Aleph (high-end uncountable infinity))]} { Hyper Hyper Hyper Aleph (high-end uncountable infinity) (Aleph (high-end uncountable infinity), Aleph (high-end uncountable infinity))[Hyper Aleph (high-end uncountable infinity) (Aleph (high-end uncountable infinity), Aleph (high-end uncountable infinity)), Hyper Aleph (high-end uncountable infinity) (Aleph (high-end uncountable infinity), Aleph (high-end uncountable infinity))] {Hyper Hyper Aleph (high-end uncountable infinity) (Aleph (high-end uncountable infinity), Aleph (high-end uncountable infinity))[Hyper Aleph (high-end uncountable infinity) (Aleph (high-end uncountable infinity), Aleph (high-end uncountable infinity)), Hyper Aleph (high-end uncountable infinity) (Aleph (high-end uncountable infinity), Aleph (high-end uncountable infinity))], Hyper Hyper Aleph (high-end uncountable infinity) (Aleph (high-end uncountable infinity), Aleph (high-end uncountable infinity))[Hyper Aleph (high-end uncountable infinity) (Aleph (high-end uncountable infinity), Aleph (high-end uncountable infinity)), Hyper Aleph (high-end uncountable infinity) (Aleph (high-end uncountable infinity), Aleph (high-end uncountable infinity))]}, Hyper Hyper Hyper Aleph (high-end uncountable infinity) (Aleph (high-end uncountable infinity), Aleph (high-end uncountable infinity))[Hyper Aleph (high-end uncountable infinity) (Aleph (high-end uncountable infinity), Aleph (high-end uncountable infinity)), Hyper Aleph (high-end uncountable infinity) (Aleph (high-end uncountable infinity), Aleph (high-end uncountable infinity))] {Hyper Hyper Aleph (high-end uncountable infinity) (Aleph (high-end uncountable infinity), Aleph (high-end uncountable infinity))[Hyper Aleph (high-end uncountable infinity) (Aleph (high-end uncountable infinity), Aleph (high-end uncountable infinity)), Hyper Aleph (high-end uncountable infinity) (Aleph (high-end uncountable infinity), Aleph (high-end uncountable infinity))], Hyper Hyper Aleph (high-end uncountable infinity) (Aleph (high-end uncountable infinity), Aleph (high-end uncountable infinity))[Hyper Aleph (high-end uncountable infinity) (Aleph (high-end uncountable infinity), Aleph (high-end uncountable infinity)), Hyper Aleph (high-end uncountable infinity) (Aleph (high-end uncountable infinity), Aleph (high-end uncountable infinity))]}}, Hyper Hyper Hyper Hyper Aleph (high-end uncountable infinity) (Aleph (high-end uncountable infinity), Aleph (high-end uncountable infinity))[Hyper Aleph (high-end uncountable infinity) (Aleph (high-end uncountable infinity), Aleph (high-end uncountable infinity)), Hyper Aleph (high-end uncountable infinity) (Aleph (high-end uncountable infinity), Aleph (high-end uncountable infinity))] {Hyper Hyper Aleph (high-end uncountable infinity) (Aleph (high-end uncountable infinity), Aleph (high-end uncountable infinity))[Hyper Aleph (high-end uncountable infinity) (Aleph (high-end uncountable infinity), Aleph (high-end uncountable infinity)), Hyper Aleph (high-end uncountable infinity) (Aleph (high-end uncountable infinity), Aleph (high-end uncountable infinity))], Hyper Hyper Aleph (high-end uncountable infinity) (Aleph (high-end uncountable infinity), Aleph (high-end uncountable infinity))[Hyper Aleph (high-end uncountable infinity) (Aleph (high-end uncountable infinity), Aleph (high-end uncountable infinity)), Hyper Aleph (high-end uncountable infinity) (Aleph (high-end uncountable infinity), Aleph (high-end uncountable infinity))]} { Hyper Hyper Hyper Aleph (high-end uncountable infinity) (Aleph (high-end uncountable infinity), Aleph (high-end uncountable infinity))[Hyper Aleph (high-end uncountable infinity) (Aleph (high-end uncountable infinity), Aleph (high-end uncountable infinity)), Hyper Aleph (high-end uncountable infinity) (Aleph (high-end uncountable infinity), Aleph (high-end uncountable infinity))] {Hyper Hyper Aleph (high-end uncountable infinity) (Aleph (high-end uncountable infinity), Aleph (high-end uncountable infinity))[Hyper Aleph (high-end uncountable infinity) (Aleph (high-end uncountable infinity), Aleph (high-end uncountable infinity)), Hyper Aleph (high-end uncountable infinity) (Aleph (high-end uncountable infinity), Aleph (high-end uncountable infinity))], Hyper Hyper Aleph (high-end uncountable infinity) (Aleph (high-end uncountable infinity), Aleph (high-end uncountable infinity))[Hyper Aleph (high-end uncountable infinity) (Aleph (high-end uncountable infinity), Aleph (high-end uncountable infinity)), Hyper Aleph (high-end uncountable infinity) (Aleph (high-end uncountable infinity), Aleph (high-end uncountable infinity))]}, Hyper Hyper Hyper Aleph (high-end uncountable infinity) (Aleph (high-end uncountable infinity), Aleph (high-end uncountable infinity))[Hyper Aleph (high-end uncountable infinity) (Aleph (high-end uncountable infinity), Aleph (high-end uncountable infinity)), Hyper Aleph (high-end uncountable infinity) (Aleph (high-end uncountable infinity), Aleph (high-end uncountable infinity))] {Hyper Hyper Aleph (high-end uncountable infinity) (Aleph (high-end uncountable infinity), Aleph (high-end uncountable infinity))[Hyper Aleph (high-end uncountable infinity) (Aleph (high-end uncountable infinity), Aleph (high-end uncountable infinity)), Hyper Aleph (high-end uncountable infinity) (Aleph (high-end uncountable infinity), Aleph (high-end uncountable infinity))], Hyper Hyper Aleph (high-end uncountable infinity) (Aleph (high-end uncountable infinity), Aleph (high-end uncountable infinity))[Hyper Aleph (high-end uncountable infinity) (Aleph (high-end uncountable infinity), Aleph (high-end uncountable infinity)), Hyper Aleph (high-end uncountable infinity) (Aleph (high-end uncountable infinity), Aleph (high-end uncountable infinity))]}}}{ Hyper Hyper Hyper Hyper Hyper Aleph (high-end uncountable infinity) (Aleph (high-end uncountable infinity), Aleph (high-end uncountable infinity))[Hyper Aleph (high-end uncountable infinity) (Aleph (high-end uncountable infinity), Aleph (high-end uncountable infinity)), Hyper Aleph (high-end uncountable infinity) (Aleph (high-end uncountable infinity), Aleph (high-end uncountable infinity))] {Hyper Hyper Aleph (high-end uncountable infinity) (Aleph (high-end uncountable infinity), Aleph (high-end uncountable infinity))[Hyper Aleph (high-end uncountable infinity) (Aleph (high-end uncountable infinity), Aleph (high-end uncountable infinity)), Hyper Aleph (high-end uncountable infinity) (Aleph (high-end uncountable infinity), Aleph (high-end uncountable infinity))], Hyper Hyper Aleph (high-end uncountable infinity) (Aleph (high-end uncountable infinity), Aleph (high-end uncountable infinity))[Hyper Aleph (high-end uncountable infinity) (Aleph (high-end uncountable infinity), Aleph (high-end uncountable infinity)), Hyper Aleph (high-end uncountable infinity) (Aleph (high-end uncountable infinity),

Aleph (high-end uncountable infinity))]} { Hyper Hyper Hyper Aleph (high-end uncountable infinity) (Aleph (high-end uncountable infinity), Aleph (high-end uncountable infinity))[Hyper Aleph (high-end uncountable infinity) (Aleph (high-end uncountable infinity), Aleph (high-end uncountable infinity)), Hyper Aleph (high-end uncountable infinity) (Aleph (high-end uncountable infinity), Aleph (high-end uncountable infinity))] {Hyper Hyper Aleph (high-end uncountable infinity) (Aleph (high-end uncountable infinity), Aleph (high-end uncountable infinity))[Hyper Aleph (high-end uncountable infinity) (Aleph (high-end uncountable infinity), Aleph (high-end uncountable infinity)), Hyper Aleph (high-end uncountable infinity) (Aleph (high-end uncountable infinity), Aleph (high-end uncountable infinity))], Hyper Hyper Aleph (high-end uncountable infinity) (Aleph (high-end uncountable infinity), Aleph (high-end uncountable infinity))[Hyper Aleph (high-end uncountable infinity) (Aleph (high-end uncountable infinity), Aleph (high-end uncountable infinity)), Hyper Aleph (high-end uncountable infinity) (Aleph (high-end uncountable infinity), Aleph (high-end uncountable infinity))]}, Hyper Hyper Hyper Aleph (high-end uncountable infinity) (Aleph (high-end uncountable infinity), Aleph (high-end uncountable infinity))[Hyper Aleph (high-end uncountable infinity) (Aleph (high-end uncountable infinity), Aleph (high-end uncountable infinity)), Hyper Aleph (high-end uncountable infinity) (Aleph (high-end uncountable infinity), Aleph (high-end uncountable infinity))] {Hyper Hyper Aleph (high-end uncountable infinity) (Aleph (high-end uncountable infinity), Aleph (high-end uncountable infinity))[Hyper Aleph (high-end uncountable infinity) (Aleph (high-end uncountable infinity), Aleph (high-end uncountable infinity)), Hyper Aleph (high-end uncountable infinity) (Aleph (high-end uncountable infinity), Aleph (high-end uncountable infinity))], Hyper Hyper Aleph (high-end uncountable infinity) (Aleph (high-end uncountable infinity), Aleph (high-end uncountable infinity))[Hyper Aleph (high-end uncountable infinity) (Aleph (high-end uncountable infinity), Aleph (high-end uncountable infinity)), Hyper Aleph (high-end uncountable infinity) (Aleph (high-end uncountable infinity), Aleph (high-end uncountable infinity))]}}{ Hyper Hyper Hyper Hyper Aleph (high-end uncountable infinity) (Aleph (high-end uncountable infinity), Aleph (high-end uncountable infinity))[Hyper Aleph (high-end uncountable infinity) (Aleph (high-end uncountable infinity), Aleph (high-end uncountable infinity)), Hyper Aleph (high-end uncountable infinity) (Aleph (high-end uncountable infinity), Aleph (high-end uncountable infinity))] {Hyper Hyper Aleph (high-end uncountable infinity) (Aleph (high-end uncountable infinity), Aleph (high-end uncountable infinity))[Hyper Aleph (high-end uncountable infinity) (Aleph (high-end uncountable infinity), Aleph (high-end uncountable infinity)), Hyper Aleph (high-end uncountable infinity) (Aleph (high-end uncountable infinity), Aleph (high-end uncountable infinity))], Hyper Hyper Aleph (high-end uncountable infinity) (Aleph (high-end uncountable infinity), Aleph (high-end uncountable infinity))[Hyper Aleph (high-end uncountable infinity) (Aleph (high-end uncountable infinity), Aleph (high-end uncountable infinity)), Hyper Aleph (high-end uncountable infinity) (Aleph (high-end uncountable infinity), Aleph (high-end uncountable infinity))]} { Hyper Hyper Hyper Aleph (high-end uncountable infinity) (Aleph (high-end uncountable infinity), Aleph (high-end uncountable infinity))[Hyper Aleph (high-end uncountable infinity) (Aleph (high-end uncountable infinity), Aleph (high-end uncountable infinity)), Hyper Aleph (high-end uncountable infinity) (Aleph (high-end uncountable infinity), Aleph (high-end uncountable infinity))] {Hyper Hyper Aleph (high-end uncountable infinity) (Aleph (high-end uncountable infinity), Aleph (high-end uncountable infinity))[Hyper Aleph (high-end uncountable infinity) (Aleph (high-end uncountable infinity), Aleph (high-end uncountable infinity)), Hyper Aleph (high-end uncountable infinity) (Aleph (high-end uncountable infinity), Aleph (high-end uncountable infinity))], Hyper Hyper Aleph (high-end uncountable infinity) (Aleph (high-end uncountable infinity), Aleph (high-end uncountable infinity))[Hyper Aleph (high-end uncountable infinity) (Aleph (high-end uncountable infinity), Aleph (high-end uncountable infinity)), Hyper Aleph (high-end uncountable infinity) (Aleph (high-end uncountable infinity), Aleph (high-end uncountable infinity))]}, Hyper Hyper Hyper Aleph (high-end uncountable infinity) (Aleph (high-end uncountable infinity), Aleph (high-end uncountable infinity))[Hyper Aleph (high-end uncountable infinity) (Aleph (high-end uncountable infinity), Aleph (high-end uncountable infinity)), Hyper Aleph (high-end uncountable infinity) (Aleph (high-end uncountable infinity), Aleph (high-end uncountable infinity))] {Hyper Hyper Aleph (high-end uncountable infinity) (Aleph (high-end uncountable infinity), Aleph (high-end uncountable infinity))[Hyper Aleph (high-end uncountable infinity) (Aleph (high-end uncountable infinity), Aleph (high-end uncountable infinity)), Hyper Aleph (high-end uncountable infinity) (Aleph (high-end uncountable infinity), Aleph (high-end uncountable infinity))], Hyper Hyper Aleph (high-end uncountable infinity) (Aleph (high-end uncountable infinity), Aleph (high-end uncountable infinity))[Hyper Aleph (high-end uncountable infinity) (Aleph (high-end uncountable infinity), Aleph (high-end uncountable infinity)), Hyper Aleph (high-end uncountable infinity) (Aleph (high-end uncountable infinity), Aleph (high-end uncountable infinity))]}}, Hyper Hyper Hyper Hyper Aleph (high-end uncountable infinity) (Aleph (high-end uncountable infinity), Aleph (high-end uncountable infinity))[Hyper Aleph (high-end uncountable infinity) (Aleph (high-end uncountable infinity), Aleph (high-end uncountable infinity)), Hyper Aleph (high-end uncountable infinity) (Aleph (high-end uncountable infinity), Aleph (high-end uncountable infinity))] {Hyper Hyper Aleph (high-end uncountable infinity) (Aleph (high-end uncountable infinity), Aleph (high-end uncountable infinity))[Hyper Aleph (high-end uncountable infinity) (Aleph (high-end uncountable infinity), Aleph (high-end uncountable infinity)), Hyper Aleph (high-end uncountable infinity) (Aleph (high-end uncountable infinity), Aleph (high-end uncountable infinity))], Hyper Hyper Aleph (high-end uncountable infinity) (Aleph (high-end uncountable infinity), Aleph (high-end uncountable infinity))[Hyper Aleph (high-end uncountable infinity) (Aleph (high-end uncountable infinity), Aleph (high-end uncountable infinity)), Hyper Aleph (high-end uncountable infinity) (Aleph (high-end uncountable infinity), Aleph (high-end uncountable infinity))]} { Hyper Hyper Hyper Aleph (high-end uncountable infinity) (Aleph (high-end uncountable infinity), Aleph (high-end uncountable infinity))[Hyper Aleph (high-end uncountable infinity) (Aleph (high-end uncountable infinity), Aleph (high-end uncountable infinity)), Hyper Aleph (high-end uncountable infinity) (Aleph (high-end uncountable infinity), Aleph (high-end uncountable infinity))] {Hyper Hyper Aleph (high-end uncountable infinity) (Aleph (high-end uncountable infinity), Aleph (high-end uncountable infinity))[Hyper Aleph (high-end uncountable infinity) (Aleph (high-end uncountable infinity), Aleph (high-end uncountable infinity)), Hyper Aleph (high-end uncountable infinity) (Aleph (high-end uncountable infinity), Aleph (high-end uncountable infinity))], Hyper Hyper Aleph (high-end uncountable infinity) (Aleph (high-end uncountable infinity), Aleph (high-end uncountable infinity))[Hyper Aleph (high-end uncountable infinity) (Aleph (high-end uncountable infinity), Aleph (high-end uncountable infinity)), Hyper Aleph (high-end uncountable infinity) (Aleph (high-end uncountable infinity), Aleph (high-end uncountable infinity))]}, Hyper Hyper Hyper Aleph (high-end uncountable infinity) (Aleph (high-end uncountable infinity), Aleph (high-end uncountable infinity))[Hyper Aleph (high-end uncountable infinity) (Aleph (high-end uncountable infinity), Aleph (high-end uncountable infinity)), Hyper Aleph (high-end uncountable infinity) (Aleph (high-end uncountable infinity), Aleph (high-end uncountable infinity))] {Hyper Hyper Aleph (high-end uncountable infinity) (Aleph (high-end uncountable infinity), Aleph (high-end uncountable infinity))[Hyper Aleph (high-end uncountable infinity) (Aleph (high-end uncountable infinity), Aleph (high-end uncountable infinity)), Hyper Aleph (high-end uncountable infinity) (Aleph (high-end uncountable infinity), Aleph (high-end uncountable infinity))], Hyper Hyper Aleph (high-end uncountable infinity) (Aleph (high-end uncountable infinity), Aleph (high-end uncountable infinity))[Hyper Aleph (high-end uncountable infinity) (Aleph (high-end uncountable infinity), Aleph (high-end uncountable infinity)), Hyper Aleph (high-end uncountable infinity) (Aleph (high-end uncountable infinity), Aleph (high-end uncountable infinity))]]}}}, Hyper Hyper Hyper Hyper Hyper Aleph (high-end uncountable infinity) (Aleph (high-end uncountable infinity), Aleph (high-end uncountable infinity))[Hyper Aleph (high-end uncountable infinity) (Aleph (high-end uncountable infinity), Aleph (high-end uncountable infinity)), Hyper Aleph (high-end uncountable infinity) (Aleph (high-end uncountable infinity), Aleph (high-end uncountable infinity))] {Hyper Hyper Aleph (high-end uncountable infinity) (Aleph (high-end uncountable infinity), Aleph (high-end uncountable infinity))[Hyper Aleph (high-end uncountable infinity) (Aleph (high-end uncountable infinity), Aleph (high-end uncountable infinity)), Hyper Aleph (high-end uncountable infinity) (Aleph (high-end uncountable infinity), Aleph (high-end uncountable infinity))], Hyper Hyper Aleph (high-end uncountable infinity) (Aleph (high-end uncountable infinity), Aleph (high-end uncountable infinity))[Hyper Aleph (high-end uncountable infinity) (Aleph (high-end uncountable infinity), Aleph (high-end uncountable infinity)), Hyper Aleph (high-end uncountable infinity) (Aleph (high-end uncountable infinity), Aleph (high-end uncountable infinity))]} { Hyper Hyper Hyper Aleph (high-end uncountable infinity) (Aleph (high-end uncountable infinity), Aleph (high-end uncountable infinity))[Hyper Aleph (high-end uncountable infinity) (Aleph (high-end uncountable infinity), Aleph (high-end uncountable infinity)), Hyper Aleph (high-end uncountable infinity) (Aleph (high-end uncountable infinity), Aleph (high-end uncountable infinity))] {Hyper Hyper Aleph (high-end uncountable infinity) (Aleph (high-end uncountable infinity), Aleph (high-end uncountable infinity))[Hyper Aleph (high-end uncountable infinity) (Aleph (high-end uncountable infinity), Aleph (high-end uncountable infinity)), Hyper Aleph (high-end uncountable infinity) (Aleph (high-end uncountable infinity), Aleph (high-end uncountable infinity))], Hyper Hyper Aleph (high-end uncountable infinity) (Aleph (high-end uncountable infinity), Aleph (high-end uncountable infinity))[Hyper Aleph (high-end uncountable infinity) (Aleph (high-end uncountable infinity), Aleph (high-end uncountable infinity)), Hyper Aleph (high-end uncountable infinity) (Aleph (high-end uncountable infinity), Aleph (high-end uncountable infinity))]}, Hyper Hyper Hyper Aleph (high-end uncountable infinity) (Aleph (high-end uncountable infinity), Aleph (high-end uncountable infinity))[Hyper Aleph (high-end uncountable infinity) (Aleph (high-end uncountable infinity), Aleph (high-end uncountable infinity)), Hyper Aleph (high-end uncountable infinity) (Aleph (high-end uncountable infinity), Aleph (high-end uncountable infinity))] {Hyper Hyper Aleph (high-end uncountable infinity) (Aleph (high-end uncountable infinity), Aleph (high-end uncountable infinity))[Hyper Aleph (high-end uncountable infinity) (Aleph (high-end uncountable infinity), Aleph (high-end uncountable infinity)), Hyper Aleph (high-end uncountable infinity) (Aleph (high-end uncountable infinity), Aleph (high-end uncountable infinity))], Hyper Hyper Aleph (high-end uncountable infinity) (Aleph (high-end uncountable infinity), Aleph (high-end uncountable infinity))[Hyper Aleph (high-end uncountable infinity) (Aleph (high-end uncountable infinity), Aleph (high-end uncountable infinity)), Hyper Aleph (high-end uncountable infinity) (Aleph (high-end uncountable infinity), Aleph (high-end uncountable infinity))]}}{ Hyper Hyper Hyper Hyper Aleph (high-end uncountable infinity) (Aleph (high-end uncountable infinity), Aleph (high-end uncountable infinity))[Hyper Aleph (high-end uncountable infinity) (Aleph (high-end uncountable infinity), Aleph (high-end uncountable infinity)), Hyper Aleph (high-end uncountable infinity) (Aleph (high-end uncountable infinity), Aleph (high-end uncountable infinity))] {Hyper Hyper Aleph (high-end uncountable infinity) (Aleph (high-end uncountable infinity), Aleph (high-end uncountable infinity))[Hyper Aleph (high-end uncountable infinity) (Aleph (high-end uncountable infinity), Aleph (high-end uncountable infinity)), Hyper Aleph (high-end uncountable infinity) (Aleph (high-end uncountable infinity), Aleph (high-end uncountable infinity))], Hyper Hyper Aleph (high-end uncountable infinity) (Aleph (high-end uncountable infinity), Aleph (high-end uncountable infinity))[Hyper Aleph (high-end uncountable infinity) (Aleph (high-end uncountable infinity), Aleph (high-end uncountable infinity)), Hyper Aleph (high-end uncountable infinity) (Aleph (high-end uncountable infinity), Aleph (high-end uncountable infinity))]} { Hyper Hyper Hyper Aleph (high-end uncountable infinity) (Aleph (high-end uncountable infinity), Aleph (high-end uncountable infinity))[Hyper Aleph (high-end uncountable infinity) (Aleph (high-end uncountable infinity), Aleph (high-end uncountable infinity)), Hyper Aleph (high-end uncountable infinity) (Aleph (high-end uncountable infinity), Aleph (high-end uncountable infinity))] {Hyper Hyper Aleph (high-end uncountable infinity) (Aleph (high-end uncountable infinity), Aleph (high-end uncountable infinity))[Hyper Aleph (high-end uncountable infinity) (Aleph (high-end uncountable infinity), Aleph (high-end uncountable infinity)), Hyper Aleph (high-end uncountable infinity) (Aleph (high-end uncountable infinity), Aleph (high-end uncountable infinity))], Hyper Hyper Aleph (high-end uncountable infinity) (Aleph (high-end uncountable infinity), Aleph (high-end uncountable infinity))[Hyper Aleph (high-end uncountable infinity) (Aleph (high-end uncountable infinity), Aleph (high-end uncountable infinity)), Hyper Aleph (high-end uncountable infinity) (Aleph (high-end uncountable infinity), Aleph (high-end uncountable infinity))]}, Hyper Hyper Hyper Aleph (high-end uncountable infinity) (Aleph (high-end uncountable infinity), Aleph (high-end uncountable infinity))[Hyper Aleph (high-end uncountable infinity) (Aleph (high-end uncountable infinity), Aleph (high-end uncountable infinity)), Hyper Aleph (high-end uncountable infinity) (Aleph (high-end uncountable infinity), Aleph (high-end uncountable infinity))] {Hyper Hyper Aleph (high-end uncountable infinity) (Aleph (high-end uncountable infinity), Aleph (high-end uncountable infinity))[Hyper Aleph (high-end uncountable infinity) (Aleph (high-end uncountable infinity), Aleph (high-end uncountable infinity)), Hyper Aleph (high-end uncountable infinity) (Aleph (high-end uncountable infinity), Aleph (high-end uncountable infinity))], Hyper Hyper Aleph (high-end uncountable infinity) (Aleph (high-end uncountable infinity), Aleph (high-end uncountable infinity))[Hyper Aleph (high-end uncountable infinity) (Aleph (high-end uncountable infinity), Aleph (high-end uncountable infinity)), Hyper Aleph (high-end uncountable infinity) (Aleph (high-end uncountable infinity), Aleph (high-end uncountable infinity))]}}, Hyper Hyper Hyper Hyper Aleph (high-end uncountable infinity) (Aleph (high-end uncountable infinity), Aleph (high-end uncountable infinity))[Hyper Aleph (high-end uncountable infinity) (Aleph (high-end uncountable infinity), Aleph (high-end uncountable infinity)), Hyper Aleph (high-end uncountable infinity) (Aleph (high-end uncountable infinity), Aleph (high-end uncountable infinity))] {Hyper Hyper Aleph (high-end uncountable infinity) (Aleph (high-end uncountable infinity), Aleph (high-end uncountable infinity))[Hyper Aleph (high-end uncountable infinity) (Aleph (high-end uncountable infinity), Aleph (high-end uncountable infinity)), Hyper Aleph (high-end uncountable infinity) (Aleph (high-end uncountable infinity), Aleph (high-end uncountable infinity))], Hyper Hyper Aleph (high-end uncountable infinity) (Aleph (high-end uncountable infinity), Aleph (high-end uncountable infinity))[Hyper Aleph (high-end uncountable infinity) (Aleph (high-end uncountable infinity), Aleph (high-end uncountable infinity)), Hyper Aleph (high-end uncountable infinity) (Aleph (high-end uncountable infinity), Aleph (high-end uncountable infinity))]} { Hyper Hyper Hyper Aleph (high-end uncountable infinity) (Aleph (high-end uncountable infinity), Aleph (high-end uncountable infinity))[Hyper Aleph (high-end uncountable infinity) (Aleph (high-end uncountable infinity), Aleph (high-end uncountable infinity)), Hyper Aleph (high-end uncountable infinity) (Aleph (high-end uncountable infinity), Aleph (high-end uncountable infinity))] {Hyper Hyper Aleph (high-end uncountable infinity) (Aleph (high-end uncountable infinity), Aleph (high-end uncountable infinity))[Hyper Aleph (high-end uncountable infinity) (Aleph (high-end uncountable infinity), Aleph (high-end uncountable infinity)), Hyper Aleph (high-end uncountable infinity) (Aleph (high-end uncountable infinity), Aleph (high-end uncountable infinity))], Hyper Hyper Aleph (high-end uncountable infinity) (Aleph (high-end uncountable infinity), Aleph (high-end uncountable infinity))[Hyper Aleph (high-end uncountable infinity) (Aleph (high-end uncountable infinity), Aleph (high-end uncountable infinity)), Hyper Aleph (high-end uncountable infinity) (Aleph (high-end uncountable infinity), Aleph (high-end uncountable infinity))]}, Hyper Hyper Hyper Aleph (high-end uncountable infinity) (Aleph (high-end uncountable infinity), Aleph (high-end uncountable infinity))[Hyper Aleph (high-end uncountable infinity) (Aleph (high-end uncountable infinity), Aleph (high-end uncountable infinity)), Hyper Aleph (high-end uncountable infinity) (Aleph (high-end uncountable infinity), Aleph (high-end uncountable infinity))] {Hyper Hyper Aleph (high-end uncountable infinity) (Aleph (high-end uncountable infinity), Aleph (high-end uncountable infinity))[Hyper Aleph (high-end uncountable infinity) (Aleph (high-end uncountable infinity), Aleph (high-end uncountable infinity)), Hyper Aleph (high-end uncountable infinity) (Aleph (high-end uncountable infinity), Aleph (high-end uncountable infinity))], Hyper Hyper Aleph (high-end uncountable infinity) (Aleph (high-end uncountable infinity), Aleph (high-end uncountable infinity))[Hyper Aleph (high-end uncountable infinity) (Aleph (high-end uncountable infinity), Aleph (high-end uncountable infinity)), Hyper Aleph (high-end uncountable infinity) (Aleph (high-end uncountable infinity), Aleph (high-end uncountable infinity))]}}}}{ Hyper Hyper Hyper Hyper Hyper Hyper Aleph (high-end uncountable infinity) (Aleph (high-end uncountable infinity), Aleph (high-end uncountable infinity))[Hyper Aleph (high-end uncountable infinity) (Aleph (high-end uncountable infinity), Aleph (high-end uncountable infinity)), Hyper Aleph (high-end uncountable infinity) (Aleph (high-end uncountable infinity), Aleph (high-end uncountable infinity))] {Hyper Hyper Aleph (high-end uncountable infinity) (Aleph (high-end uncountable infinity), Aleph (high-end uncountable infinity))[Hyper Aleph (high-end uncountable infinity) (Aleph (high-end uncountable infinity), Aleph (high-end uncountable infinity)), Hyper Aleph (high-end uncountable infinity) (Aleph (high-end uncountable infinity), Aleph (high-end uncountable infinity))],

Hyper Hyper Aleph (high-end uncountable infinity) (Aleph (high-end uncountable infinity), Aleph (high-end uncountable infinity))[Hyper Aleph (high-end uncountable infinity) (Aleph (high-end uncountable infinity), Aleph (high-end uncountable infinity)), Hyper Aleph (high-end uncountable infinity) (Aleph (high-end uncountable infinity), Aleph (high-end uncountable infinity))]} { Hyper Hyper Hyper Aleph (high-end uncountable infinity) (Aleph (high-end uncountable infinity), Aleph (high-end uncountable infinity))[Hyper Aleph (high-end uncountable infinity) (Aleph (high-end uncountable infinity), Aleph (high-end uncountable infinity)), Hyper Aleph (high-end uncountable infinity) (Aleph (high-end uncountable infinity), Aleph (high-end uncountable infinity))] {Hyper Hyper Aleph (high-end uncountable infinity) (Aleph (high-end uncountable infinity), Aleph (high-end uncountable infinity))[Hyper Aleph (high-end uncountable infinity) (Aleph (high-end uncountable infinity), Aleph (high-end uncountable infinity)), Hyper Aleph (high-end uncountable infinity) (Aleph (high-end uncountable infinity), Aleph (high-end uncountable infinity))], Hyper Hyper Aleph (high-end uncountable infinity) (Aleph (high-end uncountable infinity), Aleph (high-end uncountable infinity))[Hyper Aleph (high-end uncountable infinity) (Aleph (high-end uncountable infinity), Aleph (high-end uncountable infinity)), Hyper Aleph (high-end uncountable infinity) (Aleph (high-end uncountable infinity), Aleph (high-end uncountable infinity))]}, Hyper Hyper Hyper Aleph (high-end uncountable infinity) (Aleph (high-end uncountable infinity), Aleph (high-end uncountable infinity))[Hyper Aleph (high-end uncountable infinity) (Aleph (high-end uncountable infinity), Aleph (high-end uncountable infinity)), Hyper Aleph (high-end uncountable infinity) (Aleph (high-end uncountable infinity), Aleph (high-end uncountable infinity))] {Hyper Hyper Aleph (high-end uncountable infinity) (Aleph (high-end uncountable infinity), Aleph (high-end uncountable infinity))[Hyper Aleph (high-end uncountable infinity) (Aleph (high-end uncountable infinity), Aleph (high-end uncountable infinity)), Hyper Aleph (high-end uncountable infinity) (Aleph (high-end uncountable infinity), Aleph (high-end uncountable infinity))], Hyper Hyper Aleph (high-end uncountable infinity) (Aleph (high-end uncountable infinity), Aleph (high-end uncountable infinity))[Hyper Aleph (high-end uncountable infinity) (Aleph (high-end uncountable infinity), Aleph (high-end uncountable infinity)), Hyper Aleph (high-end uncountable infinity) (Aleph (high-end uncountable infinity), Aleph (high-end uncountable infinity))]}}{ Hyper Hyper Hyper Hyper Aleph (high-end uncountable infinity) (Aleph (high-end uncountable infinity), Aleph (high-end uncountable infinity))[Hyper Aleph (high-end uncountable infinity) (Aleph (high-end uncountable infinity), Aleph (high-end uncountable infinity)), Hyper Aleph (high-end uncountable infinity) (Aleph (high-end uncountable infinity), Aleph (high-end uncountable infinity))] {Hyper Hyper Aleph (high-end uncountable infinity) (Aleph (high-end uncountable infinity), Aleph (high-end uncountable infinity))[Hyper Aleph (high-end uncountable infinity) (Aleph (high-end uncountable infinity), Aleph (high-end uncountable infinity)), Hyper Aleph (high-end uncountable infinity) (Aleph (high-end uncountable infinity), Aleph (high-end uncountable infinity))], Hyper Hyper Aleph (high-end uncountable infinity) (Aleph (high-end uncountable infinity), Aleph (high-end uncountable infinity))[Hyper Aleph (high-end uncountable infinity) (Aleph (high-end uncountable infinity), Aleph (high-end uncountable infinity)), Hyper Aleph (high-end uncountable infinity) (Aleph (high-end uncountable infinity), Aleph (high-end uncountable infinity))]} { Hyper Hyper Hyper Aleph (high-end uncountable infinity) (Aleph (high-end uncountable infinity), Aleph (high-end uncountable infinity))[Hyper Aleph (high-end uncountable infinity) (Aleph (high-end uncountable infinity), Aleph (high-end uncountable infinity)), Hyper Aleph (high-end uncountable infinity) (Aleph (high-end uncountable infinity), Aleph (high-end uncountable infinity))] {Hyper Hyper Aleph (high-end uncountable infinity) (Aleph (high-end uncountable infinity), Aleph (high-end uncountable infinity))[Hyper Aleph (high-end uncountable infinity) (Aleph (high-end uncountable infinity), Aleph (high-end uncountable infinity)), Hyper Aleph (high-end uncountable infinity) (Aleph (high-end uncountable infinity), Aleph (high-end uncountable infinity))], Hyper Hyper Aleph (high-end uncountable infinity) (Aleph (high-end uncountable infinity), Aleph (high-end uncountable infinity))[Hyper Aleph (high-end uncountable infinity) (Aleph (high-end uncountable infinity), Aleph (high-end uncountable infinity)), Hyper Aleph (high-end uncountable infinity) (Aleph (high-end uncountable infinity), Aleph (high-end uncountable infinity))]}, Hyper Hyper Hyper Aleph (high-end uncountable infinity) (Aleph (high-end uncountable infinity), Aleph (high-end uncountable infinity))[Hyper Aleph (high-end uncountable infinity) (Aleph (high-end uncountable infinity), Aleph (high-end uncountable infinity)), Hyper Aleph (high-end uncountable infinity) (Aleph (high-end uncountable infinity), Aleph (high-end uncountable infinity))] {Hyper Hyper Aleph (high-end uncountable infinity) (Aleph (high-end uncountable infinity), Aleph (high-end uncountable infinity))[Hyper Aleph (high-end uncountable infinity) (Aleph (high-end uncountable infinity), Aleph (high-end uncountable infinity)), Hyper Aleph (high-end uncountable infinity) (Aleph (high-end uncountable infinity), Aleph (high-end uncountable infinity))], Hyper Hyper Aleph (high-end uncountable infinity) (Aleph (high-end uncountable infinity), Aleph (high-end uncountable infinity))[Hyper Aleph (high-end uncountable infinity) (Aleph (high-end uncountable infinity), Aleph (high-end uncountable infinity)), Hyper Aleph (high-end uncountable infinity) (Aleph (high-end uncountable infinity), Aleph (high-end uncountable infinity))]}}, Hyper Hyper Hyper Hyper Aleph (high-end uncountable infinity) (Aleph (high-end uncountable infinity), Aleph (high-end uncountable infinity))[Hyper Aleph (high-end uncountable infinity) (Aleph (high-end uncountable infinity), Aleph (high-end uncountable infinity)), Hyper Aleph (high-end uncountable infinity) (Aleph (high-end uncountable infinity), Aleph (high-end uncountable infinity))] {Hyper Hyper Aleph (high-end uncountable infinity) (Aleph (high-end uncountable infinity), Aleph (high-end uncountable infinity))[Hyper Aleph (high-end uncountable infinity) (Aleph (high-end uncountable infinity), Aleph (high-end uncountable infinity)), Hyper Aleph (high-end uncountable infinity) (Aleph (high-end uncountable infinity), Aleph (high-end uncountable infinity))], Hyper Hyper Aleph (high-end uncountable infinity) (Aleph (high-end uncountable infinity), Aleph (high-end uncountable infinity))[Hyper Aleph (high-end uncountable infinity) (Aleph (high-end uncountable infinity), Aleph (high-end uncountable infinity)), Hyper Aleph (high-end uncountable infinity) (Aleph (high-end uncountable infinity), Aleph (high-end uncountable infinity))]]} { Hyper Hyper Hyper Aleph (high-end uncountable infinity) (Aleph (high-end uncountable infinity), Aleph (high-end uncountable infinity))[Hyper Aleph (high-end uncountable infinity) (Aleph (high-end uncountable infinity), Aleph (high-end uncountable infinity)), Hyper Aleph (high-end uncountable infinity) (Aleph (high-end uncountable infinity), Aleph (high-end uncountable infinity))] {Hyper Hyper Aleph (high-end uncountable infinity) (Aleph (high-end uncountable infinity), Aleph (high-end uncountable infinity))[Hyper Aleph (high-end uncountable infinity) (Aleph (high-end uncountable infinity), Aleph (high-end uncountable infinity)), Hyper Aleph (high-end uncountable infinity) (Aleph (high-end uncountable infinity), Aleph (high-end uncountable infinity))], Hyper Hyper Aleph (high-end uncountable infinity) (Aleph (high-end uncountable infinity), Aleph (high-end uncountable infinity))[Hyper Aleph (high-end uncountable infinity) (Aleph (high-end uncountable infinity), Aleph (high-end uncountable infinity)), Hyper Aleph (high-end uncountable infinity) (Aleph (high-end uncountable infinity), Aleph (high-end uncountable infinity))]}, Hyper Hyper Hyper Aleph (high-end uncountable infinity) (Aleph (high-end uncountable infinity), Aleph (high-end uncountable infinity))[Hyper Aleph (high-end uncountable infinity) (Aleph (high-end uncountable infinity), Aleph (high-end uncountable infinity)), Hyper Aleph (high-end uncountable infinity) (Aleph (high-end uncountable infinity), Aleph (high-end uncountable infinity))] {Hyper Hyper Aleph (high-end uncountable infinity) (Aleph (high-end uncountable infinity), Aleph (high-end uncountable infinity))[Hyper Aleph (high-end uncountable infinity) (Aleph (high-end uncountable infinity), Aleph (high-end uncountable infinity)), Hyper Aleph (high-end uncountable infinity) (Aleph (high-end uncountable infinity), Aleph (high-end uncountable infinity))], Hyper Hyper Aleph (high-end uncountable infinity) (Aleph (high-end uncountable infinity), Aleph (high-end uncountable infinity))[Hyper Aleph (high-end uncountable infinity) (Aleph (high-end uncountable infinity), Aleph (high-end uncountable infinity)), Hyper Aleph (high-end uncountable infinity) (Aleph (high-end uncountable infinity), Aleph (high-end uncountable infinity))]]}}}{ Hyper Hyper Hyper Hyper Hyper Aleph (high-end uncountable infinity) (Aleph (high-end uncountable infinity), Aleph (high-end uncountable infinity))[Hyper Aleph (high-end uncountable infinity) (Aleph (high-end uncountable infinity), Aleph (high-end uncountable infinity)), Hyper Aleph (high-end uncountable infinity) (Aleph (high-end uncountable infinity), Aleph (high-end uncountable infinity))] {Hyper Hyper Aleph (high-end uncountable infinity) (Aleph (high-end uncountable infinity), Aleph (high-end uncountable infinity))[Hyper Aleph (high-end uncountable infinity) (Aleph (high-end uncountable infinity), Aleph (high-end uncountable infinity)), Hyper Aleph (high-end uncountable infinity) (Aleph (high-end uncountable infinity), Aleph (high-end uncountable infinity))], Hyper Hyper Aleph (high-end uncountable infinity) (Aleph (high-end uncountable infinity), Aleph (high-end uncountable infinity))[Hyper Aleph (high-end uncountable infinity) (Aleph (high-end uncountable infinity), Aleph (high-end uncountable infinity)), Hyper Aleph (high-end uncountable infinity) (Aleph (high-end uncountable infinity), Aleph (high-end uncountable infinity))]} { Hyper Hyper Hyper Aleph (high-end uncountable infinity) (Aleph (high-end uncountable infinity), Aleph (high-end uncountable infinity))[Hyper Aleph (high-end uncountable infinity) (Aleph (high-end uncountable infinity), Aleph (high-end uncountable infinity)), Hyper Aleph (high-end uncountable infinity) (Aleph (high-end uncountable infinity), Aleph (high-end uncountable infinity))] {Hyper Hyper Aleph (high-end uncountable infinity) (Aleph (high-end uncountable infinity), Aleph (high-end uncountable infinity))[Hyper Aleph (high-end uncountable infinity) (Aleph (high-end uncountable infinity), Aleph (high-end uncountable infinity)), Hyper Aleph (high-end uncountable infinity) (Aleph (high-end uncountable infinity), Aleph (high-end uncountable infinity))], Hyper Hyper Aleph (high-end uncountable infinity) (Aleph (high-end uncountable infinity), Aleph (high-end uncountable infinity))[Hyper Aleph (high-end uncountable infinity) (Aleph (high-end uncountable infinity), Aleph (high-end uncountable infinity)), Hyper Aleph (high-end uncountable infinity) (Aleph (high-end uncountable infinity), Aleph (high-end uncountable infinity))]}, Hyper Hyper Hyper Aleph (high-end uncountable infinity) (Aleph (high-end uncountable infinity), Aleph (high-end uncountable infinity))[Hyper Aleph (high-end uncountable infinity) (Aleph (high-end uncountable infinity), Aleph (high-end uncountable infinity)), Hyper Aleph (high-end uncountable infinity) (Aleph (high-end uncountable infinity), Aleph (high-end uncountable infinity))] {Hyper Hyper Aleph (high-end uncountable infinity) (Aleph (high-end uncountable infinity), Aleph (high-end uncountable infinity))[Hyper Aleph (high-end uncountable infinity) (Aleph (high-end uncountable infinity), Aleph (high-end uncountable infinity)), Hyper Aleph (high-end uncountable infinity) (Aleph (high-end uncountable infinity), Aleph (high-end uncountable infinity))], Hyper Hyper Aleph (high-end uncountable infinity) (Aleph (high-end uncountable infinity), Aleph (high-end uncountable infinity))[Hyper Aleph (high-end uncountable infinity) (Aleph (high-end uncountable infinity), Aleph (high-end uncountable infinity)), Hyper Aleph (high-end uncountable infinity) (Aleph (high-end uncountable infinity), Aleph (high-end uncountable infinity))]}}{ Hyper Hyper Hyper Hyper Aleph (high-end uncountable infinity) (Aleph (high-end uncountable infinity), Aleph (high-end uncountable infinity))[Hyper Aleph (high-end uncountable infinity) (Aleph (high-end uncountable infinity), Aleph (high-end uncountable infinity)), Hyper Aleph (high-end uncountable infinity) (Aleph (high-end uncountable infinity), Aleph (high-end uncountable infinity))] {Hyper Hyper Aleph (high-end uncountable infinity) (Aleph (high-end uncountable infinity), Aleph (high-end uncountable infinity))[Hyper

Aleph (high-end uncountable infinity) (Aleph (high-end uncountable infinity), Aleph (high-end uncountable infinity)), Hyper Aleph (high-end uncountable infinity) (Aleph (high-end uncountable infinity), Aleph (high-end uncountable infinity))], Hyper Hyper Aleph (high-end uncountable infinity) (Aleph (high-end uncountable infinity), Aleph (high-end uncountable infinity))[Hyper Aleph (high-end uncountable infinity) (Aleph (high-end uncountable infinity), Aleph (high-end uncountable infinity)), Hyper Aleph (high-end uncountable infinity) (Aleph (high-end uncountable infinity), Aleph (high-end uncountable infinity))]} { Hyper Hyper Hyper Aleph (high-end uncountable infinity) (Aleph (high-end uncountable infinity), Aleph (high-end uncountable infinity))[Hyper Aleph (high-end uncountable infinity) (Aleph (high-end uncountable infinity), Aleph (high-end uncountable infinity)), Hyper Aleph (high-end uncountable infinity) (Aleph (high-end uncountable infinity), Aleph (high-end uncountable infinity))] {Hyper Hyper Aleph (high-end uncountable infinity) (Aleph (high-end uncountable infinity), Aleph (high-end uncountable infinity))[Hyper Aleph (high-end uncountable infinity) (Aleph (high-end uncountable infinity), Aleph (high-end uncountable infinity)), Hyper Aleph (high-end uncountable infinity) (Aleph (high-end uncountable infinity), Aleph (high-end uncountable infinity))], Hyper Hyper Aleph (high-end uncountable infinity) (Aleph (high-end uncountable infinity), Aleph (high-end uncountable infinity))[Hyper Aleph (high-end uncountable infinity) (Aleph (high-end uncountable infinity), Aleph (high-end uncountable infinity)), Hyper Aleph (high-end uncountable infinity) (Aleph (high-end uncountable infinity), Aleph (high-end uncountable infinity))]}, Hyper Hyper Hyper Aleph (high-end uncountable infinity) (Aleph (high-end uncountable infinity), Aleph (high-end uncountable infinity))[Hyper Aleph (high-end uncountable infinity) (Aleph (high-end uncountable infinity), Aleph (high-end uncountable infinity)), Hyper Aleph (high-end uncountable infinity) (Aleph (high-end uncountable infinity), Aleph (high-end uncountable infinity))] {Hyper Hyper Aleph (high-end uncountable infinity) (Aleph (high-end uncountable infinity), Aleph (high-end uncountable infinity))[Hyper Aleph (high-end uncountable infinity) (Aleph (high-end uncountable infinity), Aleph (high-end uncountable infinity)), Hyper Aleph (high-end uncountable infinity) (Aleph (high-end uncountable infinity), Aleph (high-end uncountable infinity))], Hyper Hyper Aleph (high-end uncountable infinity) (Aleph (high-end uncountable infinity), Aleph (high-end uncountable infinity))[Hyper Aleph (high-end uncountable infinity) (Aleph (high-end uncountable infinity), Aleph (high-end uncountable infinity)), Hyper Aleph (high-end uncountable infinity) (Aleph (high-end uncountable infinity), Aleph (high-end uncountable infinity))]}}, Hyper Hyper Hyper Hyper Aleph (high-end uncountable infinity) (Aleph (high-end uncountable infinity), Aleph (high-end uncountable infinity))[Hyper Aleph (high-end uncountable infinity) (Aleph (high-end uncountable infinity), Aleph (high-end uncountable infinity)), Hyper Aleph (high-end uncountable infinity) (Aleph (high-end uncountable infinity), Aleph (high-end uncountable infinity))] {Hyper Hyper Aleph (high-end uncountable infinity) (Aleph (high-end uncountable infinity), Aleph (high-end uncountable infinity))[Hyper Aleph (high-end uncountable infinity) (Aleph (high-end uncountable infinity), Aleph (high-end uncountable infinity)), Hyper Aleph (high-end uncountable infinity) (Aleph (high-end uncountable infinity), Aleph (high-end uncountable infinity))], Hyper Hyper Aleph (high-end uncountable infinity) (Aleph (high-end uncountable infinity), Aleph (high-end uncountable infinity))[Hyper Aleph (high-end uncountable infinity) (Aleph (high-end uncountable infinity), Aleph (high-end uncountable infinity)), Hyper Aleph (high-end uncountable infinity) (Aleph (high-end uncountable infinity), Aleph (high-end uncountable infinity))]} { Hyper Hyper Hyper Aleph (high-end uncountable infinity) (Aleph (high-end uncountable infinity), Aleph (high-end uncountable infinity))[Hyper Aleph (high-end uncountable infinity) (Aleph (high-end uncountable infinity), Aleph (high-end uncountable infinity)), Hyper Aleph (high-end uncountable infinity) (Aleph (high-end uncountable infinity), Aleph (high-end uncountable infinity))] {Hyper Hyper Aleph (high-end uncountable infinity) (Aleph (high-end uncountable infinity), Aleph (high-end uncountable infinity))[Hyper Aleph (high-end uncountable infinity) (Aleph (high-end uncountable infinity), Aleph (high-end uncountable infinity)), Hyper Aleph (high-end uncountable infinity) (Aleph (high-end uncountable infinity), Aleph (high-end uncountable infinity))], Hyper Hyper Aleph (high-end uncountable infinity) (Aleph (high-end uncountable infinity), Aleph (high-end uncountable infinity))[Hyper Aleph (high-end uncountable infinity) (Aleph (high-end uncountable infinity), Aleph (high-end uncountable infinity)), Hyper Aleph (high-end uncountable infinity) (Aleph (high-end uncountable infinity), Aleph (high-end uncountable infinity))]}, Hyper Hyper Hyper Aleph (high-end uncountable infinity) (Aleph (high-end uncountable infinity), Aleph (high-end uncountable infinity))[Hyper Aleph (high-end uncountable infinity) (Aleph (high-end uncountable infinity), Aleph (high-end uncountable infinity)), Hyper Aleph (high-end uncountable infinity) (Aleph (high-end uncountable infinity), Aleph (high-end uncountable infinity))] {Hyper Hyper Aleph (high-end uncountable infinity) (Aleph (high-end uncountable infinity), Aleph (high-end uncountable infinity))[Hyper Aleph (high-end uncountable infinity) (Aleph (high-end uncountable infinity), Aleph (high-end uncountable infinity)), Hyper Aleph (high-end uncountable infinity) (Aleph (high-end uncountable infinity), Aleph (high-end uncountable infinity))], Hyper Hyper Aleph (high-end uncountable infinity) (Aleph (high-end uncountable infinity), Aleph (high-end uncountable infinity))[Hyper Aleph (high-end uncountable infinity) (Aleph (high-end uncountable infinity), Aleph (high-end uncountable infinity)), Hyper Aleph (high-end uncountable infinity) (Aleph (high-end uncountable infinity), Aleph (high-end uncountable infinity))]}}}, Hyper Hyper Hyper Hyper Hyper Aleph (high-end uncountable infinity) (Aleph (high-end uncountable infinity), Aleph (high-end uncountable infinity))[Hyper Aleph (high-end uncountable infinity) (Aleph (high-end uncountable infinity), Aleph (high-end uncountable infinity)), Hyper Aleph (high-end uncountable infinity) (Aleph (high-end uncountable infinity), Aleph (high-end uncountable infinity))] {Hyper Hyper Aleph (high-end uncountable infinity) (Aleph (high-end uncountable infinity), Aleph (high-end uncountable infinity))[Hyper Aleph (high-end uncountable infinity) (Aleph (high-end uncountable infinity), Aleph (high-end uncountable infinity)), Hyper Aleph (high-end uncountable infinity) (Aleph (high-end uncountable infinity), Aleph (high-end uncountable infinity))], Hyper Hyper Aleph (high-end uncountable infinity) (Aleph (high-end uncountable infinity), Aleph (high-end uncountable infinity))[Hyper Aleph (high-end uncountable infinity) (Aleph (high-end uncountable infinity), Aleph (high-end uncountable infinity)), Hyper Aleph (high-end uncountable infinity) (Aleph (high-end uncountable infinity), Aleph (high-end uncountable infinity))]} { Hyper Hyper Hyper Aleph (high-end uncountable infinity) (Aleph (high-end uncountable infinity), Aleph (high-end uncountable infinity))[Hyper Aleph (high-end uncountable infinity) (Aleph (high-end uncountable infinity), Aleph (high-end uncountable infinity)), Hyper Aleph (high-end uncountable infinity) (Aleph (high-end uncountable infinity), Aleph (high-end uncountable infinity))] {Hyper Hyper Aleph (high-end uncountable infinity) (Aleph (high-end uncountable infinity), Aleph (high-end uncountable infinity))[Hyper Aleph (high-end uncountable infinity) (Aleph (high-end uncountable infinity), Aleph (high-end uncountable infinity)), Hyper Aleph (high-end uncountable infinity) (Aleph (high-end uncountable infinity), Aleph (high-end uncountable infinity))], Hyper Hyper Aleph (high-end uncountable infinity) (Aleph (high-end uncountable infinity), Aleph (high-end uncountable infinity))[Hyper Aleph (high-end uncountable infinity) (Aleph (high-end uncountable infinity), Aleph (high-end uncountable infinity)), Hyper Aleph (high-end uncountable infinity) (Aleph (high-end uncountable infinity), Aleph (high-end uncountable infinity))]}, Hyper Hyper Hyper Aleph (high-end uncountable infinity) (Aleph (high-end uncountable infinity), Aleph (high-end uncountable infinity))[Hyper Aleph (high-end uncountable infinity) (Aleph (high-end uncountable infinity), Aleph (high-end uncountable infinity)), Hyper Aleph (high-end uncountable infinity) (Aleph (high-end uncountable infinity), Aleph (high-end uncountable infinity))] {Hyper Hyper Aleph (high-end uncountable infinity) (Aleph (high-end uncountable infinity), Aleph (high-end uncountable infinity))[Hyper Aleph (high-end uncountable infinity) (Aleph (high-end uncountable infinity), Aleph (high-end uncountable infinity)), Hyper Aleph (high-end uncountable infinity) (Aleph (high-end uncountable infinity), Aleph (high-end uncountable infinity))], Hyper Hyper Aleph (high-end uncountable infinity) (Aleph (high-end uncountable infinity), Aleph (high-end uncountable infinity))[Hyper Aleph (high-end uncountable infinity) (Aleph (high-end uncountable infinity), Aleph (high-end uncountable infinity)), Hyper Aleph (high-end uncountable infinity) (Aleph (high-end uncountable infinity), Aleph (high-end uncountable infinity))]}}{ Hyper Hyper Hyper Hyper Aleph (high-end uncountable infinity) (Aleph (high-end uncountable infinity), Aleph (high-end uncountable infinity))[Hyper Aleph (high-end uncountable infinity) (Aleph (high-end uncountable infinity), Aleph (high-end uncountable infinity)), Hyper Aleph (high-end uncountable infinity) (Aleph (high-end uncountable infinity), Aleph (high-end uncountable infinity))] {Hyper Hyper Aleph (high-end uncountable infinity) (Aleph (high-end uncountable infinity), Aleph (high-end uncountable infinity))[Hyper Aleph (high-end uncountable infinity) (Aleph (high-end uncountable infinity), Aleph (high-end uncountable infinity)), Hyper Aleph (high-end uncountable infinity) (Aleph (high-end uncountable infinity), Aleph (high-end uncountable infinity))], Hyper Hyper Aleph (high-end uncountable infinity) (Aleph (high-end uncountable infinity), Aleph (high-end uncountable infinity))[Hyper Aleph (high-end uncountable infinity) (Aleph (high-end uncountable infinity), Aleph (high-end uncountable infinity)), Hyper Aleph (high-end uncountable infinity) (Aleph (high-end uncountable infinity), Aleph (high-end uncountable infinity))]} { Hyper Hyper Hyper Aleph (high-end uncountable infinity) (Aleph (high-end uncountable infinity), Aleph (high-end uncountable infinity))[Hyper Aleph (high-end uncountable infinity) (Aleph (high-end uncountable infinity), Aleph (high-end uncountable infinity)), Hyper Aleph (high-end uncountable infinity) (Aleph (high-end uncountable infinity), Aleph (high-end uncountable infinity))] {Hyper Hyper Aleph (high-end uncountable infinity) (Aleph (high-end uncountable infinity), Aleph (high-end uncountable infinity))[Hyper Aleph (high-end uncountable infinity) (Aleph (high-end uncountable infinity), Aleph (high-end uncountable infinity)), Hyper Aleph (high-end uncountable infinity) (Aleph (high-end uncountable infinity), Aleph (high-end uncountable infinity))], Hyper Hyper Aleph (high-end uncountable infinity) (Aleph (high-end uncountable infinity), Aleph (high-end uncountable infinity))[Hyper Aleph (high-end uncountable infinity) (Aleph (high-end uncountable infinity), Aleph (high-end uncountable infinity)), Hyper Aleph (high-end uncountable infinity) (Aleph (high-end uncountable infinity), Aleph (high-end uncountable infinity))]}, Hyper Hyper Hyper Aleph (high-end uncountable infinity) (Aleph (high-end uncountable infinity), Aleph (high-end uncountable infinity))[Hyper Aleph (high-end uncountable infinity) (Aleph (high-end uncountable infinity), Aleph (high-end uncountable infinity)), Hyper Aleph (high-end uncountable infinity) (Aleph (high-end uncountable infinity), Aleph (high-end uncountable infinity))] {Hyper Hyper Aleph (high-end uncountable infinity) (Aleph (high-end uncountable infinity), Aleph (high-end uncountable infinity))[Hyper Aleph (high-end uncountable infinity) (Aleph (high-end uncountable infinity), Aleph (high-end uncountable infinity)), Hyper Aleph (high-end uncountable infinity) (Aleph (high-end uncountable infinity), Aleph (high-end uncountable infinity))], Hyper Hyper Aleph (high-end uncountable infinity) (Aleph (high-end uncountable infinity), Aleph (high-end uncountable infinity))[Hyper Aleph (high-end uncountable infinity) (Aleph (high-end uncountable infinity), Aleph (high-end uncountable infinity)), Hyper Aleph (high-end uncountable infinity) (Aleph (high-end uncountable infinity), Aleph (high-end uncountable infinity))]}}, Hyper Hyper Hyper Hyper Aleph (high-end uncountable infinity) (Aleph (high-end uncountable infinity), Aleph (high-end uncountable infinity))[Hyper Aleph (high-end uncountable infinity) (Aleph (high-end uncountable infinity), Aleph (high-end uncountable infinity)), Hyper Aleph (high-end uncountable infinity) (Aleph (high-end uncountable infinity), Aleph (high-end uncountable infinity))] {Hyper Hyper Aleph (high-end uncountable infinity) (Aleph (high-end uncountable infinity), Aleph (high-end uncountable infinity))[Hyper Aleph (high-end uncountable infinity) (Aleph (high-end uncountable infinity), Aleph (high-end uncountable infinity)), Hyper Aleph (high-end uncountable infinity) (Aleph (high-end uncountable infinity), Aleph (high-end uncountable infinity))], Hyper Hyper Aleph (high-end uncountable infinity) (Aleph (high-end uncountable infinity), Aleph (high-end uncountable infinity))[Hyper Aleph (high-end uncountable infinity) (Aleph (high-end uncountable infinity), Aleph (high-end uncountable infinity)), Hyper Aleph (high-end uncountable infinity) (Aleph (high-end uncountable infinity), Aleph (high-end uncountable infinity))]} { Hyper Hyper Hyper Aleph (high-end uncountable infinity) (Aleph (high-end uncountable infinity), Aleph (high-end uncountable infinity))[Hyper Aleph (high-end uncountable infinity) (Aleph (high-end uncountable infinity), Aleph (high-end uncountable infinity)), Hyper Aleph (high-end uncountable infinity) (Aleph (high-end uncountable infinity), Aleph (high-end uncountable infinity))] {Hyper Hyper Aleph (high-end uncountable infinity) (Aleph (high-end uncountable infinity), Aleph (high-end uncountable infinity))[Hyper Aleph (high-end uncountable infinity) (Aleph (high-end uncountable infinity), Aleph (high-end uncountable infinity)), Hyper Aleph (high-end uncountable infinity) (Aleph (high-end uncountable infinity), Aleph (high-end uncountable infinity))], Hyper Hyper Aleph (high-end uncountable infinity) (Aleph (high-end uncountable infinity), Aleph (high-end uncountable infinity))[Hyper Aleph (high-end uncountable infinity) (Aleph (high-end uncountable infinity), Aleph (high-end uncountable infinity)), Hyper Aleph (high-end uncountable infinity) (Aleph (high-end uncountable infinity), Aleph (high-end uncountable infinity))]}, Hyper Hyper Hyper Aleph (high-end uncountable infinity) (Aleph (high-end uncountable infinity), Aleph (high-end uncountable infinity))[Hyper Aleph (high-end uncountable infinity) (Aleph (high-end uncountable infinity), Aleph (high-end uncountable infinity)), Hyper Aleph (high-end uncountable infinity) (Aleph (high-end uncountable infinity), Aleph (high-end uncountable infinity))] {Hyper Hyper Aleph (high-end uncountable infinity) (Aleph (high-end uncountable infinity), Aleph (high-end uncountable infinity))[Hyper Aleph (high-end uncountable infinity) (Aleph (high-end uncountable infinity), Aleph (high-end uncountable infinity)), Hyper Aleph (high-end uncountable infinity) (Aleph (high-end uncountable infinity), Aleph (high-end uncountable infinity))], Hyper Hyper Aleph (high-end uncountable infinity) (Aleph (high-end uncountable infinity), Aleph (high-end uncountable infinity))[Hyper Aleph (high-end uncountable infinity) (Aleph (high-end uncountable infinity), Aleph (high-end uncountable infinity)), Hyper Aleph (high-end uncountable infinity) (Aleph (high-end uncountable infinity), Aleph (high-end uncountable infinity))]}}}}, Hyper Hyper Hyper Hyper Hyper Hyper Aleph (high-end uncountable infinity) (Aleph (high-end uncountable infinity), Aleph (high-end uncountable infinity))[Hyper Aleph (high-end uncountable infinity) (Aleph (high-end uncountable infinity), Aleph (high-end uncountable infinity)),

Hyper Aleph (high-end uncountable infinity) (Aleph (high-end uncountable infinity), Aleph (high-end uncountable infinity))] {Hyper Hyper Aleph (high-end uncountable infinity) (Aleph (high-end uncountable infinity), Aleph (high-end uncountable infinity))[Hyper Aleph (high-end uncountable infinity) (Aleph (high-end uncountable infinity), Aleph (high-end uncountable infinity)), Hyper Aleph (high-end uncountable infinity) (Aleph (high-end uncountable infinity), Aleph (high-end uncountable infinity))], Hyper Hyper Aleph (high-end uncountable infinity) (Aleph (high-end uncountable infinity), Aleph (high-end uncountable infinity))[Hyper Aleph (high-end uncountable infinity) (Aleph (high-end uncountable infinity), Aleph (high-end uncountable infinity)), Hyper Aleph (high-end uncountable infinity) (Aleph (high-end uncountable infinity), Aleph (high-end uncountable infinity))]} { Hyper Hyper Hyper Aleph (high-end uncountable infinity) (Aleph (high-end uncountable infinity), Aleph (high-end uncountable infinity))[Hyper Aleph (high-end uncountable infinity) (Aleph (high-end uncountable infinity), Aleph (high-end uncountable infinity)), Hyper Aleph (high-end uncountable infinity) (Aleph (high-end uncountable infinity), Aleph (high-end uncountable infinity))] {Hyper Hyper Aleph (high-end uncountable infinity) (Aleph (high-end uncountable infinity), Aleph (high-end uncountable infinity))[Hyper Aleph (high-end uncountable infinity) (Aleph (high-end uncountable infinity), Aleph (high-end uncountable infinity)), Hyper Aleph (high-end uncountable infinity) (Aleph (high-end uncountable infinity), Aleph (high-end uncountable infinity))], Hyper Hyper Aleph (high-end uncountable infinity) (Aleph (high-end uncountable infinity), Aleph (high-end uncountable infinity))[Hyper Aleph (high-end uncountable infinity) (Aleph (high-end uncountable infinity), Aleph (high-end uncountable infinity)), Hyper Aleph (high-end uncountable infinity) (Aleph (high-end uncountable infinity), Aleph (high-end uncountable infinity))]}, Hyper Hyper Hyper Aleph (high-end uncountable infinity) (Aleph (high-end uncountable infinity), Aleph (high-end uncountable infinity))[Hyper Aleph (high-end uncountable infinity) (Aleph (high-end uncountable infinity), Aleph (high-end uncountable infinity)), Hyper Aleph (high-end uncountable infinity) (Aleph (high-end uncountable infinity), Aleph (high-end uncountable infinity))] {Hyper Hyper Aleph (high-end uncountable infinity) (Aleph (high-end uncountable infinity), Aleph (high-end uncountable infinity))[Hyper Aleph (high-end uncountable infinity) (Aleph (high-end uncountable infinity), Aleph (high-end uncountable infinity)), Hyper Aleph (high-end uncountable infinity) (Aleph (high-end uncountable infinity), Aleph (high-end uncountable infinity))], Hyper Hyper Aleph (high-end uncountable infinity) (Aleph (high-end uncountable infinity), Aleph (high-end uncountable infinity))[Hyper Aleph (high-end uncountable infinity) (Aleph (high-end uncountable infinity), Aleph (high-end uncountable infinity)), Hyper Aleph (high-end uncountable infinity) (Aleph (high-end uncountable infinity), Aleph (high-end uncountable infinity))]}}{ Hyper Hyper Hyper Hyper Aleph (high-end uncountable infinity) (Aleph (high-end uncountable infinity), Aleph (high-end uncountable infinity))[Hyper Aleph (high-end uncountable infinity)

(Aleph (high-end uncountable infinity), Aleph (high-end uncountable infinity)), Hyper Aleph (high-end uncountable infinity) (Aleph (high-end uncountable infinity), Aleph (high-end uncountable infinity))] {Hyper Hyper Aleph (high-end uncountable infinity) (Aleph (high-end uncountable infinity), Aleph (high-end uncountable infinity))[Hyper Aleph (high-end uncountable infinity) (Aleph (high-end uncountable infinity), Aleph (high-end uncountable infinity)), Hyper Aleph (high-end uncountable infinity) (Aleph (high-end uncountable infinity), Aleph (high-end uncountable infinity))], Hyper Hyper Aleph (high-end uncountable infinity) (Aleph (high-end uncountable infinity), Aleph (high-end uncountable infinity))[Hyper Aleph (high-end uncountable infinity) (Aleph (high-end uncountable infinity), Aleph (high-end uncountable infinity)), Hyper Aleph (high-end uncountable infinity) (Aleph (high-end uncountable infinity), Aleph (high-end uncountable infinity))]} { Hyper Hyper Hyper Aleph (high-end uncountable infinity) (Aleph (high-end uncountable infinity), Aleph (high-end uncountable infinity))[Hyper Aleph (high-end uncountable infinity) (Aleph (high-end uncountable infinity), Aleph (high-end uncountable infinity)), Hyper Aleph (high-end uncountable infinity) (Aleph (high-end uncountable infinity), Aleph (high-end uncountable infinity))] {Hyper Hyper Aleph (high-end uncountable infinity) (Aleph (high-end uncountable infinity), Aleph (high-end uncountable infinity))[Hyper Aleph (high-end uncountable infinity) (Aleph (high-end uncountable infinity), Aleph (high-end uncountable infinity)), Hyper Aleph (high-end uncountable infinity) (Aleph (high-end uncountable infinity), Aleph (high-end uncountable infinity))], Hyper Hyper Aleph (high-end uncountable infinity) (Aleph (high-end uncountable infinity), Aleph (high-end uncountable infinity))[Hyper Aleph (high-end uncountable infinity) (Aleph (high-end uncountable infinity), Aleph (high-end uncountable infinity)), Hyper Aleph (high-end uncountable infinity) (Aleph (high-end uncountable infinity), Aleph (high-end uncountable infinity))]}, Hyper Hyper Hyper Aleph (high-end uncountable infinity) (Aleph (high-end uncountable infinity), Aleph (high-end uncountable infinity))[Hyper Aleph (high-end uncountable infinity) (Aleph (high-end uncountable infinity), Aleph (high-end uncountable infinity)), Hyper Aleph (high-end uncountable infinity) (Aleph (high-end uncountable infinity), Aleph (high-end uncountable infinity))] {Hyper Hyper Aleph (high-end uncountable infinity) (Aleph (high-end uncountable infinity), Aleph (high-end uncountable infinity))[Hyper Aleph (high-end uncountable infinity) (Aleph (high-end uncountable infinity), Aleph (high-end uncountable infinity)), Hyper Aleph (high-end uncountable infinity) (Aleph (high-end uncountable infinity), Aleph (high-end uncountable infinity))], Hyper Hyper Aleph (high-end uncountable infinity) (Aleph (high-end uncountable infinity), Aleph (high-end uncountable infinity))[Hyper Aleph (high-end uncountable infinity) (Aleph (high-end uncountable infinity), Aleph (high-end uncountable infinity)), Hyper Aleph (high-end uncountable infinity) (Aleph (high-end uncountable infinity), Aleph (high-end uncountable infinity))]}}, Hyper Hyper Hyper Hyper Aleph (high-end uncountable infinity) (Aleph (high-end uncountable infinity), Aleph (high-end uncountable infinity))[Hyper

Aleph (high-end uncountable infinity) (Aleph (high-end uncountable infinity), Aleph (high-end uncountable infinity)), Hyper Aleph (high-end uncountable infinity) (Aleph (high-end uncountable infinity), Aleph (high-end uncountable infinity))] {Hyper Hyper Aleph (high-end uncountable infinity) (Aleph (high-end uncountable infinity), Aleph (high-end uncountable infinity))[Hyper Aleph (high-end uncountable infinity) (Aleph (high-end uncountable infinity), Aleph (high-end uncountable infinity)), Hyper Aleph (high-end uncountable infinity) (Aleph (high-end uncountable infinity), Aleph (high-end uncountable infinity))], Hyper Hyper Aleph (high-end uncountable infinity) (Aleph (high-end uncountable infinity), Aleph (high-end uncountable infinity))[Hyper Aleph (high-end uncountable infinity) (Aleph (high-end uncountable infinity), Aleph (high-end uncountable infinity)), Hyper Aleph (high-end uncountable infinity) (Aleph (high-end uncountable infinity), Aleph (high-end uncountable infinity))]} { Hyper Hyper Hyper Aleph (high-end uncountable infinity) (Aleph (high-end uncountable infinity), Aleph (high-end uncountable infinity))[Hyper Aleph (high-end uncountable infinity) (Aleph (high-end uncountable infinity), Aleph (high-end uncountable infinity)), Hyper Aleph (high-end uncountable infinity) (Aleph (high-end uncountable infinity), Aleph (high-end uncountable infinity))] {Hyper Hyper Aleph (high-end uncountable infinity) (Aleph (high-end uncountable infinity), Aleph (high-end uncountable infinity))[Hyper Aleph (high-end uncountable infinity) (Aleph (high-end uncountable infinity), Aleph (high-end uncountable infinity)), Hyper Aleph (high-end uncountable infinity) (Aleph (high-end uncountable infinity), Aleph (high-end uncountable infinity))], Hyper Hyper Aleph (high-end uncountable infinity) (Aleph (high-end uncountable infinity), Aleph (high-end uncountable infinity))[Hyper Aleph (high-end uncountable infinity) (Aleph (high-end uncountable infinity), Aleph (high-end uncountable infinity)), Hyper Aleph (high-end uncountable infinity) (Aleph (high-end uncountable infinity), Aleph (high-end uncountable infinity))]}, Hyper Hyper Hyper Aleph (high-end uncountable infinity) (Aleph (high-end uncountable infinity), Aleph (high-end uncountable infinity))[Hyper Aleph (high-end uncountable infinity) (Aleph (high-end uncountable infinity), Aleph (high-end uncountable infinity)), Hyper Aleph (high-end uncountable infinity) (Aleph (high-end uncountable infinity), Aleph (high-end uncountable infinity))] {Hyper Hyper Aleph (high-end uncountable infinity) (Aleph (high-end uncountable infinity), Aleph (high-end uncountable infinity))[Hyper Aleph (high-end uncountable infinity) (Aleph (high-end uncountable infinity), Aleph (high-end uncountable infinity)), Hyper Aleph (high-end uncountable infinity) (Aleph (high-end uncountable infinity), Aleph (high-end uncountable infinity))], Hyper Hyper Aleph (high-end uncountable infinity) (Aleph (high-end uncountable infinity), Aleph (high-end uncountable infinity))[Hyper Aleph (high-end uncountable infinity) (Aleph (high-end uncountable infinity), Aleph (high-end uncountable infinity)), Hyper Aleph (high-end uncountable infinity) (Aleph (high-end uncountable infinity), Aleph (high-end uncountable infinity))]}}}{ Hyper Hyper Hyper Hyper Hyper Aleph (high-end uncountable infinity) (Aleph (high-end uncountable infinity), Aleph (high-end uncountable infinity))[Hyper Aleph (high-end uncountable infinity) (Aleph (high-end uncountable infinity), Aleph (high-end uncountable infinity)), Hyper Aleph (high-end uncountable infinity) (Aleph (high-end uncountable infinity), Aleph (high-end uncountable infinity))] {Hyper Hyper Aleph (high-end uncountable infinity) (Aleph (high-end uncountable infinity), Aleph (high-end uncountable infinity))[Hyper Aleph (high-end uncountable infinity) (Aleph (high-end uncountable infinity), Aleph (high-end uncountable infinity)), Hyper Aleph (high-end uncountable infinity) (Aleph (high-end uncountable infinity), Aleph (high-end uncountable infinity))], Hyper Hyper Aleph (high-end uncountable infinity) (Aleph (high-end uncountable infinity), Aleph (high-end uncountable infinity))[Hyper Aleph (high-end uncountable infinity) (Aleph (high-end uncountable infinity), Aleph (high-end uncountable infinity)), Hyper Aleph (high-end uncountable infinity) (Aleph (high-end uncountable infinity), Aleph (high-end uncountable infinity))]} { Hyper Hyper Hyper Aleph (high-end uncountable infinity) (Aleph (high-end uncountable infinity), Aleph (high-end uncountable infinity))[Hyper Aleph (high-end uncountable infinity) (Aleph (high-end uncountable infinity), Aleph (high-end uncountable infinity)), Hyper Aleph (high-end uncountable infinity) (Aleph (high-end uncountable infinity), Aleph (high-end uncountable infinity))] {Hyper Hyper Aleph (high-end uncountable infinity) (Aleph (high-end uncountable infinity), Aleph (high-end uncountable infinity))[Hyper Aleph (high-end uncountable infinity) (Aleph (high-end uncountable infinity), Aleph (high-end uncountable infinity)), Hyper Aleph (high-end uncountable infinity) (Aleph (high-end uncountable infinity), Aleph (high-end uncountable infinity))], Hyper Hyper Aleph (high-end uncountable infinity) (Aleph (high-end uncountable infinity), Aleph (high-end uncountable infinity))[Hyper Aleph (high-end uncountable infinity) (Aleph (high-end uncountable infinity), Aleph (high-end uncountable infinity)), Hyper Aleph (high-end uncountable infinity) (Aleph (high-end uncountable infinity), Aleph (high-end uncountable infinity))]}, Hyper Hyper Hyper Aleph (high-end uncountable infinity) (Aleph (high-end uncountable infinity), Aleph (high-end uncountable infinity))[Hyper Aleph (high-end uncountable infinity) (Aleph (high-end uncountable infinity), Aleph (high-end uncountable infinity)), Hyper Aleph (high-end uncountable infinity) (Aleph (high-end uncountable infinity), Aleph (high-end uncountable infinity))] {Hyper Hyper Aleph (high-end uncountable infinity) (Aleph (high-end uncountable infinity), Aleph (high-end uncountable infinity))[Hyper Aleph (high-end uncountable infinity) (Aleph (high-end uncountable infinity), Aleph (high-end uncountable infinity)), Hyper Aleph (high-end uncountable infinity) (Aleph (high-end uncountable infinity), Aleph (high-end uncountable infinity))], Hyper Hyper Aleph (high-end uncountable infinity) (Aleph (high-end uncountable infinity), Aleph (high-end uncountable infinity))[Hyper Aleph (high-end uncountable infinity) (Aleph (high-end uncountable infinity), Aleph (high-end uncountable infinity)), Hyper Aleph (high-end uncountable infinity) (Aleph (high-end uncountable infinity), Aleph (high-end uncountable infinity))]}}{ Hyper Hyper Hyper

Hyper Aleph (high-end uncountable infinity) (Aleph (high-end uncountable infinity), Aleph (high-end uncountable infinity))[Hyper Aleph (high-end uncountable infinity) (Aleph (high-end uncountable infinity), Aleph (high-end uncountable infinity)), Hyper Aleph (high-end uncountable infinity) (Aleph (high-end uncountable infinity), Aleph (high-end uncountable infinity))] {Hyper Hyper Aleph (high-end uncountable infinity) (Aleph (high-end uncountable infinity), Aleph (high-end uncountable infinity))[Hyper Aleph (high-end uncountable infinity) (Aleph (high-end uncountable infinity), Aleph (high-end uncountable infinity)), Hyper Aleph (high-end uncountable infinity) (Aleph (high-end uncountable infinity), Aleph (high-end uncountable infinity))], Hyper Hyper Aleph (high-end uncountable infinity) (Aleph (high-end uncountable infinity), Aleph (high-end uncountable infinity))[Hyper Aleph (high-end uncountable infinity) (Aleph (high-end uncountable infinity), Aleph (high-end uncountable infinity)), Hyper Aleph (high-end uncountable infinity) (Aleph (high-end uncountable infinity), Aleph (high-end uncountable infinity))]} { Hyper Hyper Hyper Aleph (high-end uncountable infinity) (Aleph (high-end uncountable infinity), Aleph (high-end uncountable infinity))[Hyper Aleph (high-end uncountable infinity) (Aleph (high-end uncountable infinity), Aleph (high-end uncountable infinity)), Hyper Aleph (high-end uncountable infinity) (Aleph (high-end uncountable infinity), Aleph (high-end uncountable infinity))] {Hyper Hyper Aleph (high-end uncountable infinity) (Aleph (high-end uncountable infinity), Aleph (high-end uncountable infinity))[Hyper Aleph (high-end uncountable infinity) (Aleph (high-end uncountable infinity), Aleph (high-end uncountable infinity)), Hyper Aleph (high-end uncountable infinity) (Aleph (high-end uncountable infinity), Aleph (high-end uncountable infinity))], Hyper Hyper Aleph (high-end uncountable infinity) (Aleph (high-end uncountable infinity), Aleph (high-end uncountable infinity))[Hyper Aleph (high-end uncountable infinity) (Aleph (high-end uncountable infinity), Aleph (high-end uncountable infinity)), Hyper Aleph (high-end uncountable infinity) (Aleph (high-end uncountable infinity), Aleph (high-end uncountable infinity))]}, Hyper Hyper Hyper Aleph (high-end uncountable infinity) (Aleph (high-end uncountable infinity), Aleph (high-end uncountable infinity))[Hyper Aleph (high-end uncountable infinity) (Aleph (high-end uncountable infinity), Aleph (high-end uncountable infinity)), Hyper Aleph (high-end uncountable infinity) (Aleph (high-end uncountable infinity), Aleph (high-end uncountable infinity))] {Hyper Hyper Aleph (high-end uncountable infinity) (Aleph (high-end uncountable infinity), Aleph (high-end uncountable infinity))[Hyper Aleph (high-end uncountable infinity) (Aleph (high-end uncountable infinity), Aleph (high-end uncountable infinity)), Hyper Aleph (high-end uncountable infinity) (Aleph (high-end uncountable infinity), Aleph (high-end uncountable infinity))], Hyper Hyper Aleph (high-end uncountable infinity) (Aleph (high-end uncountable infinity), Aleph (high-end uncountable infinity))[Hyper Aleph (high-end uncountable infinity) (Aleph (high-end uncountable infinity), Aleph (high-end uncountable infinity)), Hyper Aleph (high-end uncountable infinity) (Aleph (high-end uncountable infinity), Aleph (high-end uncountable infinity))]}}, Hyper Hyper Hyper Hyper Aleph (high-end uncountable infinity) (Aleph (high-end uncountable infinity), Aleph (high-end uncountable infinity))[Hyper Aleph (high-end uncountable infinity) (Aleph (high-end uncountable infinity), Aleph (high-end uncountable infinity)), Hyper Aleph (high-end uncountable infinity) (Aleph (high-end uncountable infinity), Aleph (high-end uncountable infinity))] {Hyper Hyper Aleph (high-end uncountable infinity) (Aleph (high-end uncountable infinity), Aleph (high-end uncountable infinity))[Hyper Aleph (high-end uncountable infinity) (Aleph (high-end uncountable infinity), Aleph (high-end uncountable infinity)), Hyper Aleph (high-end uncountable infinity) (Aleph (high-end uncountable infinity), Aleph (high-end uncountable infinity))], Hyper Hyper Aleph (high-end uncountable infinity) (Aleph (high-end uncountable infinity), Aleph (high-end uncountable infinity))[Hyper Aleph (high-end uncountable infinity) (Aleph (high-end uncountable infinity), Aleph (high-end uncountable infinity)), Hyper Aleph (high-end uncountable infinity) (Aleph (high-end uncountable infinity), Aleph (high-end uncountable infinity))]} { Hyper Hyper Hyper Aleph (high-end uncountable infinity) (Aleph (high-end uncountable infinity), Aleph (high-end uncountable infinity))[Hyper Aleph (high-end uncountable infinity) (Aleph (high-end uncountable infinity), Aleph (high-end uncountable infinity)), Hyper Aleph (high-end uncountable infinity) (Aleph (high-end uncountable infinity), Aleph (high-end uncountable infinity))] {Hyper Hyper Aleph (high-end uncountable infinity) (Aleph (high-end uncountable infinity), Aleph (high-end uncountable infinity))[Hyper Aleph (high-end uncountable infinity) (Aleph (high-end uncountable infinity), Aleph (high-end uncountable infinity)), Hyper Aleph (high-end uncountable infinity) (Aleph (high-end uncountable infinity), Aleph (high-end uncountable infinity))], Hyper Hyper Aleph (high-end uncountable infinity) (Aleph (high-end uncountable infinity), Aleph (high-end uncountable infinity))[Hyper Aleph (high-end uncountable infinity) (Aleph (high-end uncountable infinity), Aleph (high-end uncountable infinity)), Hyper Aleph (high-end uncountable infinity) (Aleph (high-end uncountable infinity), Aleph (high-end uncountable infinity))]}, Hyper Hyper Hyper Aleph (high-end uncountable infinity) (Aleph (high-end uncountable infinity), Aleph (high-end uncountable infinity))[Hyper Aleph (high-end uncountable infinity) (Aleph (high-end uncountable infinity), Aleph (high-end uncountable infinity)), Hyper Aleph (high-end uncountable infinity) (Aleph (high-end uncountable infinity), Aleph (high-end uncountable infinity))] {Hyper Hyper Aleph (high-end uncountable infinity) (Aleph (high-end uncountable infinity), Aleph (high-end uncountable infinity))[Hyper Aleph (high-end uncountable infinity) (Aleph (high-end uncountable infinity), Aleph (high-end uncountable infinity)), Hyper Aleph (high-end uncountable infinity) (Aleph (high-end uncountable infinity), Aleph (high-end uncountable infinity))], Hyper Hyper Aleph (high-end uncountable infinity) (Aleph (high-end uncountable infinity), Aleph (high-end uncountable infinity))[Hyper Aleph (high-end uncountable infinity) (Aleph (high-end uncountable infinity), Aleph (high-end uncountable infinity)), Hyper Aleph (high-end uncountable infinity) (Aleph (high-end uncountable infinity), Aleph (high-end uncountable infinity))]}}}, Hyper Hyper Hyper Hyper Hyper Aleph (high-end uncountable infinity) (Aleph (high-end uncountable infinity), Aleph (high-end uncountable infinity))[Hyper Aleph (high-end uncountable infinity) (Aleph (high-end uncountable infinity), Aleph (high-end uncountable infinity)), Hyper Aleph (high-end uncountable infinity) (Aleph (high-end uncountable infinity), Aleph (high-end uncountable infinity))] {Hyper Hyper Aleph (high-end uncountable infinity) (Aleph (high-end uncountable infinity), Aleph (high-end uncountable infinity))[Hyper Aleph (high-end uncountable infinity) (Aleph (high-end uncountable infinity), Aleph (high-end uncountable infinity)), Hyper Aleph (high-end uncountable infinity) (Aleph (high-end uncountable infinity), Aleph (high-end uncountable infinity))], Hyper Hyper Aleph (high-end uncountable infinity) (Aleph (high-end uncountable infinity), Aleph (high-end uncountable infinity))[Hyper Aleph (high-end uncountable infinity) (Aleph (high-end uncountable infinity), Aleph (high-end uncountable infinity)), Hyper Aleph (high-end uncountable infinity) (Aleph (high-end uncountable infinity), Aleph (high-end uncountable infinity))]} { Hyper Hyper Hyper Aleph (high-end uncountable infinity) (Aleph (high-end uncountable infinity), Aleph (high-end uncountable infinity))[Hyper Aleph (high-end uncountable infinity) (Aleph (high-end uncountable infinity), Aleph (high-end uncountable infinity)), Hyper Aleph (high-end uncountable infinity) (Aleph (high-end uncountable infinity), Aleph (high-end uncountable infinity))] {Hyper Hyper Aleph (high-end uncountable infinity) (Aleph (high-end uncountable infinity), Aleph (high-end uncountable infinity))[Hyper Aleph (high-end uncountable infinity) (Aleph (high-end uncountable infinity), Aleph (high-end uncountable infinity)), Hyper Aleph (high-end uncountable infinity) (Aleph (high-end uncountable infinity), Aleph (high-end uncountable infinity))], Hyper Hyper Aleph (high-end uncountable infinity) (Aleph (high-end uncountable infinity), Aleph (high-end uncountable infinity))[Hyper Aleph (high-end uncountable infinity) (Aleph (high-end uncountable infinity), Aleph (high-end uncountable infinity)), Hyper Aleph (high-end uncountable infinity) (Aleph (high-end uncountable infinity), Aleph (high-end uncountable infinity))]}, Hyper Hyper Hyper Aleph (high-end uncountable infinity) (Aleph (high-end uncountable infinity), Aleph (high-end uncountable infinity))[Hyper Aleph (high-end uncountable infinity) (Aleph (high-end uncountable infinity), Aleph (high-end uncountable infinity)), Hyper Aleph (high-end uncountable infinity) (Aleph (high-end uncountable infinity), Aleph (high-end uncountable infinity))] {Hyper Hyper Aleph (high-end uncountable infinity) (Aleph (high-end uncountable infinity), Aleph (high-end uncountable infinity))[Hyper Aleph (high-end uncountable infinity) (Aleph (high-end uncountable infinity), Aleph (high-end uncountable infinity)), Hyper Aleph (high-end uncountable infinity) (Aleph (high-end uncountable infinity), Aleph (high-end uncountable infinity))], Hyper Hyper Aleph (high-end uncountable infinity) (Aleph (high-end uncountable infinity), Aleph (high-end uncountable infinity))[Hyper Aleph (high-end uncountable infinity) (Aleph (high-end uncountable infinity), Aleph (high-end uncountable infinity)), Hyper Aleph (high-end uncountable infinity) (Aleph (high-end uncountable infinity), Aleph (high-end uncountable infinity))]}}{ Hyper Hyper Hyper Hyper Aleph (high-end uncountable infinity) (Aleph (high-end uncountable infinity), Aleph (high-end uncountable infinity))[Hyper Aleph (high-end uncountable infinity) (Aleph (high-end uncountable infinity), Aleph (high-end uncountable infinity)), Hyper Aleph (high-end uncountable infinity) (Aleph (high-end uncountable infinity), Aleph (high-end uncountable infinity))] {Hyper Hyper Aleph (high-end uncountable infinity) (Aleph (high-end uncountable infinity), Aleph (high-end uncountable infinity))[Hyper Aleph (high-end uncountable infinity) (Aleph (high-end uncountable infinity), Aleph (high-end uncountable infinity)), Hyper Aleph (high-end uncountable infinity) (Aleph (high-end uncountable infinity), Aleph (high-end uncountable infinity))], Hyper Hyper Aleph (high-end uncountable infinity) (Aleph (high-end uncountable infinity), Aleph (high-end uncountable infinity))[Hyper Aleph (high-end uncountable infinity) (Aleph (high-end uncountable infinity), Aleph (high-end uncountable infinity)), Hyper Aleph (high-end uncountable infinity) (Aleph (high-end uncountable infinity), Aleph (high-end uncountable infinity))]} { Hyper Hyper Hyper Aleph (high-end uncountable infinity) (Aleph (high-end uncountable infinity), Aleph (high-end uncountable infinity))[Hyper Aleph (high-end uncountable infinity) (Aleph (high-end uncountable infinity), Aleph (high-end uncountable infinity)), Hyper Aleph (high-end uncountable infinity) (Aleph (high-end uncountable infinity), Aleph (high-end uncountable infinity))] {Hyper Hyper Aleph (high-end uncountable infinity) (Aleph (high-end uncountable infinity), Aleph (high-end uncountable infinity))[Hyper Aleph (high-end uncountable infinity) (Aleph (high-end uncountable infinity), Aleph (high-end uncountable infinity)), Hyper Aleph (high-end uncountable infinity) (Aleph (high-end uncountable infinity), Aleph (high-end uncountable infinity))], Hyper Hyper Aleph (high-end uncountable infinity) (Aleph (high-end uncountable infinity), Aleph (high-end uncountable infinity))[Hyper Aleph (high-end uncountable infinity) (Aleph (high-end uncountable infinity), Aleph (high-end uncountable infinity)), Hyper Aleph (high-end uncountable infinity) (Aleph (high-end uncountable infinity), Aleph (high-end uncountable infinity))]}, Hyper Hyper Hyper Aleph (high-end uncountable infinity) (Aleph (high-end uncountable infinity), Aleph (high-end uncountable infinity))[Hyper Aleph (high-end uncountable infinity) (Aleph (high-end uncountable infinity), Aleph (high-end uncountable infinity)), Hyper Aleph (high-end uncountable infinity) (Aleph (high-end uncountable infinity), Aleph (high-end uncountable infinity))] {Hyper Hyper Aleph (high-end uncountable infinity) (Aleph (high-end uncountable infinity), Aleph (high-end uncountable infinity))[Hyper Aleph (high-end uncountable infinity) (Aleph (high-end uncountable infinity), Aleph (high-end uncountable infinity)), Hyper Aleph (high-end uncountable infinity) (Aleph (high-end uncountable infinity), Aleph (high-end uncountable infinity))], Hyper Hyper Aleph (high-end uncountable infinity) (Aleph (high-end uncountable infinity), Aleph (high-end uncountable infinity))[Hyper Aleph (high-end uncountable infinity) (Aleph (high-end uncountable infinity), Aleph (high-end uncountable infinity)), Hyper Aleph (high-end uncountable infinity) (Aleph (high-end uncountable infinity), Aleph (high-end uncountable infinity))]}}, Hyper Hyper Hyper Hyper Aleph (high-end uncountable infinity) (Aleph (high-end uncountable infinity), Aleph (high-end uncountable infinity))[Hyper Aleph (high-end uncountable infinity) (Aleph (high-end uncountable infinity), Aleph (high-end uncountable infinity)), Hyper Aleph (high-end uncountable infinity) (Aleph (high-end uncountable infinity), Aleph (high-end uncountable infinity))] {Hyper Hyper Aleph (high-end uncountable infinity) (Aleph (high-end uncountable infinity), Aleph (high-end uncountable infinity))[Hyper Aleph (high-end uncountable infinity) (Aleph (high-end uncountable infinity), Aleph (high-end uncountable infinity)), Hyper Aleph (high-end uncountable infinity) (Aleph (high-end uncountable infinity), Aleph (high-end uncountable infinity))], Hyper Hyper Aleph (high-end uncountable infinity) (Aleph (high-end uncountable infinity), Aleph (high-end uncountable infinity))[Hyper Aleph (high-end uncountable infinity) (Aleph (high-end uncountable infinity), Aleph (high-end uncountable infinity)), Hyper Aleph (high-end uncountable infinity) (Aleph (high-end uncountable infinity), Aleph (high-end uncountable infinity))]} { Hyper Hyper Hyper Aleph (high-end uncountable infinity) (Aleph (high-end uncountable infinity), Aleph (high-end uncountable infinity))[Hyper Aleph (high-end uncountable infinity) (Aleph (high-end uncountable infinity), Aleph (high-end uncountable infinity)), Hyper Aleph (high-end uncountable infinity) (Aleph (high-end uncountable infinity), Aleph (high-end uncountable infinity))] {Hyper Hyper Aleph (high-end uncountable infinity) (Aleph (high-end uncountable infinity), Aleph (high-end uncountable infinity))[Hyper Aleph (high-end uncountable infinity) (Aleph (high-end uncountable infinity), Aleph (high-end uncountable infinity)), Hyper Aleph (high-end uncountable infinity) (Aleph (high-end uncountable infinity), Aleph (high-end uncountable infinity))], Hyper Hyper Aleph (high-end uncountable infinity) (Aleph (high-end uncountable infinity), Aleph (high-end uncountable infinity))[Hyper Aleph (high-end uncountable infinity) (Aleph (high-end uncountable infinity), Aleph (high-end uncountable infinity)), Hyper Aleph (high-end uncountable infinity) (Aleph (high-end uncountable infinity), Aleph (high-end uncountable infinity))]}, Hyper Hyper Hyper Aleph (high-end uncountable infinity) (Aleph (high-end uncountable infinity), Aleph (high-end uncountable infinity))[Hyper Aleph (high-end uncountable infinity) (Aleph (high-end uncountable infinity), Aleph (high-end uncountable infinity)), Hyper Aleph (high-end uncountable infinity) (Aleph (high-end uncountable infinity), Aleph (high-end uncountable infinity))] {Hyper Hyper Aleph (high-end uncountable infinity) (Aleph (high-end uncountable infinity), Aleph (high-end uncountable infinity))[Hyper Aleph (high-end uncountable infinity) (Aleph (high-end uncountable infinity), Aleph (high-end uncountable infinity)), Hyper Aleph (high-end uncountable infinity) (Aleph (high-end uncountable infinity), Aleph (high-end uncountable infinity))], Hyper Hyper Aleph (high-end uncountable infinity) (Aleph (high-end uncountable infinity), Aleph (high-end uncountable infinity))[Hyper Aleph (high-end uncountable infinity) (Aleph (high-end uncountable infinity), Aleph (high-end uncountable infinity)), Hyper Aleph (high-end uncountable infinity) (Aleph (high-end uncountable infinity), Aleph (high-end uncountable infinity))]}}}}}

We have just considered a huge infinity in only a few minutes of inspection.

We could approximately double the size of the latest expression above every Planck Time Unit from now onward for ever to perpetually build ever greater infinities. Imagine how large of any infinity we could build after and Hyper Aleph Ω (Aleph Ω, Aleph Ω) aeons have gone by in the future. An aeon is one current universal age unit or about 13.8 billion years.

Now, we can intelligibly consider the abstraction of a set that has so many members that the number of members goes beyond any infinite value.

Accordingly, we can consider sets that are so big that the set contains a non-ordinal or non-numeric multiplicity of elements. We will consider such a set as having non-ordinal-1- multiplicity of elements.

We can go still further to consider sets of non-ordinal-2-multiplicity of elements which are to sets of non-ordinal-1- multiplicity of elements as sets of non-ordinal-1-multiplicity of elements are to infinite sets.

We can go still further to consider to consider sets of non-ordinal-3-multiplicity of elements which are to sets of non-ordinal-2-multiplicity of elements as sets of non-ordinal-2-multiplicity of elements are to sets of non-ordinal-1- multiplicity of elements as sets of non-ordinal-1-multiplicity of elements are to infinite sets.

We can go further to sets of non-ordinal-4-multiplicity of elements, sets of non-ordinal-5-multiplicity of elements, all the way to sets of non-ordinal-(Aleph 0)-multiplicity of elements, all the way to sets of non-ordinal-(Aleph 1)-multiplicity of elements, all the way to sets of non-ordinal-(Aleph 2)-multiplicity of elements, all the way to sets of non-ordinal-(Aleph 3)-multiplicity of elements, and so-on to, all the way to sets of non-ordinal-(Aleph Ω)-multiplicity of elements, and so-on and so-on further to sets of sets of non-ordinal-[Hyper Aleph Ω (Aleph Ω, Aleph Ω)]-multiplicity of elements and continue on from there.

Now here is a real zinger.

So we can produce the abstraction of Hyper Aleph (non-ordinal-[Hyper Aleph Ω (Aleph Ω, Aleph Ω)]-multiplicity) (Aleph (non-ordinal-[Hyper Aleph Ω (Aleph Ω, Aleph Ω)]-multiplicity), Aleph (non-ordinal-[Hyper Aleph Ω (Aleph Ω, Aleph Ω)]-multiplicity)).

Infinities can be as big as we would like. For example, consider the following infinity.

Hyper Hyper Aleph (non-ordinal-[Hyper Aleph Ω (Aleph Ω, Aleph Ω)]-multiplicity) (Aleph (non-ordinal-[Hyper Aleph Ω (Aleph Ω, Aleph Ω)]-multiplicity), Aleph (non-ordinal-[Hyper Aleph Ω (Aleph Ω, Aleph Ω)]-multiplicity))[Hyper Aleph (non-ordinal-[Hyper Aleph Ω (Aleph Ω, Aleph Ω)]-multiplicity) (Aleph (non-ordinal-[Hyper Aleph Ω (Aleph Ω, Aleph Ω)]-multiplicity), Aleph (non-ordinal-[Hyper Aleph Ω (Aleph Ω, Aleph Ω)]-multiplicity)), Hyper Aleph (non-ordinal-[Hyper Aleph Ω (Aleph Ω, Aleph Ω)]-multiplicity) (Aleph (non-ordinal-[Hyper Aleph Ω (Aleph Ω, Aleph Ω)]-multiplicity), Aleph (non-ordinal-[Hyper Aleph Ω (Aleph Ω, Aleph Ω)]-multiplicity))]

We can go still larger to develop the following infinity.

Hyper Hyper Hyper Aleph (non-ordinal-[Hyper Aleph Ω (Aleph Ω, Aleph Ω)]-multiplicity) (Aleph (non-ordinal-[Hyper Aleph Ω (Aleph Ω, Aleph Ω)]-multiplicity), Aleph (non-ordinal-[Hyper Aleph Ω (Aleph Ω, Aleph Ω)]-multiplicity))[Hyper Aleph (non-ordinal-[Hyper Aleph Ω (Aleph Ω, Aleph Ω)]-multiplicity) (Aleph (non-ordinal-[Hyper Aleph Ω (Aleph Ω, Aleph Ω)]-multiplicity), Aleph (non-ordinal-[Hyper Aleph Ω (Aleph Ω, Aleph Ω)]-multiplicity)), Hyper Aleph (non-ordinal-[Hyper Aleph Ω (Aleph Ω, Aleph Ω)]-multiplicity) (Aleph (non-ordinal-[Hyper Aleph Ω (Aleph Ω, Aleph Ω)]-multiplicity), Aleph (non-ordinal-[Hyper Aleph Ω (Aleph Ω, Aleph Ω)]-multiplicity))] {Hyper Hyper Aleph (non-ordinal-[Hyper Aleph Ω (Aleph Ω, Aleph Ω)]-multiplicity) (Aleph (non-ordinal-[Hyper Aleph Ω (Aleph Ω, Aleph Ω)]-multiplicity), Aleph (non-ordinal-[Hyper Aleph Ω (Aleph Ω, Aleph Ω)]-multiplicity))[Hyper Aleph (non-ordinal-[Hyper Aleph Ω (Aleph Ω, Aleph Ω)]-multiplicity) (Aleph (non-ordinal-[Hyper Aleph Ω (Aleph Ω, Aleph Ω)]-multiplicity), Aleph (non-ordinal-[Hyper Aleph Ω (Aleph Ω, Aleph Ω)]-multiplicity)), Hyper Aleph (non-ordinal-[Hyper Aleph Ω (Aleph Ω, Aleph Ω)]-multiplicity) (Aleph (non-ordinal-[Hyper Aleph Ω (Aleph Ω, Aleph Ω)]-multiplicity), Aleph (non-ordinal-[Hyper Aleph Ω (Aleph Ω, Aleph Ω)]-multiplicity))], Hyper Hyper Aleph (non-ordinal-[Hyper Aleph Ω (Aleph Ω, Aleph Ω)]-multiplicity) (Aleph (non-ordinal-[Hyper Aleph Ω (Aleph Ω, Aleph Ω)]-multiplicity), Aleph (non-ordinal-[Hyper Aleph Ω (Aleph Ω, Aleph Ω)]-multiplicity))[Hyper Aleph (non-ordinal-[Hyper Aleph Ω (Aleph Ω, Aleph Ω)]-multiplicity) (Aleph (non-ordinal-[Hyper Aleph Ω (Aleph Ω, Aleph Ω)]-multiplicity), Aleph (non-ordinal-[Hyper Aleph Ω (Aleph Ω, Aleph Ω)]-multiplicity)), Hyper Aleph (non-ordinal-[Hyper Aleph Ω (Aleph Ω, Aleph Ω)]-

multiplicity) (Aleph (non-ordinal-[Hyper Aleph Ω (Aleph Ω, Aleph Ω)]-multiplicity), Aleph (non-ordinal-[Hyper Aleph Ω (Aleph Ω, Aleph Ω)]-multiplicity))]}

We can go still larger to develop the following infinity.

Hyper Hyper Hyper Hyper Aleph (non-ordinal-[Hyper Aleph Ω (Aleph Ω, Aleph Ω)]-multiplicity) (Aleph (non-ordinal-[Hyper Aleph Ω (Aleph Ω, Aleph Ω)]-multiplicity), Aleph (non-ordinal-[Hyper Aleph Ω (Aleph Ω, Aleph Ω)]-multiplicity))[Hyper Aleph (non-ordinal-[Hyper Aleph Ω (Aleph Ω, Aleph Ω)]-multiplicity) (Aleph (non-ordinal-[Hyper Aleph Ω (Aleph Ω, Aleph Ω)]-multiplicity), Aleph (non-ordinal-[Hyper Aleph Ω (Aleph Ω, Aleph Ω)]-multiplicity)), Hyper Aleph (non-ordinal-[Hyper Aleph Ω (Aleph Ω, Aleph Ω)]-multiplicity) (Aleph (non-ordinal-[Hyper Aleph Ω (Aleph Ω, Aleph Ω)]-multiplicity), Aleph (non-ordinal-[Hyper Aleph Ω (Aleph Ω, Aleph Ω)]-multiplicity))] {Hyper Hyper Aleph (non-ordinal-[Hyper Aleph Ω (Aleph Ω, Aleph Ω)]-multiplicity) (Aleph (non-ordinal-[Hyper Aleph Ω (Aleph Ω, Aleph Ω)]-multiplicity), Aleph (non-ordinal-[Hyper Aleph Ω (Aleph Ω, Aleph Ω)]-multiplicity))[Hyper Aleph (non-ordinal-[Hyper Aleph Ω (Aleph Ω, Aleph Ω)]-multiplicity) (Aleph (non-ordinal-[Hyper Aleph Ω (Aleph Ω, Aleph Ω)]-multiplicity), Aleph (non-ordinal-[Hyper Aleph Ω (Aleph Ω, Aleph Ω)]-multiplicity)), Hyper Aleph (non-ordinal-[Hyper Aleph Ω (Aleph Ω, Aleph Ω)]-multiplicity) (Aleph (non-ordinal-[Hyper Aleph Ω (Aleph Ω, Aleph Ω)]-multiplicity), Aleph (non-ordinal-[Hyper Aleph Ω (Aleph Ω, Aleph Ω)]-multiplicity))], Hyper Hyper Aleph (non-ordinal-[Hyper Aleph Ω (Aleph Ω, Aleph Ω)]-multiplicity) (Aleph (non-ordinal-[Hyper Aleph Ω (Aleph Ω, Aleph Ω)]-multiplicity), Aleph (non-ordinal-[Hyper Aleph Ω (Aleph Ω, Aleph Ω)]-multiplicity))[Hyper Aleph (non-ordinal-[Hyper Aleph Ω (Aleph Ω, Aleph Ω)]-multiplicity) (Aleph (non-ordinal-[Hyper Aleph Ω (Aleph Ω, Aleph Ω)]-multiplicity), Aleph (non-ordinal-[Hyper Aleph Ω (Aleph Ω, Aleph Ω)]-multiplicity)), Hyper Aleph (non-ordinal-[Hyper Aleph Ω (Aleph Ω, Aleph Ω)]-multiplicity) (Aleph (non-ordinal-[Hyper Aleph Ω (Aleph Ω, Aleph Ω)]-multiplicity), Aleph (non-ordinal-[Hyper Aleph Ω (Aleph Ω, Aleph Ω)]-multiplicity))]} { Hyper Hyper Hyper Aleph (non-ordinal-[Hyper Aleph Ω (Aleph Ω, Aleph Ω)]-multiplicity) (Aleph (non-ordinal-[Hyper Aleph Ω (Aleph Ω, Aleph Ω)]-multiplicity), Aleph (non-ordinal-[Hyper Aleph Ω (Aleph Ω, Aleph Ω)]-multiplicity))[Hyper Aleph (non-ordinal-[Hyper Aleph Ω (Aleph Ω, Aleph Ω)]-multiplicity) (Aleph (non-ordinal-[Hyper Aleph Ω (Aleph Ω, Aleph Ω)]-multiplicity), Aleph (non-ordinal-[Hyper Aleph Ω (Aleph Ω, Aleph Ω)]-multiplicity)), Hyper Aleph (non-ordinal-[Hyper Aleph Ω (Aleph Ω, Aleph Ω)]-multiplicity) (Aleph (non-ordinal-[Hyper Aleph Ω (Aleph Ω, Aleph Ω)]-multiplicity), Aleph (non-ordinal-[Hyper Aleph Ω (Aleph Ω, Aleph Ω)]-multiplicity))] {Hyper Hyper Aleph (non-ordinal-[Hyper Aleph Ω (Aleph Ω, Aleph Ω)]-multiplicity) (Aleph (non-ordinal-[Hyper Aleph Ω (Aleph Ω, Aleph Ω)]-multiplicity), Aleph (non-ordinal-[Hyper Aleph Ω (Aleph Ω, Aleph Ω)]-

multiplicity))[Hyper Aleph (non-ordinal-[Hyper Aleph Ω (Aleph Ω, Aleph Ω)]-multiplicity) (Aleph (non-ordinal-[Hyper Aleph Ω (Aleph Ω, Aleph Ω)]-multiplicity), Aleph (non-ordinal-[Hyper Aleph Ω (Aleph Ω, Aleph Ω)]-multiplicity)), Hyper Aleph (non-ordinal-[Hyper Aleph Ω (Aleph Ω, Aleph Ω)]-multiplicity) (Aleph (non-ordinal-[Hyper Aleph Ω (Aleph Ω, Aleph Ω)]-multiplicity), Aleph (non-ordinal-[Hyper Aleph Ω (Aleph Ω, Aleph Ω)]-multiplicity))], Hyper Hyper Aleph (non-ordinal-[Hyper Aleph Ω (Aleph Ω, Aleph Ω)]-multiplicity) (Aleph (non-ordinal-[Hyper Aleph Ω (Aleph Ω, Aleph Ω)]-multiplicity), Aleph (non-ordinal-[Hyper Aleph Ω (Aleph Ω, Aleph Ω)]-multiplicity))[Hyper Aleph (non-ordinal-[Hyper Aleph Ω (Aleph Ω, Aleph Ω)]-multiplicity) (Aleph (non-ordinal-[Hyper Aleph Ω (Aleph Ω, Aleph Ω)]-multiplicity), Aleph (non-ordinal-[Hyper Aleph Ω (Aleph Ω, Aleph Ω)]-multiplicity)), Hyper Aleph (non-ordinal-[Hyper Aleph Ω (Aleph Ω, Aleph Ω)]-multiplicity) (Aleph (non-ordinal-[Hyper Aleph Ω (Aleph Ω, Aleph Ω)]-multiplicity), Aleph (non-ordinal-[Hyper Aleph Ω (Aleph Ω, Aleph Ω)]-multiplicity))]}, Hyper Hyper Hyper Aleph (non-ordinal-[Hyper Aleph Ω (Aleph Ω, Aleph Ω)]-multiplicity) (Aleph (non-ordinal-[Hyper Aleph Ω (Aleph Ω, Aleph Ω)]-multiplicity), Aleph (non-ordinal-[Hyper Aleph Ω (Aleph Ω, Aleph Ω)]-multiplicity))[Hyper Aleph (non-ordinal-[Hyper Aleph Ω (Aleph Ω, Aleph Ω)]-multiplicity) (Aleph (non-ordinal-[Hyper Aleph Ω (Aleph Ω, Aleph Ω)]-multiplicity), Aleph (non-ordinal-[Hyper Aleph Ω (Aleph Ω, Aleph Ω)]-multiplicity)), Hyper Aleph (non-ordinal-[Hyper Aleph Ω (Aleph Ω, Aleph Ω)]-multiplicity) (Aleph (non-ordinal-[Hyper Aleph Ω (Aleph Ω, Aleph Ω)]-multiplicity), Aleph (non-ordinal-[Hyper Aleph Ω (Aleph Ω, Aleph Ω)]-multiplicity))] {Hyper Hyper Aleph (non-ordinal-[Hyper Aleph Ω (Aleph Ω, Aleph Ω)]-multiplicity) (Aleph (non-ordinal-[Hyper Aleph Ω (Aleph Ω, Aleph Ω)]-multiplicity), Aleph (non-ordinal-[Hyper Aleph Ω (Aleph Ω, Aleph Ω)]-multiplicity))[Hyper Aleph (non-ordinal-[Hyper Aleph Ω (Aleph Ω, Aleph Ω)]-multiplicity) (Aleph (non-ordinal-[Hyper Aleph Ω (Aleph Ω, Aleph Ω)]-multiplicity), Aleph (non-ordinal-[Hyper Aleph Ω (Aleph Ω, Aleph Ω)]-multiplicity)), Hyper Aleph (non-ordinal-[Hyper Aleph Ω (Aleph Ω, Aleph Ω)]-multiplicity) (Aleph (non-ordinal-[Hyper Aleph Ω (Aleph Ω, Aleph Ω)]-multiplicity), Aleph (non-ordinal-[Hyper Aleph Ω (Aleph Ω, Aleph Ω)]-multiplicity))], Hyper Hyper Aleph (non-ordinal-[Hyper Aleph Ω (Aleph Ω, Aleph Ω)]-multiplicity) (Aleph (non-ordinal-[Hyper Aleph Ω (Aleph Ω, Aleph Ω)]-multiplicity), Aleph (non-ordinal-[Hyper Aleph Ω (Aleph Ω, Aleph Ω)]-multiplicity))[Hyper Aleph (non-ordinal-[Hyper Aleph Ω (Aleph Ω, Aleph Ω)]-multiplicity) (Aleph (non-ordinal-[Hyper Aleph Ω (Aleph Ω, Aleph Ω)]-multiplicity), Aleph (non-ordinal-[Hyper Aleph Ω (Aleph Ω, Aleph Ω)]-multiplicity)), Hyper Aleph (non-ordinal-[Hyper Aleph Ω (Aleph Ω, Aleph Ω)]-multiplicity) (Aleph (non-ordinal-[Hyper Aleph Ω (Aleph Ω, Aleph Ω)]-multiplicity), Aleph (non-ordinal-[Hyper Aleph Ω (Aleph Ω, Aleph Ω)]-multiplicity))]}}

We can go still larger to develop the following infinity.

Hyper Hyper Hyper Hyper Hyper Aleph (non-ordinal-[Hyper Aleph Ω (Aleph Ω, Aleph Ω)]-multiplicity) (Aleph (non-ordinal-[Hyper Aleph Ω (Aleph Ω, Aleph Ω)]-multiplicity), Aleph (non-ordinal-[Hyper Aleph Ω (Aleph Ω, Aleph Ω)]-multiplicity))[Hyper Aleph (non-ordinal-[Hyper Aleph Ω (Aleph Ω, Aleph Ω)]-multiplicity) (Aleph (non-ordinal-[Hyper Aleph Ω (Aleph Ω, Aleph Ω)]-multiplicity), Aleph (non-ordinal-[Hyper Aleph Ω (Aleph Ω, Aleph Ω)]-multiplicity)), Hyper Aleph (non-ordinal-[Hyper Aleph Ω (Aleph Ω, Aleph Ω)]-multiplicity) (Aleph (non-ordinal-[Hyper Aleph Ω (Aleph Ω, Aleph Ω)]-multiplicity), Aleph (non-ordinal-[Hyper Aleph Ω (Aleph Ω, Aleph Ω)]-multiplicity))] {Hyper Hyper Aleph (non-ordinal-[Hyper Aleph Ω (Aleph Ω, Aleph Ω)]-multiplicity) (Aleph (non-ordinal-[Hyper Aleph Ω (Aleph Ω, Aleph Ω)]-multiplicity), Aleph (non-ordinal-[Hyper Aleph Ω (Aleph Ω, Aleph Ω)]-multiplicity))[Hyper Aleph (non-ordinal-[Hyper Aleph Ω (Aleph Ω, Aleph Ω)]-multiplicity) (Aleph (non-ordinal-[Hyper Aleph Ω (Aleph Ω, Aleph Ω)]-multiplicity), Aleph (non-ordinal-[Hyper Aleph Ω (Aleph Ω, Aleph Ω)]-multiplicity)), Hyper Aleph (non-ordinal-[Hyper Aleph Ω (Aleph Ω, Aleph Ω)]-multiplicity) (Aleph (non-ordinal-[Hyper Aleph Ω (Aleph Ω, Aleph Ω)]-multiplicity), Aleph (non-ordinal-[Hyper Aleph Ω (Aleph Ω, Aleph Ω)]-multiplicity))], Hyper Hyper Aleph (non-ordinal-[Hyper Aleph Ω (Aleph Ω, Aleph Ω)]-multiplicity) (Aleph (non-ordinal-[Hyper Aleph Ω (Aleph Ω, Aleph Ω)]-multiplicity), Aleph (non-ordinal-[Hyper Aleph Ω (Aleph Ω, Aleph Ω)]-multiplicity))[Hyper Aleph (non-ordinal-[Hyper Aleph Ω (Aleph Ω, Aleph Ω)]-multiplicity) (Aleph (non-ordinal-[Hyper Aleph Ω (Aleph Ω, Aleph Ω)]-multiplicity), Aleph (non-ordinal-[Hyper Aleph Ω (Aleph Ω, Aleph Ω)]-multiplicity)), Hyper Aleph (non-ordinal-[Hyper Aleph Ω (Aleph Ω, Aleph Ω)]-multiplicity) (Aleph (non-ordinal-[Hyper Aleph Ω (Aleph Ω, Aleph Ω)]-multiplicity), Aleph (non-ordinal-[Hyper Aleph Ω (Aleph Ω, Aleph Ω)]-multiplicity))]} { Hyper Hyper Hyper Aleph (non-ordinal-[Hyper Aleph Ω (Aleph Ω, Aleph Ω)]-multiplicity) (Aleph (non-ordinal-[Hyper Aleph Ω (Aleph Ω, Aleph Ω)]-multiplicity), Aleph (non-ordinal-[Hyper Aleph Ω (Aleph Ω, Aleph Ω)]-multiplicity))[Hyper Aleph (non-ordinal-[Hyper Aleph Ω (Aleph Ω, Aleph Ω)]-multiplicity) (Aleph (non-ordinal-[Hyper Aleph Ω (Aleph Ω, Aleph Ω)]-multiplicity), Aleph (non-ordinal-[Hyper Aleph Ω (Aleph Ω, Aleph Ω)]-multiplicity)), Hyper Aleph (non-ordinal-[Hyper Aleph Ω (Aleph Ω, Aleph Ω)]-multiplicity) (Aleph (non-ordinal-[Hyper Aleph Ω (Aleph Ω, Aleph Ω)]-multiplicity), Aleph (non-ordinal-[Hyper Aleph Ω (Aleph Ω, Aleph Ω)]-multiplicity))] {Hyper Hyper Aleph (non-ordinal-[Hyper Aleph Ω (Aleph Ω, Aleph Ω)]-multiplicity) (Aleph (non-ordinal-[Hyper Aleph Ω (Aleph Ω, Aleph Ω)]-multiplicity), Aleph (non-ordinal-[Hyper Aleph Ω (Aleph Ω, Aleph Ω)]-multiplicity))[Hyper Aleph (non-ordinal-[Hyper Aleph Ω (Aleph Ω, Aleph Ω)]-multiplicity) (Aleph (non-ordinal-[Hyper Aleph Ω (Aleph Ω, Aleph Ω)]-multiplicity), Aleph (non-ordinal-[Hyper Aleph Ω (Aleph Ω, Aleph Ω)]-multiplicity)), Hyper Aleph (non-ordinal-[Hyper Aleph Ω (Aleph Ω, Aleph Ω)]-multiplicity) (Aleph (non-ordinal-[Hyper Aleph Ω (Aleph Ω, Aleph Ω)]-multiplicity), Aleph (non-ordinal-[Hyper Aleph Ω (Aleph Ω, Aleph Ω)]-multiplicity))], Hyper Hyper Aleph (non-ordinal-[Hyper Aleph Ω (Aleph Ω, Aleph Ω)]-multiplicity) (Aleph (non-ordinal-[Hyper Aleph Ω (Aleph Ω, Aleph Ω)]-multiplicity), Aleph (non-ordinal-[Hyper Aleph Ω (Aleph Ω, Aleph Ω)]-multiplicity))[Hyper Aleph (non-ordinal-[Hyper Aleph Ω (Aleph Ω, Aleph Ω)]-multiplicity) (Aleph (non-ordinal-[Hyper Aleph Ω (Aleph Ω, Aleph Ω)]-multiplicity), Aleph (non-ordinal-[Hyper Aleph Ω (Aleph Ω, Aleph Ω)]-multiplicity)), Hyper Aleph (non-ordinal-[Hyper Aleph Ω (Aleph Ω, Aleph Ω)]-multiplicity) (Aleph (non-ordinal-[Hyper Aleph Ω (Aleph Ω, Aleph Ω)]-multiplicity), Aleph (non-ordinal-[Hyper Aleph Ω (Aleph Ω, Aleph Ω)]-multiplicity))]}, Hyper Hyper Hyper Aleph (non-ordinal-[Hyper Aleph Ω (Aleph Ω, Aleph Ω)]-multiplicity) (Aleph (non-ordinal-[Hyper Aleph Ω (Aleph Ω, Aleph Ω)]-multiplicity), Aleph (non-ordinal-[Hyper Aleph Ω (Aleph Ω, Aleph Ω)]-multiplicity))[Hyper Aleph (non-ordinal-[Hyper Aleph Ω (Aleph Ω, Aleph Ω)]-multiplicity) (Aleph (non-ordinal-[Hyper Aleph Ω (Aleph Ω, Aleph Ω)]-multiplicity), Aleph (non-ordinal-[Hyper Aleph Ω (Aleph Ω, Aleph Ω)]-multiplicity)), Hyper Aleph (non-ordinal-[Hyper Aleph Ω (Aleph Ω, Aleph Ω)]-multiplicity) (Aleph (non-ordinal-[Hyper Aleph Ω (Aleph Ω, Aleph Ω)]-multiplicity), Aleph (non-ordinal-[Hyper Aleph Ω (Aleph Ω, Aleph Ω)]-multiplicity))] {Hyper Hyper Aleph (non-ordinal-[Hyper Aleph Ω (Aleph Ω, Aleph Ω)]-multiplicity) (Aleph (non-ordinal-[Hyper Aleph Ω (Aleph Ω, Aleph Ω)]-multiplicity), Aleph (non-ordinal-[Hyper Aleph Ω (Aleph Ω, Aleph Ω)]-multiplicity))[Hyper Aleph (non-ordinal-[Hyper Aleph Ω (Aleph Ω, Aleph Ω)]-multiplicity) (Aleph (non-ordinal-[Hyper Aleph Ω (Aleph Ω, Aleph Ω)]-multiplicity), Aleph (non-ordinal-[Hyper Aleph Ω (Aleph Ω, Aleph Ω)]-multiplicity)), Hyper Aleph (non-ordinal-[Hyper Aleph Ω (Aleph Ω, Aleph Ω)]-multiplicity) (Aleph (non-ordinal-[Hyper Aleph Ω (Aleph Ω, Aleph Ω)]-multiplicity), Aleph (non-ordinal-[Hyper Aleph Ω (Aleph Ω, Aleph Ω)]-multiplicity))], Hyper Hyper Aleph (non-ordinal-[Hyper Aleph Ω (Aleph Ω, Aleph Ω)]-multiplicity) (Aleph (non-ordinal-[Hyper Aleph Ω (Aleph Ω, Aleph Ω)]-multiplicity), Aleph (non-ordinal-[Hyper Aleph Ω (Aleph Ω, Aleph Ω)]-multiplicity))[Hyper Aleph (non-ordinal-[Hyper Aleph Ω (Aleph Ω, Aleph Ω)]-multiplicity) (Aleph (non-ordinal-[Hyper Aleph Ω (Aleph Ω, Aleph Ω)]-multiplicity), Aleph (non-ordinal-[Hyper Aleph Ω (Aleph Ω, Aleph Ω)]-multiplicity)), Hyper Aleph (non-ordinal-[Hyper Aleph Ω (Aleph Ω, Aleph Ω)]-multiplicity) (Aleph (non-ordinal-[Hyper Aleph Ω (Aleph Ω, Aleph Ω)]-multiplicity), Aleph (non-ordinal-[Hyper Aleph Ω (Aleph Ω, Aleph Ω)]-multiplicity))]}}{ Hyper Hyper Hyper Hyper Aleph (non-ordinal-[Hyper Aleph Ω (Aleph Ω, Aleph Ω)]-multiplicity) (Aleph (non-ordinal-[Hyper Aleph Ω (Aleph Ω, Aleph Ω)]-multiplicity), Aleph (non-ordinal-[Hyper Aleph Ω (Aleph Ω, Aleph Ω)]-multiplicity))[Hyper Aleph (non-ordinal-[Hyper Aleph Ω (Aleph Ω, Aleph Ω)]-multiplicity) (Aleph (non-ordinal-[Hyper Aleph Ω (Aleph Ω, Aleph Ω)]-multiplicity), Aleph (non-ordinal-[Hyper Aleph Ω (Aleph Ω, Aleph Ω)]-multiplicity)), Hyper Aleph (non-ordinal-[Hyper Aleph Ω (Aleph Ω, Aleph Ω)]-multiplicity) (Aleph (non-ordinal-[Hyper Aleph Ω (Aleph Ω, Aleph Ω)]-multiplicity), Aleph (non-ordinal-[Hyper Aleph Ω (Aleph Ω, Aleph Ω)]-multiplicity))] {Hyper Hyper Aleph (non-ordinal-[Hyper Aleph Ω (Aleph Ω, Aleph Ω)]-multiplicity) (Aleph (non-ordinal-[Hyper Aleph Ω (Aleph Ω, Aleph Ω)]-multiplicity), Aleph (non-ordinal-[Hyper Aleph Ω (Aleph Ω, Aleph Ω)]-multiplicity))[Hyper Aleph (non-ordinal-[Hyper Aleph Ω (Aleph Ω, Aleph Ω)]-multiplicity)

(Aleph (non-ordinal-[Hyper Aleph Ω (Aleph Ω, Aleph Ω)]-multiplicity), Aleph (non-ordinal-[Hyper Aleph Ω (Aleph Ω, Aleph Ω)]-multiplicity)), Hyper Aleph (non-ordinal-[Hyper Aleph Ω (Aleph Ω, Aleph Ω)]-multiplicity) (Aleph (non-ordinal-[Hyper Aleph Ω (Aleph Ω, Aleph Ω)]-multiplicity), Aleph (non-ordinal-[Hyper Aleph Ω (Aleph Ω, Aleph Ω)]-multiplicity))], Hyper Hyper Aleph (non-ordinal-[Hyper Aleph Ω (Aleph Ω, Aleph Ω)]-multiplicity) (Aleph (non-ordinal-[Hyper Aleph Ω (Aleph Ω, Aleph Ω)]-multiplicity), Aleph (non-ordinal-[Hyper Aleph Ω (Aleph Ω, Aleph Ω)]-multiplicity))[Hyper Aleph (non-ordinal-[Hyper Aleph Ω (Aleph Ω, Aleph Ω)]-multiplicity) (Aleph (non-ordinal-[Hyper Aleph Ω (Aleph Ω, Aleph Ω)]-multiplicity), Aleph (non-ordinal-[Hyper Aleph Ω (Aleph Ω, Aleph Ω)]-multiplicity)), Hyper Aleph (non-ordinal-[Hyper Aleph Ω (Aleph Ω, Aleph Ω)]-multiplicity) (Aleph (non-ordinal-[Hyper Aleph Ω (Aleph Ω, Aleph Ω)]-multiplicity), Aleph (non-ordinal-[Hyper Aleph Ω (Aleph Ω, Aleph Ω)]-multiplicity))]]} { Hyper Hyper Hyper Aleph (non-ordinal-[Hyper Aleph Ω (Aleph Ω, Aleph Ω)]-multiplicity) (Aleph (non-ordinal-[Hyper Aleph Ω (Aleph Ω, Aleph Ω)]-multiplicity), Aleph (non-ordinal-[Hyper Aleph Ω (Aleph Ω, Aleph Ω)]-multiplicity))[Hyper Aleph (non-ordinal-[Hyper Aleph Ω (Aleph Ω, Aleph Ω)]-multiplicity) (Aleph (non-ordinal-[Hyper Aleph Ω (Aleph Ω, Aleph Ω)]-multiplicity), Aleph (non-ordinal-[Hyper Aleph Ω (Aleph Ω, Aleph Ω)]-multiplicity)), Hyper Aleph (non-ordinal-[Hyper Aleph Ω (Aleph Ω, Aleph Ω)]-multiplicity) (Aleph (non-ordinal-[Hyper Aleph Ω (Aleph Ω, Aleph Ω)]-multiplicity), Aleph (non-ordinal-[Hyper Aleph Ω (Aleph Ω, Aleph Ω)]-multiplicity))] {Hyper Hyper Aleph (non-ordinal-[Hyper Aleph Ω (Aleph Ω, Aleph Ω)]-multiplicity) (Aleph (non-ordinal-[Hyper Aleph Ω (Aleph Ω, Aleph Ω)]-multiplicity), Aleph (non-ordinal-[Hyper Aleph Ω (Aleph Ω, Aleph Ω)]-multiplicity))[Hyper Aleph (non-ordinal-[Hyper Aleph Ω (Aleph Ω, Aleph Ω)]-multiplicity) (Aleph (non-ordinal-[Hyper Aleph Ω (Aleph Ω, Aleph Ω)]-multiplicity), Aleph (non-ordinal-[Hyper Aleph Ω (Aleph Ω, Aleph Ω)]-multiplicity)), Hyper Aleph (non-ordinal-[Hyper Aleph Ω (Aleph Ω, Aleph Ω)]-multiplicity) (Aleph (non-ordinal-[Hyper Aleph Ω (Aleph Ω, Aleph Ω)]-multiplicity), Aleph (non-ordinal-[Hyper Aleph Ω (Aleph Ω, Aleph Ω)]-multiplicity))], Hyper Hyper Aleph (non-ordinal-[Hyper Aleph Ω (Aleph Ω, Aleph Ω)]-multiplicity) (Aleph (non-ordinal-[Hyper Aleph Ω (Aleph Ω, Aleph Ω)]-multiplicity), Aleph (non-ordinal-[Hyper Aleph Ω (Aleph Ω, Aleph Ω)]-multiplicity))[Hyper Aleph (non-ordinal-[Hyper Aleph Ω (Aleph Ω, Aleph Ω)]-multiplicity) (Aleph (non-ordinal-[Hyper Aleph Ω (Aleph Ω, Aleph Ω)]-multiplicity), Aleph (non-ordinal-[Hyper Aleph Ω (Aleph Ω, Aleph Ω)]-multiplicity)), Hyper Aleph (non-ordinal-[Hyper Aleph Ω (Aleph Ω, Aleph Ω)]-multiplicity) (Aleph (non-ordinal-[Hyper Aleph Ω (Aleph Ω, Aleph Ω)]-multiplicity), Aleph (non-ordinal-[Hyper Aleph Ω (Aleph Ω, Aleph Ω)]-multiplicity))]]}, Hyper Hyper Hyper Aleph (non-ordinal-[Hyper Aleph Ω (Aleph Ω, Aleph Ω)]-multiplicity) (Aleph (non-ordinal-[Hyper Aleph Ω (Aleph Ω, Aleph Ω)]-multiplicity), Aleph (non-ordinal-[Hyper Aleph Ω (Aleph Ω, Aleph Ω)]-multiplicity))[Hyper Aleph (non-ordinal-[Hyper Aleph Ω (Aleph Ω, Aleph Ω)]-multiplicity) (Aleph (non-ordinal-[Hyper Aleph Ω (Aleph Ω, Aleph Ω)]-multiplicity), Aleph (non-ordinal-[Hyper Aleph Ω (Aleph Ω, Aleph Ω)]-multiplicity)), Hyper

Aleph (non-ordinal-[Hyper Aleph Ω (Aleph Ω, Aleph Ω)]-multiplicity) (Aleph (non-ordinal-[Hyper Aleph Ω (Aleph Ω, Aleph Ω)]-multiplicity), Aleph (non-ordinal-[Hyper Aleph Ω (Aleph Ω, Aleph Ω)]-multiplicity))] {Hyper Hyper Aleph (non-ordinal-[Hyper Aleph Ω (Aleph Ω, Aleph Ω)]-multiplicity) (Aleph (non-ordinal-[Hyper Aleph Ω (Aleph Ω, Aleph Ω)]-multiplicity), Aleph (non-ordinal-[Hyper Aleph Ω (Aleph Ω, Aleph Ω)]-multiplicity))[Hyper Aleph (non-ordinal-[Hyper Aleph Ω (Aleph Ω, Aleph Ω)]-multiplicity) (Aleph (non-ordinal-[Hyper Aleph Ω (Aleph Ω, Aleph Ω)]-multiplicity), Aleph (non-ordinal-[Hyper Aleph Ω (Aleph Ω, Aleph Ω)]-multiplicity)), Hyper Aleph (non-ordinal-[Hyper Aleph Ω (Aleph Ω, Aleph Ω)]-multiplicity) (Aleph (non-ordinal-[Hyper Aleph Ω (Aleph Ω, Aleph Ω)]-multiplicity), Aleph (non-ordinal-[Hyper Aleph Ω (Aleph Ω, Aleph Ω)]-multiplicity))], Hyper Hyper Aleph (non-ordinal-[Hyper Aleph Ω (Aleph Ω, Aleph Ω)]-multiplicity) (Aleph (non-ordinal-[Hyper Aleph Ω (Aleph Ω, Aleph Ω)]-multiplicity), Aleph (non-ordinal-[Hyper Aleph Ω (Aleph Ω, Aleph Ω)]-multiplicity))[Hyper Aleph (non-ordinal-[Hyper Aleph Ω (Aleph Ω, Aleph Ω)]-multiplicity) (Aleph (non-ordinal-[Hyper Aleph Ω (Aleph Ω, Aleph Ω)]-multiplicity), Aleph (non-ordinal-[Hyper Aleph Ω (Aleph Ω, Aleph Ω)]-multiplicity)), Hyper Aleph (non-ordinal-[Hyper Aleph Ω (Aleph Ω, Aleph Ω)]-multiplicity) (Aleph (non-ordinal-[Hyper Aleph Ω (Aleph Ω, Aleph Ω)]-multiplicity), Aleph (non-ordinal-[Hyper Aleph Ω (Aleph Ω, Aleph Ω)]-multiplicity))]}}, Hyper Hyper Hyper Hyper Aleph (non-ordinal-[Hyper Aleph Ω (Aleph Ω, Aleph Ω)]-multiplicity) (Aleph (non-ordinal-[Hyper Aleph Ω (Aleph Ω, Aleph Ω)]-multiplicity), Aleph (non-ordinal-[Hyper Aleph Ω (Aleph Ω, Aleph Ω)]-multiplicity))[Hyper Aleph (non-ordinal-[Hyper Aleph Ω (Aleph Ω, Aleph Ω)]-multiplicity) (Aleph (non-ordinal-[Hyper Aleph Ω (Aleph Ω, Aleph Ω)]-multiplicity), Aleph (non-ordinal-[Hyper Aleph Ω (Aleph Ω, Aleph Ω)]-multiplicity)), Hyper Aleph (non-ordinal-[Hyper Aleph Ω (Aleph Ω, Aleph Ω)]-multiplicity) (Aleph (non-ordinal-[Hyper Aleph Ω (Aleph Ω, Aleph Ω)]-multiplicity), Aleph (non-ordinal-[Hyper Aleph Ω (Aleph Ω, Aleph Ω)]-multiplicity))] {Hyper Hyper Aleph (non-ordinal-[Hyper Aleph Ω (Aleph Ω, Aleph Ω)]-multiplicity) (Aleph (non-ordinal-[Hyper Aleph Ω (Aleph Ω, Aleph Ω)]-multiplicity), Aleph (non-ordinal-[Hyper Aleph Ω (Aleph Ω, Aleph Ω)]-multiplicity))[Hyper Aleph (non-ordinal-[Hyper Aleph Ω (Aleph Ω, Aleph Ω)]-multiplicity) (Aleph (non-ordinal-[Hyper Aleph Ω (Aleph Ω, Aleph Ω)]-multiplicity), Aleph (non-ordinal-[Hyper Aleph Ω (Aleph Ω, Aleph Ω)]-multiplicity)), Hyper Aleph (non-ordinal-[Hyper Aleph Ω (Aleph Ω, Aleph Ω)]-multiplicity) (Aleph (non-ordinal-[Hyper Aleph Ω (Aleph Ω, Aleph Ω)]-multiplicity), Aleph (non-ordinal-[Hyper Aleph Ω (Aleph Ω, Aleph Ω)]-multiplicity))], Hyper Hyper Aleph (non-ordinal-[Hyper Aleph Ω (Aleph Ω, Aleph Ω)]-multiplicity) (Aleph (non-ordinal-[Hyper Aleph Ω (Aleph Ω, Aleph Ω)]-multiplicity), Aleph (non-ordinal-[Hyper Aleph Ω (Aleph Ω, Aleph Ω)]-multiplicity))[Hyper Aleph (non-ordinal-[Hyper Aleph Ω (Aleph Ω, Aleph Ω)]-multiplicity) (Aleph (non-ordinal-[Hyper Aleph Ω (Aleph Ω, Aleph Ω)]-multiplicity), Aleph (non-ordinal-[Hyper Aleph Ω (Aleph Ω, Aleph Ω)]-multiplicity)), Hyper Aleph (non-ordinal-[Hyper Aleph Ω (Aleph Ω, Aleph Ω)]-multiplicity) (Aleph (non-ordinal-[Hyper Aleph Ω (Aleph Ω, Aleph Ω)]-multiplicity), Aleph (non-ordinal-[Hyper Aleph Ω (Aleph Ω, Aleph Ω)]-multiplicity))]} { Hyper Hyper Hyper Aleph (non-ordinal-[Hyper Aleph Ω (Aleph Ω, Aleph Ω)]-multiplicity) (Aleph (non-ordinal-[Hyper Aleph Ω (Aleph Ω, Aleph Ω)]-multiplicity), Aleph (non-ordinal-[Hyper Aleph Ω (Aleph Ω, Aleph Ω)]-multiplicity))[Hyper Aleph (non-ordinal-[Hyper Aleph Ω (Aleph Ω, Aleph Ω)]-multiplicity) (Aleph (non-ordinal-[Hyper Aleph Ω (Aleph Ω, Aleph Ω)]-multiplicity), Aleph (non-ordinal-[Hyper Aleph Ω (Aleph Ω, Aleph Ω)]-multiplicity)), Hyper Aleph (non-ordinal-[Hyper Aleph Ω (Aleph Ω, Aleph Ω)]-multiplicity) (Aleph (non-ordinal-[Hyper Aleph Ω (Aleph Ω, Aleph Ω)]-multiplicity), Aleph (non-ordinal-[Hyper Aleph Ω (Aleph Ω, Aleph Ω)]-multiplicity))] {Hyper Hyper Aleph (non-ordinal-[Hyper Aleph Ω (Aleph Ω, Aleph Ω)]-multiplicity) (Aleph (non-ordinal-[Hyper Aleph Ω (Aleph Ω, Aleph Ω)]-multiplicity), Aleph (non-ordinal-[Hyper Aleph Ω (Aleph Ω, Aleph Ω)]-multiplicity))[Hyper Aleph (non-ordinal-[Hyper Aleph Ω (Aleph Ω, Aleph Ω)]-multiplicity) (Aleph (non-ordinal-[Hyper Aleph Ω (Aleph Ω, Aleph Ω)]-multiplicity), Aleph (non-ordinal-[Hyper Aleph Ω (Aleph Ω, Aleph Ω)]-multiplicity)), Hyper Aleph (non-ordinal-[Hyper Aleph Ω (Aleph Ω, Aleph Ω)]-multiplicity) (Aleph (non-ordinal-[Hyper Aleph Ω (Aleph Ω, Aleph Ω)]-multiplicity), Aleph (non-ordinal-[Hyper Aleph Ω (Aleph Ω, Aleph Ω)]-multiplicity))], Hyper Hyper Aleph (non-ordinal-[Hyper Aleph Ω (Aleph Ω, Aleph Ω)]-multiplicity) (Aleph (non-ordinal-[Hyper Aleph Ω (Aleph Ω, Aleph Ω)]-multiplicity), Aleph (non-ordinal-[Hyper Aleph Ω (Aleph Ω, Aleph Ω)]-multiplicity))[Hyper Aleph (non-ordinal-[Hyper Aleph Ω (Aleph Ω, Aleph Ω)]-multiplicity) (Aleph (non-ordinal-[Hyper Aleph Ω (Aleph Ω, Aleph Ω)]-multiplicity), Aleph (non-ordinal-[Hyper Aleph Ω (Aleph Ω, Aleph Ω)]-multiplicity)), Hyper Aleph (non-ordinal-[Hyper Aleph Ω (Aleph Ω, Aleph Ω)]-multiplicity) (Aleph (non-ordinal-[Hyper Aleph Ω (Aleph Ω, Aleph Ω)]-multiplicity), Aleph (non-ordinal-[Hyper Aleph Ω (Aleph Ω, Aleph Ω)]-multiplicity))]}, Hyper Hyper Hyper Aleph (non-ordinal-[Hyper Aleph Ω (Aleph Ω, Aleph Ω)]-multiplicity) (Aleph (non-ordinal-[Hyper Aleph Ω (Aleph Ω, Aleph Ω)]-multiplicity), Aleph (non-ordinal-[Hyper Aleph Ω (Aleph Ω, Aleph Ω)]-multiplicity))[Hyper Aleph (non-ordinal-[Hyper Aleph Ω (Aleph Ω, Aleph Ω)]-multiplicity) (Aleph (non-ordinal-[Hyper Aleph Ω (Aleph Ω, Aleph Ω)]-multiplicity), Aleph (non-ordinal-[Hyper Aleph Ω (Aleph Ω, Aleph Ω)]-multiplicity)), Hyper Aleph (non-ordinal-[Hyper Aleph Ω (Aleph Ω, Aleph Ω)]-multiplicity) (Aleph (non-ordinal-[Hyper Aleph Ω (Aleph Ω, Aleph Ω)]-multiplicity), Aleph (non-ordinal-[Hyper Aleph Ω (Aleph Ω, Aleph Ω)]-multiplicity))] {Hyper Hyper Aleph (non-ordinal-[Hyper Aleph Ω (Aleph Ω, Aleph Ω)]-multiplicity) (Aleph (non-ordinal-[Hyper Aleph Ω (Aleph Ω, Aleph Ω)]-multiplicity), Aleph (non-ordinal-[Hyper Aleph Ω (Aleph Ω, Aleph Ω)]-multiplicity))[Hyper Aleph (non-ordinal-[Hyper Aleph Ω (Aleph Ω, Aleph Ω)]-multiplicity) (Aleph (non-ordinal-[Hyper Aleph Ω (Aleph Ω, Aleph Ω)]-multiplicity), Aleph (non-ordinal-[Hyper Aleph Ω (Aleph Ω, Aleph Ω)]-multiplicity)), Hyper Aleph (non-ordinal-[Hyper Aleph Ω (Aleph Ω, Aleph Ω)]-multiplicity) (Aleph (non-ordinal-[Hyper Aleph Ω (Aleph Ω, Aleph Ω)]-multiplicity), Aleph (non-ordinal-[Hyper Aleph Ω (Aleph Ω, Aleph Ω)]-multiplicity))], Hyper Hyper Aleph (non-ordinal-[Hyper Aleph Ω (Aleph Ω, Aleph Ω)]-

multiplicity) (Aleph (non-ordinal-[Hyper Aleph Ω (Aleph Ω, Aleph Ω)]-multiplicity), Aleph (non-ordinal-[Hyper Aleph Ω (Aleph Ω, Aleph Ω)]-multiplicity))[Hyper Aleph (non-ordinal-[Hyper Aleph Ω (Aleph Ω, Aleph Ω)]-multiplicity) (Aleph (non-ordinal-[Hyper Aleph Ω (Aleph Ω, Aleph Ω)]-multiplicity), Aleph (non-ordinal-[Hyper Aleph Ω (Aleph Ω, Aleph Ω)]-multiplicity)), Hyper Aleph (non-ordinal-[Hyper Aleph Ω (Aleph Ω, Aleph Ω)]-multiplicity) (Aleph (non-ordinal-[Hyper Aleph Ω (Aleph Ω, Aleph Ω)]-multiplicity), Aleph (non-ordinal-[Hyper Aleph Ω (Aleph Ω, Aleph Ω)]-multiplicity))]}}}

We can go still larger to develop the following infinity.

Hyper Hyper Hyper Hyper Hyper Hyper Aleph (non-ordinal-[Hyper Aleph Ω (Aleph Ω, Aleph Ω)]-multiplicity) (Aleph (non-ordinal-[Hyper Aleph Ω (Aleph Ω, Aleph Ω)]-multiplicity), Aleph (non-ordinal-[Hyper Aleph Ω (Aleph Ω, Aleph Ω)]-multiplicity))[Hyper Aleph (non-ordinal-[Hyper Aleph Ω (Aleph Ω, Aleph Ω)]-multiplicity) (Aleph (non-ordinal-[Hyper Aleph Ω (Aleph Ω, Aleph Ω)]-multiplicity), Aleph (non-ordinal-[Hyper Aleph Ω (Aleph Ω, Aleph Ω)]-multiplicity)), Hyper Aleph (non-ordinal-[Hyper Aleph Ω (Aleph Ω, Aleph Ω)]-multiplicity) (Aleph (non-ordinal-[Hyper Aleph Ω (Aleph Ω, Aleph Ω)]-multiplicity), Aleph (non-ordinal-[Hyper Aleph Ω (Aleph Ω, Aleph Ω)]-multiplicity))] {Hyper Hyper Aleph (non-ordinal-[Hyper Aleph Ω (Aleph Ω, Aleph Ω)]-multiplicity) (Aleph (non-ordinal-[Hyper Aleph Ω (Aleph Ω, Aleph Ω)]-multiplicity), Aleph (non-ordinal-[Hyper Aleph Ω (Aleph Ω, Aleph Ω)]-multiplicity))[Hyper Aleph (non-ordinal-[Hyper Aleph Ω (Aleph Ω, Aleph Ω)]-multiplicity) (Aleph (non-ordinal-[Hyper Aleph Ω (Aleph Ω, Aleph Ω)]-multiplicity), Aleph (non-ordinal-[Hyper Aleph Ω (Aleph Ω, Aleph Ω)]-multiplicity)), Hyper Aleph (non-ordinal-[Hyper Aleph Ω (Aleph Ω, Aleph Ω)]-multiplicity) (Aleph (non-ordinal-[Hyper Aleph Ω (Aleph Ω, Aleph Ω)]-multiplicity), Aleph (non-ordinal-[Hyper Aleph Ω (Aleph Ω, Aleph Ω)]-multiplicity))], Hyper Hyper Aleph (non-ordinal-[Hyper Aleph Ω (Aleph Ω, Aleph Ω)]-multiplicity) (Aleph (non-ordinal-[Hyper Aleph Ω (Aleph Ω, Aleph Ω)]-multiplicity), Aleph (non-ordinal-[Hyper Aleph Ω (Aleph Ω, Aleph Ω)]-multiplicity))[Hyper Aleph (non-ordinal-[Hyper Aleph Ω (Aleph Ω, Aleph Ω)]-multiplicity) (Aleph (non-ordinal-[Hyper Aleph Ω (Aleph Ω, Aleph Ω)]-multiplicity), Aleph (non-ordinal-[Hyper Aleph Ω (Aleph Ω, Aleph Ω)]-multiplicity)), Hyper Aleph (non-ordinal-[Hyper Aleph Ω (Aleph Ω, Aleph Ω)]-multiplicity) (Aleph (non-ordinal-[Hyper Aleph Ω (Aleph Ω, Aleph Ω)]-multiplicity), Aleph (non-ordinal-[Hyper Aleph Ω (Aleph Ω, Aleph Ω)]-multiplicity))]} { Hyper Hyper Hyper Aleph (non-ordinal-[Hyper Aleph Ω (Aleph Ω, Aleph Ω)]-multiplicity) (Aleph (non-ordinal-[Hyper Aleph Ω (Aleph Ω, Aleph Ω)]-multiplicity), Aleph (non-ordinal-[Hyper Aleph Ω (Aleph Ω, Aleph Ω)]-multiplicity))[Hyper Aleph (non-ordinal-[Hyper Aleph Ω (Aleph Ω, Aleph Ω)]-multiplicity) (Aleph (non-ordinal-[Hyper Aleph Ω (Aleph Ω, Aleph Ω)]-multiplicity), Aleph (non-ordinal-[Hyper Aleph Ω (Aleph Ω,

Aleph Ω)]-multiplicity)), Hyper Aleph (non-ordinal-[Hyper Aleph Ω (Aleph Ω, Aleph Ω)]-multiplicity) (Aleph (non-ordinal-[Hyper Aleph Ω (Aleph Ω, Aleph Ω)]-multiplicity), Aleph (non-ordinal-[Hyper Aleph Ω (Aleph Ω, Aleph Ω)]-multiplicity))] {Hyper Hyper Aleph (non-ordinal-[Hyper Aleph Ω (Aleph Ω, Aleph Ω)]-multiplicity) (Aleph (non-ordinal-[Hyper Aleph Ω (Aleph Ω, Aleph Ω)]-multiplicity), Aleph (non-ordinal-[Hyper Aleph Ω (Aleph Ω, Aleph Ω)]-multiplicity))[Hyper Aleph (non-ordinal-[Hyper Aleph Ω (Aleph Ω, Aleph Ω)]-multiplicity) (Aleph (non-ordinal-[Hyper Aleph Ω (Aleph Ω, Aleph Ω)]-multiplicity), Aleph (non-ordinal-[Hyper Aleph Ω (Aleph Ω, Aleph Ω)]-multiplicity)), Hyper Aleph (non-ordinal-[Hyper Aleph Ω (Aleph Ω, Aleph Ω)]-multiplicity) (Aleph (non-ordinal-[Hyper Aleph Ω (Aleph Ω, Aleph Ω)]-multiplicity), Aleph (non-ordinal-[Hyper Aleph Ω (Aleph Ω, Aleph Ω)]-multiplicity))], Hyper Hyper Aleph (non-ordinal-[Hyper Aleph Ω (Aleph Ω, Aleph Ω)]-multiplicity) (Aleph (non-ordinal-[Hyper Aleph Ω (Aleph Ω, Aleph Ω)]-multiplicity), Aleph (non-ordinal-[Hyper Aleph Ω (Aleph Ω, Aleph Ω)]-multiplicity))[Hyper Aleph (non-ordinal-[Hyper Aleph Ω (Aleph Ω, Aleph Ω)]-multiplicity) (Aleph (non-ordinal-[Hyper Aleph Ω (Aleph Ω, Aleph Ω)]-multiplicity), Aleph (non-ordinal-[Hyper Aleph Ω (Aleph Ω, Aleph Ω)]-multiplicity)), Hyper Aleph (non-ordinal-[Hyper Aleph Ω (Aleph Ω, Aleph Ω)]-multiplicity) (Aleph (non-ordinal-[Hyper Aleph Ω (Aleph Ω, Aleph Ω)]-multiplicity), Aleph (non-ordinal-[Hyper Aleph Ω (Aleph Ω, Aleph Ω)]-multiplicity))]}, Hyper Hyper Hyper Aleph (non-ordinal-[Hyper Aleph Ω (Aleph Ω, Aleph Ω)]-multiplicity) (Aleph (non-ordinal-[Hyper Aleph Ω (Aleph Ω, Aleph Ω)]-multiplicity), Aleph (non-ordinal-[Hyper Aleph Ω (Aleph Ω, Aleph Ω)]-multiplicity))[Hyper Aleph (non-ordinal-[Hyper Aleph Ω (Aleph Ω, Aleph Ω)]-multiplicity) (Aleph (non-ordinal-[Hyper Aleph Ω (Aleph Ω, Aleph Ω)]-multiplicity), Aleph (non-ordinal-[Hyper Aleph Ω (Aleph Ω, Aleph Ω)]-multiplicity)), Hyper Aleph (non-ordinal-[Hyper Aleph Ω (Aleph Ω, Aleph Ω)]-multiplicity) (Aleph (non-ordinal-[Hyper Aleph Ω (Aleph Ω, Aleph Ω)]-multiplicity), Aleph (non-ordinal-[Hyper Aleph Ω (Aleph Ω, Aleph Ω)]-multiplicity))] {Hyper Hyper Aleph (non-ordinal-[Hyper Aleph Ω (Aleph Ω, Aleph Ω)]-multiplicity) (Aleph (non-ordinal-[Hyper Aleph Ω (Aleph Ω, Aleph Ω)]-multiplicity), Aleph (non-ordinal-[Hyper Aleph Ω (Aleph Ω, Aleph Ω)]-multiplicity))[Hyper Aleph (non-ordinal-[Hyper Aleph Ω (Aleph Ω, Aleph Ω)]-multiplicity) (Aleph (non-ordinal-[Hyper Aleph Ω (Aleph Ω, Aleph Ω)]-multiplicity), Aleph (non-ordinal-[Hyper Aleph Ω (Aleph Ω, Aleph Ω)]-multiplicity)), Hyper Aleph (non-ordinal-[Hyper Aleph Ω (Aleph Ω, Aleph Ω)]-multiplicity) (Aleph (non-ordinal-[Hyper Aleph Ω (Aleph Ω, Aleph Ω)]-multiplicity), Aleph (non-ordinal-[Hyper Aleph Ω (Aleph Ω, Aleph Ω)]-multiplicity))], Hyper Hyper Aleph (non-ordinal-[Hyper Aleph Ω (Aleph Ω, Aleph Ω)]-multiplicity) (Aleph (non-ordinal-[Hyper Aleph Ω (Aleph Ω, Aleph Ω)]-multiplicity), Aleph (non-ordinal-[Hyper Aleph Ω (Aleph Ω, Aleph Ω)]-multiplicity))[Hyper Aleph (non-ordinal-[Hyper Aleph Ω (Aleph Ω, Aleph Ω)]-multiplicity) (Aleph (non-ordinal-[Hyper Aleph Ω (Aleph Ω, Aleph Ω)]-multiplicity), Aleph (non-ordinal-[Hyper Aleph Ω (Aleph Ω, Aleph Ω)]-multiplicity)), Hyper Aleph (non-ordinal-[Hyper Aleph Ω (Aleph Ω, Aleph Ω)]-multiplicity) (Aleph (non-ordinal-[Hyper Aleph Ω (Aleph Ω, Aleph Ω)]-multiplicity), Aleph (non-ordinal-[Hyper Aleph Ω (Aleph Ω, Aleph Ω)]-multiplicity))]}}{ Hyper Hyper Hyper Hyper Aleph (non-ordinal-[Hyper Aleph Ω (Aleph Ω, Aleph Ω)]-multiplicity) (Aleph (non-ordinal-[Hyper Aleph Ω (Aleph Ω, Aleph Ω)]-multiplicity), Aleph (non-ordinal-[Hyper Aleph Ω (Aleph Ω, Aleph Ω)]-multiplicity))[Hyper Aleph (non-ordinal-[Hyper Aleph Ω (Aleph Ω, Aleph Ω)]-multiplicity) (Aleph (non-ordinal-[Hyper Aleph Ω (Aleph Ω, Aleph Ω)]-multiplicity), Aleph (non-ordinal-[Hyper Aleph Ω (Aleph Ω, Aleph Ω)]-multiplicity)), Hyper Aleph (non-ordinal-[Hyper Aleph Ω (Aleph Ω, Aleph Ω)]-multiplicity) (Aleph (non-ordinal-[Hyper Aleph Ω (Aleph Ω, Aleph Ω)]-multiplicity), Aleph (non-ordinal-[Hyper Aleph Ω (Aleph Ω, Aleph Ω)]-multiplicity))] {Hyper Hyper Aleph (non-ordinal-[Hyper Aleph Ω (Aleph Ω, Aleph Ω)]-multiplicity) (Aleph (non-ordinal-[Hyper Aleph Ω (Aleph Ω, Aleph Ω)]-multiplicity), Aleph (non-ordinal-[Hyper Aleph Ω (Aleph Ω, Aleph Ω)]-multiplicity))[Hyper Aleph (non-ordinal-[Hyper Aleph Ω (Aleph Ω, Aleph Ω)]-multiplicity) (Aleph (non-ordinal-[Hyper Aleph Ω (Aleph Ω, Aleph Ω)]-multiplicity), Aleph (non-ordinal-[Hyper Aleph Ω (Aleph Ω, Aleph Ω)]-multiplicity)), Hyper Aleph (non-ordinal-[Hyper Aleph Ω (Aleph Ω, Aleph Ω)]-multiplicity) (Aleph (non-ordinal-[Hyper Aleph Ω (Aleph Ω, Aleph Ω)]-multiplicity), Aleph (non-ordinal-[Hyper Aleph Ω (Aleph Ω, Aleph Ω)]-multiplicity))], Hyper Hyper Aleph (non-ordinal-[Hyper Aleph Ω (Aleph Ω, Aleph Ω)]-multiplicity) (Aleph (non-ordinal-[Hyper Aleph Ω (Aleph Ω, Aleph Ω)]-multiplicity), Aleph (non-ordinal-[Hyper Aleph Ω (Aleph Ω, Aleph Ω)]-multiplicity))[Hyper Aleph (non-ordinal-[Hyper Aleph Ω (Aleph Ω, Aleph Ω)]-multiplicity) (Aleph (non-ordinal-[Hyper Aleph Ω (Aleph Ω, Aleph Ω)]-multiplicity), Aleph (non-ordinal-[Hyper Aleph Ω (Aleph Ω, Aleph Ω)]-multiplicity)), Hyper Aleph (non-ordinal-[Hyper Aleph Ω (Aleph Ω, Aleph Ω)]-multiplicity) (Aleph (non-ordinal-[Hyper Aleph Ω (Aleph Ω, Aleph Ω)]-multiplicity), Aleph (non-ordinal-[Hyper Aleph Ω (Aleph Ω, Aleph Ω)]-multiplicity))]} { Hyper Hyper Hyper Aleph (non-ordinal-[Hyper Aleph Ω (Aleph Ω, Aleph Ω)]-multiplicity) (Aleph (non-ordinal-[Hyper Aleph Ω (Aleph Ω, Aleph Ω)]-multiplicity), Aleph (non-ordinal-[Hyper Aleph Ω (Aleph Ω, Aleph Ω)]-multiplicity))[Hyper Aleph (non-ordinal-[Hyper Aleph Ω (Aleph Ω, Aleph Ω)]-multiplicity) (Aleph (non-ordinal-[Hyper Aleph Ω (Aleph Ω, Aleph Ω)]-multiplicity), Aleph (non-ordinal-[Hyper Aleph Ω (Aleph Ω, Aleph Ω)]-multiplicity)), Hyper Aleph (non-ordinal-[Hyper Aleph Ω (Aleph Ω, Aleph Ω)]-multiplicity) (Aleph (non-ordinal-[Hyper Aleph Ω (Aleph Ω, Aleph Ω)]-multiplicity), Aleph (non-ordinal-[Hyper Aleph Ω (Aleph Ω, Aleph Ω)]-multiplicity))] {Hyper Hyper Aleph (non-ordinal-[Hyper Aleph Ω (Aleph Ω, Aleph Ω)]-multiplicity) (Aleph (non-ordinal-[Hyper Aleph Ω (Aleph Ω, Aleph Ω)]-multiplicity), Aleph (non-ordinal-[Hyper Aleph Ω (Aleph Ω, Aleph Ω)]-multiplicity))[Hyper Aleph (non-ordinal-[Hyper Aleph Ω (Aleph Ω, Aleph Ω)]-multiplicity) (Aleph (non-ordinal-[Hyper Aleph Ω (Aleph Ω, Aleph Ω)]-multiplicity), Aleph (non-ordinal-[Hyper Aleph Ω (Aleph Ω, Aleph Ω)]-multiplicity)), Hyper Aleph (non-ordinal-[Hyper Aleph Ω (Aleph Ω, Aleph Ω)]-multiplicity) (Aleph (non-ordinal-[Hyper Aleph Ω (Aleph Ω, Aleph Ω)]-multiplicity), Aleph (non-ordinal-[Hyper Aleph Ω (Aleph Ω, Aleph Ω)]-multiplicity))], Hyper Hyper Aleph (non-ordinal-[Hyper Aleph Ω (Aleph Ω, Aleph Ω)]-

multiplicity) (Aleph (non-ordinal-[Hyper Aleph Ω (Aleph Ω, Aleph Ω)]-multiplicity), Aleph (non-ordinal-[Hyper Aleph Ω (Aleph Ω, Aleph Ω)]-multiplicity))[Hyper Aleph (non-ordinal-[Hyper Aleph Ω (Aleph Ω, Aleph Ω)]-multiplicity) (Aleph (non-ordinal-[Hyper Aleph Ω (Aleph Ω, Aleph Ω)]-multiplicity), Aleph (non-ordinal-[Hyper Aleph Ω (Aleph Ω, Aleph Ω)]-multiplicity)), Hyper Aleph (non-ordinal-[Hyper Aleph Ω (Aleph Ω, Aleph Ω)]-multiplicity) (Aleph (non-ordinal-[Hyper Aleph Ω (Aleph Ω, Aleph Ω)]-multiplicity), Aleph (non-ordinal-[Hyper Aleph Ω (Aleph Ω, Aleph Ω)]-multiplicity))]}, Hyper Hyper Hyper Aleph (non-ordinal-[Hyper Aleph Ω (Aleph Ω, Aleph Ω)]-multiplicity) (Aleph (non-ordinal-[Hyper Aleph Ω (Aleph Ω, Aleph Ω)]-multiplicity), Aleph (non-ordinal-[Hyper Aleph Ω (Aleph Ω, Aleph Ω)]-multiplicity))[Hyper Aleph (non-ordinal-[Hyper Aleph Ω (Aleph Ω, Aleph Ω)]-multiplicity) (Aleph (non-ordinal-[Hyper Aleph Ω (Aleph Ω, Aleph Ω)]-multiplicity), Aleph (non-ordinal-[Hyper Aleph Ω (Aleph Ω, Aleph Ω)]-multiplicity)), Hyper Aleph (non-ordinal-[Hyper Aleph Ω (Aleph Ω, Aleph Ω)]-multiplicity) (Aleph (non-ordinal-[Hyper Aleph Ω (Aleph Ω, Aleph Ω)]-multiplicity), Aleph (non-ordinal-[Hyper Aleph Ω (Aleph Ω, Aleph Ω)]-multiplicity))] {Hyper Hyper Aleph (non-ordinal-[Hyper Aleph Ω (Aleph Ω, Aleph Ω)]-multiplicity) (Aleph (non-ordinal-[Hyper Aleph Ω (Aleph Ω, Aleph Ω)]-multiplicity), Aleph (non-ordinal-[Hyper Aleph Ω (Aleph Ω, Aleph Ω)]-multiplicity))[Hyper Aleph (non-ordinal-[Hyper Aleph Ω (Aleph Ω, Aleph Ω)]-multiplicity) (Aleph (non-ordinal-[Hyper Aleph Ω (Aleph Ω, Aleph Ω)]-multiplicity), Aleph (non-ordinal-[Hyper Aleph Ω (Aleph Ω, Aleph Ω)]-multiplicity)), Hyper Aleph (non-ordinal-[Hyper Aleph Ω (Aleph Ω, Aleph Ω)]-multiplicity) (Aleph (non-ordinal-[Hyper Aleph Ω (Aleph Ω, Aleph Ω)]-multiplicity), Aleph (non-ordinal-[Hyper Aleph Ω (Aleph Ω, Aleph Ω)]-multiplicity))], Hyper Hyper Aleph (non-ordinal-[Hyper Aleph Ω (Aleph Ω, Aleph Ω)]-multiplicity) (Aleph (non-ordinal-[Hyper Aleph Ω (Aleph Ω, Aleph Ω)]-multiplicity), Aleph (non-ordinal-[Hyper Aleph Ω (Aleph Ω, Aleph Ω)]-multiplicity))[Hyper Aleph (non-ordinal-[Hyper Aleph Ω (Aleph Ω, Aleph Ω)]-multiplicity) (Aleph (non-ordinal-[Hyper Aleph Ω (Aleph Ω, Aleph Ω)]-multiplicity), Aleph (non-ordinal-[Hyper Aleph Ω (Aleph Ω, Aleph Ω)]-multiplicity)), Hyper Aleph (non-ordinal-[Hyper Aleph Ω (Aleph Ω, Aleph Ω)]-multiplicity) (Aleph (non-ordinal-[Hyper Aleph Ω (Aleph Ω, Aleph Ω)]-multiplicity), Aleph (non-ordinal-[Hyper Aleph Ω (Aleph Ω, Aleph Ω)]-multiplicity))]}}, Hyper Hyper Hyper Hyper Aleph (non-ordinal-[Hyper Aleph Ω (Aleph Ω, Aleph Ω)]-multiplicity) (Aleph (non-ordinal-[Hyper Aleph Ω (Aleph Ω, Aleph Ω)]-multiplicity), Aleph (non-ordinal-[Hyper Aleph Ω (Aleph Ω, Aleph Ω)]-multiplicity))[Hyper Aleph (non-ordinal-[Hyper Aleph Ω (Aleph Ω, Aleph Ω)]-multiplicity) (Aleph (non-ordinal-[Hyper Aleph Ω (Aleph Ω, Aleph Ω)]-multiplicity), Aleph (non-ordinal-[Hyper Aleph Ω (Aleph Ω, Aleph Ω)]-multiplicity)), Hyper Aleph (non-ordinal-[Hyper Aleph Ω (Aleph Ω, Aleph Ω)]-multiplicity) (Aleph (non-ordinal-[Hyper Aleph Ω (Aleph Ω, Aleph Ω)]-multiplicity), Aleph (non-ordinal-[Hyper Aleph Ω (Aleph Ω, Aleph Ω)]-multiplicity))] {Hyper Hyper Aleph (non-ordinal-[Hyper Aleph Ω (Aleph Ω, Aleph Ω)]-multiplicity) (Aleph (non-ordinal-[Hyper Aleph Ω (Aleph Ω, Aleph Ω)]-multiplicity), Aleph (non-ordinal-[Hyper Aleph Ω (Aleph Ω, Aleph Ω)]-

multiplicity))[Hyper Aleph (non-ordinal-[Hyper Aleph Ω (Aleph Ω, Aleph Ω)]-multiplicity) (Aleph (non-ordinal-[Hyper Aleph Ω (Aleph Ω, Aleph Ω)]-multiplicity), Aleph (non-ordinal-[Hyper Aleph Ω (Aleph Ω, Aleph Ω)]-multiplicity)), Hyper Aleph (non-ordinal-[Hyper Aleph Ω (Aleph Ω, Aleph Ω)]-multiplicity) (Aleph (non-ordinal-[Hyper Aleph Ω (Aleph Ω, Aleph Ω)]-multiplicity), Aleph (non-ordinal-[Hyper Aleph Ω (Aleph Ω, Aleph Ω)]-multiplicity))], Hyper Hyper Aleph (non-ordinal-[Hyper Aleph Ω (Aleph Ω, Aleph Ω)]-multiplicity) (Aleph (non-ordinal-[Hyper Aleph Ω (Aleph Ω, Aleph Ω)]-multiplicity), Aleph (non-ordinal-[Hyper Aleph Ω (Aleph Ω, Aleph Ω)]-multiplicity))[Hyper Aleph (non-ordinal-[Hyper Aleph Ω (Aleph Ω, Aleph Ω)]-multiplicity) (Aleph (non-ordinal-[Hyper Aleph Ω (Aleph Ω, Aleph Ω)]-multiplicity), Aleph (non-ordinal-[Hyper Aleph Ω (Aleph Ω, Aleph Ω)]-multiplicity)), Hyper Aleph (non-ordinal-[Hyper Aleph Ω (Aleph Ω, Aleph Ω)]-multiplicity) (Aleph (non-ordinal-[Hyper Aleph Ω (Aleph Ω, Aleph Ω)]-multiplicity), Aleph (non-ordinal-[Hyper Aleph Ω (Aleph Ω, Aleph Ω)]-multiplicity))]} { Hyper Hyper Hyper Aleph (non-ordinal-[Hyper Aleph Ω (Aleph Ω, Aleph Ω)]-multiplicity) (Aleph (non-ordinal-[Hyper Aleph Ω (Aleph Ω, Aleph Ω)]-multiplicity), Aleph (non-ordinal-[Hyper Aleph Ω (Aleph Ω, Aleph Ω)]-multiplicity))[Hyper Aleph (non-ordinal-[Hyper Aleph Ω (Aleph Ω, Aleph Ω)]-multiplicity) (Aleph (non-ordinal-[Hyper Aleph Ω (Aleph Ω, Aleph Ω)]-multiplicity), Aleph (non-ordinal-[Hyper Aleph Ω (Aleph Ω, Aleph Ω)]-multiplicity)), Hyper Aleph (non-ordinal-[Hyper Aleph Ω (Aleph Ω, Aleph Ω)]-multiplicity) (Aleph (non-ordinal-[Hyper Aleph Ω (Aleph Ω, Aleph Ω)]-multiplicity), Aleph (non-ordinal-[Hyper Aleph Ω (Aleph Ω, Aleph Ω)]-multiplicity))] {Hyper Hyper Aleph (non-ordinal-[Hyper Aleph Ω (Aleph Ω, Aleph Ω)]-multiplicity) (Aleph (non-ordinal-[Hyper Aleph Ω (Aleph Ω, Aleph Ω)]-multiplicity), Aleph (non-ordinal-[Hyper Aleph Ω (Aleph Ω, Aleph Ω)]-multiplicity))[Hyper Aleph (non-ordinal-[Hyper Aleph Ω (Aleph Ω, Aleph Ω)]-multiplicity) (Aleph (non-ordinal-[Hyper Aleph Ω (Aleph Ω, Aleph Ω)]-multiplicity), Aleph (non-ordinal-[Hyper Aleph Ω (Aleph Ω, Aleph Ω)]-multiplicity)), Hyper Aleph (non-ordinal-[Hyper Aleph Ω (Aleph Ω, Aleph Ω)]-multiplicity) (Aleph (non-ordinal-[Hyper Aleph Ω (Aleph Ω, Aleph Ω)]-multiplicity), Aleph (non-ordinal-[Hyper Aleph Ω (Aleph Ω, Aleph Ω)]-multiplicity))], Hyper Hyper Aleph (non-ordinal-[Hyper Aleph Ω (Aleph Ω, Aleph Ω)]-multiplicity) (Aleph (non-ordinal-[Hyper Aleph Ω (Aleph Ω, Aleph Ω)]-multiplicity), Aleph (non-ordinal-[Hyper Aleph Ω (Aleph Ω, Aleph Ω)]-multiplicity))[Hyper Aleph (non-ordinal-[Hyper Aleph Ω (Aleph Ω, Aleph Ω)]-multiplicity) (Aleph (non-ordinal-[Hyper Aleph Ω (Aleph Ω, Aleph Ω)]-multiplicity), Aleph (non-ordinal-[Hyper Aleph Ω (Aleph Ω, Aleph Ω)]-multiplicity)), Hyper Aleph (non-ordinal-[Hyper Aleph Ω (Aleph Ω, Aleph Ω)]-multiplicity) (Aleph (non-ordinal-[Hyper Aleph Ω (Aleph Ω, Aleph Ω)]-multiplicity), Aleph (non-ordinal-[Hyper Aleph Ω (Aleph Ω, Aleph Ω)]-multiplicity))]}, Hyper Hyper Hyper Aleph (non-ordinal-[Hyper Aleph Ω (Aleph Ω, Aleph Ω)]-multiplicity) (Aleph (non-ordinal-[Hyper Aleph Ω (Aleph Ω, Aleph Ω)]-multiplicity), Aleph (non-ordinal-[Hyper Aleph Ω (Aleph Ω, Aleph Ω)]-multiplicity))[Hyper Aleph (non-ordinal-[Hyper Aleph Ω (Aleph Ω, Aleph Ω)]-multiplicity) (Aleph (non-ordinal-[Hyper Aleph Ω (Aleph Ω, Aleph Ω)]-

multiplicity), Aleph (non-ordinal-[Hyper Aleph Ω (Aleph Ω, Aleph Ω)]-multiplicity)), Hyper Aleph (non-ordinal-[Hyper Aleph Ω (Aleph Ω, Aleph Ω)]-multiplicity) (Aleph (non-ordinal-[Hyper Aleph Ω (Aleph Ω, Aleph Ω)]-multiplicity), Aleph (non-ordinal-[Hyper Aleph Ω (Aleph Ω, Aleph Ω)]-multiplicity))] {Hyper Hyper Aleph (non-ordinal-[Hyper Aleph Ω (Aleph Ω, Aleph Ω)]-multiplicity) (Aleph (non-ordinal-[Hyper Aleph Ω (Aleph Ω, Aleph Ω)]-multiplicity), Aleph (non-ordinal-[Hyper Aleph Ω (Aleph Ω, Aleph Ω)]-multiplicity))[Hyper Aleph (non-ordinal-[Hyper Aleph Ω (Aleph Ω, Aleph Ω)]-multiplicity) (Aleph (non-ordinal-[Hyper Aleph Ω (Aleph Ω, Aleph Ω)]-multiplicity), Aleph (non-ordinal-[Hyper Aleph Ω (Aleph Ω, Aleph Ω)]-multiplicity)), Hyper Aleph (non-ordinal-[Hyper Aleph Ω (Aleph Ω, Aleph Ω)]-multiplicity) (Aleph (non-ordinal-[Hyper Aleph Ω (Aleph Ω, Aleph Ω)]-multiplicity), Aleph (non-ordinal-[Hyper Aleph Ω (Aleph Ω, Aleph Ω)]-multiplicity))], Hyper Hyper Aleph (non-ordinal-[Hyper Aleph Ω (Aleph Ω, Aleph Ω)]-multiplicity) (Aleph (non-ordinal-[Hyper Aleph Ω (Aleph Ω, Aleph Ω)]-multiplicity), Aleph (non-ordinal-[Hyper Aleph Ω (Aleph Ω, Aleph Ω)]-multiplicity))[Hyper Aleph (non-ordinal-[Hyper Aleph Ω (Aleph Ω, Aleph Ω)]-multiplicity) (Aleph (non-ordinal-[Hyper Aleph Ω (Aleph Ω, Aleph Ω)]-multiplicity), Aleph (non-ordinal-[Hyper Aleph Ω (Aleph Ω, Aleph Ω)]-multiplicity)), Hyper Aleph (non-ordinal-[Hyper Aleph Ω (Aleph Ω, Aleph Ω)]-multiplicity) (Aleph (non-ordinal-[Hyper Aleph Ω (Aleph Ω, Aleph Ω)]-multiplicity), Aleph (non-ordinal-[Hyper Aleph Ω (Aleph Ω, Aleph Ω)]-multiplicity))]}}}{ Hyper Hyper Hyper Hyper Hyper Aleph (non-ordinal-[Hyper Aleph Ω (Aleph Ω, Aleph Ω)]-multiplicity) (Aleph (non-ordinal-[Hyper Aleph Ω (Aleph Ω, Aleph Ω)]-multiplicity), Aleph (non-ordinal-[Hyper Aleph Ω (Aleph Ω, Aleph Ω)]-multiplicity))[Hyper Aleph (non-ordinal-[Hyper Aleph Ω (Aleph Ω, Aleph Ω)]-multiplicity) (Aleph (non-ordinal-[Hyper Aleph Ω (Aleph Ω, Aleph Ω)]-multiplicity), Aleph (non-ordinal-[Hyper Aleph Ω (Aleph Ω, Aleph Ω)]-multiplicity)), Hyper Aleph (non-ordinal-[Hyper Aleph Ω (Aleph Ω, Aleph Ω)]-multiplicity) (Aleph (non-ordinal-[Hyper Aleph Ω (Aleph Ω, Aleph Ω)]-multiplicity), Aleph (non-ordinal-[Hyper Aleph Ω (Aleph Ω, Aleph Ω)]-multiplicity))] {Hyper Hyper Aleph (non-ordinal-[Hyper Aleph Ω (Aleph Ω, Aleph Ω)]-multiplicity) (Aleph (non-ordinal-[Hyper Aleph Ω (Aleph Ω, Aleph Ω)]-multiplicity), Aleph (non-ordinal-[Hyper Aleph Ω (Aleph Ω, Aleph Ω)]-multiplicity))[Hyper Aleph (non-ordinal-[Hyper Aleph Ω (Aleph Ω, Aleph Ω)]-multiplicity) (Aleph (non-ordinal-[Hyper Aleph Ω (Aleph Ω, Aleph Ω)]-multiplicity), Aleph (non-ordinal-[Hyper Aleph Ω (Aleph Ω, Aleph Ω)]-multiplicity)), Hyper Aleph (non-ordinal-[Hyper Aleph Ω (Aleph Ω, Aleph Ω)]-multiplicity) (Aleph (non-ordinal-[Hyper Aleph Ω (Aleph Ω, Aleph Ω)]-multiplicity), Aleph (non-ordinal-[Hyper Aleph Ω (Aleph Ω, Aleph Ω)]-multiplicity))], Hyper Hyper Aleph (non-ordinal-[Hyper Aleph Ω (Aleph Ω, Aleph Ω)]-multiplicity) (Aleph (non-ordinal-[Hyper Aleph Ω (Aleph Ω, Aleph Ω)]-multiplicity), Aleph (non-ordinal-[Hyper Aleph Ω (Aleph Ω, Aleph Ω)]-multiplicity))[Hyper Aleph (non-ordinal-[Hyper Aleph Ω (Aleph Ω, Aleph Ω)]-multiplicity) (Aleph (non-ordinal-[Hyper Aleph Ω (Aleph Ω, Aleph Ω)]-multiplicity), Aleph (non-ordinal-[Hyper Aleph Ω (Aleph Ω, Aleph Ω)]-multiplicity)), Hyper Aleph (non-ordinal-[Hyper Aleph Ω (Aleph Ω, Aleph Ω)]-

multiplicity) (Aleph (non-ordinal-[Hyper Aleph Ω (Aleph Ω, Aleph Ω)]-multiplicity), Aleph (non-ordinal-[Hyper Aleph Ω (Aleph Ω, Aleph Ω)]-multiplicity))]} { Hyper Hyper Hyper Aleph (non-ordinal-[Hyper Aleph Ω (Aleph Ω, Aleph Ω)]-multiplicity) (Aleph (non-ordinal-[Hyper Aleph Ω (Aleph Ω, Aleph Ω)]-multiplicity), Aleph (non-ordinal-[Hyper Aleph Ω (Aleph Ω, Aleph Ω)]-multiplicity))[Hyper Aleph (non-ordinal-[Hyper Aleph Ω (Aleph Ω, Aleph Ω)]-multiplicity) (Aleph (non-ordinal-[Hyper Aleph Ω (Aleph Ω, Aleph Ω)]-multiplicity), Aleph (non-ordinal-[Hyper Aleph Ω (Aleph Ω, Aleph Ω)]-multiplicity)), Hyper Aleph (non-ordinal-[Hyper Aleph Ω (Aleph Ω, Aleph Ω)]-multiplicity) (Aleph (non-ordinal-[Hyper Aleph Ω (Aleph Ω, Aleph Ω)]-multiplicity), Aleph (non-ordinal-[Hyper Aleph Ω (Aleph Ω, Aleph Ω)]-multiplicity))] {Hyper Hyper Aleph (non-ordinal-[Hyper Aleph Ω (Aleph Ω, Aleph Ω)]-multiplicity) (Aleph (non-ordinal-[Hyper Aleph Ω (Aleph Ω, Aleph Ω)]-multiplicity), Aleph (non-ordinal-[Hyper Aleph Ω (Aleph Ω, Aleph Ω)]-multiplicity))[Hyper Aleph (non-ordinal-[Hyper Aleph Ω (Aleph Ω, Aleph Ω)]-multiplicity) (Aleph (non-ordinal-[Hyper Aleph Ω (Aleph Ω, Aleph Ω)]-multiplicity), Aleph (non-ordinal-[Hyper Aleph Ω (Aleph Ω, Aleph Ω)]-multiplicity)), Hyper Aleph (non-ordinal-[Hyper Aleph Ω (Aleph Ω, Aleph Ω)]-multiplicity) (Aleph (non-ordinal-[Hyper Aleph Ω (Aleph Ω, Aleph Ω)]-multiplicity), Aleph (non-ordinal-[Hyper Aleph Ω (Aleph Ω, Aleph Ω)]-multiplicity))], Hyper Hyper Aleph (non-ordinal-[Hyper Aleph Ω (Aleph Ω, Aleph Ω)]-multiplicity) (Aleph (non-ordinal-[Hyper Aleph Ω (Aleph Ω, Aleph Ω)]-multiplicity), Aleph (non-ordinal-[Hyper Aleph Ω (Aleph Ω, Aleph Ω)]-multiplicity))[Hyper Aleph (non-ordinal-[Hyper Aleph Ω (Aleph Ω, Aleph Ω)]-multiplicity) (Aleph (non-ordinal-[Hyper Aleph Ω (Aleph Ω, Aleph Ω)]-multiplicity), Aleph (non-ordinal-[Hyper Aleph Ω (Aleph Ω, Aleph Ω)]-multiplicity)), Hyper Aleph (non-ordinal-[Hyper Aleph Ω (Aleph Ω, Aleph Ω)]-multiplicity) (Aleph (non-ordinal-[Hyper Aleph Ω (Aleph Ω, Aleph Ω)]-multiplicity), Aleph (non-ordinal-[Hyper Aleph Ω (Aleph Ω, Aleph Ω)]-multiplicity))]}, Hyper Hyper Hyper Aleph (non-ordinal-[Hyper Aleph Ω (Aleph Ω, Aleph Ω)]-multiplicity) (Aleph (non-ordinal-[Hyper Aleph Ω (Aleph Ω, Aleph Ω)]-multiplicity), Aleph (non-ordinal-[Hyper Aleph Ω (Aleph Ω, Aleph Ω)]-multiplicity))[Hyper Aleph (non-ordinal-[Hyper Aleph Ω (Aleph Ω, Aleph Ω)]-multiplicity) (Aleph (non-ordinal-[Hyper Aleph Ω (Aleph Ω, Aleph Ω)]-multiplicity), Aleph (non-ordinal-[Hyper Aleph Ω (Aleph Ω, Aleph Ω)]-multiplicity)), Hyper Aleph (non-ordinal-[Hyper Aleph Ω (Aleph Ω, Aleph Ω)]-multiplicity) (Aleph (non-ordinal-[Hyper Aleph Ω (Aleph Ω, Aleph Ω)]-multiplicity), Aleph (non-ordinal-[Hyper Aleph Ω (Aleph Ω, Aleph Ω)]-multiplicity))] {Hyper Hyper Aleph (non-ordinal-[Hyper Aleph Ω (Aleph Ω, Aleph Ω)]-multiplicity) (Aleph (non-ordinal-[Hyper Aleph Ω (Aleph Ω, Aleph Ω)]-multiplicity), Aleph (non-ordinal-[Hyper Aleph Ω (Aleph Ω, Aleph Ω)]-multiplicity))[Hyper Aleph (non-ordinal-[Hyper Aleph Ω (Aleph Ω, Aleph Ω)]-multiplicity) (Aleph (non-ordinal-[Hyper Aleph Ω (Aleph Ω, Aleph Ω)]-multiplicity), Aleph (non-ordinal-[Hyper Aleph Ω (Aleph Ω, Aleph Ω)]-multiplicity)), Hyper Aleph (non-ordinal-[Hyper Aleph Ω (Aleph Ω, Aleph Ω)]-multiplicity) (Aleph (non-ordinal-[Hyper Aleph Ω (Aleph Ω, Aleph Ω)]-multiplicity), Aleph (non-ordinal-[Hyper Aleph Ω (Aleph Ω, Aleph Ω)]-

multiplicity))], Hyper Hyper Aleph (non-ordinal-[Hyper Aleph Ω (Aleph Ω, Aleph Ω)]-multiplicity) (Aleph (non-ordinal-[Hyper Aleph Ω (Aleph Ω, Aleph Ω)]-multiplicity), Aleph (non-ordinal-[Hyper Aleph Ω (Aleph Ω, Aleph Ω)]-multiplicity))[Hyper Aleph (non-ordinal-[Hyper Aleph Ω (Aleph Ω, Aleph Ω)]-multiplicity) (Aleph (non-ordinal-[Hyper Aleph Ω (Aleph Ω, Aleph Ω)]-multiplicity), Aleph (non-ordinal-[Hyper Aleph Ω (Aleph Ω, Aleph Ω)]-multiplicity)), Hyper Aleph (non-ordinal-[Hyper Aleph Ω (Aleph Ω, Aleph Ω)]-multiplicity) (Aleph (non-ordinal-[Hyper Aleph Ω (Aleph Ω, Aleph Ω)]-multiplicity), Aleph (non-ordinal-[Hyper Aleph Ω (Aleph Ω, Aleph Ω)]-multiplicity))]}}{ Hyper Hyper Hyper Hyper Aleph (non-ordinal-[Hyper Aleph Ω (Aleph Ω, Aleph Ω)]-multiplicity) (Aleph (non-ordinal-[Hyper Aleph Ω (Aleph Ω, Aleph Ω)]-multiplicity), Aleph (non-ordinal-[Hyper Aleph Ω (Aleph Ω, Aleph Ω)]-multiplicity))[Hyper Aleph (non-ordinal-[Hyper Aleph Ω (Aleph Ω, Aleph Ω)]-multiplicity) (Aleph (non-ordinal-[Hyper Aleph Ω (Aleph Ω, Aleph Ω)]-multiplicity), Aleph (non-ordinal-[Hyper Aleph Ω (Aleph Ω, Aleph Ω)]-multiplicity)), Hyper Aleph (non-ordinal-[Hyper Aleph Ω (Aleph Ω, Aleph Ω)]-multiplicity) (Aleph (non-ordinal-[Hyper Aleph Ω (Aleph Ω, Aleph Ω)]-multiplicity), Aleph (non-ordinal-[Hyper Aleph Ω (Aleph Ω, Aleph Ω)]-multiplicity))] {Hyper Hyper Aleph (non-ordinal-[Hyper Aleph Ω (Aleph Ω, Aleph Ω)]-multiplicity) (Aleph (non-ordinal-[Hyper Aleph Ω (Aleph Ω, Aleph Ω)]-multiplicity), Aleph (non-ordinal-[Hyper Aleph Ω (Aleph Ω, Aleph Ω)]-multiplicity))[Hyper Aleph (non-ordinal-[Hyper Aleph Ω (Aleph Ω, Aleph Ω)]-multiplicity) (Aleph (non-ordinal-[Hyper Aleph Ω (Aleph Ω, Aleph Ω)]-multiplicity), Aleph (non-ordinal-[Hyper Aleph Ω (Aleph Ω, Aleph Ω)]-multiplicity)), Hyper Aleph (non-ordinal-[Hyper Aleph Ω (Aleph Ω, Aleph Ω)]-multiplicity) (Aleph (non-ordinal-[Hyper Aleph Ω (Aleph Ω, Aleph Ω)]-multiplicity), Aleph (non-ordinal-[Hyper Aleph Ω (Aleph Ω, Aleph Ω)]-multiplicity))], Hyper Hyper Aleph (non-ordinal-[Hyper Aleph Ω (Aleph Ω, Aleph Ω)]-multiplicity) (Aleph (non-ordinal-[Hyper Aleph Ω (Aleph Ω, Aleph Ω)]-multiplicity), Aleph (non-ordinal-[Hyper Aleph Ω (Aleph Ω, Aleph Ω)]-multiplicity))[Hyper Aleph (non-ordinal-[Hyper Aleph Ω (Aleph Ω, Aleph Ω)]-multiplicity) (Aleph (non-ordinal-[Hyper Aleph Ω (Aleph Ω, Aleph Ω)]-multiplicity), Aleph (non-ordinal-[Hyper Aleph Ω (Aleph Ω, Aleph Ω)]-multiplicity)), Hyper Aleph (non-ordinal-[Hyper Aleph Ω (Aleph Ω, Aleph Ω)]-multiplicity) (Aleph (non-ordinal-[Hyper Aleph Ω (Aleph Ω, Aleph Ω)]-multiplicity), Aleph (non-ordinal-[Hyper Aleph Ω (Aleph Ω, Aleph Ω)]-multiplicity))]} { Hyper Hyper Hyper Aleph (non-ordinal-[Hyper Aleph Ω (Aleph Ω, Aleph Ω)]-multiplicity) (Aleph (non-ordinal-[Hyper Aleph Ω (Aleph Ω, Aleph Ω)]-multiplicity), Aleph (non-ordinal-[Hyper Aleph Ω (Aleph Ω, Aleph Ω)]-multiplicity))[Hyper Aleph (non-ordinal-[Hyper Aleph Ω (Aleph Ω, Aleph Ω)]-multiplicity) (Aleph (non-ordinal-[Hyper Aleph Ω (Aleph Ω, Aleph Ω)]-multiplicity), Aleph (non-ordinal-[Hyper Aleph Ω (Aleph Ω, Aleph Ω)]-multiplicity)), Hyper Aleph (non-ordinal-[Hyper Aleph Ω (Aleph Ω, Aleph Ω)]-multiplicity) (Aleph (non-ordinal-[Hyper Aleph Ω (Aleph Ω, Aleph Ω)]-multiplicity), Aleph (non-ordinal-[Hyper Aleph Ω (Aleph Ω, Aleph Ω)]-multiplicity))] {Hyper Hyper Aleph (non-ordinal-[Hyper Aleph Ω (Aleph Ω, Aleph Ω)]-multiplicity) (Aleph (non-ordinal-[Hyper Aleph Ω (Aleph Ω, Aleph

Ω)]-multiplicity), Aleph (non-ordinal-[Hyper Aleph Ω (Aleph Ω, Aleph Ω)]-multiplicity))[Hyper Aleph (non-ordinal-[Hyper Aleph Ω (Aleph Ω, Aleph Ω)]-multiplicity) (Aleph (non-ordinal-[Hyper Aleph Ω (Aleph Ω, Aleph Ω)]-multiplicity), Aleph (non-ordinal-[Hyper Aleph Ω (Aleph Ω, Aleph Ω)]-multiplicity)), Hyper Aleph (non-ordinal-[Hyper Aleph Ω (Aleph Ω, Aleph Ω)]-multiplicity) (Aleph (non-ordinal-[Hyper Aleph Ω (Aleph Ω, Aleph Ω)]-multiplicity), Aleph (non-ordinal-[Hyper Aleph Ω (Aleph Ω, Aleph Ω)]-multiplicity))], Hyper Hyper Aleph (non-ordinal-[Hyper Aleph Ω (Aleph Ω, Aleph Ω)]-multiplicity) (Aleph (non-ordinal-[Hyper Aleph Ω (Aleph Ω, Aleph Ω)]-multiplicity), Aleph (non-ordinal-[Hyper Aleph Ω (Aleph Ω, Aleph Ω)]-multiplicity))[Hyper Aleph (non-ordinal-[Hyper Aleph Ω (Aleph Ω, Aleph Ω)]-multiplicity) (Aleph (non-ordinal-[Hyper Aleph Ω (Aleph Ω, Aleph Ω)]-multiplicity), Aleph (non-ordinal-[Hyper Aleph Ω (Aleph Ω, Aleph Ω)]-multiplicity)), Hyper Aleph (non-ordinal-[Hyper Aleph Ω (Aleph Ω, Aleph Ω)]-multiplicity) (Aleph (non-ordinal-[Hyper Aleph Ω (Aleph Ω, Aleph Ω)]-multiplicity), Aleph (non-ordinal-[Hyper Aleph Ω (Aleph Ω, Aleph Ω)]-multiplicity))]}, Hyper Hyper Hyper Aleph (non-ordinal-[Hyper Aleph Ω (Aleph Ω, Aleph Ω)]-multiplicity) (Aleph (non-ordinal-[Hyper Aleph Ω (Aleph Ω, Aleph Ω)]-multiplicity), Aleph (non-ordinal-[Hyper Aleph Ω (Aleph Ω, Aleph Ω)]-multiplicity))[Hyper Aleph (non-ordinal-[Hyper Aleph Ω (Aleph Ω, Aleph Ω)]-multiplicity) (Aleph (non-ordinal-[Hyper Aleph Ω (Aleph Ω, Aleph Ω)]-multiplicity), Aleph (non-ordinal-[Hyper Aleph Ω (Aleph Ω, Aleph Ω)]-multiplicity)), Hyper Aleph (non-ordinal-[Hyper Aleph Ω (Aleph Ω, Aleph Ω)]-multiplicity) (Aleph (non-ordinal-[Hyper Aleph Ω (Aleph Ω, Aleph Ω)]-multiplicity), Aleph (non-ordinal-[Hyper Aleph Ω (Aleph Ω, Aleph Ω)]-multiplicity))] {Hyper Hyper Aleph (non-ordinal-[Hyper Aleph Ω (Aleph Ω, Aleph Ω)]-multiplicity) (Aleph (non-ordinal-[Hyper Aleph Ω (Aleph Ω, Aleph Ω)]-multiplicity), Aleph (non-ordinal-[Hyper Aleph Ω (Aleph Ω, Aleph Ω)]-multiplicity))[Hyper Aleph (non-ordinal-[Hyper Aleph Ω (Aleph Ω, Aleph Ω)]-multiplicity) (Aleph (non-ordinal-[Hyper Aleph Ω (Aleph Ω, Aleph Ω)]-multiplicity), Aleph (non-ordinal-[Hyper Aleph Ω (Aleph Ω, Aleph Ω)]-multiplicity)), Hyper Aleph (non-ordinal-[Hyper Aleph Ω (Aleph Ω, Aleph Ω)]-multiplicity) (Aleph (non-ordinal-[Hyper Aleph Ω (Aleph Ω, Aleph Ω)]-multiplicity), Aleph (non-ordinal-[Hyper Aleph Ω (Aleph Ω, Aleph Ω)]-multiplicity))], Hyper Hyper Aleph (non-ordinal-[Hyper Aleph Ω (Aleph Ω, Aleph Ω)]-multiplicity) (Aleph (non-ordinal-[Hyper Aleph Ω (Aleph Ω, Aleph Ω)]-multiplicity), Aleph (non-ordinal-[Hyper Aleph Ω (Aleph Ω, Aleph Ω)]-multiplicity))[Hyper Aleph (non-ordinal-[Hyper Aleph Ω (Aleph Ω, Aleph Ω)]-multiplicity) (Aleph (non-ordinal-[Hyper Aleph Ω (Aleph Ω, Aleph Ω)]-multiplicity), Aleph (non-ordinal-[Hyper Aleph Ω (Aleph Ω, Aleph Ω)]-multiplicity)), Hyper Aleph (non-ordinal-[Hyper Aleph Ω (Aleph Ω, Aleph Ω)]-multiplicity) (Aleph (non-ordinal-[Hyper Aleph Ω (Aleph Ω, Aleph Ω)]-multiplicity), Aleph (non-ordinal-[Hyper Aleph Ω (Aleph Ω, Aleph Ω)]-multiplicity))]}}, Hyper Hyper Hyper Hyper Aleph (non-ordinal-[Hyper Aleph Ω (Aleph Ω, Aleph Ω)]-multiplicity) (Aleph (non-ordinal-[Hyper Aleph Ω (Aleph Ω, Aleph Ω)]-multiplicity), Aleph (non-ordinal-[Hyper Aleph Ω (Aleph Ω, Aleph Ω)]-multiplicity))[Hyper Aleph (non-ordinal-[Hyper Aleph Ω

(Aleph Ω, Aleph Ω)]-multiplicity) (Aleph (non-ordinal-[Hyper Aleph Ω (Aleph Ω, Aleph Ω)]-multiplicity), Aleph (non-ordinal-[Hyper Aleph Ω (Aleph Ω, Aleph Ω)]-multiplicity)), Hyper Aleph (non-ordinal-[Hyper Aleph Ω (Aleph Ω, Aleph Ω)]-multiplicity) (Aleph (non-ordinal-[Hyper Aleph Ω (Aleph Ω, Aleph Ω)]-multiplicity), Aleph (non-ordinal-[Hyper Aleph Ω (Aleph Ω, Aleph Ω)]-multiplicity))] {Hyper Hyper Aleph (non-ordinal-[Hyper Aleph Ω (Aleph Ω, Aleph Ω)]-multiplicity) (Aleph (non-ordinal-[Hyper Aleph Ω (Aleph Ω, Aleph Ω)]-multiplicity), Aleph (non-ordinal-[Hyper Aleph Ω (Aleph Ω, Aleph Ω)]-multiplicity))[Hyper Aleph (non-ordinal-[Hyper Aleph Ω (Aleph Ω, Aleph Ω)]-multiplicity) (Aleph (non-ordinal-[Hyper Aleph Ω (Aleph Ω, Aleph Ω)]-multiplicity), Aleph (non-ordinal-[Hyper Aleph Ω (Aleph Ω, Aleph Ω)]-multiplicity)), Hyper Aleph (non-ordinal-[Hyper Aleph Ω (Aleph Ω, Aleph Ω)]-multiplicity) (Aleph (non-ordinal-[Hyper Aleph Ω (Aleph Ω, Aleph Ω)]-multiplicity), Aleph (non-ordinal-[Hyper Aleph Ω (Aleph Ω, Aleph Ω)]-multiplicity))], Hyper Hyper Aleph (non-ordinal-[Hyper Aleph Ω (Aleph Ω, Aleph Ω)]-multiplicity) (Aleph (non-ordinal-[Hyper Aleph Ω (Aleph Ω, Aleph Ω)]-multiplicity), Aleph (non-ordinal-[Hyper Aleph Ω (Aleph Ω, Aleph Ω)]-multiplicity))[Hyper Aleph (non-ordinal-[Hyper Aleph Ω (Aleph Ω, Aleph Ω)]-multiplicity) (Aleph (non-ordinal-[Hyper Aleph Ω (Aleph Ω, Aleph Ω)]-multiplicity), Aleph (non-ordinal-[Hyper Aleph Ω (Aleph Ω, Aleph Ω)]-multiplicity)), Hyper Aleph (non-ordinal-[Hyper Aleph Ω (Aleph Ω, Aleph Ω)]-multiplicity) (Aleph (non-ordinal-[Hyper Aleph Ω (Aleph Ω, Aleph Ω)]-multiplicity), Aleph (non-ordinal-[Hyper Aleph Ω (Aleph Ω, Aleph Ω)]-multiplicity))]} { Hyper Hyper Hyper Aleph (non-ordinal-[Hyper Aleph Ω (Aleph Ω, Aleph Ω)]-multiplicity) (Aleph (non-ordinal-[Hyper Aleph Ω (Aleph Ω, Aleph Ω)]-multiplicity), Aleph (non-ordinal-[Hyper Aleph Ω (Aleph Ω, Aleph Ω)]-multiplicity))[Hyper Aleph (non-ordinal-[Hyper Aleph Ω (Aleph Ω, Aleph Ω)]-multiplicity) (Aleph (non-ordinal-[Hyper Aleph Ω (Aleph Ω, Aleph Ω)]-multiplicity), Aleph (non-ordinal-[Hyper Aleph Ω (Aleph Ω, Aleph Ω)]-multiplicity)), Hyper Aleph (non-ordinal-[Hyper Aleph Ω (Aleph Ω, Aleph Ω)]-multiplicity) (Aleph (non-ordinal-[Hyper Aleph Ω (Aleph Ω, Aleph Ω)]-multiplicity), Aleph (non-ordinal-[Hyper Aleph Ω (Aleph Ω, Aleph Ω)]-multiplicity))] {Hyper Hyper Aleph (non-ordinal-[Hyper Aleph Ω (Aleph Ω, Aleph Ω)]-multiplicity) (Aleph (non-ordinal-[Hyper Aleph Ω (Aleph Ω, Aleph Ω)]-multiplicity), Aleph (non-ordinal-[Hyper Aleph Ω (Aleph Ω, Aleph Ω)]-multiplicity))[Hyper Aleph (non-ordinal-[Hyper Aleph Ω (Aleph Ω, Aleph Ω)]-multiplicity) (Aleph (non-ordinal-[Hyper Aleph Ω (Aleph Ω, Aleph Ω)]-multiplicity), Aleph (non-ordinal-[Hyper Aleph Ω (Aleph Ω, Aleph Ω)]-multiplicity)), Hyper Aleph (non-ordinal-[Hyper Aleph Ω (Aleph Ω, Aleph Ω)]-multiplicity) (Aleph (non-ordinal-[Hyper Aleph Ω (Aleph Ω, Aleph Ω)]-multiplicity), Aleph (non-ordinal-[Hyper Aleph Ω (Aleph Ω, Aleph Ω)]-multiplicity))], Hyper Hyper Aleph (non-ordinal-[Hyper Aleph Ω (Aleph Ω, Aleph Ω)]-multiplicity) (Aleph (non-ordinal-[Hyper Aleph Ω (Aleph Ω, Aleph Ω)]-multiplicity), Aleph (non-ordinal-[Hyper Aleph Ω (Aleph Ω, Aleph Ω)]-multiplicity))[Hyper Aleph (non-ordinal-[Hyper Aleph Ω (Aleph Ω, Aleph Ω)]-multiplicity) (Aleph (non-ordinal-[Hyper Aleph Ω (Aleph Ω, Aleph Ω)]-multiplicity), Aleph (non-ordinal-[Hyper Aleph Ω (Aleph Ω, Aleph

Ω)]-multiplicity)), Hyper Aleph (non-ordinal-[Hyper Aleph Ω (Aleph Ω, Aleph Ω)]-multiplicity) (Aleph (non-ordinal-[Hyper Aleph Ω (Aleph Ω, Aleph Ω)]-multiplicity), Aleph (non-ordinal-[Hyper Aleph Ω (Aleph Ω, Aleph Ω)]-multiplicity))]}, Hyper Hyper Hyper Aleph (non-ordinal-[Hyper Aleph Ω (Aleph Ω, Aleph Ω)]-multiplicity) (Aleph (non-ordinal-[Hyper Aleph Ω (Aleph Ω, Aleph Ω)]-multiplicity), Aleph (non-ordinal-[Hyper Aleph Ω (Aleph Ω, Aleph Ω)]-multiplicity))[Hyper Aleph (non-ordinal-[Hyper Aleph Ω (Aleph Ω, Aleph Ω)]-multiplicity) (Aleph (non-ordinal-[Hyper Aleph Ω (Aleph Ω, Aleph Ω)]-multiplicity), Aleph (non-ordinal-[Hyper Aleph Ω (Aleph Ω, Aleph Ω)]-multiplicity)), Hyper Aleph (non-ordinal-[Hyper Aleph Ω (Aleph Ω, Aleph Ω)]-multiplicity) (Aleph (non-ordinal-[Hyper Aleph Ω (Aleph Ω, Aleph Ω)]-multiplicity), Aleph (non-ordinal-[Hyper Aleph Ω (Aleph Ω, Aleph Ω)]-multiplicity))] {Hyper Hyper Aleph (non-ordinal-[Hyper Aleph Ω (Aleph Ω, Aleph Ω)]-multiplicity) (Aleph (non-ordinal-[Hyper Aleph Ω (Aleph Ω, Aleph Ω)]-multiplicity), Aleph (non-ordinal-[Hyper Aleph Ω (Aleph Ω, Aleph Ω)]-multiplicity))[Hyper Aleph (non-ordinal-[Hyper Aleph Ω (Aleph Ω, Aleph Ω)]-multiplicity) (Aleph (non-ordinal-[Hyper Aleph Ω (Aleph Ω, Aleph Ω)]-multiplicity), Aleph (non-ordinal-[Hyper Aleph Ω (Aleph Ω, Aleph Ω)]-multiplicity)), Hyper Aleph (non-ordinal-[Hyper Aleph Ω (Aleph Ω, Aleph Ω)]-multiplicity) (Aleph (non-ordinal-[Hyper Aleph Ω (Aleph Ω, Aleph Ω)]-multiplicity), Aleph (non-ordinal-[Hyper Aleph Ω (Aleph Ω, Aleph Ω)]-multiplicity))], Hyper Hyper Aleph (non-ordinal-[Hyper Aleph Ω (Aleph Ω, Aleph Ω)]-multiplicity) (Aleph (non-ordinal-[Hyper Aleph Ω (Aleph Ω, Aleph Ω)]-multiplicity), Aleph (non-ordinal-[Hyper Aleph Ω (Aleph Ω, Aleph Ω)]-multiplicity))[Hyper Aleph (non-ordinal-[Hyper Aleph Ω (Aleph Ω, Aleph Ω)]-multiplicity) (Aleph (non-ordinal-[Hyper Aleph Ω (Aleph Ω, Aleph Ω)]-multiplicity), Aleph (non-ordinal-[Hyper Aleph Ω (Aleph Ω, Aleph Ω)]-multiplicity)), Hyper Aleph (non-ordinal-[Hyper Aleph Ω (Aleph Ω, Aleph Ω)]-multiplicity) (Aleph (non-ordinal-[Hyper Aleph Ω (Aleph Ω, Aleph Ω)]-multiplicity), Aleph (non-ordinal-[Hyper Aleph Ω (Aleph Ω, Aleph Ω)]-multiplicity))]}}}, Hyper Hyper Hyper Hyper Hyper Aleph (non-ordinal-[Hyper Aleph Ω (Aleph Ω, Aleph Ω)]-multiplicity) (Aleph (non-ordinal-[Hyper Aleph Ω (Aleph Ω, Aleph Ω)]-multiplicity), Aleph (non-ordinal-[Hyper Aleph Ω (Aleph Ω, Aleph Ω)]-multiplicity))[Hyper Aleph (non-ordinal-[Hyper Aleph Ω (Aleph Ω, Aleph Ω)]-multiplicity) (Aleph (non-ordinal-[Hyper Aleph Ω (Aleph Ω, Aleph Ω)]-multiplicity), Aleph (non-ordinal-[Hyper Aleph Ω (Aleph Ω, Aleph Ω)]-multiplicity)), Hyper Aleph (non-ordinal-[Hyper Aleph Ω (Aleph Ω, Aleph Ω)]-multiplicity) (Aleph (non-ordinal-[Hyper Aleph Ω (Aleph Ω, Aleph Ω)]-multiplicity), Aleph (non-ordinal-[Hyper Aleph Ω (Aleph Ω, Aleph Ω)]-multiplicity))] {Hyper Hyper Aleph (non-ordinal-[Hyper Aleph Ω (Aleph Ω, Aleph Ω)]-multiplicity) (Aleph (non-ordinal-[Hyper Aleph Ω (Aleph Ω, Aleph Ω)]-multiplicity), Aleph (non-ordinal-[Hyper Aleph Ω (Aleph Ω, Aleph Ω)]-multiplicity))[Hyper Aleph (non-ordinal-[Hyper Aleph Ω (Aleph Ω, Aleph Ω)]-multiplicity) (Aleph (non-ordinal-[Hyper Aleph Ω (Aleph Ω, Aleph Ω)]-multiplicity), Aleph (non-ordinal-[Hyper Aleph Ω (Aleph Ω, Aleph Ω)]-multiplicity)), Hyper Aleph (non-ordinal-[Hyper Aleph Ω (Aleph Ω, Aleph Ω)]-multiplicity) (Aleph (non-ordinal-[Hyper Aleph Ω (Aleph Ω,

Aleph Ω)]-multiplicity), Aleph (non-ordinal-[Hyper Aleph Ω (Aleph Ω, Aleph Ω)]-multiplicity))], Hyper Hyper Aleph (non-ordinal-[Hyper Aleph Ω (Aleph Ω, Aleph Ω)]-multiplicity) (Aleph (non-ordinal-[Hyper Aleph Ω (Aleph Ω, Aleph Ω)]-multiplicity), Aleph (non-ordinal-[Hyper Aleph Ω (Aleph Ω, Aleph Ω)]-multiplicity))[Hyper Aleph (non-ordinal-[Hyper Aleph Ω (Aleph Ω, Aleph Ω)]-multiplicity) (Aleph (non-ordinal-[Hyper Aleph Ω (Aleph Ω, Aleph Ω)]-multiplicity), Aleph (non-ordinal-[Hyper Aleph Ω (Aleph Ω, Aleph Ω)]-multiplicity)), Hyper Aleph (non-ordinal-[Hyper Aleph Ω (Aleph Ω, Aleph Ω)]-multiplicity) (Aleph (non-ordinal-[Hyper Aleph Ω (Aleph Ω, Aleph Ω)]-multiplicity), Aleph (non-ordinal-[Hyper Aleph Ω (Aleph Ω, Aleph Ω)]-multiplicity))]} { Hyper Hyper Hyper Aleph (non-ordinal-[Hyper Aleph Ω (Aleph Ω, Aleph Ω)]-multiplicity) (Aleph (non-ordinal-[Hyper Aleph Ω (Aleph Ω, Aleph Ω)]-multiplicity), Aleph (non-ordinal-[Hyper Aleph Ω (Aleph Ω, Aleph Ω)]-multiplicity))[Hyper Aleph (non-ordinal-[Hyper Aleph Ω (Aleph Ω, Aleph Ω)]-multiplicity) (Aleph (non-ordinal-[Hyper Aleph Ω (Aleph Ω, Aleph Ω)]-multiplicity), Aleph (non-ordinal-[Hyper Aleph Ω (Aleph Ω, Aleph Ω)]-multiplicity)), Hyper Aleph (non-ordinal-[Hyper Aleph Ω (Aleph Ω, Aleph Ω)]-multiplicity) (Aleph (non-ordinal-[Hyper Aleph Ω (Aleph Ω, Aleph Ω)]-multiplicity), Aleph (non-ordinal-[Hyper Aleph Ω (Aleph Ω, Aleph Ω)]-multiplicity))] {Hyper Hyper Aleph (non-ordinal-[Hyper Aleph Ω (Aleph Ω, Aleph Ω)]-multiplicity) (Aleph (non-ordinal-[Hyper Aleph Ω (Aleph Ω, Aleph Ω)]-multiplicity), Aleph (non-ordinal-[Hyper Aleph Ω (Aleph Ω, Aleph Ω)]-multiplicity))[Hyper Aleph (non-ordinal-[Hyper Aleph Ω (Aleph Ω, Aleph Ω)]-multiplicity) (Aleph (non-ordinal-[Hyper Aleph Ω (Aleph Ω, Aleph Ω)]-multiplicity), Aleph (non-ordinal-[Hyper Aleph Ω (Aleph Ω, Aleph Ω)]-multiplicity)), Hyper Aleph (non-ordinal-[Hyper Aleph Ω (Aleph Ω, Aleph Ω)]-multiplicity) (Aleph (non-ordinal-[Hyper Aleph Ω (Aleph Ω, Aleph Ω)]-multiplicity), Aleph (non-ordinal-[Hyper Aleph Ω (Aleph Ω, Aleph Ω)]-multiplicity))], Hyper Hyper Aleph (non-ordinal-[Hyper Aleph Ω (Aleph Ω, Aleph Ω)]-multiplicity) (Aleph (non-ordinal-[Hyper Aleph Ω (Aleph Ω, Aleph Ω)]-multiplicity), Aleph (non-ordinal-[Hyper Aleph Ω (Aleph Ω, Aleph Ω)]-multiplicity))[Hyper Aleph (non-ordinal-[Hyper Aleph Ω (Aleph Ω, Aleph Ω)]-multiplicity) (Aleph (non-ordinal-[Hyper Aleph Ω (Aleph Ω, Aleph Ω)]-multiplicity), Aleph (non-ordinal-[Hyper Aleph Ω (Aleph Ω, Aleph Ω)]-multiplicity)), Hyper Aleph (non-ordinal-[Hyper Aleph Ω (Aleph Ω, Aleph Ω)]-multiplicity) (Aleph (non-ordinal-[Hyper Aleph Ω (Aleph Ω, Aleph Ω)]-multiplicity), Aleph (non-ordinal-[Hyper Aleph Ω (Aleph Ω, Aleph Ω)]-multiplicity))]}, Hyper Hyper Hyper Aleph (non-ordinal-[Hyper Aleph Ω (Aleph Ω, Aleph Ω)]-multiplicity) (Aleph (non-ordinal-[Hyper Aleph Ω (Aleph Ω, Aleph Ω)]-multiplicity), Aleph (non-ordinal-[Hyper Aleph Ω (Aleph Ω, Aleph Ω)]-multiplicity))[Hyper Aleph (non-ordinal-[Hyper Aleph Ω (Aleph Ω, Aleph Ω)]-multiplicity) (Aleph (non-ordinal-[Hyper Aleph Ω (Aleph Ω, Aleph Ω)]-multiplicity), Aleph (non-ordinal-[Hyper Aleph Ω (Aleph Ω, Aleph Ω)]-multiplicity)), Hyper Aleph (non-ordinal-[Hyper Aleph Ω (Aleph Ω, Aleph Ω)]-multiplicity) (Aleph (non-ordinal-[Hyper Aleph Ω (Aleph Ω, Aleph Ω)]-multiplicity), Aleph (non-ordinal-[Hyper Aleph Ω (Aleph Ω, Aleph Ω)]-multiplicity))] {Hyper Hyper Aleph (non-ordinal-[Hyper Aleph Ω

(Aleph Ω, Aleph Ω)]-multiplicity) (Aleph (non-ordinal-[Hyper Aleph Ω (Aleph Ω, Aleph Ω)]-multiplicity), Aleph (non-ordinal-[Hyper Aleph Ω (Aleph Ω, Aleph Ω)]-multiplicity))[Hyper Aleph (non-ordinal-[Hyper Aleph Ω (Aleph Ω, Aleph Ω)]-multiplicity) (Aleph (non-ordinal-[Hyper Aleph Ω (Aleph Ω, Aleph Ω)]-multiplicity), Aleph (non-ordinal-[Hyper Aleph Ω (Aleph Ω, Aleph Ω)]-multiplicity)), Hyper Aleph (non-ordinal-[Hyper Aleph Ω (Aleph Ω, Aleph Ω)]-multiplicity) (Aleph (non-ordinal-[Hyper Aleph Ω (Aleph Ω, Aleph Ω)]-multiplicity), Aleph (non-ordinal-[Hyper Aleph Ω (Aleph Ω, Aleph Ω)]-multiplicity))], Hyper Hyper Aleph (non-ordinal-[Hyper Aleph Ω (Aleph Ω, Aleph Ω)]-multiplicity) (Aleph (non-ordinal-[Hyper Aleph Ω (Aleph Ω, Aleph Ω)]-multiplicity), Aleph (non-ordinal-[Hyper Aleph Ω (Aleph Ω, Aleph Ω)]-multiplicity))[Hyper Aleph (non-ordinal-[Hyper Aleph Ω (Aleph Ω, Aleph Ω)]-multiplicity) (Aleph (non-ordinal-[Hyper Aleph Ω (Aleph Ω, Aleph Ω)]-multiplicity), Aleph (non-ordinal-[Hyper Aleph Ω (Aleph Ω, Aleph Ω)]-multiplicity)), Hyper Aleph (non-ordinal-[Hyper Aleph Ω (Aleph Ω, Aleph Ω)]-multiplicity) (Aleph (non-ordinal-[Hyper Aleph Ω (Aleph Ω, Aleph Ω)]-multiplicity), Aleph (non-ordinal-[Hyper Aleph Ω (Aleph Ω, Aleph Ω)]-multiplicity))]}}{ Hyper Hyper Hyper Hyper Aleph (non-ordinal-[Hyper Aleph Ω (Aleph Ω, Aleph Ω)]-multiplicity) (Aleph (non-ordinal-[Hyper Aleph Ω (Aleph Ω, Aleph Ω)]-multiplicity), Aleph (non-ordinal-[Hyper Aleph Ω (Aleph Ω, Aleph Ω)]-multiplicity))[Hyper Aleph (non-ordinal-[Hyper Aleph Ω (Aleph Ω, Aleph Ω)]-multiplicity) (Aleph (non-ordinal-[Hyper Aleph Ω (Aleph Ω, Aleph Ω)]-multiplicity), Aleph (non-ordinal-[Hyper Aleph Ω (Aleph Ω, Aleph Ω)]-multiplicity)), Hyper Aleph (non-ordinal-[Hyper Aleph Ω (Aleph Ω, Aleph Ω)]-multiplicity) (Aleph (non-ordinal-[Hyper Aleph Ω (Aleph Ω, Aleph Ω)]-multiplicity), Aleph (non-ordinal-[Hyper Aleph Ω (Aleph Ω, Aleph Ω)]-multiplicity))] {Hyper Hyper Aleph (non-ordinal-[Hyper Aleph Ω (Aleph Ω, Aleph Ω)]-multiplicity) (Aleph (non-ordinal-[Hyper Aleph Ω (Aleph Ω, Aleph Ω)]-multiplicity), Aleph (non-ordinal-[Hyper Aleph Ω (Aleph Ω, Aleph Ω)]-multiplicity))[Hyper Aleph (non-ordinal-[Hyper Aleph Ω (Aleph Ω, Aleph Ω)]-multiplicity) (Aleph (non-ordinal-[Hyper Aleph Ω (Aleph Ω, Aleph Ω)]-multiplicity), Aleph (non-ordinal-[Hyper Aleph Ω (Aleph Ω, Aleph Ω)]-multiplicity)), Hyper Aleph (non-ordinal-[Hyper Aleph Ω (Aleph Ω, Aleph Ω)]-multiplicity) (Aleph (non-ordinal-[Hyper Aleph Ω (Aleph Ω, Aleph Ω)]-multiplicity), Aleph (non-ordinal-[Hyper Aleph Ω (Aleph Ω, Aleph Ω)]-multiplicity))], Hyper Hyper Aleph (non-ordinal-[Hyper Aleph Ω (Aleph Ω, Aleph Ω)]-multiplicity) (Aleph (non-ordinal-[Hyper Aleph Ω (Aleph Ω, Aleph Ω)]-multiplicity), Aleph (non-ordinal-[Hyper Aleph Ω (Aleph Ω, Aleph Ω)]-multiplicity))[Hyper Aleph (non-ordinal-[Hyper Aleph Ω (Aleph Ω, Aleph Ω)]-multiplicity) (Aleph (non-ordinal-[Hyper Aleph Ω (Aleph Ω, Aleph Ω)]-multiplicity), Aleph (non-ordinal-[Hyper Aleph Ω (Aleph Ω, Aleph Ω)]-multiplicity)), Hyper Aleph (non-ordinal-[Hyper Aleph Ω (Aleph Ω, Aleph Ω)]-multiplicity) (Aleph (non-ordinal-[Hyper Aleph Ω (Aleph Ω, Aleph Ω)]-multiplicity), Aleph (non-ordinal-[Hyper Aleph Ω (Aleph Ω, Aleph Ω)]-multiplicity))]} { Hyper Hyper Hyper Aleph (non-ordinal-[Hyper Aleph Ω (Aleph Ω, Aleph Ω)]-multiplicity) (Aleph (non-ordinal-[Hyper Aleph Ω (Aleph Ω, Aleph Ω)]-multiplicity), Aleph (non-ordinal-[Hyper Aleph Ω

(Aleph Ω, Aleph Ω)]-multiplicity))[Hyper Aleph (non-ordinal-[Hyper Aleph Ω (Aleph Ω, Aleph Ω)]-multiplicity) (Aleph (non-ordinal-[Hyper Aleph Ω (Aleph Ω, Aleph Ω)]-multiplicity), Aleph (non-ordinal-[Hyper Aleph Ω (Aleph Ω, Aleph Ω)]-multiplicity)), Hyper Aleph (non-ordinal-[Hyper Aleph Ω (Aleph Ω, Aleph Ω)]-multiplicity) (Aleph (non-ordinal-[Hyper Aleph Ω (Aleph Ω, Aleph Ω)]-multiplicity), Aleph (non-ordinal-[Hyper Aleph Ω (Aleph Ω, Aleph Ω)]-multiplicity))] {Hyper Hyper Aleph (non-ordinal-[Hyper Aleph Ω (Aleph Ω, Aleph Ω)]-multiplicity) (Aleph (non-ordinal-[Hyper Aleph Ω (Aleph Ω, Aleph Ω)]-multiplicity), Aleph (non-ordinal-[Hyper Aleph Ω (Aleph Ω, Aleph Ω)]-multiplicity))[Hyper Aleph (non-ordinal-[Hyper Aleph Ω (Aleph Ω, Aleph Ω)]-multiplicity) (Aleph (non-ordinal-[Hyper Aleph Ω (Aleph Ω, Aleph Ω)]-multiplicity), Aleph (non-ordinal-[Hyper Aleph Ω (Aleph Ω, Aleph Ω)]-multiplicity)), Hyper Aleph (non-ordinal-[Hyper Aleph Ω (Aleph Ω, Aleph Ω)]-multiplicity) (Aleph (non-ordinal-[Hyper Aleph Ω (Aleph Ω, Aleph Ω)]-multiplicity), Aleph (non-ordinal-[Hyper Aleph Ω (Aleph Ω, Aleph Ω)]-multiplicity))], Hyper Hyper Aleph (non-ordinal-[Hyper Aleph Ω (Aleph Ω, Aleph Ω)]-multiplicity) (Aleph (non-ordinal-[Hyper Aleph Ω (Aleph Ω, Aleph Ω)]-multiplicity), Aleph (non-ordinal-[Hyper Aleph Ω (Aleph Ω, Aleph Ω)]-multiplicity))[Hyper Aleph (non-ordinal-[Hyper Aleph Ω (Aleph Ω, Aleph Ω)]-multiplicity) (Aleph (non-ordinal-[Hyper Aleph Ω (Aleph Ω, Aleph Ω)]-multiplicity), Aleph (non-ordinal-[Hyper Aleph Ω (Aleph Ω, Aleph Ω)]-multiplicity)), Hyper Aleph (non-ordinal-[Hyper Aleph Ω (Aleph Ω, Aleph Ω)]-multiplicity) (Aleph (non-ordinal-[Hyper Aleph Ω (Aleph Ω, Aleph Ω)]-multiplicity), Aleph (non-ordinal-[Hyper Aleph Ω (Aleph Ω, Aleph Ω)]-multiplicity))]}, Hyper Hyper Hyper Aleph (non-ordinal-[Hyper Aleph Ω (Aleph Ω, Aleph Ω)]-multiplicity) (Aleph (non-ordinal-[Hyper Aleph Ω (Aleph Ω, Aleph Ω)]-multiplicity), Aleph (non-ordinal-[Hyper Aleph Ω (Aleph Ω, Aleph Ω)]-multiplicity))[Hyper Aleph (non-ordinal-[Hyper Aleph Ω (Aleph Ω, Aleph Ω)]-multiplicity) (Aleph (non-ordinal-[Hyper Aleph Ω (Aleph Ω, Aleph Ω)]-multiplicity), Aleph (non-ordinal-[Hyper Aleph Ω (Aleph Ω, Aleph Ω)]-multiplicity)), Hyper Aleph (non-ordinal-[Hyper Aleph Ω (Aleph Ω, Aleph Ω)]-multiplicity) (Aleph (non-ordinal-[Hyper Aleph Ω (Aleph Ω, Aleph Ω)]-multiplicity), Aleph (non-ordinal-[Hyper Aleph Ω (Aleph Ω, Aleph Ω)]-multiplicity))] {Hyper Hyper Aleph (non-ordinal-[Hyper Aleph Ω (Aleph Ω, Aleph Ω)]-multiplicity) (Aleph (non-ordinal-[Hyper Aleph Ω (Aleph Ω, Aleph Ω)]-multiplicity), Aleph (non-ordinal-[Hyper Aleph Ω (Aleph Ω, Aleph Ω)]-multiplicity))[Hyper Aleph (non-ordinal-[Hyper Aleph Ω (Aleph Ω, Aleph Ω)]-multiplicity) (Aleph (non-ordinal-[Hyper Aleph Ω (Aleph Ω, Aleph Ω)]-multiplicity), Aleph (non-ordinal-[Hyper Aleph Ω (Aleph Ω, Aleph Ω)]-multiplicity)), Hyper Aleph (non-ordinal-[Hyper Aleph Ω (Aleph Ω, Aleph Ω)]-multiplicity) (Aleph (non-ordinal-[Hyper Aleph Ω (Aleph Ω, Aleph Ω)]-multiplicity), Aleph (non-ordinal-[Hyper Aleph Ω (Aleph Ω, Aleph Ω)]-multiplicity))], Hyper Hyper Aleph (non-ordinal-[Hyper Aleph Ω (Aleph Ω, Aleph Ω)]-multiplicity) (Aleph (non-ordinal-[Hyper Aleph Ω (Aleph Ω, Aleph Ω)]-multiplicity), Aleph (non-ordinal-[Hyper Aleph Ω (Aleph Ω, Aleph Ω)]-multiplicity))[Hyper Aleph (non-ordinal-[Hyper Aleph Ω (Aleph Ω, Aleph Ω)]-multiplicity) (Aleph (non-ordinal-[Hyper Aleph Ω

(Aleph Ω, Aleph Ω)]-multiplicity), Aleph (non-ordinal-[Hyper Aleph Ω (Aleph Ω, Aleph Ω)]-multiplicity)), Hyper Aleph (non-ordinal-[Hyper Aleph Ω (Aleph Ω, Aleph Ω)]-multiplicity) (Aleph (non-ordinal-[Hyper Aleph Ω (Aleph Ω, Aleph Ω)]-multiplicity), Aleph (non-ordinal-[Hyper Aleph Ω (Aleph Ω, Aleph Ω)]-multiplicity))]}}, Hyper Hyper Hyper Hyper Aleph (non-ordinal-[Hyper Aleph Ω (Aleph Ω, Aleph Ω)]-multiplicity) (Aleph (non-ordinal-[Hyper Aleph Ω (Aleph Ω, Aleph Ω)]-multiplicity), Aleph (non-ordinal-[Hyper Aleph Ω (Aleph Ω, Aleph Ω)]-multiplicity))[Hyper Aleph (non-ordinal-[Hyper Aleph Ω (Aleph Ω, Aleph Ω)]-multiplicity) (Aleph (non-ordinal-[Hyper Aleph Ω (Aleph Ω, Aleph Ω)]-multiplicity), Aleph (non-ordinal-[Hyper Aleph Ω (Aleph Ω, Aleph Ω)]-multiplicity)), Hyper Aleph (non-ordinal-[Hyper Aleph Ω (Aleph Ω, Aleph Ω)]-multiplicity) (Aleph (non-ordinal-[Hyper Aleph Ω (Aleph Ω, Aleph Ω)]-multiplicity), Aleph (non-ordinal-[Hyper Aleph Ω (Aleph Ω, Aleph Ω)]-multiplicity))] {Hyper Hyper Aleph (non-ordinal-[Hyper Aleph Ω (Aleph Ω, Aleph Ω)]-multiplicity) (Aleph (non-ordinal-[Hyper Aleph Ω (Aleph Ω, Aleph Ω)]-multiplicity), Aleph (non-ordinal-[Hyper Aleph Ω (Aleph Ω, Aleph Ω)]-multiplicity))[Hyper Aleph (non-ordinal-[Hyper Aleph Ω (Aleph Ω, Aleph Ω)]-multiplicity) (Aleph (non-ordinal-[Hyper Aleph Ω (Aleph Ω, Aleph Ω)]-multiplicity), Aleph (non-ordinal-[Hyper Aleph Ω (Aleph Ω, Aleph Ω)]-multiplicity)), Hyper Aleph (non-ordinal-[Hyper Aleph Ω (Aleph Ω, Aleph Ω)]-multiplicity) (Aleph (non-ordinal-[Hyper Aleph Ω (Aleph Ω, Aleph Ω)]-multiplicity), Aleph (non-ordinal-[Hyper Aleph Ω (Aleph Ω, Aleph Ω)]-multiplicity))], Hyper Hyper Aleph (non-ordinal-[Hyper Aleph Ω (Aleph Ω, Aleph Ω)]-multiplicity) (Aleph (non-ordinal-[Hyper Aleph Ω (Aleph Ω, Aleph Ω)]-multiplicity), Aleph (non-ordinal-[Hyper Aleph Ω (Aleph Ω, Aleph Ω)]-multiplicity))[Hyper Aleph (non-ordinal-[Hyper Aleph Ω (Aleph Ω, Aleph Ω)]-multiplicity) (Aleph (non-ordinal-[Hyper Aleph Ω (Aleph Ω, Aleph Ω)]-multiplicity), Aleph (non-ordinal-[Hyper Aleph Ω (Aleph Ω, Aleph Ω)]-multiplicity)), Hyper Aleph (non-ordinal-[Hyper Aleph Ω (Aleph Ω, Aleph Ω)]-multiplicity) (Aleph (non-ordinal-[Hyper Aleph Ω (Aleph Ω, Aleph Ω)]-multiplicity), Aleph (non-ordinal-[Hyper Aleph Ω (Aleph Ω, Aleph Ω)]-multiplicity))]} { Hyper Hyper Hyper Aleph (non-ordinal-[Hyper Aleph Ω (Aleph Ω, Aleph Ω)]-multiplicity) (Aleph (non-ordinal-[Hyper Aleph Ω (Aleph Ω, Aleph Ω)]-multiplicity), Aleph (non-ordinal-[Hyper Aleph Ω (Aleph Ω, Aleph Ω)]-multiplicity))[Hyper Aleph (non-ordinal-[Hyper Aleph Ω (Aleph Ω, Aleph Ω)]-multiplicity) (Aleph (non-ordinal-[Hyper Aleph Ω (Aleph Ω, Aleph Ω)]-multiplicity), Aleph (non-ordinal-[Hyper Aleph Ω (Aleph Ω, Aleph Ω)]-multiplicity)), Hyper Aleph (non-ordinal-[Hyper Aleph Ω (Aleph Ω, Aleph Ω)]-multiplicity) (Aleph (non-ordinal-[Hyper Aleph Ω (Aleph Ω, Aleph Ω)]-multiplicity), Aleph (non-ordinal-[Hyper Aleph Ω (Aleph Ω, Aleph Ω)]-multiplicity))] {Hyper Hyper Aleph (non-ordinal-[Hyper Aleph Ω (Aleph Ω, Aleph Ω)]-multiplicity) (Aleph (non-ordinal-[Hyper Aleph Ω (Aleph Ω, Aleph Ω)]-multiplicity), Aleph (non-ordinal-[Hyper Aleph Ω (Aleph Ω, Aleph Ω)]-multiplicity))[Hyper Aleph (non-ordinal-[Hyper Aleph Ω (Aleph Ω, Aleph Ω)]-multiplicity) (Aleph (non-ordinal-[Hyper Aleph Ω (Aleph Ω, Aleph Ω)]-multiplicity), Aleph (non-ordinal-[Hyper Aleph Ω (Aleph Ω, Aleph Ω)]-multiplicity)), Hyper Aleph (non-ordinal-[Hyper

Aleph Ω (Aleph Ω, Aleph Ω)]-multiplicity) (Aleph (non-ordinal-[Hyper Aleph Ω (Aleph Ω, Aleph Ω)]-multiplicity), Aleph (non-ordinal-[Hyper Aleph Ω (Aleph Ω, Aleph Ω)]-multiplicity))], Hyper Hyper Aleph (non-ordinal-[Hyper Aleph Ω (Aleph Ω, Aleph Ω)]-multiplicity) (Aleph (non-ordinal-[Hyper Aleph Ω (Aleph Ω, Aleph Ω)]-multiplicity), Aleph (non-ordinal-[Hyper Aleph Ω (Aleph Ω, Aleph Ω)]-multiplicity))[Hyper Aleph (non-ordinal-[Hyper Aleph Ω (Aleph Ω, Aleph Ω)]-multiplicity) (Aleph (non-ordinal-[Hyper Aleph Ω (Aleph Ω, Aleph Ω)]-multiplicity), Aleph (non-ordinal-[Hyper Aleph Ω (Aleph Ω, Aleph Ω)]-multiplicity)), Hyper Aleph (non-ordinal-[Hyper Aleph Ω (Aleph Ω, Aleph Ω)]-multiplicity) (Aleph (non-ordinal-[Hyper Aleph Ω (Aleph Ω, Aleph Ω)]-multiplicity), Aleph (non-ordinal-[Hyper Aleph Ω (Aleph Ω, Aleph Ω)]-multiplicity))]}, Hyper Hyper Hyper Aleph (non-ordinal-[Hyper Aleph Ω (Aleph Ω, Aleph Ω)]-multiplicity) (Aleph (non-ordinal-[Hyper Aleph Ω (Aleph Ω, Aleph Ω)]-multiplicity), Aleph (non-ordinal-[Hyper Aleph Ω (Aleph Ω, Aleph Ω)]-multiplicity))[Hyper Aleph (non-ordinal-[Hyper Aleph Ω (Aleph Ω, Aleph Ω)]-multiplicity) (Aleph (non-ordinal-[Hyper Aleph Ω (Aleph Ω, Aleph Ω)]-multiplicity), Aleph (non-ordinal-[Hyper Aleph Ω (Aleph Ω, Aleph Ω)]-multiplicity)), Hyper Aleph (non-ordinal-[Hyper Aleph Ω (Aleph Ω, Aleph Ω)]-multiplicity) (Aleph (non-ordinal-[Hyper Aleph Ω (Aleph Ω, Aleph Ω)]-multiplicity), Aleph (non-ordinal-[Hyper Aleph Ω (Aleph Ω, Aleph Ω)]-multiplicity))] {Hyper Hyper Aleph (non-ordinal-[Hyper Aleph Ω (Aleph Ω, Aleph Ω)]-multiplicity) (Aleph (non-ordinal-[Hyper Aleph Ω (Aleph Ω, Aleph Ω)]-multiplicity), Aleph (non-ordinal-[Hyper Aleph Ω (Aleph Ω, Aleph Ω)]-multiplicity))[Hyper Aleph (non-ordinal-[Hyper Aleph Ω (Aleph Ω, Aleph Ω)]-multiplicity) (Aleph (non-ordinal-[Hyper Aleph Ω (Aleph Ω, Aleph Ω)]-multiplicity), Aleph (non-ordinal-[Hyper Aleph Ω (Aleph Ω, Aleph Ω)]-multiplicity)), Hyper Aleph (non-ordinal-[Hyper Aleph Ω (Aleph Ω, Aleph Ω)]-multiplicity) (Aleph (non-ordinal-[Hyper Aleph Ω (Aleph Ω, Aleph Ω)]-multiplicity), Aleph (non-ordinal-[Hyper Aleph Ω (Aleph Ω, Aleph Ω)]-multiplicity))], Hyper Hyper Aleph (non-ordinal-[Hyper Aleph Ω (Aleph Ω, Aleph Ω)]-multiplicity) (Aleph (non-ordinal-[Hyper Aleph Ω (Aleph Ω, Aleph Ω)]-multiplicity), Aleph (non-ordinal-[Hyper Aleph Ω (Aleph Ω, Aleph Ω)]-multiplicity))[Hyper Aleph (non-ordinal-[Hyper Aleph Ω (Aleph Ω, Aleph Ω)]-multiplicity) (Aleph (non-ordinal-[Hyper Aleph Ω (Aleph Ω, Aleph Ω)]-multiplicity), Aleph (non-ordinal-[Hyper Aleph Ω (Aleph Ω, Aleph Ω)]-multiplicity)), Hyper Aleph (non-ordinal-[Hyper Aleph Ω (Aleph Ω, Aleph Ω)]-multiplicity) (Aleph (non-ordinal-[Hyper Aleph Ω (Aleph Ω, Aleph Ω)]-multiplicity), Aleph (non-ordinal-[Hyper Aleph Ω (Aleph Ω, Aleph Ω)]-multiplicity))]}}}}

We can go still yet further to develop the following infinity.

Hyper Hyper Hyper Hyper Hyper Hyper Hyper Aleph (non-ordinal-[Hyper Aleph Ω (Aleph Ω, Aleph Ω)]-multiplicity) (Aleph (non-ordinal-[Hyper Aleph Ω (Aleph Ω, Aleph

Ω)]-multiplicity), Aleph (non-ordinal-[Hyper Aleph Ω (Aleph Ω, Aleph Ω)]-multiplicity))[Hyper Aleph (non-ordinal-[Hyper Aleph Ω (Aleph Ω, Aleph Ω)]-multiplicity) (Aleph (non-ordinal-[Hyper Aleph Ω (Aleph Ω, Aleph Ω)]-multiplicity), Aleph (non-ordinal-[Hyper Aleph Ω (Aleph Ω, Aleph Ω)]-multiplicity)), Hyper Aleph (non-ordinal-[Hyper Aleph Ω (Aleph Ω, Aleph Ω)]-multiplicity) (Aleph (non-ordinal-[Hyper Aleph Ω (Aleph Ω, Aleph Ω)]-multiplicity), Aleph (non-ordinal-[Hyper Aleph Ω (Aleph Ω, Aleph Ω)]-multiplicity))] {Hyper Hyper Aleph (non-ordinal-[Hyper Aleph Ω (Aleph Ω, Aleph Ω)]-multiplicity) (Aleph (non-ordinal-[Hyper Aleph Ω (Aleph Ω, Aleph Ω)]-multiplicity), Aleph (non-ordinal-[Hyper Aleph Ω (Aleph Ω, Aleph Ω)]-multiplicity))[Hyper Aleph (non-ordinal-[Hyper Aleph Ω (Aleph Ω, Aleph Ω)]-multiplicity) (Aleph (non-ordinal-[Hyper Aleph Ω (Aleph Ω, Aleph Ω)]-multiplicity), Aleph (non-ordinal-[Hyper Aleph Ω (Aleph Ω, Aleph Ω)]-multiplicity)), Hyper Aleph (non-ordinal-[Hyper Aleph Ω (Aleph Ω, Aleph Ω)]-multiplicity) (Aleph (non-ordinal-[Hyper Aleph Ω (Aleph Ω, Aleph Ω)]-multiplicity), Aleph (non-ordinal-[Hyper Aleph Ω (Aleph Ω, Aleph Ω)]-multiplicity))], Hyper Hyper Aleph (non-ordinal-[Hyper Aleph Ω (Aleph Ω, Aleph Ω)]-multiplicity) (Aleph (non-ordinal-[Hyper Aleph Ω (Aleph Ω, Aleph Ω)]-multiplicity), Aleph (non-ordinal-[Hyper Aleph Ω (Aleph Ω, Aleph Ω)]-multiplicity))[Hyper Aleph (non-ordinal-[Hyper Aleph Ω (Aleph Ω, Aleph Ω)]-multiplicity) (Aleph (non-ordinal-[Hyper Aleph Ω (Aleph Ω, Aleph Ω)]-multiplicity), Aleph (non-ordinal-[Hyper Aleph Ω (Aleph Ω, Aleph Ω)]-multiplicity)), Hyper Aleph (non-ordinal-[Hyper Aleph Ω (Aleph Ω, Aleph Ω)]-multiplicity) (Aleph (non-ordinal-[Hyper Aleph Ω (Aleph Ω, Aleph Ω)]-multiplicity), Aleph (non-ordinal-[Hyper Aleph Ω (Aleph Ω, Aleph Ω)]-multiplicity))]} { Hyper Hyper Hyper Aleph (non-ordinal-[Hyper Aleph Ω (Aleph Ω, Aleph Ω)]-multiplicity) (Aleph (non-ordinal-[Hyper Aleph Ω (Aleph Ω, Aleph Ω)]-multiplicity), Aleph (non-ordinal-[Hyper Aleph Ω (Aleph Ω, Aleph Ω)]-multiplicity))[Hyper Aleph (non-ordinal-[Hyper Aleph Ω (Aleph Ω, Aleph Ω)]-multiplicity) (Aleph (non-ordinal-[Hyper Aleph Ω (Aleph Ω, Aleph Ω)]-multiplicity), Aleph (non-ordinal-[Hyper Aleph Ω (Aleph Ω, Aleph Ω)]-multiplicity)), Hyper Aleph (non-ordinal-[Hyper Aleph Ω (Aleph Ω, Aleph Ω)]-multiplicity) (Aleph (non-ordinal-[Hyper Aleph Ω (Aleph Ω, Aleph Ω)]-multiplicity), Aleph (non-ordinal-[Hyper Aleph Ω (Aleph Ω, Aleph Ω)]-multiplicity))] {Hyper Hyper Aleph (non-ordinal-[Hyper Aleph Ω (Aleph Ω, Aleph Ω)]-multiplicity) (Aleph (non-ordinal-[Hyper Aleph Ω (Aleph Ω, Aleph Ω)]-multiplicity), Aleph (non-ordinal-[Hyper Aleph Ω (Aleph Ω, Aleph Ω)]-multiplicity))[Hyper Aleph (non-ordinal-[Hyper Aleph Ω (Aleph Ω, Aleph Ω)]-multiplicity) (Aleph (non-ordinal-[Hyper Aleph Ω (Aleph Ω, Aleph Ω)]-multiplicity), Aleph (non-ordinal-[Hyper Aleph Ω (Aleph Ω, Aleph Ω)]-multiplicity)), Hyper Aleph (non-ordinal-[Hyper Aleph Ω (Aleph Ω, Aleph Ω)]-multiplicity) (Aleph (non-ordinal-[Hyper Aleph Ω (Aleph Ω, Aleph Ω)]-multiplicity), Aleph (non-ordinal-[Hyper Aleph Ω (Aleph Ω, Aleph Ω)]-multiplicity))], Hyper Hyper Aleph (non-ordinal-[Hyper Aleph Ω (Aleph Ω, Aleph Ω)]-multiplicity) (Aleph (non-ordinal-[Hyper Aleph Ω (Aleph Ω, Aleph Ω)]-multiplicity), Aleph (non-ordinal-[Hyper Aleph Ω (Aleph Ω, Aleph Ω)]-multiplicity))[Hyper Aleph (non-ordinal-[Hyper Aleph Ω (Aleph Ω, Aleph Ω)]-multiplicity)

(Aleph (non-ordinal-[Hyper Aleph Ω (Aleph Ω, Aleph Ω)]-multiplicity), Aleph (non-ordinal-[Hyper Aleph Ω (Aleph Ω, Aleph Ω)]-multiplicity)), Hyper Aleph (non-ordinal-[Hyper Aleph Ω (Aleph Ω, Aleph Ω)]-multiplicity) (Aleph (non-ordinal-[Hyper Aleph Ω (Aleph Ω, Aleph Ω)]-multiplicity), Aleph (non-ordinal-[Hyper Aleph Ω (Aleph Ω, Aleph Ω)]-multiplicity))]}, Hyper Hyper Hyper Aleph (non-ordinal-[Hyper Aleph Ω (Aleph Ω, Aleph Ω)]-multiplicity) (Aleph (non-ordinal-[Hyper Aleph Ω (Aleph Ω, Aleph Ω)]-multiplicity), Aleph (non-ordinal-[Hyper Aleph Ω (Aleph Ω, Aleph Ω)]-multiplicity))[Hyper Aleph (non-ordinal-[Hyper Aleph Ω (Aleph Ω, Aleph Ω)]-multiplicity) (Aleph (non-ordinal-[Hyper Aleph Ω (Aleph Ω, Aleph Ω)]-multiplicity), Aleph (non-ordinal-[Hyper Aleph Ω (Aleph Ω, Aleph Ω)]-multiplicity)), Hyper Aleph (non-ordinal-[Hyper Aleph Ω (Aleph Ω, Aleph Ω)]-multiplicity) (Aleph (non-ordinal-[Hyper Aleph Ω (Aleph Ω, Aleph Ω)]-multiplicity), Aleph (non-ordinal-[Hyper Aleph Ω (Aleph Ω, Aleph Ω)]-multiplicity))] {Hyper Hyper Aleph (non-ordinal-[Hyper Aleph Ω (Aleph Ω, Aleph Ω)]-multiplicity) (Aleph (non-ordinal-[Hyper Aleph Ω (Aleph Ω, Aleph Ω)]-multiplicity), Aleph (non-ordinal-[Hyper Aleph Ω (Aleph Ω, Aleph Ω)]-multiplicity))[Hyper Aleph (non-ordinal-[Hyper Aleph Ω (Aleph Ω, Aleph Ω)]-multiplicity) (Aleph (non-ordinal-[Hyper Aleph Ω (Aleph Ω, Aleph Ω)]-multiplicity), Aleph (non-ordinal-[Hyper Aleph Ω (Aleph Ω, Aleph Ω)]-multiplicity)), Hyper Aleph (non-ordinal-[Hyper Aleph Ω (Aleph Ω, Aleph Ω)]-multiplicity) (Aleph (non-ordinal-[Hyper Aleph Ω (Aleph Ω, Aleph Ω)]-multiplicity), Aleph (non-ordinal-[Hyper Aleph Ω (Aleph Ω, Aleph Ω)]-multiplicity))], Hyper Hyper Aleph (non-ordinal-[Hyper Aleph Ω (Aleph Ω, Aleph Ω)]-multiplicity) (Aleph (non-ordinal-[Hyper Aleph Ω (Aleph Ω, Aleph Ω)]-multiplicity), Aleph (non-ordinal-[Hyper Aleph Ω (Aleph Ω, Aleph Ω)]-multiplicity))[Hyper Aleph (non-ordinal-[Hyper Aleph Ω (Aleph Ω, Aleph Ω)]-multiplicity) (Aleph (non-ordinal-[Hyper Aleph Ω (Aleph Ω, Aleph Ω)]-multiplicity), Aleph (non-ordinal-[Hyper Aleph Ω (Aleph Ω, Aleph Ω)]-multiplicity)), Hyper Aleph (non-ordinal-[Hyper Aleph Ω (Aleph Ω, Aleph Ω)]-multiplicity) (Aleph (non-ordinal-[Hyper Aleph Ω (Aleph Ω, Aleph Ω)]-multiplicity), Aleph (non-ordinal-[Hyper Aleph Ω (Aleph Ω, Aleph Ω)]-multiplicity))]}}{ Hyper Hyper Hyper Hyper Aleph (non-ordinal-[Hyper Aleph Ω (Aleph Ω, Aleph Ω)]-multiplicity) (Aleph (non-ordinal-[Hyper Aleph Ω (Aleph Ω, Aleph Ω)]-multiplicity), Aleph (non-ordinal-[Hyper Aleph Ω (Aleph Ω, Aleph Ω)]-multiplicity))[Hyper Aleph (non-ordinal-[Hyper Aleph Ω (Aleph Ω, Aleph Ω)]-multiplicity) (Aleph (non-ordinal-[Hyper Aleph Ω (Aleph Ω, Aleph Ω)]-multiplicity), Aleph (non-ordinal-[Hyper Aleph Ω (Aleph Ω, Aleph Ω)]-multiplicity)), Hyper Aleph (non-ordinal-[Hyper Aleph Ω (Aleph Ω, Aleph Ω)]-multiplicity) (Aleph (non-ordinal-[Hyper Aleph Ω (Aleph Ω, Aleph Ω)]-multiplicity), Aleph (non-ordinal-[Hyper Aleph Ω (Aleph Ω, Aleph Ω)]-multiplicity))] {Hyper Hyper Aleph (non-ordinal-[Hyper Aleph Ω (Aleph Ω, Aleph Ω)]-multiplicity) (Aleph (non-ordinal-[Hyper Aleph Ω (Aleph Ω, Aleph Ω)]-multiplicity), Aleph (non-ordinal-[Hyper Aleph Ω (Aleph Ω, Aleph Ω)]-multiplicity))[Hyper Aleph (non-ordinal-[Hyper Aleph Ω (Aleph Ω, Aleph Ω)]-multiplicity) (Aleph (non-ordinal-[Hyper Aleph Ω (Aleph Ω, Aleph Ω)]-multiplicity), Aleph (non-ordinal-[Hyper Aleph Ω (Aleph Ω, Aleph Ω)]-multiplicity)), Hyper Aleph (non-ordinal-[Hyper Aleph Ω (Aleph Ω, Aleph Ω)]-multiplicity) (Aleph (non-ordinal-[Hyper Aleph Ω (Aleph Ω, Aleph Ω)]-multiplicity), Aleph (non-ordinal-[Hyper Aleph Ω (Aleph Ω, Aleph Ω)]-multiplicity))], Hyper Hyper Aleph (non-ordinal-[Hyper Aleph Ω (Aleph Ω, Aleph Ω)]-multiplicity) (Aleph (non-ordinal-[Hyper Aleph Ω (Aleph Ω, Aleph Ω)]-multiplicity), Aleph (non-ordinal-[Hyper Aleph Ω (Aleph Ω, Aleph Ω)]-multiplicity))[Hyper Aleph (non-ordinal-[Hyper Aleph Ω (Aleph Ω, Aleph Ω)]-multiplicity) (Aleph (non-ordinal-[Hyper Aleph Ω (Aleph Ω, Aleph Ω)]-multiplicity), Aleph (non-ordinal-[Hyper Aleph Ω (Aleph Ω, Aleph Ω)]-multiplicity)), Hyper Aleph (non-ordinal-[Hyper Aleph Ω (Aleph Ω, Aleph Ω)]-multiplicity) (Aleph (non-ordinal-[Hyper Aleph Ω (Aleph Ω, Aleph Ω)]-multiplicity), Aleph (non-ordinal-[Hyper Aleph Ω (Aleph Ω, Aleph Ω)]-multiplicity))]]} { Hyper Hyper Hyper Aleph (non-ordinal-[Hyper Aleph Ω (Aleph Ω, Aleph Ω)]-multiplicity) (Aleph (non-ordinal-[Hyper Aleph Ω (Aleph Ω, Aleph Ω)]-multiplicity), Aleph (non-ordinal-[Hyper Aleph Ω (Aleph Ω, Aleph Ω)]-multiplicity))[Hyper Aleph (non-ordinal-[Hyper Aleph Ω (Aleph Ω, Aleph Ω)]-multiplicity) (Aleph (non-ordinal-[Hyper Aleph Ω (Aleph Ω, Aleph Ω)]-multiplicity), Aleph (non-ordinal-[Hyper Aleph Ω (Aleph Ω, Aleph Ω)]-multiplicity)), Hyper Aleph (non-ordinal-[Hyper Aleph Ω (Aleph Ω, Aleph Ω)]-multiplicity) (Aleph (non-ordinal-[Hyper Aleph Ω (Aleph Ω, Aleph Ω)]-multiplicity), Aleph (non-ordinal-[Hyper Aleph Ω (Aleph Ω, Aleph Ω)]-multiplicity))] {Hyper Hyper Aleph (non-ordinal-[Hyper Aleph Ω (Aleph Ω, Aleph Ω)]-multiplicity) (Aleph (non-ordinal-[Hyper Aleph Ω (Aleph Ω, Aleph Ω)]-multiplicity), Aleph (non-ordinal-[Hyper Aleph Ω (Aleph Ω, Aleph Ω)]-multiplicity))[Hyper Aleph (non-ordinal-[Hyper Aleph Ω (Aleph Ω, Aleph Ω)]-multiplicity) (Aleph (non-ordinal-[Hyper Aleph Ω (Aleph Ω, Aleph Ω)]-multiplicity), Aleph (non-ordinal-[Hyper Aleph Ω (Aleph Ω, Aleph Ω)]-multiplicity)), Hyper Aleph (non-ordinal-[Hyper Aleph Ω (Aleph Ω, Aleph Ω)]-multiplicity) (Aleph (non-ordinal-[Hyper Aleph Ω (Aleph Ω, Aleph Ω)]-multiplicity), Aleph (non-ordinal-[Hyper Aleph Ω (Aleph Ω, Aleph Ω)]-multiplicity))], Hyper Hyper Aleph (non-ordinal-[Hyper Aleph Ω (Aleph Ω, Aleph Ω)]-multiplicity) (Aleph (non-ordinal-[Hyper Aleph Ω (Aleph Ω, Aleph Ω)]-multiplicity), Aleph (non-ordinal-[Hyper Aleph Ω (Aleph Ω, Aleph Ω)]-multiplicity))[Hyper Aleph (non-ordinal-[Hyper Aleph Ω (Aleph Ω, Aleph Ω)]-multiplicity) (Aleph (non-ordinal-[Hyper Aleph Ω (Aleph Ω, Aleph Ω)]-multiplicity), Aleph (non-ordinal-[Hyper Aleph Ω (Aleph Ω, Aleph Ω)]-multiplicity)), Hyper Aleph (non-ordinal-[Hyper Aleph Ω (Aleph Ω, Aleph Ω)]-multiplicity) (Aleph (non-ordinal-[Hyper Aleph Ω (Aleph Ω, Aleph Ω)]-multiplicity), Aleph (non-ordinal-[Hyper Aleph Ω (Aleph Ω, Aleph Ω)]-multiplicity))]}, Hyper Hyper Hyper Aleph (non-ordinal-[Hyper Aleph Ω (Aleph Ω, Aleph Ω)]-multiplicity) (Aleph (non-ordinal-[Hyper Aleph Ω (Aleph Ω, Aleph Ω)]-multiplicity), Aleph (non-ordinal-[Hyper Aleph Ω (Aleph Ω, Aleph Ω)]-multiplicity))[Hyper Aleph (non-ordinal-[Hyper Aleph Ω (Aleph Ω, Aleph Ω)]-multiplicity) (Aleph (non-ordinal-[Hyper Aleph Ω (Aleph Ω, Aleph Ω)]-multiplicity), Aleph (non-ordinal-[Hyper Aleph Ω (Aleph Ω, Aleph Ω)]-multiplicity)), Hyper Aleph (non-ordinal-[Hyper Aleph Ω (Aleph Ω, Aleph Ω)]-multiplicity) (Aleph (non-ordinal-[Hyper Aleph Ω (Aleph Ω, Aleph Ω)]-multiplicity), Aleph (non-ordinal-[Hyper Aleph Ω

(Aleph Ω, Aleph Ω)]-multiplicity))] {Hyper Hyper Aleph (non-ordinal-[Hyper Aleph Ω (Aleph Ω, Aleph Ω)]-multiplicity) (Aleph (non-ordinal-[Hyper Aleph Ω (Aleph Ω, Aleph Ω)]-multiplicity), Aleph (non-ordinal-[Hyper Aleph Ω (Aleph Ω, Aleph Ω)]-multiplicity))[Hyper Aleph (non-ordinal-[Hyper Aleph Ω (Aleph Ω, Aleph Ω)]-multiplicity) (Aleph (non-ordinal-[Hyper Aleph Ω (Aleph Ω, Aleph Ω)]-multiplicity), Aleph (non-ordinal-[Hyper Aleph Ω (Aleph Ω, Aleph Ω)]-multiplicity)), Hyper Aleph (non-ordinal-[Hyper Aleph Ω (Aleph Ω, Aleph Ω)]-multiplicity) (Aleph (non-ordinal-[Hyper Aleph Ω (Aleph Ω, Aleph Ω)]-multiplicity), Aleph (non-ordinal-[Hyper Aleph Ω (Aleph Ω, Aleph Ω)]-multiplicity))], Hyper Hyper Aleph (non-ordinal-[Hyper Aleph Ω (Aleph Ω, Aleph Ω)]-multiplicity) (Aleph (non-ordinal-[Hyper Aleph Ω (Aleph Ω, Aleph Ω)]-multiplicity), Aleph (non-ordinal-[Hyper Aleph Ω (Aleph Ω, Aleph Ω)]-multiplicity))[Hyper Aleph (non-ordinal-[Hyper Aleph Ω (Aleph Ω, Aleph Ω)]-multiplicity) (Aleph (non-ordinal-[Hyper Aleph Ω (Aleph Ω, Aleph Ω)]-multiplicity), Aleph (non-ordinal-[Hyper Aleph Ω (Aleph Ω, Aleph Ω)]-multiplicity)), Hyper Aleph (non-ordinal-[Hyper Aleph Ω (Aleph Ω, Aleph Ω)]-multiplicity) (Aleph (non-ordinal-[Hyper Aleph Ω (Aleph Ω, Aleph Ω)]-multiplicity), Aleph (non-ordinal-[Hyper Aleph Ω (Aleph Ω, Aleph Ω)]-multiplicity))]}}, Hyper Hyper Hyper Hyper Aleph (non-ordinal-[Hyper Aleph Ω (Aleph Ω, Aleph Ω)]-multiplicity) (Aleph (non-ordinal-[Hyper Aleph Ω (Aleph Ω, Aleph Ω)]-multiplicity), Aleph (non-ordinal-[Hyper Aleph Ω (Aleph Ω, Aleph Ω)]-multiplicity))[Hyper Aleph (non-ordinal-[Hyper Aleph Ω (Aleph Ω, Aleph Ω)]-multiplicity) (Aleph (non-ordinal-[Hyper Aleph Ω (Aleph Ω, Aleph Ω)]-multiplicity), Aleph (non-ordinal-[Hyper Aleph Ω (Aleph Ω, Aleph Ω)]-multiplicity)), Hyper Aleph (non-ordinal-[Hyper Aleph Ω (Aleph Ω, Aleph Ω)]-multiplicity) (Aleph (non-ordinal-[Hyper Aleph Ω (Aleph Ω, Aleph Ω)]-multiplicity), Aleph (non-ordinal-[Hyper Aleph Ω (Aleph Ω, Aleph Ω)]-multiplicity))] {Hyper Hyper Aleph (non-ordinal-[Hyper Aleph Ω (Aleph Ω, Aleph Ω)]-multiplicity) (Aleph (non-ordinal-[Hyper Aleph Ω (Aleph Ω, Aleph Ω)]-multiplicity), Aleph (non-ordinal-[Hyper Aleph Ω (Aleph Ω, Aleph Ω)]-multiplicity))[Hyper Aleph (non-ordinal-[Hyper Aleph Ω (Aleph Ω, Aleph Ω)]-multiplicity) (Aleph (non-ordinal-[Hyper Aleph Ω (Aleph Ω, Aleph Ω)]-multiplicity), Aleph (non-ordinal-[Hyper Aleph Ω (Aleph Ω, Aleph Ω)]-multiplicity)), Hyper Aleph (non-ordinal-[Hyper Aleph Ω (Aleph Ω, Aleph Ω)]-multiplicity) (Aleph (non-ordinal-[Hyper Aleph Ω (Aleph Ω, Aleph Ω)]-multiplicity), Aleph (non-ordinal-[Hyper Aleph Ω (Aleph Ω, Aleph Ω)]-multiplicity))], Hyper Hyper Aleph (non-ordinal-[Hyper Aleph Ω (Aleph Ω, Aleph Ω)]-multiplicity) (Aleph (non-ordinal-[Hyper Aleph Ω (Aleph Ω, Aleph Ω)]-multiplicity), Aleph (non-ordinal-[Hyper Aleph Ω (Aleph Ω, Aleph Ω)]-multiplicity))[Hyper Aleph (non-ordinal-[Hyper Aleph Ω (Aleph Ω, Aleph Ω)]-multiplicity) (Aleph (non-ordinal-[Hyper Aleph Ω (Aleph Ω, Aleph Ω)]-multiplicity), Aleph (non-ordinal-[Hyper Aleph Ω (Aleph Ω, Aleph Ω)]-multiplicity)), Hyper Aleph (non-ordinal-[Hyper Aleph Ω (Aleph Ω, Aleph Ω)]-multiplicity) (Aleph (non-ordinal-[Hyper Aleph Ω (Aleph Ω, Aleph Ω)]-multiplicity), Aleph (non-ordinal-[Hyper Aleph Ω (Aleph Ω, Aleph Ω)]-multiplicity))]} { Hyper Hyper Hyper Aleph (non-ordinal-[Hyper Aleph Ω (Aleph Ω, Aleph Ω)]-multiplicity) (Aleph (non-ordinal-

[Hyper Aleph Ω (Aleph Ω, Aleph Ω)]-multiplicity), Aleph (non-ordinal-[Hyper Aleph Ω (Aleph Ω, Aleph Ω)]-multiplicity))[Hyper Aleph (non-ordinal-[Hyper Aleph Ω (Aleph Ω, Aleph Ω)]-multiplicity) (Aleph (non-ordinal-[Hyper Aleph Ω (Aleph Ω, Aleph Ω)]-multiplicity), Aleph (non-ordinal-[Hyper Aleph Ω (Aleph Ω, Aleph Ω)]-multiplicity)), Hyper Aleph (non-ordinal-[Hyper Aleph Ω (Aleph Ω, Aleph Ω)]-multiplicity) (Aleph (non-ordinal-[Hyper Aleph Ω (Aleph Ω, Aleph Ω)]-multiplicity), Aleph (non-ordinal-[Hyper Aleph Ω (Aleph Ω, Aleph Ω)]-multiplicity))] {Hyper Hyper Aleph (non-ordinal-[Hyper Aleph Ω (Aleph Ω, Aleph Ω)]-multiplicity) (Aleph (non-ordinal-[Hyper Aleph Ω (Aleph Ω, Aleph Ω)]-multiplicity), Aleph (non-ordinal-[Hyper Aleph Ω (Aleph Ω, Aleph Ω)]-multiplicity))[Hyper Aleph (non-ordinal-[Hyper Aleph Ω (Aleph Ω, Aleph Ω)]-multiplicity) (Aleph (non-ordinal-[Hyper Aleph Ω (Aleph Ω, Aleph Ω)]-multiplicity), Aleph (non-ordinal-[Hyper Aleph Ω (Aleph Ω, Aleph Ω)]-multiplicity)), Hyper Aleph (non-ordinal-[Hyper Aleph Ω (Aleph Ω, Aleph Ω)]-multiplicity) (Aleph (non-ordinal-[Hyper Aleph Ω (Aleph Ω, Aleph Ω)]-multiplicity), Aleph (non-ordinal-[Hyper Aleph Ω (Aleph Ω, Aleph Ω)]-multiplicity))], Hyper Hyper Aleph (non-ordinal-[Hyper Aleph Ω (Aleph Ω, Aleph Ω)]-multiplicity) (Aleph (non-ordinal-[Hyper Aleph Ω (Aleph Ω, Aleph Ω)]-multiplicity), Aleph (non-ordinal-[Hyper Aleph Ω (Aleph Ω, Aleph Ω)]-multiplicity))[Hyper Aleph (non-ordinal-[Hyper Aleph Ω (Aleph Ω, Aleph Ω)]-multiplicity) (Aleph (non-ordinal-[Hyper Aleph Ω (Aleph Ω, Aleph Ω)]-multiplicity), Aleph (non-ordinal-[Hyper Aleph Ω (Aleph Ω, Aleph Ω)]-multiplicity)), Hyper Aleph (non-ordinal-[Hyper Aleph Ω (Aleph Ω, Aleph Ω)]-multiplicity) (Aleph (non-ordinal-[Hyper Aleph Ω (Aleph Ω, Aleph Ω)]-multiplicity), Aleph (non-ordinal-[Hyper Aleph Ω (Aleph Ω, Aleph Ω)]-multiplicity))]}, Hyper Hyper Hyper Aleph (non-ordinal-[Hyper Aleph Ω (Aleph Ω, Aleph Ω)]-multiplicity) (Aleph (non-ordinal-[Hyper Aleph Ω (Aleph Ω, Aleph Ω)]-multiplicity), Aleph (non-ordinal-[Hyper Aleph Ω (Aleph Ω, Aleph Ω)]-multiplicity))[Hyper Aleph (non-ordinal-[Hyper Aleph Ω (Aleph Ω, Aleph Ω)]-multiplicity) (Aleph (non-ordinal-[Hyper Aleph Ω (Aleph Ω, Aleph Ω)]-multiplicity), Aleph (non-ordinal-[Hyper Aleph Ω (Aleph Ω, Aleph Ω)]-multiplicity)), Hyper Aleph (non-ordinal-[Hyper Aleph Ω (Aleph Ω, Aleph Ω)]-multiplicity) (Aleph (non-ordinal-[Hyper Aleph Ω (Aleph Ω, Aleph Ω)]-multiplicity), Aleph (non-ordinal-[Hyper Aleph Ω (Aleph Ω, Aleph Ω)]-multiplicity))] {Hyper Hyper Aleph (non-ordinal-[Hyper Aleph Ω (Aleph Ω, Aleph Ω)]-multiplicity) (Aleph (non-ordinal-[Hyper Aleph Ω (Aleph Ω, Aleph Ω)]-multiplicity), Aleph (non-ordinal-[Hyper Aleph Ω (Aleph Ω, Aleph Ω)]-multiplicity))[Hyper Aleph (non-ordinal-[Hyper Aleph Ω (Aleph Ω, Aleph Ω)]-multiplicity) (Aleph (non-ordinal-[Hyper Aleph Ω (Aleph Ω, Aleph Ω)]-multiplicity), Aleph (non-ordinal-[Hyper Aleph Ω (Aleph Ω, Aleph Ω)]-multiplicity)), Hyper Aleph (non-ordinal-[Hyper Aleph Ω (Aleph Ω, Aleph Ω)]-multiplicity) (Aleph (non-ordinal-[Hyper Aleph Ω (Aleph Ω, Aleph Ω)]-multiplicity), Aleph (non-ordinal-[Hyper Aleph Ω (Aleph Ω, Aleph Ω)]-multiplicity))], Hyper Hyper Aleph (non-ordinal-[Hyper Aleph Ω (Aleph Ω, Aleph Ω)]-multiplicity) (Aleph (non-ordinal-[Hyper Aleph Ω (Aleph Ω, Aleph Ω)]-multiplicity), Aleph (non-ordinal-[Hyper Aleph Ω (Aleph Ω, Aleph Ω)]-multiplicity))[Hyper Aleph (non-ordinal-

[Hyper Aleph Ω (Aleph Ω, Aleph Ω)]-multiplicity) (Aleph (non-ordinal-[Hyper Aleph Ω (Aleph Ω, Aleph Ω)]-multiplicity), Aleph (non-ordinal-[Hyper Aleph Ω (Aleph Ω, Aleph Ω)]-multiplicity)), Hyper Aleph (non-ordinal-[Hyper Aleph Ω (Aleph Ω, Aleph Ω)]-multiplicity) (Aleph (non-ordinal-[Hyper Aleph Ω (Aleph Ω, Aleph Ω)]-multiplicity), Aleph (non-ordinal-[Hyper Aleph Ω (Aleph Ω, Aleph Ω)]-multiplicity))]}}}{ Hyper Hyper Hyper Hyper Hyper Aleph (non-ordinal-[Hyper Aleph Ω (Aleph Ω, Aleph Ω)]-multiplicity) (Aleph (non-ordinal-[Hyper Aleph Ω (Aleph Ω, Aleph Ω)]-multiplicity), Aleph (non-ordinal-[Hyper Aleph Ω (Aleph Ω, Aleph Ω)]-multiplicity))[Hyper Aleph (non-ordinal-[Hyper Aleph Ω (Aleph Ω, Aleph Ω)]-multiplicity) (Aleph (non-ordinal-[Hyper Aleph Ω (Aleph Ω, Aleph Ω)]-multiplicity), Aleph (non-ordinal-[Hyper Aleph Ω (Aleph Ω, Aleph Ω)]-multiplicity)), Hyper Aleph (non-ordinal-[Hyper Aleph Ω (Aleph Ω, Aleph Ω)]-multiplicity) (Aleph (non-ordinal-[Hyper Aleph Ω (Aleph Ω, Aleph Ω)]-multiplicity), Aleph (non-ordinal-[Hyper Aleph Ω (Aleph Ω, Aleph Ω)]-multiplicity))] {Hyper Hyper Aleph (non-ordinal-[Hyper Aleph Ω (Aleph Ω, Aleph Ω)]-multiplicity) (Aleph (non-ordinal-[Hyper Aleph Ω (Aleph Ω, Aleph Ω)]-multiplicity), Aleph (non-ordinal-[Hyper Aleph Ω (Aleph Ω, Aleph Ω)]-multiplicity))[Hyper Aleph (non-ordinal-[Hyper Aleph Ω (Aleph Ω, Aleph Ω)]-multiplicity) (Aleph (non-ordinal-[Hyper Aleph Ω (Aleph Ω, Aleph Ω)]-multiplicity), Aleph (non-ordinal-[Hyper Aleph Ω (Aleph Ω, Aleph Ω)]-multiplicity)), Hyper Aleph (non-ordinal-[Hyper Aleph Ω (Aleph Ω, Aleph Ω)]-multiplicity) (Aleph (non-ordinal-[Hyper Aleph Ω (Aleph Ω, Aleph Ω)]-multiplicity), Aleph (non-ordinal-[Hyper Aleph Ω (Aleph Ω, Aleph Ω)]-multiplicity))], Hyper Hyper Aleph (non-ordinal-[Hyper Aleph Ω (Aleph Ω, Aleph Ω)]-multiplicity) (Aleph (non-ordinal-[Hyper Aleph Ω (Aleph Ω, Aleph Ω)]-multiplicity), Aleph (non-ordinal-[Hyper Aleph Ω (Aleph Ω, Aleph Ω)]-multiplicity))[Hyper Aleph (non-ordinal-[Hyper Aleph Ω (Aleph Ω, Aleph Ω)]-multiplicity) (Aleph (non-ordinal-[Hyper Aleph Ω (Aleph Ω, Aleph Ω)]-multiplicity), Aleph (non-ordinal-[Hyper Aleph Ω (Aleph Ω, Aleph Ω)]-multiplicity)), Hyper Aleph (non-ordinal-[Hyper Aleph Ω (Aleph Ω, Aleph Ω)]-multiplicity) (Aleph (non-ordinal-[Hyper Aleph Ω (Aleph Ω, Aleph Ω)]-multiplicity), Aleph (non-ordinal-[Hyper Aleph Ω (Aleph Ω, Aleph Ω)]-multiplicity))]} { Hyper Hyper Hyper Aleph (non-ordinal-[Hyper Aleph Ω (Aleph Ω, Aleph Ω)]-multiplicity) (Aleph (non-ordinal-[Hyper Aleph Ω (Aleph Ω, Aleph Ω)]-multiplicity), Aleph (non-ordinal-[Hyper Aleph Ω (Aleph Ω, Aleph Ω)]-multiplicity))[Hyper Aleph (non-ordinal-[Hyper Aleph Ω (Aleph Ω, Aleph Ω)]-multiplicity) (Aleph (non-ordinal-[Hyper Aleph Ω (Aleph Ω, Aleph Ω)]-multiplicity), Aleph (non-ordinal-[Hyper Aleph Ω (Aleph Ω, Aleph Ω)]-multiplicity)), Hyper Aleph (non-ordinal-[Hyper Aleph Ω (Aleph Ω, Aleph Ω)]-multiplicity) (Aleph (non-ordinal-[Hyper Aleph Ω (Aleph Ω, Aleph Ω)]-multiplicity), Aleph (non-ordinal-[Hyper Aleph Ω (Aleph Ω, Aleph Ω)]-multiplicity))] {Hyper Hyper Aleph (non-ordinal-[Hyper Aleph Ω (Aleph Ω, Aleph Ω)]-multiplicity) (Aleph (non-ordinal-[Hyper Aleph Ω (Aleph Ω, Aleph Ω)]-multiplicity), Aleph (non-ordinal-[Hyper Aleph Ω (Aleph Ω, Aleph Ω)]-multiplicity))[Hyper Aleph (non-ordinal-[Hyper Aleph Ω (Aleph Ω, Aleph Ω)]-multiplicity) (Aleph (non-ordinal-[Hyper Aleph Ω (Aleph Ω, Aleph Ω)]-multiplicity), Aleph (non-ordinal-

[Hyper Aleph Ω (Aleph Ω, Aleph Ω)]-multiplicity)), Hyper Aleph (non-ordinal-[Hyper Aleph Ω (Aleph Ω, Aleph Ω)]-multiplicity) (Aleph (non-ordinal-[Hyper Aleph Ω (Aleph Ω, Aleph Ω)]-multiplicity), Aleph (non-ordinal-[Hyper Aleph Ω (Aleph Ω, Aleph Ω)]-multiplicity))], Hyper Hyper Aleph (non-ordinal-[Hyper Aleph Ω (Aleph Ω, Aleph Ω)]-multiplicity) (Aleph (non-ordinal-[Hyper Aleph Ω (Aleph Ω, Aleph Ω)]-multiplicity), Aleph (non-ordinal-[Hyper Aleph Ω (Aleph Ω, Aleph Ω)]-multiplicity))[Hyper Aleph (non-ordinal-[Hyper Aleph Ω (Aleph Ω, Aleph Ω)]-multiplicity) (Aleph (non-ordinal-[Hyper Aleph Ω (Aleph Ω, Aleph Ω)]-multiplicity), Aleph (non-ordinal-[Hyper Aleph Ω (Aleph Ω, Aleph Ω)]-multiplicity)), Hyper Aleph (non-ordinal-[Hyper Aleph Ω (Aleph Ω, Aleph Ω)]-multiplicity) (Aleph (non-ordinal-[Hyper Aleph Ω (Aleph Ω, Aleph Ω)]-multiplicity), Aleph (non-ordinal-[Hyper Aleph Ω (Aleph Ω, Aleph Ω)]-multiplicity))]}, Hyper Hyper Hyper Aleph (non-ordinal-[Hyper Aleph Ω (Aleph Ω, Aleph Ω)]-multiplicity) (Aleph (non-ordinal-[Hyper Aleph Ω (Aleph Ω, Aleph Ω)]-multiplicity), Aleph (non-ordinal-[Hyper Aleph Ω (Aleph Ω, Aleph Ω)]-multiplicity))[Hyper Aleph (non-ordinal-[Hyper Aleph Ω (Aleph Ω, Aleph Ω)]-multiplicity) (Aleph (non-ordinal-[Hyper Aleph Ω (Aleph Ω, Aleph Ω)]-multiplicity), Aleph (non-ordinal-[Hyper Aleph Ω (Aleph Ω, Aleph Ω)]-multiplicity)), Hyper Aleph (non-ordinal-[Hyper Aleph Ω (Aleph Ω, Aleph Ω)]-multiplicity) (Aleph (non-ordinal-[Hyper Aleph Ω (Aleph Ω, Aleph Ω)]-multiplicity), Aleph (non-ordinal-[Hyper Aleph Ω (Aleph Ω, Aleph Ω)]-multiplicity))] {Hyper Hyper Aleph (non-ordinal-[Hyper Aleph Ω (Aleph Ω, Aleph Ω)]-multiplicity) (Aleph (non-ordinal-[Hyper Aleph Ω (Aleph Ω, Aleph Ω)]-multiplicity), Aleph (non-ordinal-[Hyper Aleph Ω (Aleph Ω, Aleph Ω)]-multiplicity))[Hyper Aleph (non-ordinal-[Hyper Aleph Ω (Aleph Ω, Aleph Ω)]-multiplicity) (Aleph (non-ordinal-[Hyper Aleph Ω (Aleph Ω, Aleph Ω)]-multiplicity), Aleph (non-ordinal-[Hyper Aleph Ω (Aleph Ω, Aleph Ω)]-multiplicity)), Hyper Aleph (non-ordinal-[Hyper Aleph Ω (Aleph Ω, Aleph Ω)]-multiplicity) (Aleph (non-ordinal-[Hyper Aleph Ω (Aleph Ω, Aleph Ω)]-multiplicity), Aleph (non-ordinal-[Hyper Aleph Ω (Aleph Ω, Aleph Ω)]-multiplicity))], Hyper Hyper Aleph (non-ordinal-[Hyper Aleph Ω (Aleph Ω, Aleph Ω)]-multiplicity) (Aleph (non-ordinal-[Hyper Aleph Ω (Aleph Ω, Aleph Ω)]-multiplicity), Aleph (non-ordinal-[Hyper Aleph Ω (Aleph Ω, Aleph Ω)]-multiplicity))[Hyper Aleph (non-ordinal-[Hyper Aleph Ω (Aleph Ω, Aleph Ω)]-multiplicity) (Aleph (non-ordinal-[Hyper Aleph Ω (Aleph Ω, Aleph Ω)]-multiplicity), Aleph (non-ordinal-[Hyper Aleph Ω (Aleph Ω, Aleph Ω)]-multiplicity)), Hyper Aleph (non-ordinal-[Hyper Aleph Ω (Aleph Ω, Aleph Ω)]-multiplicity) (Aleph (non-ordinal-[Hyper Aleph Ω (Aleph Ω, Aleph Ω)]-multiplicity), Aleph (non-ordinal-[Hyper Aleph Ω (Aleph Ω, Aleph Ω)]-multiplicity))]}}{ Hyper Hyper Hyper Hyper Aleph (non-ordinal-[Hyper Aleph Ω (Aleph Ω, Aleph Ω)]-multiplicity) (Aleph (non-ordinal-[Hyper Aleph Ω (Aleph Ω, Aleph Ω)]-multiplicity), Aleph (non-ordinal-[Hyper Aleph Ω (Aleph Ω, Aleph Ω)]-multiplicity))[Hyper Aleph (non-ordinal-[Hyper Aleph Ω (Aleph Ω, Aleph Ω)]-multiplicity) (Aleph (non-ordinal-[Hyper Aleph Ω (Aleph Ω, Aleph Ω)]-multiplicity), Aleph (non-ordinal-[Hyper Aleph Ω (Aleph Ω, Aleph Ω)]-multiplicity)), Hyper Aleph (non-ordinal-[Hyper Aleph Ω (Aleph Ω, Aleph Ω)]-multiplicity) (Aleph (non-ordinal-[Hyper Aleph Ω (Aleph Ω, Aleph Ω)]-multiplicity), Aleph (non-ordinal-[Hyper Aleph Ω (Aleph Ω, Aleph Ω)]-multiplicity))] {Hyper Hyper Aleph (non-ordinal-[Hyper Aleph Ω (Aleph Ω, Aleph Ω)]-multiplicity) (Aleph (non-ordinal-[Hyper Aleph Ω (Aleph Ω, Aleph Ω)]-multiplicity), Aleph (non-ordinal-[Hyper Aleph Ω (Aleph Ω, Aleph Ω)]-multiplicity))[Hyper Aleph (non-ordinal-[Hyper Aleph Ω (Aleph Ω, Aleph Ω)]-multiplicity) (Aleph (non-ordinal-[Hyper Aleph Ω (Aleph Ω, Aleph Ω)]-multiplicity), Aleph (non-ordinal-[Hyper Aleph Ω (Aleph Ω, Aleph Ω)]-multiplicity)), Hyper Aleph (non-ordinal-[Hyper Aleph Ω (Aleph Ω, Aleph Ω)]-multiplicity) (Aleph (non-ordinal-[Hyper Aleph Ω (Aleph Ω, Aleph Ω)]-multiplicity), Aleph (non-ordinal-[Hyper Aleph Ω (Aleph Ω, Aleph Ω)]-multiplicity))], Hyper Hyper Aleph (non-ordinal-[Hyper Aleph Ω (Aleph Ω, Aleph Ω)]-multiplicity) (Aleph (non-ordinal-[Hyper Aleph Ω (Aleph Ω, Aleph Ω)]-multiplicity), Aleph (non-ordinal-[Hyper Aleph Ω (Aleph Ω, Aleph Ω)]-multiplicity))[Hyper Aleph (non-ordinal-[Hyper Aleph Ω (Aleph Ω, Aleph Ω)]-multiplicity) (Aleph (non-ordinal-[Hyper Aleph Ω (Aleph Ω, Aleph Ω)]-multiplicity), Aleph (non-ordinal-[Hyper Aleph Ω (Aleph Ω, Aleph Ω)]-multiplicity)), Hyper Aleph (non-ordinal-[Hyper Aleph Ω (Aleph Ω, Aleph Ω)]-multiplicity) (Aleph (non-ordinal-[Hyper Aleph Ω (Aleph Ω, Aleph Ω)]-multiplicity), Aleph (non-ordinal-[Hyper Aleph Ω (Aleph Ω, Aleph Ω)]-multiplicity))]} { Hyper Hyper Hyper Aleph (non-ordinal-[Hyper Aleph Ω (Aleph Ω, Aleph Ω)]-multiplicity) (Aleph (non-ordinal-[Hyper Aleph Ω (Aleph Ω, Aleph Ω)]-multiplicity), Aleph (non-ordinal-[Hyper Aleph Ω (Aleph Ω, Aleph Ω)]-multiplicity))[Hyper Aleph (non-ordinal-[Hyper Aleph Ω (Aleph Ω, Aleph Ω)]-multiplicity) (Aleph (non-ordinal-[Hyper Aleph Ω (Aleph Ω, Aleph Ω)]-multiplicity), Aleph (non-ordinal-[Hyper Aleph Ω (Aleph Ω, Aleph Ω)]-multiplicity)), Hyper Aleph (non-ordinal-[Hyper Aleph Ω (Aleph Ω, Aleph Ω)]-multiplicity) (Aleph (non-ordinal-[Hyper Aleph Ω (Aleph Ω, Aleph Ω)]-multiplicity), Aleph (non-ordinal-[Hyper Aleph Ω (Aleph Ω, Aleph Ω)]-multiplicity))] {Hyper Hyper Aleph (non-ordinal-[Hyper Aleph Ω (Aleph Ω, Aleph Ω)]-multiplicity) (Aleph (non-ordinal-[Hyper Aleph Ω (Aleph Ω, Aleph Ω)]-multiplicity), Aleph (non-ordinal-[Hyper Aleph Ω (Aleph Ω, Aleph Ω)]-multiplicity))[Hyper Aleph (non-ordinal-[Hyper Aleph Ω (Aleph Ω, Aleph Ω)]-multiplicity) (Aleph (non-ordinal-[Hyper Aleph Ω (Aleph Ω, Aleph Ω)]-multiplicity), Aleph (non-ordinal-[Hyper Aleph Ω (Aleph Ω, Aleph Ω)]-multiplicity)), Hyper Aleph (non-ordinal-[Hyper Aleph Ω (Aleph Ω, Aleph Ω)]-multiplicity) (Aleph (non-ordinal-[Hyper Aleph Ω (Aleph Ω, Aleph Ω)]-multiplicity), Aleph (non-ordinal-[Hyper Aleph Ω (Aleph Ω, Aleph Ω)]-multiplicity))], Hyper Hyper Aleph (non-ordinal-[Hyper Aleph Ω (Aleph Ω, Aleph Ω)]-multiplicity) (Aleph (non-ordinal-[Hyper Aleph Ω (Aleph Ω, Aleph Ω)]-multiplicity), Aleph (non-ordinal-[Hyper Aleph Ω (Aleph Ω, Aleph Ω)]-multiplicity))[Hyper Aleph (non-ordinal-[Hyper Aleph Ω (Aleph Ω, Aleph Ω)]-multiplicity) (Aleph (non-ordinal-[Hyper Aleph Ω (Aleph Ω, Aleph Ω)]-multiplicity), Aleph (non-ordinal-[Hyper Aleph Ω (Aleph Ω, Aleph Ω)]-multiplicity)), Hyper Aleph (non-ordinal-[Hyper Aleph Ω (Aleph Ω, Aleph Ω)]-multiplicity) (Aleph (non-ordinal-[Hyper Aleph Ω (Aleph Ω, Aleph Ω)]-multiplicity), Aleph (non-ordinal-[Hyper Aleph Ω (Aleph Ω, Aleph Ω)]-multiplicity))]}, Hyper Hyper Hyper

Aleph (non-ordinal-[Hyper Aleph Ω (Aleph Ω, Aleph Ω)]-multiplicity) (Aleph (non-ordinal-[Hyper Aleph Ω (Aleph Ω, Aleph Ω)]-multiplicity), Aleph (non-ordinal-[Hyper Aleph Ω (Aleph Ω, Aleph Ω)]-multiplicity))[Hyper Aleph (non-ordinal-[Hyper Aleph Ω (Aleph Ω, Aleph Ω)]-multiplicity) (Aleph (non-ordinal-[Hyper Aleph Ω (Aleph Ω, Aleph Ω)]-multiplicity), Aleph (non-ordinal-[Hyper Aleph Ω (Aleph Ω, Aleph Ω)]-multiplicity)), Hyper Aleph (non-ordinal-[Hyper Aleph Ω (Aleph Ω, Aleph Ω)]-multiplicity) (Aleph (non-ordinal-[Hyper Aleph Ω (Aleph Ω, Aleph Ω)]-multiplicity), Aleph (non-ordinal-[Hyper Aleph Ω (Aleph Ω, Aleph Ω)]-multiplicity))] {Hyper Hyper Aleph (non-ordinal-[Hyper Aleph Ω (Aleph Ω, Aleph Ω)]-multiplicity) (Aleph (non-ordinal-[Hyper Aleph Ω (Aleph Ω, Aleph Ω)]-multiplicity), Aleph (non-ordinal-[Hyper Aleph Ω (Aleph Ω, Aleph Ω)]-multiplicity))[Hyper Aleph (non-ordinal-[Hyper Aleph Ω (Aleph Ω, Aleph Ω)]-multiplicity) (Aleph (non-ordinal-[Hyper Aleph Ω (Aleph Ω, Aleph Ω)]-multiplicity), Aleph (non-ordinal-[Hyper Aleph Ω (Aleph Ω, Aleph Ω)]-multiplicity)), Hyper Aleph (non-ordinal-[Hyper Aleph Ω (Aleph Ω, Aleph Ω)]-multiplicity) (Aleph (non-ordinal-[Hyper Aleph Ω (Aleph Ω, Aleph Ω)]-multiplicity), Aleph (non-ordinal-[Hyper Aleph Ω (Aleph Ω, Aleph Ω)]-multiplicity))], Hyper Hyper Aleph (non-ordinal-[Hyper Aleph Ω (Aleph Ω, Aleph Ω)]-multiplicity) (Aleph (non-ordinal-[Hyper Aleph Ω (Aleph Ω, Aleph Ω)]-multiplicity), Aleph (non-ordinal-[Hyper Aleph Ω (Aleph Ω, Aleph Ω)]-multiplicity))[Hyper Aleph (non-ordinal-[Hyper Aleph Ω (Aleph Ω, Aleph Ω)]-multiplicity) (Aleph (non-ordinal-[Hyper Aleph Ω (Aleph Ω, Aleph Ω)]-multiplicity), Aleph (non-ordinal-[Hyper Aleph Ω (Aleph Ω, Aleph Ω)]-multiplicity)), Hyper Aleph (non-ordinal-[Hyper Aleph Ω (Aleph Ω, Aleph Ω)]-multiplicity) (Aleph (non-ordinal-[Hyper Aleph Ω (Aleph Ω, Aleph Ω)]-multiplicity), Aleph (non-ordinal-[Hyper Aleph Ω (Aleph Ω, Aleph Ω)]-multiplicity))]}}, Hyper Hyper Hyper Hyper Aleph (non-ordinal-[Hyper Aleph Ω (Aleph Ω, Aleph Ω)]-multiplicity) (Aleph (non-ordinal-[Hyper Aleph Ω (Aleph Ω, Aleph Ω)]-multiplicity), Aleph (non-ordinal-[Hyper Aleph Ω (Aleph Ω, Aleph Ω)]-multiplicity))[Hyper Aleph (non-ordinal-[Hyper Aleph Ω (Aleph Ω, Aleph Ω)]-multiplicity) (Aleph (non-ordinal-[Hyper Aleph Ω (Aleph Ω, Aleph Ω)]-multiplicity), Aleph (non-ordinal-[Hyper Aleph Ω (Aleph Ω, Aleph Ω)]-multiplicity)), Hyper Aleph (non-ordinal-[Hyper Aleph Ω (Aleph Ω, Aleph Ω)]-multiplicity) (Aleph (non-ordinal-[Hyper Aleph Ω (Aleph Ω, Aleph Ω)]-multiplicity), Aleph (non-ordinal-[Hyper Aleph Ω (Aleph Ω, Aleph Ω)]-multiplicity))] {Hyper Hyper Aleph (non-ordinal-[Hyper Aleph Ω (Aleph Ω, Aleph Ω)]-multiplicity) (Aleph (non-ordinal-[Hyper Aleph Ω (Aleph Ω, Aleph Ω)]-multiplicity), Aleph (non-ordinal-[Hyper Aleph Ω (Aleph Ω, Aleph Ω)]-multiplicity))[Hyper Aleph (non-ordinal-[Hyper Aleph Ω (Aleph Ω, Aleph Ω)]-multiplicity) (Aleph (non-ordinal-[Hyper Aleph Ω (Aleph Ω, Aleph Ω)]-multiplicity), Aleph (non-ordinal-[Hyper Aleph Ω (Aleph Ω, Aleph Ω)]-multiplicity)), Hyper Aleph (non-ordinal-[Hyper Aleph Ω (Aleph Ω, Aleph Ω)]-multiplicity) (Aleph (non-ordinal-[Hyper Aleph Ω (Aleph Ω, Aleph Ω)]-multiplicity), Aleph (non-ordinal-[Hyper Aleph Ω (Aleph Ω, Aleph Ω)]-multiplicity))], Hyper Hyper Aleph (non-ordinal-[Hyper Aleph Ω (Aleph Ω, Aleph Ω)]-multiplicity) (Aleph (non-ordinal-[Hyper Aleph Ω (Aleph Ω, Aleph Ω)]-multiplicity), Aleph (non-ordinal-[Hyper Aleph Ω (Aleph Ω, Aleph Ω)]-multiplicity))[Hyper Aleph (non-ordinal-[Hyper Aleph Ω (Aleph Ω, Aleph Ω)]-multiplicity) (Aleph (non-ordinal-[Hyper Aleph Ω (Aleph Ω, Aleph Ω)]-multiplicity), Aleph (non-ordinal-[Hyper Aleph Ω (Aleph Ω, Aleph Ω)]-multiplicity)), Hyper Aleph (non-ordinal-[Hyper Aleph Ω (Aleph Ω, Aleph Ω)]-multiplicity) (Aleph (non-ordinal-[Hyper Aleph Ω (Aleph Ω, Aleph Ω)]-multiplicity), Aleph (non-ordinal-[Hyper Aleph Ω (Aleph Ω, Aleph Ω)]-multiplicity))]} { Hyper Hyper Hyper Aleph (non-ordinal-[Hyper Aleph Ω (Aleph Ω, Aleph Ω)]-multiplicity) (Aleph (non-ordinal-[Hyper Aleph Ω (Aleph Ω, Aleph Ω)]-multiplicity), Aleph (non-ordinal-[Hyper Aleph Ω (Aleph Ω, Aleph Ω)]-multiplicity))[Hyper Aleph (non-ordinal-[Hyper Aleph Ω (Aleph Ω, Aleph Ω)]-multiplicity) (Aleph (non-ordinal-[Hyper Aleph Ω (Aleph Ω, Aleph Ω)]-multiplicity), Aleph (non-ordinal-[Hyper Aleph Ω (Aleph Ω, Aleph Ω)]-multiplicity)), Hyper Aleph (non-ordinal-[Hyper Aleph Ω (Aleph Ω, Aleph Ω)]-multiplicity) (Aleph (non-ordinal-[Hyper Aleph Ω (Aleph Ω, Aleph Ω)]-multiplicity), Aleph (non-ordinal-[Hyper Aleph Ω (Aleph Ω, Aleph Ω)]-multiplicity))] {Hyper Hyper Aleph (non-ordinal-[Hyper Aleph Ω (Aleph Ω, Aleph Ω)]-multiplicity) (Aleph (non-ordinal-[Hyper Aleph Ω (Aleph Ω, Aleph Ω)]-multiplicity), Aleph (non-ordinal-[Hyper Aleph Ω (Aleph Ω, Aleph Ω)]-multiplicity))[Hyper Aleph (non-ordinal-[Hyper Aleph Ω (Aleph Ω, Aleph Ω)]-multiplicity) (Aleph (non-ordinal-[Hyper Aleph Ω (Aleph Ω, Aleph Ω)]-multiplicity), Aleph (non-ordinal-[Hyper Aleph Ω (Aleph Ω, Aleph Ω)]-multiplicity)), Hyper Aleph (non-ordinal-[Hyper Aleph Ω (Aleph Ω, Aleph Ω)]-multiplicity) (Aleph (non-ordinal-[Hyper Aleph Ω (Aleph Ω, Aleph Ω)]-multiplicity), Aleph (non-ordinal-[Hyper Aleph Ω (Aleph Ω, Aleph Ω)]-multiplicity))], Hyper Hyper Aleph (non-ordinal-[Hyper Aleph Ω (Aleph Ω, Aleph Ω)]-multiplicity) (Aleph (non-ordinal-[Hyper Aleph Ω (Aleph Ω, Aleph Ω)]-multiplicity), Aleph (non-ordinal-[Hyper Aleph Ω (Aleph Ω, Aleph Ω)]-multiplicity))[Hyper Aleph (non-ordinal-[Hyper Aleph Ω (Aleph Ω, Aleph Ω)]-multiplicity) (Aleph (non-ordinal-[Hyper Aleph Ω (Aleph Ω, Aleph Ω)]-multiplicity), Aleph (non-ordinal-[Hyper Aleph Ω (Aleph Ω, Aleph Ω)]-multiplicity)), Hyper Aleph (non-ordinal-[Hyper Aleph Ω (Aleph Ω, Aleph Ω)]-multiplicity) (Aleph (non-ordinal-[Hyper Aleph Ω (Aleph Ω, Aleph Ω)]-multiplicity), Aleph (non-ordinal-[Hyper Aleph Ω (Aleph Ω, Aleph Ω)]-multiplicity))]}, Hyper Hyper Hyper Aleph (non-ordinal-[Hyper Aleph Ω (Aleph Ω, Aleph Ω)]-multiplicity) (Aleph (non-ordinal-[Hyper Aleph Ω (Aleph Ω, Aleph Ω)]-multiplicity), Aleph (non-ordinal-[Hyper Aleph Ω (Aleph Ω, Aleph Ω)]-multiplicity))[Hyper Aleph (non-ordinal-[Hyper Aleph Ω (Aleph Ω, Aleph Ω)]-multiplicity) (Aleph (non-ordinal-[Hyper Aleph Ω (Aleph Ω, Aleph Ω)]-multiplicity), Aleph (non-ordinal-[Hyper Aleph Ω (Aleph Ω, Aleph Ω)]-multiplicity)), Hyper Aleph (non-ordinal-[Hyper Aleph Ω (Aleph Ω, Aleph Ω)]-multiplicity) (Aleph (non-ordinal-[Hyper Aleph Ω (Aleph Ω, Aleph Ω)]-multiplicity), Aleph (non-ordinal-[Hyper Aleph Ω (Aleph Ω, Aleph Ω)]-multiplicity))] {Hyper Hyper Aleph (non-ordinal-[Hyper Aleph Ω (Aleph Ω, Aleph Ω)]-multiplicity) (Aleph (non-ordinal-[Hyper Aleph Ω (Aleph Ω, Aleph Ω)]-multiplicity), Aleph (non-ordinal-[Hyper Aleph Ω (Aleph Ω, Aleph Ω)]-multiplicity))[Hyper Aleph (non-ordinal-[Hyper Aleph Ω (Aleph Ω, Aleph Ω)]-multiplicity)

(Aleph (non-ordinal-[Hyper Aleph Ω (Aleph Ω, Aleph Ω)]-multiplicity), Aleph (non-ordinal-[Hyper Aleph Ω (Aleph Ω, Aleph Ω)]-multiplicity)), Hyper Aleph (non-ordinal-[Hyper Aleph Ω (Aleph Ω, Aleph Ω)]-multiplicity) (Aleph (non-ordinal-[Hyper Aleph Ω (Aleph Ω, Aleph Ω)]-multiplicity), Aleph (non-ordinal-[Hyper Aleph Ω (Aleph Ω, Aleph Ω)]-multiplicity))], Hyper Hyper Aleph (non-ordinal-[Hyper Aleph Ω (Aleph Ω, Aleph Ω)]-multiplicity) (Aleph (non-ordinal-[Hyper Aleph Ω (Aleph Ω, Aleph Ω)]-multiplicity), Aleph (non-ordinal-[Hyper Aleph Ω (Aleph Ω, Aleph Ω)]-multiplicity))[Hyper Aleph (non-ordinal-[Hyper Aleph Ω (Aleph Ω, Aleph Ω)]-multiplicity) (Aleph (non-ordinal-[Hyper Aleph Ω (Aleph Ω, Aleph Ω)]-multiplicity), Aleph (non-ordinal-[Hyper Aleph Ω (Aleph Ω, Aleph Ω)]-multiplicity)), Hyper Aleph (non-ordinal-[Hyper Aleph Ω (Aleph Ω, Aleph Ω)]-multiplicity) (Aleph (non-ordinal-[Hyper Aleph Ω (Aleph Ω, Aleph Ω)]-multiplicity), Aleph (non-ordinal-[Hyper Aleph Ω (Aleph Ω, Aleph Ω)]-multiplicity))]}}}, Hyper Hyper Hyper Hyper Hyper Aleph (non-ordinal-[Hyper Aleph Ω (Aleph Ω, Aleph Ω)]-multiplicity) (Aleph (non-ordinal-[Hyper Aleph Ω (Aleph Ω, Aleph Ω)]-multiplicity), Aleph (non-ordinal-[Hyper Aleph Ω (Aleph Ω, Aleph Ω)]-multiplicity))[Hyper Aleph (non-ordinal-[Hyper Aleph Ω (Aleph Ω, Aleph Ω)]-multiplicity) (Aleph (non-ordinal-[Hyper Aleph Ω (Aleph Ω, Aleph Ω)]-multiplicity), Aleph (non-ordinal-[Hyper Aleph Ω (Aleph Ω, Aleph Ω)]-multiplicity)), Hyper Aleph (non-ordinal-[Hyper Aleph Ω (Aleph Ω, Aleph Ω)]-multiplicity) (Aleph (non-ordinal-[Hyper Aleph Ω (Aleph Ω, Aleph Ω)]-multiplicity), Aleph (non-ordinal-[Hyper Aleph Ω (Aleph Ω, Aleph Ω)]-multiplicity))] {Hyper Hyper Aleph (non-ordinal-[Hyper Aleph Ω (Aleph Ω, Aleph Ω)]-multiplicity) (Aleph (non-ordinal-[Hyper Aleph Ω (Aleph Ω, Aleph Ω)]-multiplicity), Aleph (non-ordinal-[Hyper Aleph Ω (Aleph Ω, Aleph Ω)]-multiplicity))[Hyper Aleph (non-ordinal-[Hyper Aleph Ω (Aleph Ω, Aleph Ω)]-multiplicity) (Aleph (non-ordinal-[Hyper Aleph Ω (Aleph Ω, Aleph Ω)]-multiplicity), Aleph (non-ordinal-[Hyper Aleph Ω (Aleph Ω, Aleph Ω)]-multiplicity)), Hyper Aleph (non-ordinal-[Hyper Aleph Ω (Aleph Ω, Aleph Ω)]-multiplicity) (Aleph (non-ordinal-[Hyper Aleph Ω (Aleph Ω, Aleph Ω)]-multiplicity), Aleph (non-ordinal-[Hyper Aleph Ω (Aleph Ω, Aleph Ω)]-multiplicity))], Hyper Hyper Aleph (non-ordinal-[Hyper Aleph Ω (Aleph Ω, Aleph Ω)]-multiplicity) (Aleph (non-ordinal-[Hyper Aleph Ω (Aleph Ω, Aleph Ω)]-multiplicity), Aleph (non-ordinal-[Hyper Aleph Ω (Aleph Ω, Aleph Ω)]-multiplicity))[Hyper Aleph (non-ordinal-[Hyper Aleph Ω (Aleph Ω, Aleph Ω)]-multiplicity) (Aleph (non-ordinal-[Hyper Aleph Ω (Aleph Ω, Aleph Ω)]-multiplicity), Aleph (non-ordinal-[Hyper Aleph Ω (Aleph Ω, Aleph Ω)]-multiplicity)), Hyper Aleph (non-ordinal-[Hyper Aleph Ω (Aleph Ω, Aleph Ω)]-multiplicity) (Aleph (non-ordinal-[Hyper Aleph Ω (Aleph Ω, Aleph Ω)]-multiplicity), Aleph (non-ordinal-[Hyper Aleph Ω (Aleph Ω, Aleph Ω)]-multiplicity))]} { Hyper Hyper Hyper Aleph (non-ordinal-[Hyper Aleph Ω (Aleph Ω, Aleph Ω)]-multiplicity) (Aleph (non-ordinal-[Hyper Aleph Ω (Aleph Ω, Aleph Ω)]-multiplicity), Aleph (non-ordinal-[Hyper Aleph Ω (Aleph Ω, Aleph Ω)]-multiplicity))[Hyper Aleph (non-ordinal-[Hyper Aleph Ω (Aleph Ω, Aleph Ω)]-multiplicity) (Aleph (non-ordinal-[Hyper Aleph Ω (Aleph Ω, Aleph Ω)]-multiplicity), Aleph (non-ordinal-[Hyper Aleph Ω (Aleph Ω, Aleph Ω)]-multiplicity)), Hyper

Aleph (non-ordinal-[Hyper Aleph Ω (Aleph Ω, Aleph Ω)]-multiplicity) (Aleph (non-ordinal-[Hyper Aleph Ω (Aleph Ω, Aleph Ω)]-multiplicity), Aleph (non-ordinal-[Hyper Aleph Ω (Aleph Ω, Aleph Ω)]-multiplicity))] {Hyper Hyper Aleph (non-ordinal-[Hyper Aleph Ω (Aleph Ω, Aleph Ω)]-multiplicity) (Aleph (non-ordinal-[Hyper Aleph Ω (Aleph Ω, Aleph Ω)]-multiplicity), Aleph (non-ordinal-[Hyper Aleph Ω (Aleph Ω, Aleph Ω)]-multiplicity))[Hyper Aleph (non-ordinal-[Hyper Aleph Ω (Aleph Ω, Aleph Ω)]-multiplicity) (Aleph (non-ordinal-[Hyper Aleph Ω (Aleph Ω, Aleph Ω)]-multiplicity), Aleph (non-ordinal-[Hyper Aleph Ω (Aleph Ω, Aleph Ω)]-multiplicity)), Hyper Aleph (non-ordinal-[Hyper Aleph Ω (Aleph Ω, Aleph Ω)]-multiplicity) (Aleph (non-ordinal-[Hyper Aleph Ω (Aleph Ω, Aleph Ω)]-multiplicity), Aleph (non-ordinal-[Hyper Aleph Ω (Aleph Ω, Aleph Ω)]-multiplicity))], Hyper Hyper Aleph (non-ordinal-[Hyper Aleph Ω (Aleph Ω, Aleph Ω)]-multiplicity) (Aleph (non-ordinal-[Hyper Aleph Ω (Aleph Ω, Aleph Ω)]-multiplicity), Aleph (non-ordinal-[Hyper Aleph Ω (Aleph Ω, Aleph Ω)]-multiplicity))[Hyper Aleph (non-ordinal-[Hyper Aleph Ω (Aleph Ω, Aleph Ω)]-multiplicity) (Aleph (non-ordinal-[Hyper Aleph Ω (Aleph Ω, Aleph Ω)]-multiplicity), Aleph (non-ordinal-[Hyper Aleph Ω (Aleph Ω, Aleph Ω)]-multiplicity)), Hyper Aleph (non-ordinal-[Hyper Aleph Ω (Aleph Ω, Aleph Ω)]-multiplicity) (Aleph (non-ordinal-[Hyper Aleph Ω (Aleph Ω, Aleph Ω)]-multiplicity), Aleph (non-ordinal-[Hyper Aleph Ω (Aleph Ω, Aleph Ω)]-multiplicity))]}, Hyper Hyper Hyper Aleph (non-ordinal-[Hyper Aleph Ω (Aleph Ω, Aleph Ω)]-multiplicity) (Aleph (non-ordinal-[Hyper Aleph Ω (Aleph Ω, Aleph Ω)]-multiplicity), Aleph (non-ordinal-[Hyper Aleph Ω (Aleph Ω, Aleph Ω)]-multiplicity))[Hyper Aleph (non-ordinal-[Hyper Aleph Ω (Aleph Ω, Aleph Ω)]-multiplicity) (Aleph (non-ordinal-[Hyper Aleph Ω (Aleph Ω, Aleph Ω)]-multiplicity), Aleph (non-ordinal-[Hyper Aleph Ω (Aleph Ω, Aleph Ω)]-multiplicity)), Hyper Aleph (non-ordinal-[Hyper Aleph Ω (Aleph Ω, Aleph Ω)]-multiplicity) (Aleph (non-ordinal-[Hyper Aleph Ω (Aleph Ω, Aleph Ω)]-multiplicity), Aleph (non-ordinal-[Hyper Aleph Ω (Aleph Ω, Aleph Ω)]-multiplicity))] {Hyper Hyper Aleph (non-ordinal-[Hyper Aleph Ω (Aleph Ω, Aleph Ω)]-multiplicity) (Aleph (non-ordinal-[Hyper Aleph Ω (Aleph Ω, Aleph Ω)]-multiplicity), Aleph (non-ordinal-[Hyper Aleph Ω (Aleph Ω, Aleph Ω)]-multiplicity))[Hyper Aleph (non-ordinal-[Hyper Aleph Ω (Aleph Ω, Aleph Ω)]-multiplicity) (Aleph (non-ordinal-[Hyper Aleph Ω (Aleph Ω, Aleph Ω)]-multiplicity), Aleph (non-ordinal-[Hyper Aleph Ω (Aleph Ω, Aleph Ω)]-multiplicity)), Hyper Aleph (non-ordinal-[Hyper Aleph Ω (Aleph Ω, Aleph Ω)]-multiplicity) (Aleph (non-ordinal-[Hyper Aleph Ω (Aleph Ω, Aleph Ω)]-multiplicity), Aleph (non-ordinal-[Hyper Aleph Ω (Aleph Ω, Aleph Ω)]-multiplicity))], Hyper Hyper Aleph (non-ordinal-[Hyper Aleph Ω (Aleph Ω, Aleph Ω)]-multiplicity) (Aleph (non-ordinal-[Hyper Aleph Ω (Aleph Ω, Aleph Ω)]-multiplicity), Aleph (non-ordinal-[Hyper Aleph Ω (Aleph Ω, Aleph Ω)]-multiplicity))[Hyper Aleph (non-ordinal-[Hyper Aleph Ω (Aleph Ω, Aleph Ω)]-multiplicity) (Aleph (non-ordinal-[Hyper Aleph Ω (Aleph Ω, Aleph Ω)]-multiplicity), Aleph (non-ordinal-[Hyper Aleph Ω (Aleph Ω, Aleph Ω)]-multiplicity)), Hyper Aleph (non-ordinal-[Hyper Aleph Ω (Aleph Ω, Aleph Ω)]-multiplicity) (Aleph (non-ordinal-[Hyper Aleph Ω (Aleph Ω, Aleph Ω)]-multiplicity), Aleph (non-ordinal-[Hyper Aleph Ω (Aleph Ω, Aleph Ω)]-multiplicity))]}}{ Hyper Hyper Hyper Hyper Aleph (non-ordinal-[Hyper Aleph Ω (Aleph Ω, Aleph Ω)]-multiplicity) (Aleph (non-ordinal-[Hyper Aleph Ω (Aleph Ω, Aleph Ω)]-multiplicity), Aleph (non-ordinal-[Hyper Aleph Ω (Aleph Ω, Aleph Ω)]-multiplicity))[Hyper Aleph (non-ordinal-[Hyper Aleph Ω (Aleph Ω, Aleph Ω)]-multiplicity) (Aleph (non-ordinal-[Hyper Aleph Ω (Aleph Ω, Aleph Ω)]-multiplicity), Aleph (non-ordinal-[Hyper Aleph Ω (Aleph Ω, Aleph Ω)]-multiplicity)), Hyper Aleph (non-ordinal-[Hyper Aleph Ω (Aleph Ω, Aleph Ω)]-multiplicity) (Aleph (non-ordinal-[Hyper Aleph Ω (Aleph Ω, Aleph Ω)]-multiplicity), Aleph (non-ordinal-[Hyper Aleph Ω (Aleph Ω, Aleph Ω)]-multiplicity))] {Hyper Hyper Aleph (non-ordinal-[Hyper Aleph Ω (Aleph Ω, Aleph Ω)]-multiplicity) (Aleph (non-ordinal-[Hyper Aleph Ω (Aleph Ω, Aleph Ω)]-multiplicity), Aleph (non-ordinal-[Hyper Aleph Ω (Aleph Ω, Aleph Ω)]-multiplicity))[Hyper Aleph (non-ordinal-[Hyper Aleph Ω (Aleph Ω, Aleph Ω)]-multiplicity) (Aleph (non-ordinal-[Hyper Aleph Ω (Aleph Ω, Aleph Ω)]-multiplicity), Aleph (non-ordinal-[Hyper Aleph Ω (Aleph Ω, Aleph Ω)]-multiplicity)), Hyper Aleph (non-ordinal-[Hyper Aleph Ω (Aleph Ω, Aleph Ω)]-multiplicity) (Aleph (non-ordinal-[Hyper Aleph Ω (Aleph Ω, Aleph Ω)]-multiplicity), Aleph (non-ordinal-[Hyper Aleph Ω (Aleph Ω, Aleph Ω)]-multiplicity))], Hyper Hyper Aleph (non-ordinal-[Hyper Aleph Ω (Aleph Ω, Aleph Ω)]-multiplicity) (Aleph (non-ordinal-[Hyper Aleph Ω (Aleph Ω, Aleph Ω)]-multiplicity), Aleph (non-ordinal-[Hyper Aleph Ω (Aleph Ω, Aleph Ω)]-multiplicity))[Hyper Aleph (non-ordinal-[Hyper Aleph Ω (Aleph Ω, Aleph Ω)]-multiplicity) (Aleph (non-ordinal-[Hyper Aleph Ω (Aleph Ω, Aleph Ω)]-multiplicity), Aleph (non-ordinal-[Hyper Aleph Ω (Aleph Ω, Aleph Ω)]-multiplicity)), Hyper Aleph (non-ordinal-[Hyper Aleph Ω (Aleph Ω, Aleph Ω)]-multiplicity) (Aleph (non-ordinal-[Hyper Aleph Ω (Aleph Ω, Aleph Ω)]-multiplicity), Aleph (non-ordinal-[Hyper Aleph Ω (Aleph Ω, Aleph Ω)]-multiplicity))]} { Hyper Hyper Hyper Aleph (non-ordinal-[Hyper Aleph Ω (Aleph Ω, Aleph Ω)]-multiplicity) (Aleph (non-ordinal-[Hyper Aleph Ω (Aleph Ω, Aleph Ω)]-multiplicity), Aleph (non-ordinal-[Hyper Aleph Ω (Aleph Ω, Aleph Ω)]-multiplicity))[Hyper Aleph (non-ordinal-[Hyper Aleph Ω (Aleph Ω, Aleph Ω)]-multiplicity) (Aleph (non-ordinal-[Hyper Aleph Ω (Aleph Ω, Aleph Ω)]-multiplicity), Aleph (non-ordinal-[Hyper Aleph Ω (Aleph Ω, Aleph Ω)]-multiplicity)), Hyper Aleph (non-ordinal-[Hyper Aleph Ω (Aleph Ω, Aleph Ω)]-multiplicity) (Aleph (non-ordinal-[Hyper Aleph Ω (Aleph Ω, Aleph Ω)]-multiplicity), Aleph (non-ordinal-[Hyper Aleph Ω (Aleph Ω, Aleph Ω)]-multiplicity))] {Hyper Hyper Aleph (non-ordinal-[Hyper Aleph Ω (Aleph Ω, Aleph Ω)]-multiplicity) (Aleph (non-ordinal-[Hyper Aleph Ω (Aleph Ω, Aleph Ω)]-multiplicity), Aleph (non-ordinal-[Hyper Aleph Ω (Aleph Ω, Aleph Ω)]-multiplicity))[Hyper Aleph (non-ordinal-[Hyper Aleph Ω (Aleph Ω, Aleph Ω)]-multiplicity) (Aleph (non-ordinal-[Hyper Aleph Ω (Aleph Ω, Aleph Ω)]-multiplicity), Aleph (non-ordinal-[Hyper Aleph Ω (Aleph Ω, Aleph Ω)]-multiplicity)), Hyper Aleph (non-ordinal-[Hyper Aleph Ω (Aleph Ω, Aleph Ω)]-multiplicity) (Aleph (non-ordinal-[Hyper Aleph Ω (Aleph Ω, Aleph Ω)]-multiplicity), Aleph (non-ordinal-[Hyper Aleph Ω (Aleph Ω, Aleph Ω)]-multiplicity))], Hyper Hyper Aleph (non-ordinal-[Hyper Aleph Ω (Aleph Ω, Aleph Ω)]-

multiplicity) (Aleph (non-ordinal-[Hyper Aleph Ω (Aleph Ω, Aleph Ω)]-multiplicity), Aleph (non-ordinal-[Hyper Aleph Ω (Aleph Ω, Aleph Ω)]-multiplicity))[Hyper Aleph (non-ordinal-[Hyper Aleph Ω (Aleph Ω, Aleph Ω)]-multiplicity) (Aleph (non-ordinal-[Hyper Aleph Ω (Aleph Ω, Aleph Ω)]-multiplicity), Aleph (non-ordinal-[Hyper Aleph Ω (Aleph Ω, Aleph Ω)]-multiplicity)), Hyper Aleph (non-ordinal-[Hyper Aleph Ω (Aleph Ω, Aleph Ω)]-multiplicity) (Aleph (non-ordinal-[Hyper Aleph Ω (Aleph Ω, Aleph Ω)]-multiplicity), Aleph (non-ordinal-[Hyper Aleph Ω (Aleph Ω, Aleph Ω)]-multiplicity))]}, Hyper Hyper Hyper Aleph (non-ordinal-[Hyper Aleph Ω (Aleph Ω, Aleph Ω)]-multiplicity) (Aleph (non-ordinal-[Hyper Aleph Ω (Aleph Ω, Aleph Ω)]-multiplicity), Aleph (non-ordinal-[Hyper Aleph Ω (Aleph Ω, Aleph Ω)]-multiplicity))[Hyper Aleph (non-ordinal-[Hyper Aleph Ω (Aleph Ω, Aleph Ω)]-multiplicity) (Aleph (non-ordinal-[Hyper Aleph Ω (Aleph Ω, Aleph Ω)]-multiplicity), Aleph (non-ordinal-[Hyper Aleph Ω (Aleph Ω, Aleph Ω)]-multiplicity)), Hyper Aleph (non-ordinal-[Hyper Aleph Ω (Aleph Ω, Aleph Ω)]-multiplicity) (Aleph (non-ordinal-[Hyper Aleph Ω (Aleph Ω, Aleph Ω)]-multiplicity), Aleph (non-ordinal-[Hyper Aleph Ω (Aleph Ω, Aleph Ω)]-multiplicity))] {Hyper Hyper Aleph (non-ordinal-[Hyper Aleph Ω (Aleph Ω, Aleph Ω)]-multiplicity) (Aleph (non-ordinal-[Hyper Aleph Ω (Aleph Ω, Aleph Ω)]-multiplicity), Aleph (non-ordinal-[Hyper Aleph Ω (Aleph Ω, Aleph Ω)]-multiplicity))[Hyper Aleph (non-ordinal-[Hyper Aleph Ω (Aleph Ω, Aleph Ω)]-multiplicity) (Aleph (non-ordinal-[Hyper Aleph Ω (Aleph Ω, Aleph Ω)]-multiplicity), Aleph (non-ordinal-[Hyper Aleph Ω (Aleph Ω, Aleph Ω)]-multiplicity)), Hyper Aleph (non-ordinal-[Hyper Aleph Ω (Aleph Ω, Aleph Ω)]-multiplicity) (Aleph (non-ordinal-[Hyper Aleph Ω (Aleph Ω, Aleph Ω)]-multiplicity), Aleph (non-ordinal-[Hyper Aleph Ω (Aleph Ω, Aleph Ω)]-multiplicity))], Hyper Hyper Aleph (non-ordinal-[Hyper Aleph Ω (Aleph Ω, Aleph Ω)]-multiplicity) (Aleph (non-ordinal-[Hyper Aleph Ω (Aleph Ω, Aleph Ω)]-multiplicity), Aleph (non-ordinal-[Hyper Aleph Ω (Aleph Ω, Aleph Ω)]-multiplicity))[Hyper Aleph (non-ordinal-[Hyper Aleph Ω (Aleph Ω, Aleph Ω)]-multiplicity) (Aleph (non-ordinal-[Hyper Aleph Ω (Aleph Ω, Aleph Ω)]-multiplicity), Aleph (non-ordinal-[Hyper Aleph Ω (Aleph Ω, Aleph Ω)]-multiplicity)), Hyper Aleph (non-ordinal-[Hyper Aleph Ω (Aleph Ω, Aleph Ω)]-multiplicity) (Aleph (non-ordinal-[Hyper Aleph Ω (Aleph Ω, Aleph Ω)]-multiplicity), Aleph (non-ordinal-[Hyper Aleph Ω (Aleph Ω, Aleph Ω)]-multiplicity))]}}, Hyper Hyper Hyper Hyper Aleph (non-ordinal-[Hyper Aleph Ω (Aleph Ω, Aleph Ω)]-multiplicity) (Aleph (non-ordinal-[Hyper Aleph Ω (Aleph Ω, Aleph Ω)]-multiplicity), Aleph (non-ordinal-[Hyper Aleph Ω (Aleph Ω, Aleph Ω)]-multiplicity))[Hyper Aleph (non-ordinal-[Hyper Aleph Ω (Aleph Ω, Aleph Ω)]-multiplicity) (Aleph (non-ordinal-[Hyper Aleph Ω (Aleph Ω, Aleph Ω)]-multiplicity), Aleph (non-ordinal-[Hyper Aleph Ω (Aleph Ω, Aleph Ω)]-multiplicity)), Hyper Aleph (non-ordinal-[Hyper Aleph Ω (Aleph Ω, Aleph Ω)]-multiplicity) (Aleph (non-ordinal-[Hyper Aleph Ω (Aleph Ω, Aleph Ω)]-multiplicity), Aleph (non-ordinal-[Hyper Aleph Ω (Aleph Ω, Aleph Ω)]-multiplicity))] {Hyper Hyper Aleph (non-ordinal-[Hyper Aleph Ω (Aleph Ω, Aleph Ω)]-multiplicity) (Aleph (non-ordinal-[Hyper Aleph Ω (Aleph Ω, Aleph Ω)]-multiplicity), Aleph (non-ordinal-[Hyper Aleph Ω (Aleph Ω, Aleph Ω)]-

multiplicity))[Hyper Aleph (non-ordinal-[Hyper Aleph Ω (Aleph Ω, Aleph Ω)]-multiplicity) (Aleph (non-ordinal-[Hyper Aleph Ω (Aleph Ω, Aleph Ω)]-multiplicity), Aleph (non-ordinal-[Hyper Aleph Ω (Aleph Ω, Aleph Ω)]-multiplicity)), Hyper Aleph (non-ordinal-[Hyper Aleph Ω (Aleph Ω, Aleph Ω)]-multiplicity) (Aleph (non-ordinal-[Hyper Aleph Ω (Aleph Ω, Aleph Ω)]-multiplicity), Aleph (non-ordinal-[Hyper Aleph Ω (Aleph Ω, Aleph Ω)]-multiplicity))], Hyper Hyper Aleph (non-ordinal-[Hyper Aleph Ω (Aleph Ω, Aleph Ω)]-multiplicity) (Aleph (non-ordinal-[Hyper Aleph Ω (Aleph Ω, Aleph Ω)]-multiplicity), Aleph (non-ordinal-[Hyper Aleph Ω (Aleph Ω, Aleph Ω)]-multiplicity))[Hyper Aleph (non-ordinal-[Hyper Aleph Ω (Aleph Ω, Aleph Ω)]-multiplicity) (Aleph (non-ordinal-[Hyper Aleph Ω (Aleph Ω, Aleph Ω)]-multiplicity), Aleph (non-ordinal-[Hyper Aleph Ω (Aleph Ω, Aleph Ω)]-multiplicity)), Hyper Aleph (non-ordinal-[Hyper Aleph Ω (Aleph Ω, Aleph Ω)]-multiplicity) (Aleph (non-ordinal-[Hyper Aleph Ω (Aleph Ω, Aleph Ω)]-multiplicity), Aleph (non-ordinal-[Hyper Aleph Ω (Aleph Ω, Aleph Ω)]-multiplicity))]} { Hyper Hyper Hyper Aleph (non-ordinal-[Hyper Aleph Ω (Aleph Ω, Aleph Ω)]-multiplicity) (Aleph (non-ordinal-[Hyper Aleph Ω (Aleph Ω, Aleph Ω)]-multiplicity), Aleph (non-ordinal-[Hyper Aleph Ω (Aleph Ω, Aleph Ω)]-multiplicity))[Hyper Aleph (non-ordinal-[Hyper Aleph Ω (Aleph Ω, Aleph Ω)]-multiplicity) (Aleph (non-ordinal-[Hyper Aleph Ω (Aleph Ω, Aleph Ω)]-multiplicity), Aleph (non-ordinal-[Hyper Aleph Ω (Aleph Ω, Aleph Ω)]-multiplicity)), Hyper Aleph (non-ordinal-[Hyper Aleph Ω (Aleph Ω, Aleph Ω)]-multiplicity) (Aleph (non-ordinal-[Hyper Aleph Ω (Aleph Ω, Aleph Ω)]-multiplicity), Aleph (non-ordinal-[Hyper Aleph Ω (Aleph Ω, Aleph Ω)]-multiplicity))] {Hyper Hyper Aleph (non-ordinal-[Hyper Aleph Ω (Aleph Ω, Aleph Ω)]-multiplicity) (Aleph (non-ordinal-[Hyper Aleph Ω (Aleph Ω, Aleph Ω)]-multiplicity), Aleph (non-ordinal-[Hyper Aleph Ω (Aleph Ω, Aleph Ω)]-multiplicity))[Hyper Aleph (non-ordinal-[Hyper Aleph Ω (Aleph Ω, Aleph Ω)]-multiplicity) (Aleph (non-ordinal-[Hyper Aleph Ω (Aleph Ω, Aleph Ω)]-multiplicity), Aleph (non-ordinal-[Hyper Aleph Ω (Aleph Ω, Aleph Ω)]-multiplicity)), Hyper Aleph (non-ordinal-[Hyper Aleph Ω (Aleph Ω, Aleph Ω)]-multiplicity) (Aleph (non-ordinal-[Hyper Aleph Ω (Aleph Ω, Aleph Ω)]-multiplicity), Aleph (non-ordinal-[Hyper Aleph Ω (Aleph Ω, Aleph Ω)]-multiplicity))], Hyper Hyper Aleph (non-ordinal-[Hyper Aleph Ω (Aleph Ω, Aleph Ω)]-multiplicity) (Aleph (non-ordinal-[Hyper Aleph Ω (Aleph Ω, Aleph Ω)]-multiplicity), Aleph (non-ordinal-[Hyper Aleph Ω (Aleph Ω, Aleph Ω)]-multiplicity))[Hyper Aleph (non-ordinal-[Hyper Aleph Ω (Aleph Ω, Aleph Ω)]-multiplicity) (Aleph (non-ordinal-[Hyper Aleph Ω (Aleph Ω, Aleph Ω)]-multiplicity), Aleph (non-ordinal-[Hyper Aleph Ω (Aleph Ω, Aleph Ω)]-multiplicity)), Hyper Aleph (non-ordinal-[Hyper Aleph Ω (Aleph Ω, Aleph Ω)]-multiplicity) (Aleph (non-ordinal-[Hyper Aleph Ω (Aleph Ω, Aleph Ω)]-multiplicity), Aleph (non-ordinal-[Hyper Aleph Ω (Aleph Ω, Aleph Ω)]-multiplicity))]}, Hyper Hyper Hyper Aleph (non-ordinal-[Hyper Aleph Ω (Aleph Ω, Aleph Ω)]-multiplicity) (Aleph (non-ordinal-[Hyper Aleph Ω (Aleph Ω, Aleph Ω)]-multiplicity), Aleph (non-ordinal-[Hyper Aleph Ω (Aleph Ω, Aleph Ω)]-multiplicity))[Hyper Aleph (non-ordinal-[Hyper Aleph Ω (Aleph Ω, Aleph Ω)]-multiplicity) (Aleph (non-ordinal-[Hyper Aleph Ω (Aleph Ω, Aleph Ω)]-

multiplicity), Aleph (non-ordinal-[Hyper Aleph Ω (Aleph Ω, Aleph Ω)]-multiplicity)), Hyper Aleph (non-ordinal-[Hyper Aleph Ω (Aleph Ω, Aleph Ω)]-multiplicity) (Aleph (non-ordinal-[Hyper Aleph Ω (Aleph Ω, Aleph Ω)]-multiplicity), Aleph (non-ordinal-[Hyper Aleph Ω (Aleph Ω, Aleph Ω)]-multiplicity))] {Hyper Hyper Aleph (non-ordinal-[Hyper Aleph Ω (Aleph Ω, Aleph Ω)]-multiplicity) (Aleph (non-ordinal-[Hyper Aleph Ω (Aleph Ω, Aleph Ω)]-multiplicity), Aleph (non-ordinal-[Hyper Aleph Ω (Aleph Ω, Aleph Ω)]-multiplicity))[Hyper Aleph (non-ordinal-[Hyper Aleph Ω (Aleph Ω, Aleph Ω)]-multiplicity) (Aleph (non-ordinal-[Hyper Aleph Ω (Aleph Ω, Aleph Ω)]-multiplicity), Aleph (non-ordinal-[Hyper Aleph Ω (Aleph Ω, Aleph Ω)]-multiplicity)), Hyper Aleph (non-ordinal-[Hyper Aleph Ω (Aleph Ω, Aleph Ω)]-multiplicity) (Aleph (non-ordinal-[Hyper Aleph Ω (Aleph Ω, Aleph Ω)]-multiplicity), Aleph (non-ordinal-[Hyper Aleph Ω (Aleph Ω, Aleph Ω)]-multiplicity))], Hyper Hyper Aleph (non-ordinal-[Hyper Aleph Ω (Aleph Ω, Aleph Ω)]-multiplicity) (Aleph (non-ordinal-[Hyper Aleph Ω (Aleph Ω, Aleph Ω)]-multiplicity), Aleph (non-ordinal-[Hyper Aleph Ω (Aleph Ω, Aleph Ω)]-multiplicity))[Hyper Aleph (non-ordinal-[Hyper Aleph Ω (Aleph Ω, Aleph Ω)]-multiplicity) (Aleph (non-ordinal-[Hyper Aleph Ω (Aleph Ω, Aleph Ω)]-multiplicity), Aleph (non-ordinal-[Hyper Aleph Ω (Aleph Ω, Aleph Ω)]-multiplicity)), Hyper Aleph (non-ordinal-[Hyper Aleph Ω (Aleph Ω, Aleph Ω)]-multiplicity) (Aleph (non-ordinal-[Hyper Aleph Ω (Aleph Ω, Aleph Ω)]-multiplicity), Aleph (non-ordinal-[Hyper Aleph Ω (Aleph Ω, Aleph Ω)]-multiplicity))]}}}}{ Hyper Hyper Hyper Hyper Hyper Hyper Aleph (non-ordinal-[Hyper Aleph Ω (Aleph Ω, Aleph Ω)]-multiplicity) (Aleph (non-ordinal-[Hyper Aleph Ω (Aleph Ω, Aleph Ω)]-multiplicity), Aleph (non-ordinal-[Hyper Aleph Ω (Aleph Ω, Aleph Ω)]-multiplicity))[Hyper Aleph (non-ordinal-[Hyper Aleph Ω (Aleph Ω, Aleph Ω)]-multiplicity) (Aleph (non-ordinal-[Hyper Aleph Ω (Aleph Ω, Aleph Ω)]-multiplicity), Aleph (non-ordinal-[Hyper Aleph Ω (Aleph Ω, Aleph Ω)]-multiplicity)), Hyper Aleph (non-ordinal-[Hyper Aleph Ω (Aleph Ω, Aleph Ω)]-multiplicity) (Aleph (non-ordinal-[Hyper Aleph Ω (Aleph Ω, Aleph Ω)]-multiplicity), Aleph (non-ordinal-[Hyper Aleph Ω (Aleph Ω, Aleph Ω)]-multiplicity))] {Hyper Hyper Aleph (non-ordinal-[Hyper Aleph Ω (Aleph Ω, Aleph Ω)]-multiplicity) (Aleph (non-ordinal-[Hyper Aleph Ω (Aleph Ω, Aleph Ω)]-multiplicity), Aleph (non-ordinal-[Hyper Aleph Ω (Aleph Ω, Aleph Ω)]-multiplicity))[Hyper Aleph (non-ordinal-[Hyper Aleph Ω (Aleph Ω, Aleph Ω)]-multiplicity) (Aleph (non-ordinal-[Hyper Aleph Ω (Aleph Ω, Aleph Ω)]-multiplicity), Aleph (non-ordinal-[Hyper Aleph Ω (Aleph Ω, Aleph Ω)]-multiplicity)), Hyper Aleph (non-ordinal-[Hyper Aleph Ω (Aleph Ω, Aleph Ω)]-multiplicity) (Aleph (non-ordinal-[Hyper Aleph Ω (Aleph Ω, Aleph Ω)]-multiplicity), Aleph (non-ordinal-[Hyper Aleph Ω (Aleph Ω, Aleph Ω)]-multiplicity))], Hyper Hyper Aleph (non-ordinal-[Hyper Aleph Ω (Aleph Ω, Aleph Ω)]-multiplicity) (Aleph (non-ordinal-[Hyper Aleph Ω (Aleph Ω, Aleph Ω)]-multiplicity), Aleph (non-ordinal-[Hyper Aleph Ω (Aleph Ω, Aleph Ω)]-multiplicity))[Hyper Aleph (non-ordinal-[Hyper Aleph Ω (Aleph Ω, Aleph Ω)]-multiplicity) (Aleph (non-ordinal-[Hyper Aleph Ω (Aleph Ω, Aleph Ω)]-multiplicity), Aleph (non-ordinal-[Hyper Aleph Ω (Aleph Ω, Aleph Ω)]-multiplicity)), Hyper Aleph (non-ordinal-[Hyper Aleph Ω (Aleph Ω, Aleph Ω)]-

multiplicity) (Aleph (non-ordinal-[Hyper Aleph Ω (Aleph Ω, Aleph Ω)]-multiplicity), Aleph (non-ordinal-[Hyper Aleph Ω (Aleph Ω, Aleph Ω)]-multiplicity))]} { Hyper Hyper Hyper Aleph (non-ordinal-[Hyper Aleph Ω (Aleph Ω, Aleph Ω)]-multiplicity) (Aleph (non-ordinal-[Hyper Aleph Ω (Aleph Ω, Aleph Ω)]-multiplicity), Aleph (non-ordinal-[Hyper Aleph Ω (Aleph Ω, Aleph Ω)]-multiplicity))[Hyper Aleph (non-ordinal-[Hyper Aleph Ω (Aleph Ω, Aleph Ω)]-multiplicity) (Aleph (non-ordinal-[Hyper Aleph Ω (Aleph Ω, Aleph Ω)]-multiplicity), Aleph (non-ordinal-[Hyper Aleph Ω (Aleph Ω, Aleph Ω)]-multiplicity)), Hyper Aleph (non-ordinal-[Hyper Aleph Ω (Aleph Ω, Aleph Ω)]-multiplicity) (Aleph (non-ordinal-[Hyper Aleph Ω (Aleph Ω, Aleph Ω)]-multiplicity), Aleph (non-ordinal-[Hyper Aleph Ω (Aleph Ω, Aleph Ω)]-multiplicity))] {Hyper Hyper Aleph (non-ordinal-[Hyper Aleph Ω (Aleph Ω, Aleph Ω)]-multiplicity) (Aleph (non-ordinal-[Hyper Aleph Ω (Aleph Ω, Aleph Ω)]-multiplicity), Aleph (non-ordinal-[Hyper Aleph Ω (Aleph Ω, Aleph Ω)]-multiplicity))[Hyper Aleph (non-ordinal-[Hyper Aleph Ω (Aleph Ω, Aleph Ω)]-multiplicity) (Aleph (non-ordinal-[Hyper Aleph Ω (Aleph Ω, Aleph Ω)]-multiplicity), Aleph (non-ordinal-[Hyper Aleph Ω (Aleph Ω, Aleph Ω)]-multiplicity)), Hyper Aleph (non-ordinal-[Hyper Aleph Ω (Aleph Ω, Aleph Ω)]-multiplicity) (Aleph (non-ordinal-[Hyper Aleph Ω (Aleph Ω, Aleph Ω)]-multiplicity), Aleph (non-ordinal-[Hyper Aleph Ω (Aleph Ω, Aleph Ω)]-multiplicity))], Hyper Hyper Aleph (non-ordinal-[Hyper Aleph Ω (Aleph Ω, Aleph Ω)]-multiplicity) (Aleph (non-ordinal-[Hyper Aleph Ω (Aleph Ω, Aleph Ω)]-multiplicity), Aleph (non-ordinal-[Hyper Aleph Ω (Aleph Ω, Aleph Ω)]-multiplicity))[Hyper Aleph (non-ordinal-[Hyper Aleph Ω (Aleph Ω, Aleph Ω)]-multiplicity) (Aleph (non-ordinal-[Hyper Aleph Ω (Aleph Ω, Aleph Ω)]-multiplicity), Aleph (non-ordinal-[Hyper Aleph Ω (Aleph Ω, Aleph Ω)]-multiplicity)), Hyper Aleph (non-ordinal-[Hyper Aleph Ω (Aleph Ω, Aleph Ω)]-multiplicity) (Aleph (non-ordinal-[Hyper Aleph Ω (Aleph Ω, Aleph Ω)]-multiplicity), Aleph (non-ordinal-[Hyper Aleph Ω (Aleph Ω, Aleph Ω)]-multiplicity))]}, Hyper Hyper Hyper Aleph (non-ordinal-[Hyper Aleph Ω (Aleph Ω, Aleph Ω)]-multiplicity) (Aleph (non-ordinal-[Hyper Aleph Ω (Aleph Ω, Aleph Ω)]-multiplicity), Aleph (non-ordinal-[Hyper Aleph Ω (Aleph Ω, Aleph Ω)]-multiplicity))[Hyper Aleph (non-ordinal-[Hyper Aleph Ω (Aleph Ω, Aleph Ω)]-multiplicity) (Aleph (non-ordinal-[Hyper Aleph Ω (Aleph Ω, Aleph Ω)]-multiplicity), Aleph (non-ordinal-[Hyper Aleph Ω (Aleph Ω, Aleph Ω)]-multiplicity)), Hyper Aleph (non-ordinal-[Hyper Aleph Ω (Aleph Ω, Aleph Ω)]-multiplicity) (Aleph (non-ordinal-[Hyper Aleph Ω (Aleph Ω, Aleph Ω)]-multiplicity), Aleph (non-ordinal-[Hyper Aleph Ω (Aleph Ω, Aleph Ω)]-multiplicity))] {Hyper Hyper Aleph (non-ordinal-[Hyper Aleph Ω (Aleph Ω, Aleph Ω)]-multiplicity) (Aleph (non-ordinal-[Hyper Aleph Ω (Aleph Ω, Aleph Ω)]-multiplicity), Aleph (non-ordinal-[Hyper Aleph Ω (Aleph Ω, Aleph Ω)]-multiplicity))[Hyper Aleph (non-ordinal-[Hyper Aleph Ω (Aleph Ω, Aleph Ω)]-multiplicity) (Aleph (non-ordinal-[Hyper Aleph Ω (Aleph Ω, Aleph Ω)]-multiplicity), Aleph (non-ordinal-[Hyper Aleph Ω (Aleph Ω, Aleph Ω)]-multiplicity)), Hyper Aleph (non-ordinal-[Hyper Aleph Ω (Aleph Ω, Aleph Ω)]-multiplicity) (Aleph (non-ordinal-[Hyper Aleph Ω (Aleph Ω, Aleph Ω)]-multiplicity), Aleph (non-ordinal-[Hyper Aleph Ω (Aleph Ω, Aleph Ω)]-

multiplicity))], Hyper Hyper Aleph (non-ordinal-[Hyper Aleph Ω (Aleph Ω, Aleph Ω)]-multiplicity) (Aleph (non-ordinal-[Hyper Aleph Ω (Aleph Ω, Aleph Ω)]-multiplicity), Aleph (non-ordinal-[Hyper Aleph Ω (Aleph Ω, Aleph Ω)]-multiplicity))[Hyper Aleph (non-ordinal-[Hyper Aleph Ω (Aleph Ω, Aleph Ω)]-multiplicity) (Aleph (non-ordinal-[Hyper Aleph Ω (Aleph Ω, Aleph Ω)]-multiplicity), Aleph (non-ordinal-[Hyper Aleph Ω (Aleph Ω, Aleph Ω)]-multiplicity)), Hyper Aleph (non-ordinal-[Hyper Aleph Ω (Aleph Ω, Aleph Ω)]-multiplicity) (Aleph (non-ordinal-[Hyper Aleph Ω (Aleph Ω, Aleph Ω)]-multiplicity), Aleph (non-ordinal-[Hyper Aleph Ω (Aleph Ω, Aleph Ω)]-multiplicity))]}}{ Hyper Hyper Hyper Hyper Aleph (non-ordinal-[Hyper Aleph Ω (Aleph Ω, Aleph Ω)]-multiplicity) (Aleph (non-ordinal-[Hyper Aleph Ω (Aleph Ω, Aleph Ω)]-multiplicity), Aleph (non-ordinal-[Hyper Aleph Ω (Aleph Ω, Aleph Ω)]-multiplicity))[Hyper Aleph (non-ordinal-[Hyper Aleph Ω (Aleph Ω, Aleph Ω)]-multiplicity) (Aleph (non-ordinal-[Hyper Aleph Ω (Aleph Ω, Aleph Ω)]-multiplicity), Aleph (non-ordinal-[Hyper Aleph Ω (Aleph Ω, Aleph Ω)]-multiplicity)), Hyper Aleph (non-ordinal-[Hyper Aleph Ω (Aleph Ω, Aleph Ω)]-multiplicity) (Aleph (non-ordinal-[Hyper Aleph Ω (Aleph Ω, Aleph Ω)]-multiplicity), Aleph (non-ordinal-[Hyper Aleph Ω (Aleph Ω, Aleph Ω)]-multiplicity))] {Hyper Hyper Aleph (non-ordinal-[Hyper Aleph Ω (Aleph Ω, Aleph Ω)]-multiplicity) (Aleph (non-ordinal-[Hyper Aleph Ω (Aleph Ω, Aleph Ω)]-multiplicity), Aleph (non-ordinal-[Hyper Aleph Ω (Aleph Ω, Aleph Ω)]-multiplicity))[Hyper Aleph (non-ordinal-[Hyper Aleph Ω (Aleph Ω, Aleph Ω)]-multiplicity) (Aleph (non-ordinal-[Hyper Aleph Ω (Aleph Ω, Aleph Ω)]-multiplicity), Aleph (non-ordinal-[Hyper Aleph Ω (Aleph Ω, Aleph Ω)]-multiplicity)), Hyper Aleph (non-ordinal-[Hyper Aleph Ω (Aleph Ω, Aleph Ω)]-multiplicity) (Aleph (non-ordinal-[Hyper Aleph Ω (Aleph Ω, Aleph Ω)]-multiplicity), Aleph (non-ordinal-[Hyper Aleph Ω (Aleph Ω, Aleph Ω)]-multiplicity))], Hyper Hyper Aleph (non-ordinal-[Hyper Aleph Ω (Aleph Ω, Aleph Ω)]-multiplicity) (Aleph (non-ordinal-[Hyper Aleph Ω (Aleph Ω, Aleph Ω)]-multiplicity), Aleph (non-ordinal-[Hyper Aleph Ω (Aleph Ω, Aleph Ω)]-multiplicity))[Hyper Aleph (non-ordinal-[Hyper Aleph Ω (Aleph Ω, Aleph Ω)]-multiplicity) (Aleph (non-ordinal-[Hyper Aleph Ω (Aleph Ω, Aleph Ω)]-multiplicity), Aleph (non-ordinal-[Hyper Aleph Ω (Aleph Ω, Aleph Ω)]-multiplicity)), Hyper Aleph (non-ordinal-[Hyper Aleph Ω (Aleph Ω, Aleph Ω)]-multiplicity) (Aleph (non-ordinal-[Hyper Aleph Ω (Aleph Ω, Aleph Ω)]-multiplicity), Aleph (non-ordinal-[Hyper Aleph Ω (Aleph Ω, Aleph Ω)]-multiplicity))]} { Hyper Hyper Hyper Aleph (non-ordinal-[Hyper Aleph Ω (Aleph Ω, Aleph Ω)]-multiplicity) (Aleph (non-ordinal-[Hyper Aleph Ω (Aleph Ω, Aleph Ω)]-multiplicity), Aleph (non-ordinal-[Hyper Aleph Ω (Aleph Ω, Aleph Ω)]-multiplicity))[Hyper Aleph (non-ordinal-[Hyper Aleph Ω (Aleph Ω, Aleph Ω)]-multiplicity) (Aleph (non-ordinal-[Hyper Aleph Ω (Aleph Ω, Aleph Ω)]-multiplicity), Aleph (non-ordinal-[Hyper Aleph Ω (Aleph Ω, Aleph Ω)]-multiplicity)), Hyper Aleph (non-ordinal-[Hyper Aleph Ω (Aleph Ω, Aleph Ω)]-multiplicity) (Aleph (non-ordinal-[Hyper Aleph Ω (Aleph Ω, Aleph Ω)]-multiplicity), Aleph (non-ordinal-[Hyper Aleph Ω (Aleph Ω, Aleph Ω)]-multiplicity))] {Hyper Hyper Aleph (non-ordinal-[Hyper Aleph Ω (Aleph Ω, Aleph Ω)]-multiplicity) (Aleph (non-ordinal-[Hyper Aleph Ω (Aleph Ω, Aleph

Ω)]-multiplicity), Aleph (non-ordinal-[Hyper Aleph Ω (Aleph Ω, Aleph Ω)]-multiplicity))[Hyper Aleph (non-ordinal-[Hyper Aleph Ω (Aleph Ω, Aleph Ω)]-multiplicity) (Aleph (non-ordinal-[Hyper Aleph Ω (Aleph Ω, Aleph Ω)]-multiplicity), Aleph (non-ordinal-[Hyper Aleph Ω (Aleph Ω, Aleph Ω)]-multiplicity)), Hyper Aleph (non-ordinal-[Hyper Aleph Ω (Aleph Ω, Aleph Ω)]-multiplicity) (Aleph (non-ordinal-[Hyper Aleph Ω (Aleph Ω, Aleph Ω)]-multiplicity), Aleph (non-ordinal-[Hyper Aleph Ω (Aleph Ω, Aleph Ω)]-multiplicity))], Hyper Hyper Aleph (non-ordinal-[Hyper Aleph Ω (Aleph Ω, Aleph Ω)]-multiplicity) (Aleph (non-ordinal-[Hyper Aleph Ω (Aleph Ω, Aleph Ω)]-multiplicity), Aleph (non-ordinal-[Hyper Aleph Ω (Aleph Ω, Aleph Ω)]-multiplicity))[Hyper Aleph (non-ordinal-[Hyper Aleph Ω (Aleph Ω, Aleph Ω)]-multiplicity) (Aleph (non-ordinal-[Hyper Aleph Ω (Aleph Ω, Aleph Ω)]-multiplicity), Aleph (non-ordinal-[Hyper Aleph Ω (Aleph Ω, Aleph Ω)]-multiplicity)), Hyper Aleph (non-ordinal-[Hyper Aleph Ω (Aleph Ω, Aleph Ω)]-multiplicity) (Aleph (non-ordinal-[Hyper Aleph Ω (Aleph Ω, Aleph Ω)]-multiplicity), Aleph (non-ordinal-[Hyper Aleph Ω (Aleph Ω, Aleph Ω)]-multiplicity))]}, Hyper Hyper Hyper Aleph (non-ordinal-[Hyper Aleph Ω (Aleph Ω, Aleph Ω)]-multiplicity) (Aleph (non-ordinal-[Hyper Aleph Ω (Aleph Ω, Aleph Ω)]-multiplicity), Aleph (non-ordinal-[Hyper Aleph Ω (Aleph Ω, Aleph Ω)]-multiplicity))[Hyper Aleph (non-ordinal-[Hyper Aleph Ω (Aleph Ω, Aleph Ω)]-multiplicity) (Aleph (non-ordinal-[Hyper Aleph Ω (Aleph Ω, Aleph Ω)]-multiplicity), Aleph (non-ordinal-[Hyper Aleph Ω (Aleph Ω, Aleph Ω)]-multiplicity)), Hyper Aleph (non-ordinal-[Hyper Aleph Ω (Aleph Ω, Aleph Ω)]-multiplicity) (Aleph (non-ordinal-[Hyper Aleph Ω (Aleph Ω, Aleph Ω)]-multiplicity), Aleph (non-ordinal-[Hyper Aleph Ω (Aleph Ω, Aleph Ω)]-multiplicity))] {Hyper Hyper Aleph (non-ordinal-[Hyper Aleph Ω (Aleph Ω, Aleph Ω)]-multiplicity) (Aleph (non-ordinal-[Hyper Aleph Ω (Aleph Ω, Aleph Ω)]-multiplicity), Aleph (non-ordinal-[Hyper Aleph Ω (Aleph Ω, Aleph Ω)]-multiplicity))[Hyper Aleph (non-ordinal-[Hyper Aleph Ω (Aleph Ω, Aleph Ω)]-multiplicity) (Aleph (non-ordinal-[Hyper Aleph Ω (Aleph Ω, Aleph Ω)]-multiplicity), Aleph (non-ordinal-[Hyper Aleph Ω (Aleph Ω, Aleph Ω)]-multiplicity)), Hyper Aleph (non-ordinal-[Hyper Aleph Ω (Aleph Ω, Aleph Ω)]-multiplicity) (Aleph (non-ordinal-[Hyper Aleph Ω (Aleph Ω, Aleph Ω)]-multiplicity), Aleph (non-ordinal-[Hyper Aleph Ω (Aleph Ω, Aleph Ω)]-multiplicity))], Hyper Hyper Aleph (non-ordinal-[Hyper Aleph Ω (Aleph Ω, Aleph Ω)]-multiplicity) (Aleph (non-ordinal-[Hyper Aleph Ω (Aleph Ω, Aleph Ω)]-multiplicity), Aleph (non-ordinal-[Hyper Aleph Ω (Aleph Ω, Aleph Ω)]-multiplicity))[Hyper Aleph (non-ordinal-[Hyper Aleph Ω (Aleph Ω, Aleph Ω)]-multiplicity) (Aleph (non-ordinal-[Hyper Aleph Ω (Aleph Ω, Aleph Ω)]-multiplicity), Aleph (non-ordinal-[Hyper Aleph Ω (Aleph Ω, Aleph Ω)]-multiplicity)), Hyper Aleph (non-ordinal-[Hyper Aleph Ω (Aleph Ω, Aleph Ω)]-multiplicity) (Aleph (non-ordinal-[Hyper Aleph Ω (Aleph Ω, Aleph Ω)]-multiplicity), Aleph (non-ordinal-[Hyper Aleph Ω (Aleph Ω, Aleph Ω)]-multiplicity))]}}, Hyper Hyper Hyper Hyper Aleph (non-ordinal-[Hyper Aleph Ω (Aleph Ω, Aleph Ω)]-multiplicity) (Aleph (non-ordinal-[Hyper Aleph Ω (Aleph Ω, Aleph Ω)]-multiplicity), Aleph (non-ordinal-[Hyper Aleph Ω (Aleph Ω, Aleph Ω)]-multiplicity))[Hyper Aleph (non-ordinal-[Hyper Aleph Ω

(Aleph Ω, Aleph Ω)]-multiplicity) (Aleph (non-ordinal-[Hyper Aleph Ω (Aleph Ω, Aleph Ω)]-multiplicity), Aleph (non-ordinal-[Hyper Aleph Ω (Aleph Ω, Aleph Ω)]-multiplicity)), Hyper Aleph (non-ordinal-[Hyper Aleph Ω (Aleph Ω, Aleph Ω)]-multiplicity) (Aleph (non-ordinal-[Hyper Aleph Ω (Aleph Ω, Aleph Ω)]-multiplicity), Aleph (non-ordinal-[Hyper Aleph Ω (Aleph Ω, Aleph Ω)]-multiplicity))] {Hyper Hyper Aleph (non-ordinal-[Hyper Aleph Ω (Aleph Ω, Aleph Ω)]-multiplicity) (Aleph (non-ordinal-[Hyper Aleph Ω (Aleph Ω, Aleph Ω)]-multiplicity), Aleph (non-ordinal-[Hyper Aleph Ω (Aleph Ω, Aleph Ω)]-multiplicity))[Hyper Aleph (non-ordinal-[Hyper Aleph Ω (Aleph Ω, Aleph Ω)]-multiplicity) (Aleph (non-ordinal-[Hyper Aleph Ω (Aleph Ω, Aleph Ω)]-multiplicity), Aleph (non-ordinal-[Hyper Aleph Ω (Aleph Ω, Aleph Ω)]-multiplicity)), Hyper Aleph (non-ordinal-[Hyper Aleph Ω (Aleph Ω, Aleph Ω)]-multiplicity) (Aleph (non-ordinal-[Hyper Aleph Ω (Aleph Ω, Aleph Ω)]-multiplicity), Aleph (non-ordinal-[Hyper Aleph Ω (Aleph Ω, Aleph Ω)]-multiplicity))], Hyper Hyper Aleph (non-ordinal-[Hyper Aleph Ω (Aleph Ω, Aleph Ω)]-multiplicity) (Aleph (non-ordinal-[Hyper Aleph Ω (Aleph Ω, Aleph Ω)]-multiplicity), Aleph (non-ordinal-[Hyper Aleph Ω (Aleph Ω, Aleph Ω)]-multiplicity))[Hyper Aleph (non-ordinal-[Hyper Aleph Ω (Aleph Ω, Aleph Ω)]-multiplicity) (Aleph (non-ordinal-[Hyper Aleph Ω (Aleph Ω, Aleph Ω)]-multiplicity), Aleph (non-ordinal-[Hyper Aleph Ω (Aleph Ω, Aleph Ω)]-multiplicity)), Hyper Aleph (non-ordinal-[Hyper Aleph Ω (Aleph Ω, Aleph Ω)]-multiplicity) (Aleph (non-ordinal-[Hyper Aleph Ω (Aleph Ω, Aleph Ω)]-multiplicity), Aleph (non-ordinal-[Hyper Aleph Ω (Aleph Ω, Aleph Ω)]-multiplicity))]} { Hyper Hyper Hyper Aleph (non-ordinal-[Hyper Aleph Ω (Aleph Ω, Aleph Ω)]-multiplicity) (Aleph (non-ordinal-[Hyper Aleph Ω (Aleph Ω, Aleph Ω)]-multiplicity), Aleph (non-ordinal-[Hyper Aleph Ω (Aleph Ω, Aleph Ω)]-multiplicity))[Hyper Aleph (non-ordinal-[Hyper Aleph Ω (Aleph Ω, Aleph Ω)]-multiplicity) (Aleph (non-ordinal-[Hyper Aleph Ω (Aleph Ω, Aleph Ω)]-multiplicity), Aleph (non-ordinal-[Hyper Aleph Ω (Aleph Ω, Aleph Ω)]-multiplicity)), Hyper Aleph (non-ordinal-[Hyper Aleph Ω (Aleph Ω, Aleph Ω)]-multiplicity) (Aleph (non-ordinal-[Hyper Aleph Ω (Aleph Ω, Aleph Ω)]-multiplicity), Aleph (non-ordinal-[Hyper Aleph Ω (Aleph Ω, Aleph Ω)]-multiplicity))] {Hyper Hyper Aleph (non-ordinal-[Hyper Aleph Ω (Aleph Ω, Aleph Ω)]-multiplicity) (Aleph (non-ordinal-[Hyper Aleph Ω (Aleph Ω, Aleph Ω)]-multiplicity), Aleph (non-ordinal-[Hyper Aleph Ω (Aleph Ω, Aleph Ω)]-multiplicity))[Hyper Aleph (non-ordinal-[Hyper Aleph Ω (Aleph Ω, Aleph Ω)]-multiplicity) (Aleph (non-ordinal-[Hyper Aleph Ω (Aleph Ω, Aleph Ω)]-multiplicity), Aleph (non-ordinal-[Hyper Aleph Ω (Aleph Ω, Aleph Ω)]-multiplicity)), Hyper Aleph (non-ordinal-[Hyper Aleph Ω (Aleph Ω, Aleph Ω)]-multiplicity) (Aleph (non-ordinal-[Hyper Aleph Ω (Aleph Ω, Aleph Ω)]-multiplicity), Aleph (non-ordinal-[Hyper Aleph Ω (Aleph Ω, Aleph Ω)]-multiplicity))], Hyper Hyper Aleph (non-ordinal-[Hyper Aleph Ω (Aleph Ω, Aleph Ω)]-multiplicity) (Aleph (non-ordinal-[Hyper Aleph Ω (Aleph Ω, Aleph Ω)]-multiplicity), Aleph (non-ordinal-[Hyper Aleph Ω (Aleph Ω, Aleph Ω)]-multiplicity))[Hyper Aleph (non-ordinal-[Hyper Aleph Ω (Aleph Ω, Aleph Ω)]-multiplicity) (Aleph (non-ordinal-[Hyper Aleph Ω (Aleph Ω, Aleph Ω)]-multiplicity), Aleph (non-ordinal-[Hyper Aleph Ω (Aleph Ω, Aleph

Ω)]-multiplicity)), Hyper Aleph (non-ordinal-[Hyper Aleph Ω (Aleph Ω, Aleph Ω)]-multiplicity) (Aleph (non-ordinal-[Hyper Aleph Ω (Aleph Ω, Aleph Ω)]-multiplicity), Aleph (non-ordinal-[Hyper Aleph Ω (Aleph Ω, Aleph Ω)]-multiplicity))]}, Hyper Hyper Hyper Aleph (non-ordinal-[Hyper Aleph Ω (Aleph Ω, Aleph Ω)]-multiplicity) (Aleph (non-ordinal-[Hyper Aleph Ω (Aleph Ω, Aleph Ω)]-multiplicity), Aleph (non-ordinal-[Hyper Aleph Ω (Aleph Ω, Aleph Ω)]-multiplicity))[Hyper Aleph (non-ordinal-[Hyper Aleph Ω (Aleph Ω, Aleph Ω)]-multiplicity) (Aleph (non-ordinal-[Hyper Aleph Ω (Aleph Ω, Aleph Ω)]-multiplicity), Aleph (non-ordinal-[Hyper Aleph Ω (Aleph Ω, Aleph Ω)]-multiplicity)), Hyper Aleph (non-ordinal-[Hyper Aleph Ω (Aleph Ω, Aleph Ω)]-multiplicity) (Aleph (non-ordinal-[Hyper Aleph Ω (Aleph Ω, Aleph Ω)]-multiplicity), Aleph (non-ordinal-[Hyper Aleph Ω (Aleph Ω, Aleph Ω)]-multiplicity))] {Hyper Hyper Aleph (non-ordinal-[Hyper Aleph Ω (Aleph Ω, Aleph Ω)]-multiplicity) (Aleph (non-ordinal-[Hyper Aleph Ω (Aleph Ω, Aleph Ω)]-multiplicity), Aleph (non-ordinal-[Hyper Aleph Ω (Aleph Ω, Aleph Ω)]-multiplicity))[Hyper Aleph (non-ordinal-[Hyper Aleph Ω (Aleph Ω, Aleph Ω)]-multiplicity) (Aleph (non-ordinal-[Hyper Aleph Ω (Aleph Ω, Aleph Ω)]-multiplicity), Aleph (non-ordinal-[Hyper Aleph Ω (Aleph Ω, Aleph Ω)]-multiplicity)), Hyper Aleph (non-ordinal-[Hyper Aleph Ω (Aleph Ω, Aleph Ω)]-multiplicity) (Aleph (non-ordinal-[Hyper Aleph Ω (Aleph Ω, Aleph Ω)]-multiplicity), Aleph (non-ordinal-[Hyper Aleph Ω (Aleph Ω, Aleph Ω)]-multiplicity))], Hyper Hyper Aleph (non-ordinal-[Hyper Aleph Ω (Aleph Ω, Aleph Ω)]-multiplicity) (Aleph (non-ordinal-[Hyper Aleph Ω (Aleph Ω, Aleph Ω)]-multiplicity), Aleph (non-ordinal-[Hyper Aleph Ω (Aleph Ω, Aleph Ω)]-multiplicity))[Hyper Aleph (non-ordinal-[Hyper Aleph Ω (Aleph Ω, Aleph Ω)]-multiplicity) (Aleph (non-ordinal-[Hyper Aleph Ω (Aleph Ω, Aleph Ω)]-multiplicity), Aleph (non-ordinal-[Hyper Aleph Ω (Aleph Ω, Aleph Ω)]-multiplicity)), Hyper Aleph (non-ordinal-[Hyper Aleph Ω (Aleph Ω, Aleph Ω)]-multiplicity) (Aleph (non-ordinal-[Hyper Aleph Ω (Aleph Ω, Aleph Ω)]-multiplicity), Aleph (non-ordinal-[Hyper Aleph Ω (Aleph Ω, Aleph Ω)]-multiplicity))]}}}{ Hyper Hyper Hyper Hyper Hyper Aleph (non-ordinal-[Hyper Aleph Ω (Aleph Ω, Aleph Ω)]-multiplicity) (Aleph (non-ordinal-[Hyper Aleph Ω (Aleph Ω, Aleph Ω)]-multiplicity), Aleph (non-ordinal-[Hyper Aleph Ω (Aleph Ω, Aleph Ω)]-multiplicity))[Hyper Aleph (non-ordinal-[Hyper Aleph Ω (Aleph Ω, Aleph Ω)]-multiplicity) (Aleph (non-ordinal-[Hyper Aleph Ω (Aleph Ω, Aleph Ω)]-multiplicity), Aleph (non-ordinal-[Hyper Aleph Ω (Aleph Ω, Aleph Ω)]-multiplicity)), Hyper Aleph (non-ordinal-[Hyper Aleph Ω (Aleph Ω, Aleph Ω)]-multiplicity) (Aleph (non-ordinal-[Hyper Aleph Ω (Aleph Ω, Aleph Ω)]-multiplicity), Aleph (non-ordinal-[Hyper Aleph Ω (Aleph Ω, Aleph Ω)]-multiplicity))] {Hyper Hyper Aleph (non-ordinal-[Hyper Aleph Ω (Aleph Ω, Aleph Ω)]-multiplicity) (Aleph (non-ordinal-[Hyper Aleph Ω (Aleph Ω, Aleph Ω)]-multiplicity), Aleph (non-ordinal-[Hyper Aleph Ω (Aleph Ω, Aleph Ω)]-multiplicity))[Hyper Aleph (non-ordinal-[Hyper Aleph Ω (Aleph Ω, Aleph Ω)]-multiplicity) (Aleph (non-ordinal-[Hyper Aleph Ω (Aleph Ω, Aleph Ω)]-multiplicity), Aleph (non-ordinal-[Hyper Aleph Ω (Aleph Ω, Aleph Ω)]-multiplicity)), Hyper Aleph (non-ordinal-[Hyper Aleph Ω (Aleph Ω, Aleph Ω)]-multiplicity) (Aleph (non-ordinal-[Hyper Aleph Ω (Aleph Ω,

Aleph Ω)]-multiplicity), Aleph (non-ordinal-[Hyper Aleph Ω (Aleph Ω, Aleph Ω)]-multiplicity))], Hyper Hyper Aleph (non-ordinal-[Hyper Aleph Ω (Aleph Ω, Aleph Ω)]-multiplicity) (Aleph (non-ordinal-[Hyper Aleph Ω (Aleph Ω, Aleph Ω)]-multiplicity), Aleph (non-ordinal-[Hyper Aleph Ω (Aleph Ω, Aleph Ω)]-multiplicity))[Hyper Aleph (non-ordinal-[Hyper Aleph Ω (Aleph Ω, Aleph Ω)]-multiplicity) (Aleph (non-ordinal-[Hyper Aleph Ω (Aleph Ω, Aleph Ω)]-multiplicity), Aleph (non-ordinal-[Hyper Aleph Ω (Aleph Ω, Aleph Ω)]-multiplicity)), Hyper Aleph (non-ordinal-[Hyper Aleph Ω (Aleph Ω, Aleph Ω)]-multiplicity) (Aleph (non-ordinal-[Hyper Aleph Ω (Aleph Ω, Aleph Ω)]-multiplicity), Aleph (non-ordinal-[Hyper Aleph Ω (Aleph Ω, Aleph Ω)]-multiplicity))]} { Hyper Hyper Hyper Aleph (non-ordinal-[Hyper Aleph Ω (Aleph Ω, Aleph Ω)]-multiplicity) (Aleph (non-ordinal-[Hyper Aleph Ω (Aleph Ω, Aleph Ω)]-multiplicity), Aleph (non-ordinal-[Hyper Aleph Ω (Aleph Ω, Aleph Ω)]-multiplicity))[Hyper Aleph (non-ordinal-[Hyper Aleph Ω (Aleph Ω, Aleph Ω)]-multiplicity) (Aleph (non-ordinal-[Hyper Aleph Ω (Aleph Ω, Aleph Ω)]-multiplicity), Aleph (non-ordinal-[Hyper Aleph Ω (Aleph Ω, Aleph Ω)]-multiplicity)), Hyper Aleph (non-ordinal-[Hyper Aleph Ω (Aleph Ω, Aleph Ω)]-multiplicity) (Aleph (non-ordinal-[Hyper Aleph Ω (Aleph Ω, Aleph Ω)]-multiplicity), Aleph (non-ordinal-[Hyper Aleph Ω (Aleph Ω, Aleph Ω)]-multiplicity))] {Hyper Hyper Aleph (non-ordinal-[Hyper Aleph Ω (Aleph Ω, Aleph Ω)]-multiplicity) (Aleph (non-ordinal-[Hyper Aleph Ω (Aleph Ω, Aleph Ω)]-multiplicity), Aleph (non-ordinal-[Hyper Aleph Ω (Aleph Ω, Aleph Ω)]-multiplicity))[Hyper Aleph (non-ordinal-[Hyper Aleph Ω (Aleph Ω, Aleph Ω)]-multiplicity) (Aleph (non-ordinal-[Hyper Aleph Ω (Aleph Ω, Aleph Ω)]-multiplicity), Aleph (non-ordinal-[Hyper Aleph Ω (Aleph Ω, Aleph Ω)]-multiplicity)), Hyper Aleph (non-ordinal-[Hyper Aleph Ω (Aleph Ω, Aleph Ω)]-multiplicity) (Aleph (non-ordinal-[Hyper Aleph Ω (Aleph Ω, Aleph Ω)]-multiplicity), Aleph (non-ordinal-[Hyper Aleph Ω (Aleph Ω, Aleph Ω)]-multiplicity))], Hyper Hyper Aleph (non-ordinal-[Hyper Aleph Ω (Aleph Ω, Aleph Ω)]-multiplicity) (Aleph (non-ordinal-[Hyper Aleph Ω (Aleph Ω, Aleph Ω)]-multiplicity), Aleph (non-ordinal-[Hyper Aleph Ω (Aleph Ω, Aleph Ω)]-multiplicity))[Hyper Aleph (non-ordinal-[Hyper Aleph Ω (Aleph Ω, Aleph Ω)]-multiplicity) (Aleph (non-ordinal-[Hyper Aleph Ω (Aleph Ω, Aleph Ω)]-multiplicity), Aleph (non-ordinal-[Hyper Aleph Ω (Aleph Ω, Aleph Ω)]-multiplicity)), Hyper Aleph (non-ordinal-[Hyper Aleph Ω (Aleph Ω, Aleph Ω)]-multiplicity) (Aleph (non-ordinal-[Hyper Aleph Ω (Aleph Ω, Aleph Ω)]-multiplicity), Aleph (non-ordinal-[Hyper Aleph Ω (Aleph Ω, Aleph Ω)]-multiplicity))]}, Hyper Hyper Hyper Aleph (non-ordinal-[Hyper Aleph Ω (Aleph Ω, Aleph Ω)]-multiplicity) (Aleph (non-ordinal-[Hyper Aleph Ω (Aleph Ω, Aleph Ω)]-multiplicity), Aleph (non-ordinal-[Hyper Aleph Ω (Aleph Ω, Aleph Ω)]-multiplicity))[Hyper Aleph (non-ordinal-[Hyper Aleph Ω (Aleph Ω, Aleph Ω)]-multiplicity) (Aleph (non-ordinal-[Hyper Aleph Ω (Aleph Ω, Aleph Ω)]-multiplicity), Aleph (non-ordinal-[Hyper Aleph Ω (Aleph Ω, Aleph Ω)]-multiplicity)), Hyper Aleph (non-ordinal-[Hyper Aleph Ω (Aleph Ω, Aleph Ω)]-multiplicity) (Aleph (non-ordinal-[Hyper Aleph Ω (Aleph Ω, Aleph Ω)]-multiplicity), Aleph (non-ordinal-[Hyper Aleph Ω (Aleph Ω, Aleph Ω)]-multiplicity))] {Hyper Hyper Aleph (non-ordinal-[Hyper Aleph Ω

(Aleph Ω, Aleph Ω)]-multiplicity) (Aleph (non-ordinal-[Hyper Aleph Ω (Aleph Ω, Aleph Ω)]-multiplicity), Aleph (non-ordinal-[Hyper Aleph Ω (Aleph Ω, Aleph Ω)]-multiplicity))[Hyper Aleph (non-ordinal-[Hyper Aleph Ω (Aleph Ω, Aleph Ω)]-multiplicity) (Aleph (non-ordinal-[Hyper Aleph Ω (Aleph Ω, Aleph Ω)]-multiplicity), Aleph (non-ordinal-[Hyper Aleph Ω (Aleph Ω, Aleph Ω)]-multiplicity)), Hyper Aleph (non-ordinal-[Hyper Aleph Ω (Aleph Ω, Aleph Ω)]-multiplicity) (Aleph (non-ordinal-[Hyper Aleph Ω (Aleph Ω, Aleph Ω)]-multiplicity), Aleph (non-ordinal-[Hyper Aleph Ω (Aleph Ω, Aleph Ω)]-multiplicity))], Hyper Hyper Aleph (non-ordinal-[Hyper Aleph Ω (Aleph Ω, Aleph Ω)]-multiplicity) (Aleph (non-ordinal-[Hyper Aleph Ω (Aleph Ω, Aleph Ω)]-multiplicity), Aleph (non-ordinal-[Hyper Aleph Ω (Aleph Ω, Aleph Ω)]-multiplicity))[Hyper Aleph (non-ordinal-[Hyper Aleph Ω (Aleph Ω, Aleph Ω)]-multiplicity) (Aleph (non-ordinal-[Hyper Aleph Ω (Aleph Ω, Aleph Ω)]-multiplicity), Aleph (non-ordinal-[Hyper Aleph Ω (Aleph Ω, Aleph Ω)]-multiplicity)), Hyper Aleph (non-ordinal-[Hyper Aleph Ω (Aleph Ω, Aleph Ω)]-multiplicity) (Aleph (non-ordinal-[Hyper Aleph Ω (Aleph Ω, Aleph Ω)]-multiplicity), Aleph (non-ordinal-[Hyper Aleph Ω (Aleph Ω, Aleph Ω)]-multiplicity))]}}{ Hyper Hyper Hyper Hyper Aleph (non-ordinal-[Hyper Aleph Ω (Aleph Ω, Aleph Ω)]-multiplicity) (Aleph (non-ordinal-[Hyper Aleph Ω (Aleph Ω, Aleph Ω)]-multiplicity), Aleph (non-ordinal-[Hyper Aleph Ω (Aleph Ω, Aleph Ω)]-multiplicity))[Hyper Aleph (non-ordinal-[Hyper Aleph Ω (Aleph Ω, Aleph Ω)]-multiplicity) (Aleph (non-ordinal-[Hyper Aleph Ω (Aleph Ω, Aleph Ω)]-multiplicity), Aleph (non-ordinal-[Hyper Aleph Ω (Aleph Ω, Aleph Ω)]-multiplicity)), Hyper Aleph (non-ordinal-[Hyper Aleph Ω (Aleph Ω, Aleph Ω)]-multiplicity) (Aleph (non-ordinal-[Hyper Aleph Ω (Aleph Ω, Aleph Ω)]-multiplicity), Aleph (non-ordinal-[Hyper Aleph Ω (Aleph Ω, Aleph Ω)]-multiplicity))] {Hyper Hyper Aleph (non-ordinal-[Hyper Aleph Ω (Aleph Ω, Aleph Ω)]-multiplicity) (Aleph (non-ordinal-[Hyper Aleph Ω (Aleph Ω, Aleph Ω)]-multiplicity), Aleph (non-ordinal-[Hyper Aleph Ω (Aleph Ω, Aleph Ω)]-multiplicity))[Hyper Aleph (non-ordinal-[Hyper Aleph Ω (Aleph Ω, Aleph Ω)]-multiplicity) (Aleph (non-ordinal-[Hyper Aleph Ω (Aleph Ω, Aleph Ω)]-multiplicity), Aleph (non-ordinal-[Hyper Aleph Ω (Aleph Ω, Aleph Ω)]-multiplicity)), Hyper Aleph (non-ordinal-[Hyper Aleph Ω (Aleph Ω, Aleph Ω)]-multiplicity) (Aleph (non-ordinal-[Hyper Aleph Ω (Aleph Ω, Aleph Ω)]-multiplicity), Aleph (non-ordinal-[Hyper Aleph Ω (Aleph Ω, Aleph Ω)]-multiplicity))], Hyper Hyper Aleph (non-ordinal-[Hyper Aleph Ω (Aleph Ω, Aleph Ω)]-multiplicity) (Aleph (non-ordinal-[Hyper Aleph Ω (Aleph Ω, Aleph Ω)]-multiplicity), Aleph (non-ordinal-[Hyper Aleph Ω (Aleph Ω, Aleph Ω)]-multiplicity))[Hyper Aleph (non-ordinal-[Hyper Aleph Ω (Aleph Ω, Aleph Ω)]-multiplicity) (Aleph (non-ordinal-[Hyper Aleph Ω (Aleph Ω, Aleph Ω)]-multiplicity), Aleph (non-ordinal-[Hyper Aleph Ω (Aleph Ω, Aleph Ω)]-multiplicity)), Hyper Aleph (non-ordinal-[Hyper Aleph Ω (Aleph Ω, Aleph Ω)]-multiplicity) (Aleph (non-ordinal-[Hyper Aleph Ω (Aleph Ω, Aleph Ω)]-multiplicity), Aleph (non-ordinal-[Hyper Aleph Ω (Aleph Ω, Aleph Ω)]-multiplicity))]} { Hyper Hyper Hyper Aleph (non-ordinal-[Hyper Aleph Ω (Aleph Ω, Aleph Ω)]-multiplicity) (Aleph (non-ordinal-[Hyper Aleph Ω (Aleph Ω, Aleph Ω)]-multiplicity), Aleph (non-ordinal-[Hyper Aleph Ω

(Aleph Ω, Aleph Ω)]-multiplicity))[Hyper Aleph (non-ordinal-[Hyper Aleph Ω (Aleph Ω, Aleph Ω)]-multiplicity) (Aleph (non-ordinal-[Hyper Aleph Ω (Aleph Ω, Aleph Ω)]-multiplicity), Aleph (non-ordinal-[Hyper Aleph Ω (Aleph Ω, Aleph Ω)]-multiplicity)), Hyper Aleph (non-ordinal-[Hyper Aleph Ω (Aleph Ω, Aleph Ω)]-multiplicity) (Aleph (non-ordinal-[Hyper Aleph Ω (Aleph Ω, Aleph Ω)]-multiplicity), Aleph (non-ordinal-[Hyper Aleph Ω (Aleph Ω, Aleph Ω)]-multiplicity))] {Hyper Hyper Aleph (non-ordinal-[Hyper Aleph Ω (Aleph Ω, Aleph Ω)]-multiplicity) (Aleph (non-ordinal-[Hyper Aleph Ω (Aleph Ω, Aleph Ω)]-multiplicity), Aleph (non-ordinal-[Hyper Aleph Ω (Aleph Ω, Aleph Ω)]-multiplicity))[Hyper Aleph (non-ordinal-[Hyper Aleph Ω (Aleph Ω, Aleph Ω)]-multiplicity) (Aleph (non-ordinal-[Hyper Aleph Ω (Aleph Ω, Aleph Ω)]-multiplicity), Aleph (non-ordinal-[Hyper Aleph Ω (Aleph Ω, Aleph Ω)]-multiplicity)), Hyper Aleph (non-ordinal-[Hyper Aleph Ω (Aleph Ω, Aleph Ω)]-multiplicity) (Aleph (non-ordinal-[Hyper Aleph Ω (Aleph Ω, Aleph Ω)]-multiplicity), Aleph (non-ordinal-[Hyper Aleph Ω (Aleph Ω, Aleph Ω)]-multiplicity))], Hyper Hyper Aleph (non-ordinal-[Hyper Aleph Ω (Aleph Ω, Aleph Ω)]-multiplicity) (Aleph (non-ordinal-[Hyper Aleph Ω (Aleph Ω, Aleph Ω)]-multiplicity), Aleph (non-ordinal-[Hyper Aleph Ω (Aleph Ω, Aleph Ω)]-multiplicity))[Hyper Aleph (non-ordinal-[Hyper Aleph Ω (Aleph Ω, Aleph Ω)]-multiplicity) (Aleph (non-ordinal-[Hyper Aleph Ω (Aleph Ω, Aleph Ω)]-multiplicity), Aleph (non-ordinal-[Hyper Aleph Ω (Aleph Ω, Aleph Ω)]-multiplicity)), Hyper Aleph (non-ordinal-[Hyper Aleph Ω (Aleph Ω, Aleph Ω)]-multiplicity) (Aleph (non-ordinal-[Hyper Aleph Ω (Aleph Ω, Aleph Ω)]-multiplicity), Aleph (non-ordinal-[Hyper Aleph Ω (Aleph Ω, Aleph Ω)]-multiplicity))]}, Hyper Hyper Hyper Aleph (non-ordinal-[Hyper Aleph Ω (Aleph Ω, Aleph Ω)]-multiplicity) (Aleph (non-ordinal-[Hyper Aleph Ω (Aleph Ω, Aleph Ω)]-multiplicity), Aleph (non-ordinal-[Hyper Aleph Ω (Aleph Ω, Aleph Ω)]-multiplicity))[Hyper Aleph (non-ordinal-[Hyper Aleph Ω (Aleph Ω, Aleph Ω)]-multiplicity) (Aleph (non-ordinal-[Hyper Aleph Ω (Aleph Ω, Aleph Ω)]-multiplicity), Aleph (non-ordinal-[Hyper Aleph Ω (Aleph Ω, Aleph Ω)]-multiplicity)), Hyper Aleph (non-ordinal-[Hyper Aleph Ω (Aleph Ω, Aleph Ω)]-multiplicity) (Aleph (non-ordinal-[Hyper Aleph Ω (Aleph Ω, Aleph Ω)]-multiplicity), Aleph (non-ordinal-[Hyper Aleph Ω (Aleph Ω, Aleph Ω)]-multiplicity))] {Hyper Hyper Aleph (non-ordinal-[Hyper Aleph Ω (Aleph Ω, Aleph Ω)]-multiplicity) (Aleph (non-ordinal-[Hyper Aleph Ω (Aleph Ω, Aleph Ω)]-multiplicity), Aleph (non-ordinal-[Hyper Aleph Ω (Aleph Ω, Aleph Ω)]-multiplicity))[Hyper Aleph (non-ordinal-[Hyper Aleph Ω (Aleph Ω, Aleph Ω)]-multiplicity) (Aleph (non-ordinal-[Hyper Aleph Ω (Aleph Ω, Aleph Ω)]-multiplicity), Aleph (non-ordinal-[Hyper Aleph Ω (Aleph Ω, Aleph Ω)]-multiplicity)), Hyper Aleph (non-ordinal-[Hyper Aleph Ω (Aleph Ω, Aleph Ω)]-multiplicity) (Aleph (non-ordinal-[Hyper Aleph Ω (Aleph Ω, Aleph Ω)]-multiplicity), Aleph (non-ordinal-[Hyper Aleph Ω (Aleph Ω, Aleph Ω)]-multiplicity))], Hyper Hyper Aleph (non-ordinal-[Hyper Aleph Ω (Aleph Ω, Aleph Ω)]-multiplicity) (Aleph (non-ordinal-[Hyper Aleph Ω (Aleph Ω, Aleph Ω)]-multiplicity), Aleph (non-ordinal-[Hyper Aleph Ω (Aleph Ω, Aleph Ω)]-multiplicity))[Hyper Aleph (non-ordinal-[Hyper Aleph Ω (Aleph Ω, Aleph Ω)]-multiplicity) (Aleph (non-ordinal-[Hyper Aleph Ω

(Aleph Ω, Aleph Ω)]-multiplicity), Aleph (non-ordinal-[Hyper Aleph Ω (Aleph Ω, Aleph Ω)]-multiplicity)), Hyper Aleph (non-ordinal-[Hyper Aleph Ω (Aleph Ω, Aleph Ω)]-multiplicity) (Aleph (non-ordinal-[Hyper Aleph Ω (Aleph Ω, Aleph Ω)]-multiplicity), Aleph (non-ordinal-[Hyper Aleph Ω (Aleph Ω, Aleph Ω)]-multiplicity))]}}, Hyper Hyper Hyper Hyper Aleph (non-ordinal-[Hyper Aleph Ω (Aleph Ω, Aleph Ω)]-multiplicity) (Aleph (non-ordinal-[Hyper Aleph Ω (Aleph Ω, Aleph Ω)]-multiplicity), Aleph (non-ordinal-[Hyper Aleph Ω (Aleph Ω, Aleph Ω)]-multiplicity))[Hyper Aleph (non-ordinal-[Hyper Aleph Ω (Aleph Ω, Aleph Ω)]-multiplicity) (Aleph (non-ordinal-[Hyper Aleph Ω (Aleph Ω, Aleph Ω)]-multiplicity), Aleph (non-ordinal-[Hyper Aleph Ω (Aleph Ω, Aleph Ω)]-multiplicity)), Hyper Aleph (non-ordinal-[Hyper Aleph Ω (Aleph Ω, Aleph Ω)]-multiplicity) (Aleph (non-ordinal-[Hyper Aleph Ω (Aleph Ω, Aleph Ω)]-multiplicity), Aleph (non-ordinal-[Hyper Aleph Ω (Aleph Ω, Aleph Ω)]-multiplicity))] {Hyper Hyper Aleph (non-ordinal-[Hyper Aleph Ω (Aleph Ω, Aleph Ω)]-multiplicity) (Aleph (non-ordinal-[Hyper Aleph Ω (Aleph Ω, Aleph Ω)]-multiplicity), Aleph (non-ordinal-[Hyper Aleph Ω (Aleph Ω, Aleph Ω)]-multiplicity))[Hyper Aleph (non-ordinal-[Hyper Aleph Ω (Aleph Ω, Aleph Ω)]-multiplicity) (Aleph (non-ordinal-[Hyper Aleph Ω (Aleph Ω, Aleph Ω)]-multiplicity), Aleph (non-ordinal-[Hyper Aleph Ω (Aleph Ω, Aleph Ω)]-multiplicity)), Hyper Aleph (non-ordinal-[Hyper Aleph Ω (Aleph Ω, Aleph Ω)]-multiplicity) (Aleph (non-ordinal-[Hyper Aleph Ω (Aleph Ω, Aleph Ω)]-multiplicity), Aleph (non-ordinal-[Hyper Aleph Ω (Aleph Ω, Aleph Ω)]-multiplicity))], Hyper Hyper Aleph (non-ordinal-[Hyper Aleph Ω (Aleph Ω, Aleph Ω)]-multiplicity) (Aleph (non-ordinal-[Hyper Aleph Ω (Aleph Ω, Aleph Ω)]-multiplicity), Aleph (non-ordinal-[Hyper Aleph Ω (Aleph Ω, Aleph Ω)]-multiplicity))[Hyper Aleph (non-ordinal-[Hyper Aleph Ω (Aleph Ω, Aleph Ω)]-multiplicity) (Aleph (non-ordinal-[Hyper Aleph Ω (Aleph Ω, Aleph Ω)]-multiplicity), Aleph (non-ordinal-[Hyper Aleph Ω (Aleph Ω, Aleph Ω)]-multiplicity)), Hyper Aleph (non-ordinal-[Hyper Aleph Ω (Aleph Ω, Aleph Ω)]-multiplicity) (Aleph (non-ordinal-[Hyper Aleph Ω (Aleph Ω, Aleph Ω)]-multiplicity), Aleph (non-ordinal-[Hyper Aleph Ω (Aleph Ω, Aleph Ω)]-multiplicity))]} { Hyper Hyper Hyper Aleph (non-ordinal-[Hyper Aleph Ω (Aleph Ω, Aleph Ω)]-multiplicity) (Aleph (non-ordinal-[Hyper Aleph Ω (Aleph Ω, Aleph Ω)]-multiplicity), Aleph (non-ordinal-[Hyper Aleph Ω (Aleph Ω, Aleph Ω)]-multiplicity))[Hyper Aleph (non-ordinal-[Hyper Aleph Ω (Aleph Ω, Aleph Ω)]-multiplicity) (Aleph (non-ordinal-[Hyper Aleph Ω (Aleph Ω, Aleph Ω)]-multiplicity), Aleph (non-ordinal-[Hyper Aleph Ω (Aleph Ω, Aleph Ω)]-multiplicity)), Hyper Aleph (non-ordinal-[Hyper Aleph Ω (Aleph Ω, Aleph Ω)]-multiplicity) (Aleph (non-ordinal-[Hyper Aleph Ω (Aleph Ω, Aleph Ω)]-multiplicity), Aleph (non-ordinal-[Hyper Aleph Ω (Aleph Ω, Aleph Ω)]-multiplicity))] {Hyper Hyper Aleph (non-ordinal-[Hyper Aleph Ω (Aleph Ω, Aleph Ω)]-multiplicity) (Aleph (non-ordinal-[Hyper Aleph Ω (Aleph Ω, Aleph Ω)]-multiplicity), Aleph (non-ordinal-[Hyper Aleph Ω (Aleph Ω, Aleph Ω)]-multiplicity))[Hyper Aleph (non-ordinal-[Hyper Aleph Ω (Aleph Ω, Aleph Ω)]-multiplicity) (Aleph (non-ordinal-[Hyper Aleph Ω (Aleph Ω, Aleph Ω)]-multiplicity), Aleph (non-ordinal-[Hyper Aleph Ω (Aleph Ω, Aleph Ω)]-multiplicity)), Hyper Aleph (non-ordinal-[Hyper

Aleph Ω (Aleph Ω, Aleph Ω)]-multiplicity) (Aleph (non-ordinal-[Hyper Aleph Ω (Aleph Ω, Aleph Ω)]-multiplicity), Aleph (non-ordinal-[Hyper Aleph Ω (Aleph Ω, Aleph Ω)]-multiplicity))], Hyper Hyper Aleph (non-ordinal-[Hyper Aleph Ω (Aleph Ω, Aleph Ω)]-multiplicity) (Aleph (non-ordinal-[Hyper Aleph Ω (Aleph Ω, Aleph Ω)]-multiplicity), Aleph (non-ordinal-[Hyper Aleph Ω (Aleph Ω, Aleph Ω)]-multiplicity))[Hyper Aleph (non-ordinal-[Hyper Aleph Ω (Aleph Ω, Aleph Ω)]-multiplicity) (Aleph (non-ordinal-[Hyper Aleph Ω (Aleph Ω, Aleph Ω)]-multiplicity), Aleph (non-ordinal-[Hyper Aleph Ω (Aleph Ω, Aleph Ω)]-multiplicity)), Hyper Aleph (non-ordinal-[Hyper Aleph Ω (Aleph Ω, Aleph Ω)]-multiplicity) (Aleph (non-ordinal-[Hyper Aleph Ω (Aleph Ω, Aleph Ω)]-multiplicity), Aleph (non-ordinal-[Hyper Aleph Ω (Aleph Ω, Aleph Ω)]-multiplicity))]}, Hyper Hyper Hyper Aleph (non-ordinal-[Hyper Aleph Ω (Aleph Ω, Aleph Ω)]-multiplicity) (Aleph (non-ordinal-[Hyper Aleph Ω (Aleph Ω, Aleph Ω)]-multiplicity), Aleph (non-ordinal-[Hyper Aleph Ω (Aleph Ω, Aleph Ω)]-multiplicity))[Hyper Aleph (non-ordinal-[Hyper Aleph Ω (Aleph Ω, Aleph Ω)]-multiplicity) (Aleph (non-ordinal-[Hyper Aleph Ω (Aleph Ω, Aleph Ω)]-multiplicity), Aleph (non-ordinal-[Hyper Aleph Ω (Aleph Ω, Aleph Ω)]-multiplicity)), Hyper Aleph (non-ordinal-[Hyper Aleph Ω (Aleph Ω, Aleph Ω)]-multiplicity) (Aleph (non-ordinal-[Hyper Aleph Ω (Aleph Ω, Aleph Ω)]-multiplicity), Aleph (non-ordinal-[Hyper Aleph Ω (Aleph Ω, Aleph Ω)]-multiplicity))] {Hyper Hyper Aleph (non-ordinal-[Hyper Aleph Ω (Aleph Ω, Aleph Ω)]-multiplicity) (Aleph (non-ordinal-[Hyper Aleph Ω (Aleph Ω, Aleph Ω)]-multiplicity), Aleph (non-ordinal-[Hyper Aleph Ω (Aleph Ω, Aleph Ω)]-multiplicity))[Hyper Aleph (non-ordinal-[Hyper Aleph Ω (Aleph Ω, Aleph Ω)]-multiplicity) (Aleph (non-ordinal-[Hyper Aleph Ω (Aleph Ω, Aleph Ω)]-multiplicity), Aleph (non-ordinal-[Hyper Aleph Ω (Aleph Ω, Aleph Ω)]-multiplicity)), Hyper Aleph (non-ordinal-[Hyper Aleph Ω (Aleph Ω, Aleph Ω)]-multiplicity) (Aleph (non-ordinal-[Hyper Aleph Ω (Aleph Ω, Aleph Ω)]-multiplicity), Aleph (non-ordinal-[Hyper Aleph Ω (Aleph Ω, Aleph Ω)]-multiplicity))], Hyper Hyper Aleph (non-ordinal-[Hyper Aleph Ω (Aleph Ω, Aleph Ω)]-multiplicity) (Aleph (non-ordinal-[Hyper Aleph Ω (Aleph Ω, Aleph Ω)]-multiplicity), Aleph (non-ordinal-[Hyper Aleph Ω (Aleph Ω, Aleph Ω)]-multiplicity))[Hyper Aleph (non-ordinal-[Hyper Aleph Ω (Aleph Ω, Aleph Ω)]-multiplicity) (Aleph (non-ordinal-[Hyper Aleph Ω (Aleph Ω, Aleph Ω)]-multiplicity), Aleph (non-ordinal-[Hyper Aleph Ω (Aleph Ω, Aleph Ω)]-multiplicity)), Hyper Aleph (non-ordinal-[Hyper Aleph Ω (Aleph Ω, Aleph Ω)]-multiplicity) (Aleph (non-ordinal-[Hyper Aleph Ω (Aleph Ω, Aleph Ω)]-multiplicity), Aleph (non-ordinal-[Hyper Aleph Ω (Aleph Ω, Aleph Ω)]-multiplicity))]}}}, Hyper Hyper Hyper Hyper Hyper Aleph (non-ordinal-[Hyper Aleph Ω (Aleph Ω, Aleph Ω)]-multiplicity) (Aleph (non-ordinal-[Hyper Aleph Ω (Aleph Ω, Aleph Ω)]-multiplicity), Aleph (non-ordinal-[Hyper Aleph Ω (Aleph Ω, Aleph Ω)]-multiplicity))[Hyper Aleph (non-ordinal-[Hyper Aleph Ω (Aleph Ω, Aleph Ω)]-multiplicity) (Aleph (non-ordinal-[Hyper Aleph Ω (Aleph Ω, Aleph Ω)]-multiplicity), Aleph (non-ordinal-[Hyper Aleph Ω (Aleph Ω, Aleph Ω)]-multiplicity)), Hyper Aleph (non-ordinal-[Hyper Aleph Ω (Aleph Ω, Aleph Ω)]-multiplicity) (Aleph (non-ordinal-[Hyper Aleph Ω (Aleph Ω, Aleph Ω)]-multiplicity), Aleph (non-ordinal-[Hyper

Aleph Ω (Aleph Ω, Aleph Ω)]-multiplicity))] {Hyper Hyper Aleph (non-ordinal-[Hyper Aleph Ω (Aleph Ω, Aleph Ω)]-multiplicity) (Aleph (non-ordinal-[Hyper Aleph Ω (Aleph Ω, Aleph Ω)]-multiplicity), Aleph (non-ordinal-[Hyper Aleph Ω (Aleph Ω, Aleph Ω)]-multiplicity))[Hyper Aleph (non-ordinal-[Hyper Aleph Ω (Aleph Ω, Aleph Ω)]-multiplicity) (Aleph (non-ordinal-[Hyper Aleph Ω (Aleph Ω, Aleph Ω)]-multiplicity), Aleph (non-ordinal-[Hyper Aleph Ω (Aleph Ω, Aleph Ω)]-multiplicity)), Hyper Aleph (non-ordinal-[Hyper Aleph Ω (Aleph Ω, Aleph Ω)]-multiplicity) (Aleph (non-ordinal-[Hyper Aleph Ω (Aleph Ω, Aleph Ω)]-multiplicity), Aleph (non-ordinal-[Hyper Aleph Ω (Aleph Ω, Aleph Ω)]-multiplicity))], Hyper Hyper Aleph (non-ordinal-[Hyper Aleph Ω (Aleph Ω, Aleph Ω)]-multiplicity) (Aleph (non-ordinal-[Hyper Aleph Ω (Aleph Ω, Aleph Ω)]-multiplicity), Aleph (non-ordinal-[Hyper Aleph Ω (Aleph Ω, Aleph Ω)]-multiplicity))[Hyper Aleph (non-ordinal-[Hyper Aleph Ω (Aleph Ω, Aleph Ω)]-multiplicity) (Aleph (non-ordinal-[Hyper Aleph Ω (Aleph Ω, Aleph Ω)]-multiplicity), Aleph (non-ordinal-[Hyper Aleph Ω (Aleph Ω, Aleph Ω)]-multiplicity)), Hyper Aleph (non-ordinal-[Hyper Aleph Ω (Aleph Ω, Aleph Ω)]-multiplicity) (Aleph (non-ordinal-[Hyper Aleph Ω (Aleph Ω, Aleph Ω)]-multiplicity), Aleph (non-ordinal-[Hyper Aleph Ω (Aleph Ω, Aleph Ω)]-multiplicity))]} { Hyper Hyper Hyper Aleph (non-ordinal-[Hyper Aleph Ω (Aleph Ω, Aleph Ω)]-multiplicity) (Aleph (non-ordinal-[Hyper Aleph Ω (Aleph Ω, Aleph Ω)]-multiplicity), Aleph (non-ordinal-[Hyper Aleph Ω (Aleph Ω, Aleph Ω)]-multiplicity))[Hyper Aleph (non-ordinal-[Hyper Aleph Ω (Aleph Ω, Aleph Ω)]-multiplicity) (Aleph (non-ordinal-[Hyper Aleph Ω (Aleph Ω, Aleph Ω)]-multiplicity), Aleph (non-ordinal-[Hyper Aleph Ω (Aleph Ω, Aleph Ω)]-multiplicity)), Hyper Aleph (non-ordinal-[Hyper Aleph Ω (Aleph Ω, Aleph Ω)]-multiplicity) (Aleph (non-ordinal-[Hyper Aleph Ω (Aleph Ω, Aleph Ω)]-multiplicity), Aleph (non-ordinal-[Hyper Aleph Ω (Aleph Ω, Aleph Ω)]-multiplicity))] {Hyper Hyper Aleph (non-ordinal-[Hyper Aleph Ω (Aleph Ω, Aleph Ω)]-multiplicity) (Aleph (non-ordinal-[Hyper Aleph Ω (Aleph Ω, Aleph Ω)]-multiplicity), Aleph (non-ordinal-[Hyper Aleph Ω (Aleph Ω, Aleph Ω)]-multiplicity))[Hyper Aleph (non-ordinal-[Hyper Aleph Ω (Aleph Ω, Aleph Ω)]-multiplicity) (Aleph (non-ordinal-[Hyper Aleph Ω (Aleph Ω, Aleph Ω)]-multiplicity), Aleph (non-ordinal-[Hyper Aleph Ω (Aleph Ω, Aleph Ω)]-multiplicity)), Hyper Aleph (non-ordinal-[Hyper Aleph Ω (Aleph Ω, Aleph Ω)]-multiplicity) (Aleph (non-ordinal-[Hyper Aleph Ω (Aleph Ω, Aleph Ω)]-multiplicity), Aleph (non-ordinal-[Hyper Aleph Ω (Aleph Ω, Aleph Ω)]-multiplicity))], Hyper Hyper Aleph (non-ordinal-[Hyper Aleph Ω (Aleph Ω, Aleph Ω)]-multiplicity) (Aleph (non-ordinal-[Hyper Aleph Ω (Aleph Ω, Aleph Ω)]-multiplicity), Aleph (non-ordinal-[Hyper Aleph Ω (Aleph Ω, Aleph Ω)]-multiplicity))[Hyper Aleph (non-ordinal-[Hyper Aleph Ω (Aleph Ω, Aleph Ω)]-multiplicity) (Aleph (non-ordinal-[Hyper Aleph Ω (Aleph Ω, Aleph Ω)]-multiplicity), Aleph (non-ordinal-[Hyper Aleph Ω (Aleph Ω, Aleph Ω)]-multiplicity)), Hyper Aleph (non-ordinal-[Hyper Aleph Ω (Aleph Ω, Aleph Ω)]-multiplicity) (Aleph (non-ordinal-[Hyper Aleph Ω (Aleph Ω, Aleph Ω)]-multiplicity), Aleph (non-ordinal-[Hyper Aleph Ω (Aleph Ω, Aleph Ω)]-multiplicity))]}, Hyper Hyper Hyper Aleph (non-ordinal-[Hyper Aleph Ω (Aleph Ω, Aleph Ω)]-multiplicity) (Aleph (non-ordinal-

[Hyper Aleph Ω (Aleph Ω, Aleph Ω)]-multiplicity), Aleph (non-ordinal-[Hyper Aleph Ω (Aleph Ω, Aleph Ω)]-multiplicity))[Hyper Aleph (non-ordinal-[Hyper Aleph Ω (Aleph Ω, Aleph Ω)]-multiplicity) (Aleph (non-ordinal-[Hyper Aleph Ω (Aleph Ω, Aleph Ω)]-multiplicity), Aleph (non-ordinal-[Hyper Aleph Ω (Aleph Ω, Aleph Ω)]-multiplicity)), Hyper Aleph (non-ordinal-[Hyper Aleph Ω (Aleph Ω, Aleph Ω)]-multiplicity) (Aleph (non-ordinal-[Hyper Aleph Ω (Aleph Ω, Aleph Ω)]-multiplicity), Aleph (non-ordinal-[Hyper Aleph Ω (Aleph Ω, Aleph Ω)]-multiplicity))] {Hyper Hyper Aleph (non-ordinal-[Hyper Aleph Ω (Aleph Ω, Aleph Ω)]-multiplicity) (Aleph (non-ordinal-[Hyper Aleph Ω (Aleph Ω, Aleph Ω)]-multiplicity), Aleph (non-ordinal-[Hyper Aleph Ω (Aleph Ω, Aleph Ω)]-multiplicity))[Hyper Aleph (non-ordinal-[Hyper Aleph Ω (Aleph Ω, Aleph Ω)]-multiplicity) (Aleph (non-ordinal-[Hyper Aleph Ω (Aleph Ω, Aleph Ω)]-multiplicity), Aleph (non-ordinal-[Hyper Aleph Ω (Aleph Ω, Aleph Ω)]-multiplicity)), Hyper Aleph (non-ordinal-[Hyper Aleph Ω (Aleph Ω, Aleph Ω)]-multiplicity) (Aleph (non-ordinal-[Hyper Aleph Ω (Aleph Ω, Aleph Ω)]-multiplicity), Aleph (non-ordinal-[Hyper Aleph Ω (Aleph Ω, Aleph Ω)]-multiplicity))], Hyper Hyper Aleph (non-ordinal-[Hyper Aleph Ω (Aleph Ω, Aleph Ω)]-multiplicity) (Aleph (non-ordinal-[Hyper Aleph Ω (Aleph Ω, Aleph Ω)]-multiplicity), Aleph (non-ordinal-[Hyper Aleph Ω (Aleph Ω, Aleph Ω)]-multiplicity))[Hyper Aleph (non-ordinal-[Hyper Aleph Ω (Aleph Ω, Aleph Ω)]-multiplicity) (Aleph (non-ordinal-[Hyper Aleph Ω (Aleph Ω, Aleph Ω)]-multiplicity), Aleph (non-ordinal-[Hyper Aleph Ω (Aleph Ω, Aleph Ω)]-multiplicity)), Hyper Aleph (non-ordinal-[Hyper Aleph Ω (Aleph Ω, Aleph Ω)]-multiplicity) (Aleph (non-ordinal-[Hyper Aleph Ω (Aleph Ω, Aleph Ω)]-multiplicity), Aleph (non-ordinal-[Hyper Aleph Ω (Aleph Ω, Aleph Ω)]-multiplicity))]}}{ Hyper Hyper Hyper Hyper Aleph (non-ordinal-[Hyper Aleph Ω (Aleph Ω, Aleph Ω)]-multiplicity) (Aleph (non-ordinal-[Hyper Aleph Ω (Aleph Ω, Aleph Ω)]-multiplicity), Aleph (non-ordinal-[Hyper Aleph Ω (Aleph Ω, Aleph Ω)]-multiplicity))[Hyper Aleph (non-ordinal-[Hyper Aleph Ω (Aleph Ω, Aleph Ω)]-multiplicity) (Aleph (non-ordinal-[Hyper Aleph Ω (Aleph Ω, Aleph Ω)]-multiplicity), Aleph (non-ordinal-[Hyper Aleph Ω (Aleph Ω, Aleph Ω)]-multiplicity)), Hyper Aleph (non-ordinal-[Hyper Aleph Ω (Aleph Ω, Aleph Ω)]-multiplicity) (Aleph (non-ordinal-[Hyper Aleph Ω (Aleph Ω, Aleph Ω)]-multiplicity), Aleph (non-ordinal-[Hyper Aleph Ω (Aleph Ω, Aleph Ω)]-multiplicity))] {Hyper Hyper Aleph (non-ordinal-[Hyper Aleph Ω (Aleph Ω, Aleph Ω)]-multiplicity) (Aleph (non-ordinal-[Hyper Aleph Ω (Aleph Ω, Aleph Ω)]-multiplicity), Aleph (non-ordinal-[Hyper Aleph Ω (Aleph Ω, Aleph Ω)]-multiplicity))[Hyper Aleph (non-ordinal-[Hyper Aleph Ω (Aleph Ω, Aleph Ω)]-multiplicity) (Aleph (non-ordinal-[Hyper Aleph Ω (Aleph Ω, Aleph Ω)]-multiplicity), Aleph (non-ordinal-[Hyper Aleph Ω (Aleph Ω, Aleph Ω)]-multiplicity)), Hyper Aleph (non-ordinal-[Hyper Aleph Ω (Aleph Ω, Aleph Ω)]-multiplicity) (Aleph (non-ordinal-[Hyper Aleph Ω (Aleph Ω, Aleph Ω)]-multiplicity), Aleph (non-ordinal-[Hyper Aleph Ω (Aleph Ω, Aleph Ω)]-multiplicity))], Hyper Hyper Aleph (non-ordinal-[Hyper Aleph Ω (Aleph Ω, Aleph Ω)]-multiplicity) (Aleph (non-ordinal-[Hyper Aleph Ω (Aleph Ω, Aleph Ω)]-multiplicity), Aleph (non-ordinal-[Hyper Aleph Ω (Aleph Ω, Aleph Ω)]-multiplicity))[Hyper Aleph (non-ordinal-

[Hyper Aleph Ω (Aleph Ω, Aleph Ω)]-multiplicity) (Aleph (non-ordinal-[Hyper Aleph Ω (Aleph Ω, Aleph Ω)]-multiplicity), Aleph (non-ordinal-[Hyper Aleph Ω (Aleph Ω, Aleph Ω)]-multiplicity)), Hyper Aleph (non-ordinal-[Hyper Aleph Ω (Aleph Ω, Aleph Ω)]-multiplicity) (Aleph (non-ordinal-[Hyper Aleph Ω (Aleph Ω, Aleph Ω)]-multiplicity), Aleph (non-ordinal-[Hyper Aleph Ω (Aleph Ω, Aleph Ω)]-multiplicity))]} { Hyper Hyper Hyper Aleph (non-ordinal-[Hyper Aleph Ω (Aleph Ω, Aleph Ω)]-multiplicity) (Aleph (non-ordinal-[Hyper Aleph Ω (Aleph Ω, Aleph Ω)]-multiplicity), Aleph (non-ordinal-[Hyper Aleph Ω (Aleph Ω, Aleph Ω)]-multiplicity))[Hyper Aleph (non-ordinal-[Hyper Aleph Ω (Aleph Ω, Aleph Ω)]-multiplicity) (Aleph (non-ordinal-[Hyper Aleph Ω (Aleph Ω, Aleph Ω)]-multiplicity), Aleph (non-ordinal-[Hyper Aleph Ω (Aleph Ω, Aleph Ω)]-multiplicity)), Hyper Aleph (non-ordinal-[Hyper Aleph Ω (Aleph Ω, Aleph Ω)]-multiplicity) (Aleph (non-ordinal-[Hyper Aleph Ω (Aleph Ω, Aleph Ω)]-multiplicity), Aleph (non-ordinal-[Hyper Aleph Ω (Aleph Ω, Aleph Ω)]-multiplicity))] {Hyper Hyper Aleph (non-ordinal-[Hyper Aleph Ω (Aleph Ω, Aleph Ω)]-multiplicity) (Aleph (non-ordinal-[Hyper Aleph Ω (Aleph Ω, Aleph Ω)]-multiplicity), Aleph (non-ordinal-[Hyper Aleph Ω (Aleph Ω, Aleph Ω)]-multiplicity))[Hyper Aleph (non-ordinal-[Hyper Aleph Ω (Aleph Ω, Aleph Ω)]-multiplicity) (Aleph (non-ordinal-[Hyper Aleph Ω (Aleph Ω, Aleph Ω)]-multiplicity), Aleph (non-ordinal-[Hyper Aleph Ω (Aleph Ω, Aleph Ω)]-multiplicity)), Hyper Aleph (non-ordinal-[Hyper Aleph Ω (Aleph Ω, Aleph Ω)]-multiplicity) (Aleph (non-ordinal-[Hyper Aleph Ω (Aleph Ω, Aleph Ω)]-multiplicity), Aleph (non-ordinal-[Hyper Aleph Ω (Aleph Ω, Aleph Ω)]-multiplicity))], Hyper Hyper Aleph (non-ordinal-[Hyper Aleph Ω (Aleph Ω, Aleph Ω)]-multiplicity) (Aleph (non-ordinal-[Hyper Aleph Ω (Aleph Ω, Aleph Ω)]-multiplicity), Aleph (non-ordinal-[Hyper Aleph Ω (Aleph Ω, Aleph Ω)]-multiplicity))[Hyper Aleph (non-ordinal-[Hyper Aleph Ω (Aleph Ω, Aleph Ω)]-multiplicity) (Aleph (non-ordinal-[Hyper Aleph Ω (Aleph Ω, Aleph Ω)]-multiplicity), Aleph (non-ordinal-[Hyper Aleph Ω (Aleph Ω, Aleph Ω)]-multiplicity)), Hyper Aleph (non-ordinal-[Hyper Aleph Ω (Aleph Ω, Aleph Ω)]-multiplicity) (Aleph (non-ordinal-[Hyper Aleph Ω (Aleph Ω, Aleph Ω)]-multiplicity), Aleph (non-ordinal-[Hyper Aleph Ω (Aleph Ω, Aleph Ω)]-multiplicity))]}, Hyper Hyper Hyper Aleph (non-ordinal-[Hyper Aleph Ω (Aleph Ω, Aleph Ω)]-multiplicity) (Aleph (non-ordinal-[Hyper Aleph Ω (Aleph Ω, Aleph Ω)]-multiplicity), Aleph (non-ordinal-[Hyper Aleph Ω (Aleph Ω, Aleph Ω)]-multiplicity))[Hyper Aleph (non-ordinal-[Hyper Aleph Ω (Aleph Ω, Aleph Ω)]-multiplicity) (Aleph (non-ordinal-[Hyper Aleph Ω (Aleph Ω, Aleph Ω)]-multiplicity), Aleph (non-ordinal-[Hyper Aleph Ω (Aleph Ω, Aleph Ω)]-multiplicity)), Hyper Aleph (non-ordinal-[Hyper Aleph Ω (Aleph Ω, Aleph Ω)]-multiplicity) (Aleph (non-ordinal-[Hyper Aleph Ω (Aleph Ω, Aleph Ω)]-multiplicity), Aleph (non-ordinal-[Hyper Aleph Ω (Aleph Ω, Aleph Ω)]-multiplicity))] {Hyper Hyper Aleph (non-ordinal-[Hyper Aleph Ω (Aleph Ω, Aleph Ω)]-multiplicity) (Aleph (non-ordinal-[Hyper Aleph Ω (Aleph Ω, Aleph Ω)]-multiplicity), Aleph (non-ordinal-[Hyper Aleph Ω (Aleph Ω, Aleph Ω)]-multiplicity))[Hyper Aleph (non-ordinal-[Hyper Aleph Ω (Aleph Ω, Aleph Ω)]-multiplicity) (Aleph (non-ordinal-[Hyper Aleph Ω (Aleph Ω, Aleph Ω)]-multiplicity), Aleph (non-ordinal-

[Hyper Aleph Ω (Aleph Ω, Aleph Ω)]-multiplicity)), Hyper Aleph (non-ordinal-[Hyper Aleph Ω (Aleph Ω, Aleph Ω)]-multiplicity) (Aleph (non-ordinal-[Hyper Aleph Ω (Aleph Ω, Aleph Ω)]-multiplicity), Aleph (non-ordinal-[Hyper Aleph Ω (Aleph Ω, Aleph Ω)]-multiplicity))], Hyper Hyper Aleph (non-ordinal-[Hyper Aleph Ω (Aleph Ω, Aleph Ω)]-multiplicity) (Aleph (non-ordinal-[Hyper Aleph Ω (Aleph Ω, Aleph Ω)]-multiplicity), Aleph (non-ordinal-[Hyper Aleph Ω (Aleph Ω, Aleph Ω)]-multiplicity))[Hyper Aleph (non-ordinal-[Hyper Aleph Ω (Aleph Ω, Aleph Ω)]-multiplicity) (Aleph (non-ordinal-[Hyper Aleph Ω (Aleph Ω, Aleph Ω)]-multiplicity), Aleph (non-ordinal-[Hyper Aleph Ω (Aleph Ω, Aleph Ω)]-multiplicity)), Hyper Aleph (non-ordinal-[Hyper Aleph Ω (Aleph Ω, Aleph Ω)]-multiplicity) (Aleph (non-ordinal-[Hyper Aleph Ω (Aleph Ω, Aleph Ω)]-multiplicity), Aleph (non-ordinal-[Hyper Aleph Ω (Aleph Ω, Aleph Ω)]-multiplicity))]}}, Hyper Hyper Hyper Hyper Aleph (non-ordinal-[Hyper Aleph Ω (Aleph Ω, Aleph Ω)]-multiplicity) (Aleph (non-ordinal-[Hyper Aleph Ω (Aleph Ω, Aleph Ω)]-multiplicity), Aleph (non-ordinal-[Hyper Aleph Ω (Aleph Ω, Aleph Ω)]-multiplicity))[Hyper Aleph (non-ordinal-[Hyper Aleph Ω (Aleph Ω, Aleph Ω)]-multiplicity) (Aleph (non-ordinal-[Hyper Aleph Ω (Aleph Ω, Aleph Ω)]-multiplicity), Aleph (non-ordinal-[Hyper Aleph Ω (Aleph Ω, Aleph Ω)]-multiplicity)), Hyper Aleph (non-ordinal-[Hyper Aleph Ω (Aleph Ω, Aleph Ω)]-multiplicity) (Aleph (non-ordinal-[Hyper Aleph Ω (Aleph Ω, Aleph Ω)]-multiplicity), Aleph (non-ordinal-[Hyper Aleph Ω (Aleph Ω, Aleph Ω)]-multiplicity))] {Hyper Hyper Aleph (non-ordinal-[Hyper Aleph Ω (Aleph Ω, Aleph Ω)]-multiplicity) (Aleph (non-ordinal-[Hyper Aleph Ω (Aleph Ω, Aleph Ω)]-multiplicity), Aleph (non-ordinal-[Hyper Aleph Ω (Aleph Ω, Aleph Ω)]-multiplicity))[Hyper Aleph (non-ordinal-[Hyper Aleph Ω (Aleph Ω, Aleph Ω)]-multiplicity) (Aleph (non-ordinal-[Hyper Aleph Ω (Aleph Ω, Aleph Ω)]-multiplicity), Aleph (non-ordinal-[Hyper Aleph Ω (Aleph Ω, Aleph Ω)]-multiplicity)), Hyper Aleph (non-ordinal-[Hyper Aleph Ω (Aleph Ω, Aleph Ω)]-multiplicity) (Aleph (non-ordinal-[Hyper Aleph Ω (Aleph Ω, Aleph Ω)]-multiplicity), Aleph (non-ordinal-[Hyper Aleph Ω (Aleph Ω, Aleph Ω)]-multiplicity))], Hyper Hyper Aleph (non-ordinal-[Hyper Aleph Ω (Aleph Ω, Aleph Ω)]-multiplicity) (Aleph (non-ordinal-[Hyper Aleph Ω (Aleph Ω, Aleph Ω)]-multiplicity), Aleph (non-ordinal-[Hyper Aleph Ω (Aleph Ω, Aleph Ω)]-multiplicity))[Hyper Aleph (non-ordinal-[Hyper Aleph Ω (Aleph Ω, Aleph Ω)]-multiplicity) (Aleph (non-ordinal-[Hyper Aleph Ω (Aleph Ω, Aleph Ω)]-multiplicity), Aleph (non-ordinal-[Hyper Aleph Ω (Aleph Ω, Aleph Ω)]-multiplicity)), Hyper Aleph (non-ordinal-[Hyper Aleph Ω (Aleph Ω, Aleph Ω)]-multiplicity) (Aleph (non-ordinal-[Hyper Aleph Ω (Aleph Ω, Aleph Ω)]-multiplicity), Aleph (non-ordinal-[Hyper Aleph Ω (Aleph Ω, Aleph Ω)]-multiplicity))]} { Hyper Hyper Hyper Aleph (non-ordinal-[Hyper Aleph Ω (Aleph Ω, Aleph Ω)]-multiplicity) (Aleph (non-ordinal-[Hyper Aleph Ω (Aleph Ω, Aleph Ω)]-multiplicity), Aleph (non-ordinal-[Hyper Aleph Ω (Aleph Ω, Aleph Ω)]-multiplicity))[Hyper Aleph (non-ordinal-[Hyper Aleph Ω (Aleph Ω, Aleph Ω)]-multiplicity) (Aleph (non-ordinal-[Hyper Aleph Ω (Aleph Ω, Aleph Ω)]-multiplicity), Aleph (non-ordinal-[Hyper Aleph Ω (Aleph Ω, Aleph Ω)]-multiplicity)), Hyper Aleph (non-ordinal-[Hyper Aleph Ω (Aleph Ω, Aleph Ω)]-multiplicity) (Aleph (non-ordinal-

[Hyper Aleph Ω (Aleph Ω, Aleph Ω)]-multiplicity), Aleph (non-ordinal-[Hyper Aleph Ω (Aleph Ω, Aleph Ω)]-multiplicity))] {Hyper Hyper Aleph (non-ordinal-[Hyper Aleph Ω (Aleph Ω, Aleph Ω)]-multiplicity) (Aleph (non-ordinal-[Hyper Aleph Ω (Aleph Ω, Aleph Ω)]-multiplicity), Aleph (non-ordinal-[Hyper Aleph Ω (Aleph Ω, Aleph Ω)]-multiplicity))[Hyper Aleph (non-ordinal-[Hyper Aleph Ω (Aleph Ω, Aleph Ω)]-multiplicity) (Aleph (non-ordinal-[Hyper Aleph Ω (Aleph Ω, Aleph Ω)]-multiplicity), Aleph (non-ordinal-[Hyper Aleph Ω (Aleph Ω, Aleph Ω)]-multiplicity)), Hyper Aleph (non-ordinal-[Hyper Aleph Ω (Aleph Ω, Aleph Ω)]-multiplicity) (Aleph (non-ordinal-[Hyper Aleph Ω (Aleph Ω, Aleph Ω)]-multiplicity), Aleph (non-ordinal-[Hyper Aleph Ω (Aleph Ω, Aleph Ω)]-multiplicity))], Hyper Hyper Aleph (non-ordinal-[Hyper Aleph Ω (Aleph Ω, Aleph Ω)]-multiplicity) (Aleph (non-ordinal-[Hyper Aleph Ω (Aleph Ω, Aleph Ω)]-multiplicity), Aleph (non-ordinal-[Hyper Aleph Ω (Aleph Ω, Aleph Ω)]-multiplicity))[Hyper Aleph (non-ordinal-[Hyper Aleph Ω (Aleph Ω, Aleph Ω)]-multiplicity) (Aleph (non-ordinal-[Hyper Aleph Ω (Aleph Ω, Aleph Ω)]-multiplicity), Aleph (non-ordinal-[Hyper Aleph Ω (Aleph Ω, Aleph Ω)]-multiplicity)), Hyper Aleph (non-ordinal-[Hyper Aleph Ω (Aleph Ω, Aleph Ω)]-multiplicity) (Aleph (non-ordinal-[Hyper Aleph Ω (Aleph Ω, Aleph Ω)]-multiplicity), Aleph (non-ordinal-[Hyper Aleph Ω (Aleph Ω, Aleph Ω)]-multiplicity))]}, Hyper Hyper Hyper Aleph (non-ordinal-[Hyper Aleph Ω (Aleph Ω, Aleph Ω)]-multiplicity) (Aleph (non-ordinal-[Hyper Aleph Ω (Aleph Ω, Aleph Ω)]-multiplicity), Aleph (non-ordinal-[Hyper Aleph Ω (Aleph Ω, Aleph Ω)]-multiplicity))[Hyper Aleph (non-ordinal-[Hyper Aleph Ω (Aleph Ω, Aleph Ω)]-multiplicity) (Aleph (non-ordinal-[Hyper Aleph Ω (Aleph Ω, Aleph Ω)]-multiplicity), Aleph (non-ordinal-[Hyper Aleph Ω (Aleph Ω, Aleph Ω)]-multiplicity)), Hyper Aleph (non-ordinal-[Hyper Aleph Ω (Aleph Ω, Aleph Ω)]-multiplicity) (Aleph (non-ordinal-[Hyper Aleph Ω (Aleph Ω, Aleph Ω)]-multiplicity), Aleph (non-ordinal-[Hyper Aleph Ω (Aleph Ω, Aleph Ω)]-multiplicity))] {Hyper Hyper Aleph (non-ordinal-[Hyper Aleph Ω (Aleph Ω, Aleph Ω)]-multiplicity) (Aleph (non-ordinal-[Hyper Aleph Ω (Aleph Ω, Aleph Ω)]-multiplicity), Aleph (non-ordinal-[Hyper Aleph Ω (Aleph Ω, Aleph Ω)]-multiplicity))[Hyper Aleph (non-ordinal-[Hyper Aleph Ω (Aleph Ω, Aleph Ω)]-multiplicity) (Aleph (non-ordinal-[Hyper Aleph Ω (Aleph Ω, Aleph Ω)]-multiplicity), Aleph (non-ordinal-[Hyper Aleph Ω (Aleph Ω, Aleph Ω)]-multiplicity)), Hyper Aleph (non-ordinal-[Hyper Aleph Ω (Aleph Ω, Aleph Ω)]-multiplicity) (Aleph (non-ordinal-[Hyper Aleph Ω (Aleph Ω, Aleph Ω)]-multiplicity), Aleph (non-ordinal-[Hyper Aleph Ω (Aleph Ω, Aleph Ω)]-multiplicity))], Hyper Hyper Aleph (non-ordinal-[Hyper Aleph Ω (Aleph Ω, Aleph Ω)]-multiplicity) (Aleph (non-ordinal-[Hyper Aleph Ω (Aleph Ω, Aleph Ω)]-multiplicity), Aleph (non-ordinal-[Hyper Aleph Ω (Aleph Ω, Aleph Ω)]-multiplicity))[Hyper Aleph (non-ordinal-[Hyper Aleph Ω (Aleph Ω, Aleph Ω)]-multiplicity) (Aleph (non-ordinal-[Hyper Aleph Ω (Aleph Ω, Aleph Ω)]-multiplicity), Aleph (non-ordinal-[Hyper Aleph Ω (Aleph Ω, Aleph Ω)]-multiplicity)), Hyper Aleph (non-ordinal-[Hyper Aleph Ω (Aleph Ω, Aleph Ω)]-multiplicity) (Aleph (non-ordinal-[Hyper Aleph Ω (Aleph Ω, Aleph Ω)]-multiplicity), Aleph (non-ordinal-[Hyper Aleph Ω (Aleph Ω, Aleph Ω)]-multiplicity))]}}}}, Hyper Hyper Hyper

Hyper Hyper Hyper Aleph (non-ordinal-[Hyper Aleph Ω (Aleph Ω, Aleph Ω)]-multiplicity) (Aleph (non-ordinal-[Hyper Aleph Ω (Aleph Ω, Aleph Ω)]-multiplicity), Aleph (non-ordinal-[Hyper Aleph Ω (Aleph Ω, Aleph Ω)]-multiplicity))[Hyper Aleph (non-ordinal-[Hyper Aleph Ω (Aleph Ω, Aleph Ω)]-multiplicity) (Aleph (non-ordinal-[Hyper Aleph Ω (Aleph Ω, Aleph Ω)]-multiplicity), Aleph (non-ordinal-[Hyper Aleph Ω (Aleph Ω, Aleph Ω)]-multiplicity)), Hyper Aleph (non-ordinal-[Hyper Aleph Ω (Aleph Ω, Aleph Ω)]-multiplicity) (Aleph (non-ordinal-[Hyper Aleph Ω (Aleph Ω, Aleph Ω)]-multiplicity), Aleph (non-ordinal-[Hyper Aleph Ω (Aleph Ω, Aleph Ω)]-multiplicity))] {Hyper Hyper Aleph (non-ordinal-[Hyper Aleph Ω (Aleph Ω, Aleph Ω)]-multiplicity) (Aleph (non-ordinal-[Hyper Aleph Ω (Aleph Ω, Aleph Ω)]-multiplicity), Aleph (non-ordinal-[Hyper Aleph Ω (Aleph Ω, Aleph Ω)]-multiplicity))[Hyper Aleph (non-ordinal-[Hyper Aleph Ω (Aleph Ω, Aleph Ω)]-multiplicity) (Aleph (non-ordinal-[Hyper Aleph Ω (Aleph Ω, Aleph Ω)]-multiplicity), Aleph (non-ordinal-[Hyper Aleph Ω (Aleph Ω, Aleph Ω)]-multiplicity)), Hyper Aleph (non-ordinal-[Hyper Aleph Ω (Aleph Ω, Aleph Ω)]-multiplicity) (Aleph (non-ordinal-[Hyper Aleph Ω (Aleph Ω, Aleph Ω)]-multiplicity), Aleph (non-ordinal-[Hyper Aleph Ω (Aleph Ω, Aleph Ω)]-multiplicity))], Hyper Hyper Aleph (non-ordinal-[Hyper Aleph Ω (Aleph Ω, Aleph Ω)]-multiplicity) (Aleph (non-ordinal-[Hyper Aleph Ω (Aleph Ω, Aleph Ω)]-multiplicity), Aleph (non-ordinal-[Hyper Aleph Ω (Aleph Ω, Aleph Ω)]-multiplicity))[Hyper Aleph (non-ordinal-[Hyper Aleph Ω (Aleph Ω, Aleph Ω)]-multiplicity) (Aleph (non-ordinal-[Hyper Aleph Ω (Aleph Ω, Aleph Ω)]-multiplicity), Aleph (non-ordinal-[Hyper Aleph Ω (Aleph Ω, Aleph Ω)]-multiplicity)), Hyper Aleph (non-ordinal-[Hyper Aleph Ω (Aleph Ω, Aleph Ω)]-multiplicity) (Aleph (non-ordinal-[Hyper Aleph Ω (Aleph Ω, Aleph Ω)]-multiplicity), Aleph (non-ordinal-[Hyper Aleph Ω (Aleph Ω, Aleph Ω)]-multiplicity))]} { Hyper Hyper Hyper Aleph (non-ordinal-[Hyper Aleph Ω (Aleph Ω, Aleph Ω)]-multiplicity) (Aleph (non-ordinal-[Hyper Aleph Ω (Aleph Ω, Aleph Ω)]-multiplicity), Aleph (non-ordinal-[Hyper Aleph Ω (Aleph Ω, Aleph Ω)]-multiplicity))[Hyper Aleph (non-ordinal-[Hyper Aleph Ω (Aleph Ω, Aleph Ω)]-multiplicity) (Aleph (non-ordinal-[Hyper Aleph Ω (Aleph Ω, Aleph Ω)]-multiplicity), Aleph (non-ordinal-[Hyper Aleph Ω (Aleph Ω, Aleph Ω)]-multiplicity)), Hyper Aleph (non-ordinal-[Hyper Aleph Ω (Aleph Ω, Aleph Ω)]-multiplicity) (Aleph (non-ordinal-[Hyper Aleph Ω (Aleph Ω, Aleph Ω)]-multiplicity), Aleph (non-ordinal-[Hyper Aleph Ω (Aleph Ω, Aleph Ω)]-multiplicity))] {Hyper Hyper Aleph (non-ordinal-[Hyper Aleph Ω (Aleph Ω, Aleph Ω)]-multiplicity) (Aleph (non-ordinal-[Hyper Aleph Ω (Aleph Ω, Aleph Ω)]-multiplicity), Aleph (non-ordinal-[Hyper Aleph Ω (Aleph Ω, Aleph Ω)]-multiplicity))[Hyper Aleph (non-ordinal-[Hyper Aleph Ω (Aleph Ω, Aleph Ω)]-multiplicity) (Aleph (non-ordinal-[Hyper Aleph Ω (Aleph Ω, Aleph Ω)]-multiplicity), Aleph (non-ordinal-[Hyper Aleph Ω (Aleph Ω, Aleph Ω)]-multiplicity)), Hyper Aleph (non-ordinal-[Hyper Aleph Ω (Aleph Ω, Aleph Ω)]-multiplicity) (Aleph (non-ordinal-[Hyper Aleph Ω (Aleph Ω, Aleph Ω)]-multiplicity), Aleph (non-ordinal-[Hyper Aleph Ω (Aleph Ω, Aleph Ω)]-multiplicity))], Hyper Hyper Aleph (non-ordinal-[Hyper Aleph Ω (Aleph Ω, Aleph Ω)]-multiplicity) (Aleph (non-ordinal-[Hyper Aleph Ω (Aleph Ω, Aleph Ω)]-multiplicity), Aleph (non-ordinal-[Hyper Aleph Ω (Aleph Ω, Aleph Ω)]-multiplicity))[Hyper Aleph (non-ordinal-[Hyper Aleph Ω (Aleph Ω, Aleph Ω)]-multiplicity) (Aleph (non-ordinal-[Hyper Aleph Ω (Aleph Ω, Aleph Ω)]-multiplicity), Aleph (non-ordinal-[Hyper Aleph Ω (Aleph Ω, Aleph Ω)]-multiplicity)), Hyper Aleph (non-ordinal-[Hyper Aleph Ω (Aleph Ω, Aleph Ω)]-multiplicity) (Aleph (non-ordinal-[Hyper Aleph Ω (Aleph Ω, Aleph Ω)]-multiplicity), Aleph (non-ordinal-[Hyper Aleph Ω (Aleph Ω, Aleph Ω)]-multiplicity))]}, Hyper Hyper Hyper Aleph (non-ordinal-[Hyper Aleph Ω (Aleph Ω, Aleph Ω)]-multiplicity) (Aleph (non-ordinal-[Hyper Aleph Ω (Aleph Ω, Aleph Ω)]-multiplicity), Aleph (non-ordinal-[Hyper Aleph Ω (Aleph Ω, Aleph Ω)]-multiplicity))[Hyper Aleph (non-ordinal-[Hyper Aleph Ω (Aleph Ω, Aleph Ω)]-multiplicity) (Aleph (non-ordinal-[Hyper Aleph Ω (Aleph Ω, Aleph Ω)]-multiplicity), Aleph (non-ordinal-[Hyper Aleph Ω (Aleph Ω, Aleph Ω)]-multiplicity)), Hyper Aleph (non-ordinal-[Hyper Aleph Ω (Aleph Ω, Aleph Ω)]-multiplicity) (Aleph (non-ordinal-[Hyper Aleph Ω (Aleph Ω, Aleph Ω)]-multiplicity), Aleph (non-ordinal-[Hyper Aleph Ω (Aleph Ω, Aleph Ω)]-multiplicity))] {Hyper Hyper Aleph (non-ordinal-[Hyper Aleph Ω (Aleph Ω, Aleph Ω)]-multiplicity) (Aleph (non-ordinal-[Hyper Aleph Ω (Aleph Ω, Aleph Ω)]-multiplicity), Aleph (non-ordinal-[Hyper Aleph Ω (Aleph Ω, Aleph Ω)]-multiplicity))[Hyper Aleph (non-ordinal-[Hyper Aleph Ω (Aleph Ω, Aleph Ω)]-multiplicity) (Aleph (non-ordinal-[Hyper Aleph Ω (Aleph Ω, Aleph Ω)]-multiplicity), Aleph (non-ordinal-[Hyper Aleph Ω (Aleph Ω, Aleph Ω)]-multiplicity)), Hyper Aleph (non-ordinal-[Hyper Aleph Ω (Aleph Ω, Aleph Ω)]-multiplicity) (Aleph (non-ordinal-[Hyper Aleph Ω (Aleph Ω, Aleph Ω)]-multiplicity), Aleph (non-ordinal-[Hyper Aleph Ω (Aleph Ω, Aleph Ω)]-multiplicity))], Hyper Hyper Aleph (non-ordinal-[Hyper Aleph Ω (Aleph Ω, Aleph Ω)]-multiplicity) (Aleph (non-ordinal-[Hyper Aleph Ω (Aleph Ω, Aleph Ω)]-multiplicity), Aleph (non-ordinal-[Hyper Aleph Ω (Aleph Ω, Aleph Ω)]-multiplicity))[Hyper Aleph (non-ordinal-[Hyper Aleph Ω (Aleph Ω, Aleph Ω)]-multiplicity) (Aleph (non-ordinal-[Hyper Aleph Ω (Aleph Ω, Aleph Ω)]-multiplicity), Aleph (non-ordinal-[Hyper Aleph Ω (Aleph Ω, Aleph Ω)]-multiplicity)), Hyper Aleph (non-ordinal-[Hyper Aleph Ω (Aleph Ω, Aleph Ω)]-multiplicity) (Aleph (non-ordinal-[Hyper Aleph Ω (Aleph Ω, Aleph Ω)]-multiplicity), Aleph (non-ordinal-[Hyper Aleph Ω (Aleph Ω, Aleph Ω)]-multiplicity))]}}{ Hyper Hyper Hyper Hyper Aleph (non-ordinal-[Hyper Aleph Ω (Aleph Ω, Aleph Ω)]-multiplicity) (Aleph (non-ordinal-[Hyper Aleph Ω (Aleph Ω, Aleph Ω)]-multiplicity), Aleph (non-ordinal-[Hyper Aleph Ω (Aleph Ω, Aleph Ω)]-multiplicity))[Hyper Aleph (non-ordinal-[Hyper Aleph Ω (Aleph Ω, Aleph Ω)]-multiplicity) (Aleph (non-ordinal-[Hyper Aleph Ω (Aleph Ω, Aleph Ω)]-multiplicity), Aleph (non-ordinal-[Hyper Aleph Ω (Aleph Ω, Aleph Ω)]-multiplicity)), Hyper Aleph (non-ordinal-[Hyper Aleph Ω (Aleph Ω, Aleph Ω)]-multiplicity) (Aleph (non-ordinal-[Hyper Aleph Ω (Aleph Ω, Aleph Ω)]-multiplicity), Aleph (non-ordinal-[Hyper Aleph Ω (Aleph Ω, Aleph Ω)]-multiplicity))] {Hyper Hyper Aleph (non-ordinal-[Hyper Aleph Ω (Aleph Ω, Aleph Ω)]-multiplicity) (Aleph (non-ordinal-[Hyper Aleph Ω (Aleph Ω, Aleph Ω)]-multiplicity), Aleph (non-ordinal-[Hyper Aleph Ω (Aleph Ω, Aleph Ω)]-multiplicity))[Hyper Aleph (non-ordinal-[Hyper Aleph Ω (Aleph Ω, Aleph Ω)]-multiplicity)

(Aleph (non-ordinal-[Hyper Aleph Ω (Aleph Ω, Aleph Ω)]-multiplicity), Aleph (non-ordinal-[Hyper Aleph Ω (Aleph Ω, Aleph Ω)]-multiplicity)), Hyper Aleph (non-ordinal-[Hyper Aleph Ω (Aleph Ω, Aleph Ω)]-multiplicity) (Aleph (non-ordinal-[Hyper Aleph Ω (Aleph Ω, Aleph Ω)]-multiplicity), Aleph (non-ordinal-[Hyper Aleph Ω (Aleph Ω, Aleph Ω)]-multiplicity))], Hyper Hyper Aleph (non-ordinal-[Hyper Aleph Ω (Aleph Ω, Aleph Ω)]-multiplicity) (Aleph (non-ordinal-[Hyper Aleph Ω (Aleph Ω, Aleph Ω)]-multiplicity), Aleph (non-ordinal-[Hyper Aleph Ω (Aleph Ω, Aleph Ω)]-multiplicity))[Hyper Aleph (non-ordinal-[Hyper Aleph Ω (Aleph Ω, Aleph Ω)]-multiplicity) (Aleph (non-ordinal-[Hyper Aleph Ω (Aleph Ω, Aleph Ω)]-multiplicity), Aleph (non-ordinal-[Hyper Aleph Ω (Aleph Ω, Aleph Ω)]-multiplicity)), Hyper Aleph (non-ordinal-[Hyper Aleph Ω (Aleph Ω, Aleph Ω)]-multiplicity) (Aleph (non-ordinal-[Hyper Aleph Ω (Aleph Ω, Aleph Ω)]-multiplicity), Aleph (non-ordinal-[Hyper Aleph Ω (Aleph Ω, Aleph Ω)]-multiplicity))]} { Hyper Hyper Hyper Aleph (non-ordinal-[Hyper Aleph Ω (Aleph Ω, Aleph Ω)]-multiplicity) (Aleph (non-ordinal-[Hyper Aleph Ω (Aleph Ω, Aleph Ω)]-multiplicity), Aleph (non-ordinal-[Hyper Aleph Ω (Aleph Ω, Aleph Ω)]-multiplicity))[Hyper Aleph (non-ordinal-[Hyper Aleph Ω (Aleph Ω, Aleph Ω)]-multiplicity) (Aleph (non-ordinal-[Hyper Aleph Ω (Aleph Ω, Aleph Ω)]-multiplicity), Aleph (non-ordinal-[Hyper Aleph Ω (Aleph Ω, Aleph Ω)]-multiplicity)), Hyper Aleph (non-ordinal-[Hyper Aleph Ω (Aleph Ω, Aleph Ω)]-multiplicity) (Aleph (non-ordinal-[Hyper Aleph Ω (Aleph Ω, Aleph Ω)]-multiplicity), Aleph (non-ordinal-[Hyper Aleph Ω (Aleph Ω, Aleph Ω)]-multiplicity))] {Hyper Hyper Aleph (non-ordinal-[Hyper Aleph Ω (Aleph Ω, Aleph Ω)]-multiplicity) (Aleph (non-ordinal-[Hyper Aleph Ω (Aleph Ω, Aleph Ω)]-multiplicity), Aleph (non-ordinal-[Hyper Aleph Ω (Aleph Ω, Aleph Ω)]-multiplicity))[Hyper Aleph (non-ordinal-[Hyper Aleph Ω (Aleph Ω, Aleph Ω)]-multiplicity) (Aleph (non-ordinal-[Hyper Aleph Ω (Aleph Ω, Aleph Ω)]-multiplicity), Aleph (non-ordinal-[Hyper Aleph Ω (Aleph Ω, Aleph Ω)]-multiplicity)), Hyper Aleph (non-ordinal-[Hyper Aleph Ω (Aleph Ω, Aleph Ω)]-multiplicity) (Aleph (non-ordinal-[Hyper Aleph Ω (Aleph Ω, Aleph Ω)]-multiplicity), Aleph (non-ordinal-[Hyper Aleph Ω (Aleph Ω, Aleph Ω)]-multiplicity))], Hyper Hyper Aleph (non-ordinal-[Hyper Aleph Ω (Aleph Ω, Aleph Ω)]-multiplicity) (Aleph (non-ordinal-[Hyper Aleph Ω (Aleph Ω, Aleph Ω)]-multiplicity), Aleph (non-ordinal-[Hyper Aleph Ω (Aleph Ω, Aleph Ω)]-multiplicity))[Hyper Aleph (non-ordinal-[Hyper Aleph Ω (Aleph Ω, Aleph Ω)]-multiplicity) (Aleph (non-ordinal-[Hyper Aleph Ω (Aleph Ω, Aleph Ω)]-multiplicity), Aleph (non-ordinal-[Hyper Aleph Ω (Aleph Ω, Aleph Ω)]-multiplicity)), Hyper Aleph (non-ordinal-[Hyper Aleph Ω (Aleph Ω, Aleph Ω)]-multiplicity) (Aleph (non-ordinal-[Hyper Aleph Ω (Aleph Ω, Aleph Ω)]-multiplicity), Aleph (non-ordinal-[Hyper Aleph Ω (Aleph Ω, Aleph Ω)]-multiplicity))]}, Hyper Hyper Hyper Aleph (non-ordinal-[Hyper Aleph Ω (Aleph Ω, Aleph Ω)]-multiplicity) (Aleph (non-ordinal-[Hyper Aleph Ω (Aleph Ω, Aleph Ω)]-multiplicity), Aleph (non-ordinal-[Hyper Aleph Ω (Aleph Ω, Aleph Ω)]-multiplicity))[Hyper Aleph (non-ordinal-[Hyper Aleph Ω (Aleph Ω, Aleph Ω)]-multiplicity) (Aleph (non-ordinal-[Hyper Aleph Ω (Aleph Ω, Aleph Ω)]-multiplicity), Aleph (non-ordinal-[Hyper Aleph Ω (Aleph Ω, Aleph Ω)]-multiplicity)), Hyper

Aleph (non-ordinal-[Hyper Aleph Ω (Aleph Ω, Aleph Ω)]-multiplicity) (Aleph (non-ordinal-[Hyper Aleph Ω (Aleph Ω, Aleph Ω)]-multiplicity), Aleph (non-ordinal-[Hyper Aleph Ω (Aleph Ω, Aleph Ω)]-multiplicity))] {Hyper Hyper Aleph (non-ordinal-[Hyper Aleph Ω (Aleph Ω, Aleph Ω)]-multiplicity) (Aleph (non-ordinal-[Hyper Aleph Ω (Aleph Ω, Aleph Ω)]-multiplicity), Aleph (non-ordinal-[Hyper Aleph Ω (Aleph Ω, Aleph Ω)]-multiplicity))[Hyper Aleph (non-ordinal-[Hyper Aleph Ω (Aleph Ω, Aleph Ω)]-multiplicity) (Aleph (non-ordinal-[Hyper Aleph Ω (Aleph Ω, Aleph Ω)]-multiplicity), Aleph (non-ordinal-[Hyper Aleph Ω (Aleph Ω, Aleph Ω)]-multiplicity)), Hyper Aleph (non-ordinal-[Hyper Aleph Ω (Aleph Ω, Aleph Ω)]-multiplicity) (Aleph (non-ordinal-[Hyper Aleph Ω (Aleph Ω, Aleph Ω)]-multiplicity), Aleph (non-ordinal-[Hyper Aleph Ω (Aleph Ω, Aleph Ω)]-multiplicity))], Hyper Hyper Aleph (non-ordinal-[Hyper Aleph Ω (Aleph Ω, Aleph Ω)]-multiplicity) (Aleph (non-ordinal-[Hyper Aleph Ω (Aleph Ω, Aleph Ω)]-multiplicity), Aleph (non-ordinal-[Hyper Aleph Ω (Aleph Ω, Aleph Ω)]-multiplicity))[Hyper Aleph (non-ordinal-[Hyper Aleph Ω (Aleph Ω, Aleph Ω)]-multiplicity) (Aleph (non-ordinal-[Hyper Aleph Ω (Aleph Ω, Aleph Ω)]-multiplicity), Aleph (non-ordinal-[Hyper Aleph Ω (Aleph Ω, Aleph Ω)]-multiplicity)), Hyper Aleph (non-ordinal-[Hyper Aleph Ω (Aleph Ω, Aleph Ω)]-multiplicity) (Aleph (non-ordinal-[Hyper Aleph Ω (Aleph Ω, Aleph Ω)]-multiplicity), Aleph (non-ordinal-[Hyper Aleph Ω (Aleph Ω, Aleph Ω)]-multiplicity))]}}, Hyper Hyper Hyper Hyper Aleph (non-ordinal-[Hyper Aleph Ω (Aleph Ω, Aleph Ω)]-multiplicity) (Aleph (non-ordinal-[Hyper Aleph Ω (Aleph Ω, Aleph Ω)]-multiplicity), Aleph (non-ordinal-[Hyper Aleph Ω (Aleph Ω, Aleph Ω)]-multiplicity))[Hyper Aleph (non-ordinal-[Hyper Aleph Ω (Aleph Ω, Aleph Ω)]-multiplicity) (Aleph (non-ordinal-[Hyper Aleph Ω (Aleph Ω, Aleph Ω)]-multiplicity), Aleph (non-ordinal-[Hyper Aleph Ω (Aleph Ω, Aleph Ω)]-multiplicity)), Hyper Aleph (non-ordinal-[Hyper Aleph Ω (Aleph Ω, Aleph Ω)]-multiplicity) (Aleph (non-ordinal-[Hyper Aleph Ω (Aleph Ω, Aleph Ω)]-multiplicity), Aleph (non-ordinal-[Hyper Aleph Ω (Aleph Ω, Aleph Ω)]-multiplicity))] {Hyper Hyper Aleph (non-ordinal-[Hyper Aleph Ω (Aleph Ω, Aleph Ω)]-multiplicity) (Aleph (non-ordinal-[Hyper Aleph Ω (Aleph Ω, Aleph Ω)]-multiplicity), Aleph (non-ordinal-[Hyper Aleph Ω (Aleph Ω, Aleph Ω)]-multiplicity))[Hyper Aleph (non-ordinal-[Hyper Aleph Ω (Aleph Ω, Aleph Ω)]-multiplicity) (Aleph (non-ordinal-[Hyper Aleph Ω (Aleph Ω, Aleph Ω)]-multiplicity), Aleph (non-ordinal-[Hyper Aleph Ω (Aleph Ω, Aleph Ω)]-multiplicity)), Hyper Aleph (non-ordinal-[Hyper Aleph Ω (Aleph Ω, Aleph Ω)]-multiplicity) (Aleph (non-ordinal-[Hyper Aleph Ω (Aleph Ω, Aleph Ω)]-multiplicity), Aleph (non-ordinal-[Hyper Aleph Ω (Aleph Ω, Aleph Ω)]-multiplicity))], Hyper Hyper Aleph (non-ordinal-[Hyper Aleph Ω (Aleph Ω, Aleph Ω)]-multiplicity) (Aleph (non-ordinal-[Hyper Aleph Ω (Aleph Ω, Aleph Ω)]-multiplicity), Aleph (non-ordinal-[Hyper Aleph Ω (Aleph Ω, Aleph Ω)]-multiplicity))[Hyper Aleph (non-ordinal-[Hyper Aleph Ω (Aleph Ω, Aleph Ω)]-multiplicity) (Aleph (non-ordinal-[Hyper Aleph Ω (Aleph Ω, Aleph Ω)]-multiplicity), Aleph (non-ordinal-[Hyper Aleph Ω (Aleph Ω, Aleph Ω)]-multiplicity)), Hyper Aleph (non-ordinal-[Hyper Aleph Ω (Aleph Ω, Aleph Ω)]-multiplicity) (Aleph (non-ordinal-[Hyper Aleph Ω (Aleph Ω, Aleph Ω)]-multiplicity), Aleph (non-ordinal-[Hyper Aleph Ω (Aleph Ω, Aleph Ω)]-multiplicity))]} { Hyper Hyper Hyper Aleph (non-ordinal-[Hyper Aleph Ω (Aleph Ω, Aleph Ω)]-multiplicity) (Aleph (non-ordinal-[Hyper Aleph Ω (Aleph Ω, Aleph Ω)]-multiplicity), Aleph (non-ordinal-[Hyper Aleph Ω (Aleph Ω, Aleph Ω)]-multiplicity))[Hyper Aleph (non-ordinal-[Hyper Aleph Ω (Aleph Ω, Aleph Ω)]-multiplicity) (Aleph (non-ordinal-[Hyper Aleph Ω (Aleph Ω, Aleph Ω)]-multiplicity), Aleph (non-ordinal-[Hyper Aleph Ω (Aleph Ω, Aleph Ω)]-multiplicity)), Hyper Aleph (non-ordinal-[Hyper Aleph Ω (Aleph Ω, Aleph Ω)]-multiplicity) (Aleph (non-ordinal-[Hyper Aleph Ω (Aleph Ω, Aleph Ω)]-multiplicity), Aleph (non-ordinal-[Hyper Aleph Ω (Aleph Ω, Aleph Ω)]-multiplicity))] {Hyper Hyper Aleph (non-ordinal-[Hyper Aleph Ω (Aleph Ω, Aleph Ω)]-multiplicity) (Aleph (non-ordinal-[Hyper Aleph Ω (Aleph Ω, Aleph Ω)]-multiplicity), Aleph (non-ordinal-[Hyper Aleph Ω (Aleph Ω, Aleph Ω)]-multiplicity))[Hyper Aleph (non-ordinal-[Hyper Aleph Ω (Aleph Ω, Aleph Ω)]-multiplicity) (Aleph (non-ordinal-[Hyper Aleph Ω (Aleph Ω, Aleph Ω)]-multiplicity), Aleph (non-ordinal-[Hyper Aleph Ω (Aleph Ω, Aleph Ω)]-multiplicity)), Hyper Aleph (non-ordinal-[Hyper Aleph Ω (Aleph Ω, Aleph Ω)]-multiplicity) (Aleph (non-ordinal-[Hyper Aleph Ω (Aleph Ω, Aleph Ω)]-multiplicity), Aleph (non-ordinal-[Hyper Aleph Ω (Aleph Ω, Aleph Ω)]-multiplicity))], Hyper Hyper Aleph (non-ordinal-[Hyper Aleph Ω (Aleph Ω, Aleph Ω)]-multiplicity) (Aleph (non-ordinal-[Hyper Aleph Ω (Aleph Ω, Aleph Ω)]-multiplicity), Aleph (non-ordinal-[Hyper Aleph Ω (Aleph Ω, Aleph Ω)]-multiplicity))[Hyper Aleph (non-ordinal-[Hyper Aleph Ω (Aleph Ω, Aleph Ω)]-multiplicity) (Aleph (non-ordinal-[Hyper Aleph Ω (Aleph Ω, Aleph Ω)]-multiplicity), Aleph (non-ordinal-[Hyper Aleph Ω (Aleph Ω, Aleph Ω)]-multiplicity)), Hyper Aleph (non-ordinal-[Hyper Aleph Ω (Aleph Ω, Aleph Ω)]-multiplicity) (Aleph (non-ordinal-[Hyper Aleph Ω (Aleph Ω, Aleph Ω)]-multiplicity), Aleph (non-ordinal-[Hyper Aleph Ω (Aleph Ω, Aleph Ω)]-multiplicity))]}, Hyper Hyper Hyper Aleph (non-ordinal-[Hyper Aleph Ω (Aleph Ω, Aleph Ω)]-multiplicity) (Aleph (non-ordinal-[Hyper Aleph Ω (Aleph Ω, Aleph Ω)]-multiplicity), Aleph (non-ordinal-[Hyper Aleph Ω (Aleph Ω, Aleph Ω)]-multiplicity))[Hyper Aleph (non-ordinal-[Hyper Aleph Ω (Aleph Ω, Aleph Ω)]-multiplicity) (Aleph (non-ordinal-[Hyper Aleph Ω (Aleph Ω, Aleph Ω)]-multiplicity), Aleph (non-ordinal-[Hyper Aleph Ω (Aleph Ω, Aleph Ω)]-multiplicity)), Hyper Aleph (non-ordinal-[Hyper Aleph Ω (Aleph Ω, Aleph Ω)]-multiplicity) (Aleph (non-ordinal-[Hyper Aleph Ω (Aleph Ω, Aleph Ω)]-multiplicity), Aleph (non-ordinal-[Hyper Aleph Ω (Aleph Ω, Aleph Ω)]-multiplicity))] {Hyper Hyper Aleph (non-ordinal-[Hyper Aleph Ω (Aleph Ω, Aleph Ω)]-multiplicity) (Aleph (non-ordinal-[Hyper Aleph Ω (Aleph Ω, Aleph Ω)]-multiplicity), Aleph (non-ordinal-[Hyper Aleph Ω (Aleph Ω, Aleph Ω)]-multiplicity))[Hyper Aleph (non-ordinal-[Hyper Aleph Ω (Aleph Ω, Aleph Ω)]-multiplicity) (Aleph (non-ordinal-[Hyper Aleph Ω (Aleph Ω, Aleph Ω)]-multiplicity), Aleph (non-ordinal-[Hyper Aleph Ω (Aleph Ω, Aleph Ω)]-multiplicity)), Hyper Aleph (non-ordinal-[Hyper Aleph Ω (Aleph Ω, Aleph Ω)]-multiplicity) (Aleph (non-ordinal-[Hyper Aleph Ω (Aleph Ω, Aleph Ω)]-multiplicity), Aleph (non-ordinal-[Hyper Aleph Ω (Aleph Ω, Aleph Ω)]-multiplicity))], Hyper Hyper Aleph (non-ordinal-[Hyper Aleph Ω (Aleph Ω, Aleph Ω)]-

multiplicity) (Aleph (non-ordinal-[Hyper Aleph Ω (Aleph Ω, Aleph Ω)]-multiplicity), Aleph (non-ordinal-[Hyper Aleph Ω (Aleph Ω, Aleph Ω)]-multiplicity))[Hyper Aleph (non-ordinal-[Hyper Aleph Ω (Aleph Ω, Aleph Ω)]-multiplicity) (Aleph (non-ordinal-[Hyper Aleph Ω (Aleph Ω, Aleph Ω)]-multiplicity), Aleph (non-ordinal-[Hyper Aleph Ω (Aleph Ω, Aleph Ω)]-multiplicity)), Hyper Aleph (non-ordinal-[Hyper Aleph Ω (Aleph Ω, Aleph Ω)]-multiplicity) (Aleph (non-ordinal-[Hyper Aleph Ω (Aleph Ω, Aleph Ω)]-multiplicity), Aleph (non-ordinal-[Hyper Aleph Ω (Aleph Ω, Aleph Ω)]-multiplicity))]}}}{ Hyper Hyper Hyper Hyper Hyper Aleph (non-ordinal-[Hyper Aleph Ω (Aleph Ω, Aleph Ω)]-multiplicity) (Aleph (non-ordinal-[Hyper Aleph Ω (Aleph Ω, Aleph Ω)]-multiplicity), Aleph (non-ordinal-[Hyper Aleph Ω (Aleph Ω, Aleph Ω)]-multiplicity))[Hyper Aleph (non-ordinal-[Hyper Aleph Ω (Aleph Ω, Aleph Ω)]-multiplicity) (Aleph (non-ordinal-[Hyper Aleph Ω (Aleph Ω, Aleph Ω)]-multiplicity), Aleph (non-ordinal-[Hyper Aleph Ω (Aleph Ω, Aleph Ω)]-multiplicity)), Hyper Aleph (non-ordinal-[Hyper Aleph Ω (Aleph Ω, Aleph Ω)]-multiplicity) (Aleph (non-ordinal-[Hyper Aleph Ω (Aleph Ω, Aleph Ω)]-multiplicity), Aleph (non-ordinal-[Hyper Aleph Ω (Aleph Ω, Aleph Ω)]-multiplicity))] {Hyper Hyper Aleph (non-ordinal-[Hyper Aleph Ω (Aleph Ω, Aleph Ω)]-multiplicity) (Aleph (non-ordinal-[Hyper Aleph Ω (Aleph Ω, Aleph Ω)]-multiplicity), Aleph (non-ordinal-[Hyper Aleph Ω (Aleph Ω, Aleph Ω)]-multiplicity))[Hyper Aleph (non-ordinal-[Hyper Aleph Ω (Aleph Ω, Aleph Ω)]-multiplicity) (Aleph (non-ordinal-[Hyper Aleph Ω (Aleph Ω, Aleph Ω)]-multiplicity), Aleph (non-ordinal-[Hyper Aleph Ω (Aleph Ω, Aleph Ω)]-multiplicity)), Hyper Aleph (non-ordinal-[Hyper Aleph Ω (Aleph Ω, Aleph Ω)]-multiplicity) (Aleph (non-ordinal-[Hyper Aleph Ω (Aleph Ω, Aleph Ω)]-multiplicity), Aleph (non-ordinal-[Hyper Aleph Ω (Aleph Ω, Aleph Ω)]-multiplicity))], Hyper Hyper Aleph (non-ordinal-[Hyper Aleph Ω (Aleph Ω, Aleph Ω)]-multiplicity) (Aleph (non-ordinal-[Hyper Aleph Ω (Aleph Ω, Aleph Ω)]-multiplicity), Aleph (non-ordinal-[Hyper Aleph Ω (Aleph Ω, Aleph Ω)]-multiplicity))[Hyper Aleph (non-ordinal-[Hyper Aleph Ω (Aleph Ω, Aleph Ω)]-multiplicity) (Aleph (non-ordinal-[Hyper Aleph Ω (Aleph Ω, Aleph Ω)]-multiplicity), Aleph (non-ordinal-[Hyper Aleph Ω (Aleph Ω, Aleph Ω)]-multiplicity)), Hyper Aleph (non-ordinal-[Hyper Aleph Ω (Aleph Ω, Aleph Ω)]-multiplicity) (Aleph (non-ordinal-[Hyper Aleph Ω (Aleph Ω, Aleph Ω)]-multiplicity), Aleph (non-ordinal-[Hyper Aleph Ω (Aleph Ω, Aleph Ω)]-multiplicity))]} { Hyper Hyper Hyper Aleph (non-ordinal-[Hyper Aleph Ω (Aleph Ω, Aleph Ω)]-multiplicity) (Aleph (non-ordinal-[Hyper Aleph Ω (Aleph Ω, Aleph Ω)]-multiplicity), Aleph (non-ordinal-[Hyper Aleph Ω (Aleph Ω, Aleph Ω)]-multiplicity))[Hyper Aleph (non-ordinal-[Hyper Aleph Ω (Aleph Ω, Aleph Ω)]-multiplicity) (Aleph (non-ordinal-[Hyper Aleph Ω (Aleph Ω, Aleph Ω)]-multiplicity), Aleph (non-ordinal-[Hyper Aleph Ω (Aleph Ω, Aleph Ω)]-multiplicity)), Hyper Aleph (non-ordinal-[Hyper Aleph Ω (Aleph Ω, Aleph Ω)]-multiplicity) (Aleph (non-ordinal-[Hyper Aleph Ω (Aleph Ω, Aleph Ω)]-multiplicity), Aleph (non-ordinal-[Hyper Aleph Ω (Aleph Ω, Aleph Ω)]-multiplicity))] {Hyper Hyper Aleph (non-ordinal-[Hyper Aleph Ω (Aleph Ω, Aleph Ω)]-multiplicity) (Aleph (non-ordinal-[Hyper Aleph Ω (Aleph Ω, Aleph Ω)]-multiplicity), Aleph (non-ordinal-[Hyper Aleph Ω (Aleph Ω, Aleph Ω)]-

multiplicity))[Hyper Aleph (non-ordinal-[Hyper Aleph Ω (Aleph Ω, Aleph Ω)]-multiplicity) (Aleph (non-ordinal-[Hyper Aleph Ω (Aleph Ω, Aleph Ω)]-multiplicity), Aleph (non-ordinal-[Hyper Aleph Ω (Aleph Ω, Aleph Ω)]-multiplicity)), Hyper Aleph (non-ordinal-[Hyper Aleph Ω (Aleph Ω, Aleph Ω)]-multiplicity) (Aleph (non-ordinal-[Hyper Aleph Ω (Aleph Ω, Aleph Ω)]-multiplicity), Aleph (non-ordinal-[Hyper Aleph Ω (Aleph Ω, Aleph Ω)]-multiplicity))], Hyper Hyper Aleph (non-ordinal-[Hyper Aleph Ω (Aleph Ω, Aleph Ω)]-multiplicity) (Aleph (non-ordinal-[Hyper Aleph Ω (Aleph Ω, Aleph Ω)]-multiplicity), Aleph (non-ordinal-[Hyper Aleph Ω (Aleph Ω, Aleph Ω)]-multiplicity))[Hyper Aleph (non-ordinal-[Hyper Aleph Ω (Aleph Ω, Aleph Ω)]-multiplicity) (Aleph (non-ordinal-[Hyper Aleph Ω (Aleph Ω, Aleph Ω)]-multiplicity), Aleph (non-ordinal-[Hyper Aleph Ω (Aleph Ω, Aleph Ω)]-multiplicity)), Hyper Aleph (non-ordinal-[Hyper Aleph Ω (Aleph Ω, Aleph Ω)]-multiplicity) (Aleph (non-ordinal-[Hyper Aleph Ω (Aleph Ω, Aleph Ω)]-multiplicity), Aleph (non-ordinal-[Hyper Aleph Ω (Aleph Ω, Aleph Ω)]-multiplicity))]}, Hyper Hyper Hyper Aleph (non-ordinal-[Hyper Aleph Ω (Aleph Ω, Aleph Ω)]-multiplicity) (Aleph (non-ordinal-[Hyper Aleph Ω (Aleph Ω, Aleph Ω)]-multiplicity), Aleph (non-ordinal-[Hyper Aleph Ω (Aleph Ω, Aleph Ω)]-multiplicity))[Hyper Aleph (non-ordinal-[Hyper Aleph Ω (Aleph Ω, Aleph Ω)]-multiplicity) (Aleph (non-ordinal-[Hyper Aleph Ω (Aleph Ω, Aleph Ω)]-multiplicity), Aleph (non-ordinal-[Hyper Aleph Ω (Aleph Ω, Aleph Ω)]-multiplicity)), Hyper Aleph (non-ordinal-[Hyper Aleph Ω (Aleph Ω, Aleph Ω)]-multiplicity) (Aleph (non-ordinal-[Hyper Aleph Ω (Aleph Ω, Aleph Ω)]-multiplicity), Aleph (non-ordinal-[Hyper Aleph Ω (Aleph Ω, Aleph Ω)]-multiplicity))] {Hyper Hyper Aleph (non-ordinal-[Hyper Aleph Ω (Aleph Ω, Aleph Ω)]-multiplicity) (Aleph (non-ordinal-[Hyper Aleph Ω (Aleph Ω, Aleph Ω)]-multiplicity), Aleph (non-ordinal-[Hyper Aleph Ω (Aleph Ω, Aleph Ω)]-multiplicity))[Hyper Aleph (non-ordinal-[Hyper Aleph Ω (Aleph Ω, Aleph Ω)]-multiplicity) (Aleph (non-ordinal-[Hyper Aleph Ω (Aleph Ω, Aleph Ω)]-multiplicity), Aleph (non-ordinal-[Hyper Aleph Ω (Aleph Ω, Aleph Ω)]-multiplicity)), Hyper Aleph (non-ordinal-[Hyper Aleph Ω (Aleph Ω, Aleph Ω)]-multiplicity) (Aleph (non-ordinal-[Hyper Aleph Ω (Aleph Ω, Aleph Ω)]-multiplicity), Aleph (non-ordinal-[Hyper Aleph Ω (Aleph Ω, Aleph Ω)]-multiplicity))], Hyper Hyper Aleph (non-ordinal-[Hyper Aleph Ω (Aleph Ω, Aleph Ω)]-multiplicity) (Aleph (non-ordinal-[Hyper Aleph Ω (Aleph Ω, Aleph Ω)]-multiplicity), Aleph (non-ordinal-[Hyper Aleph Ω (Aleph Ω, Aleph Ω)]-multiplicity))[Hyper Aleph (non-ordinal-[Hyper Aleph Ω (Aleph Ω, Aleph Ω)]-multiplicity) (Aleph (non-ordinal-[Hyper Aleph Ω (Aleph Ω, Aleph Ω)]-multiplicity), Aleph (non-ordinal-[Hyper Aleph Ω (Aleph Ω, Aleph Ω)]-multiplicity)), Hyper Aleph (non-ordinal-[Hyper Aleph Ω (Aleph Ω, Aleph Ω)]-multiplicity) (Aleph (non-ordinal-[Hyper Aleph Ω (Aleph Ω, Aleph Ω)]-multiplicity), Aleph (non-ordinal-[Hyper Aleph Ω (Aleph Ω, Aleph Ω)]-multiplicity))]}}{ Hyper Hyper Hyper Hyper Aleph (non-ordinal-[Hyper Aleph Ω (Aleph Ω, Aleph Ω)]-multiplicity) (Aleph (non-ordinal-[Hyper Aleph Ω (Aleph Ω, Aleph Ω)]-multiplicity), Aleph (non-ordinal-[Hyper Aleph Ω (Aleph Ω, Aleph Ω)]-multiplicity))[Hyper Aleph (non-ordinal-[Hyper Aleph Ω (Aleph Ω, Aleph Ω)]-multiplicity) (Aleph (non-ordinal-[Hyper Aleph Ω (Aleph Ω, Aleph

Ω)]-multiplicity), Aleph (non-ordinal-[Hyper Aleph Ω (Aleph Ω, Aleph Ω)]-multiplicity)), Hyper Aleph (non-ordinal-[Hyper Aleph Ω (Aleph Ω, Aleph Ω)]-multiplicity) (Aleph (non-ordinal-[Hyper Aleph Ω (Aleph Ω, Aleph Ω)]-multiplicity), Aleph (non-ordinal-[Hyper Aleph Ω (Aleph Ω, Aleph Ω)]-multiplicity))] {Hyper Hyper Aleph (non-ordinal-[Hyper Aleph Ω (Aleph Ω, Aleph Ω)]-multiplicity) (Aleph (non-ordinal-[Hyper Aleph Ω (Aleph Ω, Aleph Ω)]-multiplicity), Aleph (non-ordinal-[Hyper Aleph Ω (Aleph Ω, Aleph Ω)]-multiplicity))[Hyper Aleph (non-ordinal-[Hyper Aleph Ω (Aleph Ω, Aleph Ω)]-multiplicity) (Aleph (non-ordinal-[Hyper Aleph Ω (Aleph Ω, Aleph Ω)]-multiplicity), Aleph (non-ordinal-[Hyper Aleph Ω (Aleph Ω, Aleph Ω)]-multiplicity)), Hyper Aleph (non-ordinal-[Hyper Aleph Ω (Aleph Ω, Aleph Ω)]-multiplicity) (Aleph (non-ordinal-[Hyper Aleph Ω (Aleph Ω, Aleph Ω)]-multiplicity), Aleph (non-ordinal-[Hyper Aleph Ω (Aleph Ω, Aleph Ω)]-multiplicity))], Hyper Hyper Aleph (non-ordinal-[Hyper Aleph Ω (Aleph Ω, Aleph Ω)]-multiplicity) (Aleph (non-ordinal-[Hyper Aleph Ω (Aleph Ω, Aleph Ω)]-multiplicity), Aleph (non-ordinal-[Hyper Aleph Ω (Aleph Ω, Aleph Ω)]-multiplicity))[Hyper Aleph (non-ordinal-[Hyper Aleph Ω (Aleph Ω, Aleph Ω)]-multiplicity) (Aleph (non-ordinal-[Hyper Aleph Ω (Aleph Ω, Aleph Ω)]-multiplicity), Aleph (non-ordinal-[Hyper Aleph Ω (Aleph Ω, Aleph Ω)]-multiplicity)), Hyper Aleph (non-ordinal-[Hyper Aleph Ω (Aleph Ω, Aleph Ω)]-multiplicity) (Aleph (non-ordinal-[Hyper Aleph Ω (Aleph Ω, Aleph Ω)]-multiplicity), Aleph (non-ordinal-[Hyper Aleph Ω (Aleph Ω, Aleph Ω)]-multiplicity))]} { Hyper Hyper Hyper Aleph (non-ordinal-[Hyper Aleph Ω (Aleph Ω, Aleph Ω)]-multiplicity) (Aleph (non-ordinal-[Hyper Aleph Ω (Aleph Ω, Aleph Ω)]-multiplicity), Aleph (non-ordinal-[Hyper Aleph Ω (Aleph Ω, Aleph Ω)]-multiplicity))[Hyper Aleph (non-ordinal-[Hyper Aleph Ω (Aleph Ω, Aleph Ω)]-multiplicity) (Aleph (non-ordinal-[Hyper Aleph Ω (Aleph Ω, Aleph Ω)]-multiplicity), Aleph (non-ordinal-[Hyper Aleph Ω (Aleph Ω, Aleph Ω)]-multiplicity)), Hyper Aleph (non-ordinal-[Hyper Aleph Ω (Aleph Ω, Aleph Ω)]-multiplicity) (Aleph (non-ordinal-[Hyper Aleph Ω (Aleph Ω, Aleph Ω)]-multiplicity), Aleph (non-ordinal-[Hyper Aleph Ω (Aleph Ω, Aleph Ω)]-multiplicity))] {Hyper Hyper Aleph (non-ordinal-[Hyper Aleph Ω (Aleph Ω, Aleph Ω)]-multiplicity) (Aleph (non-ordinal-[Hyper Aleph Ω (Aleph Ω, Aleph Ω)]-multiplicity), Aleph (non-ordinal-[Hyper Aleph Ω (Aleph Ω, Aleph Ω)]-multiplicity))[Hyper Aleph (non-ordinal-[Hyper Aleph Ω (Aleph Ω, Aleph Ω)]-multiplicity) (Aleph (non-ordinal-[Hyper Aleph Ω (Aleph Ω, Aleph Ω)]-multiplicity), Aleph (non-ordinal-[Hyper Aleph Ω (Aleph Ω, Aleph Ω)]-multiplicity)), Hyper Aleph (non-ordinal-[Hyper Aleph Ω (Aleph Ω, Aleph Ω)]-multiplicity) (Aleph (non-ordinal-[Hyper Aleph Ω (Aleph Ω, Aleph Ω)]-multiplicity), Aleph (non-ordinal-[Hyper Aleph Ω (Aleph Ω, Aleph Ω)]-multiplicity))], Hyper Hyper Aleph (non-ordinal-[Hyper Aleph Ω (Aleph Ω, Aleph Ω)]-multiplicity) (Aleph (non-ordinal-[Hyper Aleph Ω (Aleph Ω, Aleph Ω)]-multiplicity), Aleph (non-ordinal-[Hyper Aleph Ω (Aleph Ω, Aleph Ω)]-multiplicity))[Hyper Aleph (non-ordinal-[Hyper Aleph Ω (Aleph Ω, Aleph Ω)]-multiplicity) (Aleph (non-ordinal-[Hyper Aleph Ω (Aleph Ω, Aleph Ω)]-multiplicity), Aleph (non-ordinal-[Hyper Aleph Ω (Aleph Ω, Aleph Ω)]-multiplicity)), Hyper Aleph (non-ordinal-[Hyper Aleph Ω (Aleph Ω, Aleph Ω)]-

multiplicity) (Aleph (non-ordinal-[Hyper Aleph Ω (Aleph Ω, Aleph Ω)]-multiplicity), Aleph (non-ordinal-[Hyper Aleph Ω (Aleph Ω, Aleph Ω)]-multiplicity))]}, Hyper Hyper Hyper Aleph (non-ordinal-[Hyper Aleph Ω (Aleph Ω, Aleph Ω)]-multiplicity) (Aleph (non-ordinal-[Hyper Aleph Ω (Aleph Ω, Aleph Ω)]-multiplicity), Aleph (non-ordinal-[Hyper Aleph Ω (Aleph Ω, Aleph Ω)]-multiplicity))[Hyper Aleph (non-ordinal-[Hyper Aleph Ω (Aleph Ω, Aleph Ω)]-multiplicity) (Aleph (non-ordinal-[Hyper Aleph Ω (Aleph Ω, Aleph Ω)]-multiplicity), Aleph (non-ordinal-[Hyper Aleph Ω (Aleph Ω, Aleph Ω)]-multiplicity)), Hyper Aleph (non-ordinal-[Hyper Aleph Ω (Aleph Ω, Aleph Ω)]-multiplicity) (Aleph (non-ordinal-[Hyper Aleph Ω (Aleph Ω, Aleph Ω)]-multiplicity), Aleph (non-ordinal-[Hyper Aleph Ω (Aleph Ω, Aleph Ω)]-multiplicity))] {Hyper Hyper Aleph (non-ordinal-[Hyper Aleph Ω (Aleph Ω, Aleph Ω)]-multiplicity) (Aleph (non-ordinal-[Hyper Aleph Ω (Aleph Ω, Aleph Ω)]-multiplicity), Aleph (non-ordinal-[Hyper Aleph Ω (Aleph Ω, Aleph Ω)]-multiplicity))[Hyper Aleph (non-ordinal-[Hyper Aleph Ω (Aleph Ω, Aleph Ω)]-multiplicity) (Aleph (non-ordinal-[Hyper Aleph Ω (Aleph Ω, Aleph Ω)]-multiplicity), Aleph (non-ordinal-[Hyper Aleph Ω (Aleph Ω, Aleph Ω)]-multiplicity)), Hyper Aleph (non-ordinal-[Hyper Aleph Ω (Aleph Ω, Aleph Ω)]-multiplicity) (Aleph (non-ordinal-[Hyper Aleph Ω (Aleph Ω, Aleph Ω)]-multiplicity), Aleph (non-ordinal-[Hyper Aleph Ω (Aleph Ω, Aleph Ω)]-multiplicity))], Hyper Hyper Aleph (non-ordinal-[Hyper Aleph Ω (Aleph Ω, Aleph Ω)]-multiplicity) (Aleph (non-ordinal-[Hyper Aleph Ω (Aleph Ω, Aleph Ω)]-multiplicity), Aleph (non-ordinal-[Hyper Aleph Ω (Aleph Ω, Aleph Ω)]-multiplicity))[Hyper Aleph (non-ordinal-[Hyper Aleph Ω (Aleph Ω, Aleph Ω)]-multiplicity) (Aleph (non-ordinal-[Hyper Aleph Ω (Aleph Ω, Aleph Ω)]-multiplicity), Aleph (non-ordinal-[Hyper Aleph Ω (Aleph Ω, Aleph Ω)]-multiplicity)), Hyper Aleph (non-ordinal-[Hyper Aleph Ω (Aleph Ω, Aleph Ω)]-multiplicity) (Aleph (non-ordinal-[Hyper Aleph Ω (Aleph Ω, Aleph Ω)]-multiplicity), Aleph (non-ordinal-[Hyper Aleph Ω (Aleph Ω, Aleph Ω)]-multiplicity))]}}, Hyper Hyper Hyper Hyper Aleph (non-ordinal-[Hyper Aleph Ω (Aleph Ω, Aleph Ω)]-multiplicity) (Aleph (non-ordinal-[Hyper Aleph Ω (Aleph Ω, Aleph Ω)]-multiplicity), Aleph (non-ordinal-[Hyper Aleph Ω (Aleph Ω, Aleph Ω)]-multiplicity))[Hyper Aleph (non-ordinal-[Hyper Aleph Ω (Aleph Ω, Aleph Ω)]-multiplicity) (Aleph (non-ordinal-[Hyper Aleph Ω (Aleph Ω, Aleph Ω)]-multiplicity), Aleph (non-ordinal-[Hyper Aleph Ω (Aleph Ω, Aleph Ω)]-multiplicity)), Hyper Aleph (non-ordinal-[Hyper Aleph Ω (Aleph Ω, Aleph Ω)]-multiplicity) (Aleph (non-ordinal-[Hyper Aleph Ω (Aleph Ω, Aleph Ω)]-multiplicity), Aleph (non-ordinal-[Hyper Aleph Ω (Aleph Ω, Aleph Ω)]-multiplicity))] {Hyper Hyper Aleph (non-ordinal-[Hyper Aleph Ω (Aleph Ω, Aleph Ω)]-multiplicity) (Aleph (non-ordinal-[Hyper Aleph Ω (Aleph Ω, Aleph Ω)]-multiplicity), Aleph (non-ordinal-[Hyper Aleph Ω (Aleph Ω, Aleph Ω)]-multiplicity))[Hyper Aleph (non-ordinal-[Hyper Aleph Ω (Aleph Ω, Aleph Ω)]-multiplicity) (Aleph (non-ordinal-[Hyper Aleph Ω (Aleph Ω, Aleph Ω)]-multiplicity), Aleph (non-ordinal-[Hyper Aleph Ω (Aleph Ω, Aleph Ω)]-multiplicity)), Hyper Aleph (non-ordinal-[Hyper Aleph Ω (Aleph Ω, Aleph Ω)]-multiplicity) (Aleph (non-ordinal-[Hyper Aleph Ω (Aleph Ω, Aleph Ω)]-multiplicity), Aleph (non-ordinal-[Hyper Aleph Ω (Aleph Ω, Aleph Ω)]-

multiplicity))], Hyper Hyper Aleph (non-ordinal-[Hyper Aleph Ω (Aleph Ω, Aleph Ω)]-multiplicity) (Aleph (non-ordinal-[Hyper Aleph Ω (Aleph Ω, Aleph Ω)]-multiplicity), Aleph (non-ordinal-[Hyper Aleph Ω (Aleph Ω, Aleph Ω)]-multiplicity))[Hyper Aleph (non-ordinal-[Hyper Aleph Ω (Aleph Ω, Aleph Ω)]-multiplicity) (Aleph (non-ordinal-[Hyper Aleph Ω (Aleph Ω, Aleph Ω)]-multiplicity), Aleph (non-ordinal-[Hyper Aleph Ω (Aleph Ω, Aleph Ω)]-multiplicity)), Hyper Aleph (non-ordinal-[Hyper Aleph Ω (Aleph Ω, Aleph Ω)]-multiplicity) (Aleph (non-ordinal-[Hyper Aleph Ω (Aleph Ω, Aleph Ω)]-multiplicity), Aleph (non-ordinal-[Hyper Aleph Ω (Aleph Ω, Aleph Ω)]-multiplicity))]} { Hyper Hyper Hyper Aleph (non-ordinal-[Hyper Aleph Ω (Aleph Ω, Aleph Ω)]-multiplicity) (Aleph (non-ordinal-[Hyper Aleph Ω (Aleph Ω, Aleph Ω)]-multiplicity), Aleph (non-ordinal-[Hyper Aleph Ω (Aleph Ω, Aleph Ω)]-multiplicity))[Hyper Aleph (non-ordinal-[Hyper Aleph Ω (Aleph Ω, Aleph Ω)]-multiplicity) (Aleph (non-ordinal-[Hyper Aleph Ω (Aleph Ω, Aleph Ω)]-multiplicity), Aleph (non-ordinal-[Hyper Aleph Ω (Aleph Ω, Aleph Ω)]-multiplicity)), Hyper Aleph (non-ordinal-[Hyper Aleph Ω (Aleph Ω, Aleph Ω)]-multiplicity) (Aleph (non-ordinal-[Hyper Aleph Ω (Aleph Ω, Aleph Ω)]-multiplicity), Aleph (non-ordinal-[Hyper Aleph Ω (Aleph Ω, Aleph Ω)]-multiplicity))] {Hyper Hyper Aleph (non-ordinal-[Hyper Aleph Ω (Aleph Ω, Aleph Ω)]-multiplicity) (Aleph (non-ordinal-[Hyper Aleph Ω (Aleph Ω, Aleph Ω)]-multiplicity), Aleph (non-ordinal-[Hyper Aleph Ω (Aleph Ω, Aleph Ω)]-multiplicity))[Hyper Aleph (non-ordinal-[Hyper Aleph Ω (Aleph Ω, Aleph Ω)]-multiplicity) (Aleph (non-ordinal-[Hyper Aleph Ω (Aleph Ω, Aleph Ω)]-multiplicity), Aleph (non-ordinal-[Hyper Aleph Ω (Aleph Ω, Aleph Ω)]-multiplicity)), Hyper Aleph (non-ordinal-[Hyper Aleph Ω (Aleph Ω, Aleph Ω)]-multiplicity) (Aleph (non-ordinal-[Hyper Aleph Ω (Aleph Ω, Aleph Ω)]-multiplicity), Aleph (non-ordinal-[Hyper Aleph Ω (Aleph Ω, Aleph Ω)]-multiplicity))], Hyper Hyper Aleph (non-ordinal-[Hyper Aleph Ω (Aleph Ω, Aleph Ω)]-multiplicity) (Aleph (non-ordinal-[Hyper Aleph Ω (Aleph Ω, Aleph Ω)]-multiplicity), Aleph (non-ordinal-[Hyper Aleph Ω (Aleph Ω, Aleph Ω)]-multiplicity))[Hyper Aleph (non-ordinal-[Hyper Aleph Ω (Aleph Ω, Aleph Ω)]-multiplicity) (Aleph (non-ordinal-[Hyper Aleph Ω (Aleph Ω, Aleph Ω)]-multiplicity), Aleph (non-ordinal-[Hyper Aleph Ω (Aleph Ω, Aleph Ω)]-multiplicity)), Hyper Aleph (non-ordinal-[Hyper Aleph Ω (Aleph Ω, Aleph Ω)]-multiplicity) (Aleph (non-ordinal-[Hyper Aleph Ω (Aleph Ω, Aleph Ω)]-multiplicity), Aleph (non-ordinal-[Hyper Aleph Ω (Aleph Ω, Aleph Ω)]-multiplicity))]}, Hyper Hyper Hyper Aleph (non-ordinal-[Hyper Aleph Ω (Aleph Ω, Aleph Ω)]-multiplicity) (Aleph (non-ordinal-[Hyper Aleph Ω (Aleph Ω, Aleph Ω)]-multiplicity), Aleph (non-ordinal-[Hyper Aleph Ω (Aleph Ω, Aleph Ω)]-multiplicity))[Hyper Aleph (non-ordinal-[Hyper Aleph Ω (Aleph Ω, Aleph Ω)]-multiplicity) (Aleph (non-ordinal-[Hyper Aleph Ω (Aleph Ω, Aleph Ω)]-multiplicity), Aleph (non-ordinal-[Hyper Aleph Ω (Aleph Ω, Aleph Ω)]-multiplicity)), Hyper Aleph (non-ordinal-[Hyper Aleph Ω (Aleph Ω, Aleph Ω)]-multiplicity) (Aleph (non-ordinal-[Hyper Aleph Ω (Aleph Ω, Aleph Ω)]-multiplicity), Aleph (non-ordinal-[Hyper Aleph Ω (Aleph Ω, Aleph Ω)]-multiplicity))] {Hyper Hyper Aleph (non-ordinal-[Hyper Aleph Ω (Aleph Ω, Aleph Ω)]-multiplicity) (Aleph (non-ordinal-[Hyper Aleph Ω (Aleph Ω, Aleph

Ω)]-multiplicity), Aleph (non-ordinal-[Hyper Aleph Ω (Aleph Ω, Aleph Ω)]-multiplicity))[Hyper Aleph (non-ordinal-[Hyper Aleph Ω (Aleph Ω, Aleph Ω)]-multiplicity) (Aleph (non-ordinal-[Hyper Aleph Ω (Aleph Ω, Aleph Ω)]-multiplicity), Aleph (non-ordinal-[Hyper Aleph Ω (Aleph Ω, Aleph Ω)]-multiplicity)), Hyper Aleph (non-ordinal-[Hyper Aleph Ω (Aleph Ω, Aleph Ω)]-multiplicity) (Aleph (non-ordinal-[Hyper Aleph Ω (Aleph Ω, Aleph Ω)]-multiplicity), Aleph (non-ordinal-[Hyper Aleph Ω (Aleph Ω, Aleph Ω)]-multiplicity))], Hyper Hyper Aleph (non-ordinal-[Hyper Aleph Ω (Aleph Ω, Aleph Ω)]-multiplicity) (Aleph (non-ordinal-[Hyper Aleph Ω (Aleph Ω, Aleph Ω)]-multiplicity), Aleph (non-ordinal-[Hyper Aleph Ω (Aleph Ω, Aleph Ω)]-multiplicity))[Hyper Aleph (non-ordinal-[Hyper Aleph Ω (Aleph Ω, Aleph Ω)]-multiplicity) (Aleph (non-ordinal-[Hyper Aleph Ω (Aleph Ω, Aleph Ω)]-multiplicity), Aleph (non-ordinal-[Hyper Aleph Ω (Aleph Ω, Aleph Ω)]-multiplicity)), Hyper Aleph (non-ordinal-[Hyper Aleph Ω (Aleph Ω, Aleph Ω)]-multiplicity) (Aleph (non-ordinal-[Hyper Aleph Ω (Aleph Ω, Aleph Ω)]-multiplicity), Aleph (non-ordinal-[Hyper Aleph Ω (Aleph Ω, Aleph Ω)]-multiplicity))]}}}, Hyper Hyper Hyper Hyper Hyper Aleph (non-ordinal-[Hyper Aleph Ω (Aleph Ω, Aleph Ω)]-multiplicity) (Aleph (non-ordinal-[Hyper Aleph Ω (Aleph Ω, Aleph Ω)]-multiplicity), Aleph (non-ordinal-[Hyper Aleph Ω (Aleph Ω, Aleph Ω)]-multiplicity))[Hyper Aleph (non-ordinal-[Hyper Aleph Ω (Aleph Ω, Aleph Ω)]-multiplicity) (Aleph (non-ordinal-[Hyper Aleph Ω (Aleph Ω, Aleph Ω)]-multiplicity), Aleph (non-ordinal-[Hyper Aleph Ω (Aleph Ω, Aleph Ω)]-multiplicity)), Hyper Aleph (non-ordinal-[Hyper Aleph Ω (Aleph Ω, Aleph Ω)]-multiplicity) (Aleph (non-ordinal-[Hyper Aleph Ω (Aleph Ω, Aleph Ω)]-multiplicity), Aleph (non-ordinal-[Hyper Aleph Ω (Aleph Ω, Aleph Ω)]-multiplicity))] {Hyper Hyper Aleph (non-ordinal-[Hyper Aleph Ω (Aleph Ω, Aleph Ω)]-multiplicity) (Aleph (non-ordinal-[Hyper Aleph Ω (Aleph Ω, Aleph Ω)]-multiplicity), Aleph (non-ordinal-[Hyper Aleph Ω (Aleph Ω, Aleph Ω)]-multiplicity))[Hyper Aleph (non-ordinal-[Hyper Aleph Ω (Aleph Ω, Aleph Ω)]-multiplicity) (Aleph (non-ordinal-[Hyper Aleph Ω (Aleph Ω, Aleph Ω)]-multiplicity), Aleph (non-ordinal-[Hyper Aleph Ω (Aleph Ω, Aleph Ω)]-multiplicity)), Hyper Aleph (non-ordinal-[Hyper Aleph Ω (Aleph Ω, Aleph Ω)]-multiplicity) (Aleph (non-ordinal-[Hyper Aleph Ω (Aleph Ω, Aleph Ω)]-multiplicity), Aleph (non-ordinal-[Hyper Aleph Ω (Aleph Ω, Aleph Ω)]-multiplicity))], Hyper Hyper Aleph (non-ordinal-[Hyper Aleph Ω (Aleph Ω, Aleph Ω)]-multiplicity) (Aleph (non-ordinal-[Hyper Aleph Ω (Aleph Ω, Aleph Ω)]-multiplicity), Aleph (non-ordinal-[Hyper Aleph Ω (Aleph Ω, Aleph Ω)]-multiplicity))[Hyper Aleph (non-ordinal-[Hyper Aleph Ω (Aleph Ω, Aleph Ω)]-multiplicity) (Aleph (non-ordinal-[Hyper Aleph Ω (Aleph Ω, Aleph Ω)]-multiplicity), Aleph (non-ordinal-[Hyper Aleph Ω (Aleph Ω, Aleph Ω)]-multiplicity)), Hyper Aleph (non-ordinal-[Hyper Aleph Ω (Aleph Ω, Aleph Ω)]-multiplicity) (Aleph (non-ordinal-[Hyper Aleph Ω (Aleph Ω, Aleph Ω)]-multiplicity), Aleph (non-ordinal-[Hyper Aleph Ω (Aleph Ω, Aleph Ω)]-multiplicity))]} { Hyper Hyper Hyper Aleph (non-ordinal-[Hyper Aleph Ω (Aleph Ω, Aleph Ω)]-multiplicity) (Aleph (non-ordinal-[Hyper Aleph Ω (Aleph Ω, Aleph Ω)]-multiplicity), Aleph (non-ordinal-[Hyper Aleph Ω (Aleph Ω, Aleph Ω)]-multiplicity))[Hyper Aleph (non-ordinal-[Hyper Aleph Ω (Aleph Ω,

Aleph Ω)]-multiplicity) (Aleph (non-ordinal-[Hyper Aleph Ω (Aleph Ω, Aleph Ω)]-multiplicity), Aleph (non-ordinal-[Hyper Aleph Ω (Aleph Ω, Aleph Ω)]-multiplicity)), Hyper Aleph (non-ordinal-[Hyper Aleph Ω (Aleph Ω, Aleph Ω)]-multiplicity) (Aleph (non-ordinal-[Hyper Aleph Ω (Aleph Ω, Aleph Ω)]-multiplicity), Aleph (non-ordinal-[Hyper Aleph Ω (Aleph Ω, Aleph Ω)]-multiplicity))] {Hyper Hyper Aleph (non-ordinal-[Hyper Aleph Ω (Aleph Ω, Aleph Ω)]-multiplicity) (Aleph (non-ordinal-[Hyper Aleph Ω (Aleph Ω, Aleph Ω)]-multiplicity), Aleph (non-ordinal-[Hyper Aleph Ω (Aleph Ω, Aleph Ω)]-multiplicity))[Hyper Aleph (non-ordinal-[Hyper Aleph Ω (Aleph Ω, Aleph Ω)]-multiplicity) (Aleph (non-ordinal-[Hyper Aleph Ω (Aleph Ω, Aleph Ω)]-multiplicity), Aleph (non-ordinal-[Hyper Aleph Ω (Aleph Ω, Aleph Ω)]-multiplicity)), Hyper Aleph (non-ordinal-[Hyper Aleph Ω (Aleph Ω, Aleph Ω)]-multiplicity) (Aleph (non-ordinal-[Hyper Aleph Ω (Aleph Ω, Aleph Ω)]-multiplicity), Aleph (non-ordinal-[Hyper Aleph Ω (Aleph Ω, Aleph Ω)]-multiplicity))], Hyper Hyper Aleph (non-ordinal-[Hyper Aleph Ω (Aleph Ω, Aleph Ω)]-multiplicity) (Aleph (non-ordinal-[Hyper Aleph Ω (Aleph Ω, Aleph Ω)]-multiplicity), Aleph (non-ordinal-[Hyper Aleph Ω (Aleph Ω, Aleph Ω)]-multiplicity))[Hyper Aleph (non-ordinal-[Hyper Aleph Ω (Aleph Ω, Aleph Ω)]-multiplicity) (Aleph (non-ordinal-[Hyper Aleph Ω (Aleph Ω, Aleph Ω)]-multiplicity), Aleph (non-ordinal-[Hyper Aleph Ω (Aleph Ω, Aleph Ω)]-multiplicity)), Hyper Aleph (non-ordinal-[Hyper Aleph Ω (Aleph Ω, Aleph Ω)]-multiplicity) (Aleph (non-ordinal-[Hyper Aleph Ω (Aleph Ω, Aleph Ω)]-multiplicity), Aleph (non-ordinal-[Hyper Aleph Ω (Aleph Ω, Aleph Ω)]-multiplicity))]}, Hyper Hyper Hyper Aleph (non-ordinal-[Hyper Aleph Ω (Aleph Ω, Aleph Ω)]-multiplicity) (Aleph (non-ordinal-[Hyper Aleph Ω (Aleph Ω, Aleph Ω)]-multiplicity), Aleph (non-ordinal-[Hyper Aleph Ω (Aleph Ω, Aleph Ω)]-multiplicity))[Hyper Aleph (non-ordinal-[Hyper Aleph Ω (Aleph Ω, Aleph Ω)]-multiplicity) (Aleph (non-ordinal-[Hyper Aleph Ω (Aleph Ω, Aleph Ω)]-multiplicity), Aleph (non-ordinal-[Hyper Aleph Ω (Aleph Ω, Aleph Ω)]-multiplicity)), Hyper Aleph (non-ordinal-[Hyper Aleph Ω (Aleph Ω, Aleph Ω)]-multiplicity) (Aleph (non-ordinal-[Hyper Aleph Ω (Aleph Ω, Aleph Ω)]-multiplicity), Aleph (non-ordinal-[Hyper Aleph Ω (Aleph Ω, Aleph Ω)]-multiplicity))] {Hyper Hyper Aleph (non-ordinal-[Hyper Aleph Ω (Aleph Ω, Aleph Ω)]-multiplicity) (Aleph (non-ordinal-[Hyper Aleph Ω (Aleph Ω, Aleph Ω)]-multiplicity), Aleph (non-ordinal-[Hyper Aleph Ω (Aleph Ω, Aleph Ω)]-multiplicity))[Hyper Aleph (non-ordinal-[Hyper Aleph Ω (Aleph Ω, Aleph Ω)]-multiplicity) (Aleph (non-ordinal-[Hyper Aleph Ω (Aleph Ω, Aleph Ω)]-multiplicity), Aleph (non-ordinal-[Hyper Aleph Ω (Aleph Ω, Aleph Ω)]-multiplicity)), Hyper Aleph (non-ordinal-[Hyper Aleph Ω (Aleph Ω, Aleph Ω)]-multiplicity) (Aleph (non-ordinal-[Hyper Aleph Ω (Aleph Ω, Aleph Ω)]-multiplicity), Aleph (non-ordinal-[Hyper Aleph Ω (Aleph Ω, Aleph Ω)]-multiplicity))], Hyper Hyper Aleph (non-ordinal-[Hyper Aleph Ω (Aleph Ω, Aleph Ω)]-multiplicity) (Aleph (non-ordinal-[Hyper Aleph Ω (Aleph Ω, Aleph Ω)]-multiplicity), Aleph (non-ordinal-[Hyper Aleph Ω (Aleph Ω, Aleph Ω)]-multiplicity))[Hyper Aleph (non-ordinal-[Hyper Aleph Ω (Aleph Ω, Aleph Ω)]-multiplicity) (Aleph (non-ordinal-[Hyper Aleph Ω (Aleph Ω, Aleph Ω)]-multiplicity), Aleph (non-ordinal-[Hyper Aleph Ω (Aleph Ω, Aleph

Ω)]-multiplicity)), Hyper Aleph (non-ordinal-[Hyper Aleph Ω (Aleph Ω, Aleph Ω)]-multiplicity) (Aleph (non-ordinal-[Hyper Aleph Ω (Aleph Ω, Aleph Ω)]-multiplicity), Aleph (non-ordinal-[Hyper Aleph Ω (Aleph Ω, Aleph Ω)]-multiplicity))]}}{ Hyper Hyper Hyper Hyper Aleph (non-ordinal-[Hyper Aleph Ω (Aleph Ω, Aleph Ω)]-multiplicity) (Aleph (non-ordinal-[Hyper Aleph Ω (Aleph Ω, Aleph Ω)]-multiplicity), Aleph (non-ordinal-[Hyper Aleph Ω (Aleph Ω, Aleph Ω)]-multiplicity))[Hyper Aleph (non-ordinal-[Hyper Aleph Ω (Aleph Ω, Aleph Ω)]-multiplicity) (Aleph (non-ordinal-[Hyper Aleph Ω (Aleph Ω, Aleph Ω)]-multiplicity), Aleph (non-ordinal-[Hyper Aleph Ω (Aleph Ω, Aleph Ω)]-multiplicity)), Hyper Aleph (non-ordinal-[Hyper Aleph Ω (Aleph Ω, Aleph Ω)]-multiplicity) (Aleph (non-ordinal-[Hyper Aleph Ω (Aleph Ω, Aleph Ω)]-multiplicity), Aleph (non-ordinal-[Hyper Aleph Ω (Aleph Ω, Aleph Ω)]-multiplicity))] {Hyper Hyper Aleph (non-ordinal-[Hyper Aleph Ω (Aleph Ω, Aleph Ω)]-multiplicity) (Aleph (non-ordinal-[Hyper Aleph Ω (Aleph Ω, Aleph Ω)]-multiplicity), Aleph (non-ordinal-[Hyper Aleph Ω (Aleph Ω, Aleph Ω)]-multiplicity))[Hyper Aleph (non-ordinal-[Hyper Aleph Ω (Aleph Ω, Aleph Ω)]-multiplicity) (Aleph (non-ordinal-[Hyper Aleph Ω (Aleph Ω, Aleph Ω)]-multiplicity), Aleph (non-ordinal-[Hyper Aleph Ω (Aleph Ω, Aleph Ω)]-multiplicity)), Hyper Aleph (non-ordinal-[Hyper Aleph Ω (Aleph Ω, Aleph Ω)]-multiplicity) (Aleph (non-ordinal-[Hyper Aleph Ω (Aleph Ω, Aleph Ω)]-multiplicity), Aleph (non-ordinal-[Hyper Aleph Ω (Aleph Ω, Aleph Ω)]-multiplicity))], Hyper Hyper Aleph (non-ordinal-[Hyper Aleph Ω (Aleph Ω, Aleph Ω)]-multiplicity) (Aleph (non-ordinal-[Hyper Aleph Ω (Aleph Ω, Aleph Ω)]-multiplicity), Aleph (non-ordinal-[Hyper Aleph Ω (Aleph Ω, Aleph Ω)]-multiplicity))[Hyper Aleph (non-ordinal-[Hyper Aleph Ω (Aleph Ω, Aleph Ω)]-multiplicity) (Aleph (non-ordinal-[Hyper Aleph Ω (Aleph Ω, Aleph Ω)]-multiplicity), Aleph (non-ordinal-[Hyper Aleph Ω (Aleph Ω, Aleph Ω)]-multiplicity)), Hyper Aleph (non-ordinal-[Hyper Aleph Ω (Aleph Ω, Aleph Ω)]-multiplicity) (Aleph (non-ordinal-[Hyper Aleph Ω (Aleph Ω, Aleph Ω)]-multiplicity), Aleph (non-ordinal-[Hyper Aleph Ω (Aleph Ω, Aleph Ω)]-multiplicity))]} { Hyper Hyper Hyper Aleph (non-ordinal-[Hyper Aleph Ω (Aleph Ω, Aleph Ω)]-multiplicity) (Aleph (non-ordinal-[Hyper Aleph Ω (Aleph Ω, Aleph Ω)]-multiplicity), Aleph (non-ordinal-[Hyper Aleph Ω (Aleph Ω, Aleph Ω)]-multiplicity))[Hyper Aleph (non-ordinal-[Hyper Aleph Ω (Aleph Ω, Aleph Ω)]-multiplicity) (Aleph (non-ordinal-[Hyper Aleph Ω (Aleph Ω, Aleph Ω)]-multiplicity), Aleph (non-ordinal-[Hyper Aleph Ω (Aleph Ω, Aleph Ω)]-multiplicity)), Hyper Aleph (non-ordinal-[Hyper Aleph Ω (Aleph Ω, Aleph Ω)]-multiplicity) (Aleph (non-ordinal-[Hyper Aleph Ω (Aleph Ω, Aleph Ω)]-multiplicity), Aleph (non-ordinal-[Hyper Aleph Ω (Aleph Ω, Aleph Ω)]-multiplicity))] {Hyper Hyper Aleph (non-ordinal-[Hyper Aleph Ω (Aleph Ω, Aleph Ω)]-multiplicity) (Aleph (non-ordinal-[Hyper Aleph Ω (Aleph Ω, Aleph Ω)]-multiplicity), Aleph (non-ordinal-[Hyper Aleph Ω (Aleph Ω, Aleph Ω)]-multiplicity))[Hyper Aleph (non-ordinal-[Hyper Aleph Ω (Aleph Ω, Aleph Ω)]-multiplicity) (Aleph (non-ordinal-[Hyper Aleph Ω (Aleph Ω, Aleph Ω)]-multiplicity), Aleph (non-ordinal-[Hyper Aleph Ω (Aleph Ω, Aleph Ω)]-multiplicity)), Hyper Aleph (non-ordinal-[Hyper Aleph Ω (Aleph Ω, Aleph Ω)]-multiplicity) (Aleph (non-ordinal-[Hyper Aleph Ω (Aleph Ω,

Aleph Ω)]-multiplicity), Aleph (non-ordinal-[Hyper Aleph Ω (Aleph Ω, Aleph Ω)]-multiplicity))], Hyper Hyper Aleph (non-ordinal-[Hyper Aleph Ω (Aleph Ω, Aleph Ω)]-multiplicity) (Aleph (non-ordinal-[Hyper Aleph Ω (Aleph Ω, Aleph Ω)]-multiplicity), Aleph (non-ordinal-[Hyper Aleph Ω (Aleph Ω, Aleph Ω)]-multiplicity))[Hyper Aleph (non-ordinal-[Hyper Aleph Ω (Aleph Ω, Aleph Ω)]-multiplicity) (Aleph (non-ordinal-[Hyper Aleph Ω (Aleph Ω, Aleph Ω)]-multiplicity), Aleph (non-ordinal-[Hyper Aleph Ω (Aleph Ω, Aleph Ω)]-multiplicity)), Hyper Aleph (non-ordinal-[Hyper Aleph Ω (Aleph Ω, Aleph Ω)]-multiplicity) (Aleph (non-ordinal-[Hyper Aleph Ω (Aleph Ω, Aleph Ω)]-multiplicity), Aleph (non-ordinal-[Hyper Aleph Ω (Aleph Ω, Aleph Ω)]-multiplicity))]}, Hyper Hyper Hyper Aleph (non-ordinal-[Hyper Aleph Ω (Aleph Ω, Aleph Ω)]-multiplicity) (Aleph (non-ordinal-[Hyper Aleph Ω (Aleph Ω, Aleph Ω)]-multiplicity), Aleph (non-ordinal-[Hyper Aleph Ω (Aleph Ω, Aleph Ω)]-multiplicity))[Hyper Aleph (non-ordinal-[Hyper Aleph Ω (Aleph Ω, Aleph Ω)]-multiplicity) (Aleph (non-ordinal-[Hyper Aleph Ω (Aleph Ω, Aleph Ω)]-multiplicity), Aleph (non-ordinal-[Hyper Aleph Ω (Aleph Ω, Aleph Ω)]-multiplicity)), Hyper Aleph (non-ordinal-[Hyper Aleph Ω (Aleph Ω, Aleph Ω)]-multiplicity) (Aleph (non-ordinal-[Hyper Aleph Ω (Aleph Ω, Aleph Ω)]-multiplicity), Aleph (non-ordinal-[Hyper Aleph Ω (Aleph Ω, Aleph Ω)]-multiplicity))] {Hyper Hyper Aleph (non-ordinal-[Hyper Aleph Ω (Aleph Ω, Aleph Ω)]-multiplicity) (Aleph (non-ordinal-[Hyper Aleph Ω (Aleph Ω, Aleph Ω)]-multiplicity), Aleph (non-ordinal-[Hyper Aleph Ω (Aleph Ω, Aleph Ω)]-multiplicity))[Hyper Aleph (non-ordinal-[Hyper Aleph Ω (Aleph Ω, Aleph Ω)]-multiplicity) (Aleph (non-ordinal-[Hyper Aleph Ω (Aleph Ω, Aleph Ω)]-multiplicity), Aleph (non-ordinal-[Hyper Aleph Ω (Aleph Ω, Aleph Ω)]-multiplicity)), Hyper Aleph (non-ordinal-[Hyper Aleph Ω (Aleph Ω, Aleph Ω)]-multiplicity) (Aleph (non-ordinal-[Hyper Aleph Ω (Aleph Ω, Aleph Ω)]-multiplicity), Aleph (non-ordinal-[Hyper Aleph Ω (Aleph Ω, Aleph Ω)]-multiplicity))], Hyper Hyper Aleph (non-ordinal-[Hyper Aleph Ω (Aleph Ω, Aleph Ω)]-multiplicity) (Aleph (non-ordinal-[Hyper Aleph Ω (Aleph Ω, Aleph Ω)]-multiplicity), Aleph (non-ordinal-[Hyper Aleph Ω (Aleph Ω, Aleph Ω)]-multiplicity))[Hyper Aleph (non-ordinal-[Hyper Aleph Ω (Aleph Ω, Aleph Ω)]-multiplicity) (Aleph (non-ordinal-[Hyper Aleph Ω (Aleph Ω, Aleph Ω)]-multiplicity), Aleph (non-ordinal-[Hyper Aleph Ω (Aleph Ω, Aleph Ω)]-multiplicity)), Hyper Aleph (non-ordinal-[Hyper Aleph Ω (Aleph Ω, Aleph Ω)]-multiplicity) (Aleph (non-ordinal-[Hyper Aleph Ω (Aleph Ω, Aleph Ω)]-multiplicity), Aleph (non-ordinal-[Hyper Aleph Ω (Aleph Ω, Aleph Ω)]-multiplicity))]}}, Hyper Hyper Hyper Hyper Aleph (non-ordinal-[Hyper Aleph Ω (Aleph Ω, Aleph Ω)]-multiplicity) (Aleph (non-ordinal-[Hyper Aleph Ω (Aleph Ω, Aleph Ω)]-multiplicity), Aleph (non-ordinal-[Hyper Aleph Ω (Aleph Ω, Aleph Ω)]-multiplicity))[Hyper Aleph (non-ordinal-[Hyper Aleph Ω (Aleph Ω, Aleph Ω)]-multiplicity) (Aleph (non-ordinal-[Hyper Aleph Ω (Aleph Ω, Aleph Ω)]-multiplicity), Aleph (non-ordinal-[Hyper Aleph Ω (Aleph Ω, Aleph Ω)]-multiplicity)), Hyper Aleph (non-ordinal-[Hyper Aleph Ω (Aleph Ω, Aleph Ω)]-multiplicity) (Aleph (non-ordinal-[Hyper Aleph Ω (Aleph Ω, Aleph Ω)]-multiplicity), Aleph (non-ordinal-[Hyper Aleph Ω (Aleph Ω, Aleph Ω)]-multiplicity))] {Hyper Hyper Aleph (non-ordinal-[Hyper

Aleph Ω (Aleph Ω, Aleph Ω)]-multiplicity) (Aleph (non-ordinal-[Hyper Aleph Ω (Aleph Ω, Aleph Ω)]-multiplicity), Aleph (non-ordinal-[Hyper Aleph Ω (Aleph Ω, Aleph Ω)]-multiplicity))[Hyper Aleph (non-ordinal-[Hyper Aleph Ω (Aleph Ω, Aleph Ω)]-multiplicity) (Aleph (non-ordinal-[Hyper Aleph Ω (Aleph Ω, Aleph Ω)]-multiplicity), Aleph (non-ordinal-[Hyper Aleph Ω (Aleph Ω, Aleph Ω)]-multiplicity)), Hyper Aleph (non-ordinal-[Hyper Aleph Ω (Aleph Ω, Aleph Ω)]-multiplicity) (Aleph (non-ordinal-[Hyper Aleph Ω (Aleph Ω, Aleph Ω)]-multiplicity), Aleph (non-ordinal-[Hyper Aleph Ω (Aleph Ω, Aleph Ω)]-multiplicity))], Hyper Hyper Aleph (non-ordinal-[Hyper Aleph Ω (Aleph Ω, Aleph Ω)]-multiplicity) (Aleph (non-ordinal-[Hyper Aleph Ω (Aleph Ω, Aleph Ω)]-multiplicity), Aleph (non-ordinal-[Hyper Aleph Ω (Aleph Ω, Aleph Ω)]-multiplicity))[Hyper Aleph (non-ordinal-[Hyper Aleph Ω (Aleph Ω, Aleph Ω)]-multiplicity) (Aleph (non-ordinal-[Hyper Aleph Ω (Aleph Ω, Aleph Ω)]-multiplicity), Aleph (non-ordinal-[Hyper Aleph Ω (Aleph Ω, Aleph Ω)]-multiplicity)), Hyper Aleph (non-ordinal-[Hyper Aleph Ω (Aleph Ω, Aleph Ω)]-multiplicity) (Aleph (non-ordinal-[Hyper Aleph Ω (Aleph Ω, Aleph Ω)]-multiplicity), Aleph (non-ordinal-[Hyper Aleph Ω (Aleph Ω, Aleph Ω)]-multiplicity))]} { Hyper Hyper Hyper Aleph (non-ordinal-[Hyper Aleph Ω (Aleph Ω, Aleph Ω)]-multiplicity) (Aleph (non-ordinal-[Hyper Aleph Ω (Aleph Ω, Aleph Ω)]-multiplicity), Aleph (non-ordinal-[Hyper Aleph Ω (Aleph Ω, Aleph Ω)]-multiplicity))[Hyper Aleph (non-ordinal-[Hyper Aleph Ω (Aleph Ω, Aleph Ω)]-multiplicity) (Aleph (non-ordinal-[Hyper Aleph Ω (Aleph Ω, Aleph Ω)]-multiplicity), Aleph (non-ordinal-[Hyper Aleph Ω (Aleph Ω, Aleph Ω)]-multiplicity)), Hyper Aleph (non-ordinal-[Hyper Aleph Ω (Aleph Ω, Aleph Ω)]-multiplicity) (Aleph (non-ordinal-[Hyper Aleph Ω (Aleph Ω, Aleph Ω)]-multiplicity), Aleph (non-ordinal-[Hyper Aleph Ω (Aleph Ω, Aleph Ω)]-multiplicity))] {Hyper Hyper Aleph (non-ordinal-[Hyper Aleph Ω (Aleph Ω, Aleph Ω)]-multiplicity) (Aleph (non-ordinal-[Hyper Aleph Ω (Aleph Ω, Aleph Ω)]-multiplicity), Aleph (non-ordinal-[Hyper Aleph Ω (Aleph Ω, Aleph Ω)]-multiplicity))[Hyper Aleph (non-ordinal-[Hyper Aleph Ω (Aleph Ω, Aleph Ω)]-multiplicity) (Aleph (non-ordinal-[Hyper Aleph Ω (Aleph Ω, Aleph Ω)]-multiplicity), Aleph (non-ordinal-[Hyper Aleph Ω (Aleph Ω, Aleph Ω)]-multiplicity)), Hyper Aleph (non-ordinal-[Hyper Aleph Ω (Aleph Ω, Aleph Ω)]-multiplicity) (Aleph (non-ordinal-[Hyper Aleph Ω (Aleph Ω, Aleph Ω)]-multiplicity), Aleph (non-ordinal-[Hyper Aleph Ω (Aleph Ω, Aleph Ω)]-multiplicity))], Hyper Hyper Aleph (non-ordinal-[Hyper Aleph Ω (Aleph Ω, Aleph Ω)]-multiplicity) (Aleph (non-ordinal-[Hyper Aleph Ω (Aleph Ω, Aleph Ω)]-multiplicity), Aleph (non-ordinal-[Hyper Aleph Ω (Aleph Ω, Aleph Ω)]-multiplicity))[Hyper Aleph (non-ordinal-[Hyper Aleph Ω (Aleph Ω, Aleph Ω)]-multiplicity) (Aleph (non-ordinal-[Hyper Aleph Ω (Aleph Ω, Aleph Ω)]-multiplicity), Aleph (non-ordinal-[Hyper Aleph Ω (Aleph Ω, Aleph Ω)]-multiplicity)), Hyper Aleph (non-ordinal-[Hyper Aleph Ω (Aleph Ω, Aleph Ω)]-multiplicity) (Aleph (non-ordinal-[Hyper Aleph Ω (Aleph Ω, Aleph Ω)]-multiplicity), Aleph (non-ordinal-[Hyper Aleph Ω (Aleph Ω, Aleph Ω)]-multiplicity))]}, Hyper Hyper Hyper Aleph (non-ordinal-[Hyper Aleph Ω (Aleph Ω, Aleph Ω)]-multiplicity) (Aleph (non-ordinal-[Hyper Aleph Ω (Aleph Ω, Aleph Ω)]-multiplicity), Aleph (non-ordinal-[Hyper Aleph Ω

(Aleph Ω, Aleph Ω)]-multiplicity))[Hyper Aleph (non-ordinal-[Hyper Aleph Ω (Aleph Ω, Aleph Ω)]-multiplicity) (Aleph (non-ordinal-[Hyper Aleph Ω (Aleph Ω, Aleph Ω)]-multiplicity), Aleph (non-ordinal-[Hyper Aleph Ω (Aleph Ω, Aleph Ω)]-multiplicity)), Hyper Aleph (non-ordinal-[Hyper Aleph Ω (Aleph Ω, Aleph Ω)]-multiplicity) (Aleph (non-ordinal-[Hyper Aleph Ω (Aleph Ω, Aleph Ω)]-multiplicity), Aleph (non-ordinal-[Hyper Aleph Ω (Aleph Ω, Aleph Ω)]-multiplicity))] {Hyper Hyper Aleph (non-ordinal-[Hyper Aleph Ω (Aleph Ω, Aleph Ω)]-multiplicity) (Aleph (non-ordinal-[Hyper Aleph Ω (Aleph Ω, Aleph Ω)]-multiplicity), Aleph (non-ordinal-[Hyper Aleph Ω (Aleph Ω, Aleph Ω)]-multiplicity))[Hyper Aleph (non-ordinal-[Hyper Aleph Ω (Aleph Ω, Aleph Ω)]-multiplicity) (Aleph (non-ordinal-[Hyper Aleph Ω (Aleph Ω, Aleph Ω)]-multiplicity), Aleph (non-ordinal-[Hyper Aleph Ω (Aleph Ω, Aleph Ω)]-multiplicity)), Hyper Aleph (non-ordinal-[Hyper Aleph Ω (Aleph Ω, Aleph Ω)]-multiplicity) (Aleph (non-ordinal-[Hyper Aleph Ω (Aleph Ω, Aleph Ω)]-multiplicity), Aleph (non-ordinal-[Hyper Aleph Ω (Aleph Ω, Aleph Ω)]-multiplicity))], Hyper Hyper Aleph (non-ordinal-[Hyper Aleph Ω (Aleph Ω, Aleph Ω)]-multiplicity) (Aleph (non-ordinal-[Hyper Aleph Ω (Aleph Ω, Aleph Ω)]-multiplicity), Aleph (non-ordinal-[Hyper Aleph Ω (Aleph Ω, Aleph Ω)]-multiplicity))[Hyper Aleph (non-ordinal-[Hyper Aleph Ω (Aleph Ω, Aleph Ω)]-multiplicity) (Aleph (non-ordinal-[Hyper Aleph Ω (Aleph Ω, Aleph Ω)]-multiplicity), Aleph (non-ordinal-[Hyper Aleph Ω (Aleph Ω, Aleph Ω)]-multiplicity)), Hyper Aleph (non-ordinal-[Hyper Aleph Ω (Aleph Ω, Aleph Ω)]-multiplicity) (Aleph (non-ordinal-[Hyper Aleph Ω (Aleph Ω, Aleph Ω)]-multiplicity), Aleph (non-ordinal-[Hyper Aleph Ω (Aleph Ω, Aleph Ω)]-multiplicity))]}}}}}

We may in principle approximately double the length of the above latest expression every future Planck Time Unit for all eternity.

A fascinating construct is to first consider the empty set which has no elements. Then we go to something totally fascinating to consider a set that is completed and absolutely the opposite of the empty set which we will refer to as the anti-empty set.

The anti-empty set would contain everything. We denote the anti-empty-set as defined above as the anti-1-empty-set.

The anti-2-empty-set is defined as being to the anti-1-empty-set as the anti-1-empty-set is to the empty set.

The anti-3-empty-set is defined as being to the anti-2-empty-set as the anti-2-empty-set is to the anti-1-empty-set as the anti-1-empty-set is to the empty set empty set.

We can go on to consider the anti-4-empty-set, the anti-5-empty-set, and so-on to the anti-(Aleph 0)-empty-set, then to the anti-(Aleph 1)-empty-set, then to the anti-(Aleph 2)-empty-set, then to the anti-(Aleph 3)-empty-set, and so-on and so-on to the anti-(Aleph Ω)-empty-set, and so-on and so-on to the anti-[Hyper Aleph Ω (Aleph Ω, Aleph Ω)]-empty-set, and continue on from there.

The future of space travel, technological progress, scientific, theological, philosophical, medical, cultural, social, and spiritual progress is graced with possibilities defined by the anti-[Hyper Aleph Ω (Aleph Ω, Aleph Ω)]-empty-set, and beyond.

Now, we considered unlimited buildouts of the expressions [Hyper Aleph Ω (Aleph Ω, Aleph Ω)], [Hyper Aleph (high-end uncountable infinity) (Aleph (high-end uncountable infinity), Aleph (high-end uncountable infinity))], and Hyper Aleph (non-ordinal-[Hyper Aleph Ω (Aleph Ω, Aleph Ω)]-multiplicity) (Aleph (non-ordinal-[Hyper Aleph Ω (Aleph Ω, Aleph Ω)]-multiplicity), Aleph (non-ordinal-[Hyper Aleph Ω (Aleph Ω, Aleph Ω)]-multiplicity)).

Accordingly, we considered unending progressive production of ever far-reaching reality by roughly doubling the size of each of the above built out expressions for the passage of each Planck Time Unit that goes by into future eternity.

Note that we thus conjectured build-out scenarios in a 3-fold progression.

We can consider as much as a [Hyper Aleph Ω (Aleph Ω, Aleph Ω)]-fold progression, even a [Hyper Aleph (high-end uncountable infinity) (Aleph (high-end uncountable infinity), Aleph (high-end uncountable infinity))]-fold progression, and perhaps even a Hyper Aleph (non-ordinal-[Hyper Aleph Ω (Aleph Ω, Aleph Ω)]-multiplicity) (Aleph (non-ordinal-[Hyper Aleph Ω (Aleph Ω, Aleph Ω)]-multiplicity), Aleph (non-ordinal-[Hyper Aleph Ω (Aleph Ω, Aleph Ω)]-multiplicity))-fold progression and so-on in a never ending digression.

End of my latest stuff in the epilog of "Breaching The Light-Barrier. Volume 8.".

My readership by now understands how I enjoy conjecture about so-called super-infinities of various levels and the like.

Well, here I desire to return to the subject of infinities to give infinities their glorious due and earned honor.

For example, a number line in a geometric extends from negative infinity to positive infinity for which Set Theory states that the number of positive integers is the first Aleph number known as Aleph 0. Aleph 1 is according to the perhaps unprovable and thus unfalsifiable Continuum Hypothesis equal to the number of real numbers but either way is equal to 2 EXP (Aleph 0). The number of Aleph numbers is infinite and in general Aleph n = 2 EXP [Aleph (n-1)] where n is any finite or infinite positive integer.

Now, we consider another construct of a hyper-extended number line or a number line that ironically extends beyond the traditional number line.

A first level of hyper-extension would extend into the realm of a first level of hyper-integers and a first level of hyper-real-numbers. As such, these ordinals would indeed be greater than the traditional infinite integers and infinite real numbers.

A spacecraft having attained hyper-real-number Lorentz factors may enter an exotic state with respect to the background that is not Lorentz invariant and may even enter an exotic state that is not Lorentz invariant in the spacecraft's reference frame.

A second level of hyper-extension would extend into the realm of a second level of hyper-integers and a second level of hyper-real-numbers. As such, these ordinals would indeed be greater than the first level of hyper-real-integers and a first level of hyper-real-numbers.

A third level of hyper-extension would extend into the realm of a third level of hyper-integers and a third level of hyper-real-numbers. As such, these ordinals would indeed be greater than the second level of hyper-real-integers and a second level of hyper-real-numbers.

We refer to these first three levels of these hyper-integers and hyper-real-numbers as hyper-1-integers and hyper-1-real-numbers, hyper-2-integers and hyper-2-real-numbers, and hyper-3-integers and hyper-3-real-numbers.

We can progress to hyper-4-integers and hyper-4-real-numbers, hyper-5-integers and hyper-5-real-numbers, all the way up to and beyond hyper-Ω-integers and hyper-Ω-real-numbers where Ω is the least transfinite or least infinite ordinal.

We can continue to and beyond hyper-(Aleph Ω)-integers and hyper-(Aleph Ω)-real-numbers.

These numbers would become progressively more bazaar and exotic thus implying not only new and bazaar spacecraft travel Lorentz factors, thermodynamics, kinematics, and topology, but also new, exotic, and bazaar metaphysical aspects, some of which would affect the being or substance of the material composition of the spacecraft and the crew members bodies and their consciousness.

We may affix the following operator to formulas for gamma and related parameters to denote the above paradigms of hyper-extended number lines and associated values for spacecraft Lorentz factors.

[w(Up to and beyond hyper-(Aleph Ω)-integers and hyper-(Aleph Ω)-real-numbers values of new and bazaar spacecraft travel Lorentz factors, thermodynamics, kinematics, and topology, but also new, exotic, and bazaar metaphysical aspects, some of which would affect the being or substance of the material composition of the spacecraft and the crew members bodies and their consciousness)]

We can also consider arbitrary orders of differentiation or integration of the above operator values with respect to ship-time, gamma, acceleration, spacecraft kinetic energy etc.. So we denote such operation by the following operator.

{w{Arbitrary orders of differentiation or integration of the following operator values with respect to ship-time, gamma, acceleration, spacecraft kinetic energy etc. [w(Up to and beyond hyper-(Aleph Ω)-integers and hyper-(Aleph Ω)-real-numbers values of new and bazaar spacecraft travel Lorentz factors, thermodynamics, kinematics, and topology, but also new, exotic, and bazaar metaphysical aspects, some of which would affect the being or substance of the material composition of the spacecraft and the crew members bodies and their consciousness)]}}

We thus may affix the following operator to formulas for gamma and related parameters to denote these possibilities.

[w(Up to and beyond hyper-(Aleph Ω)-integers and hyper-(Aleph Ω)-real-numbers values of new and bazaar spacecraft travel Lorentz factors, thermodynamics, kinematics, and topology, but also new, exotic, and bazaar metaphysical aspects, some of which would affect the being or substance of the material composition of the spacecraft and the crew members bodies and their consciousness)]:{w{Arbitrary orders of differentiation or integration of the following operator values with respect to ship-time, gamma, acceleration, spacecraft kinetic energy etc. [w(Up to and beyond hyper-(Aleph Ω)-integers and hyper-(Aleph Ω)-real-numbers values of new and bazaar spacecraft travel Lorentz factors, thermodynamics, kinematics, and topology, but also new, exotic, and bazaar metaphysical aspects, some of which would affect the being or substance of the material composition of the spacecraft and the crew members bodies and their consciousness)]}}

Here we consider the future as an ever distant but ever more rapidly progressed in light-speed spacecraft and the spacecraft crew.

In a sense, a sufficiently extreme Lorentz factor spacecraft even if finite is moved forward in time at a rapidly increased rate.

The ultimate future will never be arrived at. None-the-less, it will always beckon us forward as if an eternally ever distance dim but sublime beacon.

For light-speed spacecraft at infinite Lorentz factors, travel into the future becomes infinitely fast compared to the background. Yet, there is an ultimate future of a first level that will always be eternally temporally far-off into the future. Pun both not intended and intended! RLOL!

In a sense, a first level eternally far-off future can and will provide a great sense of mystique for both folks who stay back at home and folks who travel in spacecraft having achieved the speed of light but which progresses to ever greater Lorentz factors.

No matter how infinite the Lorentz factors of a spacecraft and no matter how long it will travel, even if traveling infinite time spans in the craft's own reference frame time, a first level eternally far off future will remain forever out of reach to the spacecraft and its crew but will always serve as a subtly distant dim but wonderfully mystical far-out beacon to draw us forward to each new cosmic phase change and cosmic era to follow our current temporal locations.

These cosmic eras I believe will be available for the souls in Heaven and all souls who go to Hell. So, in a sense, we can all by our solidarity as personal creations have an unending series of future cosmic eras to look forward to guided by an eternally ever distant beautiful, mystical, beacon of light which is a first level eternally far off future.

However, GOD is so big that HE provides and manifests not just one first level eternally far off future, but also a second, a third, and so-on in a never ending series of levels of eternally far off futures.

We will never attain these additional futures and may never even see them but we will have the wondrous knowledge that they exist and are so far-out existentially that we can delight in their evermore mysterious and wondrous nature. All the while, we will have one cosmic era after another cosmic era to enjoy knowing that we will never exhaust the number of new cosmic era to be entered. The awesome beacon of the ever so far-out first level of ultimate future will always provide beauty, mystique, and wonder as we develop along with the rest of creation.

In the end, could GOD change HIS mind and then let us see the second level of ultimate future, then eventually a third level of ultimate future, and so-on? I will hazard a guess of saying never say never and that it might actually be true that GOD can and will change HIS MIND to let us accordingly proceed to see higher levels of ultimate futures.

For now, let us be content to look forward to a return to the Moon by a woman and a man planned by NASA for year 2025. This is close at only about three years away.

Further into this Century, I believe the first human starships will be launched with living human crew members.

These starships will plod along at perhaps 10 to 20 percent of the speed of light and will easily travel from star system to star system in under one ordinary human familial generation.

Later in this millennium, we will master extremely relativistic space flight and be able to attain spacecraft Lorentz factors in the range of 1,000.

Over the next 10 Eons, we will have traveled to near the edge of the then current light cone and will have spacecraft undergoing missions having Lorentz factors on the order of a billion to a trillion.

In some impossibly distant future, we will have worked out impulse travel in light-speed and associated infinite Lorentz factors. This, I believe will and can be accomplished even if it takes a googol or a googol-plex number of background years to develop the required technology. A googol is 10 EXP 100 and a googol-plex is (10 EXP 100) EXP (10 EXP 100). From there, we will simply carry on with developing spacecraft capable of ever greater infinite Lorentz factors. As such, eternity simply goes on, forever, without end.

We continue to consider analogs of first, second, third and the like levels of ultimate future.

Accordingly, these analogs would on average be roughly as bold if not bolder that the levels of ultimate future. However, each set of such analogs would be completely distinct from the set of levels of ultimate future.

Each set of analogs of ultimate future may have infinitely many levels. Moreover, the number of species of sets of analogs may be arbitrarily infinite.

By infinite, I even consider ordinals on hyper-extended number lines in the range of infinite numbers.

Each set of analogs may be as if a positive coordinate axis in an orthogonal systems of axes each mutually orthogonal where the number of axes is as infinite as GOD desires and GOD may know no bounds to HIS creative abilities and creative desires.

There may even be hybrids among the sets of analogs thus the reason for including said infinite dimensional coordinate systems conjectures and the associated points in the coordinate systems. Each point not strictly placed on a given axis ray can be view as the head of a ray defining a hybrid analog and sets thereof.

So, fasten your seat belts as we are in for a real never ending fun and awesomely enjoyable ride.

With the huge number of; realms of travel and entrance, possible time lines and numbers of temporal dimensions, infinite Lorentz factors, super-luminal velocities, and the entire host of other parameters maxed to and beyond countable infinity numbers in value, we are getting close to technologies so advanced that the huge infinity of infinite realms and scenarios available to spacecraft and crew are opportunities seeming as if requiring GOD's creative intervention all the while manifesting as scientific technological developments. We will label the huge infinite number of infinite sets of parameters of opportunities as such as "para-Divine-opportunities". We are a people meant for GOD and each other so I am convinced the para-Divine-opportunities will always superabound and become increasingly infinite in number and in extremity.

We may affix the following operator to formulas for gamma and related parameters to denote the above conjectures and associated mechanisms.

[w(Para-Divine-opportunities always superabounding and becoming increasingly infinite in number and in extremity)]

With the huge number of; realms of travel and entrance, possible time lines and numbers of temporal dimensions, infinite Lorentz factors, super-luminal velocities, and the entire host of other parameters maxed to and beyond countable infinity numbers in value, we are getting close to technologies so advanced that the huge infinity of infinite realms and scenarios available to spacecraft and crew are opportunities seeming as if requiring GOD's creative intervention all the while manifesting as scientific technological developments. We will label the huge infinite number of infinite sets of parameters of opportunities as such as "para-Divine-opportunities". We are a people meant for GOD and each other so I am convinced the para-Divine-opportunities will always superabound and become increasingly infinite in number and in extremity.

We may affix the following operator to formulas for gamma and related parameters to denote the above conjectures and associated mechanisms.

[w(Para-Divine-opportunities always superabounding and becoming increasingly infinite in number and in extremity)]

We can form the set of derivatives and integrals of Para-Divine-opportunities where the variable(s) of derivatives and/or integrals is any ones of the hugely infinite number of para-Divine-opportunities.

We affix the following operator to the lengthy formulas for gamma and related parameters in this part of this section of the book to denote the above conjectures and associated mechanisms of derivatives and integrals.

{u{[First and/or higher mth-Para-Divine-opportunity derivatives and integrals of [Σ(k = 1; k = ∞↑):kth-Para-Divine-opportunity]]| m = 1, 2, 3, …}}

Note that the variables of ship-time, acceleration, gamma, kinetic energy and the like are assumed part of the set of but not all of the set of para-Divine-opportunities.

Here, we explore light-speed effects related to chaos, super-chaos, and super-causality-chaos as opportunities for created persons to do extremely determinism violating acts of free will and to experience spontaneous creative spells, emotions and feelings, and intellectual and spiritual insights.

Note that I have never experienced anything like such causality chaos but would welcome an opportunity for such for spontaneous personal growth.

A cosmic order of chaos, super-chaos, super-causality-chaos and the like may be a sink for enabling thermo-info-dynamic order to grow to increasingly extreme extents.

The order as such can not only affect and be associated with cosmic stability, but also affect and be associated the composition created personal lifeforms, and/or technological applications. Such degrees of order can become so great that they morph into hyper-order or hyper-1-order.

Still more extreme scenarios can enable hyper-2-order which is to hyper-1-order as hyper-1-order is to ordinary order.

Still more extreme scenarios can enable hyper-3-order which is to hyper-2-order as hyper-2-order is to hyper-1-order as hyper-1-order is to ordinary order.

We can continue to hyper-4-order, hyper-5-order all the way up to and beyond hyper-∞↑-order. The infinity sign followed by the vertical arrow indicates unlimited potential growth.

It could be that such super-causality-chaos is necessary for reality to develop and maintain its integrity. This latter statement may be a form and/or analog of the concepts that “Nature abhors a vacuum.”.

Folks experiencing super-causality-chaos may enjoy the ride and randomness in terms of their souls in cases where they would go to Hell. Accordingly, they may seem to suffer and actually do suffer but prefer Hell to Heaven. I am a firm believer that GOD has something fantastic planned for the souls and angels in Hell in the utter depths of future eternity. However, I will not ever expound nor highly publicize my conjectures on such since the Church authorities would almost certainly censor me. Here, I am speaking from my interior sense of empathy and not for the Church nor for the sake of faith-based counseling.

However, trust me! You do not want to go to Hell as you will experience grave discomfort there. So do not use my speculation as a means to be lax in opportunities to grow in grace and get to Heaven. I merely offer these speculations and personal beliefs of mine as a way to share ideas and to promote a sense of peace in the hearts and minds of all folks like me who are extremely saddened by the reality of Hell.

Hell may have pockets of extreme order and levels of hyper-order too.

For some residents of Hell, these pockets of order and levels of hyper-order may be really, really, boring and this boredom may be an aspect of their locations in Hell.

I would hazard a guess that some souls in Hell although they suffer may be very kind and offer up their being consigned to Hell so that their family members, relatives, friends, spouses and the like can avoid going to Hell.

Regardless, spacecraft having attained the speed of light may enter a state where the cause and effect within the spacecraft and of the spacecraft may be scrambled relative to the background reference frame.

Still more infinite spacecraft Lorentz factors may enable the spacecraft cause and effect within the spacecraft in its own reference frame to be scrambled. As I mentioned previously, such scrambling may be associated with wild indeterminism thus promoting and enhancing the effects of human free will, and the extents of indeterminism of acts of free will by the spacecraft crew members.

We may affix the following operator to formulas for gamma and related parameters to denote the above light-speed mechanisms of indeterminism, levels of hyper-order, and levels of super-causality-chaos and applications for technological and existential evolution of created persons.

[u(Light-speed mechanisms of indeterminism, for promoting levels of hyper-order, and levels of super-causality-chaos, and applications for technological and existential evolution of created persons)].

We can take first thru infinite order derivatives and integrals of the above operator mechanisms with respect to ship-time, gamma, acceleration, ship kinetic energy, or any other relevant parameters.

We may affix the following operator to formulas for gamma and related parameters to denote first thru infinite order derivatives and integrals of the above operator mechanisms with respect to ship-time, gamma, acceleration, ship kinetic energy, or any other relevant parameters.

{1st thru ∞↑ arbitrary der & ∫ of [u(Light-speed mechanisms of indeterminism, for promoting levels of hyper-order, and levels of super-causality-chaos, and applications for technological and existential evolution of created persons)]}

I've produced the following summary of some really cool concepts regarding aspects of the spiritual and immortal soul which at the time of this writing I just added to my work-in-progress book,

"THE SOUL. OUR INNERMOST, IMMORTAL, FIRE OF LIFE. VOLUME 1."

Here, we consider infinite Lorentz factor light-speed reference frames which develop secondary co-present reference frames.

Accordingly, a light-speed inertial spacecraft may attain such a great infinite Lorentz factor such that the spacecraft not only has its own reference frame, but also one and progressively more reference frames in an ancillary or primitive sense.

Accordingly, there would exist or become real spacecraft reference frames that are distinct from the ordinarily visible accidental attributes of the spacecraft.

Such additional spacecraft reference frames may include semi-substantial and substantial aspects as well as the essence of the prime-matter composition of the spacecraft.

For crew members, the existence and/or onset of additional reference frames may include non-Lorentz invariant processes. Such processes can include non-Lorentz invariant aspects of the human soul such as its accidents, faculties, substance, and essence.

However, there may also at the same time exist Lorentz invariant perception, mobility, consciousness, and other psychodynamic processes for crew members. Accordingly, the crew members would experience nothing different nor out of the ordinary even while at the same time Lorentz varying existential processes or states occur in the human soul such as its accidents, faculties, substance, and essence.

The reality that Lorentz varying and Lorentz invariant processes can occur for the crew members, their souls, and the spacecraft affirm that there are higher or transcendent levels of physical and created spiritual realities.

A spacecraft having attained suitably infinite Lorentz factors may even begin to experience enhanced reference frames of human crew members' states of grace and Sacramental seals were applicable, yet notice nothing unusual in their fields of sensory perception and conscious activities. Accordingly, the life of grace may have more than one reality including Lorentz invariant and Lorentz varying aspects.

The number of distinct reference frame modes of the spacecraft and crew members is likely only a few, but perhaps could grow to infinite numbers thus providing infinite numbers of ship-frame and crew-members-frames of reference.

Oddly enough, the reality that the crew members in extremely great finite and infinite Lorentz factors reference frames would not have any Lorentz variant experiences affirms the existence of transcendent aspects of not only the physical composition of the spacecraft, but also of the spiritual composition of human souls including accidents, faculties, substances, and essences of the human soul. Oddly enough, this would also affirm that human conscious and even self-aware psychodynamic and sensory processes would have invisible or non-observable aspects, even for the persons under consideration.

We may affix the following operator to the following lengthy formulas for the existential size and significance of the human soul to denote the possible Lorentz-Variant-Invariant mechanisms of the human soul and the implied extremity of the visible and invisible aspects of the accidents, faculties, substance, and essence of the human soul.

Since the substance and essence of human souls underwrites existentially the faculties and accidents of the human soul, the reality that in theory, light-speed travel would not alter human crew members' waking consciousness affirms that there are also associated hidden or veiled aspects of the substance and essence of the human soul that have not previously been considered in philosophy and religious belief systems. The reason for this is the existential coupling between the substance and essence of the human soul and its faculties and accidental aspects.

[u(Lorentz-Variant-Invariant mechanisms of the human soul and the implied extremity of the visible and invisible aspects of the accidents, faculties, substance, and essence of the human soul)].

The proposed reality of Lorentz varying acquisition and/or enhancement of unknown numbers of spacecraft reference frames suggest a commensurately multi-faceted set of aspects of the human soul and a soul with commensurately many transcendent properties.

Now, the acquisition of additional reference frames as considered above is not simply a matter of a spacecraft having light-speed secondary motions in its own translational reference frame, but instead also with respect to each and every differential volumetric element of the spacecraft.

So, we take the above provided operator and modify it as follows to denote the above aspects of the spacecraft materiality also.

[u(Lorentz-Variant-Invariant mechanisms of the human soul and fiat material composition and the implied extremity of the visible and invisible aspects of the accidents, faculties, substance, and essence of the human soul and also of mattergy)].

We may affix another operator as follows to denote the wealth of the acquisition of an unknown number of species of additional spacecraft and crew member reference frames for suitably infinite Lorentz factor spacecraft translational motion reference frames.

[w(Wealth of the acquisition of an unknown number of species of additional spacecraft and crew member reference frames for suitably infinite Lorentz factor spacecraft translational motion reference frames)]

We actually combine both operators as the following single compound operator.

{[u(Lorentz-Variant-Invariant mechanisms of the human soul and fiat material composition and the implied extremity of the visible and invisible aspects of the accidents, faculties, substance, and essence of the human soul and also of mattergy)]:
[w(Wealth of the acquisition of an unknown number of species of additional spacecraft and crew member reference frames for suitably infinite Lorentz factor spacecraft translational motion reference frames)]}

Back in the days while I was a teenager, I attended a private school. The school psychologist was a consecrated Catholic religious brother with dark hair and a dark beard who used to let me ride in his fancy Ford Thunderbird. The car had a black exterior and interior.

Well, at about the same time, the sitcom, “The Jeffersons” was popular and the show theme song had a refrain that went like “Well we're movin' on up. (Movin' on up). To the east side.”.

Even back then I was interested in interstellar travel concepts.

To make a long story short, I associated the school psychologist and rides in his Thunderbird with my internalized mantra of Movin’ on up, to the future, at near light-speed. Thus, I became more hooked on special relativistic space travel and time dilation. The fire of my imagination for near light-speed travel was lit just as assuredly as the black Ford Thunderbird resembled the black cosmic void. I knew then the ramifications of infinite time dilation and infinite forward time travel and distances through space mathematically plausible for light-speed impulse travel.

So, if you have the courage to delve further into this book, or even only study select portions thereof, you will likely if not already also become intrigued with Movin’ on up into the future with Special Relativity. As we now have a space travel industry, we have set before us the seas of infinity.

In a way, we are all movin' on up into the future special relativistically. Such a future is unbounded in terms of forward temporal progression and the eternal duration of reality.

Even for spacecraft having attained light-speed in a least infinite Lorentz factor, there is no mathematical reason why the light-speed spacecraft cannot continue having increasing infinite Lorentz factors.

Since the human soul is immortal, even after a first eternity comes to a conclusion, there is eternity after eternity to follow without end. The never ending series of eternities, the first of which ironically will never end, are a grand, bold, prolife, joy-filled, wondrous future in special relativistic movin' on up.

Essentially, we will always be movin' on up into the future.

Giving sufficiently infinite spacecraft Lorentz factors, we may leave a universe, multiverse, forest, biosphere, and the like of travel altogether and enter another type of realm and temporal or non-temporal path of eternity or eternity-like progression. Given sufficiently lengthy additional periods of positive acceleration, the craft may enter yet another type of realm and temporal or non-temporal path of eternity-like progression.

In plausibility, there may by a never ending and always growing set of eternities or eternity-like realms of progression by which any created personal life-forms can continue movin on up into. Even the eternity that we inhabit, per the assumed infinite future duration of the cosmos we live in, we will always be movin' on up within in some profound ways even if we attain spacecraft Lorentz factors sufficient to move us into other eternities.

We might affix the following operator to formulas for gamma and related parameters to denote these movin' on up scenarios.

{[u(kth special relativistic mode of mov'n on up opportunities)] | k is any finite or infinite positive integer}

However, we can also take arbitrary orders of derivatives and integrals of the rates of the movin' on up opportunities with respect to arbitrary parameters. So, in practice, we will affix the following operator instead.

{[u(kth special relativistic mode of mov'n on up opportunities)] | k is any finite or infinite positive integer} and/or arbitrary orders of der and ∫ of the rates of progression thereof with respect to arbitrary parameters}

These arbitrary parameters include ship-time, acceleration, gamma, kinetic energy and perhaps infinitely many others.

Here, we consider light-speed accelerating spacecraft as experiencing ontological growth in the value and existence of the craft Lorentz factors but contraction-like intra-burrowing to denser and denser compositions, real and/or mathematically abstract-wise.

Accordingly, just as a gravitational mass in space obtains sufficient density to cause a huge star to form, with the eventual fate of the star being a neutron star and/or core collapse into a black hole state, we accordingly conjecture that the suitably large finite and infinite Lorentz factors of a spacecraft will ontologically and really collapse into a metaphorical infinite density state.

Such metaphorical collapse into an infinite density state is a akin to a mechanism for which each finite factoral increase in infinite Lorentz factors, the number of possible Lorentz factors in the interval of increase of gamma is infinitely greater than in cases for finite factoral increases of finite Lorentz factors.

Another way to consider this general mechanism is that just as the center of a black hole may collapse to infinite density, or a huge finite Planck Mass Density core, analogous real and abstract collapse of suitably great finite or infinite spacecraft Lorentz factors may occur.

A consequence of such an occurance might be a runaway increase in the actual value of gamma to ever greater infinite values.

Some additional fascinating prospects consider that the collapsing and/or but runaway values of gamma may result in gamma developing an onion-like core with layer upon layer developing as the value of gamma increases. Thus, a runaway increasing infinite value of gamma may be analogous to an ocean that becomes deeper and deeper without depths limits.

We may ordinarily affix the following operator to formulas for gamma and related parameters to denote the above light-speed, Lorentz factors collapse, obtainment of infinite real and abstract densities and values, infinite runaway increases in gamma, and the fathomless increase in oceanic depth of gamma and increases in number of Lorentz factor layers.

[q(light-speed, γ collapse, obtainment of ∞ real and abstract densities and values, ∞ runaway increases in gamma, and the fathomless increase in oceanic depth of gamma and increases in number of γ layers)]

However, we can also consider arbitrary order derivatives and integrals of the rate of progression in the mechanisms implied in the above operator with respect to arbitrary parameters.

So, instead, we actually affix the following operator to formulas for gamma and related parameters as such.

{[q(c, γ collapse, obtainment of ∞ real & abstract densities & values, ∞ runaway increases in γ, & the fathomless increase in oceanic depth of γ & increases in number of γ layers)] &/or arbitrary (f) der &/or ∫ thereof}

Here (f) stands for function as parameter of differentiation and/or integration.

A fascinating construct is the notion that due to Special and General Relativity, each person lives his own world-line for which their own reference frame is relative to all of the other reference frames with respect to velocity, momentum, kinetic energy, temperature, acceleration and the like even though accelerated reference frames are considered under General Relativity.

This means that in a sense, each human person and any other species of created persons is in his or her own cocoon, all nestled safe and snug, answering ultimately only to GOD.

The manifestation of this relativity becomes all the more extreme for crew members in a relativistic spacecraft. Were the spacecraft to attain a suitably infinite Lorentz factor, the relativity of the spacecraft would max out to some extreme boundary condition, but perhaps only one of untold additional boundary conditions as the spacecraft Lorentz factor would actually continue to increase.

Now, at some point, a spacecraft may attain such an extreme infinite Lorentz factor such that is acquires a large fraction of the total mass-energy content of the universe, multiverse, forest, biosphere and the like, hyperspace, bulk, and/or hyperbulk of travel. Thus, the spacecraft in the associated realm of travel may no longer be truly relativized as it would have strong distinction of incorporating a large fraction of the available and visible energy in the given realm of travel.

However, relativity may yet in some sense still exist or perhaps re-emerge.

For example, other universes, multiverses, forests, biospheres and the like, hyperspaces, bulks, and/or hyperbulks that are non-casually coupled to the one of travel may not even be existentially present to the realm of travel. In time and in the first few additional more broadly manifested time-like dimensions, these non-casually coupled realms may not exist relative to the realm of travel and the spacecraft considered.

So, in a sense, existence can be a much deeper relative construct. If GOD has manifested such non-relatively existing realms, it behooves us to consider that existence may be just one reality parameter among many perhaps in the same class as existence but distinct from existence.

So, another aspect of relativity may be relativity of existence. There could even manifest other non-existence relativities in the same or different class as existence. Note, however, that there may be more than one level of existential relativity.

For example, as a spacecraft would continue to uptake all the more cosmic energy, the spacecraft may supersede one level of relative existence after another and thus become more and more absolute.

These constructs are fascinating to consider and are highly plausible per the above analysis and toy models.

Now, the positive energy in the universe may be almost or of identical magnitude as the universe negative energy. Thus, there is likely a physical mechanism that is pre-ordinate and more primitive than energy, in both positive and negative forms.

Accordingly, a spacecraft having achieved sufficient positive kinetic energy may need to produce more cosmic negative energy in order to continue its acquisition of kinetic energy. If balancing negative energy can be produced in sufficiently unlimited quantities, then the spacecraft should be able to super-abound in ever greater infinite kinetic energy and thus ever greater infinite Lorentz factors.

Another option which may have aspects related to or not to the concepts in the previous paragraph is the manipulation for the physical entity stated as preordinate or more primitive than energy. Accordingly, manipulation of this mechanism may be used to grow more positive and negative energy. The spacecraft may then incorporate the increased grown positive energy at least in part.

A second consideration of the above conjectured option is that this ephemeral meta-energy principle of preordination may be directly harnessed to enhance spacecraft locomotion and thus transport without the production of additional positive and requisite negative energy. As such, the spacecraft may have Lorentz factors that continue to super-abound in ever increasing infinite values without addition of more kinetic energy.

So, in a sense, there may exist physical principles that are more primitive than and preordinate to physical energy. In some cases, it may be possible to convert some of these more primitive physical principles directly into energy but not in the same sense as mass energy equivalence.

Accordingly, these more primitive principles may be screened from directly observable accidental space-time-mass-energy compositions thus enabling in principle a relativistic rocket with an effective mass ratio just slightly in excess of one being able to attain infinite Lorentz factors in light-speed travel. Such screening may also be applicable to infinite beam energy production for which a high powered beam could be directed onto a light-sail, photovoltaic, photo-electric, thermoelectric, or turboelectric power generation system where the electrical power generated would energize electrical propulsion systems. Such systems would be indefinitely operable for a given spacecraft as long as a light-cone in a given universe, multiverse, forest, biosphere and the like, hyperspace, bulk, or hyperbulk of travel expands faster than the speed of light in terms of its light-cone on centered radial coordinate of extension.

Some such electrical propulsion systems include:

1) Ion rockets,

2) Electron rockets,

3) Muon rockets,

4) Tauon rockets,

5) Exotic baryon rockets,

6) Meson rockets,

7) Photon rockets,

8) Magneto-hydrodynamic-plasma-drives,

9) Electro-hydrodynamic-plasma-drives,

10) Electro-magneto-hydrodynamic-plasma-drives,

11) Electromagnetic-hydrodynamic-plasma-drives,

12) Magnetic plasma sails,

13) Magnetospheric-plasma-bottle sails,

14) Mini-Magnetospheric-Plasma-Propulsion systems,

15) Magnetic field-effect propulsion systems,

16) Electric field-effect propulsion systems,

17) Relativistic mass driver rockets,

18) Nuclear fission and/or fusion powered interstellar ramjets,

19) Neutralization of background ions systems that intake emission photons from neutralized ions,

20) Other method not listed or as may become known in the future.

We can take the ship-time, gamma, acceleration, kinetic energy, and the like derivatives and integrals of arbitrary order of the above conjectured physical mechanisms.

We may affix the following operator to formulas for gamma and related parameters to denote the above relativity and energy derivation concepts and applicability to spacecraft Lorentz factors and other related parameters.

{[q(k,1,th spacecraft existential absolutizement mech)]:[q(k,2,th + energy creation mech)]:[q(k,3,th preordinate principle direct harnessing and/or energy conversion)] &/or arbitrary (f) der &/or ∫ thereof where: k,1; k,2; k,3 = 1, 2, 3, ...}

SO TRULY, YOU MUST NOT BE BORED OF ANY LIGHT-SPEED TRAVEL LIMITS!

Now, we see manifestations of numbers everywhere from science and math, philosophy and theology, medicine and social science, trades and blue collar work, industry and manufacturing, energy and food production, and everywhere else.

However, how often do we consider realities that are in a limited sense, more numbers than numbers?

Accordingly, I am not referring to mathematical functions and relations, nor the rules and operators of symbolic login, nor indices, nor qualitative realities, nor probability amplitudes or components thereof of quantum wave-functions.

Instead, I am referring to some analog of numbers that is to numbers as numbers are to non-numbers.

Here, there may be an analog of the number, 1, that is entirely distinct from unity.

There would also be analogs of the concepts of countable and uncountable infinities. Here, I am not referring to qualitative infinities, but instead, realities that loosely translate into forms that are existentially on the opposite side of numbers than as are qualities.

There may exist additional levels or hierarchies of so-called meta-numbers.

For example, meta-1-numbers may be to numbers what numbers are to qualities or other common non-numerical constructs while meta-2-numbers are to meta-1-numbers what meta-1-numbers are to numbers as numbers are to qualities or other common non-numerical constructs.

We can likewise consider still higher levels such as meta-3-numbers, meta-4-numbers, and so-on in a never ending ascending series.

I am not sure what realities the various levels of meta-k-numbers would apply to. Perhaps such meta-k-numbers would apply to Divine realities.

Future intrepid human space travelers who will have long severed ties with what occurs back on Earth may end up being in the presence of the Lord upon which the Lord will begin to reveal to them the various mysteries of realities defined by the meta-k-numbers.

Assuming there is only one GOD, perhaps the one GOD is also the meta-k-realities META-K-GOD. The latter reality would belong to the one GOD who would also be the meta-k-realities META-K-GOD. It is also note worthy to realize that what we know above GOD is very limited even in the sense of abstract knowledge.

We may affix the following operator to formulas for gamma and related parameters to denote the above mechanisms and manifestations of the revelation of the various mysteries of realities defined by the meta-k-numbers.

[u(Mechanisms and manifestations of the revelation of the various mysteries of realities defined by the meta-k-numbers)]

Now, consider the hyper-operator notation that was designed to express huge values not otherwise expressible.

For example, Note that Hyper4(a, n) is equal to a tetrated n or a raised to the power of itself n-1 times. The latter value is symbolically written as n subscript a. For example 3 EXP 4 = 81, but 4 subscript3 is approximately equal to 10 EXP (1,000,000,000,000).

Alternatively 4 subscript 2 = 2 EXP 2 EXP 2 EXP 2 = 2 EXP [2 EXP [2 EXP 2]] = 2 EXP (2 EXP 4) = 2 EXP 16 = 65,536.

For example, Hyper5(4, 4)is equal to 4 tetrated 4 tetrated 4 tetrated 4. This value is commonly referred to as 4 pentated 4.

Hyper 6, (4,4) is 4 pentated 4 pentated 4 pentated 4 and is also referred to as 4 hexataed 4.

Hyper 7, (4,4) is 4 hexated 4 hexated 4 hexated 4 and so-on

Aleph 0 is the infinite number of integers.

Aleph 1 according to the perhaps unprovable and thus unfalsifiable Continuum Hypotheses is the number of real numbers which is greater than Aleph 0 by a multiplicative factor of infinity.

Aleph 2 is similarly greater than Aleph 1.

Aleph 3 is similarly greater than Aleph 2.

Aleph 4 is similarly greater than Aleph 3.

and so-on

In general, Aleph n = 2 EXP [Aleph (n-1)], where n is any finite or infinite counting number.

The number Ω is commonly stated as the least infinite positive integer or ordinal.

Now here is a real zinger.

So we can produce the abstraction of Hyper Aleph Ω (Aleph Ω, Aleph Ω).

We can go on to produce the abstraction of,

Hyper [Hyper Aleph Ω (Aleph Ω, Aleph Ω)] {[Hyper Aleph Ω (Aleph Ω, Aleph Ω)],[Hyper Aleph Ω (Aleph Ω, Aleph Ω)]}

We can denote this value as Hyper,primed,2: Hyper Aleph Ω (Aleph Ω, Aleph Ω).

We can go further to produce the following huge infinity.

Hyper {Hyper [Hyper Aleph Ω (Aleph Ω, Aleph Ω)] {[Hyper Aleph Ω (Aleph Ω, Aleph Ω)],[Hyper Aleph Ω (Aleph Ω, Aleph Ω)]}}{{Hyper [Hyper Aleph Ω (Aleph Ω, Aleph Ω)] {[Hyper Aleph Ω (Aleph Ω, Aleph Ω)],[Hyper Aleph Ω (Aleph Ω, Aleph Ω)]}},{Hyper [Hyper Aleph Ω (Aleph Ω, Aleph Ω)] {[Hyper Aleph Ω (Aleph Ω, Aleph Ω)],[Hyper Aleph Ω (Aleph Ω, Aleph Ω)]}}

We can denote this value as Hyper,primed,3: Hyper Aleph Ω (Aleph Ω, Aleph Ω)

We can go further to produce the following huge infinity.

Hyper { Hyper {Hyper [Hyper Aleph Ω (Aleph Ω, Aleph Ω)] {[Hyper Aleph Ω (Aleph Ω, Aleph Ω)],[Hyper Aleph Ω (Aleph Ω, Aleph Ω)]}}{{Hyper [Hyper Aleph Ω (Aleph Ω, Aleph Ω)] {[Hyper Aleph Ω (Aleph Ω, Aleph Ω)],[Hyper Aleph Ω (Aleph Ω, Aleph

Ω)]}},{Hyper [Hyper Aleph Ω (Aleph Ω, Aleph Ω)] {[Hyper Aleph Ω (Aleph Ω, Aleph Ω)],[Hyper Aleph Ω (Aleph Ω, Aleph Ω)]}}}{{ Hyper {Hyper [Hyper Aleph Ω (Aleph Ω, Aleph Ω)] {[Hyper Aleph Ω (Aleph Ω, Aleph Ω)],[Hyper Aleph Ω (Aleph Ω, Aleph Ω)]}}{{Hyper [Hyper Aleph Ω (Aleph Ω, Aleph Ω)] {[Hyper Aleph Ω (Aleph Ω, Aleph Ω)],[Hyper Aleph Ω (Aleph Ω, Aleph Ω)]}},{Hyper [Hyper Aleph Ω (Aleph Ω, Aleph Ω)] {[Hyper Aleph Ω (Aleph Ω, Aleph Ω)],[Hyper Aleph Ω (Aleph Ω, Aleph Ω)]}}},{ Hyper {Hyper [Hyper Aleph Ω (Aleph Ω, Aleph Ω)] {[Hyper Aleph Ω (Aleph Ω, Aleph Ω)],[Hyper Aleph Ω (Aleph Ω, Aleph Ω)]}}{{Hyper [Hyper Aleph Ω (Aleph Ω, Aleph Ω)] {[Hyper Aleph Ω (Aleph Ω, Aleph Ω)],[Hyper Aleph Ω (Aleph Ω, Aleph Ω)]}},{Hyper [Hyper Aleph Ω (Aleph Ω, Aleph Ω)] {[Hyper Aleph Ω (Aleph Ω, Aleph Ω)],[Hyper Aleph Ω (Aleph Ω, Aleph Ω)]}}}

We can denote this value as Hyper,primed,4: Hyper Aleph Ω (Aleph Ω, Aleph Ω)

We can likewise continue to produce

Hyper,primed,5: Hyper Aleph Ω (Aleph Ω, Aleph Ω)

Hyper,primed,6: Hyper Aleph Ω (Aleph Ω, Aleph Ω)

and so-on and so-on to,

Hyper,primed,(Aleph 0): Hyper Aleph Ω (Aleph Ω, Aleph Ω)

and so-on and so-on to,

Hyper,primed,(Aleph 1): Hyper Aleph Ω (Aleph Ω, Aleph Ω)

and so-on and so-on to,

Hyper,primed,(Aleph 2): Hyper Aleph Ω (Aleph Ω, Aleph Ω)

and so-on and so-on to,

Hyper,primed,(Aleph 3): Hyper Aleph Ω (Aleph Ω, Aleph Ω)

and so-on and so-on to,

Hyper,primed,(Aleph Ω): Hyper Aleph Ω (Aleph Ω, Aleph Ω)

and so-on and so-on to,

Hyper,primed,[Hyper Aleph Ω (Aleph Ω, Aleph Ω)]: Hyper Aleph Ω (Aleph Ω, Aleph Ω)

and so-on and so-on without end.

These expressions can provide us with a sense of the whimsical when considering spacecraft Lorentz factors.

For example, imagine a spacecraft at light-speed having Lorentz factor of Hyper,primed,[Hyper Aleph Ω (Aleph Ω, Aleph Ω)]: Hyper Aleph Ω (Aleph Ω, Aleph Ω).

Well, this bad boy will travel about Hyper,primed,[Hyper Aleph Ω (Aleph Ω, Aleph Ω)]: Hyper Aleph Ω (Aleph Ω, Aleph Ω) light-years thru space in one year ship time and Hyper,primed,[Hyper Aleph Ω (Aleph Ω, Aleph Ω)]: Hyper Aleph Ω (Aleph Ω, Aleph Ω) years into the future in one year ship time.

We may affix the following operator to formulas for gamma and related parameters to denote such tremendous travel parameters.

{w{Gamma ≥ {Hyper,primed,[Hyper Aleph Ω (Aleph Ω, Aleph Ω)]: Hyper Aleph Ω (Aleph Ω, Aleph Ω)}}}

Now, a spacecraft traveling at light-speed in a least transfinite Lorentz factor will likely still perceive the passage of ship-time. One second ship time will correspond with Ω seconds background time.

More than likely according to my paradigm, a spacecraft can climb to much qualitatively greater infinite quantity Lorentz factors.

Eventually, such a light-speed spacecraft will no longer experience the passage of ship-time but may continue having perpetually runaway infinite Lorentz factors.

Accordingly, the spacecraft and her live crew would experience an incredible calm or state of changelessness.

A first state of changelessness for the crew would include some sort of altered consciousness while also providing extraordinary natural inner peace.

As the spacecraft Lorentz factor continued to climb to ever greater infinite values, the state of changelessness would be continually enhanced to promote continued levels of altered consciousness and growing of natural inner peace.

Other novel psychological phenomenon can manifest such as the complete stillness of emotions, feelings, mental imagery, emotional imagery, state of volition and the like.

At still greater infinite spacecraft Lorentz factors, new and completely currently unknown cognitive, affective, instinctive, and volitional states will manifest.

At still yet greater infinite spacecraft Lorentz factors, the human psyche may develop additional unknown faculties as different from and distinct from the faculties of sensibility, intellect, will, and memory as these latter faculties are to each other.

Eventually, the psychodynamic aspects of the human personality will grow beyond the three modes of consciousness, sub-conscious, and unconscious, or beyond the three modes of the ID, ego, and superego.

Thus, we arrive at new definitions of velocity.

Accordingly, the rate of development of ever more extreme changelessness and the like of the spacecraft and her crew can be expressed in a first or higher orders of gamma derivatives and integrals. The derivatives indicate more of speeds and the integrals indicate more of distance traveled where distance is associated with the metric of growth in changelessness. The referenced speeds are a different class of speeds relative to the common meaning associated with motion thru space.

Similar derivatives can be taken with respect to spacecraft kinetic energy, spacecraft acceleration, and any other useful arbitrary but applicable variables.

We may affix the following lengthy operator to formulas for gamma and related parameters to denote taking all of the above mechanisms into account.

{{[w(Rate of altering and temporally freezing of human cognition, sensory perceptions, affect, volition, and memory as craft obtains Lorentz factors which are suitably infinite)]:[w(Rate of development of new psychodynamic faculties distinct from the sensibility, intellect, will, and memory as craft obtains Lorentz factors which are suitably infinite)]:[w(Rate of development of new psychodynamic structures distinct from the ID, ego, and superego as craft obtains Lorentz factors which are suitably infinite)]:[w(Rate of development of new experiential psychodynamic levels distinct from the consciousness, sub-conscious, unconscious, and ultra-unconscious as craft obtains Lorentz factors which are suitably infinite)]:[w(Rate of development of extraordinary natural inner peace as craft obtains Lorentz factors which are suitably infinite)]} and/or first and higher orders of differentiation and/or integration thereof with respect to gamma, spacecraft kinetic energy, spacecraft acceleration, and any other useful arbitrary but applicable variables}

Thus, light-speed travel may become so extreme that it becomes a real mind-trip, a real head-trip, or whatever. So, new meanings of the concept of travel and velocity are possible.

Here, we consider ever-so-slightly faster than light travel such that the craft does not violate chronological protection no-go theorems but is able to react off of the early birth energy states of our; universe, multiverse, forest, biosphere and the like as a limited version of what we will refer to as pseudo-backward time travel or connection.

So, we are not actually considering that the craft travels back in time but simply just reacts of the temporally backward energy states of our universe, multiverse, forest, biosphere and the like.

In a way, such backward reactance can be explained simply as reacting off backward information which is conserved in the present by conservative information laws.

A real blessing would be the ability of a craft or fiat to react off of the very instants of creation, such as perhaps, those commonly attributed to GOD's creative acts of will. As such, and Divinely permitting, craft or fiats can experience in principle enormously infinite jumps in Lorentz factors and kinetic energies as well as ship-frame acceleration.

I often have a mental image of GOD blowing the universe or greater cosmos into existence as if by a horn shaped abstract arrangement. The horn is typically very abstracted but with a mental image as a black colored speaker, with an array of sub-emitters that is three emitters tall and four emitters long for a total of 12 sub-units. I have this mental auditory image of the horn-like apparatus sounding like a tuba.

The above mental image is simply an abstraction but the horn-like contraptions is imagined as coming out of nowhere and surrounded by a black void.

The three horizontal rows represent the three families of quarks, charged leptons, and neutrinos, while the four columns represent the four known forces. In line with Pope John Paul II The Great, the Divine blowing process is the state of affairs during and at the very beginning of the first Planck Time Unit, a region that may be forever veiled to scientific observation because it is very holy. Pope John Paul II stated that we should not probe back before the end of the first Planck Time Unit because this era or aspect of creation is holy.

However, I am willing to be proved wrong in this by observational and theoretical cosmology as already there are models folks have come up with that explain the pre-Planck time era of our universe.

We may affix the following operator to formulas for gamma and related parameters to denote the above retroactive thermodynamics and infodynamic states of our cosmic orders for spacecraft propulsion already at or to light-speed and perhaps to velocities ever-so-slightly greater than that of light.

[w(Retroactive thermodynamics and infodynamic states of our cosmic orders for spacecraft propulsion already at or to light-speed and perhaps to velocities ever-so-slightly greater than that of light)]

I have decided I want to paint or draw an expression of this mental image of mine.

Another mechanism of suitably infinite Lorentz factor manifestations in light-speed or ever-so-slightly faster than light travel involves the change of a present region in the same Planck Time Unit that the change takes place or is desired to take place.

Accordingly, such a technology would be one step removed from changing the past and might actually be possible.

Here, the suitably infinite Lorentz factors light-speed or ever-so-slightly faster than light travel would be an intermediary state between actually fully faster than light travel and backward time travel and travel into the future at an accelerated ship-frame rate for extreme finite Lorentz factor or moderately infinite Lorentz factor light-speed travel.

We may affix the following operator to formulas for gamma and related parameters to denote the above present change concepts and associated ramifications.

[w(Suitably infinite Lorentz factor manifestations in light-speed or ever-so-slightly faster than light travel involving the change of a present region in the same Planck Time Unit that the change takes place or is desired to take place)]

A fascinating consideration involves the notion of the cause of a created entity in the sense of the cause it stands for.

For example, contemporary TV series on the American Civil War had a segment on the cause of Union recruitment of troops displaying a poster that had the heading “Sharp Shooters Wanted For The Cause” or something similar.

Well, the cause I am presenting you has nothing to do with warfare, at least not directly, and is on another subject such as the cause of a given soul in the sense that Divine providence creates every soul that comes into being as part of the Divine Cause for creation and the subset of creatures known as human souls and then so for absolutely each and every soul, even the ones that go to Hell.

Another closely related but distinct concept of a cause is related to what each and every human soul ever greated is intended for.

The first cause mentioned above is more or less related to the immediate actions of the Divine Will but with eternal ramifications while the second concept of cause is more of an avolitional purpose or reason for a creature, and in our cases, our souls.

So, each and every soul ever created has both causes even the souls that go to Hell.

Since, these two causes of the soul transcend its composition and its existential properties, we can say that the two causes of the soul are irrevocable hyper-immortalities of the soul.

These irrevocable dicausal aspects of the soul are so extreme that these aspects transcend the very immortality and spirituality of the human soul.

Even considering the space-time of ordinary 3-D-Space-1-D-time of our universe, the structure and division of our universal space-time is kind of really whimsically weird and this is really cool.

For example, at the edge of our current light-cone is a boundary that we cannot see past even with our best most advanced telescopes. However, beyond this boundary is in a sense a region causally decoupled from our light-cones. So, in a sense, such decoupling makes the space-time structure of our universe more complicated than simply a 4-D space-time template that we can imagine abstractly as a 3-D coordinate axis within which objects move commensurate with a 4th dimension known as time.

We may even scrunch up or deform of imaginary 4-D system to account for warpage of space-time.

However, the reality that every person has their own cosmic light-cone boundary while at the same time, all the residents of a given planet or within a common galaxy are causally decoupled from the space-time regions outside of their common light-cones or averaged light-cones, indicates a profound structure to our universe that is likely well under-appreciated and not well considered by cosmologists.

A fascinating take on these ideas includes the conjecture that a cosmic habitat might be assembled for which some, much, or almost all of the associated structure is not co-located into a single cosmic light-cone width or radius.

Accordingly, such a habitat may include mass-driver travel tubes, gravitating road or rail transport pathways, and/or electrical conduits for carrying electrical pulses formed by the electrical current analogs of the beaming up processes in the famous science fiction series of TV episodes and movies commonly known as Star Trek.

All of the above considerations for super-light-cone habitat extensions and travel highways in a sense, although limited as such, manifest as super-luminal or faster-than-light conditions.

For example, a nano-technology self-assembly mechanism may be accelerated to near the velocity of light at which the mechanism drops off molecular or even micron scale machines that self-replicate into copies of themselves all the while assembling a mass-driver, road or railway, and/or an electrical conduit.

To the extent that these transportation infrastructures eventually extend so far out that they cross into another light-cone relative to the region of origin is a proxy for superluminal information transfer and superluminal causal effects.

Also note that tube-like habs can be likewise fabricated and which as continuous living quarters may extend or span more than one cosmic light-cone with respect to the origin of construction.

So, we have some interesting really weird stuff that may be possible.

First, is the complex geometry, and I will further generalize, the complex topology of the space-time structure of super-cosmic-light-cone sized habitats.

Second is the effective super-luminal propagation of the far edge of the habitat and road-way system from its point of origin.

Third, is the prospect of enabling superluminal recessionary velocities of fiats and/or residents of the system with respect to others located within the overall system.

And fourth, is the causal decoupling in a direct sense between locations and persons that are more than one cosmic light-cone radius distant from each other.

Regarding the space-time topology of the system relative to a given person, the topology of the super-cosmic-light-cone system may be as if cellular or uniformly tiled.

Accordingly, the light-cone in which a given person is located may be as if a tile, albeit one of 3 spatial dimensions and one time dimension.

Relative to the above said given person, the next cosmic-light-cone intervals or regions would be as if surrounding tiles.

Relative to persons situated in the center of these adjacent cosmic-light-cones and on the far sides of these adjacent cosmic-light-cones relative to the original person of consideration would begin more cosmic-light-cone cells, and the pattern would simply continue with associated causal breaks in repeated patterns relative to the person originally considered above.

Another topological manifestation which need not be contrary to the above uniform tiling convention is a division of super-light-cone regions relative to said original person considered in the form of tiles of progressively distorted geometry relative to said original person considered. Such distortion can be considered slightly analogous to the non-spherically symmetric nature associated with relativistic aberration of the space-time background relative to a near light-speed inertial reference frame.

Another manifestation may include the partitioning of extra-cosmic-light-cone sized portions of the universe as a concentric set of ring or spheres relative to said original observer. Once again, such a model may entail gaps between cosmic-light-cone distance boundaries that are uniform and/or radially skewed.

All of the above models may be existentially manifest simultaneously while still allowing for such varied differentiations.

We may affix the following operator to formulas for gamma and related parameters to denote the above super-luminal tiling mechanisms based on finite cosmic light-cone radii where the multiple tile models are valid including ones for uniform and distorted tiles relative to an observer.

[w(Super-luminal tiling mechanisms based on finite cosmic light-cone radii where the multiple tile models are valid including ones for uniform and distorted tiles relative to an observer)]